T0399116

FOOD, MEDICAL, AND ENVIRONMENTAL APPLICATIONS OF NANOMATERIALS

FOOD, MEDICAL, AND ENVIRONMENTAL APPLICATIONS OF NANOMATERIALS

Edited by

KUNAL PAL
Department of Biotechnology and Medical Engineering, National Institute of Technology, Rourkela, Odisha, India

ANGANA SARKAR
Department of Biotechnology and Medical Engineering, National Institute of Technology, Rourkela, Odisha, India

PREETAM SARKAR
Department of Food Process Engineering, National Institute of Technology Rourkela, Rourkela, India

NANDIKA BANDARA
Department of Food and Human Nutritional Sciences, Richardson Centre for Food Technology and Research, University of Manitoba, Winnipeg, Manitoba, Canada

VEERIAH JEGATHEESAN
Water: Effective Technologies and Tools (WETT) Research Centre, School of Engineering, RMIT University, Melbourne, Australia

ELSEVIER

Elsevier
Radarweg 29, PO Box 211, 1000 AE Amsterdam, Netherlands
The Boulevard, Langford Lane, Kidlington, Oxford OX5 1GB, United Kingdom
50 Hampshire Street, 5th Floor, Cambridge, MA 02139, United States

Notices
Knowledge and best practice in this field are constantly changing. As new research and experience broaden our
understanding, changes in research methods, professional practices, or medical treatment may become
necessary.

Practitioners and researchers must always rely on their own experience and knowledge in evaluating and using
any information, methods, compounds, or experiments described herein. In using such information or methods
they should be mindful of their own safety and the safety of others, including parties for whom they have a
professional responsibility.

To the fullest extent of the law, neither the Publisher nor the authors, contributors, or editors, assume any liability
for any injury and/or damage to persons or property as a matter of products liability, negligence or otherwise, or
from any use or operation of any methods, products, instructions, or ideas contained in the material herein.

ISBN: 978-0-12-822858-6

For information on all Elsevier publications
visit our website at https://www.elsevier.com/books-and-journals

Publisher: Matthew Deans
Acquisitions Editor: Sabrina Webber
Editorial Project Manager: Aleksandra Packowska
Production Project Manager: Stalin Viswanathan
Cover Designer: Miles Hitchen

Typeset by STRAIVE, India

Working together
to grow libraries in
developing countries

www.elsevier.com • www.bookaid.org

Contents

18. Nano-formulations in drug delivery 473

Melissa Garcia-Carrasco, Itzel F. Parra-Aguilar,
Erick P. Gutiérrez-Grijalva, Angel Licea-Claverie, and
J. Basilio Heredia

19. Nano-materials as biosensor for heavy metal detection 493

Samprit Bose, Sourav Maity, and Angana Sarkar

20. Smart nano-biosensors in sustainable agriculture and environmental applications 527

Rani Puthukulangara Ramachandran, Chelladurai Vellaichamy, and
Chyngyz Erkinbaev

Contributors

Tarun Agarwal Department of Biotechnology, Indian Institute of Technology, Kharagpur, West Bengal, India

Ibrahim N. Amirrah Centre for Tissue Engineering and Regenerative Medicine, Faculty of Medicine, National University of Malaysia, Kuala Lumpur, Malaysia

C. Anandharamakrishnan Computational Modeling and Nanoscale Processing Unit, National Institute of Food Technology, Entrepreneurship and Management—Thanjavur, Ministry of Food Processing Industries, Govt. of India, Thanjavur, Tamil Nadu, India

Aafreen Ansari Department of Biotechnology and Medical Engineering, National Institute of Technology Rourkela, Rourkela, Orissa, India

Mahreen Arooj Department of Chemistry, College of Sciences, University of Sharjah, Sharjah, United Arab Emirates

Nandika Bandara Department of Food and Human Nutritional Sciences, Richardson Centre for Food Technology and Research, University of Manitoba, Winnipeg, MB, Canada

Nabaraj Banjara Department of Biological and Physical Science, University of Holy Cross, New Orleans, LA, United States

Ananya Barui Centre for Healthcare Science and Technology, Indian Institute of Engineering Science and Technology, Shibpur, Howrah, West Bengal, India

J. Basilio Heredia Research Center for Food and Development (CIAD), Nutraceuticals and Functional Foods Laboratory, Culiacán, Sinaloa, Mexico

Kanishka Bhunia Agricultural and Food Engineering Department, Indian Institute of Technology Kharagpur, Kharagpur, WB, India

Samprit Bose Department of Biotechnology and Medical Engineering, National Institute of Technology, Rourkela, Odisha, India

Rashmi Chawdhary All India Institute of Medical Sciences (AIIMS), Bhopal, Madhya Pradesh, India

Rahul Chetri Department of Food Process Engineering, National Institute of Technology Rourkela, Rourkela, India

Nandita Dasgupta Department of Biotechnology, Institute of Engineering and Technology, Dr. A.P.J. Abdul Kalam Technical University, Lucknow, Uttar Pradesh, India

Shalini Dasgupta Centre for Healthcare Science and Technology, Indian Institute of Engineering Science and Technology, Shibpur, Howrah, West Bengal, India

Pallab Datta Department of Pharmaceutics, National Institute of Pharmaceutical Education and Research (NIPER), Kolkata, West Bengal, India

Sangeetha Dharmalingam Department of Mechanical Engineering, Anna University, Chennai, Tamil Nadu, India

Dipanjan Dwari Centre for Healthcare Science and Technology, Indian Institute of Engineering Science and Technology (IIEST), Howrah, West Bengal, India

Chyngyz Erkinbaev Department of Biosystems Engineering, University of Manitoba, Winnipeg, MB, Canada

Mh Busra Fauzi Centre for Tissue Engineering and Regenerative Medicine, Faculty of Medicine, National University of Malaysia, Kuala Lumpur, Malaysia

Melissa Garcia-Carrasco Research Center for Food and Development (CIAD), Nutraceuticals and Functional Foods Laboratory, Culiacán, Sinaloa, Mexico

Anujit Ghosal   Department of Food and Human Nutritional Sciences, Faculty of Agricultural and Food Sciences, University of Manitoba; Richardson Centre for Functional Foods & Nutraceuticals, Winnipeg, MB, Canada

Nikhil Gorhe   Council of Scientific and Industrial Research—Advanced Materials and Processes Research Institute; Academy of Council Scientific and Industrial Research (AcSIR)—Advanced Materials and Processes Research Institute (AMPRI), Hoshangabad Road, Bhopal, Madhya Pradesh, India

Erick P. Gutiérrez-Grijalva   CATEDRAS CONACYT-Research Center for Food and Development (CIAD), Nutraceuticals and Functional Foods Laboratory, Culiacán, Sinaloa, Mexico

Vaishnavi Hada   Council of Scientific and Industrial Research—Advanced Materials and Processes Research Institute, Bhopal, Madhya Pradesh, India

Hushnaara Hadem   Centre for Nano Sciences and Technology, Pondicherry University, Chinna Kalapet, Kalapet, Puducherry, India

Mehavesh Hameed   Department of Chemistry, College of Sciences, University of Sharjah, Sharjah, United Arab Emirates

S.A.R. Hashmi   Council of Scientific and Industrial Research—Advanced Materials and Processes Research Institute; Academy of Council Scientific and Industrial Research (AcSIR)—Advanced Materials and Processes Research Institute (AMPRI), Hoshangabad Road, Bhopal, Madhya Pradesh, India

Navam Hettiarachchy   Department of Food Science, University of Arkansas, Fayetteville, AR, United States

Darryl L. Holliday   Department of Biological and Physical Science, University of Holy Cross, New Orleans, LA, United States

Nishant Rachayya Swami Hulle   Department of Food Engineering and Technology, Tezpur University, Tezpur, Assam, India

K. Jagajjanani Rao   Department of Biotechnology, Vel Tech Rangarajan Dr. Sagunthala R & D Institute of Science and Technology, Chennai, India

Lily Jaiswal   Department of Food and Nutrition, BioNanocomposite Research Institute, Kyung Hee University, Seoul, Republic of Korea

Law Xia Jian   Centre for Tissue Engineering and Regenerative Medicine, Faculty of Medicine, National University of Malaysia, Kuala Lumpur, Malaysia

Nurkhuzaiah Kamaruzaman   Centre for Tissue Engineering and Regenerative Medicine, Faculty of Medicine, National University of Malaysia, Kuala Lumpur, Malaysia

Manal Khan   All India Institute of Medical Sciences (AIIMS), Bhopal, Madhya Pradesh, India

Tarangini Korumilli   Department of Biotechnology, Vel Tech Rangarajan Dr. Sagunthala R & D Institute of Science and Technology, Chennai, India

Vaidhegi Kugarajah   Department of Mechanical Engineering, Anna University, Chennai, Tamil Nadu, India

Jia Xian Law   Centre for Tissue Engineering and Regenerative Medicine, Faculty of Medicine, National University of Malaysia, Kuala Lumpur, Malaysia

Angel Licea-Claverie   TNM, Tijuana Technological Institute, Graduate Studies and Chemistry Research Center, Tijuana, Baja California, México

Alya Limayem   Research Department of Pharmaceutical Sciences, College of Pharmacy, University of South Florida Centre for Research & Education in Nanobioengineering, USF Health, Tampa, FL, United States

Yogeswaran Lokanathan   Centre for Tissue Engineering and Regenerative Medicine, Faculty of Medicine, National University of Malaysia, Kuala Lumpur, Malaysia

L. Mahalakshmi Computational Modeling and Nanoscale Processing Unit, National Institute of Food Technology, Entrepreneurship and Management—Thanjavur, Ministry of Food Processing Industries, Govt. of India, Thanjavur, Tamil Nadu, India

Tapas Kumar Maiti Department of Biotechnology, Indian Institute of Technology, Kharagpur, West Bengal, India

Sourav Maity Department of Biotechnology and Medical Engineering, National Institute of Technology, Rourkela, Odisha, India

Zawani Mazlan Centre for Tissue Engineering and Regenerative Medicine, Faculty of Medicine, National University of Malaysia, Kuala Lumpur, Malaysia

Medha Mili Council of Scientific and Industrial Research—Advanced Materials and Processes Research Institute; Academy of Council Scientific and Industrial Research (AcSIR)—Advanced Materials and Processes Research Institute (AMPRI), Hoshangabad Road, Bhopal, Madhya Pradesh, India

Bhartendu Nath Mishra Department of Biotechnology, Institute of Engineering and Technology, Dr. A.P.J. Abdul Kalam Technical University, Lucknow, Uttar Pradesh, India

Ahmed A. Mohamed Department of Chemistry, College of Sciences, University of Sharjah, Sharjah, United Arab Emirates

J.A. Moses Computational Modeling and Nanoscale Processing Unit, National Institute of Food Technology, Entrepreneurship and Management—Thanjavur, Ministry of Food Processing Industries, Govt. of India, Thanjavur, Tamil Nadu, India

Soma Mukherjee Department of Biological and Physical Science, University of Holy Cross, New Orleans, LA, United States

Ajay Naik Council of Scientific and Industrial Research—Advanced Materials and Processes Research Institute; Academy of Council Scientific and Industrial Research (AcSIR)—Advanced Materials and Processes Research Institute (AMPRI), Hoshangabad Road, Bhopal, Madhya Pradesh, India

Debarshi Nath Department of Food Process Engineering, National Institute of Technology Rourkela, Rourkela, India

Min Hwei Ng Centre for Tissue Engineering and Regenerative Medicine, Faculty of Medicine, National University of Malaysia, Kuala Lumpur, Malaysia

Lei Nie College of Life Sciences, Xinyang Normal University, Xinyang, China

Sophia Devi Nongmaithem Department of Food Engineering and Technology, Tezpur University, Tezpur, Assam, India

Fatimah Mohd Nor KPJ Ampang Puteri Specialist Hospital, Ampang, Selangor, Malaysia

Atul Kumar Ojha Centre for Nano Sciences and Technology, Pondicherry University, Chinna Kalapet, Kalapet, Puducherry, India

Abhinandan Pal Agricultural and Food Engineering Department, Indian Institute of Technology Kharagpur, Kharagpur, WB, India

Kunal Pal Department of Biotechnology and Medical Engineering, National Institute of Technology, Rourkela, Odisha, India

Seema Panicker Department of Chemistry, College of Sciences, University of Sharjah, Sharjah, United Arab Emirates

Itzel F. Parra-Aguilar Research Center for Food and Development (CIAD), Nutraceuticals and Functional Foods Laboratory, Culiacán, Sinaloa, Mexico

N. Prashant Council of Scientific and Industrial Research—Advanced Materials and Processes Research Institute; Academy of Council Scientific and Industrial Research (AcSIR)—Advanced Materials and Processes Research Institute (AMPRI), Hoshangabad Road, Bhopal, Madhya Pradesh, India

Niloofar Khoshdel Rad Department of Stem Cells and Developmental Biology, Cell Science Research Center, Royan Institute for Stem Cell Biology and Technology, ACECR, Tehran, Iran

Mukesh Kumar Ram Centre for Healthcare Science and Technology, Indian Institute of Engineering Science and Technology (IIEST), Howrah, West Bengal, India

Rani Puthukulangara Ramachandran
Department of Biosystems Engineering,
University of Manitoba, Winnipeg, MB,
Canada

Shivendu Ranjan Faculty of Engineering and
the Built Environment, University of
Johannesburg, Johannesburg, South Africa;
Animal Cell and Tissue Culture Lab, Gujarat
Biotechnology Research Centre, Department of
Science and Technology, Government of
Gujarat, Gandhinagar, Gujarat, India

Akash Roy Centre for Healthcare Science and
Technology, Indian Institute of Engineering
Science and Technology (IIEST), Howrah, West
Bengal, India

Sai Sateesh Sagiri Agro-Nanotechnology and
Advanced Materials Research Center, Institute
of Postharvest and Food Sciences, Agricultural
Research Organization, The Volcani Center,
Rishon LeZion, Israel

Sayandeep Saha Centre for Healthcare Science
and Technology, Indian Institute of
Engineering Science and Technology, Shibpur,
Howrah, West Bengal, India

Atiqah Salleh Centre for Tissue Engineering
and Regenerative Medicine, Faculty of
Medicine, National University of Malaysia,
Kuala Lumpur, Malaysia

Nusaibah Sallehuddin Centre for Tissue
Engineering and Regenerative Medicine,
Faculty of Medicine, National University of
Malaysia, Kuala Lumpur, Malaysia

R. Santhosh Department of Food Process
Engineering, National Institute of Technology
Rourkela, Rourkela, India

Angana Sarkar Department of Biotechnology
and Medical Engineering, National Institute of
Technology, Rourkela, Odisha, India

Preetam Sarkar Department of Food Process
Engineering, National Institute of Technology
Rourkela, Rourkela, India

Shiv Shankar Department of Food and
Nutrition, BioNanocomposite Research
Institute, Kyung Hee University, Seoul,
Republic of Korea; Research Laboratories in
Sciences Applied to Food, INRS-Institute
Armand-Frappier, Laval, QC, Canada

Ihsan Shehadi Department of Chemistry,
College of Sciences, University of Sharjah,
Sharjah, United Arab Emirates

Ali Smandri Centre for Tissue Engineering and
Regenerative Medicine, Faculty of Medicine,
National University of Malaysia, Kuala
Lumpur, Malaysia

A.K. Srivastava Council of Scientific and
Industrial Research—Advanced Materials
and Processes Research Institute; Academy of
Council Scientific and Industrial Research
(AcSIR)—Advanced Materials and
Processes Research Institute (AMPRI),
Hoshangabad Road, Bhopal, Madhya
Pradesh, India

Nadiah Sulaiman Centre for Tissue
Engineering and Regenerative Medicine,
Faculty of Medicine, National University of
Malaysia, Kuala Lumpur, Malaysia

Sheri-Ann Tan Department of Bioscience,
Faculty of Applied Sciences, Tunku Abdul
Rahman University College, Kuala Lumpur,
Malaysia

Chelladurai Vellaichamy Department of
Agricultural Engineering, Bannariamman
Institute of Technology, Sathyamangalam,
Tamil Nadu, India

Sarika Verma Council of Scientific and
Industrial Research—Advanced Materials and
Processes Research Institute; Academy of
Council Scientific and Industrial Research
(AcSIR)—Advanced Materials and
Processes Research Institute (AMPRI),
Hoshangabad Road, Bhopal,
Madhya Pradesh, India

Massoud Vosough Department of Stem Cells
and Developmental Biology; Department of
Regenerative Medicine, Cell Science
Research Centre, Royan Institute for Stem
Cell Biology and Technology, ACECR,
Tehran, Iran

K.S. Yoha Computational Modeling and
Nanoscale Processing Unit, National Institute
of Food Technology, Entrepreneurship and
Management—Thanjavur, Ministry of Food
Processing Industries, Govt. of India,
Thanjavur, Tamil Nadu, India

Ensieh Zahmatkesh   Department of Stem Cells
and Developmental Biology, Cell Science
Research Center, Royan Institute for Stem Cell
Biology and Technology, ACECR, Tehran, Iran

Ibrahim Zarkesh   Department of Stem Cells and
Developmental Biology, Cell Science Research
Center, Royan Institute for Stem Cell Biology
and Technology, ACECR, Tehran, Iran

Izzat Zulkiflee   Centre for Tissue Engineering
and Regenerative Medicine, Faculty of
Medicine, National University of Malaysia,
Kuala Lumpur, Malaysia

CHAPTER

1

Fabrication of nanomaterials

Vaidhegi Kugarajah[a],*, Hushnaara Hadem[b],*, Atul Kumar Ojha[b], Shivendu Ranjan[c,d], Nandita Dasgupta[e], Bhartendu Nath Mishra[e], and Sangeetha Dharmalingam[a]

[a]Department of Mechanical Engineering, Anna University, Chennai, Tamil Nadu, India [b]Centre for Nano Sciences and Technology, Pondicherry University, Chinna Kalapet, Kalapet, Puducherry, India [c]Faculty of Engineering and the Built Environment, University of Johannesburg, Johannesburg, South Africa [d]Animal Cell and Tissue Culture Lab, Gujarat Biotechnology Research Centre, Department of Science and Technology, Government of Gujarat, Gandhinagar, Gujarat, India [e]Department of Biotechnology, Institute of Engineering and Technology, Dr. A.P.J. Abdul Kalam Technical University, Lucknow, Uttar Pradesh, India

OUTLINE

*Authors contributed equally.

1 Introduction

Nanotechnology deals with materials with 1–100 nm dimension in size. The concept of nanotechnology was introduced by Richard P. Feynman (Nobel Laureate in Physics, 1965) in "There's Plenty of Room at the Bottom." He introduced the concept of the possibility to arrange the atoms in the nanoscale (Feynman, 1960). Nowadays, most electronics, optical communications, and biological systems are based on nanotechnology (Enescu et al., 2019; Mathew et al., 2019; Shang et al., 2019; Kumar et al., 2020). This is due to the unique physical, chemical, and thermal properties and high surface area to volume ratio. It was found that using nanotechnology, billions of transistors can be packed in computer chips (Srivastava and Kotov, 2008). In biomedical, nanotechnology has been used to achieve targeted drug delivery (Oroojalian et al., 2020; Saxena et al., 2020), gene replacement (Pandey et al., 2019; Cheng et al., 2020), tissue regeneration (Yang et al., 2019), etc. Optical lithography is another best application of nanotechnology (Crucho and Barros, 2017; Karaballi et al., 2020), which has been used for printing small objects (Albisetti et al., 2016; Bose et al., 2018). Also, nanotechnology has various applications such as in flat display devices (Lim, 2019), medical imaging (Röthlisberger et al., 2017; Tibbals, 2017), paint (Lutz, 2019), additives, automobile components (Werner et al., 2018), satellite components (Pourzahedi et al., 2017), high-energy storage system, fuel cells (Elumalai and Sangeetha, 2018; Kugarajah and Dharmalingam, 2020), optical devices (Wang et al., 2019), electromagnetic interference shielding, food and beverage packaging (Enescu et al., 2019), sensors (Farzin et al., 2020), aircraft components, etc.

Various methods are being used to fabricate nanomaterial-based products such as sol-gel synthesis, plasma synthesis, chemical synthesis, hydrothermal synthesis, alloying, blending, mechanical, and mechanochemical synthesis, etc. (Crucho and Barros, 2017; Dastan, 2017; Jamkhande et al., 2019; Karaballi et al., 2020). In order to explain the fabrication, design, and application of nanomaterials, nanoscience uses the basic concepts of properties and mechanisms of nanomaterials used (Zhong, 2009). The historical aspect of nanotechnology is mentioned in Table 1.

Crystallization process (nucleation and growth) has a great impact on the crystal structure and shape during nanoparticle synthesis. The LaMer theory and Sugimoto model can be used to study the kinetics of nucleation and growth mechanism of nanomaterials (Sugimoto, 2007; Mehranpour et al., 2010). According to LaMer theory, when the solute concentration reaches the critical concentration (which is the minimum concentration for nucleation), it starts nucleating. Then, the solute concentration reaches its maximum, which decides the consumption rate for the nucleation and the growth of the generated nuclei. However, a further increase in the solute concentration for the growth of the generated nuclei results in the declination of the curve. It indicates the end of nucleation (as shown in Fig. 1) of the concentration vs time curve. This theory was only proposed for monodisperse particle formation. The basic assumptions of this theory were

(1) Mass balance between the supply rate of solute and its consumption rate for nucleation and growth of the generated nuclei;
(2) The supply rate of solute is independent of the subsequent precipitation events;
(3) The nucleation rate is controlled only by the growth of the preformed nuclei at a fixed supply rate of solute when precursor solute is transferred by slow irreversible generation in a closed system or by a continuous feed from outside in an open system (LaMer and Dinegar, 1950).

TABLE 1 Historical aspects of nanotechnology (Horikoshi and Serpone, 2013).

Year	Remarks	Country/people
1200–1300 BC	Discovery of soluble gold	Egypt and China
290–325 AD	Lycurgus cup	Alexandria or Rome
1618	The first book on colloidal gold	F. Antonii
1676	Book published on drinkable gold that contains metallic gold in neutral media	J. von Löwenstern-Kunckel (Germany)
1718	Publication of a complete treatise on colloidal gold	Hans Heinrich Helcher
1857	Synthesis of colloidal gold	M. Faraday (The Royal Institution of Great Britain)
1902	Surface plasmon resonance (SPR)	R. W. Wood (Johns Hopkins University, United States)
1908	Scattering and absorption of electromagnetic fields by a nanosphere	G. Mie (University of Göttingen, Germany)
1931	Transmission electron microscope (TEM)	M. Knoll and E. Ruska (Technical University of Berlin, Germany)
1937	Scanning electron microscope (SEM)	M. von Ardenne (Forschungslaboratorium für Elektronenphysik, Germany)
1959	Feynman's Lecture on "There's Plenty of Room at the Bottom"	R. P. Feynman (California Institute of Technology, Pasadena, CA, United States)
1960	Microelectromechanical systems (MEMS)	I. Igarashi (Toyota Central R & D Labs, Japan)
1960	Successful oscillation of a laser	T. H. Maiman (Hughes Research Laboratories, United States)
1962	The Kubo effect	R. Kubo (University of Tokyo, Japan)
1965	Moore's Law	G. Moore (Fairchild Semiconductor Inc., United States)
1969	The Honda–Fujishima effect	A. Fujishima and K. Honda (University of Tokyo, Japan)
1972	Amorphous heterostructure photodiode created with bottom-up process	E. Maruyama (Hitachi Co. Ltd., Japan)
1974	Concept of nanotechnology proposed	N. Taniguchi (Tokyo University of Science, Japan)
1976	Carbon nanofiber	M. Endo (Shinshu University, Japan)
1976	Amorphous silicon solar cells	D. E. Carlson and C. R. Wronski (RCA, United States)
1980	Quantum hall effect (Nobel Prize)	K. von Klitzing (University of Würzburg, Germany)

Continued

TABLE 1   Historical aspects of nanotechnology (Horikoshi and Serpone, 2013)—cont'd

Year	Remarks	Country/people
1982	Scanning tunneling microscope (STM) (Nobel Prize)	G. Binnig and H. Rohrer (IBM Zurich Research Lab., Switzerland)
1986	Atomic force microscope (AFM)	G. Binnig (IBM Zurich Research Lab., Switzerland)
1986	Three-dimensional space manipulation of atoms demonstrated (Nobel Prize)	S. Chu (Bell Lab., United States)
1987	Gold nanoparticle catalysis	M. Haruta (Industrial Research Institute of Osaka, Japan)
1990	Atoms controlled with scanning tunneling microscope (STM)	D. M. Eigler (IBM, United States)
1991	Carbon nanotubes discovered	S. Iijima (NEC Co., Japan)
1992	Japan's National Project on Ultimate Manipulation of Atoms and Molecules begins	
1995	Nanoimprinting	S. Y. Chou (University of Minnesota, United States)
1996	Nano sheets	T. Sasaki (National Institute for Research in Inorganic Materials, Japan)
2000	National Nanotechnology Initiative (NNI), United States	
2003	21st Century Nanotechnology Research and Development Act, United States	
2005	Nanosciences and Nanotechnologies: An action plan, Europe	

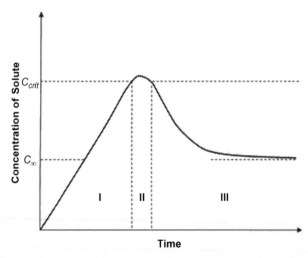

FIG. 1   LaMer diagram as a schematic explanation for the formation process of monodisperse particles, where C_∞ and C_{crit} are the equilibrium concentration of solute with the bulk solid and the critical concentration as the minimum concentration for nucleation, respectively. The regions I, II, and III represent the prenucleation, nucleation, and growth stages, respectively. *Reproduced from Sugimoto, T., 2007. Underlying mechanisms in size control of uniform nanoparticles. J. Colloid Interface Sci. 309, 106–118. https://doi.org/10.1016/j.jcis.2007.01.036.*

The size and the uniformity of the final nanoparticles can be controlled by reducing the growth rate of nuclei, which is possible by detailed knowledge about factors such as temperature, pH, adsorption of additives, and affinity of solvents (Sugimoto, 2007). Further, fluorescent biosensor light-emitting diodes can be fabricated using these quantum dots or nanoparticles (Lin et al., 2017; Mu et al., 2020). Sharp and symmetrical emission spectra, high quantum yield, good photochemical stability, and size-dependent emission-wavelength tenability are the main properties of quantum dots or nanoparticles, owing to their quantum confinement effect (Raphaël et al., 2011). Therefore, semiconductor quantum dots nanomaterials play a significant role in modern electronics light-emitting diode (Cao et al., 2017; Dong et al., 2019), digital and analog integrated circuits, transistors, solar cells (Ahmad et al., 2018), solar photovoltaic panels, wastewater treatment by adsorption of pollutants (Mustapha et al., 2020), etc. Functionalization of nanomaterials offers to achieve better performance and properties (Palit and Hussain, 2020). The functionalization of nanomaterials is carried out based on the interaction between main nanostructured materials and neighboring material systems (Iordache et al., 2011). According to the Hall-Petch equation, the nanomaterial-based composites system also shows better mechanical properties, according to which the strength of materials is inversely proportional to the root diameter of grain. Nanomaterials offer excellent barrier properties due to the increased diffusion length of the solvent gas. The main objective of this chapter is to explain the fabrication of nanomaterials.

2 Fabrication of nanomaterials

Materials can be classified according to their mechanical, electrical, magnetic/dielectric properties and their dimensions, including bulk materials, 0D nanomaterials (all three dimensions in nano-range), 1D (two dimensions in nano-range) (Wei et al., 2017), 2D (one dimension in nano-range) (Wang et al., 2017a), and 3D (polycrystalline nanomaterials) (Zhu et al., 2017). In this chapter, we will focus on the fabrication methods of nanomaterials. The technique to fabricate and process the nanomaterials is the main issue in nanotechnology. Probing their unique physical properties and understanding their possible fabrication techniques provide advantages for using the nanomaterials for their final use. Generally, physical methods or chemical methods are employed to fabricate nanomaterials. The physical methods include top-down and bottom-up approaches.

Top-down such as solid-state route, ball milling and bottom-up such as sol-gel and coprecipitation are the two methods that have been adopted to synthesize the nanomaterials. In some cases, the bottom-up and top-down methods are applied together to fabricate nanomaterials such as in Lithography process (Gregorczyk and Knez, 2016). Such an approach is called a hybrid method. A flow chart for some methods of nanomaterials fabrication is mentioned in Fig. 2. There are several methods to characterize the nanomaterials prepared. Some major characterization techniques include X-ray diffraction, scanning electron microscopy, transmission electron microscopy, energy dissipative analysis, ultraviolet-visible spectroscopy, dynamic light scattering, and Fourier transformation infrared spectroscopy (Thomas et al., 2017). Morphology, physical, chemical, and thermal properties of

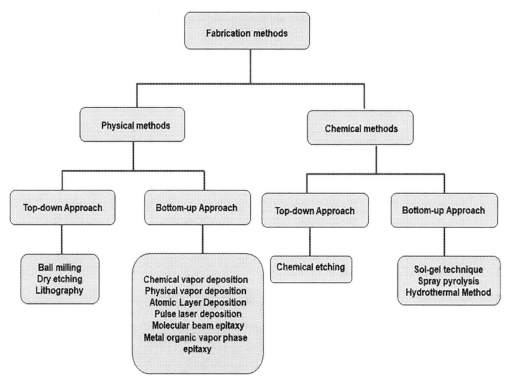

FIG. 2   A flow chart for nanomaterials fabrication.

nanomaterials depend on the fabrication techniques and precursors of the nanomaterials that were used to fabricate them. The properties of nanomaterials decide the applications of nanomaterials in various areas such as catalysis, food industry, medicine, and electronics. Hydrothermal synthesis, combustion synthesis, gas-phase methods, microwave synthesis, and sol-gel processing have been used to synthesize nano metal oxides (Cheng et al., 1995; Gopal et al., 1997; Kim et al., 1999; Wang and Ying, 1999; Watson et al., 2004; Wolf, 2008). Other technologies such as self-assembly, nontraditional lithography, template growth, and biomimetics have also been used to produce nanomaterial-based products, quantum dots, nanorods, nanotubes, etc. (Wolf, 2008). Nanomaterials have at least one dimension in the nanometer range. The nanomaterials show different physical and chemical properties from their bulk counterparts. Nanotechnology allows us to understand the size effect properties of nanostructured materials and their possible applications. Nowadays, nanotechnology is being used as an interdisciplinary science and technology, which includes nanochemistry, nanoelectronics, nanomaterials science, nanophysics, nanorobotics, nanobiotechnology, nanometrology, etc.

In addition to the above, there are several methods, including coprecipitation, hydrothermal, microwave, microemulsion, ultrasound, template synthesis, biological synthesis, electrochemical synthesis, etc. which are discussed in the following sections.

3 Top-down fabrication methods

The top-down approach is a well-developed technology commonly used to divide a massive solid into smaller particles up to the nanometer size range (Su and Chang, 2018). Dry etching, ball milling, and lithography are mainly employed for fabricating the nanoparticles through a top-down approach. There are certain characteristics of nanomaterials, which can be achieved by these fabrication techniques:

(a) Mono (uniform) size distribution, i.e., all the particles will have the same size in nano-range,
(b) Similar shape and (microstructure) morphology,
(c) Similar chemical composition, and
(d) Lower agglomeration (by surface area and various types of interaction).

However, there are also many drawbacks in these methods, including crystallographic defects, roughness, and the introduction of impurities (Su and Chang, 2018). These impurities and defects alter the surface properties of the resulting nanomaterials. For example, the conductivity of nanomaterials is altered by the surface imperfections, and excessive generation of heat is another problem of these defects. Top-down approach has another drawback, such as smaller flakes or particles with a wide size distribution (Habiba et al., 2014). The top-down approach for nanomaterials is mentioned in Fig. 3. Top-down approach is also called as solid-phase fabrication method.

3.1 Mechanical methods

Mechanical milling has been widely adopted for fabricating various kinds of materials so far. Mechanical milling is performed using the ball shaking, rotating disk, and rotating cylinder. The main aim of mechanical milling is to reduce the particle size, mechanical mixing, or alloying by applying mechanical energy and mechanical shear forces on the powder sample of the mixture (Boldyrev and Tkacova, 2000). Mechanical milling has been used to fabricate nanomaterials and nanocomposites. The principle of mechanical milling is based on the

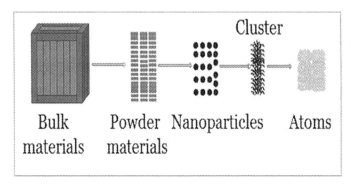

Bulk materials **Powder materials** **Nanoparticles** **Cluster** **Atoms**

FIG. 3 A schematic representation of the top-down approach for nanomaterials.

ball-powder-ball collision in the case of shaker mills, frictional and impact forces in the case of planetary ball mills (Prasad Yadav et al., 2012), and very high shear and impact forces in the case of Attrition mills (Rajput, 2015). These mechanical and shear forces cause the powder particles to undergo deformation and/or fracture to achieve the desired microstructure of the powder. The mechanical characteristics of powder component or powder mixture is a crucial factor in mechanical ball milling; there are various types of powder mixture such as ductile-ductile, brittle-brittle, ductile-brittle system (Benjamin and Volin, 1974; Davis et al., 1988; Lee and Koch, 1988). However, the temperature and the nonuniformity of particle size during milling profoundly affect the structural, microstructural changes (Su and Chang, 2018). The speed of the ball, the material properties of the powder, and milling environment can influence the temperature of powder during milling. Phase transfer can be induced by the ball milling, which depends on the diffusivity and defect concentration in the powder, while, diffusivity and defect concentration in the powder are influenced by the temperature of the powder (Sun et al., 2002; Prasad Yadav et al., 2012). A scheme for the ball milling process is shown in Fig. 4. The mechanical milling can be performed in various ways such as attrition ball mill, planetary ball mill, vibrating ball mill, low-energy tumbling mill, and high-energy ball mill.

Nanoparticle-based powders can be easily produced by high-energy ball milling (Jamkhande et al., 2019). In high-energy ball milling, powders and heavy steel or tungsten carbide are loaded together in a container, and shaking or high-speed rotation is applied to transfer the high energy on the loaded powder; the collision between balls generates the energy. However, high-energy ball milling has certain drawbacks such as contamination problems, low surface, highly polydisperse size distribution, and partially amorphous state of the powder. To resolve this issue, inert atmosphere and/or high vacuum processes are applied. The milling duration, characteristics of materials, and size of ball are the main parameters (Prasad Yadav et al., 2012). Low-energy tumbling mill is employed to fabricate nanoparticles-based alloy materials in a simple way and at low cost. Vibrating ball mills are used to produce amorphous alloys. A vibrating ball supplies a high amount of energy to the particles. The vibrating container is used to accelerate the milling process. The vibrating frequency and amplitude of oscillation are found to be around 1500–3000 oscillation/min and 2–3mm, respectively. The grinding ball is made of steel or carbide ball and is heavier than the

FIG. 4 Ball milling process.

Planetary ball milling Grinding ball   Ball-powder-ball Collision

powder particles and 10–20 mm in diameter. Planetary ball milling is performed in a small container, these small containers are fixed on to a rotating drum, and the containers rotate in opposite direction to the rotating drum. Centrifugal forces are imparted by the rotation of the supporting disk and autonomous turning of the containers. The ball and powder roll on the inner wall of the containers and are thrown off across the containers at around 360 rpm. As attrition indicates, the finer particles are obtained by the wear or rubbing. Attrition milling (Stirred Ball Mills) is performed via a vertical shaft with horizontal arms such as flat disks, disks with various geometric openings, and concentric rings. The shaft is kept in media, which contains the ball and powder slurry. When the shaft rotates, stirrer action of the horizontal arm on the powder slurry reduces the size of particles. The shaker mill reduces the size of powder by the shaking process. The powder and the ball are kept in the container; during the shaking process, balls collide with each other and the container wall, which produces high shear and impact forces on the powder. This high impact and shear force reduce the powder size (Ullah et al., 2014). In one of the studies, researchers prepared Mg-doped ZnO nanoparticles by planetary ball milling at a speed of 400 rpm and milled for 20 h. The spherical nanoparticles were agglomerated into a cluster in order to reduce the surface-free energy, as shown in Fig. 5 (Suwanboon and Amornpitoksuk, 2012). A recent study by Abu-Oqail et al. (2019) showed that the increase in milling time to 20 h improved the microstructure of the

FIG. 5 SEM images of Mg-doped ZnO nanoparticle at different Mg contents (Suwanboon and Amornpitoksuk, 2012).

Cu-ZrO$_2$ nanocomposites up to 10% higher than that of pure copper, which enabled better mechanical properties of the nanocomposites. Another study showed the influence of ball milling in starch nanoparticle production for its application in drug and food formulations. It was observed that the relative crystalline size decreased after ball milling with an average nanoparticle size in the range of 9–12 nm (Ahmad et al., 2020).

3.2 Mechanochemical synthesis

Mechanochemical synthesis is the combination of mechanical and chemical methods to get nanomaterials (Dutková et al., 2018; Galaburda et al., 2019). Mechanochemical synthesis is an entirely different process from the ball milling process (Hai Nguyen et al., 2020). It is performed by solid-state displacement reaction during the ball milling process to obtain nanoparticles embedded in by-product finally. Milling temperature, milling collision energy, volume and particle size, milling time, molar ratio of precursor, powder mixture to ball ratio are the main factors that influence the particles size and particles size distribution. The main drawbacks of this system are (1) contamination, (2) long processing time, (3) uncontrollable particle microstructure, and (4) agglomeration. Fig. 6 shows the flow chart of mechanochemical synthesis method for nanoparticles. In one study, researchers fabricated ZnO nanoparticles by mechanochemical method using ZnCl$_2$, NaCl, and Na$_2$CO$_3$ as raw materials. TEM image shows that calcination has a profound effect on the particle size. It was found that the particle size increases with an increase in the calcination temperature. As Fig. 7 shows, the average size of particles was approximately 20–30 nm (Moballegh et al., 2007). It should be pointed out that the mechanochemical synthesis is a result of milling and chemical interaction between precursors. Researchers have fabricated SiO$_2$/TDI and SiO$_2$/TDI/(PDMS-OH) hybrid particles by the mechanochemical method in the presence of TDI (2,4-diisocyanatotoluene) and hydroxyl silicone oil (PDMS-OH), as shown in Fig. 8. TEM microstructure indicates that hybrid particles (SiO$_2$/TDI and SiO$_2$/TDI/(PDMS-OH)) prepared by ball milling method exhibit much better miscibility and dispersion in PDMS matrix when

FIG. 6   Mechanochemical process.

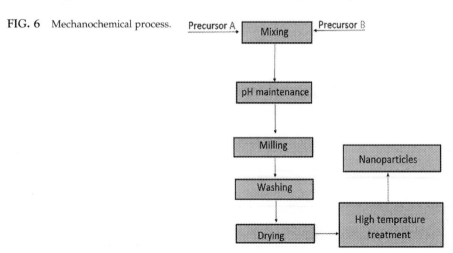

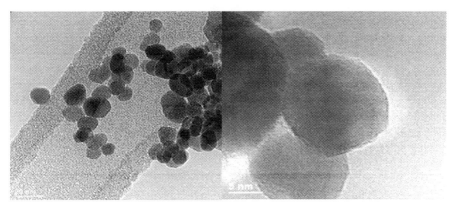

FIG. 7 TEM micrographs of ZnO nanoparticles of heat treated at 300°C (Moballegh et al., 2007).

compared with those particles prepared by a common mixing device (Lin et al., 2010). A recent report shows the preparation of InP/TiO$_2$-C nanocomposites synthesized through the mechanochemical process as an anode for its application in Li-ion batteries to improve efficiency and stability (Hai Nguyen et al., 2020).

3.3 Lithographic processes

Nanolithography has been considered as the branch of nanotechnology, which is used to fabricate nanometer-scale structures. Recently, various nano-electromechanical systems, nano-transistors, nanodiodes, nano switches, nanologic gates, semiconducting integrated circuits (ICs), etc. have been fabricated through the nanolithography technology (Chou et al., 1996). Nanolithography is derived from the Greek word "nanos, lithos, grapho," which means dwarf (nano), rock, or stone (litho) and writes (grapho), respectively. It is explained as tiny writing on stone as nano terms suggest the patterning with at least one dimension in nanoscale. Nanolithography has several advantages such a higher degree of safety, improved stability, and robustness, a higher degree of efficiency and capability, predictable properties of nanocomposites and materials, environmental competitiveness, and flexibility. Nanolithography consists of masked or mask-less, top-down or bottom-up, beam- or tip-based, resist-based or resist-less, and serial or parallel methods. Therefore, many nanolithography technologies have been used to fabricate the micro/nanopattern; those technologies are photolithography, ion beam lithography, X-ray lithography, electron beam lithography, microcontact printing, nanoimprint lithography, scanning probe lithography.

The photolithography system is performed using a light source, a mask, and an optical projection system (Wang et al., 2017b). In photolithography approach, initially, a wafer or substrate with an oxide layer is prepared; after that photoresist materials (light-sensitive materials or chemical) are coated on the substrate, a photomask of the desired pattern is fixed on to the coated substrate, which is developed onto a glass substrate. The UV light is exposed through the photomask, then the photoresist undergoes a chemical reaction during the light exposure, and depending on the characteristics of photoresist, a pattern is created, and etching was performed to get an oxide layer pattern. The photoresist is classified as a negative or

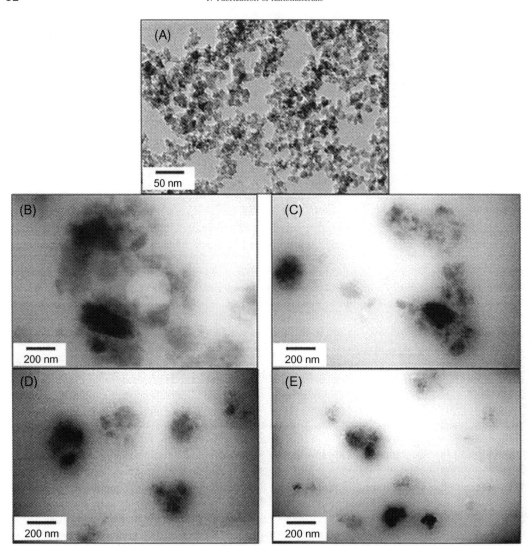

FIG. 8   TEM images of nano-SiO$_2$ (A) dispersed in D4(Octamethylcyclotetrasiloxane), hybrid particles prepared by common method dispersed in PDMS (SiO$_2$/TDI (B) and SiO$_2$/TDI/(PDMS-OH) (C)), and hybrid particles prepared by ball milling method dispersed in PDMS(SiO$_2$/TDI (D) and SiO$_2$/TDI/(PDMS-OH) (E)) (Lin et al., 2010).

positive resist, the negative resist is cheaper with low resolution and positive resist becomes polymerized under light exposure. The system of photolithography can be contact printing (photomask and photoresist show direct contact), proximity printing (proximate contact between mast and photoresist), and projection printing (projection printing by using a lens having high resolution). The photolithography process scheme is shown in Fig. 9 (Mosher et al., 2009; Khumpuang and Hara, 2015; Srivastava, 2016). In 2019, Liu et al. developed a new method of photolithography based on electron beam (Liu et al., 2019). However, electron

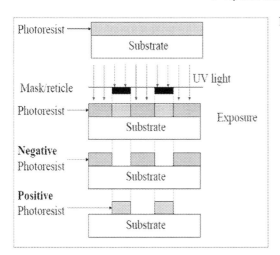

FIG. 9 Photolithography process.

beam lithography is a type of mask-less lithography employed for nanoscale patterns. Accelerated electron beams are exposed to the electron-sensitive materials. A conducting substrate is used to overcome the charging. The result of this exposure can be a positive tone resist or negative tone resist. The electron beam is scanned (layer-by-layer fashion) on the surface of the resist with the diameter in the range of nanometers. Finally, the resulting pattern is obtained by an etching process or vacuum evaporation, as shown in Fig. 10 (Pfeiffer, 2009, 2010). Similarly, extreme ultraviolet (EUV) lithography, another new technology (Torretti et al., 2020), has been utilized to fabricate electronic circuits of sub-20 nm range using low-energy electrons in tin-based nanomaterials (Bespalov et al., 2020). Also, soft lithography has been used; microcontact printing is a kind of soft lithography. Micro- or nanostructured surfaces are fabricated using microcontact printing (µCP); polymeric stamp with a relief pattern is the main component of soft lithography. The substrate surface and stamp (inked) are brought in contact to get the desired pattern; once the inked stamp is contacted with the substrate, the pattern is transferred from the stamp to the substrate contact area. Elastomer and

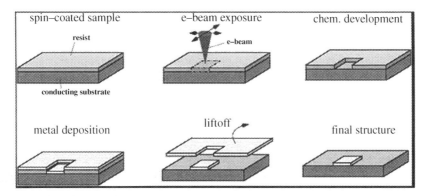

FIG. 10 Electron beam lithography process.

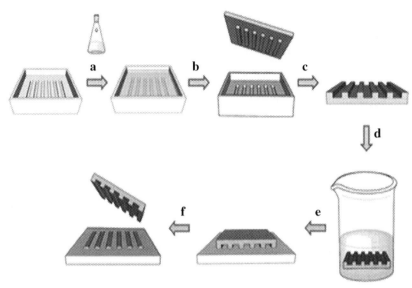

FIG. 11   Key steps in microcontact printing (μCP): (a) a prepolymer is poured on a photolithographically structured master, (b) the prepolymer is cured, and the elastomer stamp is peeled off the master, (c) the stamp is cut into smaller pieces, (d) the stamp is inked by soaking it in an ink solution, (e) the ink is printed by contacting an inked stamp with a suitable surface, and (f) a patterned substrate is obtained. Alternatively, to (a) and (b), a stamp can also be obtained by hot embossing. Alternatively, to (c), wafer-size stamps can also be used. Alternatively to (d), a stamp can be inked by spreading a drop of ink on the stamp, or by using an ink pad (Kaufmann and Ravoo, 2010).

polymer ink are used as stamp in microcontact printing because they are flexible and mechanically stable (Kaufmann and Ravoo, 2010). The key steps of microcontact printing are mentioned in Fig. 11.

Another approach, nanoimprint lithography, has been used to fabricate the nanostructured pattern (Jiang et al., 2019). Nanoimprint and soft lithography are almost similar processes. Nanoimprint lithography follows the two basic steps (Fig. 12):

(1) Imprint step produces a thickness variation pattern in the resist and
(2) Selective etching process of residual resist.

FIG. 12   Steps of nanoimprint lithography.

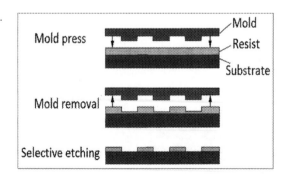

During the imprint step, the resist is heated to a temperature above its glass transition temperature. At that temperature, the resist, which is thermoplastic, becomes a viscous liquid and can flow and, therefore, can be readily deformed into the shape of the mold (Chou et al., 1996). The technique of X-ray lithography is a version of photolithography because, in the case of photolithography, UV source is used, while, in the case of X-ray lithography, the X-ray source is used in place of UV source, as shown in Fig. 13. A brief remark on the various lithography processes is mentioned in Table 2.

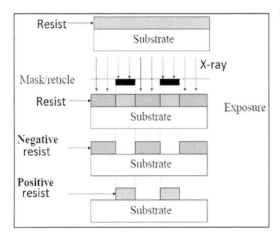

FIG. 13   X-ray lithography process.

TABLE 2   Features of various lithography processes (Srivastava, 2016).

Lithography methods	Dimension range	Remarks	Application
Photolithography	37 nm (projection printing), 2–3 μm (contact, proximity printing)	Less expensive and highly efficient; requires liquids, contaminants, and environmental hazards free room; requires completely flat substrate	Fabrication of integrated circuit and central processing unit chip
Electron beam lithography	<5 nm	High resolution, costly, time-taking, inefficient, and complex method	Integrated circuit fabrication and production of photonic crystals
Microcontact printing	30 nm	Cheaper and flexible technique; the main problem is shrinkage of stamp and stamp deformation	Fabrication of microelectromechanical systems
Nanoimprint lithography	6–40 nm	Simple process, low cost, wear of templates, difficult to make alignment between the template and the resist on the substrate	Fabrication of biosensors, MOSFET, and nanowires
X-ray lithography	15 nm	X-ray mask, long time to expose (in the case of insensitive resist), resolution is limited due to Fresnel diffraction	Fabrication of microchips and microfluidic structures

3.4 Laser ablation in the liquid synthesis

Laser ablation or pulsed laser ablation in liquid techniques is a chemical-free technique to fabricate nanoparticles (NPs) or hybrid nanosystems (Barcikowski et al., 2009). This method has several advantages such as decreased agglomeration, impurity problems, and this method is free from capping agents and chemical precursors and is environment friendly. This method consists of some basic steps to produce the nanoparticles of bulk materials in liquid. The steps in laser ablation include:

(a) Laser-matter interaction (irradiates the metal target in bulk liquid and starts the breakdown process),
(b) Plasma generation and plasma energy transfer to liquid,
(c) Cavitation, bubble formation, and bubble collapse when it reaches the maximum radius,
(d) Collapsing of the bubble and then particle release in solution.

This method has been successfully applied for fabricating metal, alloy, oxide, semiconductor, ceramics, and carbon nanoparticles. The schematic representation of laser ablation is shown in Fig. 14. A study in 2020 by Menazea showed the synthesis of silver nanoparticles using femtosecond laser ablation. It was observed that the Ag nanoparticles exhibited maximum antibacterial activity against a wide range of microbes and were chemically stable (Menazea, 2020). It was also observed that laser ablation protocol was best suited for producing metal nanoparticles such as silver, gold, copper, etc. (Sadrolhosseini et al., 2019).

3.5 Arc discharge synthesis

In a study conducted by Iijima et al., carbon nanotubes were produced by arc discharge method (Iijima, 1991). Arc discharge process includes the electrical breakdown of gas to produce plasma. The chamber contains a pair of electrodes (anode and cathode) that are attached vertically or horizontally; the anode is loaded with powdered precursor accompanying the catalyst, and the cathode is generally precursor rod. The chamber is packed with a gas or submerged in a liquid environment. After turning on the power equipment (AC or DC), the

FIG. 14   Sketch of the experimental setup (Palazzo et al., 2017).

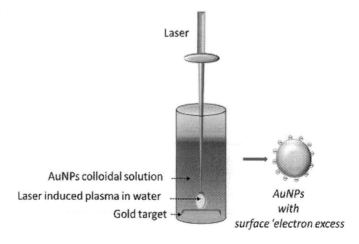

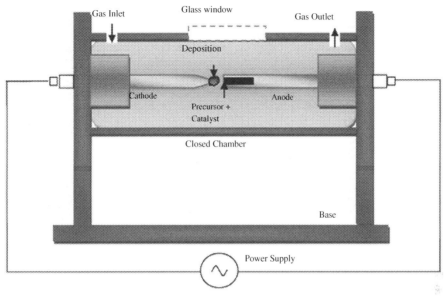

FIG. 15   Schematic of an arc discharge setup (Arora and Sharma, 2014).

electrodes are moved close to form an arc and are held at the gap of 1–2 mm to achieve a constant discharge. A steady current is carried on within the electrodes to achieve a nonfluctuating arc for which closed-loop automation is applied to set the gap automatically. A fluctuating arc can be produced in unstable plasma, and the character of the synthesized product is changed. The arc current produces the plasma of extremely high temperature ~4000–6000 K, which boosts the precursor loaded inside the anode. The precursor vapors aggregate in the gas phase and tend toward the cathode where it cools down due to the temperature gradient. After an arc application time of a few minutes, the discharge is stopped and cathodic deposit, which carries nanomaterials along with the residue, is collected from the walls of the chamber. The deposit is further purified and observed under an electron microscope to investigate their morphology (Kim and Kim, 2006; Keidar, 2007; Kim et al., 2012; Arora and Sharma, 2014). The schematic arc discharge is mentioned in Fig. 15. A recent study by Tharchanaa et al. showed the synthesis of Cu and CuO nanoparticles with the average particle size of 78 nm and 67 nm prepared by plasma arc discharge method for biomedical applications. The results showed that the prepared nanoparticles exhibited better antibacterial performance against the Gram-positive and Gram-negative strains subjected (Tharchanaa et al., 2020). Another study by Zhang et al. (2019) shows the synthesis of controllable carbon nanoparticles using the direct current arc discharge method.

4 Bottom-up fabrication methods

Bottom-up fabrication techniques are used to fabricate the nanostructures from atomic or molecular species. Sol-gel processing, chemical vapor deposition (CVD), plasma or flame

FIG. 16 Schematic representation of bottom-up approaches.

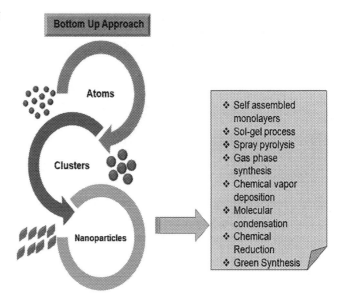

sprays synthesis, laser pyrolysis, atomic or molecular condensation have been used to fabricate the nanomaterials by bottom-up approach. Arrested precipitation and physical restriction by using templates are two general methods to control the formation and growth of the nanoparticles. Gas phase and liquid phase are two methods, which are employed to fabricate the nanostructured materials through bottom-up approach. Gas-phase synthesis includes the plasma arc, chemical vapor deposition, while liquid-phase synthesis consists of self-assembly, sol-gel, etc. The schematic representation of the bottom-up approach is represented in Fig. 16.

4.1 Self-assembled monolayers

The possibility of self-assembled monolayers (SAMs) has grown in artificial elegance and depth of explanation over the past 40 years. Zisman et al. reported the development of a monomolecular layer by adsorption (self-assembly) of surfactant onto a clean metal surface (Bigelow et al., 1946). SAMs are defined as the ordered molecular assemblies produced by the adsorption of an active surfactant on a solid surface. We can say that self-assembly is a natural fabrication tool, which helps to produce natural, organic, and inorganic materials. In a biological system such as DNA double helix, the formation of membrane cells is the best example of molecular self-assembly in nanoscale. The self-assembly is based on noncovalent interactions. Lately, several investigations in nanoscience have been dedicated to the self-assembly of nanoscale building blocks, which help produce excellent super structured nanomaterials (Fig. 17) (Polshettiwar et al., 2009; Gao et al., 2012). These nanomaterials have great potential in the field of magnetic bioseparators (Katz, 2019), targeted drug delivery (Mallick et al., 2019), magnetic resonance imaging (Akakuru et al., 2020), enzyme immobilization, etc. (Leroux, 2007; Zhu et al., 2007; Liong et al., 2008).

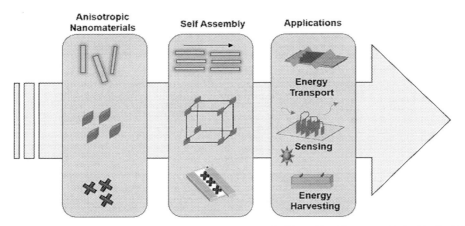

FIG. 17 Applications of nanomaterials prepared by SAM. *Modified from Thorkelsson, K., Bai, P., Xu, T., 2015. Self-assembly and applications of anisotropic nanomaterials: a review. Nano Today 10, 48–66. doi:https://doi.org/10.1016/j.nantod.2014.12.005.*

4.2 Sol-gel technology

Sol-gel synthesis can be applied to make materials with various morphology, such as porous structures, thin fibers, dense powders, and thin films (Rane et al., 2018). The level of organization of nanostructures and their characteristics depends on the character of the organic and inorganic elements of the structure that can create synergic interactions (Soler-Illia et al., 2002). The sol-gel method is used to fabricate nanostructured materials at low temperature and pressure. Sol-gel consists of two terms; one is sol, and the other is a gel; sol is defined as the stable dispersion of colloidal particles (amorphous or crystalline) or polymers (amorphous or semicrystalline) in a solvent. A gel is a three-dimensional continuous network enclosed by the liquid phase. Sol-gel process is useful to fabricate integrated network (so-called gel) of metal oxides or hybrid polymers. Sol-gel approach includes dissolving the solid in a liquid and finally obtaining the solid in a controlled way. Sol-gel is performed in various steps:

(1) Solid precursor suspension in liquid to form sol,
(2) Gelation,
(3) Drying, and
(4) Sintering.

As mentioned in the steps, inorganic metal salts such as metal chlorides and organic metal compounds such as metal alkoxide precursor are suspended in liquid. Further, hydrolysis, and poly-condensation reaction converts it into sol, where sol consists of nanoparticles dispersed in a solvent. This sol is transferred into gel by continuous network formation of particles. After solvent extraction by sublimation, we can obtain aerogel (liquid phase replaced with air). If the sol-gel solution is allowed to dry (evaporation), it will give a xerogel, which is a high-density aerogel, as shown in Fig. 18 (Owens et al., 2016). Some of the attractive features of this technique include:

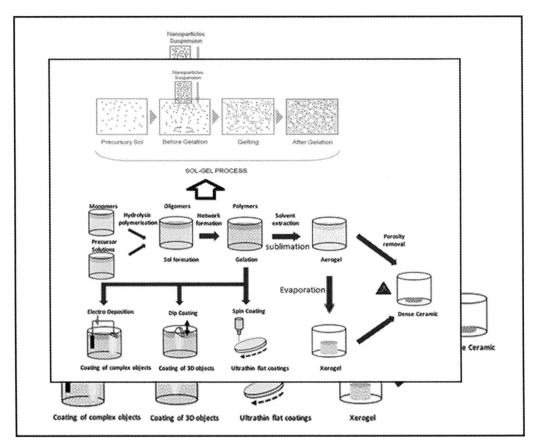

FIG. 18   Sol-gel process (Han et al., 2015; Owens et al., 2016).

- Final products obtained are homogenous with starting materials obtained from the molecular level.
- Nanoparticles are highly pure with controllable porosity.
- Synthesis at low temperature.
- Control in the nanoparticle chemical composition that is highly advantageous during multicomponent preparation.

Further, the major constraint observed in this technique includes the long reaction time and usage of organic solvents, which may be harmful to the human body (Rane et al., 2018).

4.3 Spray pyrolysis

Spray pyrolysis is a liquid-to-particle conversion, treated as "breakdown" method for aerosol processing. This method is called a liquid-phase method because sols are used to obtain nanostructured materials. In the case of spray pyrolysis, the starting solution is produced by dissolving solutes or reactants in the solvent, called as a precursor. The precursor is atomized in an atomizer into droplets; these droplets are introduced to a tubular furnace/reactor, where the carrier gas(es) have already been added. Finally, nanoparticles are produced inside the furnace by the evaporation of the solvent, drying, diffusion of solute, precipitation, the reaction between the precursors, surrounding gas, and pyrolysis. Spray drying is equivalent to spray pyrolysis except for the character of the precursor. As in the spray drying method, the colloidal particles or sols are generally adopted as precursors. This technique has the ability to develop consistently spherical particles of submicron to micron sizes. If the suspension has colloidal nanoparticles, these particles also include nanoparticles to produce a nanostructured powder. Accordingly, spray drying may be a good process for consolidating nanoparticles into macroscopic compacts and submicron spherical powders that have nanometer-scaled properties (Okuyama and Lenggoro, 2003). A recent report by Cho (2019) shows the facile synthesis of low crystalline MnO_3/carbon nanocomposites for large-scale application prepared by spray pyrolysis with higher electrical conductivity and mechanical stability as shown in Fig. 19.

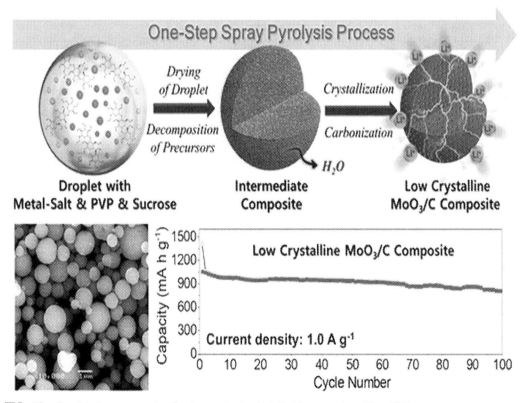

FIG. 19   Graphical representation for the synthesis of MoO_3/C composites (Cho, 2019).

4.4 Gas-phase synthesis

It is found that the gas-phase method is well suited for fabricating nanomaterials. These techniques have several advantages such as tunable shape, size, and chemical composition of the nanostructures (Schulz et al., 2019; Weyell et al., 2020). These methods follow some procedure, which differs from the other methods. The steps include:

(1) Suspending the precursor materials in a gas phase,
(2) Converting the precursor material to small clusters,
(3) Growth of these nanoclusters into nanoparticles,
(4) Finally, collection of the nanoparticles.

4.4.1 Inert gas condensation

This approach is the most fundamental of all the gas-phase fabrication methods. The method is easy as it just involves heating the material within a furnace generally filled with an inert gas such as nitrogen or helium. This approach is particularly suitable for the materials, which possess low vapor pressure. The method is considered useful in the development of metallic nanoparticles because these materials exhibit acceptable rates of evaporation at working temperatures. While commonly conducted in an inert gas, reactive gases can also be injected into the heated chamber to assist the reactions; this is especially helpful in preparing metal oxide and metal halide nanoparticles. Nanoparticle characteristics such as shape, size, and distribution are principally regulated by the rate of evaporation (heating rate/ temperature), condensation (cooling rate), and gas flow rate. It is a bottom-up approach; the individual atoms, ions, and molecules collectively condense to produce nanoparticles (Wang et al., 2020; Zhu et al., 2020). The schematic inert gas condensation instrumentation is shown in Fig. 20.

It consists of two chambers: a condensation chamber and a deposition chamber. Radio-frequency magnetron sputtering is employed as the source to create individual atoms, ions, and molecules of the target in the condensation chamber. The adjoining chamber, recognized

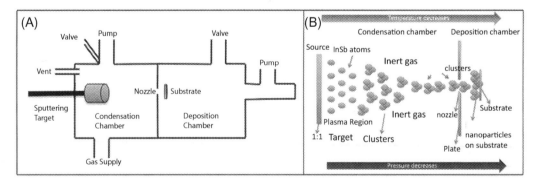

FIG. 20   (A) Schematic of the inert gas condensation instrument used in our laboratory, (B) diagram explaining the formation of the nanoparticles during the inert gas condensation process (Vishwakarma et al., 2013; Pandya and Kordesch, 2015).

as the deposition chamber, is separated from the condensation chamber by a small nozzle. The nozzle either has a hole of mm diameter or mm long and mm wide slit. The inert gas is supplied in the condensation chamber, while deposition chamber is evacuated, which helps create a high-pressure difference between these two chambers. The pressure difference helps to move away from the produced individual atoms, ions, and molecules of the target from the plasma region, which then collide with inert gas ions and lose their energy and momentum and condense to form clusters of the target. Subsequently, extraction of nanoparticles from the cluster is completed, and then nanoparticles are immediately deposited onto a substrate that is located exactly in front of the nozzle, at a distance of about cm in the deposition chamber (Vishwakarma et al., 2013; Pandya and Kordesch, 2015).

4.4.2 Pulsed laser ablation

Laser ablation is a process of eradicating material from a solid or liquid surface through irradiation with a laser beam, as shown in Fig. 21. At a lower laser flux, the material is heated with the absorbed laser energy followed by evaporation or sublimation. However, at higher laser flux, the material converts into plasma (Rane et al., 2018). Rather than just evaporating a material to create supersaturated vapor, one can apply a pulsed laser to vaporize a crest of material that is tightly constrained, both spatially and temporally. This method can usually only provide small amounts of nanoparticles. Laser ablation can vaporize materials that cannot easily be evaporated (Marine et al., 2000). The ablation is performed at the low vapor pressure of inert or reactive gas. As expansion and cooling occur, condensation occurs inside the ablated vapor, and the condensed particles experience multiple collisions with ambient gas molecules, resulting in the stabilization of the nanoclusters. Prior to that, they take place on the substrate surface. The size of the nanoclusters can be adjusted by the laser parameters: fluence, wavelength, pulse duration, and by the ambient gas conditions: pressure, nature, and flow parameters (Murakami et al., 1996; Marine et al., 1996, 2000).

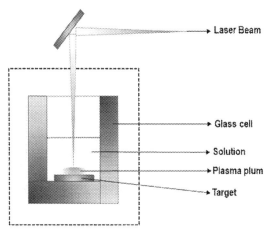

FIG. 21   Nanoparticle preparation by laser ablation.

Laser Beam

Glass cell

Solution

Plasma plum

Target

4.4.3 Spark discharge generation

A different method of evaporating metals is to charge electrodes constructed of the metal to be vaporized in the existence of an inert background gas until the breakdown voltage is achieved. The arc (spark) developed over the electrodes then evaporates a little quantity of metal. This creates very small amounts of nanoparticles, however, it is comparatively reproducible. The basic component of the spark discharge generator chamber is mentioned in Fig. 22 (Meuller et al., 2012). This method is promising due to its low cost and for industrial-scale nanofabrication of materials (Rane et al., 2018). A recent report in 2020 introduces the ablating Zn and Ti metal plates in water through ablation where the laser pulses were lower (2.0 J/pulse) and higher (5.0 J/pulse). The obtained composite nanomaterials exhibited better properties than pure ZnO and TiO_2 (Mintcheva et al., 2020). The attractive features of spark discharge include the possibility of synthesizing nanoparticles of diameter less than a nm from various materials such as metal oxides, carbon, semiconductors, etc. Further, high chemical purity with low impurities and relative simplicity of the process implementation are some of the key characteristics of this method. However, the main limiting factor of this method includes low production rate (Efimov et al., 2016; Rane et al., 2018).

4.4.4 Ion sputtering

Sputtering refers to a process in which atoms are ejected from a target through bombardment with highly energetic particles. It is often related to a momentum transfer process where the bombarding ions drive the atoms from a target. These sputtered atoms are known to travel until they strike a substrate in which they further deposit to form the desired layer (Rane et al., 2018). The common source for ions includes the utilization of the phenomenon of glow discharge by applying an electric field between two electrodes in gas at lower pressure. When a minimum level of voltage is attained, the gas breaks down to conduct electricity, which is now termed as plasma. The ions of plasma are accelerated against the target by a large electrical field. Thus, when the ions hit the target, atoms or molecules are ejected, which are further deposited on the required substrate. The process is often termed DC sputtering. Evaporation of a solid is done by sputtering it with a beam of inert gas ions to avoid any chemical reaction between the sputtered ions and the gas. The formation of nanoparticles of a various dozen metals applying magnetron sputtering of metal targets has been reported (Uppal et al.,

FIG. 22   Spark discharge generator chamber (Meuller et al., 2012).

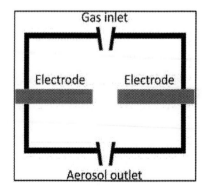

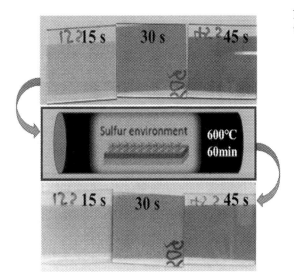

FIG. 23 Postannealing treatment for the production of the nanomaterials (Zhong et al., 2018).

2019; Nam et al., 2020). This process must be carried out at relatively low pressures because the processing of nanoparticles in aerosol form is difficult (Rexer et al., 2000; Weber et al., 2001). A report by Zhong et al. (2018) showed the synthesis of molybdenum disulfide (MoS_2) on SiO_2/Si substrate through radio frequency magnetron sputtering (RFMS) with sulfurization in two steps, thin MoS_2 were deposited by RFMS at 400°C followed by sulfurization at 600°C for 60 min (Fig. 23). In some cases, including deposition of oxides and sulfides, experiments are performed where reactive gases are purposely added to the inert gas such that the deposited film is a chemical compound where the process is termed as reactive sputtering. The efficiency of sputtering is further increased by magnetic confinement using a magnetron source where the effect of the applied magnetic field allows for the spiral of electrons which enhances the ionizing collision and improves the plasma to be operated at higher density. This process is termed as magnetron sputtering which can be used along with DC or reactive sputtering (Rane et al., 2018).

4.4.5 Chemical vapor synthesis

In this method, vapor-phase precursors are transported into a hot-wall reactor under conditions that promote nucleation of particles in the vapor phase rather than deposition of a film on the wall. It is called chemical vapor synthesis or chemical vapor condensation, and this process is similar to the chemical vapor deposition (CVD) processes; chemical vapor deposition (CVD) processes are employed to deposit thin solid films on surfaces (Ostraat et al., 2001; Wang et al., 2016; Sufian, 2020). A report by Ozkan et al. fabricated superhydrophobic, antibacterial, copper-coated polymer films by aerosol-assisted chemical vapor deposition. The prepared nanomaterials had a dual application in cytotoxic activity and limited cell adhesion, as shown in Fig. 24 (Ozkan et al., 2016).

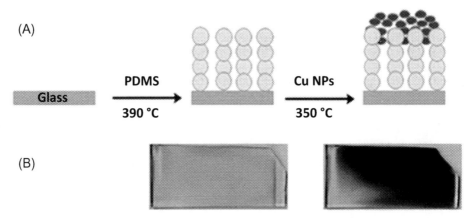

FIG. 24 Schematic representation for the preparation of copper-coated polymer films through chemical vapor deposition (Ozkan et al., 2016).

4.4.6 Spray pyrolysis

In the previous method, the transfer of nanoparticle precursors occurred in a hot reactor as a vapor. However, in this method, a nebulizer or atomizer is used to inject very small droplets of the precursor solution instantly. This is called spray pyrolysis, aerosol breakdown synthesis, droplet-to-particle conversion, etc. The reaction usually proceeds in solution in the form of droplets, accompanied by solvent evaporation (Studenikin et al., 1998; Swihart, 2003). Some other methods, such as laser pyrolysis/photothermal synthesis, thermal plasma synthesis, flame synthesis, flame spray pyrolysis, low-temperature reactive synthesis, have been used to fabricate nanomaterials from liquid or vapor precursors. Chemical precursors are heated and mixed to induce gas-phase reactions that produce a state of supersaturation in the gas phase. The supersaturation is needed to induce homogeneous nucleation of particles in a chemical reaction (Swihart, 2003; Li et al., 2017; Zahid et al., 2018). A recent study by Saha et al. (2020) reported the impact of yttrium oxide (YO_x) on zinc oxide (ZnO) thin-film transistor (TFT) based on Al_2O_3 gate insulator by spray pyrolysis to improve the stability of the prepared nanocomposites (Fig. 25).

5 Other common methods available for nanomaterials production

5.1 Coprecipitation

This method involves the simultaneous process of nucleation, growth, coarsening, and agglomeration (Rane et al., 2018). The products of this type of synthesis are often obtained under supersaturation due to chemical reactions as shown in Fig. 26. The critical step in the process is nucleation; however, during a secondary process such as aggregation, the size, shape, morphology, and properties are affected. Few advantages of utilizing this method include the simple protocol for preparation, controllable particle size, and composition, energy-efficient,

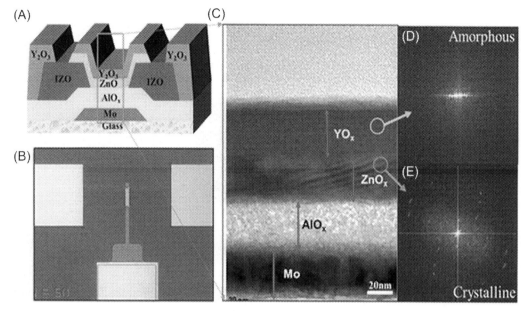

FIG. 25 (A) Schematic representation and (B) optical image representing the bottom contact ZnO TFT with Y_2O_3 passivation through spray pyrolysis. (C) Cross-sectional transmission electron microscope (TEM) image of a ZnO TFT. Fast Fourier transform (FFT) of (D) *yellow circled area* confirms the amorphous nature of Y_2O_3 layer deposited at 400°C and (E) *green circled area* confirms the hexagonal-structured ZnO deposited at 350°C by spray pyrolysis (Saha et al., 2020).

operable at lower temperatures. Some of the limitations include the process not applicable for uncharged species, reproducibility, and impurities problem, and time consumption (Pereira et al., 2018; Ajeesha et al., 2020; Kafi-Ahmadi et al., 2020).

5.2 Hydrothermal method

This technique is one of the most explored methods among scientists and technologists from various disciplines. The hydrothermal method was first introduced by Roderick Muchison, a British geologist (1792–1871), where the action of water on the elevated temperature and pressure on the earth's crust was investigated (Rane et al., 2018). The process refers to the synthesis of particles by undergoing chemical reaction above ambient temperature and pressure. For instance, the synthesis of titanium nanotubes (TNT) from TiO_2 (anatase) by the hydrothermal method was recently reported to produce tubular TNT (Kugarajah and Dharmalingam, 2020). Hydrothermal synthesis has been utilized in the preparation of nanoparticles of controlled size and shape. The synthesis particularly depends on the solubility of the material at high temperature and pressure. The salient features of the hydrothermal method include: significant improvement in the chemical reactivity of the reactant, products of intermediate metastable and specific phase can be easily produced, control over shape and size, monitoring of the phase of the material through reaction temperature, solvent, precursor type, etc. (Meng et al., 2016; Rane et al., 2018; Ghosh et al., 2020).

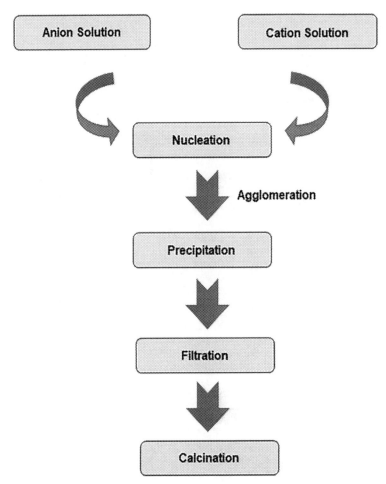

FIG. 26   Schematic representation for nanoparticle preparation through coprecipitation.

5.3 Microwave-assisted method

This type of synthesis is highly utilized in areas of the biochemical process to nanotechnology. This method offers control over reaction mixing, operation at various pressures and temperatures, and has good reproducibility. Separation of nucleation and growth stages during the preparation of nanomaterials at room temperature can be achieved through this technique. The method is important for scalability as it provides selectivity toward the activation of the nanomaterial precursor materials. The potential of this method is achieved by heat selectivity, particularly by solvent or precursors used in nanomaterial preparation (Chikan and McLaurin, 2016). Further, microwave irradiation is a method widely applied for the synthesis of hybrid (organic-inorganic), organic, and inorganic materials owing to its wide advantages over other processes (Cele, 2020). Enhanced reaction rate, improved purity, higher yield, ease of progress and eco-friendly nature, less reaction time, application in stoichiometric

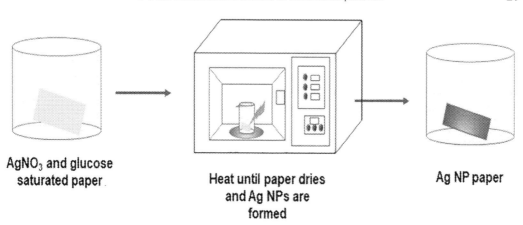

**AgNO₃ and glucose
saturated paper**

**Heat until paper dries
and Ag NPs are
formed**

Ag NP paper

FIG. 27   Schematic representation for silver nanoparticle preparation through microwave-assisted method. *Modified from Dankovich, T.A., 2014. Microwave-assisted incorporation of silver nanoparticles in paper for point-of-use water purification. Environ. Sci. Nano 1, 367–378. doi:https://doi.org/10.1039/C4EN00067F.*

composition of materials, particularly in semiconducting material preparations are common advantages for choosing this technique (Gul et al., 2020; Kannan et al., 2020). The major limitation of microwave-assisted method includes the chemical inhomogeneity for which various investigations are currently being explored. A modified schematic representation for the production of silver nanoparticles reported by Dankovich (2014) is represented in Fig. 27.

5.4 Microemulsion method

This method is one of the ideal protocols available for the preparation of inorganic nanoparticles (Rane et al., 2018); however, the mechanism of the formation through the process is still under research as represented in Fig. 28. In a typical process, the microemulsion materials, including reactants, are mixed where the exchange occurs through the collision of water droplets. The process of reaction exchange occurs extremely fast such that the precipitation occurs in nanodroplets, followed by nucleation and coagulation resulting in the end product surrounded by water/surfactants (Moradi et al., 2020; Yousuf et al., 2020). A review by Malik et al. (2012) establishes the synthesis of organic and inorganic nanoparticles through microemulsion.

5.5 Ultrasound method

Ultrasound is another tool applied for the synthesis of nanoparticles based on the ultrasonic cavitation, which occurs when a liquid is irradiated with ultrasonic radiation. The cavitation produced results in both chemical and physical effects such as variation in temperature, pressure, and cooling rate, which results in the unique condition for the chemical reactions. Ultrasound technique is used for the preparation of nanoparticles with controllable morphologies (Taheri-Ledari et al., 2020). The salient characteristics of this method include operation at ambient conditions, simple, relatively lower time required for operation.

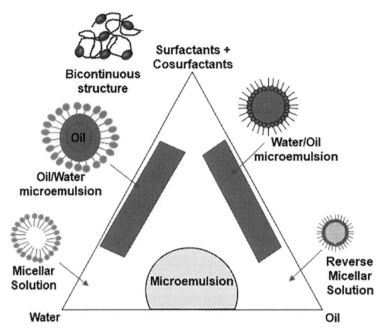

FIG. 28   Schematic representation of the microemulsion method for nanoparticles synthesis.

Further various structures, shapes, and assemblies of nanomaterials can be synthesized (Rane et al., 2018). The major limitation is the scalability, nonapplicability to heat-sensitive materials, and intensive energy requirement.

5.6 Template synthesis

With the green chemistry revolution aspect, an environmentally friendly route for the facile synthesis of nanomaterials is an attractive solution in the current scenario. Keeping that into consideration, template synthesis is considered the most promising method for the synthesis of inorganic nanoparticles with uniform void spaces (Vibulyaseak et al., 2020). The process occurs twofold; initially, the first step involves the reproduction of the structure with reproducibility to organize the skeleton for various applications in different interfaces (Rane et al., 2018). A recent report in 2020 represented the synthesis of ternary Ni/Co/N co-doped carbon film (porous) through template synthesis for its catalytic activity and electron conductivity (Qian et al., 2020). A schematic representation for template synthesis is illustrated in Fig. 29.

5.7 Biological synthesis

This technique is an interdisciplinary approach utilizing the aspects of nanotechnology and biotechnology. Despite the stability and eco-friendly approach, this method's major

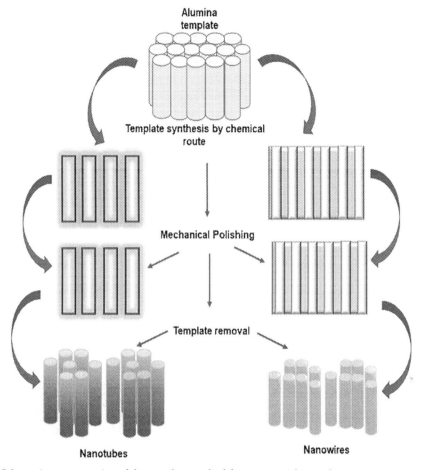

FIG. 29 Schematic representation of the template method for nanoparticles synthesis.

limitation is the slower rate of synthesis (Pantidos and Horsfall, 2014). The concentration of the molecules/components utilized in nucleation varies with respect to time, which also prolongs the nucleation period and results in polydispersity of the particles synthesized. Optimization on the microbial, plant culture utilized, photobiological methods are widely explored recently for nanoparticle synthesis through biological approach (Desai et al., 2020; Naik, 2020).

5.8 Electrochemical synthesis

Electrochemical methods are used to synthesize nanoparticles in an electrochemical cell. The major application of this method over other reactions involves the rejection of alternative wastes and the accuracy of the technique (Cele, 2020). Electrochemical synthesis has been widely explored, particularly for the synthesis of silver nanoparticles by the dissolution of the metallic anode in an aprotic solvent. Silver nanoparticles produced by the process of electroreduction ranged in size between 2 and 7 nm, which was obtained by varying the current density and utilizing different counter electrodes to evaluate the effect of the electrochemical parameters on particle size (Rodriguez-Sanchez et al., 2000). Another approach showed the synthesis of red fluorescent silicon (Si) nanoparticles, stabilized by styrene. The obtained Si nanoparticles emitted fluorescence upon UV excitation, which was applicable in optic applications. It was also observed that the emitted Si nanoparticles which interacted with styrene in ethanol resulted in Si-C substitution to Si-H, which produced red fluorescence at 365 nm (Choi et al., 2014).

6 Nanocomposites

Nanocomposites are multiphase solid materials in which one of the phases either comprises one, two, or three dimensions less than 100 nm nanostructures at repeated distances between various phases of the material. Colloids, porous media, gels, copolymers, etc., come under this category, although it usually means a solid combination of a solid bulk matrix and nano-dimensional phase, which alters the properties due to dissimilarities in structure and chemistry (Rane et al., 2018). A summarized representation portraying the fabrication techniques for ceramic, metal, and polymer nanocomposites is represented in Fig. 30.

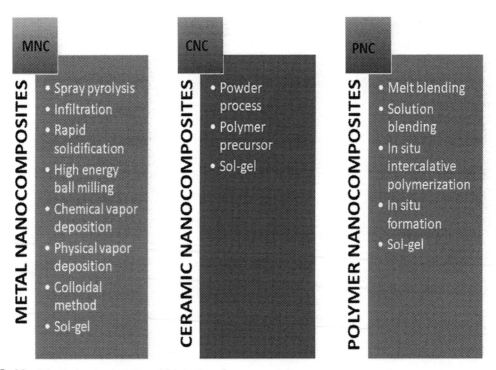

FIG. 30 Schematic representation of fabrication of nanocomposites.

7 Future trends

This chapter merely provides an overview of the various methods of fabrication of nanomaterials. At the industrial level, bigger challenges are faced during their fabrication. Fabrication methods have to be designed to be able to control: (a) particle size, (b) particle shape, (c) size distribution, (d) particle composition, and (e) degree of particle agglomeration. Therefore, interaction among scientists with different backgrounds is crucial to meet practical industrial challenges and discover new technological possibilities. Significant knowledge deficits in the basics of particle formation, challenges in reproducibility, and scalability are some of the limitations that need to be addressed in the coming future to harness nanotechnology completely. Biomimetic synthesis is a relatively new technique for nanomaterial fabrication. Still, in the research stage, it promises to provide unparalleled advantages. The basic idea behind this technology is the combination of synthetic and biological materials, architectures, and systems, respectively, and the imitation of biological processes for technological applications. It can be expected that in the near future, biomimetic synthesis will eventually be a high-level, smart, and green method for nanomaterial fabrication.

References

Abu-Oqail, A., Wagih, A., Fathy, A., et al., 2019. Effect of high energy ball milling on strengthening of Cu-ZrO$_2$ nanocomposites. Ceram. Int. 45, 5866–5875. https://doi.org/10.1016/j.ceramint.2018.12.053.

Ahmad, M.S., Pandey, A.K., Abd Rahim, N., et al., 2018. Chemical sintering of TiO$_2$ based photoanode for efficient dye sensitized solar cells using Zn nanoparticles. Ceram. Int. 44, 18444–18449.

Ahmad, M., Gani, A., Masoodi, F.A., Rizvi, S.H., 2020. Influence of ball milling on the production of starch nanoparticles and its effect on structural, thermal and functional properties. Int. J. Biol. Macromol. 151, 85–91. https://doi.org/10.1016/j.ijbiomac.2020.02.139.

Ajeesha, T.L., Anantharaman, A., Baby, J.N., George, M., 2020. Structural, magnetic, electrical and photo-Fenton properties of copper substituted strontium M-hexagonal ferrite nanomaterials via chemical coprecipitation approach. J. Nanosci. Nanotechnol. 20, 1589–1604.

Akakuru, O.U., Iqbal, M.Z., Liu, C., et al., 2020. Self-assembled, biocompatible and biodegradable TEMPO-conjugated nanoparticles enable folate-targeted tumor magnetic resonance imaging. Appl. Mater. Today 18, 100524.

Albisetti, E., Petti, D., Pancaldi, M., et al., 2016. Nanopatterning reconfigurable magnetic landscapes via thermally assisted scanning probe lithography. Nat. Nanotechnol. 11, 545–551.

Arora, N., Sharma, N.N., 2014. Arc discharge synthesis of carbon nanotubes: comprehensive review. Diam. Relat. Mater. 50, 135–150.

Barcikowski, S., Devesa, F., Moldenhauer, K., 2009. Impact and structure of literature on nanoparticle generation by laser ablation in liquids. J. Nanopart. Res. 11, 1883–1893. https://doi.org/10.1007/s11051-009-9765-0.

Benjamin, J.S., Volin, T.E., 1974. The mechanism of mechanical alloying. Metall. Trans. A 5, 1929–1934. https://doi.org/10.1007/BF02644161.

Bespalov, I., Zhang, Y., Haitjema, J., et al., 2020. Key role of very low energy electrons in tin-based molecular resists for extreme ultraviolet nanolithography. ACS Appl. Mater. Interfaces 12, 9881–9889. https://doi.org/10.1021/acsami.9b19004.

Bigelow, W.C., Pickett, D.L., Zisman, W.A., 1946. Oleophobic monolayers. Films adsorbed from solution in non-polar liquids. J. Colloid Sci. 1, 513–538. https://doi.org/10.1016/0095-8522(46)90059-1.

Boldyrev, V.V., Tkacova, K., 2000. Mechanochemistry of solids: past, present, and prospects. J. Mater. Synth. Process. 8, 121–132. https://doi.org/10.1023/A:1011347706721.

Bose, S., Cunha, J.M.V., Suresh, S., et al., 2018. Optical lithography patterning of SiO$_2$ layers for interface passivation of thin film solar cells. Sol. RRL 2, 1800212.

Cao, S., Zheng, J., Zhao, J., et al., 2017. Enhancing the performance of quantum dot light-emitting diodes using room-temperature-processed Ga-doped ZnO nanoparticles as the electron transport layer. ACS Appl. Mater. Interfaces 9, 15605–15614.

Cele, T., 2020. Preparation of nanoparticles. In: Silver Nanoparticles-Health and Safety. IntechOpen.

Cheng, H., Ma, J., Zhao, Z., Qi, L., 1995. Hydrothermal preparation of uniform nanosize rutile and anatase particles. Chem. Mater. 7, 663–671. https://doi.org/10.1021/cm00052a010.

Cheng, Q., Wei, T., Farbiak, L., et al., 2020. Selective organ targeting (SORT) nanoparticles for tissue-specific mRNA delivery and CRISPR–Cas gene editing. Nat. Nanotechnol. 15, 313–320.

Chikan, V., McLaurin, E.J., 2016. Rapid nanoparticle synthesis by magnetic and microwave heating. Nanomaterials 6, 85.

Cho, J.S., 2019. Large scale process for low crystalline MoO3-carbon composite microspheres prepared by one-step spray pyrolysis for anodes in lithium-ion batteries. Nanomaterials 9, 539.

Choi, J., Kim, K., Han, H., et al., 2014. Electrochemical synthesis of red fluorescent silicon nanoparticles. Bull. Kor. Chem. Soc. 35, 35–38.

Chou, S.Y., Krauss, P.R., Renstrom, P.J., 1996. Nanoimprint lithography. J. Vac. Sci. Technol. B Microelectron. Nanometer Struct. 14, 4129. https://doi.org/10.1116/1.588605.

Crucho, C.I.C., Barros, M.T., 2017. Polymeric nanoparticles: a study on the preparation variables and characterization methods. Mater. Sci. Eng. C 80, 771–784.

Dankovich, T.A., 2014. Microwave-assisted incorporation of silver nanoparticles in paper for point-of-use water purification. Environ. Sci. Nano 1, 367–378. https://doi.org/10.1039/C4EN00067F.

Dastan, D., 2017. Effect of preparation methods on the properties of titania nanoparticles: solvothermal versus sol–gel. Appl. Phys. A Mater. Sci. Process. 123, 699.

Davis, R.M., McDermott, B., Koch, C.C., 1988. Mechanical alloying of brittle materials. Metall. Trans. A. 19, 2867–2874. https://doi.org/10.1007/BF02647712.

Desai, M.P., Patil, R.V., Harke, S.S., Pawar, K.D., 2020. Bacterium mediated facile and green method for optimized biosynthesis of gold nanoparticles for simple and visual detection of two metal ions. J. Clust. Sci. 32, 1–10.

Dong, S., Zhang, X., Li, X., et al., 2019. SiC whiskers-reduced graphene oxide composites decorated with MnO nanoparticles for tunable microwave absorption. Chem. Eng. J. 392, 123817.

Dutková, E., Bujňáková, Z., Kováč, J., et al., 2018. Mechanochemical synthesis, structural, magnetic, optical and electrooptical properties of CuFeS$_2$ nanoparticles. Adv. Powder Technol. 29, 1820–1826. https://doi.org/10.1016/j.apt.2018.04.018.

Efimov, A.A., Lizunova, A.A., Volkov, I.A., et al., 2016. A new approach to the high-yield synthesis of nanoparticles by spark discharge. J. Phys. Conf. Ser. 741, 12035.

Elumalai, V., Sangeetha, D., 2018. Anion exchange composite membrane based on octa quaternary ammonium Polyhedral Oligomeric Silsesquioxane for alkaline fuel cells. J. Power Sources 375, 412–420.

Enescu, D., Cerqueira, M.A., Fucinos, P., Pastrana, L.M., 2019. Recent advances and challenges on applications of nanotechnology in food packaging. A literature review. Food Chem. Toxicol. 134, 110814.

Farzin, L., Shamsipur, M., Samandari, L., Sheibani, S., 2020. HIV biosensors for early diagnosis of infection: the intertwine of nanotechnology with sensing strategies. Talanta 206, 120201.

Feynman, R.P., 1960. There's plenty of room at the bottom. Calif. Inst. Technol. Eng. Sci. Mag. 183, 3.

Galaburda, M., Kovalska, E., Hogan, B.T., et al., 2019. Mechanochemical synthesis of carbon-stabilized Cu/C, Co/C and Ni/C nanocomposites with prolonged resistance to oxidation. Sci. Rep. 9, 17435. https://doi.org/10.1038/s41598-019-54007-2.

Gao, Q., Zhao, A., Gan, Z., et al., 2012. Facile fabrication and growth mechanism of 3D flower-like Fe$_3$O$_4$ nanostructures and their application as SERS substrates. CrystEngComm 14, 4834–4842. https://doi.org/10.1039/C2CE25198A.

Ghosh, R., Kundu, S., Majumder, R., Chowdhury, M.P., 2020. Hydrothermal synthesis and characterization of multifunctional ZnO nanomaterials. Mater. Today Proc. 26, 77–81.

Gopal, M., Moberly Chan, W.J., De Jonghe, L.C., 1997. Room temperature synthesis of crystalline metal oxides. J. Mater. Sci. 32, 6001–6008. https://doi.org/10.1023/a:1018671212890.

Gregorczyk, K., Knez, M., 2016. Hybrid nanomaterials through molecular and atomic layer deposition: top down, bottom up, and in-between approaches to new materials. Prog. Mater. Sci. 75, 1–37.

Gul, U., Kanwal, S., Tabassum, S., et al., 2020. Microwave-assisted synthesis of carbon dots as reductant and stabilizer for silver nanoparticles with enhanced-peroxidase like activity for colorimetric determination of hydrogen peroxide and glucose. Microchim. Acta 187, 1–8.

Habiba, K., Makarov, V.I., Weiner, B.R., Morell, G., 2014. Fabrication of nanomaterials by pulsed laser synthesis. Manufacturing Nanostructures 10, 263–292.

Hai Nguyen, Q., So, S., Hanh Nguyen, Q., et al., 2020. Mechanochemical synthesis of InP nanoparticles embedded in hybrid conductive matrix for high-performance lithium-ion batteries. Chem. Eng. J. 399. https://doi.org/10.1016/j.cej.2020.125826, 125826.

Han, X., Williamson, F., Bhaduri, G.A., et al., 2015. Synthesis and characterisation of ambient pressure dried composites of silica aerogel matrix and embedded nickel nanoparticles. J. Supercrit. Fluids 106, 140–144. https://doi.org/10.1016/j.supflu.2015.06.017.

Horikoshi, S., Serpone, N. (Eds.), 2013. Microwaves in Nanoparticle Synthesis: Fundamentals and Applications. Wiley.

Iijima, S., 1991. Helical microtubules of graphitic carbon. Nature 354, 56–58. https://doi.org/10.1038/350055a0.

Iordache, P.Z., Petrea, N., Lungu, R.M., Petre, R., Său, C., Safta, I., 2011. Nanocomposite materials with oriented functionalized structure. In: Nanomaterials. IntechOpen, pp. 69–98.

Jamkhande, P.G., Ghule, N.W., Bamer, A.H., Kalaskar, M.G., 2019. Metal nanoparticles synthesis: an overview on methods of preparation, advantages and disadvantages, and applications. J. Drug Deliv. Sci. Technol. 53, 101174.

Jiang, Y., Luo, B., Cheng, X., 2019. Enhanced thermal stability of thermoplastic polymer nanostructures for nanoimprint lithography. Materials (Basel) 12, 545.

Kafi-Ahmadi, L., Khademinia, S., Nansa, M.N., et al., 2020. Coprecipitation synthesis, characterization of $CoFe_2O_4$ nanomaterial and evaluation of its toxicity behavior on human leukemia cancer K562 cell line. J. Chil. Chem. Soc. 65, 4845–4848.

Kannan, K., Radhika, D., Vijayalakshmi, S., et al., 2020. Facile fabrication of CuO nanoparticles via microwave-assisted method: photocatalytic, antimicrobial and anticancer enhancing performance. Int. J. Environ. Anal. Chem., 1–4.

Karaballi, R.A., Monfared, Y.E., Dasog, M., 2020. Overview of synthetic methods to prepare plasmonic transition metal nitride nanoparticles. Chem. Eur. J. 26 (39), 8499–8505.

Katz, E., 2019. Synthesis, properties and applications of magnetic nanoparticles and nanowires—a brief introduction. Magnetochemistry 5, 61.

Kaufmann, T., Ravoo, B.J., 2010. Stamps, inks and substrates: polymers in microcontact printing. Polym. Chem. 1, 371. https://doi.org/10.1039/b9py00281b.

Keidar, M., 2007. Factors affecting synthesis of single wall carbon nanotubes in arc discharge. J. Phys. D. Appl. Phys. 40, 2388. https://doi.org/10.1088/0022-3727/40/8/S18.

Khumpuang, S., Hara, S., 2015. A MOSFET fabrication using a maskless lithography system in clean-localized environment of minimal fab. IEEE Trans. Semicond. Manuf. 28, 393–398. https://doi.org/10.1109/TSM.2015.2429572.

Kim, H.H., Kim, H.J., 2006. The preparation of carbon nanotubes by dc arc discharge using a carbon cathode coated with catalyst. Mater. Sci. Eng. B 130, 73–80. https://doi.org/10.1016/j.mseb.2006.02.072.

Kim, S.-J., Park, S.-D., Jeong, Y.H., Park, S., 1999. Homogeneous precipitation of TiO_2 ultrafine powders from aqueous $TiOCl_2$ solution. J. Am. Ceram. Soc. 82, 927–932. https://doi.org/10.1111/j.1151-2916.1999.tb01855.x.

Kim, Y.A., Muramatsu, H., Hayashi, T., Endo, M., 2012. Catalytic metal-free formation of multi-walled carbon nanotubes in atmospheric arc discharge. Carbon 50, 4588–4595. https://doi.org/10.1016/j.carbon.2012.05.044.

Kugarajah, V., Dharmalingam, S., 2020. Investigation of a cation exchange membrane comprising sulphonated poly ether ether ketone and sulphonated titanium nanotubes in microbial fuel cell and preliminary insights on microbial adhesion. Chem. Eng. J. 398, 125558.

Kumar, S., Nehra, M., Kedia, D., et al., 2020. Nanotechnology-based biomaterials for orthopaedic applications: recent advances and future prospects. Mater. Sci. Eng. C 106, 110154.

LaMer, V., Dinegar, R., 1950. Theory, production and mechanism of formation of monodispersed hydrosols. J. Am. Chem. 72, 4847–4854. https://doi.org/10.1021/ja01167a001.

Lee, P.Y., Koch, C.C., 1988. Formation of amorphous Ni-Zr alloy powder by mechanical alloying of intermetallic powder mixtures and mixtures of nickel or zirconium with intermetallics. J. Mater. Sci. 23, 2837–2845. https://doi.org/10.1007/BF00547458.

Leroux, J.-C., 2007. Injectable nanocarriers for biodetoxification. Nat. Nanotechnol. 2, 679–684. https://doi.org/10.1038/nnano.2007.339.

Li, H., Ci, S., Zhang, M., et al., 2017. Facile spray-pyrolysis synthesis of yolk–shell earth-abundant elemental nickel–iron-based nanohybrid electrocatalysts for full water splitting. ChemSusChem 10, 4756–4763.

Lim, J.-S., 2019. Measuring the economic impact of quantum-dot nanotechnology on display/TV industries. Asian J. Innov. Policy 8, 274–287.

Lin, J., Chen, H., Yao, L., 2010. Surface tailoring of SiO_2 nanoparticles by mechanochemical method based on simple milling. Appl. Surf. Sci. 256, 5978–5984. https://doi.org/10.1016/j.apsusc.2010.03.105.

Lin, S., Wang, Z., Zhang, Y., et al., 2017. Easy synthesis of silver nanoparticles-orange emissive carbon dots hybrids exhibiting enhanced fluorescence for white light emitting diodes. J. Alloys Compd. 700, 75–82.

Liong, M., Lu, J., Kovochich, M., et al., 2008. Multifunctional inorganic nanoparticles for imaging, targeting, and drug delivery. ACS Nano 2, 889–896. https://doi.org/10.1021/nn800072t.

Liu, P., Zhao, W., Lin, X.Y., Zhou, D.L., Zhang, C.H., Jiang, K.L., Fan, S.S., 2019. inventors; Tsinghua University, Hon Hai Precision Industry Co Ltd, assignee. Photolithography method based on electronic beam. United States Patent US 10,216,088.

Lutz, T., 2019. New optical super-resolution imaging approaches involving DNA nanotechnology.

Malik, M.A., Wani, M.Y., Hashim, M.A., 2012. Microemulsion method: a novel route to synthesize organic and inorganic nanomaterials: 1st nano update. Arab. J. Chem. 5, 397–417. https://doi.org/10.1016/j.arabjc.2010.09.027.

Mallick, S., Song, S.J., Bae, Y., Choi, J.S., 2019. Self-assembled nanoparticles composed of glycol chitosan-dequalinium for mitochondria-targeted drug delivery. Int. J. Biol. Macromol. 132, 451–460.

Marine, W., Movtchan, I., Simakine, A., et al., 1996. Pulsed laser deposition of Si nanocluster films. In: Materials Research Society Symposium-Proceedings.

Marine, W., Patrone, L., Luk'yanchuk, B., Sentis, M., 2000. Strategy of nanocluster and nanostructure synthesis by conventional pulsed laser ablation. Appl. Surf. Sci. 154–155, 345–352. https://doi.org/10.1016/S0169-4332(99)00450-X.

Mathew, J., Joy, J., George, S.C., 2019. Potential applications of nanotechnology in transportation: a review. J. King Saud Univ. 31, 586–594.

Mehranpour, H., Askari, M., Ghamsari, S.M., Farzalibeik, H., 2010. Application of Sugimoto model on particle size prediction of colloidal TiO_2 nanoparticles. Nanotechnology 1, 436–439.

Menazea, A.A., 2020. Femtosecond laser ablation-assisted synthesis of silver nanoparticles in organic and inorganic liquids medium and their antibacterial efficiency. Radiat. Phys. Chem. 168. https://doi.org/10.1016/j.radphyschem.2019.108616, 108616.

Meng, L.-Y., Wang, B., Ma, M.-G., Lin, K.-L., 2016. The progress of microwave-assisted hydrothermal method in the synthesis of functional nanomaterials. Mater. Today Chem. 1, 63–83.

Meuller, B.O., Messing, M.E., Engberg, D.L.J., et al., 2012. Review of spark discharge generators for production of nanoparticle aerosols. Aerosol Sci. Technol. 46, 1256–1270. https://doi.org/10.1080/02786826.2012.705448.

Mintcheva, N., Yamaguchi, S., Kulinich, S.A., 2020. Hybrid TiO_2-ZnO nanomaterials prepared using laser ablation in liquid. Materials (Basel) 13, 719.

Moballegh, A., Shahverdi, H.R., Aghababazadeh, R., Mirhabibi, A.R., 2007. ZnO nanoparticles obtained by mechanochemical technique and the optical properties. Surf. Sci. 601, 2850–2854. https://doi.org/10.1016/j.susc.2006.12.012.

Moradi, S., Shayesteh, K., Behbudi, G., 2020. Preparation and characterization of biodegradable lignin-sulfonate nanoparticles using the microemulsion method to enhance the acetylation efficiency of lignin-sulfonate. Int. J. Biol. Macromol. 160, 632–641. https://doi.org/10.1016/j.ijbiomac.2020.05.157.

Mosher, L., Waits, C.M., Morgan, B., Ghodssi, R., 2009. Double-exposure grayscale photolithography. J. Microelectromech. Syst. 18, 308–315. https://doi.org/10.1109/JMEMS.2008.2011703.

Mu, H., Yao, M., Wang, R., et al., 2020. White organic light emitting diodes based on localized surface plasmon resonance of Au nanoparticles and neat thermally activated delayed fluorescence and phosphorescence emission layers. J. Lumin. 220, 117022.

Murakami, K., Makimura, T., Kunii, Y., 1996. Light emission from nanometer-sized silicon particles fabricated by the laser ablation method. Jpn. J. Appl. Phys. 35, 4780.

Mustapha, S., Ndamitso, M.M., Abdulkareem, A.S., et al., 2020. Application of TiO_2 and ZnO nanoparticles immobilized on clay in wastewater treatment: a review. Appl. Water Sci. 10, 1–36.

Naik, B.S., 2020. Biosynthesis of silver nanoparticles from endophytic fungi and their role in plant disease management. In: Microbial Endophytes. Elsevier, pp. 307–321.

Nam, J.H., Jang, M.J., Jang, H.Y., et al., 2020. Room-temperature sputtered electrocatalyst WSe2 nanomaterials for hydrogen evolution reaction. J. Energy Chem. 47, 107–111.

Okuyama, K., Lenggoro, W.W., 2003. Preparation of nanoparticles via spray route. Chem. Eng. Sci. 58, 537–547. https://doi.org/10.1016/S0009-2509(02)00578-X.

Oroojalian, F., Charbgoo, F., Hashemi, M., et al., 2020. Recent advances in nanotechnology-based drug delivery systems for the kidney. J. Control. Release 321, 442–462. https://doi.org/10.1016/j.jconrel.2020.02.027.

Ostraat, M.L., De Blauwe, J.W., Green, M.L., et al., 2001. Ultraclean two-stage aerosol reactor for production of oxide-passivated silicon nanoparticles for novel memory devices. J. Electrochem. Soc. 148, G265–G270. https://doi.org/10.1149/1.1360210.

Owens, G.J., Singh, R.K., Foroutan, F., et al., 2016. Sol-gel based materials for biomedical applications. Prog. Mater. Sci. 77, 1–79. https://doi.org/10.1016/j.pmatsci.2015.12.001.

Ozkan, E., Crick, C.C., Taylor, A., et al., 2016. Copper-based water repellent and antibacterial coatings by aerosol assisted chemical vapour deposition. Chem. Sci. 7, 5126–5131.

Palazzo, G., Valenza, G., Dell'Aglio, M., De Giacomo, A., 2017. On the stability of gold nanoparticles synthesized by laser ablation in liquids. J. Colloid Interface Sci. 489, 47–56. https://doi.org/10.1016/j.jcis.2016.09.017.

Palit, S., Hussain, C.M., 2020. Functionalization of nanomaterials for industrial applications: recent and future perspectives. In: Handbook of Functionalized Nanomaterials for Industrial Applications. Elsevier, pp. 3–14.

Pandey, B., Singh, A.K., Singh, S.P., 2019. Nanoparticles mediated gene knockout through miRNA replacement: recent progress and challenges. In: Applications of Targeted Nano Drugs and Delivery Systems. Elsevier, pp. 469–497.

Pandya, S.G., Kordesch, M.E., 2015. Characterization of InSb nanoparticles synthesized using inert gas condensation. Nanoscale Res. Lett. 10, 966. https://doi.org/10.1186/s11671-015-0966-4.

Pantidos, N., Horsfall, L.E., 2014. Biological synthesis of metallic nanoparticles by bacteria, fungi and plants. J. Nanomed. Nanotechnol. 5, 1.

Pereira, C., Costa, R.S., Lopes, L., et al., 2018. Multifunctional mixed valence N-doped CNT@ MFe_2O_4 hybrid nanomaterials: from engineered one-pot coprecipitation to application in energy storage paper supercapacitors. Nanoscale 10, 12820–12840.

Pfeiffer, H.C., 2009. New prospects for electron beams as tools for semiconductor lithography. In: Scanning Microscopy. 7378. International Society for Optics and Photonics, pp. 737802–737812.

Pfeiffer, H.C., 2010. Direct write electron beam lithography: a historical overview. Proc. SPIE 7823, 782316. https://doi.org/10.1117/12.868477.

Polshettiwar, V., Baruwati, B., Varma, R.S., 2009. Self-assembly of metal oxides into three-dimensional nanostructures: synthesis and application in catalysis. ACS Nano 3, 728–736. https://doi.org/10.1021/nn800903p.

Pourzahedi, L., Zhai, P., Isaacs, J.A., Eckelman, M.J., 2017. Life cycle energy benefits of carbon nanotubes for electromagnetic interference (EMI) shielding applications. J. Clean. Prod. 142, 1971–1978.

Prasad Yadav, T., Manohar Yadav, R., Pratap Singh, D., 2012. Mechanical milling: a top down approach for the synthesis of nanomaterials and nanocomposites. Nanosci. Nanotechnol. 2, 22–48. https://doi.org/10.5923/j.nn.20120203.01.

Qian, M., Xu, M., Zhou, S., et al., 2020. Template synthesis of two-dimensional ternary nickel-cobalt-nitrogen co-doped porous carbon film: promoting the conductivity and more active sites for oxygen reduction. J. Colloid Interface Sci. 564, 276–285.

Rajput, N., 2015. Methods of preparation of nanoparticles—a review. Int. J. Adv. Eng. Technol. 7, 1806.

Rane, A.V., Kanny, K., Abitha, V.K., Thomas, S., 2018. Methods for synthesis of nanoparticles and fabrication of nanocomposites. In: Mohan Bhagyaraj, S., Oluwafemi, O.S., Kalarikkal, N., Thomas, S. (Eds.), Micro and Nano Technologies. Woodhead Publishing, pp. 121–139 (Chapter 5).

Raphaël, S., Lavinia, B., Fadi, A., 2011. Synthesis, characterization and biological applications of water-soluble ZnO quantum dots. Nanomaterials. https://doi.org/10.5772/1371.

Rexer, E.F., Wilbur, D.B., Mills, J.L., et al., 2000. Production of metal oxide thin films by pulsed arc molecular beam deposition. Rev. Sci. Instrum. 71, 2125. https://doi.org/10.1063/1.1150593.

Rodriguez-Sanchez, L., Blanco, M.C., Lopez-Quintela, M.A., 2000. Electrochemical synthesis of silver nanoparticles. J. Phys. Chem. B 104, 9683–9688.

Röthlisberger, P., Gasse, C., Hollenstein, M., 2017. Nucleic acid aptamers: emerging applications in medical imaging, nanotechnology, neurosciences, and drug delivery. Int. J. Mol. Sci. 18, 2430.

Sadrolhosseini, A.R., Mahdi, M.A., Alizadeh, F., Rashid, S.A., 2019. Laser Ablation Technique for Synthesis of Metal Nanoparticle in Liquid. IntechOpen.

Saha, J.K., Bukke, R.N., Mude, N.N., Jang, J., 2020. Remarkable stability improvement of ZnO TFT with Al_2O_3 gate insulator by yttrium passivation with spray pyrolysis. Nanomaterials 10, 976.

Saxena, S.K., Nyodu, R., Kumar, S., Maurya, V.K., 2020. Current advances in nanotechnology and medicine. In: NanoBioMedicine. Springer, pp. 3–16.

Schulz, C., Dreier, T., Fikri, M., Wiggers, H., 2019. Gas-phase synthesis of functional nanomaterials: challenges to kinetics, diagnostics, and process development. Proc. Combust. Inst. 37, 83–108.

Shang, Y., Hasan, M., Ahammed, G.J., et al., 2019. Applications of nanotechnology in plant growth and crop protection: a review. Molecules 24, 2558.

Soler-Illia, G.J.D.A.A., Sanchez, C., Lebeau, B., Patarin, J., 2002. Chemical strategies to design textured materials: from microporous and mesoporous oxides to nanonetworks and hierarchical structures. Chem. Rev. 102, 4093–4138. https://doi.org/10.1021/cr0200062.

Srivastava, R., 2016. Nanolithography: processing methods for nanofabrication development. Imp. J. Interdiscip. Res. 2454-1362. 2.

Srivastava, S., Kotov, N.A., 2008. Layer-by-layer (LBL) assembly with semiconductor nanoparticles and nanowires. In: Semiconductor Nanocrystal Quantum Dots: Synthesis, Assembly, Spectroscopy and Applications. Springer, pp. 197–216.

Studenikin, S.A., Golego, N., Cocivera, M., 1998. Fabrication of green and orange photoluminescent, undoped ZnO films using spray pyrolysis. J. Appl. Phys. 83, 2104. https://doi.org/10.1063/1.368295.

Su, S.S., Chang, I., 2018. Review of production routes of nanomaterials. In: Commercialization of Nanotechnologies—A Case Study Approach. Springer, pp. 15–29.

Sufian, S., 2020. Conversion of CO_2-CH_4 mixture into carbon nanomaterials via chemical vapour deposition. Malays. J. Microsc. 16, 1–16.

Sugimoto, T., 2007. Underlying mechanisms in size control of uniform nanoparticles. J. Colloid Interface Sci. 309, 106–118. https://doi.org/10.1016/j.jcis.2007.01.036.

Sun, J., Zhang, J.X., Fu, Y.Y., Hu, G.X., 2002. Microstructural evolution of an $Al_{67}Mn_8Ti_{24}Nb_1$ alloy during mechanical milling and subsequent annealing process. Mater. Sci. Eng. A 329–331, 703–707.

Suwanboon, S., Amornpitoksuk, P., 2012. Preparation of Mg-doped ZnO nanoparticles by mechanical milling and their optical properties. Procedia Eng. 32, 821–826.

Swihart, M.T., 2003. Vapor-phase synthesis of nanoparticles. Curr. Opin. Colloid Interface Sci. 8, 127–133.

Taheri-Ledari, R., Rahimi, J., Maleki, A., 2020. Method screening for conjugation of the small molecules onto the vinyl-coated Fe_3O_4/silica nanoparticles: highlighting the efficiency of ultrasonication. Mater. Res. Express 7, 15067.

Tharchanaa, S.B., Priyanka, K., Preethi, K., Shanmugavelayutham, G., 2020. Facile synthesis of Cu and CuO nanoparticles from copper scrap using plasma arc discharge method and evaluation of antibacterial activity. Mater. Technol. 36, 1–8.

Thomas, S., Thomas, R., Zachariah, A.K., Kumar, R., 2017. Microscopy Methods in Nanomaterials Characterization. Elsevier.

Tibbals, H.F., 2017. Medical Nanotechnology and Nanomedicine. CRC Press.

Torretti, F., Sheil, J., Schupp, R., et al., 2020. Prominent radiative contributions from multiply-excited states in laser-produced tin plasma for nanolithography. Nat. Commun. 11, 2334. https://doi.org/10.1038/s41467-020-15678-y.

Ullah, M., Ali, M., Hamid, S.B.A., 2014. Surfactant-assisted ball milling: a novel route to novel materials with controlled nanostructure—a review. Rev. Adv. Mater. Sci. 37, 1–14.

Uppal, H., Chawla, S., Joshi, A.G., et al., 2019. Facile chemical synthesis and novel application of zinc oxysulfide nanomaterial for instant and superior adsorption of arsenic from water. J. Clean. Prod. 208, 458–469.

Vibulyaseak, K., Kudo, A., Ogawa, M., 2020. Template synthesis of well-defined rutile nanoparticles by solid-state reaction at room temperature. Inorg. Chem. 59, 7934–7938.

Vishwakarma, S.R., Kumar, A., Tripathi, R.S.N., Das, S., 2013. Fabrication and characterization of n-InSb thin film of different thicknesses. Indian J. Pure Appl. Phys. 51, 260–266.

Wang, C.-C., Ying, J.Y., 1999. Sol–gel synthesis and hydrothermal processing of anatase and rutile titania nanocrystals. Chem. Mater. 11, 3113–3120. https://doi.org/10.1021/cm990180f.

Wang, B.B., Zhu, K., Feng, J., et al., 2016. Low-pressure thermal chemical vapour deposition of molybdenum oxide nanorods. J. Alloys Compd. 661, 66–71.

Wang, L., Xiong, Q., Xiao, F., Duan, H., 2017a. 2D nanomaterials based electrochemical biosensors for cancer diagnosis. Biosens. Bioelectron. 89, 136–151.

Wang, Y., Fedin, I., Zhang, H., Talapin, D.V., 2017b. Direct optical lithography of functional inorganic nanomaterials. Science 357, 385–388.

Wang, K., Chen, Y., Zheng, J., et al., 2019. Black phosphorus quantum dot based all-optical signal processing: ultrafast optical switching and wavelength converting. Nanotechnology 30, 415202.

Wang, J., Wu, S., Fu, S., et al., 2020. Ultrahigh hardness with exceptional thermal stability of a nanocrystalline CoCrFeNiMn high-entropy alloy prepared by inert gas condensation. Scr. Mater. 187, 335–339.

Watson, S., Beydoun, D., Scott, J., Amal, R., 2004. Preparation of nanosized crystalline TiO_2 particles at low temperature for photocatalysis. J. Nanopart. Res. 6, 193–207. https://doi.org/10.1023/B:Nano.0000034623.33083.71.

Weber, A.P., Seipenbusch, M., Kasper, G., 2001. Application of aerosol techniques to study the catalytic formation of methane on gasborne nickel nanoparticles. J. Phys. Chem. A 105, 8958–8963. https://doi.org/10.1021/jp0115594.

Wei, Q., Xiong, F., Tan, S., et al., 2017. Porous one-dimensional nanomaterials: design, fabrication and applications in electrochemical energy storage. Adv. Mater. 29, 1602300.

Werner, M., Wondrak, W., Johnston, C., 2018. Nanotechnology and transport: applications in the automotive industry. In: Nanoscience and Nanotechnology: Advances and Development in Nano-Sized Materials. De Gruyter, pp. 260–282 (Chapter 15).

Weyell, P., Kurland, H.-D., Hülser, T., et al., 2020. Risk and life cycle assessment of nanoparticles for medical applications prepared using safe- and benign-by-design gas-phase syntheses. Green Chem. 22, 814–827.

Wolf, E.L., 2008. Nanophysics and Nanotechnology: An Introduction to Modern Concepts in Nanoscience. Wiley.

Yang, Y., Chawla, A., Zhang, J., et al., 2019. Applications of nanotechnology for regenerative medicine; healing tissues at the nanoscale. In: Principles of Regenerative Medicine. Elsevier, pp. 485–504.

Yousuf, M.A., Jabeen, S., Shahi, M.N., et al., 2020. Magnetic and electrical properties of yttrium substituted manganese ferrite nanoparticles prepared via micro-emulsion route. Results Phys. 16, 102973.

Zahid, M.U., Pervaiz, E., Hussain, A., et al., 2018. Synthesis of carbon nanomaterials from different pyrolysis techniques: a review. Mater. Res. Express 5, 52002.

Zhang, D., Ye, K., Yao, Y., et al., 2019. Controllable synthesis of carbon nanomaterials by direct current arc discharge from the inner wall of the chamber. Carbon 142, 278–284. https://doi.org/10.1016/j.carbon.2018.10.062.

Zhong, Z.J., 2009. Optical Properties and Spectroscopy of Nanomaterials. 359 World Sci Publ Co Pte Ltd, pp. 1–383, https://doi.org/10.1017/CBO9781107415324.004.

Zhong, W., Deng, S., Wang, K., et al., 2018. Feasible route for a large area few-layer MoS2 with magnetron sputtering. Nanomaterials 8, 590.

Zhu, Y., Kaskel, S., Shi, J., et al., 2007. Immobilization of trametes versicolor laccase on magnetically separable mesoporous silica spheres. Chem. Mater. 19, 6408–6413. https://doi.org/10.1021/cm071265g.

Zhu, C., Liu, T., Qian, F., et al., 2017. 3D printed functional nanomaterials for electrochemical energy storage. Nano Today 15, 107–120.

Zhu, X., ten Brink, G.H., de Graaf, S., et al., 2020. Gas-phase synthesis of tunable-size germanium nanocrystals by inert gas condensation. Chem. Mater. 32, 1627–1635.

Nanoparticles and nanofluids: Characteristics and behavior aspects

Vaidhegi Kugarajah[a,*], Atul Kumar Ojha[b,*], Hushnaara Hadem[b], Nandita Dasgupta[c], Bhartendu Nath Mishra[c], Shivendu Ranjan[d,e], and Sangeetha Dharmalingam[a]

[a]Department of Mechanical Engineering, Anna University, Chennai, Tamil Nadu, India [b]Centre for Nano Sciences and Technology, Pondicherry University, Chinna Kalapet, Kalapet, Puducherry, India [c]Department of Biotechnology, Institute of Engineering and Technology, Dr. A.P.J. Abdul Kalam Technical University, Lucknow, Uttar Pradesh, India [d]Faculty of Engineering and the Built Environment, University of Johannesburg, Johannesburg, South Africa [e]Animal Cell and Tissue Culture Lab, Gujarat Biotechnology Research Centre, Department of Science and Technology, Government of Gujarat, Gandhinagar, Gujarat, India

OUTLINE

[*]Authors contributed equally.

1 Introduction

"Nanofluids" concern the suspension of nanoparticles in common base fluids. It is pointed out that dispersing nanoparticles, nanofibers, nanotubes, nanowires, nanorods, and nanosheets in fluids (organic and inorganic) can produce the nanofluids. These nanofluids are nanoscale colloidal suspensions, containing nanomaterials, where solid phase and liquid phase both exist in a single nanofluid. The properties of the base fluid such as thermal conductivity, thermal diffusivity, viscosity, abrasion-related properties, convective heat transfer coefficient, optical properties, solar energy absorption capability, and flow behavior are enhanced by the larger surface area and improved properties of nanoparticles. Nanoparticles not only enhance the properties of base fluid but also solve the suspension stability problem of micro-sized metallic particles in fluids (Choi and Eastman, 1995). Heat exchanger (Qi et al., 2018; Huminic and Huminic, 2012), nuclear power plant (Bang and Kim, 2010; Bafrani et al., 2020), cooling of proton exchange membrane fuel cell (Islam and Shabani, 2019; Bargal et al., 2020), alternative energy sources (Al-Waeli et al., 2018), solar systems (Qu et al., 2019; Khanafer and Vafai, 2018), and "clean" energy are the main areas (Salehin et al., 2018) where nanofluid could be frequently used. To prepare the nanofluids, one-step method and two-step method are used. In one-step method, nanoparticles are prepared in the fluid itself to produce nanofluids whereas in two-step method, the first step includes preparation of nanomaterials and then dispersing the nanomaterials into the base fluid by mechanical mixing and sonication (Devendiran and Amirtham, 2016; Babar et al., 2019). A simple schematic representation for the preparation of one-step and two-step processes is illustrated in Fig. 1.

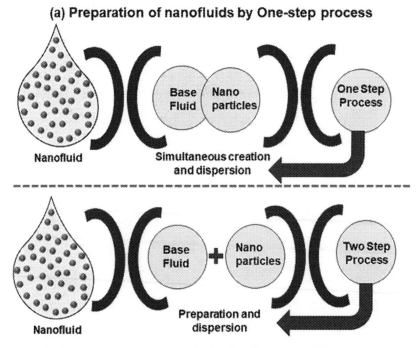

FIG. 1 Schematic representation for the preparation of nanofluids by different modes (A) one-step and (B) two-step processes.

In the two-step process, mechanical stirring and sonication are useful to reduce nanomaterials agglomeration during mixing with base fluid. Without the knowledge of the properties of nanofluid such as viscosity, mechanical, thermal, and electrical properties, investigation on the real applications of nanofluid is impossible (Manikandan et al., 2014). The type of nanoparticle, their volume fraction, temperatures, and morphology of nanoparticle have a significant impact on the properties of nanofluids (Yu and Xie, 2012). Further, Reynolds number, Nusselt number, pressure coefficient, surface vorticity, and recirculation length have an impact on the nanofluids (Bovand et al., 2015). The viscosity of nanofluids is suppressed by increasing the temperature and vice versa. Viscosity also depends on the content and size of nanoparticles, which increases on increasing the content of nanoparticles and on decreasing the size of nanoparticles (Tseng and Lin, 2003; Nguyen et al., 2008; Murshed et al., 2008). Table 1 shows the viscosity of nanofluid for various systems.

Nowadays, a lot of research work are being initiated to design thermal systems based on the nanofluids, analysis of thermal conductivity, density, heat capacity, and viscosity of nanofluids (Khullar et al., 2018; Abbas et al., 2019). Among all of the above-mentioned

TABLE 1 A literature review of the available studies on the viscosity of nanofluids containing ethylene glycol (EG).

Solid concentration	Temperature range (°C)	Base fluid	Particles	References
0–10 vol%	(−35)–50	EG:water	SiO_2	Namburu et al. (2007b)
0–6.12 vol%	(−35)–50	EG:water	CuO	Namburu et al. (2007b)
1–10 vol%	(−35)–50	EG:water	Al_2O_3	Sahoo et al. (2009)
0–1 vol%	0–50	EG:water	Fe_3O_4	Sundar et al. (2012)
1–8 vol%	15–40	EG:water	Al_2O_3 and TiO_2	Yiamsawas et al. (2013)
0–2 vol%	27.5–50	EG:water	FMWCNTs-SiO_2	Eshgarf and Afrand (2016)
0–6 vol%	20–50	EG	Al_2O_3	Anoop et al. (2009)
0.2–5 vol%	20–60	EG	ZnO	Yu et al. (2009)
0.25–5 vol%	25–50	EG	ZnO	Hemmat Esfe and Saedodin (2014)
1.75–10.5 wt%	15–55	EG	ZnO	Li et al. (2015)
0.1–1 vol%	30–60	EG	MgO-MWCNTs	Soltani and Akbari (2016)
1–7 vol%	10–50	EG	TiO_2	Khedkar et al. (2016)
0–0.1 vol%	30–60	EG	SWCNT	Baratpour et al. (2016)
0–1.2 vol%	25–50	EG	Fe_3O_4-Ag	Afrand et al. (2016)

Adopted with permission from Akbari, M., Afrand, M., Arshi, A., Karimipour, A., 2017. An experimental study on rheological behavior of ethylene glycol based nanofluid: proposing a new correlation as a function of silica concentration and temperature. J. Mol. Liq. 233, 352–57.

parameters, viscosity is considered as the most important property of nanofluids. Various heat transfer-related processes in fluids like convection heat transfer, pressure loss for turbulent and laminar flows are dependent on the viscosity of nanofluids. It was found that the viscosity and thermal conductivity of nanofluids were improved on increasing the Fe_3O_4 content for Fe_3O_4/ethylene glycol nanofluid (Sonawane and Juwar, 2016). Here, it signifies that addition of suspended nanoparticles in nanofluids attributed to the rise in heat transfer potential. In this manner, TiO_2 nanoparticles have been used in Ethylene glycol-water base fluid to increase the thermal conductivity of base Ethylene glycol-water fluid. The result showed that the thermal conductivity was a function of temperature and content of TiO_2 nanoparticles and it enhanced the thermal conductivity up to 7% (Reddy and Rao, 2013). Another study on EG/ZnO nanoparticle-based nanofluids exhibited an improved thermal conductivity up to 12% at 25°C (Li et al., 2015). The enhancement of thermal conductivity of double-walled carbon nanotube-based ethylene glycol (EG) nanofluids can be achieved up to 25% (Shamaeil et al., 2016). Hybrid filler-based nanofluids have also been found to enhance the thermal conductivity. A recent review by Jamei et al. provided insights comparing 275 works of literature for polyethylene glycol-based hybrid nanofluids for accurate thermal conductivity prediction based on artificial intelligence. From the results obtained, it was observed that the genetic programming for the dataset ($R = 0.950$, RSME $= 0.0225$) had the best performance for thermal conductivity of polyethylene glycol-based hybrid nanofluids compared to model test and model linear regression results (Jamei et al., 2020).

The kind of nanofiller used and the properties of the fluid together determine the improved properties of the nanofluid. Engine oil, ethylene glycol, acetone, water, and kerosene find many applications in microelectronics, air-conditioning, transportation, and chemical production industries as a heat transfer fluid. Low thermal conductivities and heat transfer coefficients have restricted their heat transfer ability (Yu and Xie, 2012). Titanium oxide (TiO_2) (Wei et al., 2017), alumina oxide (Al_2O_3) (Choi et al., 2019), copper oxide (CuO) (Mechiri et al., 2017), iron oxide (Fe_3O_4) and graphene oxide (GO) in water (Akbari et al., 2019), kerosene, and engine oil have been used to improve the heat capacity at a constant pressure, thermal conductivity, volumetric coefficient of thermal expansion (Hussanan et al., 2017). Nanofluids can also give better performance for solar collectors; various types of nanofluids have shown promising results for this application (Koca et al., 2017). Vitamins, minerals, and polyphenols are main nutrients in natural foods such as honey, fruits, vegetables, flowers, and tea. Prior to consumption, the removal of pesticides and the complex matrices from food is very essential. Solid-phase extraction (SPE) and liquid-liquid extraction (LLE) are generally used for removal of pesticides; however, these methods have a lot of drawbacks. ZnO and TiO_2 based nanofluids have been found to possess the ability to speed up the mass-transfer capacity in the extraction processes. These nanofluid systems are beneficial in pesticide residue analysis (Wu et al., 2017). This chapter provides a comprehensive review of the characteristics and behavior of nanoparticles/nanofluids. Table 2 shows the different types of nanofluids, and Fig. 2 shows the application of nanofluids in various fields.

TABLE 2 Possibility of different types of nanofluids preparation and their constituents (Sadeghinezhad et al., 2016).

Base fluids	Water, ethylene glycol, engine oil, decene
Metallic nanomaterials	Cu, Ag, Au, Fe, Si
Nano metal oxides	Al_2O_3, CuO, TiO_2, SiO_2, SiC, CeO_2
Carbon nanomaterials	Fullerene, graphene, carbon nanotubes
Colloidal form	Nanodroplet
Surfactants	Sodium dodecyl sulfate, gum arabic, anionic surfactant sodium dodecylbenzenesulfonate

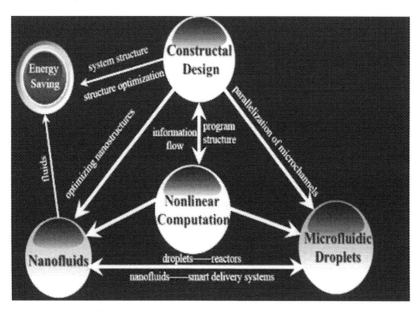

FIG. 2 Illustration showing different interconnected applications of nanofluids based on structure, delivery system, and computation.

2 Nanoparticle aggregation and dispersion behavior

Nanomaterials are a novel class of materials; their inherent properties make them well suited for various applications (Siripireddy et al., 2017; Sireesh et al., 2015, 2017; Ranjan and Ramalingam, 2016; Tammina et al., 2017; Janardan et al., 2016; Ranjan et al., 2016; Dasgupta et al., 2016a,b; Nandita et al., 2016; Walia et al., 2017; Cheng et al., 2020; Zhai and Zhou, 2019; Jin et al., 2018). Currently, many research articles are being published based on nanomaterials and related topics. These articles include the dispersion, distribution, and application of nanomaterials in various types of medium. For example, nanoparticle-based systems offer their utility in environmental clean-up technology (Khalaj et al., 2018), electromagnetic shielding technology (Kumar, 2019; Zhu et al., 2019), improving the stability of the product, enhancing the mechanical properties of the systems such as microbial fuel cells (Kugarajah and Dharmalingam, 2020b; Sugumar and Dharmalingam, 2020; Kugarajah et al., 2020), biomedical applications (Rathinavel et al., 2020; Ekambaram and Dharmalingam, 2020), water cleaning technologies (Villaseñor and Ríos, 2018; Teow and Mohammad, 2019), better energy storage systems (Cao et al., 2016), etc. However, dispersion of nanoparticles is a primary concern to obtain the desired performance of the product. The aggregation of nanoparticles always plays a crucial role in the alteration of the properties of the related system. By the aggregation, nanoparticles convert into micron size clusters (Phenrat et al., 2008). To understand the nanoparticles aggregation and dispersion behavior, colloids, and heterogeneous systems; dispersion (colloids) and dispersant must be considered. Accordingly, a particle with nm size range is well dispersed in the dispersant. However, colloidal science is not valid for all the cases of nanomaterials, particularly when particle surfaces are in contact with each other and particle-particle attachment allows aggregation of colloidal particles (Gambinossi et al., 2014; Shakib-Manesh et al., 2002). The collision between nanoparticles is controlled by long-range attractive or repulsive force between the particles; these forces are controlled by the Brownian diffusion.

Homo-aggregation and hetero-aggregation are the two main aggregation types in nanoparticles when nanoparticles are suspended in any medium (Jang and Lee, 2018). Two similar types of nanomaterials, such as carbon nanotubes (CNTs), associate via homogeneous aggregation in any medium. While two different types of nanomaterials, such as CNTs and gold nanoparticles, tend to undergo heterogeneous aggregation. The properties such as reactive surface area, reactivity, bioavailability, and toxicity have been altered by the aggregation and size of aggregation. For proper analysis, the nature of aggregation must be considered (Hansen et al., 2002, 2007), and attachment of particle and aggregation can be quantified by the net force between particles, i.e., sum of attractive and repulsive forces (Behrens et al., 2000). Van der Waals attractive and electrostatic double layer (EDL) forces can be used to determine the probability of sticking two particles together. The net force (sum of the force) will be attractive if the sum of the force is negative and vice-versa. Table 3 shows various forces and their origin. Fig. 3 shows the Derjagiun, Landau, Verway, and Overbeek (DLVO) curve; this curve gives the information about the net attraction in a primary or secondary minimum and reversible-irreversible aggregation (Hoek and Agarwal, 2006; Yu and Xie, 2012). The aggregation is called primary aggregation if the particles lie in the primary minimum, while the aggregation is irreversible. Whereas, when particles are at a secondary minimum, they can be destroyed by shear forces. The influence of other forces such as steric repulsion, magnetic forces, osmotic repulsion, magnetic attraction, and hydrophobic Lewis acid-base on the aggregation can be explained by the extended DLVO, these short forces also act along with van der Waals attractive and EDL forces. These forces are mostly found in the case of adsorbed or grafted organic macromolecular nanoparticles. There are many challenges to understand nanoparticles aggregation with DLVO and extended DLVO theory because the shape, size, structure, composition, and adsorbed or grafted organic

TABLE 3 Various types of forces and their origin source which affect aggregation behavior of nanomaterials.

Forces	Origin	References
Electrostatic double layer	Charged surfaces	Rojas Delgado et al. (2008)
Magnetic attraction	Alignment of electron spins	Phenrat et al. (2007)
Osmotic repulsion	Concentration of ions between two particles	Romero-Cano et al. (2001)
Elastic-steric repulsion	Molecules on particle surfaces resist loss of entropy due to compaction	Saleh et al. (2008)
Hydrophobic (Lewis acid-base)	Entropic penalty of separating hydrogen bonds in water	Hoek and Agarwal (2006)
Van der Waals attraction	Interaction of electron in particles	Rojas Delgado et al. (2008)

macromolecules are mentioned in these theories. When a particle size is too small, the analysis of aggregation is a challenge by the DLVO model, because electrostatic double expression does not give correct results, it can give accurate results only with proper assumption (Lowry et al., 2010; Wikipedia, n.d.; Delgado et al., 2007; Hoek and Agarwal, 2006; Yu and Xie, 2012).

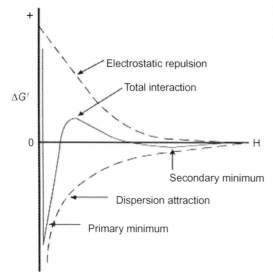

FIG. 3 A typical DLVO curve illustrating attraction-repulsion among particles relevant to nature of aggregation reversible/irreversible (Wikipedia, n.d.).

Vander Waals attractive and EDL forces tend to change with changing particles' size. Size polydispersity, size, structure, and composition affect the aggregation of nanoparticles. The smaller particles have higher aggregation tendency. Understanding the aggregation behavior of nanoparticles is also challenging through the size polydispersity concept. Magnetic materials tend to undergo rapid agglomeration because magnetic interactions can induce higher aggregation tendency. Therefore, metals and magnetic particles have higher tendency to agglomerate than the polymeric materials. Entropy also plays an important role in aggregating the nanomaterials. In the case of carbon nanomaterials present in water medium, due to the hydrophobic nature of carbon nanomaterials and hydrophilic medium, hydrophobic surface induces degree of randomness with water molecules and causes aggregation (Wang et al., 2005). Chemical composition alters the surface potential, which in turn drives the aggregation of particles. Higher aggregation tendency is related to lower surface potentials and vice versa (Evans and Wennerström, 1999). Hamaker constant, saturation magnetization, surface charge, and hydrophobicity are the primary parameters to understand aggregation through chemical composition where van der Waals attraction, magnetic attraction, EDL interaction, and hydrophobic interaction are mainly governed by the above mentioned parameters, respectively (Evans and Wennerström, 1999). The complex crystal structure of nanoparticles also affects the Hamaker constant, which means crystal structure plays a significant role in analyzing surface charge and Hamaker constant. Subsequently, it affects the aggregation properties of nanoparticles. Furthermore, aggregation of nanoparticles also depends on the pH and dissolved ionic solutes in the system. Above studies suggest that the aggregation of nanoparticles depends on the surrounding medium. Other studies involved in investigating the aggregations of nanofluids include

- Sedimentation and Centrifugation analysis (Guo, 2020)
- Zeta potential analysis (Subramanian et al., 2019)
- Spectral absorbency analysis (Yu and Xie, 2012)

3 Physicochemical characteristics of nanoparticles

Nanomaterials are a modern class of materials that consist of tiny particles. These tiny particles have at least one dimension within the size range from 1 to 100 nm. The physical, chemical, and biological properties of these nanomaterials are significantly different from the bulk materials. For example, the size of DNA double helix is around two nanometers in diameter, water molecules are around 0.1 nm, glucose is around 1 nm, the antibody is around 10 nm, and size of virus is around 100 nm. Nanomaterials can be characterized into various types according to their origin, dimensions, and structural configuration. Virus, protein molecules including antibodies are the examples of natural nanomaterials; CNTs, quantum dots are artificial nanomaterials; zero-dimensional (gold and silver nanoparticles and quantum dots) (Liang et al., 2020), one-dimensional (Nanowire and nanotubes of metals, oxides) (Pang et al., 2020), two-dimensional (Nano films, nano sheets, or nano-walls) (Su et al., 2019), and three-dimensional (All dimensions in nano-range) nanomaterials (Tian et al., 2020; Kuang et al., 2019) are the examples of dimension-based nanomaterials. Hollow spheres

(Wu et al., 2019a), ellipsoids (Wu et al., 2019b), or tubes (Kugarajah and Dharmalingam, 2020a) are the configuration-based nanomaterials.

In recent years, many electronics, energy, automobile, aerospace, medical, cosmetic, sports, and food products are produced based on nanomaterials. Nanotubes and nano-sheets such as one-dimensional (1D) CNTs and two-dimensional (2D) graphene and semiconductor nano-crystal such as quantum dots have been used for making electronics and electromagnetic shielding materials, nanosensors, chemical and physical sensors owing to their high surface-to-volume ratio and excellent thermal and mechanical properties. Cathode-ray emission-based devices such as television and computer screens can be produced by the CNTs due to their ability to emit electrons in the presence of an electric field (Xing et al., 2018; Kennedy et al., 2017). Efficient solar cells can be produced by the graphene and metallic nanostructure materials. Splitting water (H_2O) into energy in the form of hydrogen fuel is an important application of nanomaterials and was achieved by titanium dioxide (TiO_2) nanomaterials in the presence of sunlight (McCray, 2006). Nanomaterials have also been used in the automobile industries to produce smooth and thin heat-reflecting coating, self-cleaning glass, and to increase the durability of tyres. They are also used to make materials with high strength, low weight, multifunctional nature, and resistance to extreme environments and radiation. Nanoparticles-based systems such as nanorobots, prosthetic body parts, and targeted drug therapy for cancer patients are also well-established applications of nanomaterials. Nanomaterials are used in the food industry to enhance the flavor, to make the packaging of food materials, to increase the shelf life of food products, etc. (McCray, 2006; Jain et al., 2018; Rai and Bai, 2018). Application of nanomaterials depends on the electrical, catalytic, magnetic, mechanical, thermal, or imaging features of nanomaterials and their surrounding medium. Surface morphology, surface area, size, solubility, chemical composition, surface energy, surface charge, shape, agglomeration state, and crystal structure of nanomaterials manifest the performance and application of nanomaterials. Fig. 4 shows the physicochemical characteristics of nanomaterials, which have been used in various studies. Size, surface properties, crystal structures, and morphologies of nanoparticles are the key factors to decide the application of nanoparticles. The particle size and surface area influence the interaction of nanoparticles with any system; surface area-to-volume ratio tends to increase when the size of nanoparticles decreases as compared to its bulk materials. The shape of particles has the ability to influence the electromagnetic, optical, and catalytic properties of nanomaterials. It has been mentioned that smaller nanoparticles can easily affect the biological system through passive as well as active interaction (Jiang et al., 2008).

The catalytic activity of nanomaterials varies with the size of nanomaterials. Smaller size nanoparticles show higher surface-to-volume ratio, and it offers higher catalytic activity and can easily interact with the substrate (Gao et al., 2007). The nanomaterials also give size-dependent toxicity and antibacterial characteristics. Not only the size, but shapes of nanomaterials also have a significant impact on their nano-bio interfaces, cellular uptake, reactivity, antimicrobial activity, etc. (Cormode et al., 2018). The transport, delivery, and biodistribution of nanomaterials are affected by the composition of a nanomaterial. Temperature, moisture, solvents, pH, particle/molecular size, ionizing and nonionizing radiation, and enzyme can affect the stability of nanomaterials (Briscoe and Hage, 2009). Carbon nanotubes, silica, allotropies, nickel, gold, and titanium nanomaterials can show the shape-related toxicity (Ispas et al., 2009; Chithrani et al., 2006). Relatively less toxicity has been

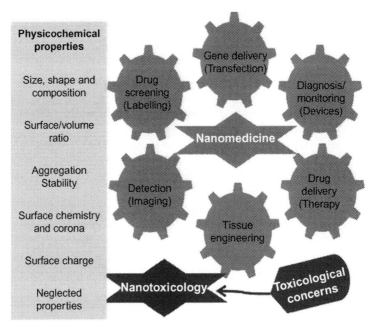

FIG. 4   Schematic representation of various physicochemical properties of nanomaterials which influence their theranostics potential in nanomedicine field (Navya and Daima, 2016).

found in the spherical nanoparticles, while biological consequences have been found to be more in nonspherical nanomaterials, because of more disposal during the flow through capillaries (Kim et al., 2012). It has been reported that the spherical carbon fullerenes are less effective to block K^+ ions than the rod-shaped SWCNT. Food additives are based on amorphous silica however, crystalline silica shows a toxic effect. Cytotoxic properties also depend on the shape of nanomaterials, spherical TiO_2 nanoparticles show less cytotoxicity than TiO_2 fibers, and gold spherical nanospheres have high uptake ability than the gold nanorods. The higher aspect ratio also produces a more toxic effect (Chen et al., 2009; Hsiao and Huang, 2011). Size, surface charge, and composition offer aggregation effect to nanoparticles, and aggregated nanoparticles show highly toxic behavior.

Nanoparticles have different colors than their bulk counterparts because in nanoparticles the surface plasmon resonance phenomena is responsible for absorbing and reflecting the light. Once bulk materials change into a nanoscale dimension, the high electron density-surface plasmons are found at the surface of nanoparticles. Under the influence of light at specific wavelengths, plasmons undergo oscillation called surface plasmon resonance. It results in the absorption and reflection of a specific color of the light spectrum. Melting point depression is another phenomenon of nanomaterials caused by the alteration of thermodynamic and thermal properties. Larger surface-to-volume ratio alters the thermodynamic and thermal properties, which is further responsible for reducing nanomaterials' melting point than bulk materials.

Crystal perfection or reduction of defects have been found in the nanomaterials and there-fore, nanomaterials offer higher mechanical properties than bulk materials. The crystal per-fection in nanomaterials enhances the volume resistivity of nanomaterials, because electron scattering is less pronounced in the case of perfect crystals. However, the surface scattering of electrons is enhanced due to high surface area, which increases the surface resistivity. The wide, discrete bandgap and modified electronic structure result in very small-sized nanoparticles (below the electron de Broglie wavelength), which increases the resistivity of nanomaterials. The chemical properties of nanomaterials are quite different than their bulk counterpart. Nanomaterials show higher energy as compared to the bulk material's atom. The high surface area-to-volume ratio increases the reactivity of nanomaterials (Vollath, 2008). Thermodynamic, electronic, and optical properties of metal nanomaterials are quite different compared to bulk materials, due to space electronic wave function confinement in metallic nanoparticles (Knight, 1992; Perenboom et al., 1981). Electron-electron interaction in materials can be understood by the femtosecond techniques, which provide information about electron-phonon, electron-surface scattering. Initially, femtosecond techniques were applied to bulk materials where, the characteristic lengths, delocalization and screening lengths or mean free paths, of charge decide the physical properties of bulk materials (Del Fatti et al., 2000; Voisin et al., 2000). It is reported that the electron-electron interactions are increased below the 5 nm metal nanoparticle size, and it increases the energy exchange rate of electron-electron, as well as accelerates the electron-electron scattering (Voisin et al., 2000). That means electron-electron interaction depends on the size of nanoparticles. Moreover, the magnetic behavior of atom can be understood from the atom with an odd number of elec-trons. The magnetic behavior of materials is measured by the electronic structure. Decreasing the particle size changes the system's boundary conditions and reduces the symmetry of the system; this in turn changes the electronic properties and hence the magnetic properties of the system get altered.

4 Interactions between nanoparticles

Interactions between nanoparticles or thin films in multilayer structures are critical aspects. The interaction of nanoparticles is essential to understand the risk and applica-tion assessment of nanomaterials. Physical properties of bulk are modified when the size of bulk materials is reduced to nanoscale. The stable dispersion of nanoparticles can be obtained by making passivation or stabilization of nanoparticle surfaces. Electrostatic steric/electro steric (Shem et al., 2014), solvent-based (Kitchens et al., 2003), or ligand-based stabilizers (Hagendorfer, 2011) are the key mechanisms to achieve stabili-zation of NPs nanoparticles as shown in Fig. 5. State of dispersion and the stabilization of nanoparticles result from internanoparticle forces from various origins. The internanostructure forces, the interactions between the nanostructures, and the environ-ment of nanostructure influence the properties of nanoparticles and nanofluids. The phase behavior of nanoparticles-based complex system also depends on the interaction between nanoparticles. Interaction of nanoparticles refers to the various forces at the nanoscale level. Here, the gravitational forces among massive objects always dominate

FIG. 5 Schematic representation of stabilization of nanoparticles via different passivation steps utilizing electric charges, ligand attachment.

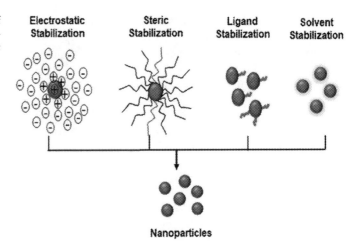

at the macroscale, while the force from electric charge dominates at the nanoscale, weak force and nuclear force dominate at the subatomic level. Also, electric charge forces mostly cause interaction, not only for nanoparticles but also for atoms and molecules. The stable complex material is produced by bringing together the entities, possibly by the attraction through the electric charge forces. The static charge produces the electric charge force and is called electrostatic force. Dipole-dipole interactions (between partial opposite charges), ionic interactions (between ions), interactions by covalent bond, and interactions by coordinate covalent bond are electrostatic interactions. A better understanding of electric forces is necessary for interaction, design, fabrication, and manipulation of nanomaterials. Adsorption of ions from the solvent, desorption of surface ligands, change in deviations in stoichiometry, dissolution of ions from the particle, and adsorption induce the charge on the nanoparticles or nanomaterials (Shevchenko et al., 2006; Yan et al., 2020; Samadi et al., 2018).

Van der Waals force is responsible for the agglomeration of nanoparticles whereas, destabilization of suspension is due to the sedimentation and flocculation mechanisms. Gravitational force on the particles results in sedimentation, while the Brownian motion is helpful for upward diffusion of particles which overcome the down motion of sedimentation. If the Brownian motion is unable to overcome the sedimentation effect, then stirring could help to prevent sedimentation. Flocculation is a result of particle-particle interaction through van der Waals forces of attraction, and then particles attract each other and form primary aggregates termed as floc. The electrostatic repulsion reduces the aggregations between particles. Attractive van der Waals interaction and repulsive electrostatic interaction are explained through the DLVO theory. At nanoscale distances, the attraction between particles is influenced by the magnetic or electric dipole-dipole interactions (Israelachvili, 1992).

Van der Waals interaction for spheres and cylinders induce dipole interaction of two materials interacting through a third medium and can be explained by the Hamaker constant.

Coupled-dipole method models can offer better accuracy for measuring the van der Waals energy of the individual atoms in the particles. This is because of particle separation which is comparable to the particle size for cluster-like particles. Moreover, the result from Hamaker theory provides inaccurate van der Waals interaction energy (Kim et al., 2007). Electrostatic repulsion and steric repulsion also exist between particles. The charge on nanoparticles or the ions in solution produces the electrostatic interaction. In brief, negative ions create the Stern layer on the particle surface and then form a diffusion layer by repulsion of negative ions due to a higher population of negative ions. The ions present in the diffuse layer tend to be mobile, so it is called as the diffusion layer. Finally, the electrical double layer around the particles is formed by the Stern and diffusion layers. This double layer and concentration of ions in double layer have significant impact on the repulsive force between two identical colloids (nanoparticles or ions in solution). There is a significant difference between solutions, colloids, and suspensions. For instance, if the particle size is in the range less than 1 nm, it is called a solution, while if the size of particles is between 1 and 1000 nm and larger than 1000 nm, it is called a colloid and suspension, respectively. Subsequently, the zeta potential is defined as the charge imparted to particles, when dispersed in water. The sign of charged particles is estimated by using the appropriate instrument; the movement of particles is used to find the charge on the particles under the influence of the electric field. The range of zeta potential is found in the magnitude of -70 to $+70$ mV. For a stable colloid, the zeta potential is found to be -30 mV. This zeta potential value is useful for obtaining the stable suspension by the appropriate mutual repulsion. Generally, the agglomeration threshold of the suspension is found higher than -15 mV (e.g., ≥ -14 mV). The zeta potential around 0 mV shows rapid coagulation or flocculation.

If two identical charged colloids are brought near each other in the presence of their double layers, they will not repel each other as strongly as they would, if there were no double layers. This phenomenon, charge screening, reduces the magnitude of the particle-particle Coulomb interaction. However, if the concentration of ions in the double layer is high enough, then the double layers of the particles will collide, increasing the electrostatic repulsion. In order to calculate the energy of the Coulomb interaction between colloids with double layers, we need a method to characterize the charge surrounding the colloids. The other kinds of repulsive interaction, steric repulsion, result from ligand molecules or polymers on the surface of a colloid. Steric repulsion is an entropic effect; as the colloids are brought closer, the number of possible ligand conformations is reduced. For nanoparticles, steric repulsion helps to prevent them from aggregating irreversibly due to van der Waals forces (Key and Maass, 2001).

5 Properties of nanofluid

Nanofluid is the best option available to enhance the thermophysical properties and heat transfer performance of fluids. For many years, the thermal conductivity of base fluid has been enhanced by adding small solid particles into a base fluid (Maxwell, 1873). Initially, particles of millimeter- or micrometer size range were added, nowadays, nanoparticles are being used into base fluid (called as nanofluids), which overcomes the limitations of micro and macro particles-based fluid such as poor suspension stability and channel clogging (Choi and Eastman, 1995). Nanofluids have a lot of advantages such as high stability, reduced particle clogging, and high heat transfer capabilities over the micro and macro particle-based fluids. However, the main drawback of nanofluid is viscosity, which is increased on increasing the content of nanoparticles (Quemada and Berli, 2002). The characterization of nanofluids has gained significance because heat transfer devices work under high-temperature conditions. Thus, the main aim to characterize the nanofluid is to evaluate the effective thermal conductivity, viscosity, heat transfer coefficient, density, and specific heat capacity.

5.1 Thermal conductivity and heat transfer in nanofluid

Most of the research and development studies have been focused on heat transfer in cooling processes. Conventional methods such as extended surface (fins) and increasing the flow rate increase the cooling rate. However, these processes also increase the additional maintenance and power consumption. One can enhance the thermal conductivity of base fluid by addition of particles to it, thereby supporting the nanofluids to overcome above-mentioned problems. Maxwell proposed that the addition of solid particles in the base fluid can enhance the physical properties of the fluid (Maxwell, 1873). In this manner, nanofluids are being used to improve the thermal and heat transfer of the base fluids. According to many researchers' studies, it was found that the particle volume concentration, types of particle materials, particle size, and particle shape, types of base fluid material, operating temperature, additives, and acidity are the main parameters, which can affect the thermal conductivity and heat transfer properties of nanofluids (Yang et al., 2019; Xu et al., 2019; Ahmadi et al., 2018). It was reported that the heat transfer of nanofluids is increased by increasing particle concentration (Wen and Ding, 2004). The researchers reported that an increase in heat transfer coefficient of ethylene glycol/water mixture can be obtained by the addition of 20 nm SiO_2 nanoparticles, as shown in Fig. 6.

Fig. 7 shows the enhancement of relative thermal conductivity of CuO-SWCNTs-EG water (40:60) with increasing temperature and volume fraction of particles (Rostamian et al., 2017).

In another work, the authors mentioned about the thermal conductivity of magnetic metal oxide $MnFe_2O_4$/water nanofluids. They found that the thermal conductivity can be altered by changing the concentration of magnetic particles, temperature, and magnetic field. The thermal conductivity can be enhanced by increasing the volume fraction of particles and the nanofluid temperature. However, this result was achieved in the absence of a magnetic field. Furthermore, once the magnetic field was applied, the thermal conductivity decreased with increase in temperature. In addition, magnetic metal oxide $MnFe_2O_4$/water nanofluids can offer better thermal conductivity under the magnetic field, if the nanofluid is used at room temperature, as shown in Figs. 8–10. Fig. 8 shows the variation of thermal conductivity of magnetic metal oxide $MnFe_2O_4$/water nanofluids with respect to magnetic metal oxide $MnFe_2O_4$ concentration and temperature. The Brownian motion and clustering of nanoparticles in the water fluid enhanced the thermal conductivity, which means clustering of nanoparticles increases the creation of paths with lower thermal resistance, which leads to the increment of heat conduction (Amani et al., 2017).

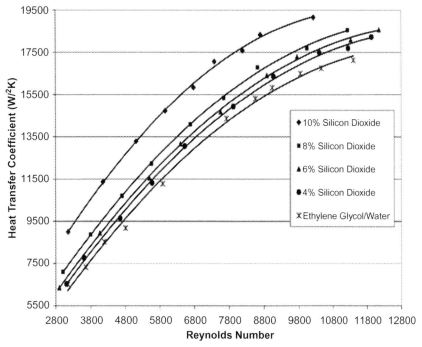

FIG. 6 The convective heat transfer coefficient of SiO₂ nanofluids (20nm diameter) in ethylene glycol/water mixture. *Adopted from Wen, D., Ding, Y., 2004. Experimental investigation into convective heat transfer of nanofluids at the entrance region under laminar flow conditions. Int. J.Heat Mass Transf. 47 (24), 5181–88. https://doi.org/10.1016/j.ijheatmasstransfer.2004.07.012.*

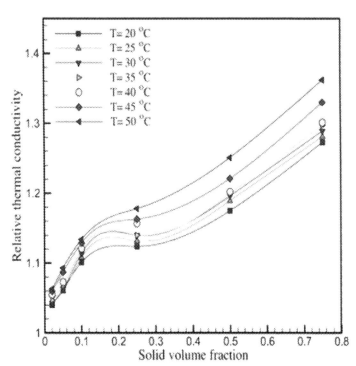

FIG. 7 Relative thermal conductivity of CuO-SWCNTs-EG water (40:60) with respect to solid volume fraction for different temperatures. *Adopted from Rostamian, S.H., Biglari, M., Saedodin, S., Hemmat Esfe, M., 2017. An inspection of thermal conductivity of CuO-SWCNTs hybrid Nanofluid versus temperature and concentration using experimental data, ANN modeling and new correlation. J. Mol. Liq. 231 (April), 364–69. https://doi.org/10.1016/j.molliq.2017.02. 015withpermission.*

FIG. 8 The thermal conductivity of MnFe$_2$O$_4$/water nanofluids against volume concentration signifies increase in thermal conductivity with rise in nanofluidic temperature (Amani et al., 2017).

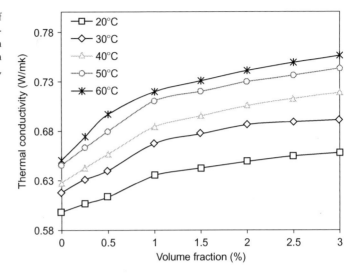

FIG. 9 The thermal conductivity of MnFe$_2$O$_4$/water nanofluid via the magnetic field intensity at different concentrations (Amani et al., 2017).

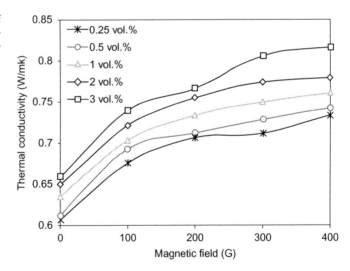

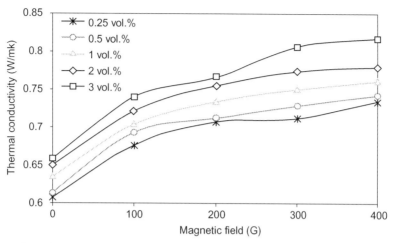

FIG. 10 Thermal conductivity via volume fraction at 20°C and 60°C under magnetic field with 200 and 400 G (Amani et al., 2017).

Volumetric concentration loading of magnetic particles in nanofluids makes a difference attributing to enhanced heat transfer capacity. Fig. 9 displays the variation of thermal conductivity of the $MnFe_2O_4$/water nanofluids with respect to volume fraction under the applied magnetic field and at 20°C. The intensity of the magnetic field enhances the thermal conductivity of $MnFe_2O_4$/water nanofluids; however, the enhancement of thermal conductivity was more pronounced at a higher concentration of magnetic particles; this is due to the distribution and formation of chain-like clusters of magnetic nanoparticles. The thermal conductivity was enhanced by the magnetic field at 20°C because the effective attraction between ferromagnetic nanoparticles depends on the magnetic moments of particles and distances; magnetic attraction is greater than the thermal energy under the magnetic field at 20°C. This helps in the formation of short nanoparticle chains oriented in the applied magnetic field (Amani et al., 2017).

The variation of thermal conductivity of $MnFe_2O_4$/water nanofluid under various temperatures and magnetic fields has been mentioned in Fig. 10. It was explained that the thermal conductivity decreases at higher temperatures under the magnetic field resulting due to absence of magnetic induction. It is a well-known fact that the magnetization of nanofluids changes inversely with temperature. The decrement in magnetization at high temperatures is due to thermomagnetic convection. Therefore, destruction of chain-like clusters due to the resulting increase in the speed of magnetic nanoparticles leads to the decrement of the nanofluid thermal conductivity (Amani et al., 2017).

The particle's shape and size have a profound effect on the thermal conductivity (Ambreen and Kim, 2020); therefore, fiber, elliptical, high aspect ratio particles can offer better thermal conductivity than spherical particles. If the conductivity of the base fluid is high, nanofluid base showed higher thermal conductivity than low thermal conducting fluids. At higher temperature, nanofluid showed better thermal conductivity due to the thermal migration and kinetic energy and thermal transfer. The thermal collision between base fluid molecules,

TABLE 4 A summary of some existing experimental studies for the thermal conductivity enhancement of different nanofluids.

Authors	Base fluid	Concentration	Dispersed particles	Maximum enhancement
Hwang et al.	Water	0.25–1 vol%	MWCNT	11.3%
Amrollahi et al.	EG	0.5–2.5 vol%	SWCNT	20.0%
Nanda et al.	EG	0–1.5 vol%	SWCNT	35.0%
Glory et al.	Water	0.01–3 wt%	MWCNT	64.0%
Jha and Ramaprabhua	Water	0.005–0.03 vol%	Ag-MWCNT	37.3%
Liu et al.	EG	0.2–1 vol%	MWCNT	12.4%
Harish et al.	EG	0.05–0.2 vol%	SWCNT	14.8%
Esfahani et al.	Water	0.01–0.5 wt%	Graphene oxide	19.9%
Sunder et al.	Water	0–1 vol%	Nanodiamond	22.8%
Hemmat Esfe et al.	Water	0–1 vol%	MWCNT	45%
Hemmat Esfe et al.	Water	0.4–1 vol%	DWCNT	7.5%

Adopted from Rostamian, S.H., Biglari, M., Saedodin, S., Hemmat Esfe, M., 2017. An inspection of thermal conductivity of CuO-SWCNTs hybrid Nanofluid versus temperature and concentration using experimental data, ANN modeling and new correlation. J. Mol. Liq. 231 (April), 364–69. https://doi.org/:10.1016/j.molliq.2017.02.015 with permission.

thermal diffusion in nanoparticles, and collision between nanoparticles enhanced the thermal conductivity. Reducing the surface area-to-volume ratio can be observed with agglomeration, which reduces the thermal interaction of particles; therefore, thermal conductivity decreases. A review of nanofluids' thermal conductivity by Ahmadi et al. showed that the increase in temperature, size, shape, and concentration leads to the increase in thermal conductivity in many nanofluids with a future scope of studying binary fluids as the base fluid or hybrid nanofluids (Ahmadi et al., 2018). Table 4 shows the enhancement in thermal conductivity with increasing particle concentration (Rostamian et al., 2017). A recent report by Wang Na et al. reported that the thermal conductivity of MgO-based nanofluids using mathematical models such as Multivariate Adaptive Regression Spline (MARS), polynomial correlation, and Group Method of Data Handling (GMDH) which showed a R^2 value of 0.9952, 0.9949, and 0.991 with the importance of variables shown in Fig. 11 (Wang et al., 2020).

5.2 Viscosity

Various science and engineering streams, especially in the fabrication of the microfluidics devices, artificial cilia, and physiological transport pumps, have motivated the research toward the nanofluids (Yang et al., 2020). Nowadays, various transportation, electronics, medical, food, defense, nuclear, and space engineering applications have been based on the application of nanofluids (Sahin et al., 2020). Advanced drug delivery vehicles and cancer treatment techniques are now more efficient in healthcare due to involvement of nanofluidic approach. Heat exchangers, improved heat transfer, reduced heat transfer fluid inventory,

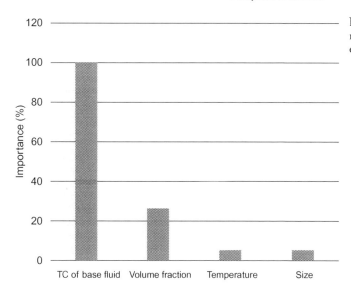

FIG. 11 Importance of variables with respect to thermal conductivity (Wang et al., 2020).

and reduced emissions are the main applications of the nanofluids in the field of heat transfer. The thermal conductivity and viscosity are the main features of the nanofluids (Kumar et al., 2012; Wang et al., 2020). Various mathematical models and experimental approaches have been used to study the viscosity of nanofluids (Bashirnezhad et al., 2016; Minakov et al., 2020; Mishra et al., 2014). These studies show the flow behavior of nanofluids with respect to particle concentration, particle types, and morphology of particles. It was reported that increasing the volume fraction of the nanoparticle enhances the viscosity of nanofluids (Murshed et al., 2008). Godson et al. showed that the viscosity of silver-water nanofluids increases with the increase in the silver nanoparticle volume concentrations of 0.3%, 0.6%, and 0.9% and decreases with an increase in the temperature, as shown in Fig. 12 (Godson et al., 2010).

The viscosity of the Al_2O_3/H_2O nanofluid has been reported. In this study, Al_2O_3 nanoparticles (nominal diameter of 43 nm) have been fabricated by the microwave-assisted chemical precipitation method, and subsequently, fabricated Al_2O_3/H_2O nanofluids were produced by using a sonicator. The viscosity of nanofluid increased with the concentration of the Al_2O_3 nanoparticles content (Chandrasekar et al., 2010). Bohlin rotational rheometer was used to measure the viscosity of TiO_2 nanoparticles dispersed in water with a volume concentration of 0.2–2 vol% nanofluids in the temperature range of 15–35°C. The result showed that the viscosity of nanofluids could be increased by increasing the particle concentrations and decreased by increasing nanofluid temperatures (Duangthongsuk and Wongwises, 2009). However, TiO_2/water nanofluids with different particle (agglomerate) sizes and concentrations can influence the viscosity of nanofluids (He et al., 2007). It was reported that the nanofluids show shear shinning behavior below the shear rate around $100 s^{-1}$, while nanofluids show the constant shear viscosity around $100 s^{-1}$. The (agglomerate) size and particle concentration increase the viscosity (He et al., 2007). Nanofluids based on the

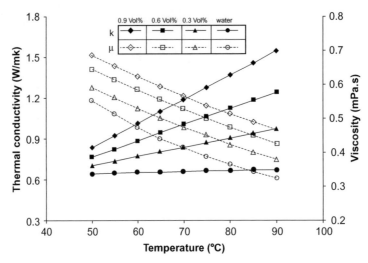

FIG. 12 Thermal conductivity and viscosity of silver-water nanofluids as a function of temperature and volume fraction (Godson et al., 2010).

Ethylene glycol (EG) containing ZnO nanoparticles show Newtonian behaviors at low concentration of ZnO nanoparticles, and shear-shinning (non-Newtonian behaviors) behavior can be observed at high concentration of ZnO nanoparticles (see Figs. 13 and 14). This means ZnO-EG nanofluids show different rheological behavior at 2vol% and above the 2vol% (Yu et al., 2009).

A recent report investigated seven nanofluids (Al_2O_3, ZrO_2, TiO_2, CuO, Fe_3O_4, Fe_2O_3, and nano diamonds) by varying the nanoparticle volume concentration from 0.25% to 8% in DI

FIG. 13 Graph signifying viscosity as a function of shear rate for 2vol% ZnO-EG nanofluid at different temperatures (Yu et al., 2009).

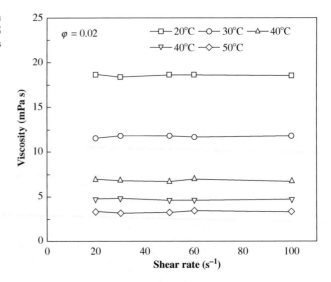

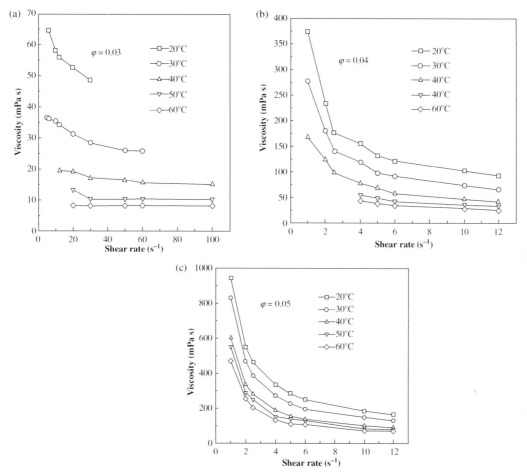

FIG. 14 Viscosity as a function of shear rate for (A) 3, (B) 4, and (C) 5 vol% ZnO-EG nanofluid at different temperatures (Yu et al., 2009).

and EG. From the results, it was observed that the viscosity of the nanofluid depended on the material used. Similarly, at high nanoparticle concentration, the viscosity was dependent on temperature, which was characterized by hysteresis as the viscosity varied depending on the heating and cooling (Minakov et al., 2020).

5.3 Specific heat

Nowadays, global warming, greenhouse effect, climate change, less efficient energy transfer in baseboard heaters, heat exchangers, automobiles, lack of energy storage, and fuel crisis are the major problems on our planet. By enhancing the heat transfer in heat recovery systems, we can overcome these problems. Energy and exergy performances of a thermal system depend on the specific heat capacity. One solution to this problem is nanofluids because of their volume fraction, temperature, types, and size of nanoparticles which enhance the specific capacity of base fluids (Shahrul et al., 2014). In the case of nanofluids, effective transfer base fluids contain solid nanoparticles with sizes of 1 to 100 nm (Eastman et al., 2001). To better understand nanofluids properties, nanofluids based on ethylene glycol and water mixture/silicon dioxide nanoparticles can show better thermophysical properties in cold regions, which is not possible alone with ethylene glycol and water base fluids. It was reported that increasing the volume concentration of SiO_2 in nanofluids decreases the specific heat of base fluid, specific heat of nanofluids containing 10% SiO_2 nanoparticle was suppressed 12% than that of the base fluid (Namburu et al., 2007a). This type of nanofluid is a good candidate for cold regions such as Alaska, Canada, and the circumpolar countries. In general, specific heat value of nanofluids depends on the volumetric concentration of nanoparticles, which decreases with increasing volume fraction of nanoparticles, and another concern is temperature; specific heat of nanofluids increases with increasing the temperature. Therefore, nanofluid has been used as heat transfer fluid, refrigerating and air-conditioning system. The specific heat of nanofluids with respect to the particle volumetric concentration and temperature have been estimated with aluminum dioxide, zinc oxide, and silicon dioxide nanoparticle systems (Vajjha and Das, 2009). In this study, nanofluids of aluminum dioxide (Al_2O_3), zinc oxide (ZnO) nanoparticles, 60:40 ethylene glycol/water mixture, silicon dioxide (SiO_2) nanoparticles and water nanofluids were fabricated. The result showed that the increasing volume concentration of Al_2O_3 decreases the specific heat of nanofluids, and specific heat increased with increasing temperature. Specific volume also increases with temperature and the same trend was reported in the case of zinc oxide, and silicon dioxide nanoparticles-based nanofluids. High-temperature nanofluids have a significant advantage for various applications. Silica nanoparticles/lithium carbonate and potassium carbonate (62:38 ratio) can be used to obtain the high-temperature nanofluids. It was mentioned that the specific heat could be enhanced up to 19%–24%. The high specific surface energies due to the high surface area of the nanoparticles is the main reason for this enhancement (Shin and Banerjee, 2011). The specific heat capacity of nanofluids varies with the size, volume concentration of nanoparticles, and particle-liquid interface. The volumetric heat capacity of nanofluid is essential to prescribe the temperature change of nanofluid. A recent report by Li et al. showed the influence of specific heat of β-cyclodextrin-modified carbon nanofluids (CD-CNT) for its application in a direct absorption solar collector (DASC). It was observed that the specific heat capacity decreased to 9.07% with increase in the rate of CD-CNT loadings (0.1 vol%). It was also observed that temperature has a major role in influencing the specific heat capacity of the nanofluids (Li et al., 2020).

6 Mass transfer in nanofluids

Unusual carrier properties of nanofluids have emerged as a theme of passionate investigation (Timofeeva et al., 2007; Stephan and Kuhnke, 2006). Transport features, such as tracer diffusion coefficient have also been described as being high, and a relationship between the heat- and mass-transport improvements has been suggested in micro convection-based approaches (Kumar et al., 2004; Patel et al., 2005). Mass diffusion refers to a molecular phenomenon describing the relative transport of the diffusive species due to concentration gradients (Ashrafmansouri and Esfahany, 2014). A considerable variety of articles also summarize that the inclusion of nanoparticles to the fluid can lead to interfacial mass-transport rates, such as the rate of gas sorption into liquid under conviction to improve much higher (Nagy et al., 2007; Zhu et al., 2008; Ashrafmansouri and Esfahany, 2014; Kebede et al., 2020). Absorption of gasses into a liquid is typically important for two- or three-phase reactions, as the diffusion of a sparingly soluble gas, like oxygen, over a gas-liquid interface usually restricts the reaction rates. By employing a third dispersed component, the mass transfer rate can be reasonably improved. The absorption rate can be explained by the existence of very fine, nanometer-sized particles or droplets. Its analytical model should consider the description of specific properties of the nanoparticles, e.g., the Brownian motion of particles, its effect on the diffusion of the bulk-phase molecules, the mass transfer rate into the nanoparticles, its relationship on the particle size, etc. (Nagy et al., 2007). Beginning from the postulated Langevin equation, it is noted that the velocity field built around a Brownian nanoparticle is identical to the velocity field foretold by Brinkman equations pointing to the relationship between dispersion in diluted fixed beds and dispersion in nanofluids. The suggested model concludes the order of magnitude of mass diffusion enhancement. The model also presents a strong relationship on the mass transfer Péclet number (Veilleux and Coulombe, 2011). Mass diffusion of rhodamine 6G (R6G) in water-based alumina nanofluids was investigated by techniques of total internal reflection fluorescence (TIRF) microscopy. They present a mass diffusivity improvement that gives an order of magnitude in a 2 vol% nanofluid when related to the value in deionized water. As analyses were conducted with positively charged R6G, interfacial complexation between the dye and the nanoparticles was not found. The influence of local density variations on mass diffusivity measurements is also addressed and enhancement of mass diffusivity was found in order of magnitude in a 2 vol% nanofluid when compared to the value in deionized water (Veilleux and Coulombe, 2010). Research on the mass diffusion of fluorescent Rhodamine B in Cu-water nanofluids has been performed to examine improvement effect of mass transfer in binary nanofluid. The test results reveal that the diffusion coefficient of fluorescent Rhodamine B in nanofluids is higher than that in de-ionized water. The Brownian motion of suspended nanoparticles and the induced micro-convection inside the suspension fluid remarkably improve mass transfer process, which is likely to improve the energy of transport inside the nanofluid. The nanoparticle concentration and fluid temperature are two principal points concerning mass transfer process inside the nanofluid. There are many general methods for estimating molecular diffusion coefficients in solutions, such as membrane cell method, multiple beam interferometry, Taylor dispersion method, nuclear magnetic resonance method, and light scattering method. With these methods, the Taylor dispersion method may be of lower cost and higher accuracy. By utilizing a solute impulse in the solution, the mass diffusion coefficient is measured by estimating the attenuation process of the impulse. However, due to limited and controversial reports on mass diffusion coefficient in nanofluids, experimental investigations on nanofluids still need to be explored. Further, reports on convective mass transfer of nanofluids have established that suspended nanoparticles enhance the mass transfer, but similar to mass diffusion the mechanism is yet unclear for practical applications (Ashrafmansouri and Esfahany, 2014).

7 Future trends

Nanoparticles and nanofluids have great potential to revolutionize almost any field, be it industrial or medical. As heat transfer (i.e., conduction, convective, boiling) can be enhanced by nanofluids, heat exchanging devices can be made energy-efficient and compact. Reduced or compact shape may result in reduced drag, for example, in automobile and similar applications. Nanofluids' stability and their production cost are major factors that hinder the commercialization of nanofluids. By solving these challenges, it is expected that nanofluids can make a substantial impact in many fields and devices. Future research needs to focus on finding out the main parameters affecting the thermal conductivity of nanofluids. The thermal conductivity of nanofluids can be a function of parameters such as particle shape, particle agglomeration, temperature, particle polydispersity, etc. Theoretical modeling and experimental works on the effective thermal conductivity and apparent diffusivity are needed to demonstrate the full potential of nanofluids for application in various fields.

References

Abbas, N., Awan, M.B., Amer, M., Ammar, S.M., Sajjad, U., Ali, H.M., Zahra, N., Hussain, M., Badshah, M.A., Jafry, A.T., 2019. Applications of nanofluids in photovoltaic thermal systems: a review of recent advances. Physica A 536, 122513.

Afrand, M., Toghraie, D., Ruhani, B., 2016. Effects of temperature and nanoparticles concentration on rheological behavior of Fe$_3$O$_4$-Ag/EG hybrid nanofluid: an experimental study. Exp. Thermal Fluid Sci. 77, 38–44.

Ahmadi, M.H., Mirlohi, A., Nazari, M.A., Ghasempour, R., 2018. A review of thermal conductivity of various nanofluids. J. Mol. Liq. 265, 181–188. https://doi.org/10.1016/j.molliq.2018.05.124.

Akbari, A., Fazel, S.A.A., Maghsoodi, S., Kootenaei, A.S., 2019. Pool boiling heat transfer characteristics of graphene-based aqueous nanofluids. J. Therm. Anal. Calorim. 135 (1), 697–711.

Al-Waeli, A.H.A., Chaichan, M.T., Kazem, H.A., Sopian, K., Ibrahim, A., Mat, S., Ruslan, M.H., 2018. Comparison study of indoor/outdoor experiments of a photovoltaic thermal PV/T system containing SiC nanofluid as a coolant. Energy 151, 33–44.

Amani, M., Amani, P., Kasaeian, A., Mahian, O., Wongwises, S., 2017. Thermal conductivity measurement of spinel-type ferrite MnFe$_2$O$_4$ nanofluids in the presence of a uniform magnetic field. J. Mol. Liq. 230 (March), 121–128. https://doi.org/10.1016/j.molliq.2016.12.013.

Ambreen, T., Kim, M.-H., 2020. Influence of particle size on the effective thermal conductivity of nanofluids: a critical review. Appl. Energy 264, 114684.

Anoop, K.B., Kabelac, S., Sundararajan, T., Das, S.K., 2009. Rheological and flow characteristics of nanofluids: influence of electroviscous effects and particle agglomeration. J. Appl. Phys. 106 (3). https://doi.org/10.1063/1.3182807.

Ashrafmansouri, S.-S., Esfahany, M.N., 2014. Mass transfer in nanofluids: a review. Int. J. Therm. Sci. 82, 84–99.

Babar, H., Sajid, M.U., Ali, H., 2019. Viscosity of hybrid nanofluids: a critical review. Therm. Sci. 23 (January), 15. https://doi.org/10.2298/TSCI181128015B.

Bafrani, H.A., Noori-Kalkhoran, O., Gei, M., Ahangari, R., Mirzaee, M.M., 2020. On the use of boundary conditions and thermophysical properties of nanoparticles for application of nanofluids as coolant in nuclear power plants; a numerical study. Prog. Nucl. Energy 126, 103417.

Bang, I.C., Kim, J.H., 2010. Thermal-fluid characterizations of ZnO and SiC nanofluids for advanced nuclear power plants. Nucl. Technol. 170 (1), 16–27.

Baratpour, M., Karimipour, A., Afrand, M., Wongwises, S., 2016. Effects of temperature and concentration on the viscosity of nanofluids made of single-wall carbon nanotubes in ethylene glycol. Int. Commun. Heat Mass Transfer 74, 108–113. https://doi.org/10.1016/j.icheatmasstransfer.2016.02.008.

Bargal, M.H.S., Abdelkareem, M.A.A., Tao, Q., Li, J., Shi, J., Wang, Y., 2020. Liquid cooling techniques in proton exchange membrane fuel cell stacks: a detailed survey. Alex. Eng. J. 59 (2), 635–655.

Bashirnezhad, K., Bazri, S., Safaei, M.R., Goodarzi, M., Dahari, M., Mahian, O., Dalkılıça, A.S., Wongwises, S., 2016. Viscosity of nanofluids: a review of recent experimental studies. Int. Commun. Heat Mass Transfer 73, 114–123.

Behrens, S.H., Christl, D.I., Emmerzael, R., Schurtenberger, P., Borkovec, M., 2000. Charging and aggregation properties of carboxyl latex particles: experiments versus DLVO theory. Langmuir 16 (6), 2566–2575. https://doi.org/10.1021/la991154z.

Bovand, M., Rashidi, S., Esfahani, J.A., 2015. Enhancement of heat transfer by nanofluids and orientations of the equilateral triangular obstacle. Energy Convers. Manag. 97, 212–223. https://doi.org/10.1016/j.enconman.2015.03.042.

Briscoe, C.J., Hage, D.S., 2009. Factors affecting the stability of drugs and drug metabolites in biological matrices. Bioanalysis 1 (1), 205–220. https://doi.org/10.4155/bio.09.20.

Cao, X., Tan, C., Zhang, X., Zhao, W., Zhang, H., 2016. Solution-processed two-dimensional metal dichalcogenide-based nanomaterials for energy storage and conversion. Adv. Mater. 28 (29), 6167–6196.

Chandrasekar, M., Suresh, S., Chandra Bose, A., 2010. Experimental investigations and theoretical determination of thermal conductivity and viscosity of Al_2O_3/water nanofluid. Exp. Thermal Fluid Sci. 34 (2), 210–216. https://doi.org/10.1016/j.expthermflusci.2009.10.022.

Chen, Y.-S., Hung, Y.-C., Liau, I., Steve Huang, G., 2009. Assessment of the in vivo toxicity of gold nanoparticles. Nanoscale Res. Lett. 4 (8), 858–864. https://doi.org/10.1007/s11671-009-9334-6.

Cheng, L., Wang, X., Gong, F., Liu, T., Liu, Z., 2020. 2D nanomaterials for cancer theranostic applications. Adv. Mater. 32 (13), 1902333.

Chithrani, B.D., Ghazani, A.A., Chan, W.C.W., 2006. Determining the size and shape dependence of gold nanoparticle uptake into mammalian cells. Nano Lett. 6 (4), 662–668. https://doi.org/10.1021/nl0523960.

Choi, S.U.S., Eastman, J.A., 1995. Enhancing thermal conductivity of fluids with nanoparticles. In: ASME International Mechanical Engineering Congress and Exposition. 66, pp. 99–105, https://doi.org/10.1115/1.1532008.

Choi, T.J., Jang, S.P., Jung, D.S., Lim, H.M., Byeon, Y.M., Choi, I.J., 2019. Effect of the freeze-thaw on the suspension stability and thermal conductivity of EG/water-based Al_2O_3 nanofluids. J. Nanomater. 2019, 2076341.

Cormode, D.P., Gao, L., Koo, H., 2018. Emerging biomedical applications of enzyme-like catalytic nanomaterials. Trends Biotechnol. 36 (1), 15–29.

Dasgupta, N., Ranjan, S., Chakraborty, A.R., Ramalingam, C., Shanker, R., Kumar, A., 2016a. Nanoagriculture and water quality management. In: Ranjan, S., Dasgupta, N., Lichtfouse, E. (Eds.), Nanoscience in Food and Agriculture 1. Springer, Berlin Heidelberg, https://doi.org/10.1007/978-3-319-39303-2_1.

Dasgupta, N., Ranjan, S., Patra, D., Srivastava, P., Kumar, A., Ramalingam, C., 2016b. Bovine serum albumin interacts with silver nanoparticles with a "side-on" or "end on" conformation. Chem. Biol. Interact. 253, 100–111. https://doi.org/10.1016/j.cbi.2016.05.018.

Del Fatti, N., Vallée, F., Flytzanis, C., Hamanaka, Y., Nakamura, A., 2000. Electron dynamics and surface plasmon resonance nonlinearities in metal nanoparticles. Chem. Phys. 251 (1–3), 215–226. https://doi.org/10.1016/S0301-0104(99)00304-3.

Delgado, A.V., González-Caballero, F., Hunter, R.J., Koopal, L.K., Lyklema, J., 2007. Measurement and interpretation of electrokinetic phenomena. J. Colloid Interface Sci. 309 (2), 194–224. https://doi.org/10.1016/j.jcis.2006.12.075.

Devendiran, D.K., Amirtham, V.A., 2016. A review on preparation, characterization, properties and applications of nanofluids. Renew. Sust. Energ. Rev. https://doi.org/10.1016/j.rser.2016.01.055.

Duangthongsuk, W., Wongwises, S., 2009. Measurement of temperature-dependent thermal conductivity and viscosity of TiO_2-water nanofluids. Exp. Thermal Fluid Sci. 33 (4), 706–714. https://doi.org/10.1016/j.expthermflusci.2009.01.005.

Eastman, J.A., Choi, S.U.S., Li, S., Yu, W., Thompson, L.J., 2001. Anomalously increased effective thermal conductivities of ethylene glycol-based nanofluids containing copper nanoparticles. Appl. Phys. Lett. 78 (6), 718–720. https://doi.org/10.1063/1.1341218.

Ekambaram, R., Dharmalingam, S., 2020. Fabrication and evaluation of electrospun biomimetic sulphonated PEEK nanofibrous scaffold for human skin cell proliferation and wound regeneration potential. Mater. Sci. Eng. C 115. https://doi.org/10.1016/j.msec.2020.111150, 111150.

Eshgarf, H., Afrand, M., 2016. An experimental study on rheological behavior of non-Newtonian hybrid nano-coolant for application in cooling and heating systems. Exp. Thermal Fluid Sci. 76, 221–227. https://doi.org/10.1016/j.expthermflusci.2016.03.015.

Evans, D.F., Wennerström, H., 1999. The Colloidal Domain: Where Physics, Chemistry, Biology and Technology Meet. Wiley-VCH, https://doi.org/10.1002/adma.19960080318.

Gambinossi, F., Mylon, S.E., Ferri, J.K., 2014. Aggregation kinetics and colloidal stability of functionalized nanoparticles. Adv. Colloid Interf. Sci. 222, 332–349. https://doi.org/10.1016/j.cis.2014.07.015.

Gao, L., Zhuang, J., Nie, L., Zhang, J., Zhang, Y., Gu, N., Wang, T., et al., 2007. Intrinsic peroxidase-like activity of ferromagnetic nanoparticles. Nat. Nanotechnol. 2 (9), 577–583. https://doi.org/10.1038/nnano.2007.260.

Godson, L., Raja, B., Mohan Lal, D., Wongwises, S., 2010. Experimental investigation on the thermal conductivity and viscosity of silver-deionized water Nanofluid. Exp. Heat Transfer 23 (4), 317–332. https://doi.org/10.1080/08916150903564796.

Guo, Z., 2020. A review on heat transfer enhancement with nanofluids. J. Enhanc. Heat Transf. 27 (1), 1–70.

Hagendorfer, H., 2011. New Analytical Methods for Size Fractionated, Quantitative, and Element Specific Analysis of Metallic Engineered Nanoparticles in Aerosols and Dispersions., https://doi.org/10.5075/epfl-thesis-5202.

Hansen, P.L., Wagner, J.B., Helveg, S., Rostrup-Nielsen, J.R., Clausen, B.S., Topsøe, H., 2002. Atom-resolved imaging of dynamic shape changes in supported copper nanocrystals. Science 295 (5562), 2053–2055. https://doi.org/10.1126/science.1069325.

Hansen, S.F., Larsen, B.H., Olsen, S.I., Baun, A., 2007. Categorization framework to aid hazard identification of nanomaterials. Nanotoxicology 1 (3), 243–250. https://doi.org/10.1080/17435390701727509.

He, Y., Jin, Y., Chen, H., Ding, Y., Cang, D., Lu, H., 2007. Heat transfer and flow behaviour of aqueous suspensions of TiO_2 nanoparticles (nanofluids) flowing upward through a vertical pipe. Int. J. Heat Mass Transf. 50 (11–12), 2272–2281. https://doi.org/10.1016/j.ijheatmasstransfer.2006.10.024.

Hemmat Esfe, M., Saedodin, S., 2014. An experimental investigation and new correlation of viscosity of ZnO-EG nanofluid at various temperatures and different solid volume fractions. Exp. Thermal Fluid Sci. 55, 1–5. https://doi.org/10.1016/j.expthermflusci.2014.02.011.

Hoek, E.M.V., Agarwal, G.K., 2006. Extended DLVO interactions between spherical particles and rough surfaces. J. Colloid Interface Sci. 298 (1), 50–58. https://doi.org/10.1016/j.jcis.2005.12.031.

Hsiao, I.-L., Huang, Y.-J., 2011. Effects of various physicochemical characteristics on the toxicities of ZnO and TiO nanoparticles toward human lung epithelial cells. Sci. Total Environ. 409 (7), 1219–1228. https://doi.org/10.1016/j.scitotenv.2010.12.033.

Huminic, G., Huminic, A., 2012. Application of nanofluids in heat exchangers: a review. Renew. Sust. Energ. Rev. 16 (8), 5625–5638.

Hussanan, A., Salleh, M.Z., Khan, I., Shafie, S., 2017. Convection heat transfer in micropolar Nanofluids with oxide nanoparticles in water, kerosene and engine oil. J. Mol. Liq. 229 (March), 482–488. https://doi.org/10.1016/j.molliq.2016.12.040.

Islam, R., Shabani, B., 2019. Prediction of electrical conductivity of TiO_2 water and ethylene glycol-based nanofluids for cooling application in low temperature PEM fuel cells. Energy Procedia 160, 550–557.

Ispas, C., Andreescu, D., Patel, A., Goia, D.V., Andreescu, S., Wallace, K.N., 2009. Toxicity and developmental defects of different sizes and shape nickel nanoparticles in zebrafish. Environ. Sci. Technol. 43 (16), 6349–6356. https://doi.org/10.1021/es9010543.

Israelachvili, J.N., 1992. Adhesion forces between surfaces in liquids and condensable vapours. Surf. Sci. Rep. 14 (3), 109–159. https://doi.org/10.1016/0167-5729(92)90015-4.

Jain, A., Ranjan, S., Dasgupta, N., Ramalingam, C., 2018. Nanomaterials in food and agriculture: an overview on their safety concerns and regulatory issues. Crit. Rev. Food Sci. Nutr. 58 (2), 297–317.

Jamei, M., Pourrajab, R., Ahmadianfar, I., Noghrehabadi, A., 2020. Accurate prediction of thermal conductivity of ethylene glycol-based hybrid nanofluids using artificial intelligence techniques. Int. Commun. Heat Mass Transfer 116, 104624.

Janardan, S., Suman, P., Ragul, G., Anjaneyulu, U., Shivendu, R., Nanditha, D., Chidambaram, R., Sasikumar, S., Vijayakrishna, K., Sivaramakrishna, A., 2016. Assessment on antibacterial activity of nanosized silica derived from hypercoordinated silicon(IV) precursors guidelines to the referees. RSC Adv. https://doi.org/10.1039/C6RA12189F.

Jang, H.-J., Lee, H.-Y., 2018. Size control of aggregations via self-assembly of amphiphilic gold nanoparticles. Colloids Surf. A Physicochem. Eng. Asp. 538, 574–582.

Jiang, W., Kim, B.Y.S., Rutka, J.T., Chan, W.C.W., 2008. Nanoparticle-mediated cellular response is size-dependent. Nat. Nanotechnol. 3 (3), 145–150. https://doi.org/10.1038/nnano.2008.30.

Jin, H., Guo, C., Liu, X., Liu, J., Vasileff, A., Jiao, Y., Zheng, Y., Qiao, S.-Z., 2018. Emerging two-dimensional nanomaterials for electrocatalysis. Chem. Rev. 118 (13), 6337–6408.

Kebede, T., Haile, E., Awgichew, G., Walelign, T., 2020. Heat and mass transfer in unsteady boundary layer flow of Williamson nanofluids. J. Appl. Math. 2020, 1890972.

Kennedy, J., Fang, F., Futter, J., Leveneur, J., Murmu, P.P., Panin, G.N., Kang, T.W., Manikandan, E., 2017. Synthesis and enhanced field emission of zinc oxide incorporated carbon nanotubes. Diam. Relat. Mater. 71, 79–84.

Key, F.S., Maass, G., 2001. Ions, atoms and charged particles. In: Silver Colloids, pp. 1–6.

Khalaj, M., Kamali, M., Khodaparast, Z., Jahanshahi, A., 2018. Copper-based nanomaterials for environmental decontamination–an overview on technical and toxicological aspects. Ecotoxicol. Environ. Saf. 148, 813–824.

Khanafer, K., Vafai, K., 2018. A review on the applications of nanofluids in solar energy field. Renew. Energy 123, 398–406.

Khedkar, R.S., Shrivastava, N., Sonawane, S.S., Wasewar, K.L., 2016. Experimental investigations and theoretical determination of thermal conductivity and viscosity of TiO_2-ethylene glycol nanofluid. Int. Commun. Heat Mass Transfer 73, 54–61. https://doi.org/10.1016/j.icheatmasstransfer.2016.02.004.

Khullar, V., Bhalla, V., Tyagi, H., 2018. Potential heat transfer fluids (nanofluids) for direct volumetric absorption-based solar thermal systems. J. Therm. Sci. Eng. Appl. 10 (1), 13.

Kim, H.Y., Sofo, J.O., Velegol, D., Cole, M.W., Lucas, A.A., 2007. Van Der Waals dispersion forces between dielectric nanoclusters. Langmuir 23 (4), 1735–1740. https://doi.org/10.1021/la061802w.

Kim, S.T., Chompoosor, A., Yeh, Y.C., Agasti, S.S., Solfiell, D.J., Rotello, V.M., 2012. Dendronized gold nanoparticles for SiRNA delivery. Small 8 (21), 3253–3256. https://doi.org/10.1002/smll.201201141.

Kitchens, C.L., Chandler McLeod, M., Roberts, C.B., 2003. Solvent effects on the growth and steric stabilization of copper metallic nanoparticles in AOT reverse micelle systems. J. Phys. Chem. B 107 (41), 11331–11338. https://doi.org/10.1021/jp0354090.

Knight, P., 1992. Progress in optics. J. Mod. Opt. 39 (7), 1599. https://doi.org/10.1080/09500349214551611.

Koca, H.D., Doganay, S., Turgut, A., 2017. Thermal characteristics and performance of Ag-water nanofluid: application to natural circulation loops. Energy Convers. Manag. 135 (March), 9–20. https://doi.org/10.1016/j.enconman.2016.12.058.

Kuang, J., Hou, X., Xiao, T., Li, Y., Wang, Q., Jiang, P., Cao, W., 2019. Three-dimensional carbon nanotube/SiC nanowire composite network structure for high-efficiency electromagnetic wave absorption. Ceram. Int. 45 (5), 6263–6267.

Kugarajah, V., Dharmalingam, S., 2020a. Investigation of a cation exchange membrane comprising sulphonated poly ether ether ketone and sulphonated titanium nanotubes in microbial fuel cell and preliminary insights on microbial adhesion. Chem. Eng. J. 125558.

Kugarajah, V., Dharmalingam, S., 2020b. Sulphonated polyhedral oligomeric silsesquioxane/sulphonated poly ether ether ketone nanocomposite membranes for microbial fuel cell: insights to the miniatures involved. Chemosphere. https://doi.org/10.1016/j.chemosphere.2020.127593, 127593.

Kugarajah, V., Sugumar, M., Dharmalingam, S., 2020. Nanocomposite membrane and microbial community analysis for improved performance in microbial fuel cell. Enzym. Microb. Technol. 140, 109606.

Kumar, P., 2019. Ultrathin 2D nanomaterials for electromagnetic interference shielding. Adv. Mater. Interfaces 6 (24), 1901454.

Kumar, D.H., Patel, H.E., Rajeev Kumar, V.R., Sundararajan, T., Pradeep, T., Das, S.K., 2004. Model for heat conduction in nanofluids. Phys. Rev. Lett. 93 (14). https://doi.org/10.1103/PhysRevLett.93.144301.

Kumar, T.A., Pradyumna, G., Jahar, S., 2012. Investigation of thermal conductivity and viscosity of nanofluids. J. Environ. Res. 7 (2), 768–777. https://doi.org/10.1007/s13204-012-0082-z.

Li, H., Wang, L., He, Y., Hu, Y., Zhu, J., Jiang, B., 2015. Experimental investigation of thermal conductivity and viscosity of ethylene glycol based ZnO nanofluids. Appl. Therm. Eng. 88, 363–368. https://doi.org/10.1016/j.applthermaleng.2014.10.071.

Li, X., Chen, W., Zou, C., 2020. An experimental study on β-cyclodextrin modified CNTs nanofluids for the direct absorption solar collector (DASC): specific heat capacity and photo-thermal conversion performance. Sol. Energy Mater. Sol. Cells 204, 110240.

Liang, W., Bunker, C.E., Sun, Y.-P., 2020. Carbon dots: zero-dimensional carbon allotrope with unique photoinduced redox characteristics. ACS Omega 5 (2), 965–971.

Lowry, G.V., Hotze, E.M., Bernhardt, E.S., Dionysiou, D.D., Pedersen, J.A., Wiesner, M.R., Xing, B., 2010. Environmental occurrences, behavior, fate, and ecological effects of nanomaterials: an introduction to the special series. J. Environ. Qual. 39 (June), 1867–1874. https://doi.org/10.2134/jeq2010.0297.

Manikandan, S., Shylaja, A., Rajan, K.S., 2014. Thermo-physical properties of engineered dispersions of nano-sand in propylene glycol. Colloids Surf. A Physicochem. Eng. Asp. 449 (1), 8–18. https://doi.org/10.1016/j.colsurfa.2014.02.040.

Maxwell, J.C., 1873. A Treatise on Electricity and Magnetism. Clarendon Press, Oxford, https://doi.org/10.1016/0016-0032(54)90053-8.

McCray, W.P.P., 2006. Nano-hype: the truth behind the nanotechnology buzz. Isis 97 (3), 586–587.

Mechiri, S.K., Vasu, V., Venu, A., Gopal., 2017. Investigation of thermal conductivity and rheological properties of vegetable oil based hybrid nanofluids containing Cu–Zn hybrid nanoparticles. Exp. Heat Transfer 30 (3), 205–217.

Minakov, A.V., Rudyak, V.Y., Pryazhnikov, M.I., 2020. Systematic experimental study of the viscosity of nanofluids. Heat Transf. Eng. 42, 1–17.

Mishra, P.C., Mukherjee, S., Kumar Nayak, S., Panda, A., 2014. A brief review on viscosity of nanofluids. Int. Nano Lett. 4 (4), 109–120.

Murshed, S.M.S., Leong, K.C., Yang, C., 2008. Investigations of thermal conductivity and viscosity of nanofluids. Int. J. Therm. Sci. 47 (5), 560–568. https://doi.org/10.1016/j.ijthermalsci.2007.05.004.

Nagy, E., Feczkó, T., Koroknai, B., 2007. Enhancement of oxygen mass transfer rate in the presence of nanosized particles. Chem. Eng. Sci. 62 (24), 7391–7398. https://doi.org/10.1016/j.ces.2007.08.064.

Namburu, P.K., Kulkarni, D.P., Dandekar, A., Das, D.K., 2007a. Experimental investigation of viscosity and specific heat of silicon dioxide nanofluids. Micro Nano Lett. 2 (4), 67–71. https://doi.org/10.1049/mnl.

Namburu, P.K., Kulkarni, D.P., Misra, D., Das, D.K., 2007b. Viscosity of copper oxide nanoparticles dispersed in ethylene glycol and water mixture. Exp. Thermal Fluid Sci. 32 (2), 397–402. https://doi.org/10.1016/j.expthermflusci.2007.05.001.

Nandita, D., Ranjan, S., Mundra, S., Ramalingam, C., Kumar, A., 2016. Fabrication of food grade vitamin E nanoemulsion by low energy approach, characterization and its application. Int. J. Food Prop. 19, 700–708. https://doi.org/10.1080/10942912.2015.1042587.

Navya, P.N., Daima, H.K., 2016. Rational engineering of physicochemical properties of nanomaterials for biomedical applications with nanotoxicological perspectives. Nano Convergence 3 (1), 1. https://doi.org/10.1186/s40580-016-0064-z.

Nguyen, C.T., Desgranges, F., Galanis, N., Roy, G., Maré, T., Boucher, S., Angue Mintsa, H., 2008. Viscosity data for Al_2O_3-water nanofluid-hysteresis: is heat transfer enhancement using nanofluids reliable? Int. J. Therm. Sci. 47 (2), 103–111. https://doi.org/10.1016/j.ijthermalsci.2007.01.033.

Pang, H., Cao, X., Zhu, L., Zheng, M., 2020. Synthesis of one-dimensional nanomaterials. In: Synthesis of Functional Nanomaterials for Electrochemical Energy Storage. Springer, pp. 31–53.

Patel, H.E., Sundararajan, T., Pradeep, T., Dasgupta, A., Dasgupta, N., Das, S.K., 2005. A micro-convection model for thermal conductivity of nanofluids. Pramana J. Phys. 65 (5), 863–869. https://doi.org/10.1007/BF02704086.

Perenboom, J.A., Wyder, P., Meier, F., 1981. Electronic properties of small metallic particles. Phys. Rep. 78 (2), 173–292. https://doi.org/10.1016/0370-1573(81)90194-0.

Phenrat, T., Saleh, N., Sirk, K., Tilton, R.D., Lowry, G.V., 2007. Aggregation and sedimentation of aqueous nanoscale zerovalent iron dispersions. Environ. Sci. Technol. 41 (1), 284–290. https://doi.org/10.1021/es061349a.

Phenrat, T., Saleh, N., Sirk, K., Kim, H.J., Tilton, R.D., Lowry, G.V., 2008. Stabilization of aqueous nanoscale zerovalent Iron dispersions by anionic polyelectrolytes: adsorbed anionic polyelectrolyte layer properties and their effect on aggregation and sedimentation. J. Nanopart. Res. 10 (5), 795–814. https://doi.org/10.1007/s11051-007-9315-6.

Qi, C., Wang, G., Yan, Y., Mei, S., Luo, T., 2018. Effect of rotating twisted tape on thermo-hydraulic performances of nanofluids in heat-exchanger systems. Energy Convers. Manag. 166, 744–757.

Qu, J., Zhang, R., Wang, Z., Wang, Q., 2019. Photo-thermal conversion properties of hybrid CuO-MWCNT/H_2O nanofluids for direct solar thermal energy harvest. Appl. Therm. Eng. 147, 390–398.

Quemada, D., Berli, C., 2002. Energy of interaction in colloids and its implications in rheological modeling. Adv. Colloid Interf. Sci. 98 (1), 51–85. https://doi.org/10.1016/S0001-8686(01)00093-8.

Rai, V.R., Bai, J.A., 2018. Nanotechnology Applications in the Food Industry. CRC Press.

Ranjan, S., Ramalingam, C., 2016. Titanium dioxide nanoparticles induce bacterial membrane rupture by reactive oxygen species generation. Environ. Chem. Lett. https://doi.org/10.1007/s10311-016-0586-y.

Ranjan, S., Dasgupta, N., Srivastava, P., Ramalingam, C., 2016. A spectroscopic study on interaction between bovine serum albumin and titanium dioxide nanoparticle synthesized from microwave-assisted hybrid chemical approach. J. Photochem. Photobiol. B Biol. 161, 472–481. https://doi.org/10.1016/j.jphotobiol.2016.06.015.

Rathinavel, S., Ekambaram, S., Korrapati, P.S., Sangeetha, D., 2020. Design and fabrication of electrospun SBA-15-incorporated PVA with curcumin: a biomimetic nanoscaffold for skin tissue engineering. Biomed. Mater. 15 (3), 35009. https://doi.org/10.1088/1748-605x/ab6b2f.

Reddy, M.C.S., Rao, V.V., 2013. Experimental studies on thermal conductivity of blends of ethylene glycol-water-based TiO_2 nanofluids. Int. Commun. Heat Mass Transfer 46, 31–36. https://doi.org/10.1016/j.icheatmasstransfer.2013.05.009.

Rojas Delgado, R., De Pauli, C.P., Barriga Carrasco, C., Avena, M.J., 2008. Influence of MII/MIII ratio in surface-charging behavior of Zn-Al layered double hydroxides. Appl. Clay Sci. 40 (1–4), 27–37. https://doi.org/10.1016/j.clay.2007.06.010.

Romero-Cano, M.S., Martín-Rodríguez, A., De Las Nieves, F.J., 2001. Electrosteric stabilization of polymer colloids with different functionality. Langmuir 17 (11), 3505–3511. https://doi.org/10.1021/la001659l.

Rostamian, S.H., Biglari, M., Saedodin, S., Hemmat Esfe, M., 2017. An inspection of thermal conductivity of CuO-SWCNTs hybrid Nanofluid versus temperature and concentration using experimental data, ANN modeling and new correlation. J. Mol. Liq. 231 (April), 364–369. https://doi.org/10.1016/j.molliq.2017.02.015.

Sadeghinezhad, E., Mehrali, M., Saidur, R., Mehrali, M., Latibari, S.T., Reza Akhiani, A., Metselaar, H.S.C., 2016. A comprehensive review on graphene nanofluids: recent research, development and applications. Energy Convers. Manag. https://doi.org/10.1016/j.enconman.2016.01.004.

Sahin, A.Z., Uddin, M.A., Yilbas, B.S., Al-Sharafi, A., 2020. Performance enhancement of solar energy systems using nanofluids: an updated review. Renew. Energy 145, 1126–1148.

Sahoo, B.C., Vajjha, R.S., Ganguli, R., Chukwu, G.A., Das, D.K., Vajjha, R.S., Ganguli, R., 2009. Determination of rheological behavior of aluminum oxide nanofluid and development of new viscosity correlations. Pet. Sci. Technol. 27 (27), 15–1757. https://doi.org/10.1080/10916460802640241.

Saleh, N., Kim, H.-J., Phenrat, T., Matyjaszewski, K., Tilton, R.D., Lowry, G.V., 2008. Ionic strength and composition affect the mobility of surface-modified Fe^0 nanoparticles in water-saturated sand columns. Environ. Sci. Technol. 42 (9), 3349–3355.

Salehin, S., Monjurul Ehsan, M., Faysal, S.R., Sadrul Islam, A.K.M., 2018. Utilization of nanofluid in various clean energy and energy efficiency applications. In: Application of Thermo-Fluid Processes in Energy Systems. Springer, pp. 3–33.

Samadi, M., Sarikhani, N., Zirak, M., Zhang, H., Zhang, H.-L., Moshfegh, A.Z., 2018. Group 6 transition metal dichalcogenide nanomaterials: synthesis, applications and future perspectives. Nanoscale Horizons 3 (2), 90–204.

Shahrul, I.M., Mahbubul, I.M., Khaleduzzaman, S.S., Saidur, R., Sabri, M.F.M., 2014. A comparative review on the specific heat of nanofluids for energy perspective. Renew. Sust. Energ. Rev. 38 (October), 88–98. https://doi.org/10.1016/j.rser.2014.05.081.

Shakib-Manesh, A., Raiskinmäki, P., Koponen, A., Kataja, M., Timonen, J., 2002. Shear stress in a Couette flow of liquid-particle suspensions. J. Stat. Phys. 107, 67–84. https://doi.org/10.1023/A:1014598201975.

Shamaeil, M., Firouzi, M., Fakhar, A., 2016. The effects of temperature and volume fraction on the thermal conductivity of functionalized DWCNTs/ethylene glycol nanofluid. J. Therm. Anal. Calorim. 126 (3), 1455–1462. https://doi.org/10.1007/s10973-016-5548-x.

Shem, P.M., Sardar, R., Shumaker-Parry, J.S., 2014. Soft ligand stabilized gold nanoparticles: incorporation of bipyridyls and two-dimensional assembly. J. Colloid Interface Sci. 426, 107–116. https://doi.org/10.1016/j.jcis.2014.03.059.

Shevchenko, E.V., Talapin, D.V., Murray, C.B., O'Brien, S., 2006. Structural characterization of self-assembled multifunctional binary nanoparticle superlattices. J. Am. Chem. Soc. 128 (11), 3620–3637. https://doi.org/10.1021/ja0564261.

Shin, D., Banerjee, D., 2011. Enhanced specific heat of silica nanofluid. J. Heat Transf. 133 (2). https://doi.org/10.1115/1.4002600, 024501.

Sireesh, B.M., Mandal, B.K., Ranjan, S., Dasgupta, N., 2015. Diastase assisted green synthesis of size-controllable gold nanoparticles. RSC Adv. https://doi.org/10.1039/c5ra03117f.

Sireesh, B.M., Mandal, B.K., Shivendu, R., Nandita, D., 2017. Diastase induced green synthesis of bilayered reduced graphene oxide and its decoration with gold nanoparticles. J. Photochem. Photobiol. B Biol. 166, 252–258.

Siripireddy, B., Mandal, B.K., Ranjan, S., Nandita, D., Ramalingam, C., 2017. Nano-zirconia – evaluation of its antioxidant and anticancer activity. J. Photochem. Photobiol. B Biol. https://doi.org/10.1016/j.jphotobiol.2017.04.004.

Soltani, O., Akbari, M., 2016. Effects of temperature and particles concentration on the dynamic viscosity of MgO-MWCNT/ethylene glycol hybrid nanofluid: experimental study. Physica E 84, 564–570. https://doi.org/10.1016/j.physe.2016.06.015.

Sonawane, S.S., Juwar, V., 2016. Optimization of conditions for an enhancement of thermal conductivity and minimization of viscosity of ethylene glycol based Fe_3O_4 nanofluid. Appl. Therm. Eng. 109, 121–129. https://doi.org/10.1016/j.applthermaleng.2016.08.066.

Stephan, K., Kuhnke, J.F., 2006. Heat transfer mechanisms in Nanofluids - experiments and theory. In: International Heat Transfer Conference.

Su, S., Sun, Q., Gu, X., Xu, Y., Shen, J., Zhu, D., Chao, J., Fan, C., Wang, L., 2019. Two-dimensional nanomaterials for biosensing applications. TrAC Trends Anal. Chem. 119, 115610.

Subramanian, K.R.V., Rao, T.N., Balakrishnan, A., 2019. Nanofluids and Their Engineering Applications. CRC Press.

Sugumar, M., Dharmalingam, S., 2020. Statistical optimization of process parameters in microbial fuel cell for enhanced power production using sulphonated polyhedral oligomeric silsesquioxane dispersed sulphonated polystyrene ethylene butylene polystyrene nanocomposite membranes. J. Power Sources 469. https://doi.org/10.1016/j.jpowsour.2020.228400, 228400.

Sundar, L.S., Venkata Ramana, E., Singh, M.K., De Sousa, A.C.M., 2012. Viscosity of low volume concentrations of magnetic Fe_3O_4 nanoparticles dispersed in ethylene glycol and water mixture. Chem. Phys. Lett. 554, 236–242.

Tammina, S.K., Mandal, B.K., Ranjan, S., Dasgupta, N., 2017. Cytotoxicity study of Piper Nigrum seed mediated synthesized SnO_2 nanoparticles towards colorectal (HCT116) and lung cancer (A549) cell lines. J. Photochem. Photobiol. B Biol. 166, 158–168. https://doi.org/10.1016/j.jphotobiol.2016.11.017.

Teow, Y.H., Mohammad, A.W., 2019. New generation nanomaterials for water desalination: a review. Desalination 451, 2–17.

Tian, Y., Lhermitte, J.R., Bai, L., Vo, T., Xin, H.L., Li, H., Li, R., Fukuto, M., Yager, K.G., Kahn, J.S., 2020. Ordered three-dimensional nanomaterials using DNA-prescribed and valence-controlled material voxels. Nat. Mater. 19, 1–8.

Timofeeva, E.V., Gavrilov, A.N., McCloskey, J.M., Tolmachev, Y.V., Sprunt, S., Lopatina, L.M., Selinger, J.V., 2007. Thermal conductivity and particle agglomeration in alumina nanofluids: experiment and theory. Phys. Rev. E Stat. Nonlinear Soft Matter Phys. 76 (6). https://doi.org/10.1103/PhysRevE.76.061203.

Tseng, W.J., Lin, K.C., 2003. Rheology and colloidal structure of aqueous TiO_2 nanoparticle suspensions. Mater. Sci. Eng. A 355 (1–2), 186–192. https://doi.org/10.1016/S0921-5093(03)00063-7.

Vajjha, R.S., Das, D.K., 2009. Specific heat measurement of three nanofluids and development of new correlations. J. Heat Transf. 131 (7). https://doi.org/10.1115/1.3090813, 071601.

Veilleux, J., Coulombe, S., 2010. A Total internal reflection fluorescence microscopy study of mass diffusion enhancement in water-based alumina nanofluids. J. Appl. Phys. 108. https://doi.org/10.1063/1.3514138.

Veilleux, J., Coulombe, S., 2011. A dispersion model of enhanced mass diffusion in nanofluids. Chem. Eng. Sci. 66 (11), 2377–2384. https://doi.org/10.1016/j.ces.2011.02.053.

Villaseñor, M.J., Ríos, Á., 2018. Nanomaterials for water cleaning and desalination, energy production, disinfection, agriculture and green chemistry. Environ. Chem. Lett. 16 (1), 11–34.

Voisin, C., Christofilos, D., Del Fatti, N., Vallée, F., Prével, B., Cottancin, E., Lermé, J., Pellarin, M., Broyer, M., 2000. Size-dependent electron-electron interactions in metal nanoparticles. Phys. Rev. Lett. 85, 2200–2203. https://doi.org/10.1103/PhysRevLett.85.2200.

Vollath, D., 2008. Nanomaterials an introduction to synthesis, properties and application. Environ. Eng. Manage. J. 7 (6), 865–870.

Walia, N., Dasgupta, N., Ranjan, S., Chen, L., Chidambaram, R., 2017. Fish oil based vitamin D nanoencapsulation by ultrasonication and bioaccessibility analysis in simulated gastro-intestinal tract. Ultrason. Sonochem. https://doi.org/10.1016/j.ultsonch.2017.05.021.

Wang, Z.W., Liu, Y., Tay, J.H., 2005. Distribution of EPS and cell surface hydrophobicity in aerobic granules. Appl. Microbiol. Biotechnol. 69 (4), 469–473. https://doi.org/10.1007/s00253-005-1991-5.

Wang, N., Maleki, A., Alhuyi Nazari, M., Tlili, I., Shadloo, M.S., 2020. Thermal conductivity modeling of nanofluids contain MgO particles by employing different approaches. Symmetry 12 (2), 206.

Wei, B., Zou, C., Li, X., 2017. Experimental investigation on stability and thermal conductivity of diathermic oil based TiO_2 nanofluids. Int. J. Heat Mass Transf. 104, 537–543.

Wen, D., Ding, Y., 2004. Experimental investigation into convective heat transfer of nanofluids at the entrance region under laminar flow conditions. Int. J. Heat Mass Transf. 47 (24), 5181–5188. https://doi.org/10.1016/j.ijheatmasstransfer.2004.07.012.

Wikipedia. n.d. "Repulsion – Electrocratic.".

Wu, X., Li, X., Yang, M., Zeng, H., Zhang, S., Lu, R., Gao, H., Xu, D., 2017. An ionic liquid-based nanofluid of titanium dioxide nanoparticles for effervescence-assisted dispersive liquid–liquid extraction for acaricide detection. J. Chromatogr. A 1497 (May), 1–8. https://doi.org/10.1016/j.chroma.2017.03.005.

Wu, M.-F., Wang, Y., Li, S., Dong, X.-X., Yang, J.-Y., Shen, Y.-D., Wang, H., Sun, Y.-M., Lei, H.-T., Xu, Z.-L., 2019a. Ultrasensitive immunosensor for acrylamide based on chitosan/SnO_2-SiC hollow sphere nanochains/gold nanomaterial as signal amplification. Anal. Chim. Acta 1049, 188–195.

Wu, Z., Li, Z., Li, H., Sun, M., Han, S., Cai, C., Shen, W., Fu, Y.Q., 2019b. Ultrafast response/recovery and high selectivity of the H2S gas sensor based on α-Fe_2O_3 nano-ellipsoids from one-step hydrothermal synthesis. ACS Appl. Mater. Interfaces 11 (13), 12761–12769.

Xing, Y., Zhang, Y., Xu, N., Huang, H., Ke, Y., Li, B., Chen, J., She, J., Deng, S., 2018. Design and realization of microwave frequency multiplier based on field emission from carbon nanotubes cold-cathode. IEEE Trans. Electron Devices 65 (3), 1146–1150.

Xu, G., Fu, J., Dong, B., Quan, Y., Song, G., 2019. A novel method to measure thermal conductivity of nanofluids. Int. J. Heat Mass Transf. 130, 978–988.

Yan, X., Zhou, M., Yu, S., Jin, Z., Zhao, K., 2020. An overview of biodegradable nanomaterials and applications in vaccines. Vaccine 38 (5), 1096–1104.

Yang, L., Ji, W., Huang, J.-N., Xu, G., 2019. An updated review on the influential parameters on thermal conductivity of Nano-fluids. J. Mol. Liq. 296, 111780.

Yang, L., Huang, J.-N., Ji, W., Mao, M., 2020. Investigations of a new combined application of nanofluids in heat recovery and air purification. Powder Technol. 360, 956–966.

Yiamsawas, T., Mahian, O., Dalkilic, A.S., Kaewnai, S., Wongwises, S., 2013. Experimental studies on the viscosity of TiO_2 and Al_2O_3 nanoparticles suspended in a mixture of ethylene glycol and water for high temperature applications. Appl. Energy 111, 40–45. https://doi.org/10.1016/j.apenergy.2013.04.068.

Yu, W., Xie, H., 2012. A review on nanofluids: preparation, stability mechanisms, and applications. Edited by Li-Hong Liu. J. Nanomater. 2012. https://doi.org/10.1155/2012/435873, 435873.

Yu, W., Xie, H., Chen, L., Li, Y., 2009. Investigation of thermal conductivity and viscosity of ethylene glycol based ZnO Nanofluid. Thermochim. Acta 491 (1–2), 92–96. https://doi.org/10.1016/j.tca.2009.03.007.

Zhai, W., Zhou, K., 2019. Nanomaterials in superlubricity. Adv. Funct. Mater. 29 (28), 1806395.

Zhu, H., Shanks, B.H., Heindel, T.J., 2008. Enhancing CO-water mass transfer by functionalized MCM41 nanoparticles. Ind. Eng. Chem. Res. 47 (20), 7881–7887. https://doi.org/10.1021/ie800238w.

Zhu, Q., Zhang, Z., Lv, Y., Chen, X., Wu, Z., Wang, S., Zou, Y., 2019. Synthesis and electromagnetic wave absorption performance of $NiCo_2O_4$ nanomaterials with different nanostructures. CrystEngComm 21 (31), 4568–4577.

Robust organometallic gold nanoparticles in nanomedicine engineering of proteins

Mahreen Arooj, Mehavesh Hameed, Seema Panicker, Ihsan Shehadi, and Ahmed A. Mohamed

Department of Chemistry, College of Sciences, University of Sharjah, Sharjah, United Arab Emirates

O U T L I N E

1 Introduction

Gold nanoparticles (AuNPs) have entrenched an inevitable role in biomedical applications over the last few decades. Synthesis of biocompatible AuNPs using biomolecules such as protein finds ubiquitous biomedical applications in nanomedicine engineering (Mahal et al., 2013; Hameed et al., 2018, 2020; AlBab et al., 2020). Conjugation of nanoparticles with proteins introduces the biocompatible functionalities and leads to act as reducing, capping, and stabilizing agents. Since the carboxyl-modified AuNPs, AuNPs-COOH, can be fabricated using manifold routes and presented outstanding resistance under physiological conditions, they specifically captured much attention in the last few years.

Our research group carried out a few studies on a few therapeutically significant proteins such as insulin, bovine serum albumin, and lysozyme to investigate their possible use as drug delivery vehicles (Hameed et al., 2018, 2020). The AuNPs-insulin interaction was studied which resulted in fibril dissociation and inhibitory effects against the proteolytic enzymatic activity of the gastrointestinal enzymes. Besides, the increase in lysozyme hydrolytic activity on the multidrug-resistant superbugs was also studied after conjugation with green-synthesized gold-aryl nanoparticles.

The physicochemical stability of protein-benzoic acid complexes was studied with BSA, zein, and lysozyme proteins experimentally and using computational methods. High stability of protein-benzoic acid complexes was concluded from molecular dynamics calculations through interactions including hydrogen bonding, hydrophobic, and electrostatic interactions.

2 BSA conjugated gold-carbon nanoparticles with outstanding robustness and hemocompatibility

Bovine serum albumin (BSA) is comparable with human serum albumin (Peter and Reed, 1978; Walsh and Knecht, 2017). Adsorption of BSA on the surface of the bare nanoparticles leads to the formation of a dense coating known as protein corona (Lynch and Dawson, 2008). Taking advantage of this advancement, studies were carried for the synthesis, characterization, and toxicity of BSA-conjugated AuNPs to prove the biocompatibility for their use as drug delivery systems (Fig. 1). AuNPs-BSA bioconjugates were fabricated by stirring the aryldiazonium gold(III) salt with BSA in water followed by the addition of a mild chemical reducing agent. The color of the solution changed from yellow to deep pink, confirming the formation of the nanoparticles. The plasmon peak of the AuNPs-BSA bioconjugate at 540 nm demonstrated the adsorption of BSA and accounted for the bioconjugate formation.

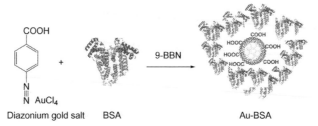

FIG. 1 Schematic diagram of AuNPs-BSA bioconjugate formation. *Credit: Elsevier 2018.*

TEM was used to picturize the core-shell structure of the AuNPs-BSA bioconjugates. Moreover, the size of the AuNPs bioconjugate was found to be the largest size in acidic media, and it was seen to be gradually decreasing when the pH was close to its isoelectric value. The size of the bioconjugates further decreased at basic pH and the smallest size was observed. Also, the isoelectric point study helped to understand the overall charge on the newly synthesized BSA conjugated AuNPs. This study also facilitated understanding the effect of pH values on the isoelectric point of AuNPs-BSA biconjugate. Interestingly, colloidal NPs immediately settled at the bottom of the tube due to rapid aggregation when the pH approached 4.5 whereas, some of the studies reported the isoelectric point of pH 4.7 (Brewer et al., 2005; Khullar et al., 2012; Pihlasalo et al., 2012). A noticeable absorbance due to SPR of the bioconjugate dispersion at 540 nm became visible at all pH values except close to the isoelectric point of BSA. Literature showing the isoelectric point of pH 4.7 for native BSA was moderately modified to a range of pH 4.1–5.0 at which the solubility declined considerably, and purple aggregates separated (Fig. 2).

Native gold-aryl nanoparticles and AuNPs-BSA bioconjugates possess remarkable hemocompatibility. Chen et al. (Chen et al., 2013) reported that the bare copper nanoparticles Cu$_2$O induce toxicity in fish RBCs and resulted in eventual hemolysis. Besides, large-sized silica-based nanomaterials showed strong hemolysis (Slowing et al., 2009; Lin and Haynes, 2010; Thomassen et al., 2011; Zhao et al., 2011). Zein protein capped gold nanoparticles synthesized by a green approach displayed membrane rupturing ability dependent on the amount of protein coating, however, a higher protein shell ratio showed less hemolysis (Mahal et al., 2013). Bare AuNPs encapsulated by BSA showed no local deformation to human RBCs membrane, however, hemolysis was carried out by encapsulation in cationic surfactants (Khullar et al., 2012). It is anticipated that an appreciable encapsulation efficiency and evenly covered surfaces of the AuNPs act as a shield to prevent hemolysis and enhance the efficiency and biocompatibility of surfaces.

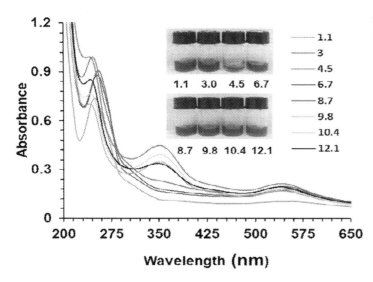

FIG. 2   UV-Vis spectra of AuNPs-BSA at different pH values. *Credit: Elsevier 2018.*

3 Green and cytocompatible carboxyl-modified gold-lysozyme antibacterial

The worldwide propagation of multidrug-resistant superbugs and the number of infections produced by such resistant strains are increasing globally. The use of nanotechnology to formulate bactericidal agents may offer a promising solution as they can obliterate bacteria and act as carriers for antibiotics and natural antimicrobial compounds (Ahmady et al., 2019). Herein, we developed green-synthesized carboxyl-functionalized gold-lysozyme antibacterial for combating multidrug-resistant superbugs. AuNPs-lysozyme antimicrobial was synthesized by stirring lysozyme with aryldiazonium gold(III) salt [HOOC-4-$C_6H_4N\equiv N$]AuCl$_4$ at pH 12. Lysozyme acted as the reducing and capping agent of AuNPs. The blue shift in the plasmon peak from 550 to 520 nm for the gold nanoparticles for the lysozyme bioconjugate is due to the basic condition of pH 12. TEM images revealed successful conjugation of AuNPs with lysozyme. The particle size was 18 ± 10 nm, however there are small sizes of 8 ± 4 nm.

The hydrolytic assay demonstrated a remarkable advancement of lysozyme hydrolytic activity with an increase of 60% measured by turbidimetric assay. It has been reported earlier that bioconjugation of lysozyme with silver, latex, and gold nanoparticles exhibited a prominent enhancement in hydrolytic activity compared to native lysozyme (Imoto et al., 1972; Gu et al., 2004; Ficai and Grumezescu, 2017; Saha et al., 2018). Furthermore, we confirmed the cell wall breakdown after 4 h of incubation. All the strains displayed genomic DNA bands whereas no bands were detected for bacteria incubated with native lysozyme (Figs. 3 and 4). MIC and MBC studies showed that AuNPs bioconjugate is highly bactericidal against Gram-positive and Gram-negative bacteria. The antibacterial activity displayed a superior improvement by 98%–99% than native lysozyme. Antibacterial activity against *P. aeruginosa* was also studied and it was found that AuNPs-lysozyme bioconjugates were more efficient than native lysozyme. Moreover, SEM was used to visualize the morphology of bacteria upon interaction with AuNPs (Fig. 5). The adsorption of lysozyme bioconjugate resulted in morphological changes of *E. coli* from rods to spheroplasts.

Hemocompatibility of AuNPs-lysozyme bioconjugate was also studied. Although the bioconjugate displayed negligible hemolysis (1%–5%) after 1 h of incubation, the hemolysis percentage increased substantially after 24 h depending upon the concentration of bioconjugate. Coating the surface of nanoparticles with a biocompatible layer could predominantly minimize the exposure to RBCs (Bakshi et al., 2011). In short, the invigorating observations of this current and related studies will enhance the replacement of antibiotics with theranostics based on AuNPs.

4 Inhibition of amyloid fibrillation at carboxyl-tailored gold-aryl nanoparticles

Amyloid fibrils are stretched protein aggregates with a rigid core, such fibrillation, in the long run, leads to neurodegenerative and neuropathic diseases (Fink, 1998; Sipe and Cohen, 2000; Tanzi and Bertram, 2005; Sipe et al., 2010; Detoma et al., 2012; Eisenberg and Jucker, 2012;

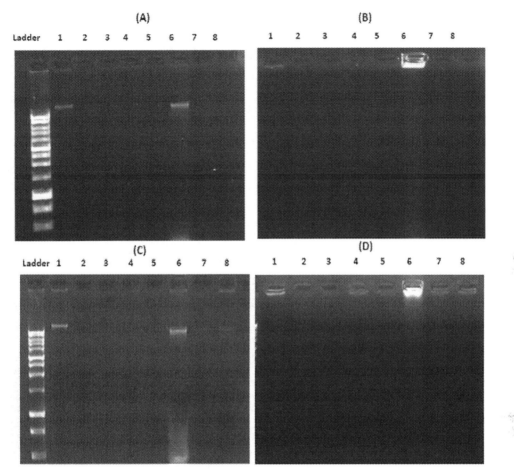

FIG. 3 Images of 1% agarose gel electrophoresis of the bacterial samples taken after 1 h of incubation. (A) With AuNPs-lysozyme bioconjugate showed DNA bands in wells 1 (*S. aureus*) and 6 (*P. aeruginosa*) and (B) with native lysozyme, no DNA bands were observed. After 3 h of incubation, (C) with AuNPs-lysozyme bioconjugate showed extra DNA bands in wells 7 (*K. pneumoniae*) and 8 (*S. typhimurium*). (D) With native lysozyme, no DNA bands were observed. *Credit: Royal Society of Chemistry 2019.*

Knowles et al., 2014; Selkoe and Hardy, 2016; Liu et al., 2019). As a result, amyloid fibrillation has gained prominence in protein-based therapeutics in cases such as misfolding disorders where the soluble states of the proteins are transformed to extremely structured fibrillar aggregates. Since they can interrupt the essential cellular functions, such fibrils are extremely deleterious to the cells (Fink, 1998).

The dissociation of the synthesized fibrils by the gold-aryl nanoparticles was investigated. Gold-insulin bioconjugates were fabricated by the addition of a mild reducing agent to a mixture of aryldiazonium gold(III) salt with insulin. The change in pH developed a sharp alteration in color and the plasmon peak underwent a major redshift from 560 to 650 nm with broadening under acidic conditions (Fig. 6). Zeta potential of the bioconjugate changed with

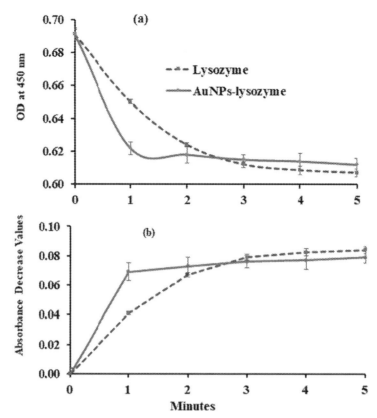

FIG. 4 (A) Linear kinetic curves show a faster reduction of substrate optical density in the first 2 min of the reaction with AuNPs-lysozyme compared to free lysozyme at the same concentration. (B) Linear graph of the reduction in optical density (ΔA450) for AuNPs-lysozyme bioconjugate and native lysozyme in μg mL^{-1}. *Credit: Royal Society of Chemistry 2019.*

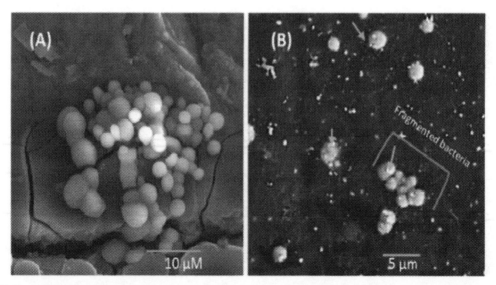

FIG. 5 SEM images of *E. coli* ESBL 1 showing integrity and morphology change upon interaction with lysozyme and gold-lysozyme bioconjugate. (A) After being treated with lysozyme for 1 h showed spheroplasts formation with negligible fragmentation. (B) Treatment with AuNPs-lysozyme for 1 h showed fragmented spherical *E. coli* cells and nanoparticles adsorbed to the *E. coli* surface as indicated by the *arrows*. *Credit: Royal Society of Chemistry 2019.*

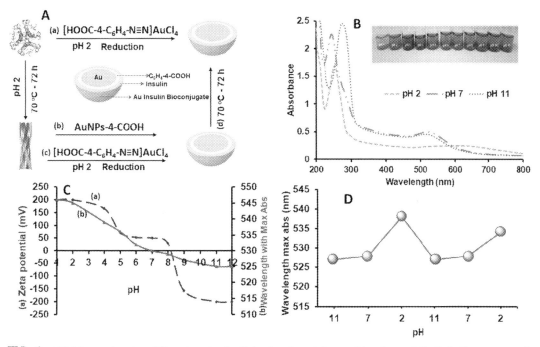

FIG. 6 (A) Scheme showing different routes (a–d) for the dissociation and breakage of fibrils in the presence of gold-aryl NPs. (B) UV-Vis examples of the change in wavelength versus pH for the insulin bioconjugate. Inset showing color change under variable pH values. (C) pH effect on the insulin bioconjugate (a) zeta potential and (b) wavelength. (D) Robustness study of the insulin bioconjugate under reversible change of pH of 2, 7, and 11. The color change is shown as *circles* matching the colors change upon pH shift between 2, 7, and 11. *Credit: Elsevier 2020.*

pH values. Gold nanoparticles/insulin can reversibly aggregate and disaggregate with pH change (Chanana et al., 2011). The quenching of insulin fibrils was studied over 120 min using thioflavin assay (Fig. 7).

Furthermore, since several proteolytic enzymes would promptly degrade peptides and proteins, the oral delivery of therapeutics is pivotal to eliminate a sudden release of protein. In the present study, it was observed that insulin release under pH 1.2 was about 30%–35%, however, under pH 7.4 was about 50% within 60 min. There was a minimal release of insulin under the gastric environment, which controlled the bioavailability of the drug. Also, to investigate the antidegradation capability of insulin-functionalized AuNPs, the formulation was reacted with pepsin and trypsin proteolytic enzymes present in the gastrointestinal tract, which are well known for the degradation of insulin and inhibit its bioavailability. It was observed that gold nanoparticles protected 26%–50% of the adsorbed insulin in the presence of pepsin and trypsin nevertheless the native insulin effectively degraded. Degradation of insulin bioconjugate occurred without a significant initial burst, thus we could achieve satisfactory oral bioavailability of insulin.

A hemocompatibility study was performed on healthy and diabetic blood samples. All samples displayed negligible hemolysis of 1%–2% even after 24 h incubation whereas

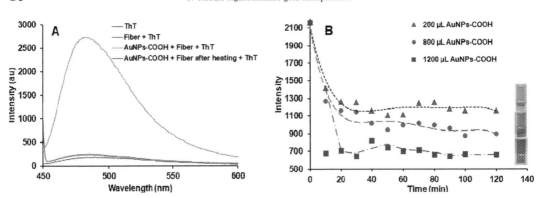

FIG. 7   (A) Insulin bioconjugate fluorescence emission in the presence of ThT dye. Fiber emission was fully quenched after the addition of bare AuNPs. (B) The intensity of insulin fibrils luminescence decreased over time by the incremental addition of AuNPs of 200, 800, and 1200 μL. *Credit: Elsevier 2020.*

samples exposed to insulin fibrils showed hemolysis of 7% and 12% in both healthy and diabetic samples, respectively. In short, the nanoparticles and their bioconjugates to both diabetic and nondiabetic exhibited a phenomenal hemocompatibility. Unquestionably, the presented nanoparticles are proficient and active systems as carriers for insulin drug delivery and to inhibit amyloid fibrillation.

5 Protein-coated gold nanoparticles: Green and chemical synthesis routes and their cellular uptake

Gold nanoparticles have shown a vast range of properties, thus considered the most dominant materials in fundamental research and applications. Several studies have shown that AuNPs can easily be functionalized using various biomolecules like DNA, proteins, antibodies, etc. (Hameed et al., 2018; Ahmad et al., 2019). Thus, biomolecules attached to AuNPs facilitate their cellular uptake. The bioconjugated AuNPs show enhanced water solubility, become physiologically compatible and nontoxic. Thereby, making them suitable for applications in a wide range of biomedical fields (Yong et al., 2009). Different routes have been developed over the years for the synthesis of AuNPs. Studies have shown that AuNPs synthesized with the carboxylic functional groups are extremely robust and can be easily bioconjugated with various biomolecules (Orefuwa et al., 2013).

The synthesis of AuNPs has been observed through two different approaches mainly by using chemicals for reduction and by green synthesis (Mohamed et al., 2015; Ahmady et al., 2019). Although both methods produce AuNPs with good efficiency, one major drawback of the chemical reduction is that the AuNPs may contain unwanted and harmful chemicals and surfactants, as part of their chemical treatment, and that may be hard to remove (Chehimi, 2012). On the other hand, the green method uses different biological reductants like plant extracts, microorganisms, and biomolecules. The advantages of green synthesis over chemical reduction were pointed out in several reports (Bakshi et al., 2011; Khullar et al., 2012; Bakshi, 2017). AuNPs synthesized by the green method exhibit slower kinetics, they are highly biocompatible, and they can be manipulated accordingly, and are more stable to aggregation. Also, these AuNPs are cost-effective, eco-friendly, and can be produced in bulk (Bakshi, 2017). To be used in biomedical applications, AuNPs along with being biofriendly must also be able to enter the cells. The uptake of AuNPs by the cells has also become one of the most widely researched areas. If the AuNPs are easily able to enter the cells, they can be very efficiently used for therapeutic purposes like drug delivery, as gene manipulating tools, intercellular messengers, etc. There are several factors, like size, shape, electrostatic interactions, and hydrophobicity of AuNPs, that determine their entry into the cells. So, the smaller the size of AuNPs, the easier their entry into the cells. Also, cells prefer uptake of spherical AuNPs compared to rod shapes. AuNPs' surface charge and hydrophobicity are also important for their interactions with cell membranes. Several studies have shown that AuNPs having a protein coating can easily enter the cells. This internalization occurs because of alteration in the physiochemical properties of AuNPs and subsequently changes the surface charge, hence enhancing their cellular uptake (Cheng et al., 2015; Saikia et al., 2016).

5.1 Synthesis of protein-coated gold nanoparticles: Green versus chemical routes

In our study, we synthesized AuNPs using both green and chemical reduction methods. In the green reduction route, proteins were used to reduce the aryl diazonium gold salt under strict conditions of pH and temperature. While in the chemical reduction we had combined the protein and diazonium gold solution and it was subjected to reduction using mild reducing agents 9-BBN and sodium borohydride (Chehimi, 2012; Orefuwa et al., 2013). This work was planned to conduct a comparative study between both types of AuNPs, for both in their characterization as well as their cellular uptake. Four different proteins were selected, BSA, lysozyme, zein, and collagen, regarding our previous work done using the first two proteins (Hameed et al., 2018; Ahmady et al., 2019) and other studies (Khullar et al., 2012; Bakshi, 2017). Thus, the synthesized protein-coated AuNPs from both methods showed the typical colors of AuNPs ranging from light pink, dark pink, purple, and brown.

5.2 Characterization of protein-coated gold nanoparticles

Characterization of the protein-coated AuNPs was studied to find out any difference in the shapes, sizes, and appearance between the two types prepared. The initial characterization was done using UV-Vis, FTIR, DLS, and TEM measurements. All the protein-coated AuNPs exhibited the typical plasmon peak in the range of 500–600 nm, thus confirming the successful synthesis of AuNPs by both methods. The AuNPs-BSA, which were synthesized using the green method at pH 3.5 gave a broad, red-shifted peak in the range of 600–650 nm, thus showing the formation of aggregation, hence forming large-sized AuNPs under acidic conditions, compared to all other AuNPs which were in the 550 nm range.

FTIR spectra for both green-synthesized and chemically reduced protein-coated AuNPs gave the absorption band stretch vibrations of groups like OH, —C=C—, —C—O, and —C—OH, thus confirming their presence in the AuNPs. The size and zeta potentials of each protein-coated AuNP were measured using DLS. The scattering intensity was represented as a function of the logarithm of the diameter of nanoparticles. All AuNPs bioconjugates

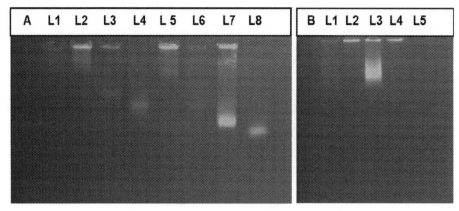

FIG. 8   (A) Electrophoretic mobility of green-reduced AuNPs bioconjugates in 0.8% agarose gel run for 40 min at 100 V in 0.5 × TBE buffer (pH 8.3). L1: AuNPs-COOH, L2: AuNPs-BSA (pH 3.5), L3: AuNPs-BSA (pH 4), L4: AuNPs-collagen (pH 7), L5: AuNPs-zein (pH 6), L6–L8: AuNPs-lysozyme (pH 9.8, 10.4, and 12). (B) Electrophoretic mobility of chemically reduced AuNPs in 0.8% agarose gel run for 40 min at 100 V in 0.5 × TBE buffer (pH 8.3); L1: uncoated AuNPs, L2: AuNPs-BSA, L3: AuNPs-lysozyme, L4: AuNPs-zein, L5: AuNPs-collagen. *Credit: American Chemical Society 2020.*

synthesized by the green method showed a smaller size except the BSA-AuNPs at a pH of 3.5. All chemically reduced AuNPs showed smaller sizes. Earlier reports also showed the relation between the size of the nanoparticles and pH, the concentration of the solution, temperature, and time of incubation (Mott et al., 2007; Wu et al., 2016). The UV-Vis and DLS results were further supported by zeta potential values.

Studies reported that the electrophoretic mobility of AuNPs in the gel is related to surface charge (Wangoo et al., 2008; Zare et al., 2012). Accordingly, our gel electrophoresis results showed the same pattern of movement. The protein-coated AuNPs, which were negatively charged, moved faster toward the positive pole, while the mobility was retarded for the ones with a positive charge that included the green-synthesized AuNPs-BSA and the chemically reduced AuNPs-lysozyme, AuNPs-zein, and AuNPs-collagen (Fig. 8).

The size and shape of the synthesized protein-coated AuNPs were further studied with TEM. The images exhibited spherical-shaped AuNPs synthesized by both methods. Green-synthesized AuNPs-zein showed different shapes like oval, triangle, and hexagonal. The nanoparticles showed lattice spacing of (200) and (111) planes, thus confirming the presence of metallic gold in all the protein-coated AuNPs.

5.3 Biological studies

For using nanoparticles in medical applications, they must be readily uptaken by the cells. Previous studies showed that chemically reduced AuNPs-BSA and native AuNPs were nontoxic as there were no RBCs hemolysis observed (Hameed et al., 2018). MG-63 osteosarcoma cell lines were used for cellular uptake studies and were assessed using confocal microscopy and flow cytometry (Figs. 9 and 10). In this study, all protein-coated AuNPs, as well as native AuNPs, were tagged with FITC fluorescent probes and incubated with the cells

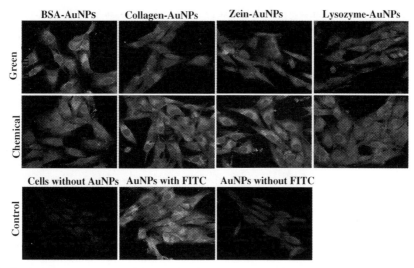

FIG. 9 Confocal microscopy images of the uptake of protein-coated gold-carbon nanoparticles synthesized with green- and chemical-reduced methods. The more intense the fluorescence, the higher uptake of the AuNPs by MG-63 cells. *Credit: American Chemical Society 2020.*

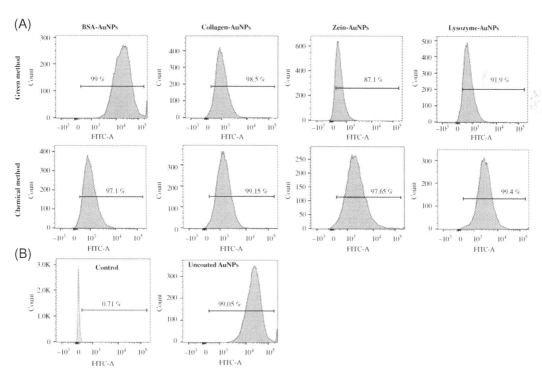

FIG. 10 MG-63 cells were treated with AuNPs-BSA, AuNPs-collagen, AuNPs-zein, and AuNPs-lysozyme and un-coated AuNPs synthesized by green- and chemical-reduced methods. All AuNPs were labeled with FITC and the cells were treated for 24 h. (A) Flow cytometric analysis measuring the mean of fluorescence intensity AuNPs for both green- and chemical-reduced methods. (B) Histograms with mean intensity for the four types of the coated AuNPs, a plot of a single parameter (FITC intensity, horizontal axis) versus the number of events detected (vertical axis) for both green- and chemical-reduced methods. *Credit: American Chemical Society 2020.*

for 24 h. Flow cytometry showed contradictory results in the case of both green-synthesized and chemically reduced AuNPs. The green-reduced AuNPs-BSA were uptaken more compared to other green-synthesized AuNPs (Fig. 10). While in the chemically reduced, the AuNPs-lysozyme entered more efficiently, followed by AuNPs-collagen. The least entry was shown by AuNPs-BSA. Similar results were obtained by confocal microscopy. From the results that we obtained from our cell studies, it was observed that each AuNPs were taken differently. The uptake of AuNPs depends mainly upon the size, shapes, and surface charges.

6 Computational methods

The use of theoretical methods such as molecular docking, molecular dynamics simulations, quantum mechanics, etc. have important implications and applications in predicting the binding modes for proteins and their substrates (Arooj et al., 2013; Bentel et al., 2017; Harb et al., 2017; Iqbal et al., 2019; Shimamura et al., 2020). With the aid of computational methods, protein-ligand interactions can be evaluated and studied in detail to understand the binding mechanisms of protein-substrate complexes and their impact on the regulation of the biological processes. In the field of nanochemistry, computational techniques provide a theoretical basis for the design and development of nanomedicines and their carriers by exploring the binding affinity and mechanisms of medically important biomolecules with various substrates and ligands (Chetty and Singh, 2020; Samal et al., 2020; Shady et al., 2020).

Theoretical and experimental studies on medically important biomolecules such as insulin, bovine serum albumin, zein, lysozyme, and various substrates have been performed to understand the biologically important phenomena such as inhibition of protein aggregation by carboxylate-terminated gold-carbon nanoparticles, dissemination of multidrug-resistant (MDR) by green- and cytocompatible carboxyl modified gold-lysozyme nano antibacterial, physicochemical stability of protein-carboxylic acid complexes, etc. (Ahmady et al., 2019; AlBab et al., 2020).

6.1 Physicochemical stability study of protein-carboxylic acid complexes

Substrates such as carboxylic acids, amines, and hydroxyls have a wide range of important applications including drug carriers, stabilizers, and food additives (Chen et al., 2019). Carboxylic substrates with excellent chemical and physical properties are the most significant due to the high solubility in polar solvents and the absence of further reactions under oxidative conditions (Zhang et al., 2018). Due to their intrinsic properties and role in the synthesis of biomaterials, modification of substrate surfaces with carboxylic moieties gained much consideration in the field of nanomedicine and biomedical engineering. Because of surface modification, the cellular uptake of drugs increases (Wu et al., 2019). Owing to its significance in several important fields, it is imperative to elucidate the binding mechanism and physicochemical stability of carboxylic acid with various proteins of biological importance.

To study the binding mode and physicochemical stability of protein-carboxylic acid complexes, our research has performed a computational study of benzoic acid with key protein targets including lysozyme, bovine serum albumin, and zein using diverse computational approaches such as homology modeling, protein-ligand molecular docking, and molecular dynamics simulations. Due to the absence of the crystal structure of zein, a homology modeling method was applied to construct the 3D structure of zein protein. For this purpose, various web servers including I-Tasser, Phyre2, and RaptorX were utilized. The best homology model was selected using several validation checks including Procheck and ProSA web servers. The highest scoring model with the best quality was subjected to subsequent molecular docking calculations. The lowest free energy conformations of the complexes of BSA-benzoic acid, zein-benzoic acid, and lysozyme-benzoic acid exhibiting the binding free energy of -5.28, -4.83, and $-4.694\,kJ/mol$, respectively, were chosen for further study. To evaluate the physicochemical stability of benzoic acid with the selected proteins, molecular dynamics (MD) simulations were performed on final complexes retrieved from molecular docking calculations.

Stability analysis on MD trajectories for all three protein complexes was performed via Root-mean-square deviation (RMSD), and radius of gyration. Outcomes from the analyses showed that all protein-ligand complexes were stable. In comparison with other proteins, the lysozyme complexed with benzoic acid exhibited the best stability with benzoic acid having diverse interactions with the active site of lysozyme namely hydrogen bonding, hydrophobic interactions (Fig. 11). The binding conformation of benzoic acid with BSA showed that it was bound with the hydrophobic subdomain IIA of the protein which is considered as the favored and the most stable binding site in the BSA protein (Hierrezuelo and Ruiz, 2015). Benzoic acid formed various hydrophobic and electrostatic interactions with the subdomain IIA of the protein (Fig. 11). Zein protein also showed optimal binding with benzoic acid via strong hydrogen bonding and hydrophobic interactions. A key feature of benzoic acid interacting with the zein was the formation of three strong hydrogen bonds along with hydrophobic contacts as well with residues including Pro116, Gln176 Pro179, Phe180, and Leu183 (Fig. 11).

Dynamics stability of complex structures was also investigated using root-mean-square fluctuation (RMSF) flexibility check. The conformational changes in the protein structures while bonded to benzoic acid at their active sites were examined in detail. As all three proteins have compact globular structures so, generally, all protein structures showed stability with the exceptions of few regions that displayed large flexibility. The relatively more flexible regions were located at the N- and C-termini of the proteins. Coil and loop parts of the protein exhibited higher flexibility as expected. Evaluation of the solvent-accessible surface area of all three proteins revealed that binding cavities of BSA and zein are deeply contained inside the protein structures in comparison with the lysozyme-binding cavity which is more open and closer to the surface. The findings of our study provide key insights into the compactness and structural dynamic stability of interactions of benzoic acid with BSA, zein, and lysozyme proteins. This research endeavor may be helpful in the drug design and discovery process in the field of biomedical and nanomedicine research by understanding the key features crucial for the stability of therapeutic protein drugs loaded on carboxyl substrates.

An experimental and computational study was conducted to elucidate the mechanism of noncovalent interactions between phenolic acids and lysozyme (Chen et al., 2019). Phenolic acids are considered very significant owing to their important biological role in antioxidation, antiinflammatory, and antiallergy. AutoDock program was used to predict the potential

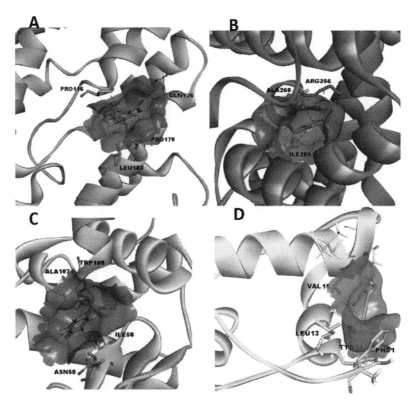

FIG. 11   Complexes of benzoic acid with proteins (A) BSA, (B) zein (C) lysozyme, and (D) insulin. Protein structures are displayed in a ribbon style. Key residues interacting with benzoic acid are represented in sticks. Benzoic acid is displayed with sticks. Nonpolar hydrogen atoms are removed for clarity. H-bonding and hydrophobic (including π-π) interactions are shown via dashed lines, respectively. *Credit: Springer 2020.*

binding conformations of four phenolic acids including chlorogenic acid (CQA), rosmarinic acid (RA), 1,3-*O*-dicaffeoylquinic acid (1,3-CQA), and 4,5-dicaffeoylquinic acid (4,5-CQA) with lysozyme. The docking results revealed the involvement of the residues Trp62, Trp63, R73, Ile98, Val99, Asp101, Gly 102, Asn103, Gly104, and Trp 108 in interactions with the four phenolic acids. To make potent binding, compounds were able to establish a variety of interactions such as hydrophobic, electrostatic, and h-bonding interactions with residues from lysozyme. The findings from this study validate the outcomes from the experimental techniques thus showing the potent binding of phenolic acid with lysozyme.

6.2 Inhibition of insulin aggregation by carboxylate-terminated nanoparticles

Protein aggregation has received widespread consideration in nanochemistry and nanomedicine because of its association with several human diseases such as Alzheimer's and type II diabetes. Especially, insulin fibril aggregation is related to various medical problems, thus, designing strong biocompatible inhibitors against such fibril assembly is crucial especially in diabetes therapy. To understand the mechanism of inhibition of insulin fibrillation by benzoic acid-terminated gold-carbon nanoparticles, various experimental techniques were used. Molecular docking calculations were applied to predict the complex of benzoic acid with insulin (AlBab et al., 2020). Blind docking is a powerful technique to calculate the binding mode of the compound for which no prior knowledge of the target active site. It is a computationally expensive technique as a docking grid is defined in such a way that the entire surface of the protein is searched for potential binding affinity for the docked ligand (Hosen et al., 2020). Therefore, the whole surface of the insulin protein was examined to calculate the possible binding pockets of insulin for the benzoic acid by setting the grid dimensions to $68 \times 66 \times 82\,\text{Å}$ thus covering the complete protein exterior.

From docking results, the lowest energy configuration of $-3.78\,\text{kJ/mol}$ from the resulting clusters was selected to explore the binding pose interactions of benzoic acid with insulin. Detailed analysis of the selected configuration of the insulin-benzoic acid complex revealed that benzoic acid interacted with insulin via various kinds of interactions including hydrogen bonding and hydrophobic interactions (Fig. 12). Benzoic acid formed key interactions with hydrophobic and aromatic residues of both chains of insulin including Phe1, and Val18 in the B chain, and Leu13, Tyr14 in the A chain. These residues belong to the fibril-forming segments (residues A13–A18) and (residues B12–B17) in protein. Previous experimental and theoretical studies have shown the significance of the residues in the aggregation of insulin (Sawaya et al., 2007; Thompson et al., 2006). A computational approach based on 3D profiling of proteins developed for identifying segments of amyloidogenic proteins forming amyloid-like fibrils predicted the above-mentioned regions of insulin prone for aggregation (Sawaya et al., 2007). So, the strong interactions of benzoic acid such as a very strong hydrogen bond

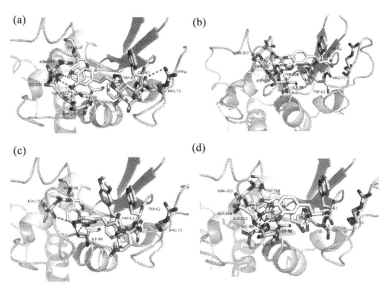

FIG. 12 Binding modes (A–D) showing the residues involved in the contacts with chlorogenic acid (CQA), rosmarinic acid (RA), 1,3-O-dicaffeoylquinic acid (1,3-CQA), and 4,5-dicaffeoylquinic acid (4,5-CQA), respectively. *Credit: Elsevier 2020.*

(1.95 Å) with the backbone of Val18 and π-π stacking interactions with Tyr14 residues belonging to the fibril-forming region of insulin may be attributed to the deterrence of conformational changes required for the fibril formation and thus aggregation of insulin. The outcomes of this work may support the progress of nanoparticle-based drug design to inhibit medical complications associated with insulin aggregation.

Research work was conducted on the use of ferulic acid for the inhibition of insulin amyloid fibril formation (Jayamani et al., 2014). Ferulic acid is an organic compound that exists in several fruits and vegetables. Results indicated that the inhibition of insulin fibrillation was dependent on the concentration of ferulic acid. Based on the blind docking method, molecular docking calculations were done to predict and evaluate the binding conformation of ferulic acid with insulin using the AutoDock Vina program. Ferulic acid was able to bind with insulin with the lowest energy conformations ranging from −5.0 to −4.3 kcal mol^{-1}. Three potential interacting sites of insulin were identified for ferulic acid as depicted in the image (Fig. 13). Ferulic acid formed several kinds of interactions including hydrogen bonding, and hydrophobic interactions using its benzene, carboxyl, and phenol hydroxyl structural

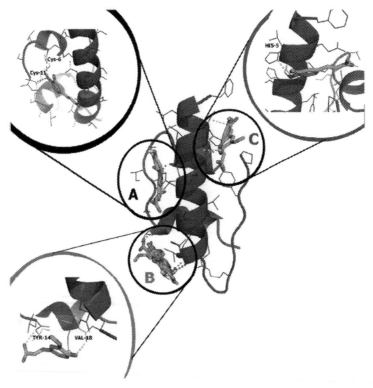

FIG. 13   Docking results. The image demonstrates the three binding regions (A, B, and C) of FA in the insulin protein. FA displays more than one configuration. The zoomed view of three binding sites shows the lowest energy modes of FA at the respective binding region. Ribbons represent the two chains of insulin. *Credit: RSC 2014.*

groups. The contacts between ferulic acid and insulin may cause conformational changes, which may further avert the unfolding of protein and hence preventing the amyloid fibril formation.

Another computational and experimental study on the use of ascorbic acid for the inhibition of human insulin aggregation revealed that ascorbic acid can also help in the prevention of aggregation of human insulin and related pathophysiology (Alam et al., 2017). To explore the binding mode of ascorbic acid with the insulin, molecular docking calculations were performed using the AutoDock program and the lowest free energy ($-4.56\,kcal\,mol^{-1}$) complex of insulin and ascorbic acid was selected for analysis. Examination of the selected binding mode of insulin-ascorbic acid complex indicated the formation of some interactions with the hydrophobic and polar residues located in the close vicinity of the aggregation-prone region. It was suggested that ascorbic acid interacting with residues located in the neighboring region of the fibril-forming section of insulin might prevent the conformational changes, which are required for fibril formation.

6.3 Carboxyl-modified gold-lysozyme antibacterial for combating multidrug-resistant superbugs

The dissemination of multidrug resistant (MDR) superbugs is considered very crucial for antibiotic resistance. AuNPs-lysozyme antibacterial was fabricated by the conjugation of AuNPs-C_6H_4-4-COOH with lysozyme (Ahmady et al., 2019). Various experimental techniques demonstrated that the lysozyme bioconjugate had excellent results against superbugs. It was reported that the use of lysozyme bioconjugates had done very little or no damage to healthy human cells. Thus, it was suggested that such antibacterial functionalities could be used to wound dressing and medical devices.

Computational studies were performed to elucidate the mechanism of benzoic acid and lysozyme binding in the carboxyl-modified gold nanoparticles-lysozyme system. Molecular electrostatic potential (MEP) calculations along with the blind docking method were applied to prepare and analyze the binding configuration of lysozyme with benzoic acid. A molecular grid with dimensions of $74 \times 104 \times 106\,\text{Å}$ was set to cover the whole surface of lysozyme so that the entire protein surface can be scanned to explore the best possible binding mode of benzoic acid. 2,500,000 energy evolutions were executed to search the minimum free energy conformation for the protein and benzoic acid complex. The best configuration with the lowest binding energy of $-4.694\,kJ/mol$ was collected from docking calculations. To explore the MEP surfaces of the lysozyme-benzoic acid complex structure, the selected conformation was subjected to a macromolecular electrostatics calculation program, Adaptive Poisson-Boltzmann Solver (APBS). Docking results revealed the binding of benzoic acid at the active site of lysozyme via strong hydrogen bonding interaction (1.93 Å) with the hydrophobic residue Ala107. Benzoic acid also formed several interactions with hydrophobic residues of the active site of lysozyme such as Ile58, Ala107, and Trp108 (Fig. 13). Key interactions with the Trp108 residue positioned in the hydrophobic matrix region of the protein were also established via π-π T-shaped stacking contacts. Trp108 is considered one of the main fluorophores in lysozyme (Imoto et al., 1972; Gu et al., 2004). Benzoic acid showed optimal binding interactions with residues at the active region such as Ile58, Asn59, Ala107, and

Trp108. Due to their position at the cleft of the protein, these residues play a key role in the potent binding of ligands at the binding pocket of lysozyme (Pushkaran et al., 2015; Saha et al., 2018; Zaman et al., 2019).

The electrostatic potential area of the binding cavity of protein bonded to the ligand in the lysozyme-benzoic acid complex was elucidated in detail. With a potential range from −5 to 5 eV, the mapping of the electrostatic potential surface onto the protein was computed. The outer surface of the lysozyme is occupied with a positive charge as depicted in the image with blue color while the inner region of the lysozyme is red-colored representing negatively charged hydrophobic cavities located in the central part of the lysozyme structure. It is quite evident that benzoic acid is well-positioned in the active site of the positively charged globular protein structure from the analysis of the electrostatic potential surface presentation of lysozyme. The outcomes from the computational methods also validate the experimental findings which report the binding of carboxyl modified gold nanoparticles with the inner region of lysozyme. Furthermore, results from MEP calculations also expose the capability of positively charged lysozyme to establish strong electrostatic contacts with the negatively charged bacterial wall.

7 Conclusion

Gold nanoparticles emerged in the last few years as the most preferred carriers for proteins and drugs. This has been proven through experimental and theoretical studies. Experimental results complied so far support the high dispersibility and biocompatibility of the synthesized bioconjugates. Besides, the ability to be uptaken by cells is critical for therapeutic and theranostics effects. Further, the theoretical studies proved the physicochemical stability of such bioconjugates on carboxylic acid surfaces. Consequently, the implementation of diverse computational techniques along with experimental approaches can be very useful to elucidate the questions of biological importance.

References

Ahmad, A.A., Panicker, S., Chehimi, M.M., Monge, M., Lopez-De-Luzuriaga, J.M., Mohamed, A.A., Bruce, A.E., Bruce, M.R., 2019. Synthesis of water-soluble gold–aryl nanoparticles with distinct catalytic performance in the reduction of the environmental pollutant 4-nitrophenol. Catal. Sci. Technol. 9 (21), 6059–6071. https://doi.org/10.1039/c9cy01402k.

Ahmady, I.M., Hameed, M.K., Almehdi, A.M., Arooj, M., Workie, B., Sahle-Demessie, E., Han, C., Mohamed, A.A., 2019. Green and cytocompatible carboxyl modified gold–lysozyme nanoantibacterial for combating multidrug-resistant superbugs. Biomater. Sci. 7 (12), 5016–5026. https://doi.org/10.1039/c9bm00935c.

Alam, P., Beg, A.Z., Siddiqi, M.K., Chaturvedi, S.K., Rajpoot, R.K., Ajmal, M.R., Zaman, M., Abdelhameed, A., Khan, R.H., 2017. Ascorbic acid inhibits human insulin aggregation and protects against amyloid induced cytotoxicity. Arch. Biochem. Biophys. 621, 54–62. https://doi.org/10.1016/j.abb.2017.04.005.

AlBab, N.D., Hameed, M.K., Maresova, A., Ahmady, I.M., Arooj, M., Han, C., Workie, B., Chehimi, M., Mohamed, A.A., 2020. Inhibition of amyloid fibrillation, enzymatic degradation and cytotoxicity of insulin at carboxyl tailored gold-aryl nanoparticles surface. Colloids Surf. A 586. https://doi.org/10.1016/j.colsurfa.2019.124279, 124279.

Arooj, M., Kim, S., Sakkiah, S., Cao, G.P., Lee, Y., Lee, K.W., 2013. Molecular modeling study for inhibition mechanism of human chymase and its application in inhibitor design. PLoS One 8 (4). https://doi.org/10.1371/journal.pone.0062740.

Bakshi, M.S., 2017. Nanotoxicity in systemic circulation and wound healing. Chem. Res. Toxicol. 30 (6), 1253–1274. https://doi.org/10.1021/acs.chemrestox.7b00068.

Bakshi, M.S., Kaur, H., Khullar, P., Banipal, T.S., Kaur, G., Singh, N., 2011. Protein films of bovine serum albumen conjugated gold nanoparticles: a synthetic route from bioconjugated nanoparticles to biodegradable protein films. J. Phys. Chem. C 115 (7), 2982–2992. https://doi.org/10.1021/jp110296y.

Bentel, J.M., Thomas, M.A., Rodgers, J.J., Arooj, M., Gray, E., Allcock, R., Fermoyle, S., Mancera, R.L., Cannell, B., Parry, J., 2017. Erdheim–Chester disease associated with a novel, complex BRAF p.Thr599_Val600delinsArgGlu mutation. BMJ Case Rep. https://doi.org/10.1136/bcr-2017-219720.

Brewer, S.H., Glomm, W.R., Johnson, M.C., Knag, M.K., Franzen, S., 2005. Probing BSA binding to citrate-coated gold nanoparticles and surfaces. Langmuir 21 (20), 9303–9307. https://doi.org/10.1021/la050588t.

Chanana, M., Correa-Duarte, M.A., Liz-Marzán, L.M., 2011. Insulin-coated gold nanoparticles: a plasmonic device for studying metal-protein interactions. Small 7 (18), 2650–2660. https://doi.org/10.1002/smll.201100735.

Chehimi, M.M. (Ed.), 2012. Aryl Diazonium Salts: New Coupling Agents in Polymer and Surface Science. Wiley Inc.

Chen, L.Q., Kang, B., Ling, J., 2013. Cytotoxicity of cuprous oxide nanoparticles to fish blood cells: hemolysis and internalization. J. Nanopart. Res. 15 (3). https://doi.org/10.1007/s11051-013-1507-7.

Chen, S., Han, Y., Wang, Y., Yang, X., Sun, C., Mao, L., Gao, Y., 2019. Zein-hyaluronic acid binary complex as a delivery vehicle of quercetagetin: fabrication, structural characterization, physicochemical stability and in vitro release property. Food Chem. 276, 322–332. https://doi.org/10.1016/j.foodchem.2018.10.034.

Cheng, X., Tian, X., Wu, A., Li, J., Tian, J., Chong, Y., Chai, Z., Zhao, Y., Chen, C., Ge, C., 2015. Protein corona influences cellular uptake of gold nanoparticles by phagocytic and nonphagocytic cells in a size-dependent manner. ACS Appl. Mater. Interfaces 7 (37), 20568–20575. https://doi.org/10.1021/acsami.5b04290.

Chetty, R., Singh, M., 2020. In-vitro interaction of cerium oxide nanoparticles with hemoglobin, insulin, and dsDNA at 310.15 K: physicochemical, spectroscopic and in-silico study. Int. J. Biol. Macromol. 156, 1022–1044. https://doi.org/10.1016/j.ijbiomac.2020.03.067.

Detoma, A.S., Salamekh, S., Ramamoorthy, A., Lim, M.H., 2012. Misfolded proteins in Alzheimer's disease and type II diabetes. Chem. Soc. Rev. 41 (2), 608–621. https://doi.org/10.1039/c1cs15112f.

Eisenberg, D., Jucker, M., 2012. The amyloid state of proteins in human diseases. Cell 148 (6), 1188–1203. https://doi.org/10.1016/j.cell.2012.02.022.

Ficai, A., Grumezescu, A. (Eds.), 2017. Nanostructures for Antimicrobial Therapy. Elsevier Inc., https://doi.org/10.1039/9781849730990-00001.

Fink, A.L., 1998. Protein aggregation: folding aggregates, inclusion bodies and amyloid. Fold. Des. 3 (1). https://doi.org/10.1016/s1359-0278(98)00002-9.

Gu, Z., Zhu, X., Ni, S., Su, Z., Zhou, H., 2004. Conformational changes of lysozyme refolding intermediates and implications for aggregation and renaturation. Int. J. Biochem. Cell Biol. 36 (5), 795–805. https://doi.org/10.1016/j.biocel.2003.08.015.

Hameed, M.K., Ahmady, I.M., Alawadhi, H., Workie, B., Sahle-Demessie, E., Han, C., Chehimi, M.M., Mohamed, A.A., 2018. Gold-carbon nanoparticles mediated delivery of BSA: remarkable robustness and hemocompatibility. Colloids Surf. A 558, 351–358. https://doi.org/10.1016/j.colsurfa.2018.09.004.

Hameed, M., Panicker, S., Abdallah, S.H., Khan, A.A., Han, C., Chehimi, M.M., Mohamed, A.A., 2020. Protein-coated aryl modified gold nanoparticles for cellular uptake study by osteosarcoma cancer cells. Langmuir 40 (36), 11765–11775. https://doi.org/10.1021/acs.langmuir.0c01443.

Harb, L.H., Arooj, M., Vrielink, A., Mancera, R.L., 2017. Computational site-directed mutagenesis studies of the role of the hydrophobic triad on substrate binding in cholesterol oxidase. Proteins 85 (9), 1645–1655. https://doi.org/10.1002/prot.25319.

Hierrezuelo, J., Ruiz, C.C., 2015. Exploring the affinity binding of alkylmaltoside surfactants to bovine serum albumin and their effect on the protein stability: a spectroscopic approach. Mater. Sci. Eng. C 53, 156–165. https://doi.org/10.1016/j.msec.2015.04.039.

Hosen, M.J., Hasan, M., Chakraborty, S., Abir, R.A., Zubaer, A., Coucke, P., 2020. Comprehensive in silico study of GLUT10: prediction of possible substrate binding sites and interacting molecules. Curr. Pharm. Biotechnol. 21 (2), 117–130. https://doi.org/10.2174/1389201020666190613152030.

Imoto, T., Forster, L.S., Rupley, J.A., Tanaka, F., 1972. Fluorescence of lysozyme: emissions from tryptophan residues 62 and 108 and energy migration. PNAS 69 (5), 1151–1155. https://doi.org/10.1073/pnas.69.5.1151.

Iqbal, Z., Morahan, G., Arooj, M., Sobolev, A.N., Hameed, S., 2019. Synthesis of new arylsulfonylspiroimidazolidine-2′,4′-diones and study of their effect on stimulation of insulin release from MIN6 cell line, inhibition of human aldose reductase, sorbitol accumulations in various tissues and oxidative stress. Eur. J. Med. Chem. 168, 154–175. https://doi.org/10.1016/j.ejmech.2019.02.036.

Jayamani, J., Shanmugam, G., Singam, E.R., 2014. Inhibition of insulin amyloid fibril formation by ferulic acid, a natural compound found in many vegetables and fruits. RSC Adv. 4 (107), 62326–62336. https://doi.org/10.1039/c4ra11291a.

Khullar, P., Singh, V., Mahal, A., Dave, P.N., Thakur, S., Kaur, G., Singh, J., Kamboj, S.S., Bakshi, M.S., 2012. Bovine serum albumin bioconjugated gold nanoparticles: synthesis, hemolysis, and cytotoxicity toward cancer cell lines. J. Phys. Chem. C 116 (15), 8834–8843. https://doi.org/10.1021/jp300585d.

Knowles, T.P., Vendruscolo, M., Dobson, C.M., 2014. The amyloid state and its association with protein misfolding diseases. Nat. Rev. Mol. Cell Biol. 15 (6), 384–396. https://doi.org/10.1038/nrm3810.

Lin, Y., Haynes, C.L., 2010. Impacts of mesoporous silica nanoparticle size, pore ordering, and pore integrity on hemolytic activity. J. Am. Chem. Soc. 132 (13), 4834–4842. https://doi.org/10.1021/ja910846q.

Liu, Y., Zhou, H., Yin, T., Gong, Y., Yuan, G., Chen, L., Liu, J., 2019. Quercetin-modified gold-palladium nanoparticles as a potential autophagy inducer for the treatment of Alzheimer's disease. J. Colloid Interface Sci. 552, 388–400. https://doi.org/10.1016/j.jcis.2019.05.066.

Lynch, I., Dawson, K.A., 2008. Protein-nanoparticle interactions. Nano Today 3 (1–2), 40–47. https://doi.org/10.1016/s1748-0132(08)70014-8.

Mahal, A., Khullar, P., Kumar, H., Kaur, G., Singh, N., Jelokhani-Niaraki, M., Bakshi, M.S., 2013. Green chemistry of zein protein toward the synthesis of bioconjugated nanoparticles: understanding unfolding, fusogenic behavior, and hemolysis. ACS Sustain. Chem. Eng. 1 (6), 627–639. https://doi.org/10.1021/sc300176r.

Mohamed, A.A., Salmi, Z., Dahoumane, S.A., Mekki, A., Carbonnier, B., Chehimi, M.M., 2015. Functionalization of nanomaterials with aryldiazonium salts. Adv. Colloid Interf. Sci. 225, 16–36. https://doi.org/10.1016/j.cis.2015.07.011.

Mott, D., Galkowski, J., Wang, L., Luo, J., Zhong, C., 2007. Synthesis of size-controlled and shaped copper nanoparticles. Langmuir 23 (10), 5740–5745. https://doi.org/10.1021/la0635092.

Orefuwa, S.A., Ravanbakhsh, M., Neal, S.N., King, J.B., Mohamed, A.A., 2013. Robust organometallic gold nanoparticles. Organometallics 33 (2), 439–442. https://doi.org/10.1021/om400927g.

Peter, T., Reed, R.G., 1978. Serum albumin as a transport protein. In: Sund, H. (Ed.), Transport by Proteins. Walter de Gruyter, Berlin, pp. 57–73.

Pihlasalo, S., Auranen, L., Hänninen, P., Härmä, H., 2012. Method for estimation of protein isoelectric point. Anal. Chem. 84 (19), 8253–8258. https://doi.org/10.1021/ac301569b.

Pushkaran, A.C., Nataraj, N., Nair, N., Götz, F., Biswas, R., Mohan, C.G., 2015. Understanding the structure–function relationship of lysozyme resistance in staphylococcus aureus by peptidoglycan O-acetylation using molecular docking, dynamics, and lysis assay. J. Chem. Inf. Model. 55 (4), 760–770. https://doi.org/10.1021/ci500734k.

Saha, B., Chowdhury, S., Sanyal, D., Chattopadhyay, K., Kumar, G.S., 2018. Comparative study of toluidine blue O and methylene blue binding to lysozyme and their inhibitory effects on protein aggregation. ACS Omega 3 (3), 2588–2601. https://doi.org/10.1021/acsomega.7b01991.

Saikia, J., Yazdimamaghani, M., Moghaddam, S.P., Ghandehari, H., 2016. Differential protein adsorption and cellular uptake of silica nanoparticles based on size and porosity. ACS Appl. Mater. Interfaces 8 (50), 34820–34832. https://doi.org/10.1021/acsami.6b09950.

Samal, R.R., Mishra, M., Subudhi, U., 2020. Differential interaction of cerium chloride with bovine liver catalase: a computational and biophysical study. Chemosphere 239. https://doi.org/10.1016/j.chemosphere.2019.124769, 124769.

Sawaya, M.R., Sambashivan, S., Nelson, R., Ivanova, M.I., Sievers, S.A., Apostol, M.I., Thompson, M.J., Balbirnie, M., Wiltzius, J.J., McFarlane, H.T., Madsen, A., Riekel, C., Eisenberg, D., 2007. Atomic structures of amyloid cross-β spines reveal varied steric zippers. Nature 447 (7143), 453–457. https://doi.org/10.1038/nature05695.

Selkoe, D.J., Hardy, J., 2016. The amyloid hypothesis of Alzheimer's disease at 25 years. EMBO Mol. Med. 8 (6), 595–608. https://doi.org/10.15252/emmm.201606210.

Shady, N.H., Khattab, A.R., Ahmed, S., Liu, M., Quinn, R.J., Fouad, M.A., Muhsinah, A.B., Krischke, M., Mueller, M.J., Abdelmohsen, U.R., 2020. Hepatitis C virus NS3 protease and helicase inhibitors from red sea sponge (Amphimedon) species in green synthesized silver nanoparticles assisted by in silico modeling and metabolic profiling. Int. J. Nanomedicine 15, 3377–3389. https://doi.org/10.2147/ijn.s233766.

Shimamura, K., Akiyama, T., Yokoyama, K., Takenoya, M., Ito, S., Sasaki, Y., Yajima, S., 2020. Structural basis of substrate recognition by the substrate binding protein (SBP) of a hydrazide transporter, obtained from microbacterium hydrocarbonoxydans. Biochem. Biophys. Res. Commun. 525 (3), 720–725. https://doi.org/10.1016/j.bbrc.2020.02.146.

Sipe, J.D., Cohen, A.S., 2000. Review: history of the amyloid fibril. J. Struct. Biol. 130 (2–3), 88–98. https://doi.org/10.1006/jsbi.2000.4221.

Sipe, J.D., Benson, M.D., Buxbaum, J.N., Ikeda, S., Merlini, G., Saraiva, M.J., Westermark, P., 2010. Amyloid fibril protein nomenclature: 2010 recommendations from the nomenclature committee of the International Society of Amyloidosis. Amyloid 17 (3–4), 101–104. https://doi.org/10.3109/13506129.2010.526812.

Slowing, I.I., Wu, C., Vivero-Escoto, J.L., Lin, V.S., 2009. Mesoporous silica nanoparticles for reducing hemolytic activity towards mammalian red blood cells. Small 5 (1), 57–62. https://doi.org/10.1002/smll.200800926.

Tanzi, R.E., Bertram, L., 2005. Twenty years of the Alzheimer's disease amyloid hypothesis: a genetic perspective. Cell 120 (4), 545–555. https://doi.org/10.1016/j.cell.2005.02.008.

Thomassen, L.C., Rabolli, V., Masschaele, K., Alberto, G., Tomatis, M., Ghiazza, M., Turci, F., Brynaert, E., Matra, G., Kirschhock, C.E., Martens, J.A., Lison, D., Fubini, B., 2011. Model system to study the influence of aggregation on the hemolytic potential of silica nanoparticles. Chem. Res. Toxicol. 24 (11), 1869–1875. https://doi.org/10.1021/tx2002178.

Thompson, M.J., Sievers, S.A., Karanicolas, J., Ivanova, M.I., Baker, D., Eisenberg, D., 2006. The 3D profile method for identifying fibril-forming segments of proteins. PNAS 103 (11), 4074–4078. https://doi.org/10.1073/pnas.0511295103.

Walsh, T.R., Knecht, M.R., 2017. Biointerface structural effects on the properties and applications of bioinspired peptide-based nanomaterials. Chem. Rev. 117 (20), 12641–12704. https://doi.org/10.1021/acs.chemrev.7b00139.

Wangoo, N., Bhasin, K., Mehta, S., Suri, C.R., 2008. Synthesis and capping of water-dispersed gold nanoparticles by an amino acid: bioconjugation and binding studies. J. Colloid Interface Sci. 323 (2), 247–254. https://doi.org/10.1016/j.jcis.2008.04.043.

Wu, Z., Yang, S., Wu, W., 2016. Shape control of inorganic nanoparticles from solution. Nanoscale 8 (3), 1237–1259. https://doi.org/10.1039/c5nr07681a.

Wu, Y., Ali, M.R., Chen, K., Fang, N., El-Sayed, M.A., 2019. Gold nanoparticles in biological optical imaging. Nano Today 24, 120–140. https://doi.org/10.1016/j.nantod.2018.12.006.

Yong, K., Roy, I., Swihart, M.T., Prasad, P.N., 2009. Multifunctional nanoparticles as biocompatible targeted probes for human cancer diagnosis and therapy. J. Mater. Chem. 19 (27), 4655. https://doi.org/10.1039/b817667c.

Zaman, M., Safdari, H.A., Khan, A.N., Zakariya, S.M., Nusrat, S., Chandel, T.I., Khan, R.H., 2019. Interaction of anticancer drug pinostrobin with lysozyme: a biophysical and molecular docking approach. J. Biomol. Struct. Dyn. 37 (16), 4338–4344. https://doi.org/10.1080/07391102.2018.1547661.

Zare, D., Akbarzadeh, A., Barkhi, M., Khoshnevisan, K., Bararpour, N., Noruzi, M., Tabatabaei, M., 2012. L-arginine and L-glutamic acid capped gold nanoparticles at physiological pH: synthesis and characterization using agarose gel electrophoresis. Synth. React. Inorg. Met.-Org. Nano-Met. Chem. 42 (2), 266–272. https://doi.org/10.1080/15533174.2011.609855.

Zhang, Y., Wu, S., Qin, Y., Liu, J., Liu, J., Wang, Q., Ren, F., Zhang, H., 2018. Interaction of phenolic acids and their derivatives with human serum albumin: structure–affinity relationships and effects on antioxidant activity. Food Chem. 240, 1072–1080. https://doi.org/10.1016/j.foodchem.2017.07.100.

Zhao, Y., Sun, X., Zhang, G., Trewyn, B.G., Slowing, I.I., Lin, V.S., 2011. Interaction of mesoporous silica nanoparticles with human red blood cell membranes: size and surface effects. ACS Nano 5 (2), 1366–1375. https://doi.org/10.1021/nn103077k.

CHAPTER

4

Polysaccharide-based nanomaterials

Lily Jaiswal[a], Alya Limayem[b], and Shiv Shankar[a,c]

[a]Department of Food and Nutrition, BioNanocomposite Research Institute, Kyung Hee University, Seoul, Republic of Korea [b]Research Department of Pharmaceutical Sciences, College of Pharmacy, University of South Florida Centre for Research & Education in Nanobioengineering, USF Health, Tampa, FL, United States [c]Research Laboratories in Sciences Applied to Food, INRS-Institute Armand-Frappier, Laval, QC, Canada

OUTLINE

1 Introduction

Nanotechnology is an interdisciplinary science for creating functional materials, devices, or systems by controlling at least one dimension of the nanometer scale and developing new phenomena and physical, chemical, and biological properties (Shankar and Rhim, 2015). The material properties change with the size of particles decreasing to the nanometer scale because of the large surface-to-volume ratio. For example, a particle of 30 nm diameter has about 5% of its atoms on its surface; however, with 10 nm diameter, the number of surface atoms changes to almost 15%; while the particle size of 3 nm in diameter have nearly 50% of its atoms on the surface. As a result, the surface area, total surface energy, and reactivity increase

(Mariano et al., 2014). Many industries have known the possible benefits of nanotechnology, and commercial products have already been fabricated in the electronics, communication, energy production, medicine, and the food industry. Nanotechnology is a rapidly growing field with multifaceted applications of new materials at the nanoscale level. Materials blended with various nanomaterials create enhanced material properties due to interfacial interactions between nanoscaled architectures, which have a great potential in optical, electrical, medical, and antibacterial applications (Drelich et al., 2011; Shankar and Rhim, 2016a). Research on biopolymer-based nanocomposite has increased drastically due to biocompatibility, biodegradability, and environmental friendliness (Shankar et al., 2016). Among various biopolymers, carbohydrate-based polymers have been widely used to prepare innovative materials for multiple applications (Rhim et al., 2013).

Moreover, particularly in drug delivery systems, the submicron size or nanosize particle offers a distinct advantage over microparticles, including relatively higher intracellular uptake than microparticles (Reis et al., 2006). Nanoaggregates, nanocapsules, and nano-spheres are nanosized systems with diameters ranging from 10 to 1000nm in size (Mora-Huertas et al., 2010). These systems can hold enzymes, drugs, and other compounds by dissolving or entrapping them or attaching them to the particle's matrix. The method used to obtain the nanoparticles determines whether nano-aggregates, nanocapsules, nano-spheres, or nanocapsules with a structured interior are received (Paques et al., 2014). Nanoaggregates can be described as nanosized colloidal systems in which the drug is physically dispersed and can have different morphologies. Nanocapsules are vesicular systems where the drug is confined to an oily or aqueous liquid core, surrounded by a polymeric membrane (Lambert et al., 2001; Lertsutthiwong et al., 2008). Nano-spheres are spherical particles with a gelled interior in which the entrapped component is physically dispersed (Mora-Huertas et al., 2010; Soppimath et al., 2001). Nanocapsules with a structured core have also been designed and combine a nanocapsule and a nano-sphere. They are often produced by first preparing a nano-sphere and subsequently an additional shell formed on the nano-sphere interface (Paques et al., 2014). Various types of polysaccharide-based nanoparticles (Fig. 1) have been developed to date using several methods (Fig. 2) and applied in numerous research (Fig. 3). This book chapter reported the types of carbohydrate-based nanoparticles, their preparation methods, and their applications.

FIG. 1 Types of polysaccharide-based nanoparticles.

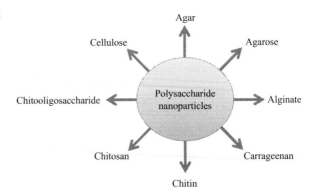

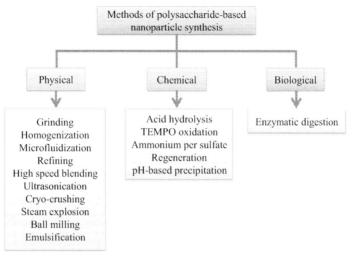

FIG. 2 Methods of polysaccharide-based nanoparticles preparation.

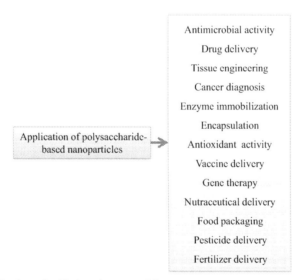

FIG. 3 Application of polysaccharide-based nanoparticles.

2 Agar nanoparticles

Agar is a polysaccharide extracted from *Rhodophyceae* class seaweed such as *Gelidiales* (*Gelidium* and *Pterocladia*) and *Gracilariales* (*Gracilaria* and *Hydropuntia*) (Lee et al., 2011; Shankar and Rhim, 2017). It is insoluble in cold water but soluble in boiling water. Agar is a mixture of agarose and agaropectin, in which the former is a linear gelling polymer

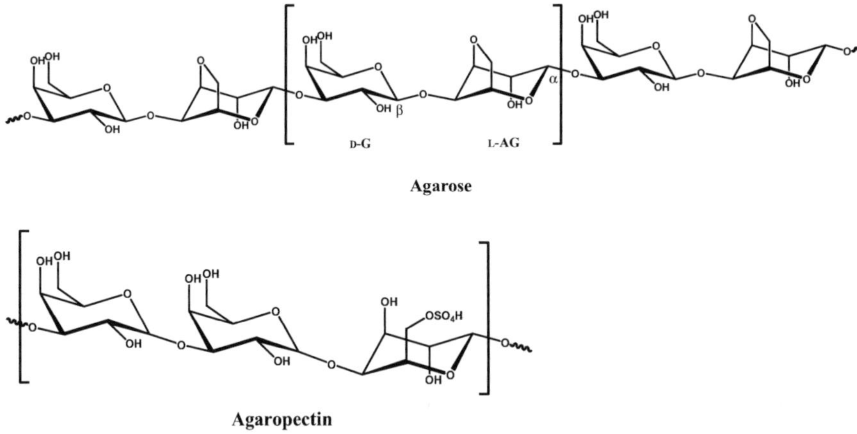

FIG. 4 Chemical structure of agarose and agaropectin (Shankar and Rhim, 2015).

composed of a repeating unit of agarobiose (composed of 1,3-linked-D-galactose and 1,4-linked 3,6-anhydro-L-galactose) and the latter is a slightly branched and sulfated nongelling polymer (Shankar and Rhim, 2015) (Fig. 4). During gelation of agar, the polysaccharide chains wrap together very tightly, and water can be trapped inside the helix. As more and more helices are formed and cross-linked, a three-dimensional network of water-containing helices is created without any net charge (Hashemi et al., 2005).

Agar has been widely used in culture media for microbiological work in the form of the agar plate. Also, agar has been used to prepare nanocomposite films for food packaging applications (Orsuwan et al., 2016; Shankar et al., 2014, 2015a, 2017; Shankar and Rhim, 2015, 2016b, 2017). In addition, agar has also been used as a bead to encapsulate microorganisms and drug delivery applications. The methods reported for preparing agar nanoparticles are usually based on water in oil emulsion due to the inert and uncharged nature of agar (Hashemi et al., 2005). A micron size of agar particle was prepared by emulsifying aqueous agar solution in paraffin oil (Manjunatha et al., 2007). The premixing of drug or bioactive materials and the aqueous agar solution can also be used for encapsulation (Moribe et al., 2008). Agar nanoparticles were prepared by water in oil emulsion using 0.5%, 1%, and 2% w/v agar emulsified in soyabean oil with an aqueous and oil phase ratio of 1:10 (Lee et al., 2011).

3 Agarose nanoparticle

Agarose is a biocompatible polysaccharide extracted from marine red algae, containing repetitions of agarobiose (disaccharide of D-galactose and 3,6-anhydro-L-galactopyranose), and can be prepared as a thermal-reversible gel. Agarose is the main component of agar, attained by the extraction of agaropectin from agar (Scionti et al., 2014; Shin et al., 2010). Depending on its molecular weight, concentration, and side groups, the agarose gelation and melting points alter from 30–40°C to 80–90°C, respectively. Moreover, it can be easily dissolved in hot water, dimethyl sulfoxide (DMSO), dimethylformamide (DMF), formamide

(FA), *N*-methylformamide (MFA), and 1-butyl-3-methylimidazolium chloride (BmimCl) (Horinaka et al., 2011; Wang et al., 2018). Interestingly, oxygen and hydrogen in the side groups of this natural carbohydrate polymer support its self-gelling feature. The agarose gelation process occurs in three steps: induction, gelation, and pseudo-equilibrium, in which the hydrogen bonding and electrostatic interaction resulted in the helical structure of agarose molecule and then formed gel (Zarrintaj et al., 2017). As a result of hydrogen bonding, agarose hydrogels can be formed without the need for toxic cross-linking agents like genipin (Campos et al., 2018), making it a biocompatible polymer (Cecilia et al., 2017). There are also some reports approving that agarose has found utility in electrophoresis due to the negative charge of pyruvate and sulfate groups that support the mobility of DNA, protein extraction, nanoparticle separation, microfluidics, micropatterning, biosensors, cell encapsulation, nano/micro drug carriers, and in vitro models for neuroscience and radiotherapy (Hlavacek et al., 2018; Johansson, 1972; Shankar and Kaur, 2018; Zhong et al., 2016).

4 Alginate nanoparticles

Alginate is a naturally occurring polyanionic polysaccharide derived from brown marine algae (Phaeophyceae), and it is commercially extracted from brown seaweed, including giant kelp (*Macrocystis pyrifera*, *Ascophyllum nodosum*) and various types of *Laminaria* species such as *Laminaria hyperborea*, *Laminaria digitata*, and *Laminaria japonica* (Rehm and Moradali, 2018). Alginate comprises a linear block copolymer of 1, 4-linked β-D-mannuronic and α-L-guluronic residues in varying proportions. As a low-cost, abundantly available, biocompatible, and environmentally friendly biopolymer, it has been used in numerous applications in the food and biotechnology industries, such as a nontoxic food additive, thickening and gelling agent, and colloidal stabilizer (Gheorghita Puscaselu et al., 2020). Alginate has been extensively reviewed concerning its physical and chemical properties and its use in the formation of microparticles, films, and hydrogels (Agüero et al., 2017; Alba and Kontogiorgos, 2018; Fernando et al., 2019; Gheorghita Puscaselu et al., 2020; Paques et al., 2014; Rehm and Moradali, 2018). Primarily alginate nanoparticle formation is based on two methods: (1) complexion and (2) emulsification.

Complexion: Complex formation can occur in an aqueous solution forming alginate nano-aggregates or on the interface of an oil droplet, creating alginate nanocapsules (Fessi et al., 1989; Lertsutthiwong et al., 2008; Mora-Huertas et al., 2010). A cross-linker such as calcium from calcium chloride is used for the complexation of alginate. Complexation can also occur through mixing alginate with an oppositely charged polyelectrolyte such as chitosan (Sæther et al., 2008). Rajaonarivony et al. (1993) described a novel method for forming alginate nano-aggregates as drug carriers with sizes ranging from 250 to 850 nm. The particles were formed in a sodium alginate solution by first adding calcium chloride and then poly-L-lysine. A so-called pregel state was formed by mixing low alginate and calcium chloride concentrations, consisting of nanosized aggregates dispersed in a continuous water phase. Addition of an aqueous polycationic solution, such as poly-L-lysine, resulted in a polyelectrolyte complex coating of these alginate nanoparticles. Since poly-L-lysine is toxic and immunogenic when injected into the human body, chitosan has been used as an alternative cationic polymer

(Sarmento et al., 2007). Alginate NPs have also been prepared by solely combining chitosan or calcium ions with a sodium alginate solution (Sæther et al., 2008; Yu et al., 2008). The alginate and cationic polymer concentration, their molecular weight, the calcium chloride concentration, and the order of addition of calcium chloride and cationic polymer to the sodium alginate solution were found to be of significant influence on the size and properties of the obtained nanoparticles (De and Robinson, 2003; Rajaonarivony et al., 1993). Alginate nano-aggregates were developed through self-assembly that did not require calcium-induced aggregation of alginate or cationic polymers to form a polyelectrolyte complex with alginate (Chang et al., 2012). They synthesized amphiphilic thiolated sodiumalginate, and sonication facilitated an oxidation reaction of the thiolgroups, inducing self-assembly of the alginate into nano-aggregates. The preparation of alginate nanocapsules consists of first mixing the drug or components that are to be encapsulated with an organic solvent and that will act as the interior oil phase of the capsules. The mixture is slowly added to an aqueous solution of alginate, containing an additional surfactant, such as Tween80. Subsequently, sonication is used to form an oil-in-water (o/w) emulsion. Calcium from a calcium chloride solution is often added to the emulsion, and alginate is deposited to form the nanoparticle membrane. Finally, the nanocapsule aqueous suspension is allowed to equilibrate for a particular time before solvent removal (Paques et al., 2014).

Emulsification: Emulsion droplets are good templates for forming spherical particles, and the mechanical power input used to prepare the emulsion determines the size of the particles. Sonication, membrane emulsification, and (self-assembled) microemulsions can also be used to form emulsion droplets. Emulsions are relatively easy to produce, and emulsion-based methods for nanoparticle formation can more easily be scaled up to industrial sizes than nozzle-based methods. Alginate-in-oil (w/o) emulsions coupled with internal or external gelation have prepared alginate micro- and nano-spheres. Alginate NPs prepared through internal gelation and external gelation have a different homogeneity. The type of gelation used, internal or external, determines the properties of the alginate particles and their applications. Alginate particles prepared with external gelation tend to have a denser structure with smaller pores at the particle's surface, compared to the particle's interior, due to a concentration gradient of the cations (high concentrations at the surface and low at the core). Alginate particles prepared through internal gelation have a more even distribution of the cations throughout the particle, resulting in a more homogenous particle (Chang et al., 2012; Paques et al., 2014).

5 Carrageenan nanoparticles

Carrageenan is the generic name for a family of high molecular weight sulfated polysaccharides obtained by extracting certain species of red seaweeds such as *Chondrus*, *Eucheuma*, *Gigartina*, and *Hypnea*. Carrageenan is mainly composed of D-galactose residues linked alternately in 3-linked-β-D-galactopyranose and 4-linked-α-D-galactopyranose units and are classified according to the degree of substitution that occurs on their free hydroxyl groups. Substitutions are generally either the addition of ester sulfate or the 3, 6-anhydride on the

4-linked residue (Alba and Kontogiorgos, 2018; Campo et al., 2009; Jiao et al., 2011; Nanaki et al., 2010). Carrageenan is widely utilized in various industries due to its excellent physical functional properties, such as gelling, thickening, emulsifying, and stabilizing abilities, and has been employed to improve the texture of cottage cheese, puddings, and dairy desserts, and as binders and stabilizers in the meat processing industry for the manufacture of sausages, patties, and low-fat hamburgers (Li et al., 2014; Reddy et al., 2011). Also, carrageenan is used in toothpaste, air freshener gels, fire fighting foam, cosmetic creams, shampoo, and shoe polish (Necas and Bartosikova, 2013). In recent years, it has been increasingly used in pharmaceutical formulations too. Carrageenan has shown several potential pharmaceutical properties, including anticoagulant, anticancer, antihyperlipidemic, and immunomodulatory activities (Udayangani et al., 2020). In vitro studies also indicate that carrageenan may also have antiviral effects, inhibiting the replication of viruses (Hans et al., 2021; Vega et al., 2020). Carrageenan also exhibits antioxidant activity and free radical scavenging activity.

Furthermore, a positive correlation between sulfate content and antioxidant activity is reported in the literature (Arunkumar et al., 2021). The nanoparticles of carrageenan can be formed in combination with chitosan. Carrageenan-chitosan complex nanoparticles can be obtained in an aqueous environment under very mild conditions based on electrostatic interaction, avoiding using organic solvents or other aggressive processes (Souza et al., 2018). Carrageenan-chitosan complex nanoparticles have potential applications in drug delivery, tissue engineering, and regenerative medicine (Manna and Jana, 2021; Pinheiro et al., 2012). It was reported that carrageenan-chitosan complex nanoparticles exhibited noncytotoxic behavior in in vitro tests using L929 fibroblasts and provided a controlled release for up to 3 weeks with ovalbumin as a model protein. The carrageenan-chitosan ratio had a significant effect on the properties of the nanoparticles (Grenha et al., 2010).

6 Chitin nanoparticles

Chitin is a linear polysaccharide made up of -(1–4)-linked 2-acetamido-2-deoxy-D-glucopyranose units which may be deacetylated to some extent (Muzzarelli, 2011). Chitin is the second most abundant biomaterial after cellulose in the world. Chitin can be obtained from the cell wall of fungi, the exoskeleton of arthropods such as crustaceans (e.g., crabs, lobsters, and shrimps) and insects, the radula of mollusks, and the beaks and the internal shells of cephalopods including squids and octopuses (Muzzarelli, 2012). Chitin nanoparticles occur in biological tissues, according to structural hierarchies, jointly with proteins and inorganic compounds. The isolation steps of chitin nanoparticles should be optimized to remove proteins and minerals present in the raw material (Danti et al., 2019). Chitin nanoparticles can be obtained by hydrochloric acid hydrolysis (Shankar et al., 2015b), TEMPO-mediated oxidation (Fan et al., 2010; Ye et al., 2021), ultrasonication (Somsak et al., 2021), electrospinning (Mallik et al., 2020; Min et al., 2004), mechanical treatment (Ifuku et al., 2010; Yang et al., 2020), and gelation (Heath et al., 2013; Kadokawa et al., 2011). The origin and isolation method of chitin determines the structure and morphology of the chitin nanoparticles or nanowhiskers (Knidri

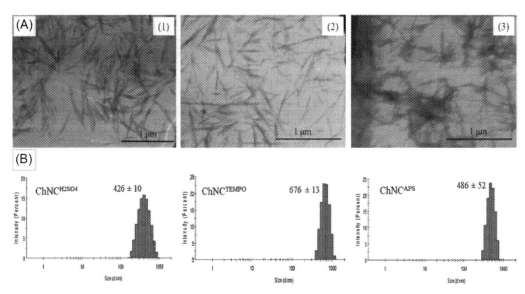

FIG. 5   (A) STEM images and (B) particle size distribution of chitin nanocrystals isolated using different methods: (1: sulfuric acid, 2: TEMPO-oxidation, and 3: ammonium persulfate) (Oun and Rhim, 2018).

et al., 2019). The chitin nanoparticles occur as rod-like or spindle-like nanowhiskers with properties comparable to perfect crystals (Shankar et al., 2015b). The dimensions for chitin whiskers obtained from crab shells are much shorter than the chitin nanowhiskers from *Riftia* tubes. The aspect ratio is an important parameter, mainly when chitin nanowhiskers reinforce polymers; a higher aspect ratio usually results in more significant reinforcement (Yang et al., 2020). Comparing with hydrochloric acid hydrolysis, the TEMPO-mediated oxidation method is more controllable by the amount of NaClO, and the yield of chitin nanowhiskers can reach 90% (Fan et al., 2008; Ye et al., 2021). Oun and Rhim (2018) isolated chitin nanowhiskers using three different methods and found that the size of chitin nanowhiskers differs from the isolation method (Fig. 5). Also, chitin nanoparticles possess higher degree of crystallinity compared to bulk counterpart. Chitin nanowhiskers possess a reactive surface covered with hydroxyl groups, which provides the possibility of modification through chemical reaction. The purpose of chemical modification is to contribute to specific functions and expand the applications of chitin nanowhiskers. Surface chemical modification of chitin nanowhiskers is a method to increase or decrease their surface energy and disperse them in aqueous and organic liquids of low polarity, respectively (Gopalan Nair et al., 2003). Since large quantities of crab and shrimp shells are produced annually as food waste, further utilization of chitins as functionalized materials is desired (Fan et al., 2008). The chitin nanowhiskers are currently obtained as aqueous suspensions, which are being studied and used as reinforcing additives for high-performance environment-friendly biodegradable nanocomposite materials, as biomedical composites for drug/gene delivery, nano scaffolds in tissue engineering, and cosmetic orthodontics (Heath et al., 2013; Kadokawa et al., 2011; Mallik et al., 2020; Yang et al., 2020).

7 Chitooligosaccharide nanoparticles

Chitosans with DP less than 20 and an average MW less than 3.9 kDa are called chitooligosaccharides, chitosan oligomers, or chitooligomers (Liang et al., 2018). Chitooligosaccharides possess a wide range of biological activities and have numerous promising applications in multiple fields such as medicine, cosmetics, food, and agriculture (Dou et al., 2009). Chitooligosaccharides, being recognized as low MW and water-soluble chitosan, have much greater demand than precursor molecules, justified by their growing commercial availability. Chitooligosaccharides generally consist of N-acetylglucosamine or glucosamine units linked by β-1,4-O-glycoside bonds. Chitooligosaccharides can be produced with a defined degree of polymerization and different degree of N-acetylation, yet, their pattern of N-acetylation is always random (Bonin et al., 2020). Homochitooligosaccharides are the oligomers of glucosamine (D unit) or N-acetylglucosamine (A unit). They are exclusively composed of glucosamine or N-acetylglucosamine units.

In contrast, heterochitooligosaccharides, comprising both glucosamine and N-acetylglucosamine units, are a combination of numerous oligomers varying in the degree of polymerization, degree of N-acetylation, degree of deacetylation, and position of N-acetyl residues in the oligomer chain. Hetero-chitooligosaccharides with a degree of polymerization less than 10 are typically water-soluble; however, the water solubility of chitooligosaccharides with a degree of polymerization more than 10 depends on the degree of N-acetylation and the pH of the solution. Food, pharmaceutical, and research scientists prefer to use hetero-chitooligosaccharides (Liaqat and Eltem, 2018; Zhai et al., 2021). Chitooligosaccharides consisting of a few monomer units are called oligomers, and the number of these monomeric units within an oligomer is designated as a degree of polymerization. Therefore, a tetra-saccharide (degree of polymerization 4) is a lower oligomer of a pentasaccharide (degree of polymerization 5). Different chitooligosaccharides may have the same degree of polymerization but a different fraction of N-acetylated residues. Oligomers of the same degree of polymerization (having the same number of monomeric units) but a different fraction of N-acetylated residues value are homologs. For example, pentasaccharide D4A1 (fraction of N-acetylated residues 0.2) is a lower homolog of pentasaccharide D2A3 (fraction of N-acetylated residues 0.6). The number of homologs comprising one oligomer is a degree of polymerization +1. Given a particular homolog, isomers may exist that differ in the sequence of D and A units but have the same degree of polymerization and fraction of N-acetylated residues (Kim, 2010). The number of all compounds comprising one oligomer increases exponentially with the degree of polymerization.

Chitooligosaccharides can be a potential drug carrier because of their unique properties such as water solubility, biodegradability, and low toxicity. Previously, chitooligosaccharides-coated nanostructured lipid carriers loaded with flurbiprofen were used for ocular drug delivery and tested for adhesion properties (Liaqat and Eltem, 2018). However, a recent study revealed the promising potential of all-trans-retinoic acid conjugated chitooligosaccharides nanoparticles as drug carriers for co-delivery of all-trans-retinoic acid, paclitaxel, and other hydrophobic therapeutic agents. In addition, lower hemolytic activity and cytotoxicity of chitooligosaccharides nanoparticles, along with the high encapsulation efficiency, made them an effective drug carrier (Zhang et al., 2015).

8 Chitosan nanoparticles

Chitosan is a linear polysaccharide that is obtained from chitin through demineralization and deproteinization and has an extensive array of medical and agricultural applications (Kim et al., 2020; Shankar and Rhim, 2018). Chitosan has drawn an undue pact of attention in different disciplines because of its exceptional biological properties, including its biocompatibility, biodegradability, nontoxicity, and antimicrobial activity (Baraketi et al., 2020; Shankar et al., 2020, 2021). Chitosan is the biopolymer that displays a cationic character due to its amino groups (Kim, 2010; Muzzarelli, 2012). Moreover, amino and hydroxyl groups of chitosan react with solutes present in the solution. However, from the adsorption point of view, the amino groups are more critical than the hydroxyl groups. Therefore, the amino groups only decide the quality of this biopolymer. Chitosan is an insoluble biopolymer in water but soluble in acidic solutions at pK_a approximately 6.3 or below. At this pH, glucosamine units ($-NH_2^+$) of chitosan convert into the protonated soluble form ($-NH_3^+$). Therefore, chitosan solubility depends on its source, molecular weight (MW), and degree of acetylation (DA). Moreover, the usage of chitosan has been restricted to its native form due to its porosity, surface area, and low solubility at neutral pH. Several chitosan derivatives have been synthesized and evaluated for pioneering applications (Bonin et al., 2020; Muzzarelli, 2011b; Rashki et al., 2021).

Chitosan NPs share features of chitosan and valuable assets of NPs, such as small size, increased surface area, and quantum size effects (Chandrasekaran et al., 2020). Chitosan NPs were produced by several procedures such as ionic gelation (Kunjachan et al., 2014), reverse micellar method (Orellano et al., 2020), microemulsion (Zhou et al., 2013), emulsion droplet coalescence (Tokumitsu et al., 1999), and spray drying (Wang et al., 2014). Among several approaches, physical cross-linking by ionic gelation was found to be the best one. The ion gelation method exploits the electrostatic communication between a positively charged group of chitosan and a negatively charged group of tripolyphosphate. Changing the ratio of chitosan to tripolyphosphate could modify the size and surface charge of NPs. Moreover, there is no chemical used for cross-linking, which reduces the toxic side effects. Different metal ions and chitosan complexes (chitosan-silver-NPs, chitosan-copper-NPs, chitosan-zinc-NPs) have been made to improve their antimicrobial activity (Polinarski et al., 2021; Vanti et al., 2020; Zhang et al., 2020b). Moreover, in further pursuit of antimicrobial efficacy, hybrid chitosan-NPs with protamine (Tamara et al., 2018), lysozyme (Wang et al., 2020; Zhang et al., 2020a), essential oil (Karimirad et al., 2020; Salehi et al., 2020), and curcumin (Bhoopathy et al., 2021) NPs, among others, have also been tested. Thus, polycationic chitosan-NPs with more surface charge density communicate with bacteria primarily than chitosan alone. Also, chitosan NPs interrupt the cell wall and membrane of bacterial cells, which leads to the efflux of intracellular molecules and bacterial cell death. Several parameters control the size of NPs and consequently affect antibacterial responses. These include the bacterial species, growth stage, concentration, pH, zeta potential, MW, and DA (Chandrasekaran et al., 2020).

While chitosan is a cost-effective nanoparticle and bioavailable polymer with well-evidenced antimicrobial properties mainly against Gram-negative bacteria, it can also act as a mucoadhesive nanodrug in the gastrointestinal tract (GIT) and withstand the harsh acidic conditions within the host cells (Vashist et al., 2013; Zhang et al., 2016). Its uniqueness arises from its delivery system property to carry other nanodrugs, including its cationic interactions with metal oxides, primarily with zinc oxide (Knorr, 1983). The synergistic effect of chitosan with zinc oxide depicted a tremendous antimicrobial effect, possessing higher potency on a broad spectrum of microbes including both Gram-positive and Gram-negative bacteria in addition to viruses, which make it an optimal nanodrug for a wide range of microbiol control from food packaging and wastewater to infectious diseases beyond hospital settings (Limayem et al., 2016, 2020).

9 Cellulose nanoparticles

Cellulose is the most abundant and sustainable natural polymer globally, with an annual output of one trillion tonnes (Oun et al., 2020). Cellulose nanofibers are isolated from various lignocellulosic sources such as wood, flax, hemp, sisal, cotton, ramie, and jute (do Nascimento et al., 2015), forest residues (Moriana et al., 2016), agricultural wastes (Oun and Rhim, 2016), marine biomass (Doh et al., 2020), and bacterial cellulose (Singhsa et al., 2017). The nanocellulose materials have several attractive characteristics such as biodegradability, sustainability, excellent mechanical properties, large aspect ratio, high surface area, low density, low environmental impact, and low cost, as well as tailorable surface properties (Oun et al., 2020). The unique physical and mechanical properties of nanocellulose materials have increased the applicability in various fields, including food packaging (Oun and Rhim, 2016), paper and nanofiber filters industry (Latifah et al., 2020), cosmetics (Almeida et al., 2021; de Amorim et al., 2020), enzyme immobilization (Tizchang et al., 2020), sensors technology (Baraketi et al., 2020; Shankar et al., 2020), and electronic device (Nyamayaro et al., 2020). Generally, two types of nanocelluloses, cellulose nanofiber (CNF) and cellulose nanocrystal (CNC), can be obtained according to the isolation method. Cellulose nanofiber is usually obtained by disrupting cellulosic fibers along an axis using mechanical means such as ultrahigh friction grinding, high-pressure homogenization, wet grinding, cryogenic cooling, and ultrasonic treatment (Adel et al., 2016; Oun et al., 2020). Mechanical methods are commonly performed together with various chemical pretreatments such as TEMPO-oxidation (Levanič et al., 2020), carboxymethylation (Tortorella et al., 2020), enzymatic hydrolysis (Zielińska et al., 2021), and phosphorylation (Helberg et al., 2021). TEMPO-oxidation is the most commonly used chemical pretreatment method for the isolation of CNF. The CNF produced nanofibrils are composed of long fibrous webs exceeding 1000 nm and generally include amorphous cellulose with a high aspect ratio and low crystallinity (Chen et al., 2011). The CNC is the crystalline portion of the CNF, which is defined as the monocrystalline cellulose region (Oun et al., 2020). The amorphous part of CNF is removed by hydrolysis using an acid hydrolysis method to produce crystalline CNC (Maciel et al., 2019). Sulfuric acid, a most widely used hydrolysis method, produces stable colloidal dispersion by grafting sulfate ester groups on the surface of the nanocellulose, producing CNC with needlelike morphology, a high crystallinity index, a low aspect ratio with 1–100 nm in diameter, and a length of several hundred nanometers (Dhali et al., 2021). The morphology, surface chemistry, dimensions, yield, and crystallinity properties of nanocelluloses are usually dependent on the source of cellulose, isolation method, isolation conditions, and pre/posttreatments

(Oun and Rhim, 2016). Several methods have been used to separate nanocellulose from the lignocellulosic sources, including mechanical method, acid hydrolysis, chemical/TEMPO-mediated oxidation, ammonium persulfate, and cellulose regeneration method (Oun et al., 2020). The characteristics of the separated nanocellulose are mainly dependent on the source of cellulose (i.e., a chemical structure of cellulose) and the methods for the isolation of nanocellulose (de Amorim et al., 2020; Oun and Rhim, 2016). Nanocellulose can be directly isolated from pure cellulose such as cotton fibers and bacterial cellulose or other lignocellulosic sources such as wood and plant grass, but pretreatment to remove noncellulosic substances such as lignin, hemicellulose, pectin, wax, and other extractive materials is needed in the latter (Dhali et al., 2021; Mariano et al., 2014).

10 Conclusion

This book chapter describes the types of polysaccharides-based nanoparticles, their source, their preparation methods, and nanotechnological application in food packaging, food preservation, drug delivery, gene delivery, tissue engineering, cancer therapy, wound dressing, biosensors, etc. Polysaccharide-based nanoparticles are highly stable, safe, biocompatible, biodegradable, nontoxic, low cost, and abundantly available. However, most of the industrial applications are still at the laboratory level. In addition, in vivo studies and clinical applications are needed to develop new commercial carbohydrate-based nanoparticle products. Polysaccharide-based nanomaterials have great potential in biomedicine, fabric, food, and pharmaceutical industries in the near future.

References

Adel, A.M., El-Gendy, A.A., Diab, M.A., Abou-Zeid, R.E., El-Zawawy, W.K., Dufresne, A., 2016. Microfibrillated cellulose from agricultural residues. Part I: papermaking application. Ind. Crop Prod. 93, 161–174.

Agüero, L., Zaldivar-Silva, D., Peña, L., Dias, M.L., 2017. Alginate microparticles as oral colon drug delivery device: a review. Carbohydr. Polym. 168, 32–43.

Alba, K., Kontogiorgos, V., 2018. Seaweed polysaccharides (agar, alginate carrageenan). In: Reference Module in Food Science. Elsevier.

Almeida, T., Silvestre, A.J., Vilela, C., Freire, C.S., 2021. Bacterial nanocellulose toward green cosmetics: recent progresses and challenges. Int. J. Mol. Sci. 22 (6), 2836.

Arunkumar, K., Raja, R., Kumar, V.S., Joseph, A., Shilpa, T., Carvalho, I.S., 2021. Antioxidant and cytotoxic activities of sulfated polysaccharides from five different edible seaweeds. J. Food Meas. Charact. 15 (1), 567–576.

Baraketi, A., D'Auria, S., Shankar, S., Fraschini, C., Salmieri, S., Menissier, J., Lacroix, M., 2020. Novel spider web trap approach based on chitosan/cellulose nanocrystals/glycerol membrane for the detection of Escherichia coli O157: H7 on food surfaces. Int. J. Biol. Macromol. 146, 1009–1014. https://doi.org/10.1016/j.ijbiomac.2019.09.225.

Bhoopathy, S., Inbakandan, D., Rajendran, T., Chandrasekaran, K., Kasilingam, R., Gopal, D., 2021. Curcumin loaded chitosan nanoparticles fortify shrimp feed pellets with enhanced antioxidant activity. Mater. Sci. Eng. C 120, 111737.

Bonin, M., Sreekumar, S., Cord-Landwehr, S., Moerschbacher, B.M., 2020. Preparation of defined chitosan oligosaccharides using chitin deacetylases. Int. J. Mol. Sci. 21 (21), 7835.

Campo, V.L., Kawano, D.F., da Silva Jr., D.B., Carvalho, I., 2009. Carrageenans: biological properties, chemical modifications and structural analysis—a review. Carbohydr. Polym. 77 (2), 167–180.

Campos, F., Bonhome-Espinosa, A.B., Vizcaino, G., Rodriguez, I.A., Durand-Herrera, D., López-López, M.T., Sánchez-Montesinos, I., Alaminos, M., Sánchez-Quevedo, M.C., Carriel, V., 2018. Generation of genipin cross-linked fibrin-agarose hydrogel tissue-like models for tissue engineering applications. Biomed. Mater. 13 (2). https://doi.org/10.1088/1748-605X/aa9ad2, 025021.

Cecilia, A., Baecker, A., Hamann, E., Rack, A., van de Kamp, T., Gruhl, F.J., Hofmann, R., Moosmann, J., Hahn, S., Kashef, J., Bauer, S., Farago, T., Helfen, L., Baumbach, T., 2017. Optimizing structural and mechanical properties of cryogel scaffolds for use in prostate cancer cell culturing. Mater. Sci. Eng. C 71, 465–472. https://doi.org/10.1016/j.msec.2016.10.038.

Chandrasekaran, M., Kim, K.D., Chun, S.C., 2020. Antibacterial activity of chitosan nanoparticles: a review. Processes 8 (9), 1173.

Chang, D., Lei, J., Cui, H., Lu, N., Sun, Y., Zhang, X., Gao, C., Zheng, H., Yin, Y., 2012. Disulfide cross-linked nanospheres from sodium alginate derivative for inflammatory bowel disease: preparation, characterization, and *in vitro* drug release behavior. Carbohydr. Polym. 88 (2), 663–669.

Chen, W., Yu, H., Liu, Y., 2011. Preparation of millimeter-long cellulose I nanofibers with diameters of 30–80 nm from bamboo fibers. Carbohydr. Polym. 86 (2), 453–461.

Danti, S., Trombi, L., Fusco, A., Azimi, B., Lazzeri, A., Morganti, P., Coltelli, M.-B., Donnarumma, G., 2019. Chitin nanofibrils and nanolignin as functional agents in skin regeneration. Int. J. Mol. Sci. 20 (11), 2669.

de Amorim, J.D.P., de Souza, K.C., Duarte, C.R., da Silva Duarte, I., de Ribeiro, F.A.S., Silva, G.S., de Farias, P.M.A., Stingl, A., Costa, A.F.S., Vinhas, G.M., 2020. Plant and bacterial nanocellulose: production, properties and applications in medicine, food, cosmetics, electronics and engineering. A review. Environ. Chem. Lett. 18 (3), 851–869.

De, S., Robinson, D., 2003. Polymer relationships during preparation of chitosan–alginate and poly-l-lysine–alginate nano-spheres. J. Control. Release 89 (1), 101–112.

Dhali, K., Ghasemlou, M., Daver, F., Cass, P., Adhikari, B., 2021. A review of nanocellulose as a new material towards environmental sustainability. Sci. Total Environ. 775, 145871.

do Nascimento, J.H.O., Luz, R.F., Galvão, F.M., Melo, J.D.D., Oliveira, F.R., Ladchumananandasivam, R., Zille, A., 2015. Extraction and characterization of cellulosic nanowhisker obtained from discarded cotton fibers. Mater. Today Proc. 2 (1), 1–7.

Doh, H., Lee, M.H., Whiteside, W.S., 2020. Physicochemical characteristics of cellulose nanocrystals isolated from seaweed biomass. Food Hydrocoll. 102, 105542.

Dou, J., Xu, Q., Tan, C., Wang, W., Du, Y., Bai, X., Ma, X., 2009. Effects of chitosan oligosaccharides on neutrophils from glycogen-induced peritonitis mice model. Carbohydr. Polym. 75 (1), 119–124.

Drelich, J., Li, B., Bowen, P., Hwang, J.-Y., Mills, O., Hoffman, D., 2011. Vermiculite decorated with copper nanoparticles: novel antibacterial hybrid material. Appl. Surf. Sci. 257 (22), 9435–9443.

Fan, Y., Saito, T., Isogai, A., 2008. Chitin nanocrystals prepared by TEMPO-mediated oxidation of α-chitin. Biomacromolecules 9 (1), 192–198.

Fan, Y., Saito, T., Isogai, A., 2010. Individual chitin nano-whiskers prepared from partially deacetylated α-chitin by fibril surface cationization. Carbohydr. Polym. 79 (4), 1046–1051.

Fernando, I.P.S., Lee, W., Han, E.J., Ahn, G., 2019. Alginate-based nanomaterials: fabrication techniques, properties, and applications. Chem. Eng. J. 391, 123823.

Fessi, H., Puisieux, F., Devissaguet, J.P., Ammoury, N., Benita, S., 1989. Nanocapsule formation by interfacial polymer deposition following solvent displacement. Int. J. Pharm. 55 (1), R1–R4.

Gheorghita Puscaselu, R., Lobiuc, A., Dimian, M., Covasa, M., 2020. Alginate: from food industry to biomedical applications and management of metabolic disorders. Polymers 12 (10), 2417.

Gopalan Nair, K., Dufresne, A., Gandini, A., Belgacem, M.N., 2003. Crab shell chitin whiskers reinforced natural rubber nanocomposites. 3. Effect of chemical modification of chitin whiskers. Biomacromolecules 4 (6), 1835–1842.

Grenha, A., Gomes, M.E., Rodrigues, M., Santo, V.E., Mano, J.F., Neves, N.M., Reis, R.L., 2010. Development of new chitosan/carrageenan nanoparticles for drug delivery applications. J. Biomed. Mater. Res. A 92 (4), 1265–1272.

Hans, N., Malik, A., Naik, S., 2021. Antiviral activity of sulfated polysaccharides from marine algae and its application in combating COVID-19: mini review. Bioresour. Technol. Rep. 13, 100623.

Hashemi, P., Rahmani, Z., Kakanejadifard, A., Niknam, E., 2005. Preparation of a fast-flow agarose-based chelating adsorbent with a novel dioxime derivative for selective column preconcentration of copper. Anal. Sci. 21 (11), 1297–1301.

Heath, L., Zhu, L., Thielemans, W., 2013. Chitin nanowhisker aerogels. ChemSusChem 6 (3), 537.

Helberg, R.M.L., Torstensen, J.Ø., Dai, Z., Janakiram, S., Chinga-Carrasco, G., Gregersen, Ø.W., Syverud, K., Deng, L., 2021. Nanocomposite membranes with high-charge and size-screened phosphorylated nanocellulose fibrils for CO_2 separation. Green Energy Environ. 6, 585–596.

Hlavacek, A., Mickert, M.J., Soukka, T., Lahtinen, S., Tallgren, T., Pizúrová, N., Król, A., Gorris, H.H., 2018. Large-scale purification of photon-upconversion nanoparticles by gel electrophoresis for analogue and digital bioassays. Anal. Chem. 91 (2), 1241–1246.

Horinaka, J., Yasuda, R., Takigawa, T., 2011. Entanglement network of agarose in various solvents. Polym. J. 43 (12), 1000–1002.

Ifuku, S., Nogi, M., Yoshioka, M., Morimoto, M., Yano, H., Saimoto, H., 2010. Fibrillation of dried chitin into 10–20 nm nanofibers by a simple grinding method under acidic conditions. Carbohydr. Polym. 81 (1), 134–139.

Jiao, G., Yu, G., Zhang, J., Ewart, H.S., 2011. Chemical structures and bioactivities of sulfated polysaccharides from marine algae. Mar. Drugs 9 (2), 196–223.

Johansson, B.G., 1972. Agarose gel electrophoresis. Scand. J. Clin. Lab. Invest. 29 (Suppl. 124), 7–19.

Kadokawa, J., Takegawa, A., Mine, S., Prasad, K., 2011. Preparation of chitin nanowhiskers using an ionic liquid and their composite materials with poly (vinyl alcohol). Carbohydr. Polym. 84 (4), 1408–1412.

Karimirad, R., Behnamian, M., Dezhsetan, S., 2020. Bitter orange oil incorporated into chitosan nanoparticles: preparation, characterization and their potential application on antioxidant and antimicrobial characteristics of white button mushroom. Food Hydrocoll. 100, 105387.

Kim, S.-K., 2010. Chitin, Chitosan, Oligosaccharides and Their Derivatives: Biological Activities and Applications. CRC Press.

Kim, Y.H., Kim, G.H., Yoon, K.S., Shankar, S., Rhim, J.-W., 2020. Comparative antibacterial and antifungal activities of sulfur nanoparticles capped with chitosan. Microb. Pathog. 144. https://doi.org/10.1016/j.micpath.2020.104178, 104178.

Knidri, H.E., Dahmani, J., Addaou, A., Laajeb, A., Lahsini, A., 2019. Rapid and efficient extraction of chitin and chitosan for scale-up production: effect of process parameters on deacetylation degree and molecular weight. Int. J. Biol. Macromol. 139, 1092–1102.

Knorr, D., 1983. Dye binding properties of chitin and chitosan. J. Food Sci. 48, 36–37.

Kunjachan, S., Jose, S., Lammers, T., 2014. Understanding the mechanism of ionic gelation for synthesis of chitosan nanoparticles using qualitative techniques. Asian J. Pharm. 4, 148–153.

Lambert, G., Fattal, E., Couvreur, P., 2001. Nanoparticulate systems for the delivery of antisense oligonucleotides. Adv. Drug Deliv. Rev. 47 (1), 99–112.

Latifah, J., Nurrul-Atika, M., Sharmiza, A., Rushdan, I., 2020. Extraction of nanofibrillated cellulose from Kelempayan (Neolamarckia cadamba) and its use as strength additive in papermaking. J. Trop. For. Sci. 32 (2), 170–178.

Lee, E.J., Park, J.K., Khan, S.A., Lim, K.-H., 2011. Preparation of agar nanoparticles by W/O emulsification. J. Chem. Eng. Jpn. 44, 502–508.

Lertsutthiwong, P., Noomun, K., Jongaroonngamsang, N., Rojsitthisak, P., Nimmannit, U., 2008. Preparation of alginate nanocapsules containing turmeric oil. Carbohydr. Polym. 74 (2), 209–214.

Levanič, J., Šenk, V.P., Nadrah, P., Poljanšek, I., Oven, P., Haapala, A., 2020. Analyzing TEMPO-oxidized cellulose fiber morphology: new insights into optimization of the oxidation process and nanocellulose dispersion quality. ACS Sustain. Chem. Eng. 8 (48), 17752–17762.

Li, L., Ni, R., Shao, Y., Mao, S., 2014. Carrageenan and its applications in drug delivery. Carbohydr. Polym. 103, 1–11.

Liang, S., Sun, Y., Dai, X., 2018. A review of the preparation, analysis and biological functions of chitooligosaccharide. Int. J. Mol. Sci. 19 (8), 2197.

Liaqat, F., Eltem, R., 2018. Chitooligosaccharides and their biological activities: a comprehensive review. Carbohydr. Polym. 184, 243–259.

Limayem, A., Gonzalez, F., Micciche, A., Haller, E., Nayak, B., Mohapatra, S., 2016. Molecular identification and nanoremediation of microbial contaminants in algal systems using untreated wastewater. J. Environ. Sci. Health B 51, 868–872.

Limayem, A., Patil, S.B., Mehta, M., Cheng, F., Nhuyen, M., 2020. A streamlined study on chitosan-zinc oxide nanomicelle properties to mitigate a drug-resistant biofilm protection mechanism. Front. Nanotechnol. 2, 592739.

Maciel, M.M.Á.D., de Carvalho Benini, K.C.C., Voorwald, H.J.C., Cioffi, M.O.H., 2019. Obtainment and characterization of nanocellulose from an unwoven industrial textile cotton waste: effect of acid hydrolysis conditions. Int. J. Biol. Macromol. 126, 496–506.

Mallik, A.K., Sakib, M.N., Shaharuzzaman, M., Haque, P., Rahman, M.M., 2020. Chitin nanomaterials: preparation and surface modifications. In: Handbook of Chitin and Chitosan: Volume 1: Preparation and Properties. Elsevier, Amsterdam, Netherlands, pp. 165–194.

Manjunatha, K.M., Ramana, M.V., Satyanarayana, D., 2007. Design and evaluation of diclofenac sodium controlled drug delivery systems. Indian J. Pharm. Sci. 69 (3), 384.

Manna, S., Jana, S., 2021. Carrageenan-based nanomaterials in drug delivery applications. In: Biopolymer-Based Nanomaterials in Drug Delivery and Biomedical Applications. Elsevier, pp. 365–382.

Mariano, M., El Kissi, N., Dufresne, A., 2014. Cellulose nanocrystals and related nanocomposites: review of some properties and challenges. J. Polym. Sci. B 52 (12), 791–806.

Min, B.-M., Lee, S.W., Lim, J.N., You, Y., Lee, T.S., Kang, P.H., Park, W.H., 2004. Chitin and chitosan nanofibers: electrospinning of chitin and deacetylation of chitin nanofibers. Polymer 45 (21), 7137–7142.

Mora-Huertas, C.E., Fessi, H., Elaissari, A., 2010. Polymer-based nanocapsules for drug delivery. Int. J. Pharm. 385 (1–2), 113–142.

Moriana, R., Vilaplana, F., Ek, M., 2016. Cellulose nanocrystals from forest residues as reinforcing agents for composites: a study from macro-to nano-dimensions. Carbohydr. Polym. 139, 139–149.

Moribe, K., Nomizu, N., Izukura, S., Yamamoto, K., Tozuka, Y., Sakurai, M., Ishida, A., Nishida, H., Miyazaki, M., 2008. Physicochemical, morphological and therapeutic evaluation of agarose hydrogel particles as a reservoir for basic fibroblast growth factor. Pharm. Dev. Technol. 13 (6), 541–547.

Muzzarelli, R.A., 2011a. Biomedical exploitation of chitin and chitosan via mechano-chemical disassembly, electrospinning, dissolution in imidazolium ionic liquids, and supercritical drying. Mar. Drugs 9 (9), 1510–1533.

Muzzarelli, R.A.A., 2012. Nanochitins and nanochitosans, paving the way to eco-friendly and energy-saving exploitation of marine resources. Polymer Science: A Comprehensive Reference. 10 Elsevier, Netherlands, pp. 153–164.

Nanaki, S., Karavas, E., Kalantzi, L., Bikiaris, D., 2010. Miscibility study of carrageenan blends and evaluation of their effectiveness as sustained release carriers. Carbohydr. Polym. 79 (4), 1157–1167.

Necas, J., Bartosikova, L., 2013. Carrageenan: a review. Vet. Med. 58 (4), 187–205.

Nyamayaro, K., Keyvani, P., D'Acierno, F., Poisson, J., Hudson, Z.M., Michal, C.A., Madden, J.D., Hatzikiriakos, S.G., Mehrkhodavandi, P., 2020. Toward biodegradable electronics: ionic diodes based on a cellulose nanocrystal-agarose hydrogel. ACS Appl. Mater. Interfaces 12 (46), 52182–52191.

Orellano, M.S., Longo, G.S., Porporatto, C., Correa, N.M., Falcone, R.D., 2020. Role of micellar interface in the synthesis of chitosan nanoparticles formulated by reverse micellar method. Colloids Surf. A Physicochem. Eng. Asp. 599, 124876.

Orsuwan, A., Shankar, S., Wang, L.-F., Sothornvit, R., Rhim, J.-W., 2016. Preparation of antimicrobial agar/banana powder blend films reinforced with silver nanoparticles. Food Hydrocoll. 60, 476–485.

Oun, A.A., Rhim, J.-W., 2016. Isolation of cellulose nanocrystals from grain straws and their use for the preparation of carboxymethyl cellulose-based nanocomposite films. Carbohydr. Polym. 150, 187–200. https://doi.org/10.1016/j.carbpol.2016.05.020.

Oun, A.A., Rhim, J.-W., 2018. I Effect of isolation methods of chitin nanocrystals on the properties of chitin-silver hybrid nanoparticles. Carbohydr. Polym. 197, 349–358.

Oun, A.A., Shankar, S., Rhim, J.-W., 2020. Multifunctional nanocellulose/metal and metal oxide nanoparticle hybrid nanomaterials. Crit. Rev. Food Sci. Nutr. 60 (3), 435–460.

Paques, J.P., van der Linden, E., van Rijn, C.J., Sagis, L.M., 2014. Preparation methods of alginate nanoparticles. Adv. Colloid Interface Sci. 209, 163–171.

Pinheiro, A.C., Bourbon, A.I., de Medeiros, B.G.S., da Silva, L.H., da Silva, M.C., Carneiro-da-Cunha, M.G., Coimbra, M.A., Vicente, A.A., 2012. Interactions between κ-carrageenan and chitosan in nanolayered coatings—structural and transport properties. Carbohydr. Polym. 87 (2), 1081–1090.

Polinarski, M.A., Beal, A.L., Silva, F.E., Bernardi-Wenzel, J., Burin, G.R., de Muniz, G.I., Alves, H.J., 2021. New perspectives of using chitosan, silver, and chitosan–silver nanoparticles against multidrug-resistant bacteria. Part. Part. Syst. Charact. 38 (4), 2100009.

Rajaonarivony, M., Vauthier, C., Couarraze, G., Puisieux, F., Couvreur, P., 1993. Development of a new drug carrier made from alginate. J. Pharm. Sci. 82 (9), 912–917.

Rashki, S., Asgarpour, K., Tarrahimofrad, H., Hashemipour, M., Ebrahimi, M.S., Fathizadeh, H., Khorshidi, A., Khan, H., Salavati-Niasari, M., Mirzaei, H., 2021. Chitosan-based nanoparticles against bacterial infections. Carbohydr. Polym. 251, 117108.

Reddy, K., Krishna Mohan, G., Satla, S., Gaikwad, S., 2011. Natural polysaccharides: versatile excipients for controlled drug delivery systems. Asian J. Pharm. Sci. 6 (6), 275–286.

Rehm, B.H., Moradali, M.F., 2018. Alginates and Their Biomedical Applications. Springer.

Reis, C.P., Neufeld, R.J., Vilela, S., Ribeiro, A.J., Veiga, F., 2006. Review and current status of emulsion/dispersion technology using an internal gelation process for the design of alginate particles. J. Microencapsul. 23 (3), 245–257.

Rhim, J.-W., Park, H.-M., Ha, C.-S., 2013. Bio-nanocomposites for food packaging applications. Prog. Polym. Sci. 38 (10−11), 1629–1652.

Sæther, H.V., Holme, H.K., Maurstad, G., Smidsrød, O., Stokke, B.T., 2008. Polyelectrolyte complex formation using alginate and chitosan. Carbohydr. Polym. 74 (4), 813–821.

Salehi, F., Behboudi, H., Kavoosi, G., Ardestani, S.K., 2020. Incorporation of Zataria multiflora essential oil into chitosan biopolymer nanoparticles: a nanoemulsion based delivery system to improve the *in-vitro* efficacy, stability and anticancer activity of ZEO against breast cancer cells. Int. J. Biol. Macromol. 143, 382–392.

Sarmento, B., Ribeiro, A., Veiga, F., Sampaio, P., Neufeld, R., Ferreira, D., 2007. Alginate/chitosan nanoparticles are effective for oral insulin delivery. Pharm. Res. 24 (12), 2198–2206.

Scionti, G., Moral, M., Toledano, M., Osorio, R., Duran, J.D., Alaminos, M., Campos, A., López-López, M.T., 2014. Effect of the hydration on the biomechanical properties in a fibrin-agarose tissue-like model. J. Biomed. Mater. Res. A 102 (8), 2573–2582.

Shankar, S., Kaur, S., 2018. Determination and distribution of cry1-type genes in Bacillus thuringiensis isolated from North India. Environ. Eng. Manage. J. 17 (3), 621–630.

Shankar, S., Rhim, J.-W., 2015. Amino acid mediated synthesis of silver nanoparticles and preparation of antimicrobial agar/silver nanoparticles composite films. Carbohydr. Polym. 130, 353–363.

Shankar, S., Rhim, J.-W., 2016a. Polymer nanocomposites for food packaging applications. In: Functional and Physical Properties of Polymer Nanocomposites. John Wiley & Sons Ltd, p. 29.

Shankar, S., Rhim, J.-W., 2016b. Preparation of nanocellulose from micro-crystalline cellulose: the effect on the performance and properties of agar-based composite films. Carbohydr. Polym. 135, 18–26.

Shankar, S., Rhim, J.-W., 2017. Preparation and characterization of agar/lignin/silver nanoparticles composite films with ultraviolet light barrier and antibacterial properties. Food Hydrocoll. 71, 76–84.

Shankar, S., Rhim, J.-W., 2018. Preparation of sulfur nanoparticle-incorporated antimicrobial chitosan films. Food Hydrocoll. 82, 116–123. https://doi.org/10.1016/j.foodhyd.2018.03.054.

Shankar, S., Teng, X., Rhim, J.-W., 2014. Properties and characterization of agar/CuNP bionanocomposite films prepared with different copper salts and reducing agents. Carbohydr. Polym. 114, 484–492.

Shankar, S., Reddy, J.P., Rhim, J.-W., 2015a. Effect of lignin on water vapor barrier, mechanical, and structural properties of agar/lignin composite films. Int. J. Biol. Macromol. 81, 267–273.

Shankar, S., Reddy, J.P., Rhim, J.-W., Kim, H.-Y., 2015b. Preparation, characterization, and antimicrobial activity of chitin nanofibrils reinforced carrageenan nanocomposite films. Carbohydr. Polym. 117, 468–475.

Shankar, S., Jaiswal, L., Rhim, J.-W., 2016. Gelatin-based nanocomposite films: potential use in antimicrobial active packaging. In: Antimicrobial Food Packaging. Elsevier, pp. 339–348.

Shankar, S., Wang, L.-F., Rhim, J.-W., 2017. Preparation and properties of carbohydrate-based composite films incorporated with CuO nanoparticles. Carbohydr. Polym. 169, 264–271.

Shankar, S., Baraketi, A., D'Auria, S., Fraschini, C., Salmieri, S., Jamshidian, M., Etty, M.C., Lacroix, M., 2020. Development of support based on chitosan and cellulose nanocrystals for the immobilization of anti-Shiga toxin 2B antibody. Carbohydr. Polym. 232. https://doi.org/10.1016/j.carbpol.2019.115785, 115785.

Shankar, S., Khodaei, D., Lacroix, M., 2021. Effect of chitosan/essential oils/silver nanoparticles composite films packaging and gamma irradiation on shelf life of strawberries. Food Hydrocoll. 117. https://doi.org/10.1016/j.foodhyd.2021.106750, 106750.

Shin, M.H., Lee, D.Y., Wohlgemuth, G., Choi, I.-G., Fiehn, O., Kim, K.H., 2010. Global metabolite profiling of agarose degradation by Saccharophagus degradans 2-40. N. Biotechnol. 27 (2), 156–168.

Singhsa, P., Narain, R., Manuspiya, H., 2017. Bacterial cellulose nanocrystals (BCNC) preparation and characterization from three bacterial cellulose sources and development of functionalized BCNCs as nucleic acid delivery systems. ACS Appl. Nano Mater. 1 (1), 209–221.

Somsak, P., Sriwattana, S., Prinyawiwatkul, W., 2021. Ultrasonic-assisted chitin nanoparticle and its application as saltiness enhancer. Int. J. Food Sci. Technol. 56 (2), 608–617.

Soppimath, K.S., Aminabhavi, T.M., Kulkarni, A.R., Rudzinski, W.E., 2001. Biodegradable polymeric nanoparticles as drug delivery devices. J. Control. Release 70 (1–2), 1–20.

Souza, M.P., Vaz, A.F., Costa, T.B., Cerqueira, M.A., De Castro, C.M., Vicente, A.A., Carneiro-da-Cunha, M.G., 2018. Construction of a biocompatible and antioxidant multilayer coating by layer-by-layer assembly of κ-carrageenan and quercetin nanoparticles. Food Bioproc. Tech. 11 (5), 1050–1060.

Tamara, F.R., Lin, C., Mi, F.-L., Ho, Y.-C., 2018. Antibacterial effects of chitosan/cationic peptide nanoparticles. Nanomaterials 8 (2), 88.

Tizchang, S., Khiabani, M.S., Mokarram, R.R., Hamishehkar, H., 2020. Bacterial cellulose nano crystal as hydrocolloid matrix in immobilized β-galactosidase onto silicon dioxide nanoparticles. LWT Food Sci. Technol. 123, 109091.

Tokumitsu, H., Ichikawa, H., Fukumori, Y., 1999. Chitosan-gadopentetic acid complex nanoparticles for gadolinium neutron-capture therapy of cancer: preparation by novel emulsion-droplet coalescence technique and characterization. Pharm. Res. 16 (12), 1830–1835.

Tortorella, S., Buratti, V.V., Maturi, M., Sambri, L., Franchini, M.C., Locatelli, E., 2020. Surface-modified nanocellulose for application in biomedical engineering and nanomedicine: a review. Int. J. Nanomedicine 15, 9909.

Udayangani, R., Somasiri, G.D.P., Wickramasinghe, I., Kim, S.-K., 2020. Potential health benefits of sulfated polysaccharides from marine algae. In: Encyclopedia of Marine Biotechnology. vol. 1. John Wiley & Sons Ltd, pp. 629–635.

Vanti, G.L., Masaphy, S., Kurjogi, M., Chakrasali, S., Nargund, V.B., 2020. Synthesis and application of chitosan-copper nanoparticles on damping off causing plant pathogenic fungi. Int. J. Biol. Macromol. 156, 1387–1395.

Vashist, A., Shahabuddin, S., Gupta, Y.K., Ahmad, S., 2013. Polyol induced interpenetrating networks: chitosan–methylmethacrylate based biocompatible and pH responsive hydrogels for drug delivery system. J. Mater. Chem. B 1, 168–178.

Vega, J.C., Bansal, S., Jonsson, C.B., Taylor, S.L., Figueroa, J.M., Dugour, A.V., Palacios, C., 2020. Iota carrageenan and xylitol inhibit SARS-CoV-2 in Vero cell culture. BioRxiv. https://doi.org/10.1101/2020.08.19.225854.

Wang, S.-L., Hiep, Đ.M., Luong, P.M., Vui, N.T., Đinh, T.M., Dzung, N.A., 2014. Preparation of chitosan nanoparticles by spray drying, and their antibacterial activity. Res. Chem. Intermed. 40 (6), 2165–2175.

Wang, S., Zhang, R., Yang, Y., Wu, S., Cao, Y., Lu, A., Zhang, L., 2018. Strength enhanced hydrogels constructed from agarose in alkali/urea aqueous solution and their application. Chem. Eng. J. 331, 177–184.

Wang, Y., Li, S., Jin, M., Han, Q., Liu, S., Chen, X., Han, Y., 2020. Enhancing the thermo-stability and anti-bacterium activity of lysozyme by immobilization on chitosan nanoparticles. Int. J. Mol. Sci. 21 (5), 1635.

Yang, X., Liu, J., Pei, Y., Zheng, X., Tang, K., 2020. Recent progress in preparation and application of nano-chitin materials. Energy Environ. Mater. 3 (4), 492–515.

Ye, W., Yokota, S., Fan, Y., Kondo, T., 2021. A combination of aqueous counter collision and TEMPO-mediated oxidation for doubled carboxyl contents of α-chitin nanofibers. Cellulose 28 (4), 2167–2181.

Yu, C.-Y., Jia, L.-H., Yin, B.-C., Zhang, X.-Z., Cheng, S.-X., Zhuo, R.-X., 2008. Fabrication of nano-spheres and vesicles as drug carriers by self-assembly of alginate. J. Phys. Chem. C 112 (43), 16774–16778.

Zarrintaj, P., Bakhshandeh, B., Rezaeian, I., Heshmatian, B., Ganjali, M.R., 2017. A novel electroactive agarose-aniline pentamer platform as a potential candidate for neural tissue engineering. Sci. Rep. 7 (1), 17187. https://doi.org/10.1038/s41598-017-17486-9.

Zhai, X., Li, C., Ren, D., Wang, J., Ma, C., Abd El-Aty, A.M., 2021. The impact of chitooligosaccharides and their derivatives on the *in vitro* and *in vivo* antitumor activity: a comprehensive review. Carbohydr. Polym. 266, 118132.

Zhang, J., Han, J., Zhang, X., Jiang, J., Xu, M., Zhang, D., Han, J., 2015. Polymeric nanoparticles based on chitooligosaccharide as drug carriers for co-delivery of all-trans-retinoic acid and paclitaxel. Carbohydr. Polym. 129, 25–34.

Zhang, H., Huang, X., Sun, Y., Xing, J., Yamamoto, A., Gao, Y., 2016. Absorption-improving effects of chitosan oligomers based on their mucoadhesive properties: a comparative study on the oral and pulmonary delivery of calcitonin. Drug Deliv. 23, 2419–2427.

Zhang, H., Feng, M., Chen, S., Shi, W., Wang, X., 2020a. Incorporation of lysozyme into cellulose nanocrystals stabilized β-chitosan nanoparticles with enhanced antibacterial activity. Carbohydr. Polym. 236, 115974.

Zhang, M., Hou, G., Hu, P., Feng, D., Wang, J., Zhu, W., 2020b. Nano chitosan-zinc complex improves the growth performance and antioxidant capacity of the small intestine in weaned piglets. Br. J. Nutr., 1–35.

Zhong, R., Yuan, M., Gao, H., Bai, Z., Guo, J., Zhao, X., Zhang, F., 2016. A subtle calculation method for nanoparticle's molar extinction coefficient: the gift from discrete protein-nanoparticle system on agarose gel electrophoresis. Funct. Mater. Lett. 9 (02), 1650029.

Zhou, Z., Jiang, F., Lee, T.-C., Yue, T., 2013. Two-step preparation of nanoscaled magnetic chitosan particles using Triton X-100 reversed-phase water-in-oil microemulsion system. J. Alloys Compd. 581, 843–848.

Zielińska, D., Szentner, K., Waśkiewicz, A., Borysiak, S., 2021. Production of nanocellulose by enzymatic treatment for application in polymer composites. Materials 14 (9), 2124.

Lipid-based nanostructures in food applications

Anujit Ghosal[a,b] *and Nandika Bandara*[c]

[a]Department of Food and Human Nutritional Sciences, Faculty of Agricultural and Food Sciences, University of Manitoba, Winnipeg, MB, Canada [b]Richardson Centre for Functional Foods & Nutraceuticals, Winnipeg, MB, Canada [c]Department of Food and Human Nutritional Sciences, Richardson Centre for Food Technology and Research, University of Manitoba, Winnipeg, MB, Canada

1 Introduction: Potential of lipid-based nanostructure

The global food industry has great value to the whole ecosystem and is one of the most crucial service providers. The smooth processing of the industry is crucial from a health and economic point of view, and hence, it is required to get support from the upcoming technologies, such as nanotechnology, to maintain the high standards. However, it has a very complex and diverse system in a way or another and touches and drives the entire living population of the world (Miwa, 2020; Maksimović et al., 2019; Costa et al., 2018). Nanotechnology has augmented the processing of food packaging, preservation against microbes, enriching of the food's shelf life, visual sensors to detect toxicity, delivery of food products/bioactive molecules, and channelizing of many other fronts of the food industry (Souza and Fernando, 2016; Gómez-Arribas et al., 2018; Go et al., 2017). Broadly, nanoparticles find application in detecting pathogens, food microbiology, taste improvement, controlling the pH, preserving food with high nutrient contents, and maintaining the shelf life of food products (Hasheminejad and Khodaiyan, 2020). In present times, the term "nanoparticle" is used in a generalized manner, defining the scale or size range of the materials used (Cao et al., 2016). The variation in the dimension of nanoparticles from their bulk counterparts generates unique physicochemical, physicomechanical, and pharmacokinetic properties. They are ubiquitously mentioned from home to healthcare applications and underwater to outer space technologies (Ghosal et al., 2013, 2019; Bhushan, 2017). Among various sectors, nanoparticles have brought a revolution in the medical sciences sector and have played a pivotal role in drug delivery, therapy, bioimaging, diagnostics, and the food sector (Ghosal et al., 2017; Puglia et al., 2019; Arya et al., 2019; Smerkova et al., 2020). They can be used to improve the state of living but can also adversely affect the living being and surroundings, as they can cross different barriers like a blood-brain barrier in our body and can travel via various alternating routes (transdermal, nasal, oral, etc.) (Tiwari et al., 2019; Ghasemiyeh and Mohammadi-Samani, 2018; Wissing et al., 2004). Further, nanoparticles can overcome the various limitations of traditional food sectors, which leads to recalling of spoiled packed food products or toxicity of the preservative ingredients at certain concentrations, thereby affecting billions of dollar economy and human life. In other words, nanoparticles are a two-sided sword that is required to be handled with utmost care. Thus, great importance is put in for the thoughts of the material selection, which is to be used in the nanoscale, particularly when the applications involve humans, like diagnostic services, pharmaceuticals, and food care industries (Sharma et al., 2012; Wang et al., 2017). A wrong choice of material can harm individuals in direct or indirect contact of the material via leaching toxins, generation of reactive oxygen species (ROS), cytotoxicity, genotoxicity, epigenicity, similar cell death processes, and mechanisms incurring foreign species (Liu et al., 2017; Alaraby et al., 2016). However, when it comes to the food sector, which directly influences humans, sustainable resource-based biosynthesized or bio-inspired nanoparticles are the most suitable candidates, among others. In this regard, lipid nanoparticle (LNP) has become one of the most efficient alternatives to metals, metal oxides, mixed ferrites, and polymer-based nanoparticles for particular applications in the food industry (Ghasemiyeh and Mohammadi-Samani, 2018; Kaczmarek et al., 2016; Khan et al., 2019; Gordillo-Galeano and Mora-Huertas, 2018). The LNPs with a balanced composition, i.e., base molecule, surfactants, emulsifier, and stabilizer, having a suitable surface charge (as per the required application) with zero toxicity, can be of utmost essence. Numerous categories in LNP can be designed by variation in the parameters, like preferred synthesis techniques, type of material used, and ultimately the desired application. The following sections will discuss these parameters along with the future perspectives for LNPs.

2 Type of lipid nanostructures used in food industries

With the help of nanotechnology, lipid biocompatibility and versatility have been explored to devise value-added products that can be utilized for bioapplications. The alteration in synthesis methodology and combination of different ingredients would result in unique nanoparticles, but most of them can be arranged in the following broad categories. Fig. 1A illustrates the wide array of applications that are explored for lipid-based nanostructures.

2.1 Solid lipid nanoparticles (SLNs)

The SLN has an average particle size ranging from 10 to 1000 nm with a solid lipid core, as shown in Fig. 1B. They are mainly prepared by oil-in-water emulsions after replacing oil with solid lipids, which get dispersed uniformly, resulting in a crystalline dispersion of lipid nanostructure in an aqueous or surfactant solution (Yousefi et al., 2019; Mukherjee et al., 2009). SLN has a large surface area, high drug-loading capacity, interactive surface for possible future modification, and the ability to increase the efficacy of delivered nutraceuticals and other active materials. The lipophilic nature of the core allows it to deliver solubilized or loaded hydrophobic drug molecules. Apart from lipids, long-chain hydrocarbon fatty acids (triglycerides, other glycerides, and vegetable oil-derived fatty acids like linoleic acid) can also be used along with the emulsifier to devise such SNPs (Li et al., 2016a).

The nanoformulation of the lipids is colloidal, but it has strategically distanced itself from general colloidal nanoparticles. Higher aqueous stability and ability to protect encapsulated molecules during application/processing procedures via different routes (oral, nasal, ocular, parenteral, dermal, or mixing with an emulsifier and other ingredients of food processing) are the key properties of SLNs that make them overcome the traditional drawbacks of polymeric or polymer-coated nanoparticles. SLNs can be prepared by ultrasonication/fast homogenization, solvent evaporation, solvent emulsification-diffusion, supercritical fluid and microemulsion, precipitation techniques, etc. (Triplett and Rathman, 2009; Ganesan and Narayanasamy, 2017; Jannin et al., 2018). SLNs have potential application in therapeutics and delivery of macromolecules within (via oral, nasal, and ocular delivery) or outside the human body. Delivery of essential bioactive compounds in food supplies, such as omega-3 fatty acids, natural pigments like carotenoids, polyphenols, resveratrol, lycopene, nisin, antioxidants, and others, has been successfully achieved by engineered SLNs (Weiss et al., 2008; Pandita et al., 2014; Zardini et al., 2018; Prombutara et al., 2012). The conversion of drug delivery vehicles like SLN, for the delivery of bioactive ingredients in foods can highlight the requirement of delivery vehicles for nutraceuticals in the food industry. The delivery mechanism is equally important as these bioactive ingredients need protection against untimely separation, pH variations, and interactions with other in vivo factors such as enzymes (Weiss et al., 2008; Aditya and Ko, 2015). Overall, these highly biocompatible nanoparticles can overcome the illness caused by low-nutrient foods, obesity, hypertension, cancer, and other related diseases. The increase in shelf life with high nutrient content would promote the health and wellness criteria of balanced dietaries. Fast-track production of SLN (10–100 nm) can be achieved by electrohydrodynamic spraying of stearic acid and

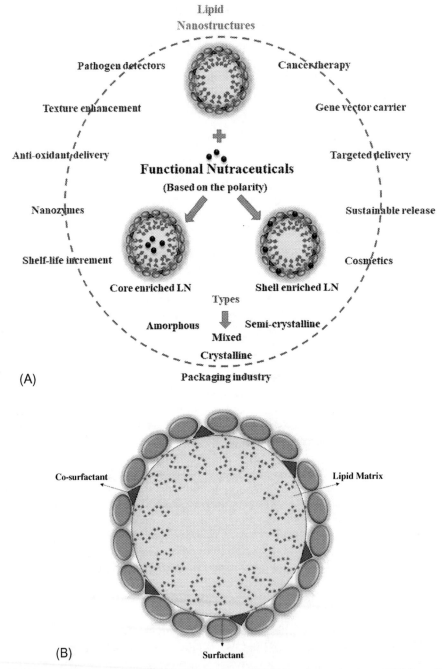

FIG. 1   (A) Schematic representation of different applications of lipid-based nanostructures. (B) Schematics of solid lipid nanoparticles made up of surfactant (emulsifier), co-surfactant, and lipophilic lipid core.

ethylcellulose encapsulating maltol flavor with an encapsulation and yield efficiency of around 69% (Eltayeb et al., 2013). A similar methodology can be further explored to deliver other lipophilic and bioactive compounds (Drosou et al., 2017; da Silva Santos et al., 2019a; Jaworek, 2016). The ability to scale up the delivery of bioactive ingredients and bioavailability of SLNs had made them the first choice for delivering such compounds or drugs in the first place. However, compared to other lipid carriers, they have lower drug-loading capacity due to drugs loaded being incorporated inside the lipid layers and lipid crystal structures, relatively higher dispersant (water: more than 75%) concentration, and drug expulsion after polymeric transition. Further, the solubility of drug within the lipid matrix majorly determines the drug encapsulation efficiency along with the lipid structure before and after polymerization. A successive advancement in nanotechnology and better understanding of lipid nanostructure indicate that the SLNs have good potential to improve and enrich the food dietaries.

2.2 Nanostructured lipid carriers (NLCs)

The NLCs were mainly explored to overcome the potential shortcoming of SLNs, such as poor drug-loading capacity and drug expulsion after lipid polymerization. The NLCs are oil-loaded SLNs, where lower crystallinity of solid lipids is combined with liquid lipids in a varied range of 70% to 30%. NLCs can be broadly classified into three categories: imperfect, amorphous, and multiple (Akhavan et al., 2018; Katouzian et al., 2017). Fig. 2A is an illustration of the key difference between SLNs and NLCs, whereas Fig. 2B represents different types of NLCs (Katouzian et al., 2017).

The preparation methodologies of NLCs are relatively similar to SLNs but contain few modifications. The drug-loading capacity was increased by using a mixture of liquid lipids and solid lipids to generate vacant spaces within the NLCs. The lipid composition has a major impact on determining the macromolecules' drug-loading capacity, release mechanism, and retention level of the encapsulated drug (da Silva Santos et al., 2019b). The NLCs fabricated using a mixture of liquid medium-chain triglycerides and solid glyceryl tristearate via homogenization and ultrasonication have influenced curcumin's loading capacity and delivery mechanism (Feng et al., 2020). An improvement of the encapsulation efficiency up to 94% was observed for astaxanthin encapsulated in NLCs prepared from caprylic/capric triglyceride and glycerol monostearate (Mao et al., 2019). In another study, a turmeric powder with a mixture of water-soluble, fat-soluble, and insoluble compounds was encapsulated with improved encapsulation efficiency in NLCs consisting of liquid lipid, medium-chain triglycerides, and solid glycerol monostearate (Park et al., 2018). Such lipid combination results in a perfect matrix that can incorporate as well as inhibit external interference with the bioactive compounds and has the ability to produce on an industrial scale. Especially, the liquid lipid used in the NLC formation should have the ability to form nanoparticles and disperse bioactive compounds and should be spatially incompatible (i.e., should not form a crystalline matrix, and the lipids should not have toxic residues) (Tamjidi et al., 2013).

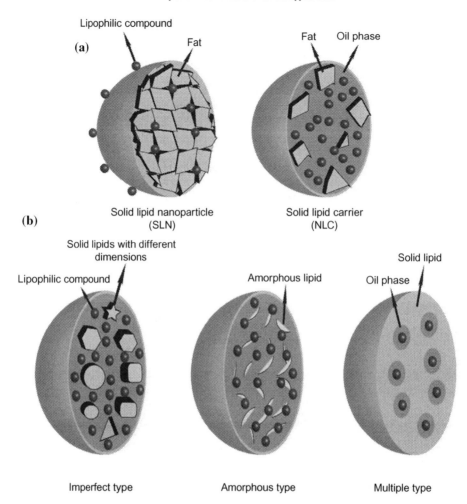

FIG. 2   (A) Schematic representations and differences between NLCs and SLNs. (B) Classification of NLCs: (a) imperfect type, (b) amorphous type, and (c) multiple types (*yellow color* (light gray in print version) specifies the solid lipid, and *orange* (light gray in print version) illustrates the liquid oil phase). *The images are reproduced with Katouzian, I., Faridi Esfanjani, A., Jafari, S.M., Akhavan, S., 2017. Formulation and application of a new generation of lipid nanocarriers for the food bioactive ingredients. Trends Food Sci. Technol. 68, 14–25. Copyright permission.*

2.3 Lipid nanogels

Hydrogels are well known for their ability to withhold a large amount of water within the polymeric skeleton and swell upon water absorption. In hydrogels, interpenetrating and cross-linked networks of polymers and copolymers will result in a three-dimensional main frame consisting of hydrophobic and hydrophilic entities, thereby delivering hydrophobic, hydrophilic, and amphiphilic active moieties. Nanoscale synthesis of these hydrogels has opened up new possibilities in delivering and safekeeping bioactive and drug molecules.

Stimulus-responsive nanogels are getting increased attention due to their unique slow release of the adsorbed drug upon changes in external stimuli, such as pH, temperature, osmotic pressure, and concentration of particular ions, or other stimuli like change in a magnetic moment or light (Ghosal et al., 2017; Molina et al., 2015; Li et al., 2020; Hou et al., 2019). Food supplements and bioactive compounds can be stored inside the lipid-based nanogels or microgels, which can be utilized for direct human consumption or developing food ingredients, such as texture modifiers or thickening agents. For example, the nanogels loaded with prebiotics were used to improve bowel movements of elderly people and to provide essential proteins for special needs (Axelos and Van de Voorde, 2017). In some cases, the nanogels were coated with the lipid layer to enable the multi-functionality and responsiveness of the lipid nanogel system (Liang and Kiick, 2014; Dabkowska et al., 2017). The lipid nanogels can be prepared by emulsification, free radical polymerization, template-assisted, ultrasonication, and photopolymerization techniques (Sun et al., 2019; Piran et al., 2020; Nejatian et al., 2018). Nanogels with a specific composition can be utilized as stabilizers for oil-in-water emulsion processing as well (Wang et al., 2020). Incorporation of various other nano-entities, like clay particles, metal/metal oxide nanoparticles, nanozymes, and catalysts, can further improve the usability in the food industry.

2.4 Nanoliposomes

Liposomes are constructed by using polar lipids or a combination of polar lipids with non-polar lipids to generate bilayer vesicle structures. They are capable of encapsulating both hydrophilic and hydrophobic drugs and other bioactive materials. The similarity in the structures of liposomes with vesicles gives an added advantage to these delivery agents in biomedicine and the food industry. However, the lower stability, burst release, and need for complex fabrication techniques limit the generalized application of liposomes (Sarabandi et al., 2019). However, the synthesis and encapsulation of liposomes in nanoscale within polymer matrixes, like pectin and chitosan, have helped improve the vesicular delivery agent's durability and drug efficacy (Shishir et al., 2019; Sohail et al., 2016; Hasan et al., 2016). Fig. 3A and B illustrates the schematics of nanoliposome preparation and the cryo-transmission electron microscopic image of nanoliposome made up of 1,2-distearoyl-sn-glycero-3-phosphoethanolamine-N-diethylenetriaminepentaacetic acid (18:0 PE DTPA (Gd)), 1,2-dimyristoyl-sn-glycero-3-phosphocholine (DMPC), 1-stearoyl-2-linoleoyl-sn-glycero-3-phosphocholine (SOPC), 1,2-dioleoyl-sn-glycero-3-ethylphosphocholine (EPC 95%), 1,2-distearoyl-sn-glycero-3-phosphocholine (DSPC), and cholesterol (Kotouček et al., 2020).

In the food sector, nanoliposomes are used to deliver a broad range of bioactive compounds, such as vitamins, proteins, enzymes, odor, pigments, antioxidants, pesticides, and many others (Shishir et al., 2019; Subramani and Ganapathyswamy, 2020). Formulating nanoliposomes involves reacting aqueous reagent with amphiphilic lipids such as phospholipids or cholesterol as lipid matrix in the presence of emulsifiers such as phosphatidylcholine. The polar ends of phospholipids will be oriented towards the aqueous phase (outside and inside of the structure), and hydrophobic hydrocarbon will be arranged within the membrane layer. Compared to other nanocarriers, these liposomes provide higher protection to sensitive drug load with possible surface modification. Traditional liposome synthesis

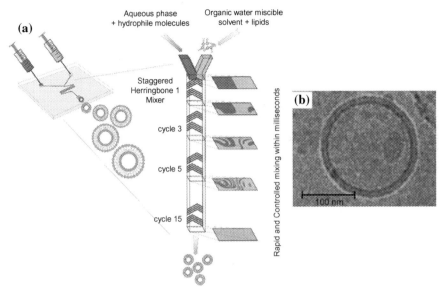

FIG. 3   (A) Principle of microfluidic mixing in herringbone channel and formation of liposomes. Organic water-miscible solvent (ethanol) contains lipids forming liposomes, while the water phase contains water-soluble components to be encapsulated. The mixing process is finished within a millisecond, and liposomes are formed via self-assembly. Various linear injector syringe pumps controlled by computers are used to drive the mixing process. (B) Cryo-TEM images of liposomes containing 5% of Gd-lipid width of phospholipid bilayer of ≈5.8nm. *The images are represented here with permission from Kotouček, J., Hubatka, F., Mašek, J., Kulich, P., Velínská, K., Bezděková, J., Fojtíková, M., Bartheldyová, E., Tomečková, A., Stráská, J., Preparation of nanoliposomes by microfluidic mixing in herring-bone channel and the role of membrane fluidity in liposomes formation. Sci. Rep. 10, 1–11. Creative Commons.*

techniques are not well enabled to produce complex nanostructured liposomes; therefore, ultrasonication, extrusion, microfluidization, freeze-thawing, and thin-film hydration are some of the techniques used for the production of nanoliposomes (Zoghi et al., 2018). The nanoscale liposomes have a higher surface area and have the potential to deliver a wide spectrum of hydrophilic, hydrophobic, and amphiphilic bioactive materials.

3 Different synthesis methodologies

3.1 Nanoemulsion

Nanoemulsion is a versatile technique, which uses two immiscible liquids (dispersed: internal phase and continuous: external phase) to generate either water-in-oil (W/O), oil-in-water (O/W), or double (O/W/O, W/O/W) emulsions as nanocarriers. This technique can be simultaneously used with some modification to create specific nanostructures, or the emulsion itself can be utilized to encapsulate drugs/bioactive ingredients (Mai et al., 2018; Colmenares et al., 2016). Surface-active agents and emulsifiers are an integral part of emulsion systems to stabilize the nonmiscible components. Emulsions can be categorized into thermodynamically unstable nanoemulsions (10–100nm), thermodynamically stable microemulsions (5 to 50 nm), and macroemulsions (0.1–5 μm) based on the size of the external phase (droplet size). The droplet size and composition (surfactants, emulsifying agents, and ingredients) will determine the emulsion system's stability, color, and shelf life. Nanoemulsions and microemulsions are far superior in terms of encapsulation and optical properties. Numerous publications support the superiority of nanoemulsions for delivery of drugs and herbal bioactive compounds (Harwansh et al., 2019). These structures are prepared mainly by either high-energy emulsification or dispersion using high shear mixing, ultrasonication, and liquid jets (Gharibzahedi and Jafari, 2018; Hörmann and Zimmer, 2016; Siddiqui et al., 2017). However, sensitive labile molecules (DNA, proteins, and enzymes) are not advised to be encapsulated by this method due to the potential heat damage, which could degrade active compounds.

3.2 Ultrasonication synthesis

Nanostructure, nanoliposomes, or nanoemulsions can be easily prepared by high-energy ultrasonication. In ultrasonication, the ultrasonic (sound energy) frequencies of more than 20 kHz are generated by a sonicator probe, which will cause pressure variation in the system. The ultrasonic transducers can generate large-intensity acoustic cavitation zones, converting the sound energy into mechanical energy. Therefore, ultrasonication can be used to fabricate SLNs, NLCs, nanoliposome, and other nanoemulsions (Jannin et al., 2018; Gharibzahedi and Jafari, 2018; Koshani and Jafari, 2019). Ultrasonication-based nanostructure formation is suitable for bioactive compounds that are highly facile and easy to scale up (Koshani and Jafari, 2019). For example, NLCs have been successfully devised using sonication with an encapsulation efficiency of 93% for curcuminoids (Park et al., 2018). Encapsulation efficiency and stability of biocompatible lipid nanostructures can be tailored by altering the sonication (bath and probe) device or variation in experimental conditions (Shah et al., 2015). Ultrasonication is also efficient in screening optimum components for the formulation of NLCs and SLNs out of natural oils and lipids. Use of probe sonication with a cycle of 0.5 s and amplitude of 70% leads to the reduction in the size of the nanocarriers, ultimately resulting in a more stable formulation, as studied for a storage period of 60 days (Babazadeh et al., 2017). A number of variations in nanostructured lipid-based entities are possible based on the intensity, frequency, number of cycles, temperature variation, and type of sonication instrument used. The technique is highly versatile, is used in various commercial product formulation centers, and has good future prospective in the field of synthesis/lipid nanostructure formulations.

3.3 Microfluidic synthesis

Microfluidic is an emerging technology that is efficient enough to work with minimal volumes of reactants. The system can be mechanical or fully automated and can generate required products continuously. High-pressure valve homogenization (HPVH)-based methods are extensively used in developing nanoemulsions. However, HPVH methods pose several challenges such as uneven droplet size and size distribution and potential damage to the bioactive compounds due to the extensive heat generation during HPVH. In contrast, microfluidization uses a capillary tube instead of a high-pressure valve for homogenization, which reduces heat generation considerably and produces nanoemulsions with smaller and uniform droplets. Due to nanoemulsions' inherent thermodynamically unstable nature, selecting a method that minimizes heat generation and develops uniform droplet size is essential in developing stable nanoemulsions. As compared to other techniques, this is the most controlled low-temperature approach to process lipid nanostructures for temperature- and environment-sensitive drugs and bioactive materials, such as DNA, RNA, proteins, enzymes, and carotenoids. The variation in flow rate, geometry of the microfluidic chip, and sonication can generate a series of different nanoparticles constructed of polyethylenimine and pH-sensitive lipid (1,2-dioleyloxy-*N,N*-dimethyl-3-aminopropane) (Li et al., 2016b). The method is highly scalable and fast and can be altered as per the demand (Maeki et al., 2018).

4 Application of lipid nanostructure in food industries

The nanoparticles of Ag, Au, Zn, ZnO, Cu, and Cu_xO_y are the most commonly used nanomaterials as nutrients, antibacterial ingredients, and preservatives. Nanoparticles in the form of catalysts (activators or deactivators) have been incorporated in food ingredients to optimize digestive needs and improve packaging and delivery systems (Rajamanickam et al., 2012; Rizwan et al., 2017). However, these metal nanoparticles could cause toxicity in certain situations. Therefore, lipid-based nanostructures are explored to avoid the toxicity associated with metal/metal oxide nanoparticles. They can be multifunctional and provide unique advantages in food systems. Most of the applications of lipid nanocarriers can be broadly categorized in the following three ways:

4.1 Delivery of preservatives and bioactive molecules

Bioactive molecules are generally a single molecule or a combination of molecules that are known to show specific activities due to available surface-reactive groups or properties generated by those groups. The active sites on such molecules are highly reactive, labile, and thermodynamically unstable. For example, anticancer drugs, antimicrobial agents, DNA, proteins, and vitamins have low thermal stability and are prone to surrounding environmental conditions. They either get deactivated with time due to unwanted reactions or cause toxicity when they accumulate. Therefore, encapsulation of bioactive compounds is considered one of the best methods to utilize these bioactive components for human benefit. The lipid source for encapsulation is selected based on low reactivity, bioavailability, and digestive stability. The delivery and adsorption of the drug/bioactive compounds increase manyfold at the nanoscale. In vitro efficacy of antibiotics (levofloxacin) against *Pseudomonas aeruginosa* biofilm has been evaluated using poly (lactic-*co*-glycolic acid) and phosphatidylcholine as lipid encapsulation systems. The constant release and long-term stability of the hybrid lipid nanoencapsulation were considered to be the main reasons behind the positive output of this system (Cheow et al., 2011). Similarly, food fortification with vitamins and minerals has been achieved using nanostructured lipid carriers and solid lipid nanoparticles (Katouzian et al., 2017).

4.2 Emulsifying agents

Emulsifiers are the substances that can stabilize an emulsion, i.e., which makes the coexistence of two or three immiscible liquids possible. The lipid-based nanostructure can be used as an emulsifier in many cases and is responsible for improving the texture and thickness of the food products. Biocompatibility allows lipid-based nanostructures to be used simultaneously as delivery agents for bioactive materials and emulsifiers. The structure of lipids determines their functional properties as well as nutritional value. Generally, mono- and diglycerides are used as emulsifiers in many food products. Lecithin, a natural emulsifier, is composed of phosphatidylcholine, phosphatidylethanolamine, phosphatidylinositol, phosphatidylserine, and phosphatidic acid. Diacetyl tartaric acid ester of monoglyceride is another ionic emulsifier that helps in stabilizing protein-rich dairy emulsions. The surface modification of SLNs using lecithin (30% to 50%) has resulted in high dispersion efficiency (Schubert and Müller-Goymann, 2005). Lecithin favors a large O/W interface and smaller overall particle size. The SLN10 and SLN20 at least require 1% emulsifier to form nanoparticles; however, nanoparticle formation was observed without any additional emulsifier in the SLN30 system (Schubert and Müller-Goymann, 2005). Similarly, the emulsifying ability has been observed in SLN with soybean phosphatidylcholine (SLN-PC) as the lipophilic emulsifier that enhances the drug delivery across the skin (Wang et al., 2009). The development of various self-emulsifying drug delivery systems using lipid vehicles, liposomes, and self-emulsifying formulations (a mixture of lipids, surfactants, and solvents) also highlights the potential of lipid nanostructures as emulsifying agents (Neslihan Gursoy and Benita, 2004; Pouton, 2000; Xie and Dunford, 2017; Rani et al., 2019).

4.3 Food structure modification agents

Apart from the delivery of the active agents and nutrients to the food system, they tend to improve eateries' texture, taste, viscosity, and other physicochemical properties. Being biocompatible and biodegradable, they are considered safe for digestion along with the encapsulated functional components. NLCs are extensively used for the development of edible beverage models. Different pH and pasteurization treatments have minimal effect on the particle size and encapsulation stability of the cinnamon essential oil within NLCs (Bashiri et al., 2020). The rheological properties of nanostructured lipid carriers are also different from traditional emulsions. Based on the composition of lipid nanostructures and interactions, they show different rheological characteristics. Further, the droplet concentration would increase the system's viscosity, and the emulsion system will exhibit solid viscoelastic-elastic properties, leading to gelation (McClements, 2011). Similar concepts can be utilized for the preparation of functional beverages, sauces, and mayonnaise. However, the use of surfactants and organic solvents may be controlled or altered to avoid possible toxicity within the gastrointestinal tract (McClements, 2013).

5 Future of lipid-based nanostructures

Lipid-based nanostructures have great potential in the agricultural and food industries. The applications are endless, where these nanocarriers have been proved to reduce the adverse effect of additives like antibacterial, antifungal, or anticancer agents by encapsulating them within the lipid matrix and delivering them to the target location. Lipid nanostructure-based edible packaging materials, delivery agents, and food modifiers can help reduce plastic wastes and improve the overall food quality. Each lipid-based nanocarrier has unique properties that can be used for a specific application like nanoliposomes to target the delivery of bioactive compounds/drugs due to their structural similarity. The chemical and structural properties will allow the SLNs and NLCs to encapsulate drugs/bioactive compounds that require long-term and sustained release. The lipid emulsion nanocarriers are easy to fabricate and highly stable and can alter or scale the amount of emulsion required. In contrast, the lipid nanogels can be used as an ingredient in food systems directly as texture/viscosity modifiers either with or without a source of bioactive compound. The use of different lipids, glycerides, or other surface-active agents derived from agricultural/food industry processing by-products and waste streams for preparing nanostructure will generate value-added applications to the by-products. Also, it will increase the biocompatibility, availability, and stability of the nanostructures. At present, many commercial lipid-based nanostructures are available in the market, such as Aquanova NovaSOL, NovaSOL Curcumin (lipid nanoemulsion), Coatsome (phospholipid-based product), Fabuless (support appetite control), and Curcosome (liposomal formulation of curcumin), and some active drugs (Norvir, Sustiva) are also delivered as lipid formulation. In this regard, many more products are expected to be projected in the future with improved delivery, minimal side effects, and higher efficacy.

Although the release mechanism of the active drug ingredients has been studied, future research should specifically study the mechanism of action, degradation of nanostructures, and alternate pathways of dispersion. In addition, advances in nanotechnology and novel technologies such as 3-D printing and computer-aided technologies such as microfluidics should be utilized to fabricate robust designs and textures with known loading capacity rapidly. Also, the co-delivery of active compounds with different chemical and physical properties needs to be studied extensively to advance this research area in the future.

References

Aditya, N.P., Ko, S., 2015. Solid lipid nanoparticles (SLNs): delivery vehicles for food bioactives. RSC Adv. 5, 30902–30911.

Akhavan, S., Assadpour, E., Katouzian, I., Jafari, S.M., 2018. Lipid nano scale cargos for the protection and delivery of food bioactive ingredients and nutraceuticals. Trends Food Sci. Technol. 74, 132–146.

Alaraby, M., Hernández, A., Marcos, R., 2016. New insights in the acute toxic/genotoxic effects of CuO nanoparticles in the in vivo Drosophila model. Nanotoxicology 10, 749–760.

Arya, G., Sharma, N., Mankamna, R., Nimesh, S., 2019. Antimicrobial silver nanoparticles: future of nanomaterials. In: Microbial Nanobionics. Springer, pp. 89–119.

Axelos, M.A., Van de Voorde, M., 2017. Nanotechnology in Agriculture and Food Science. John Wiley & Sons.

Babazadeh, A., Ghanbarzadeh, B., Hamishehkar, H., 2017. Formulation of food grade nanostructured lipid carrier (NLC) for potential applications in medicinal-functional foods. J. Drug Deliv. Sci. Technol. 39, 50–58.

Bashiri, S., Ghanbarzadeh, B., Ayaseh, A., Dehghannya, J., Ehsani, A., Ozyurt, H., 2020. Essential oil-loaded nanostructured lipid carriers: the effects of liquid lipid type on the physicochemical properties in beverage models. Food Biosci. 35, 100526.

Bhushan, B., 2017. Introduction to nanotechnology. In: Springer Handbook of Nanotechnology. Springer, pp. 1–19.

Cao, Y., Li, J., Liu, F., Li, X., Jiang, Q., Cheng, S., Gu, Y., 2016. Consideration of interaction between nanoparticles and food components for the safety assessment of nanoparticles following oral exposure: a review. Environ. Toxicol. Pharmacol. 46, 206–210.

Cheow, W.S., Chang, M.W., Hadinoto, K., 2011. The roles of lipid in anti-biofilm efficacy of lipid–polymer hybrid nanoparticles encapsulating antibiotics. Colloids Surf. A Physicochem. Eng. Asp. 389, 158–165.

Colmenares, D., Sun, Q., Shen, P., Yue, Y., McClements, D.J., Park, Y., 2016. Delivery of dietary triglycerides to *Caenorhabditis elegans* using lipid nanoparticles: nanoemulsion-based delivery systems. Food Chem. 202, 451–457.

Costa, L.B.M., Godinho Filho, M., Fredendall, L.D., Paredes, F.J.G., 2018. Lean, six sigma and lean six sigma in the food industry: a systematic literature review. Trends Food Sci. Technol. 82, 122–133.

da Silva Santos, V., Miyasaki, E.K., Cardoso, L.P., Paula, A., Ribeiro, B., Santana, M.H.A., 2019b. Crystallization, polymorphism and stability of nanostructured lipid carriers developed with soybean oil, fully hydrogenated soybean oil and free phytosterols for food applications. J. Nanotechnol. Res. 1, 1–21.

da Silva Santos, V., Ribeiro, A.P.B., Santana, M.H.A., 2019a. Solid lipid nanoparticles as carriers for lipophilic compounds for applications in foods. Food Res. Int. 122, 610–626.

Dabkowska, A.P., Valldeperas, M., Hirst, C., Montis, C., Pálsson, G.K., Wang, M., Nöjd, S., Gentile, L., Barauskas, J., Steinke, N.-J., 2017. Non-lamellar lipid assembly at interfaces: controlling layer structure by responsive nanogel particles. Interface Focus 7, 20160150.

Drosou, C.G., Krokida, M.K., Biliaderis, C.G., 2017. Encapsulation of bioactive compounds through electrospinning/electrospraying and spray drying: a comparative assessment of food-related applications. Drying Technol. 35, 139–162.

Eltayeb, M., Bakhshi, P.K., Stride, E., Edirisinghe, M., 2013. Preparation of solid lipid nanoparticles containing active compound by electrohydrodynamic spraying. Food Res. Int. 53, 88–95.

Feng, J., Huang, M., Chai, Z., Li, C., Huang, W., Cui, L., Li, Y., 2020. The influence of oil composition on the transformation, bioaccessibility, and intestinal absorption of curcumin in nanostructured lipid carriers. Food Funct. 11, 5223–5239.

Ganesan, P., Narayanasamy, D., 2017. Lipid nanoparticles: different preparation techniques, characterization, hurdles, and strategies for the production of solid lipid nanoparticles and nanostructured lipid carriers for oral drug delivery. Sustain. Chem. Pharm. 6, 37–56.

Gharibzahedi, S.M., Jafari, S.M., 2018. Fabrication of nanoemulsions by ultrasonication. In: Nanoemulsions. Elsevier, pp. 233–285.

Ghasemiyeh, P., Mohammadi-Samani, S., 2018. Solid lipid nanoparticles and nanostructured lipid carriers as novel drug delivery systems: applications, advantages and disadvantages. Res. Pharm. Sci. 13, 288.

Ghosal, A., Iqbal, S., Ahmad, S., 2019. NiO nanofiller dispersed hybrid soy epoxy anticorrosive coatings. Prog. Org. Coat. 133, 61–76.

Ghosal, A., Shah, J., Kotnala, R.K., Ahmad, S., 2013. Facile green synthesis of nickel nanostructures using natural polyol and morphology dependent dye adsorption properties. J. Mater. Chem. A 1, 12868–12878.

Ghosal, A., Tiwari, S., Mishra, A., Vashist, A., Rawat, N.K., Ahmad, S., Bhattacharya, J., 2017. Design and engineering of nanogels. In: Nanogels for Biomedical Applications, pp. 9–28.

Go, M.-R., Bae, S.-H., Kim, H.-J., Yu, J., Choi, S.-J., 2017. Interactions between food additive silica nanoparticles and food matrices. Front. Microbiol. 8, 1013.

Gómez-Arribas, L.N., Benito-Peña, E., Hurtado-Sánchez, M.D.C., Moreno-Bondi, M.C., 2018. Biosensing based on nanoparticles for food allergens detection. Sensors 18, 1087.

Gordillo-Galeano, A., Mora-Huertas, C.E., 2018. Solid lipid nanoparticles and nanostructured lipid carriers: a review emphasizing on particle structure and drug release. Eur. J. Pharm. Biopharm. 133, 285–308.

Harwansh, R.K., Deshmukh, R., Rahman, M.A., 2019. Nanoemulsion: promising nanocarrier system for delivery of herbal bioactives. J. Drug Deliv. Sci. Technol. 51, 224–233.

Hasan, M., Ben Messaoud, G., Michaux, F., Tamayol, A., Kahn, C.J.F., Belhaj, N., Linder, M., Arab-Tehrany, E., 2016. Chitosan-coated liposomes encapsulating curcumin: study of lipid–polysaccharide interactions and nanovesicle behavior. RSC Adv. 6, 45290–45304.

Hasheminejad, N., Khodaiyan, F., 2020. The effect of clove essential oil loaded chitosan nanoparticles on the shelf life and quality of pomegranate arils. Food Chem. 309, 125520.

Hörmann, K., Zimmer, A., 2016. Drug delivery and drug targeting with parenteral lipid nanoemulsions—a review. J. Control. Release 223, 85–98.

Hou, X., Pan, Y., Xiao, H., Liu, J., 2019. Controlled release of agrochemicals using pH and redox dual-responsive cellulose nanogels. J. Agric. Food Chem. 67, 6700–6707.

Jannin, V., Blas, L., Chevrier, S., Miolane, C., Demarne, F., Spitzer, D., 2018. Evaluation of the digestibility of solid lipid nanoparticles of glyceryl dibehenate produced by two techniques: ultrasonication and spray-flash evaporation. Eur. J. Pharm. Sci. 111, 91–95.

Jaworek, A., 2016. Electrohydrodynamic microencapsulation technology. In: Encapsulations. Elsevier, pp. 1–45.

Kaczmarek, J.C., Patel, A.K., Kauffman, K.J., Fenton, O.S., Webber, M.J., Heartlein, M.W., DeRosa, F., Anderson, D.G., 2016. Polymer–lipid nanoparticles for systemic delivery of mRNA to the lungs. Angew. Chem. 128, 14012–14016.

Katouzian, I., Faridi Esfanjani, A., Jafari, S.M., Akhavan, S., 2017. Formulation and application of a new generation of lipid nanocarriers for the food bioactive ingredients. Trends Food Sci. Technol. 68, 14–25.

Khan, H., Mirzaei, H.R., Amiri, A., Akkol, E.K., Haleemi, S.M.A., Mirzaei, H., 2019. Glyco-nanoparticles: new drug delivery systems in cancer therapy. Semin. Cancer Biol. 69, 24–42.

Koshani, R., Jafari, S.M., 2019. Ultrasound-assisted preparation of different nanocarriers loaded with food bioactive ingredients. Adv. Colloid Interface Sci. 270, 123–146.

Kotouček, J., Hubatka, F., Mašek, J., Kulich, P., Velínská, K., Bezděková, J., Fojtíková, M., Bartheldyová, E., Tomečková, A., Stráská, J., 2020. Preparation of nanoliposomes by microfluidic mixing in herring-bone channel and the role of membrane fluidity in liposomes formation. Sci. Rep. 10, 1–11.

Li, M., Zahi, M.R., Yuan, Q., Tian, F., Liang, H., 2016a. Preparation and stability of astaxanthin solid lipid nanoparticles based on stearic acid. Eur. J. Lipid Sci. Technol. 118, 592–602.

Li, X.-M., Li, X., Wu, Z., Wang, Y., Cheng, J.-S., Wang, T., Zhang, B., 2020. Chitosan hydrochloride/carboxymethyl starch complex nanogels stabilized Pickering emulsions for oral delivery of β-carotene: protection effect and in vitro digestion study. Food Chem. 315, 126288.

Li, Y., Huang, X., Lee, R.J., Qi, Y., Wang, K., Hao, F., Zhang, Y., Lu, J., Meng, Q., Li, S., 2016b. Synthesis of polymer-lipid nanoparticles by microfluidic focusing for siRNA delivery. Molecules 21, 1314.

Liang, Y., Kiick, K.L., 2014. Multifunctional lipid-coated polymer nanogels crosslinked by photo-triggered Michael-type addition. Polym. Chem. 5, 1728–1736.

Liu, J., Kang, Y., Yin, S., Song, B., Wei, L., Chen, L., Shao, L., 2017. Zinc oxide nanoparticles induce toxic responses in human neuroblastoma SHSY5Y cells in a size-dependent manner. Int. J. Nanomedicine 12, 8085.

Maeki, M., Kimura, N., Sato, Y., Harashima, H., Tokeshi, M., 2018. Advances in microfluidics for lipid nanoparticles and extracellular vesicles and applications in drug delivery systems. Adv. Drug Deliv. Rev. 128, 84–100.

Mai, H., Le, T., Diep, T., Le, T., Nguyen, D., Bach, L., 2018. Development of soild lipid nanoparticles of Gac (Momordica cocochinensis Spreng) oil by nano-emulsion technique. Asian J. Chem. 30, 293–297.

Maksimović, M., Omanović-Mikličanin, E., Badnjević, A., 2019. What food do we want to eat? Is nanofood food of our future? In: Nanofood and Internet of Nano Things. Springer, pp. 1–8.

Mao, X., Tian, Y., Sun, R., Wang, Q., Huang, J., Xia, Q., 2019. Stability study and in vitro evaluation of astaxanthin nanostructured lipid carriers in food industry. Integr. Ferroelectr. 200, 208–216.

McClements, D.J., 2011. Edible nanoemulsions: fabrication, properties, and functional performance. Soft Matter 7, 2297–2316.

McClements, D.J., 2013. Edible lipid nanoparticles: digestion, absorption, and potential toxicity. Prog. Lipid Res. 52, 409–423.

Miwa, N., 2020. Innovation in the food industry using microbial transglutaminase: keys to success and future prospects. Anal. Biochem. 597, 113638.

Molina, M., Asadian-Birjand, M., Balach, J., Bergueiro, J., Miceli, E., Calderón, M., 2015. Stimuli-responsive nanogel composites and their application in nanomedicine. Chem. Soc. Rev. 44, 6161–6186.

Mukherjee, S., Ray, S., Thakur, R., 2009. Solid lipid nanoparticles: a modern formulation approach in drug delivery system. Indian J. Pharm. Sci. 71, 349.

Nejatian, M., Abbasi, S., Kadkhodaee, R., 2018. Ultrasonic-assisted fabrication of concentrated triglyceride nanoemulsions and nanogels. Langmuir 34, 11433–11441.

Neslihan Gursoy, R., Benita, S., 2004. Self-emulsifying drug delivery systems (SEDDS) for improved oral delivery of lipophilic drugs. Biomed. Pharmacother. 58, 173–182.

Pandita, D., Kumar, S., Poonia, N., Lather, V., 2014. Solid lipid nanoparticles enhance oral bioavailability of resveratrol, a natural polyphenol. Food Res. Int. 62, 1165–1174.

Park, S.J., Garcia, C.V., Shin, G.H., Kim, J.T., 2018. Improvement of curcuminoid bioaccessibility from turmeric by a nanostructured lipid carrier system. Food Chem. 251, 51–57.

Piran, F., Khoshkhoo, Z., Hosseini, S., Azizi, M., 2020. Controlling the antioxidant activity of green tea extract through encapsulation in chitosan-citrate nanogel. J. Food Qual. 2020. https://doi.org/10.1155/2020/7935420, 7935420.

Pouton, C.W., 2000. Lipid formulations for oral administration of drugs: non-emulsifying, self-emulsifying and 'self-microemulsifying' drug delivery systems. Eur. J. Pharm. Sci. 11, S93–S98.

Prombutara, P., Kulwatthanasal, Y., Supaka, N., Sramala, I., Chareonpornwattana, S., 2012. Production of nisin-loaded solid lipid nanoparticles for sustained antimicrobial activity. Food Control 24, 184–190.

Puglia, C., Pignatello, R., Fuochi, V., Furneri, P.M., Lauro, M.R., Santonocito, D., Cortesi, R., Esposito, E., 2019. Lipid nanoparticles and active natural compounds: a perfect combination for pharmaceutical applications. Curr. Med. Chem. 26, 4681–4696.

Rajamanickam, U., Mylsamy, P., Viswanathan, S., Muthusamy, P., 2012. Biosynthesis of zinc nanoparticles using actinomycetes for antibacterial food packaging. In: International Conference on Nutrition and Food Sciences IPCBEE.

Rani, S., Rana, R., Saraogi, G.K., Kumar, V., Gupta, U., 2019. Self-emulsifying oral lipid drug delivery systems: advances and challenges. AAPS PharmSciTech 20, 129.

Rizwan, M., Ali, S., Qayyum, M.F., Ok, Y.S., Adrees, M., Ibrahim, M., Zia-ur-Rehman, M., Farid, M., Abbas, F., 2017. Effect of metal and metal oxide nanoparticles on growth and physiology of globally important food crops: a critical review. J. Hazard. Mater. 322, 2–16.

Sarabandi, K., Rafiee, Z., Khodaei, D., Jafari, S.M., 2019. Chapter nine—Encapsulation of food ingredients by nanoliposomes. In: Jafari, S.M. (Ed.), Lipid-Based Nanostructures for Food Encapsulation Purposes. Academic Press, pp. 347–404.

Schubert, M., Müller-Goymann, C., 2005. Characterisation of surface-modified solid lipid nanoparticles (SLN): influence of lecithin and nonionic emulsifier. Eur. J. Pharm. Biopharm. 61, 77–86.

Shah, B., Khunt, D., Bhatt, H., Misra, M., Padh, H., 2015. Application of quality by design approach for intranasal delivery of rivastigmine loaded solid lipid nanoparticles: effect on formulation and characterization parameters. Eur. J. Pharm. Sci. 78, 54–66.

Sharma, A., Madhunapantula, S.V., Robertson, G.P., 2012. Toxicological considerations when creating nanoparticle-based drugs and drug delivery systems. Expert Opin. Drug Metab. Toxicol. 8, 47–69.

Shishir, M.R.I., Karim, N., Gowd, V., Xie, J., Zheng, X., Chen, W., 2019. Pectin-chitosan conjugated nanoliposome as a promising delivery system for neohesperidin: characterization, release behavior, cellular uptake, and antioxidant property. Food Hydrocoll. 95, 432–444.

Siddiqui, S.W., Wan Mohamad, W., Rozi, M.M., Norton, I.T., 2017. Continuous, high-throughput flash-synthesis of submicron food emulsions using a confined impinging jet mixer: effect of in situ turbulence, sonication, and small surfactants. Ind. Eng. Chem. Res. 56, 12833–12847.

Smerkova, K., Dolezelikova, K., Bozdechova, L., Heger, Z., Zurek, L., Adam, V., 2020. Nanomaterials with active targeting as advanced antimicrobials. Wiley Interdiscip. Rev. Nanomed. Nanobiotechnol. 12, e1636.

Sohail, M.F., Javed, I., Hussain, S.Z., Sarwar, S., Akhtar, S., Nadhman, A., Batool, S., Bukhari, N.I., Saleem, R.S.Z., Hussain, I., 2016. Folate grafted thiolated chitosan enveloped nanoliposomes with enhanced oral bioavailability and anticancer activity of docetaxel. J. Mater. Chem. B 4, 6240–6248.

Souza, V.G.L., Fernando, A.L., 2016. Nanoparticles in food packaging: biodegradability and potential migration to food—a review. Food Packag. Shelf Life 8, 63–70.

Subramani, T., Ganapathyswamy, H., 2020. An overview of liposomal nano-encapsulation techniques and its applications in food and nutraceutical. J. Food Sci. Technol. 57, 1–11.

Sun, H., Zielinska, K., Resmini, M., Zarbakhsh, A., 2019. Interactions of NIPAM nanogels with model lipid multibilayers: a neutron reflectivity study. J. Colloid Interface Sci. 536, 598–608.

Tamjidi, F., Shahedi, M., Varshosaz, J., Nasirpour, A., 2013. Nanostructured lipid carriers (NLC): a potential delivery system for bioactive food molecules. Innov. Food Sci. Emerg. Technol. 19, 29–43.

Tiwari, S., Sharma, V., Mujawar, M., Mishra, Y.K., Kaushik, A., Ghosal, A., 2019. Biosensors for epilepsy management: state-of-art and future aspects. Sensors 19, 1525.

Triplett, M.D., Rathman, J.F., 2009. Optimization of β-carotene loaded solid lipid nanoparticles preparation using a high shear homogenization technique. J. Nanopart. Res. 11, 601–614.

Wang, J., Kaplan, J.A., Colson, Y.L., Grinstaff, M.W., 2017. Mechanoresponsive materials for drug delivery: harnessing forces for controlled release. Adv. Drug Deliv. Rev. 108, 68–82.

Wang, J.-J., Liu, K.-S., Sung, K., Tsai, C.-Y., Fang, J.-Y., 2009. Skin permeation of buprenorphine and its ester prodrugs from lipid nanoparticles: lipid emulsion, nanostructured lipid carriers and solid lipid nanoparticles. J. Microencapsul. 26, 734–747.

Wang, Z., Zhang, N., Chen, C., He, R., Ju, X., 2020. Rapeseed protein nanogels as novel Pickering stabilizers for oil-in-water emulsions. J. Agric. Food Chem. 68, 3607–3614.

Weiss, J., Decker, E.A., McClements, D.J., Kristbergsson, K., Helgason, T., Awad, T., 2008. Solid lipid nanoparticles as delivery systems for bioactive food components. Food Biophys. 3, 146–154.

Wissing, S., Kayser, O., Müller, R., 2004. Solid lipid nanoparticles for parenteral drug delivery. Adv. Drug Deliv. Rev. 56, 1257–1272.

Xie, M., Dunford, N.T., 2017. Lipid composition and emulsifying properties of canola lecithin from enzymatic degumming. Food Chem. 218, 159–164.

Yousefi, M., Ehsani, A., Jafari, S.M., 2019. Lipid-based nano delivery of antimicrobials to control foodborne bacteria. Adv. Colloid Interface Sci. 270, 263–277.

Zardini, A.A., Mohebbi, M., Farhoosh, R., Bolurian, S., 2018. Production and characterization of nanostructured lipid carriers and solid lipid nanoparticles containing lycopene for food fortification. J. Food Sci. Technol. 55, 287–298.

Zoghi, A., Khosravi-Darani, K., Omri, A., 2018. Process variables and design of experiments in liposome and nanoliposome research. Mini Rev. Med. Chem. 18, 324–344.

Bio-based multifunctional nanomaterials: Synthesis and applications

Tarangini Korumilli[a], K. Jagajjanani Rao[a], and Sai Sateesh Sagiri[b]

[a]Department of Biotechnology, Vel Tech Rangarajan Dr. Sagunthala R & D Institute of Science and Technology, Chennai, India [b]Agro-Nanotechnology and Advanced Materials Research Center, Institute of Postharvest and Food Sciences, Agricultural Research Organization, The Volcani Center, Rishon LeZion, Israel

OUTLINE

1 Introduction

Nanoparticles (NPs) are the substances of matter with 1–100 nm with one or more dimensions and can be produced from two basic approaches. One approach is "bottom-up," which involves growth and self-assembly of atoms and molecules to form large nanosized particles or structures. And the other is "top-down" approach, which involves precision-controlled slicing or cutting of a bulk material to get nanosized particles using powerful techniques like lithography and etching. Both methods have inherent advantages. Top-down approach removes building blocks in fixed crystal planes to form nanostructures, predominantly used by electronic industries. Bottom-up approach stacks building blocks to a substrate to rise to crystal planes, which are further stacked on each other to form the nanostructures, e.g., supramolecular functional materials.

At nanoscale, quantum effects can strongly modify the properties of matter like color, reactivity, and magnetic or dipolar moment; the size-dependent behavior of NPs manifests itself in markedly altered physical and chemical behavior. In nanosystems, interparticle interactions are dominated by weak van der Waals forces, strong polar or electrostatic interactions, and covalent interactions. Eyeing on the utilization of the distinctive properties (like metallic, semiconducting, magnetic, and ferroelectric) of nanomaterials, numerous synthesis methods have been explored irrespective of their scope and reliability for the large-scale production. The slogan of faster, smaller, cheaper has become an essential objective for the production of various nanomaterials by a variety of methodologies. For technology development, reliable procedures for the synthesis of nanomaterials in a safer way with controlled size and chemical composition are imperative (Ben Rogers, 2008).

1.1 Conventional methods of production

Size-quantized NPs can be prepared in bulk scale by numerous physical and chemical methods. The conventional method of synthesis involves physical processing like grinding or ball milling of larger particles and melt mixing, which involves mechanical ways and methods based on evaporation such as crystal growing, ion implantation, and molecular epitaxy. Chemical processing methods involve mainly colloids, sol-gel, Langmuir-Blodgett (L-B) films, and inversed micelles. Combination of physical and chemical synthesis routes is regarded as a hybrid method, which involves inorganic, organic, and electrochemical synthetic methodologies to construct nanostructures with distinct electronic configuration from their bulk counterparts. Preparative, polymer, and colloidal chemical techniques (like microemulsions) are employed to control the sizes and monodispersity of the incipient nanomaterials and stabilize them in solid state as well as in aqueous and nonaqueous dispersions (Busbee et al., 2003).

1.2 Greener approach as an alternative to conventional methods

Rapid growth in the design and development of nanomaterials (Fig. 1) is intriguing but poses a variety of challenges. For example, the stated physical and many chemical routes are known to be costly, use toxic chemicals, and release unwanted residuals as end products. There is indeed a need of alternative procedures for nanomaterial synthesis, which should be environmentally benign and economical. Researchers are now eying on green route synthesis for large-scale production of novel nanomaterials. The strategy involves the reduction or elimination of hazardous substances in reactions and their generation after reactions.

Application of green chemistry to nanoscience must deliver novel methodologies to synthesize various nanostructures with underpinning objectives like high precision, low waste, safer nanomaterials, energy efficiency, and commercial scale production capabilities. Coming to recent advances in green routes, we can broadly classify them into two main outlines, i.e., bio-based and non-bio-based routes of synthesis. Solution-based non-biosynthesis routes involve citrate reductions, ligand-exchange reactions, amine-based direct production, polysaccharide-assisted synthesis, tollens method, irradiation, microwave-assisted method, laser-based methods, etc. Although the stated methods are greener in approach, they suffer from certain setbacks, for example, capping difficulties in citrate reduction, use of corrosive borohydride salts as reductants, use of high temperatures and/or energy-dependent processes, and use of nonaqueous solvents (De Corte et al., 2012a). Bio-based nanomaterial synthesis has been largely under exploration in recent years and can be an effective alternative to the conventional methods. The inherent cheaper and greener pathways for nanomaterial synthesis in microorganisms and plant materials are being explored for the generation of novel inorganic nanostructures (Dhandapani et al., 2012; Gericke and Pinches, 2006a; Mohanpuria et al., 2007; Song et al., 2010; Thakkar et al., 2010).

1.3 Bio-greener ways of nanomaterial production and scope

Among all the greener synthetic routes, bio-based approaches are encouraging and exciting. The key here is the engineering ability of biomolecules to create well-defined

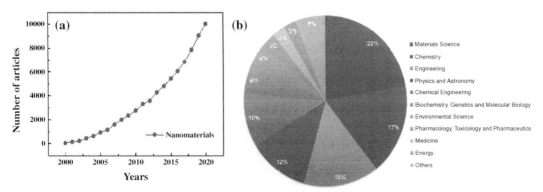

FIG. 1   Number of papers published during 2000–2020 with "Nanomaterials" as a key word (A) and published documents in percent contribution with respect to nanomaterials in various subject areas (B) from SciVerse-Scopus.

supramolecular systems (Clark, 1999; Herrera-Becerra et al., 2010). Using that inherent natural property, a variety of novel nanomaterials with well-defined size and shape can be obtained by employing suitable procedures. The synthesis of inorganic nanomaterials by green routes including biological methods has been reviewed by a few research groups (Huang and Yang, 2004; Sharma et al., 2009; Durán et al., 2011). Utilization of the naturally abundant organisms helps to have a dual advantage for the production process, making it cheaper and eco-friendly. Based on their cellular structure, metabolism, and ability, they can synthesize nanostructures via two routes, i.e., intracellular (within the cell) and extracellular (out of the cell). Several past studies have inspired many NP production methodologies in a greener way by employing natural sources, like microorganisms and plants (Mohanpuria et al., 2007).

2 Biomolecules in nanomaterial synthesis

Biosynthesis of molecules (organic/inorganic) is a highly nonequilibrium process unlike synthetic material synthesis, which is based on specific cellular functions, transformations, assembly, and biochemical translocations. Synthesis of inorganic nanomaterials alone is limited in nature but occurs at environmentally benign conditions with precise control over chemical composition and phase. Magnetic nanostructures by magnetotactic bacteria and calcium carbonate by marine organisms are certain examples of inorganic nanomaterial synthesis.

Acquaintance of biological concepts, functions, and design features of divergent biological sources enables the production of technologically important inorganic nanomaterials (Sarikaya et al., 2003). The use of biological sources for nanomaterial synthesis focuses on different aspects, such as methods based on the utilization of the entire biological system (e.g., bacteria, fungus) and/or biomolecular components, which have typical size dimensions from the lower nm size range to several μm (e.g., sugars, amino acid residues). Till now, a variety of biological systems (live and dead) and biomolecules (including plant components) have been used for various inorganic nanomaterials' synthesis (Mohanpuria et al., 2007; Roh, 2001; Song and Kim, 2008). Moreover, highly specific functionalities, metal ion interactions, and stabilization effect of biological capping agents contributed to the development of relatively new and largely unexplored area of research encompassing the bottom-up synthesis of novel inorganic nanostructures. The responsible exclusive biomolecules are mainly from two sources: (1) microbial based and (2) plant-based resources.

3 Microbial molecules in nanomaterial synthesis

This section starts with microbial machineries, which involve whole cells, viz. membrane-associated molecules (i.e., biomass aided) as well as solitary individual (extracellular) microbial compounds in inorganic NP synthesis. Research on employing biomolecules for different shape NPs with precise monodispersity is also explained. The key issues here are to explore the microbial mechanisms and enzymatic processes involved in NP production.

3.1 Nanoparticles by bacterial origin

3.1.1 Membrane-associated molecules—Intracellular synthesis

Bacteria are dominant prokaryotes that can grow in most habitats on earth. The relative ease of manipulation and progress according to different environments are key towards the evolutionary success of these organisms. Metal ion accumulation associated with leaching capacities of these microbioreactors was successfully employed to generate cellular bound nanomaterials and metal recovery (De Corte et al., 2012a; Thakkar et al., 2010). Bacterial acclimatization to improve their resistance against metal toxicity results in simultaneous metal ion reduction and nanomaterial production (Thakkar et al., 2010; Durán et al., 2011). For instance, mercury was effectively remediated by *Enterobacter* sp. via bioaccumulation and simultaneous production of uniform-sized (2–5 nm) mercury NPs (Sinha and Khare, 2011). Further, Lengke et al. showed in vitro Au NP production by *Cyanobacteria* after their ready-made acquaintance to metal deposits in (Lengke et al., 2006b).

Zaki et al. (2011) have screened some Egyptian local habituated bacterial isolates and produced spherical Ag NPs (15–50 nm) by four strains, namely *E. coli*, *Bacillus megaterium*, *Acinetobacter* sp., and *Stenotrophomonas maltophilia* (Zaki et al., 2011). Although they have not been concerned with the mechanism of particle formation, their confirmations revealed that bacterial presence played a vital role in NP synthesis. The use of single-strain *Pseudomonas aeruginosa* SM1 for multiple types of NPs (Ag, Pd, Fe, Rh, Ni, Ru, Pt, Co, and Li) was reported recently (Srivastava and Constanti, 2012) with intracellular and extracellular production capabilities. In this study, among all NPs, Co (550 ± 10 nm) and Li (950 ± 150 nm) NPs were produced intracellularly with amorphous nature. Amorphous NPs were also produced by *Shewanella putrefaciens* 200 by reducing sodium selenite. This bacterium produces amorphous selenium NPs in both aerobic and anerobic conditions as it is facultative in nature (Jiang et al., 2012). Similarly, amorphous bismuth NPs (<150 nm) were produced by a nonpigmented bismuth-reducing bacterium *Serratia marcescens*, isolated from Caspian Sea (Nazari et al., 2012).

Silver-resistant *Bacillus safensis* TEN12 strain was isolated from industrial wastewater and investigated to produce intracellular Ag NPs (Ahmed et al., 2020). Here, spherical particles of size range 7 to 31 nm (Fig. 2) exhibited significant antimicrobial and anticancer properties.

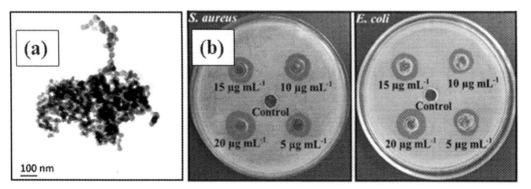

FIG. 2   TEM micrographs of Ag nanoparticles synthesized from *Bacillus safensis* TEN12 (A) and antibacterial activity of the same against *S. aureus* and *E. coli* (B) (Ahmed et al., 2020).

Furthermore, in a recent study, marine bacteria, *Halomonas* sp., were used to produce polyhydroxyalkanoate NPs intracellularly (El-malek et al., 2021), and in another study, the most common probiotic *Lactobacillus kimchicus* DCY51 bacterium was employed to produce intracellular nano-Au particles of varying sizes of 5–13 nm (Markus et al., 2016). Nanomaterials can result from bioinduced mineralization reaction by microorganisms in contrast to bio-controlled growth (usually seen in magnetotactic bacteria) indicating their tunable adaptabilities, for example, intracellular production of akaganeite (β-FeOOH) nanorods by *Anabaena* and *Calothrix* cyanobacteria (Brayner et al., 2009). Further, a list of bacteria, which can produce NPs with and without the aid of biomass, is given in Table 1.

TABLE 1   Biosynthesis of various NPs by biomass (*) and cell-free extracts (#) of certain species of bacterial kingdom.

Species or genus	Metal slat used/ particle obtained	Process/responsible compounds/capping agents	Size attained	Reference
Bacillus subtilis (FJ460362)*	Potassium hexafluorotitanate/ TiO_2	Precipitation/carboxylic group metal ion coordination/ proteins	15–20/mostly S	Dhandapani et al. (2012)
*Aeromonas hydrophila**	ZnO powder/ZnO NPs	Transformation/pH-sensitive oxidoreductases enzymes	57.72/S to oval irregular	Jayaseelan et al. (2012)
Plectonema boryanum UTEX 485 (Cyanobacterium)*	$Au(S_2O_3)_2^{3-}$/ $AuCl_4^-$/Au	Precipitation/membrane vesicles	6 µm to 10 nm/octahedral platelets	Lengke et al. (2006a)
*Klebsiella pneumonia**	Cd(NO), CdS	Precipitation/ion transport system bound by cysteine-containing cadmium-binding protein and released H_2S/ protein caps	20–200 nm	Holmes et al. (1997)
*Escherichia coli**	$PdCl_4^{2-}$/palladium	Reduction/membrane-bound reductases	50 nm	Deplanche et al. (2010)
Rhodococcus sp.*	Au	Reduction	5–15 nm	Ahmad et al. (2003b)
Bacillus cereus#	$AgNO_3$/Ag	Reduction/proteins	62.8/irregular	Silambarasan and Abraham (2012)
Streptomyces sp. ERI-3#	$AgNO_3$/Ag	Reduction	10–100/S	Faghri Zonooz and Salouti (2011)
Microbacterium sp. ARB05#	Na_2SeO_3/Se	Reduction	30–150/S	Prasad et al. (2012)

TABLE 1 Biosynthesis of various NPs by biomass (*) and cell-free extracts (#) of certain species of bacterial kingdom—cont'd

Species or genus	Metal slat used/ particle obtained	Process/responsible compounds/capping agents	Size attained	Reference
Pseudomonas aeruginosa strain BS-161R#	AgNO₃/Ag	Reduction/nitrate reductase/ rhamnolipids	13/S	Kumar and Mamidyala (2011)
B. cereus strain CM100B#	Sodium selenite/Se	Reduction/NADPH/NADH	150–200/S	Dhanjal and Cameotra (2010)
Thermomonospora sp.#	AuCl₄/Au	Reduction/enzymes/proteins	8nm	Ahmad et al. (2003c)

3.1.2 Cell-free extracellular synthesis

Even though several studies show the probability of intracellular NP production by various bacteria (i.e., membrane bound), cell-free extracellular NP production will be advantageous with the ease of separation and large-scale feasibility. Mohammad Fayaz and co-workers (Mohammed Fayaz et al., 2011) synthesized spherical Au (5–14nm) and Ag (5–35nm) NPs extracellularly using a cell-free extract of thermophilic bacterium *Geobacillus stearothermophilus* at ambient conditions with a near-neutral pH (Fig. 3). They found that among the seven extracellular proteins of the bacterium, proteins with free amine groups or cysteine residues served as capping agents and NADPH-dependent reductase class of proteins as the probable reducing agents. This study reportedly showed the advantage of cell-free extract for the large-scale production prospect with toxin-free process and ease in downstream process of the formed NPs. In contrast, a study (Nangia et al., 2009) reported that NADPH alone cannot reduce the metal salt in the absence of bacterial biomass. In this study, solution-phase Au

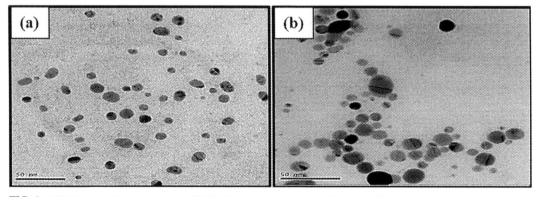

FIG. 3 TEM images of Au (A) and Ag (B) NPs from cell-free extract of thermophilic bacterium *Geobacillus stearothermophilus* (Mohammed Fayaz et al., 2011).

NPs ($\sim$40 nm) from novel *Stenotrophomonas maltophilia* were produced when the biomass was supplied with NADPH along with metal ion source, and it was proposed that NADPH served as the capping agent. From the above, we can say that the action of metal ion-reducing enzymes in the production of NPs is species specific.

Single bacterial species can be able to generate a variety of NPs. For example, iron-reducing bacterium *Shewanella oneidensis* can be used to generate Pd NPs (De Corte et al., 2012a). Further, the same metal ion-accumulating bacterial species was used to produce metal sulfide (Ag_2S) NPs at ambient conditions within 48 h (Suresh et al., 2011). The size of the particles was found to be 9 ± 3.5 nm with surface plasmon resonance exciton peak at 410 nm. Mandal and co-workers (Mandal et al., 2006) reviewed the extracellular synthesis of metal, magnetic, significant semiconductor NPs like CdS, ZnS, and PbS by different species of bacteria. Extracellular bixbyite-like Mn_2O_3 NPs were easily prepared by oxidation of $MnCl_2$ by an isolated gram-negative bacterium *Acinetobacter* sp. with sizes below 500 nm (Hosseinkhani and Emtiazi, 2011). The authors of this study quoted that the synthesis was mediated by the enzymatic pathway and noticed that biomolecules with amine groups may possibly bind to the formed NPs. Common probiotic bacteria *Lactobacillus* sp. can be used to produce ready-made extracellular TiO_2 NPs of 8–35 nm (Jha et al., 2009). The key mechanism of oxidation was mediated by pH-dependent membrane-bound oxidases to produce these oxide NPs. Cell-free extracts for NP production are always advantageous with ease in particle recovery without any interfering microbial components. At times, enzymes can be effectively utilized for metal NP production by selective bacterial species. For example, autolyzed cell-free extract of *Streptomyces* sp. can be used to generate 15–25 nm Ag NPs (Alani et al., 2011). Along with the above, a list of bacterial cell-free extracts with NP-forming ability has been given in Table 1.

3.2 Nanoparticles by fungal origin

3.2.1 Biomass-aided intracellular synthesis

Fungi, majorly due to their more protein/enzyme secretion ability, tolerance, and bioaccumulation capacity of metals, are taking the center stage of studies on biological NP production (Thakkar et al., 2010). In addition, other advantages include economic viability, ease in scale-up, and handling of biomass, prompting the extensive use of potential fungi for the large-scale production of NPs, especially metal NPs. Since many metals play a vital role in metabolic enzymatic reactions, fungi are capable of seizing and accumulating them. This property might be due to either active metal intake for metabolic drive or passive metal uptake and biosorption.

Research groups have confirmed the active role of cell-bound enzymes like NADH nitrate reductases in various metal salt reduction (Durán et al., 2011; Mukherjee et al., 2002). Contrastingly, certain research groups have showed the inability of a fungal species (e.g., *F. moniliforme*) to produce NPs, although it is having reductase enzyme (Durán et al., 2007). Recently, Salunkhe et al. have reported a wet biomass-aided silver-tolerant soil fungus *Cochliobolus lunatus* in nanosilver formation with a size range of 5–100 nm (average size being 14 nm) (Salunkhe et al., 2011). This study claims a diverse report regarding silver formation, which noticed the involvement of fungal cell wall polymer or quinones in particle formation instead of conventional enzymes like NADH-dependent reductases.

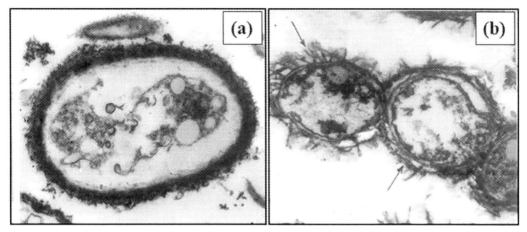

FIG. 4   TEM images of stained fungi *Trichoderma viride* cells before (A) and after (B) treatment with copper solution. *Arrows* in (B) indicate the layer of copper around the cells (Anand et al., 2006).

In some cases, fungal culture conditions are becoming crucial or deterministic in NP production and recovery. For example, Ahmad et al. (2005) synthesized biomass-mediated extracellular Au NPs under stationary condition, while the same were intracellular with shaking conditions (Ahmad et al., 2005). They found that the responsible enzymes did not get released into the medium during NP synthesis under shaking conditions. Another study by Chen and co-workers showed that pretreatment of *Phoma* sp. 3.2883 mycelia by freeze-drying can produce more Ag NPs (13.4 mg/gm) with a size of ~70 nm (Chen et al., 2003).

Metal ion accumulation by fungi is the preliminary step in processing the added metal salts to generate metal NPs. Anand et al. (Anand et al., 2006) reported the copper ion accumulation by *Trichoderma viride* as part of their growth-dependent phenomenon (Fig. 4). Here, the cell surface immobilized the metal, making it less abundant in the medium and rendering it less harmful for their growth. They also observed that the copper accumulation is dependent on added metal salt, metabolic inhibitors, temperature, and pH. Further, the cell wall's dominant role in metal uptake was referred in other strains too: *Pythium* sp. (for cadmium), *Scytalidium lignicola* (for lead), *Dictyuchus sterile* (for zinc), and wild-type *Neurospora crassa* (for copper). The stated gismo of metal accumulation can be effectively used in the generation of NPs. For instance, Mukherjee et al. (2001) synthesized ~25 nm intracellular Au NPs from the fungal biomass of *Verticillium*. Further, Castro-Longoria group observed the possibility of intra- as well as extracellular capability to synthesize metal NPs (Castro-Longoria et al., 2011). They used nonpathogenic filamentous fungus *N. crassa* suspension for the synthesis of mono- and bimetallic Au/Ag NPs with average sizes of 11 nm for silver and 32 nm for gold.

Despite the economic advantage, the downside in using fungal systems for NP synthesis is the unsolicited gene manipulation as a means of overexpressing specific enzymes (e.g., possible enzymes in metallic NP synthesis) (Thakkar et al., 2010).

3.2.2 Cell filtrate-mediated extracellular synthesis

Fungi are well-known extracellular secretors of proteins, enzymes, and other biomolecules that facilitate NP production with limited impurities. Although some studies show the dual ability (extracellular and intracellular) of certain fungus in NP production (e.g., *F. oxysporum*), many studies conveyed the extensive extracellular NP generation ability of a wide range of fungal species. Some authors claimed that in vitro synthesis by nitrate-dependent reductases along with supportive fungal molecules (e.g., phytochelatin, quinones) played a significant role in the generation of stable NPs (Chauhan et al., 2011). However, there are many studies on the role of nitrate reductase class of enzymes as sole responsible agents for various NP production (Mandal et al., 2006). For example, *Alternaria alternate*, a common fungus, utilizes NADPH-dependent nitrate reductase enzymes for the production of nanosilver (~60 nm) under ambient conditions (Gajbhiye et al., 2009). Another study (Li et al., 2012) reported stable Ag NPs of ~1 to 20 nm using cell-free extract of *Aspergillus terreus* within 24 h. Soil-borne filamentous fungus *Sclerotium rolfsii* was able to generate ~25 nm spherical Au NPs by tuning the metal ion-to-cell filtrate ratio (Narayanan and Sakthivel, 2011). The responsible compound for the formation was identified to be thermostable NADPH-dependent enzyme in this study, with proteins serving as capping agents. Sometimes, an autolyzed fungal cell-free culture can be used to produce in vitro cell-free synthesis of metal NPs like Ag NPs. Alani et al. synthesized 15–45 nm range Ag NPs using cell-free *Aspergillus fumigatus* by using the separated-out responsible intracellular metal ion reduction enzymes (Alani et al., 2011).

Mishra et al. showed the synthesis of Au NPs from various environments derived from the growth of industrially important filamentous fungus *Penicillium brevicompactum* KCCM 60390 (Mishra et al., 2011). In this study, the authors used culture supernatant broth, live cell filtrate obtained from biomass, live/dead biomass, and growth medium (potato dextrose broth) solution individually for the production of Au NPs. Among all, except by dead biomass, all other environments resulted in NPs with a prominent color change (Fig. 5i) and SPR range of ~530–538 nm. On optimizing the parameters, 20–50 nm size Au NPs resulted from 1 mM gold salt at pH-6 (Fig. 5ii). A common baker's yeast (*Saccharomyces cerevisiae*) can also be effectively utilized to produce oxide NPs like TiO_2 extracellularly (Jha et al., 2009). In this study, the mechanism of formation involves oxidation of the added $TiO(OH)_2$ by yeast cells, and the process is dependent on the partial pressure of gaseous H_2 (r-H_2), which in turn relies on the available carbon source in the culture medium. There are numerous studies available to note the intracellular and extracellular ability of fungal species that produce NPs. A list of some organisms is given in Table 2.

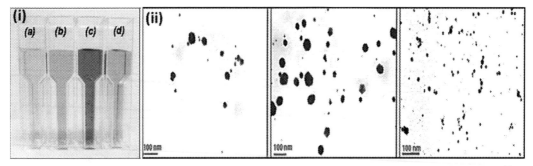

FIG. 5   (i) Color change of different mediums on exposure to 1 mM gold salt: (a) control broth, (b) live cell filtrate of *P. brevicompactum*, (c) PDB (potato dextrose broth) broth, (d) supernatant broth of *P. brevicompactum* after 12-h incubation. (ii) TEM micrographs of purified Au NPs from (b), (c), and (d) mediums *(from left to right)* (Mishra et al., 2011).

TABLE 2 Biosynthesis of various NPs by biomass (*) and cell-free extracts (#) of certain species of fungal kingdom.

Organism	Metal slat used	Process/responsible compounds and/or capping agents	Size/shape attained	Reference
*Penicillium purpurogenum**	AgNO₃	Reduction/nitrate reductase/ secreted proteins	54.6 (final)/S	Pradhan et al. (2011)
*Neurospora crassa**	AgNO₃ and HAuCl₄	Reduction/proteins and enzymes of cytoplasm/other such agents by organelles	11 (Ag), 32 (Au)/S	Castro-Longoria et al. (2011)
Verticillium sp.*	Au	Reduction/cell-bound proteins	20 nm	Mukherjee et al. (2001)
Verticillium sp.*	Ag	Reduction/enzymes in the cell wall membrane	25±12 nm	Mukherjee et al. (2001)
*V. luteoalbum**	Au	Reduction/intracellular enzymes and proteins	Few to 100 nm	Gericke and Pinches (2006b)
*Candida glabrata**	CdS	Metal sequestration-inactivation/ intracellular proteins	20 A⁰	Krumov et al. (2007)
*Schizosaccharomyces pombe**	CdS	Metal sequestration-inactivation/ intracellular proteins	1–1.5 nm	Kowshik et al. (2002)
*P. jadini**	Au	Reduction/proteins and enzymes	Few to 100 nm	Gericke and Pinches (2006a)
Geotrichum sp.	AgNO₃/Ag	Reduction/mycelia/proteins in solution	30–50/S	Jebali et al. (2011)
Fusarium solani#	AgNO₃/Ag	Reduction/reductases/proteins	~16.23 (5–35 nm)/S	Ingle et al. (2008)
Aspergillus tamarii	AgNO₃/Ag	Reduction/proteins	25–50	Rajesh Kumar et al. (2012)
Aspergillus fumigatus#	Ag	Reduction/NADH dependent reductase/proteins	5–25 nm	Bhainsa and D'Souza (2006)
Fusarium oxysporum#	Ag	Reduction/NADH dependent reductase/proteins	5–15 nm	Ahmad et al. (2003a)
F. oxysporum#	Au	Reduction/enzymes polypeptides	20–40 nm	Shankar et al. (2003)
F. oxysporum#	Zirconia	Hydrolysis/extracellular proteins	3–11 nm	Bansal et al. (2004)
F. oxysporum#	CdS	Reduction/sulfate reductase enzymes	5–20 nm	Ahmad et al. (2002)

Continued

TABLE 2   Biosynthesis of various NPs by biomass (*) and cell-free extracts (#) of certain species of fungal kingdom—cont'd

Organism	Metal slat used	Process/responsible compounds and/or capping agents	Size/shape attained	Reference
F. oxysporum#	BaTiO$_3$	Hydrolysis/hydrolyzing enzymes/proteins	4–5nm	Bansal et al. (2006)
F. oxysporum#	CdSe	Reduction/sulfate reductase enzymes		Kumar et al. (2007b)
F. oxysporum#	Silica and titanium particles (SiF2- and TiF2-)	Hydrolysis/extracellular proteins	5–15nm	Bansal et al. (2005)
Nitrate reductases (from *F. oxysporum*, a fungus)#	Ag	Reduction/NADH dependent reductase/proteins	10–25nm	Kumar et al. (2007a)
Colletotrichum sp.#	Au	Reduction/enzymes polypeptides	20–40nm	Shankar et al. (2004)
F. oxysporum and *Verticillium* sp.#	Magnetite	Hydrolysis/cationic proteins	20–50nm	Bharde et al. (2006)
Silver-tolerant yeast strain MKY3#	Ag	Reduction/proteins	2–5nm	Kowshik et al. (2003)

4 Plant resources in nanoparticle synthesis

Synthesis of nanomaterials using specific parts and/or phytochemicals extracted from plants is a hot area of research in recent years. Many plant biomolecules are antioxidants, and employing these phytochemicals enables the synthetic methods to be cost effective, simple, and risk-free by eliminating the release of unwanted products into the environment. NP synthesis by plants can be classified into different types based on the particles/material recovery sites as (1) in vivo and (2) in vitro synthesis. The former involves whole plant-mediated synthesis, and the latter involves the use of plant-derived chemicals in the nanomaterial synthesis.

4.1 In vivo synthesis

Plants act as nanofactories for different types of nanomaterials' synthesis. The use of live plants for nanomaterials was first reported by Gardea-Torresdey et al. (2003) with alfalfa sprouts uptake of silver and forming of spherical nanosilver structures (2–20nm) via coalescence and interconnections with noncrystalline silver atomic wires or clusters. Further, Bali group (Bali et al., 2006) tested different known metal hyperaccumulator plants like *Brassica juncea*, *Helianthus annus*, and *Medicago sativa* for producing NPs and noted that *B. juncea*

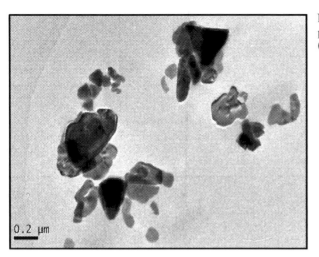

FIG. 6   TEM image of the polydisperse ZnO particles by the plant *Physalis alkekengi* L. (Qu et al., 2011a).

was the best of all to produce nanosilver from aqueous metal salt solutions. The process involved was sequestering of metals from aqueous solutions and depositing them as intracellular nanostructures (Bali et al., 2006). Some noteworthy studies provide us with valuable inferences regarding the synthesis of nanomaterials in plants through accumulation and reduction process. For instance, Qu et al. produced nanoZnO particles from zinc hyperaccumulator ornamental plant *Physalis alkekengi* L. with a mean size of ~72.5nm (Qu et al., 2011a). In this study, Zn was accumulated in the aerial parts of the plant for about 12 weeks and a polydisperse behavior was observed (Fig. 6). Furthermore, formation of NPs is dependent on the metal ion intake capacity of the specific plant species. For example, a study on two metallophytes, i.e., *B. juncea* and *M. sativa*, with Ag precursors produced Ag nanostructures of size ~50nm with varying nanosilver accumulation in terms of weight % as a part of their storage (Harris and Bali, 2007). The study also identified the negative effect of chelating agents like EDTA on metal ion uptake and reported that the rate of uptake by the used plants was 1000 times greater than previous studies on silver. Gardea-Torresdey et al. (2005) noted an increase in phytoextraction of gold by synthetic chelating agents like ammonium thiocyanate and produced 0.55nm gold nanoparticles inside the plants of *Chilopsis linearis* commonly known as desert willow. A detailed study by Haverkamp and Marshall (2009) reported the limited capacity for reducing oxidants by plants and studied various parameters like different salts of same metal ion precursor, metal uptake time, and accumulation leaf and stem of *B. juncea*. This study showed that the Ag NP formation ability of Ag-containing ions having the same strength is in the order of $AgNO_3 > Na_3Ag(S_2O_3)_2 > Ag(NH_3)_2NO_3$, which indicated the low transportation of large anions and their subsequent reduction process. The rate of silver transport into hydroponically grown *B. juncea* was noted to be the same for the entire ion species used. The study also noted that the rate of silver formation from $AgNO_3$ followed zero-order kinetics and was independent of reactant concentration.

Qu et al. (2012) produced carbon nanotubes and Cu/ZnO NPs from *B. juncea* L. isolated from a copper mine site, which has Cu (3725 ± 289 mg/kg) and Zn (2573 ± 237 mg/kg) in its shoots. The authors of this study produced carbon nanotubes of ~80 nm and Cu/ZnO NPs of sizes ~97 nm individually. In another study, Zn-hyperaccumulator plant, *Sedum alfredii* Hance, was used for synthesizing pseudo-spherical shape ZnO NPs of mean size of 53.7 nm, and it was proposed that these plant sources might be used for recycling metals (Qu et al., 2011b). Overall, studies pertaining to the in vivo or intracellular synthesis of NPs were low due to the limited potential of the plant species to accumulate metal precursors.

4.2 In vitro synthesis

Apart from live plants, nanomaterials can be effortlessly produced by active extracellular plant biomolecules present in their extracts. Abundant resource, being economical, eco-friendliness, ease in separation, and purification of nanostructures are the inherent advantages underpinning the use of plant extracts (Rao and Paria, 2015a; Jagajjanani Rao and Paria, 2017). Extracellular nanomaterials for large-scale production by various plant sources have been the focusing area of research in recent times. Use of different plant parts, such as leaf (Rao and Paria, 2015a; Manikandan et al., 2021), tuber (Cittrarasu et al., 2021), bark (Nguyen et al., 2021), and buds (Lakhan et al., 2020), has been systematically explored, and the thirst continues eying on convenient as well as easier plant synthesis routes. The methods using plant extracts involve phytochemicals, such as terpenoids (Thakkar et al., 2010), flavonoids (Raghunandan et al., 2010), phenol derivatives (Jacob et al., 2011), and plant enzymes like hydrogenases, reductases, quinones and their derivatives, and dihydric phenols (Jha et al., 2009), which act as reductants in the presence of metal salt precursors. For example, Duran et al. reviewed the mechanism of formation of Ag NPs by different plant compounds ranging from proteins to phytophenols (Durán et al., 2011).

Gold, silver, and palladium nanoparticles with interesting shapes and morphologies were successfully prepared using geranium leaves, lemon grass, neem leaves, *Aloe vera*, and tea leaf extracts (Khandel et al., 2018). NPs with varying shapes were also prepared using dead plant biomass. Gold nanorods were prepared using dead oat stalk with acidic aqueous solution of gold ions (Doyle, 2006). Independent of their ecological variations, a study (Jha et al., 2009) explored xerophyte (*Bryophyllum* sp.), mesophyte (*Cyprus* sp.), and hydrophyte (*Hydrilla* sp.) plant extracts and produced 2–5 nm Ag NPs. Moreover, the compounds responsible for the reduction process varied with different extracts and were identified as plant metabolites (organic acids and quinones in their study) or metabolic fluxes and other oxido-reductive compounds, like catechol/protocatechuic acid or ascorbate. Jae Song group produced platinum nanostructures from the leaf extract of *Diospyros kaki* using chloroplatinic acid as the reducing agent (Song et al., 2010). In this study, temperature stood as the key parameter. The maximum conversion (90%) of the given metal precursor into NPs was found to be happening at 95°C. NPs varied in size from 2 to 12 nm depending on the reaction temperature and concentrations of the leaf broth and metal ion source. At times, synthesis of multimetallic NPs is also possible by employing mixed plant extracts. The ability of aqueous extracts of *Aegle marmelos* leaf and *Syzygium aromaticum* bud extracts was studied to produce trimetallic Ag-Au-Pd NPs of size ~11 nm (Fig. 7) (Rao and Paria, 2015b). These NPs were found to be multifunctional with antimicrobial and catalytic properties.

In addition to the above, sometimes natural plant materials favor nanostructure synthesis effectively than the extracted or treated materials. For example, Satishkumar et al. produced

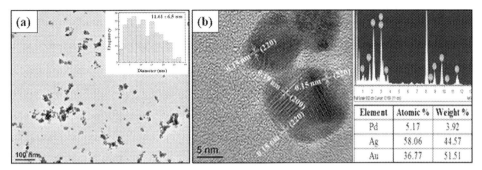

FIG. 7   (A) and (B) Correspond to TEM analysis of Ag-Au-Pd nanoparticles prepared by *Aegle marmelos* leaf and *Syzygium aromaticum* bud extracts in aqueous media. Particle size distribution is given in the inset of (A), and elemental composition is shown in (B) (Rao and Paria, 2015b).

more spherical nanosized Ag particles with crude bark powder than the bark extract of *Cinnamon zeylanicum* (Sathishkumar et al., 2009). In contrast, evidence from a study stated that more Ag particles of mixed shapes were prepared effectively by *Curcuma longa* tuber extract than its powder alone (Sathishkumar et al., 2010). Ease in the preparation of plant extracts with limited purification steps will provide large-scale applicability of the process. For instance, *Carthamus tinctorius* flower extract prepared at 50°C was used to synthesize antimicrobial Ag NPs at room temperature (Sreekanth and Lee, 2011). In another study, platinum nanostructures of irregular size and shapes were prepared from aqueous chloroplatinic acid ($H_2PtCl_6·6H_2O$) using *Ocimum sanctum* leaf broth (Soundarrajan et al., 2012). The compounds responsible for reduction were identified as ascorbic acid, gallic acid, terpenoids, certain proteins, and amino acids present in the extract. Monodisperse well-defined NPs like Ag and Au can be synthesized using same plant extracts by careful adjustment of the key parameters. Aqueous Ag and Au NPs of size below 10 nm by the extracts of vascular plant *Citrus limon* were analyzed in one such study performed at ambient temperature with tuned pH (7 for Au and 11 for Ag NPs) (Liang et al., 2012). In another study, use of a common plant extract from cypress leaves resulted in Au NPs with 94% conversion of the added metal salt within 10 min (Noruzi et al., 2012). NPs produced by various plant sources are listed in Table 3.

TABLE 3   Plant-mediated synthesis of various NPs.

Organism	Particle produced/shape	Particle occurrence/ responsible compounds	Size (nm)	Reference
Clove bud solution (*Syzygium aromaticum*)	Au irregular	Extracellular/ flavonoids	5–100	Raghunandan et al. (2010)
Azadirachta indica	Ag, Au, and Ag/Au bimetallic	Extracellular	50–100 nm	Shankar et al. (2004)

Continued

TABLE 3   Plant-mediated synthesis of various NPs—cont'd

Organism	Particle produced/shape	Particle occurrence/ responsible compounds	Size (nm)	Reference
Emblica officinalis	Ag and Au	Extracellular	(10–20 nm) and (1–25 nm)	Ankamwar et al. (2005)
Manilkara zapota	Ag/spherical and oval	Extracellular	70–140	Rajakumar and Abdul Rahuman (2012)
Pelargonium graveolens	Ag	Extracellular	16–40 nm	Shankar et al. (2003)
Catharanthus roseus Linn. G. Don	Ag	Extracellular	35–55	Ponarulselvam et al. (2012)
Cinnamomum camphora	Au and Ag	Extracellular	50–80 nm	Huang et al. (2007)
Murraya koenigii leaf extract	Ag		10–25	Christensen (2011)
Morinda citrifolia L.	Ag	Extracellular	10–60	Sathishkumar et al. (2012)
Anacardium occidentale	Pd	Extracellular	2.5–4.5	Sheny et al. (2012)
Avena sativa	KAuCl4/Au	Intracellular/biomass/ carboxylic groups and pH	5–20 25–85	Armendariz et al. (2004)
Alfalfa	Titanium and nickel complex salt or nitrate salt/Ti/Ni	Intracellular/biomass matrix	1–4 nm	Schabes-Retchkiman et al. (2006)

There is some overlapping between the microbial and plant systems regarding the modes of operation and nature of biomolecules involved in the generation of various nanostructures: for example, role of enzymes like reductases, proteins, and metabolites like quinones in the nanomaterial synthesis process. However, each system employs a unique strategy based on their cellular structure, metabolism, and ability to generate novel nanostructures as shown in Fig. 8.

5 Template-based synthesis

At times, microbial molecules provide templates for the synthesis of NPs. For instance, Katepetch and Rujiravanit (2011) used cellulose pellicles from *Acetobacter xylinum* effectively for the synthesis of magnetic NPs of a size range of 19.62 to 38.92 nm in 30 min (Fig. 9).

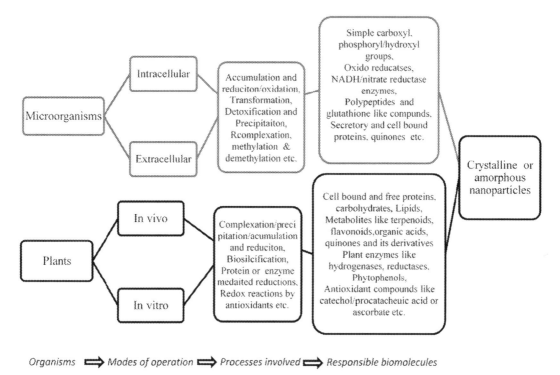

Organisms ⟹ Modes of operation ⟹ Processes involved ⟹ Responsible biomolecules

FIG. 8 Biological synthesis of NP production by plants and microbial resources.

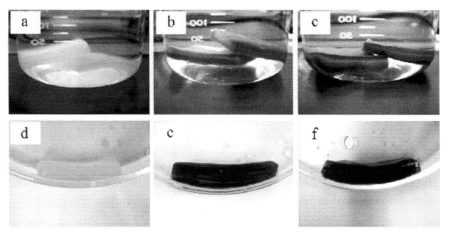

FIG. 9 Cellulose pellicles and magnetic NPs with different concentrations of iron solutions, 0.01 M (A), 0.05 M (B), and 0.1 M (C)—before reaction and 0.01 M (D), 0.05 M (E), and 0.1 M (F)—after reaction with ammonia gas devoid of air (Katepetch and Rujiravanit, 2011). The color change was due to the formation of iron NPs.

Mallick et al. used chitosan (obtained from crustaceans and some fungi) for the synthesis and stabilization of Cu NPs of size 8 ± 4 nm (Mallick et al., 2012). This chitosan-Cu NP composite was further stabilized using iodine, which can prevent oxidation of the Cu NPs and render enhanced antimicrobial property. Gelatinous polysaccharides from red algae can also be used for the generation of metallic Ag NPs. A study by Shukla et al. (2012) produced spherical Ag NPs (~6 nm) using agar from red algae *Gracilaria dura* and fabricated a nanocomposite structure. The process of reduction noted was mainly temperature dependent.

Recently, bimetallic Pd-Au and Au-Pd were synthesized using bacterial biomass of *Cupriavidus necator* as a supporting material under anerobic conditions (Hosseinkhani et al., 2012). Although they did not achieve core-shell-type structures, the bimetallic structures thus produced were effective to catalyze *p*-nitrophenol to *p*-aminophenol than their monometallic forms. Even though the authors have not produced 100% bimetallic particles, they succeeded in getting a greater number of Au-Pd (25 to 50 nm) and Pd-Au (15 to 20 nm) in their experimentation. This opens new possibilities of using economical bio-supports for novel NP synthesis. Yang et al. prepared flexible luminescent homogenous CdSe (20 nm) NPs using bacterial cellulose support and fabricated a novel quantum dot composite structure (Yang et al., 2012). The capping agent used was thioglycolic acid, and the process was done in N_2 environment with $NaBH_4$ under alkaline conditions (pH 11). The so-formed nanocomposite exhibited excellent mechanical strength with good photoluminescence properties. Electrochemically anaerobic biofilm bacteria on a conductive carbon paper produced aqueous Ag NPs of size 1–7 nm within 2 h on challenging with Ag salt, and it was separated out easily (Kalathil et al., 2011).

6 NP shape control with biomolecular systems

Recent focus of research has been on the theme of numerous size, shape, composition, and functionality of NP synthesis through greener methods. Biomolecular systems are emerging alternatives for the large-scale production of NPs and offer a convenient mode to control the said parameters in a safer and economical way. Controlled movement of electrons, holes, excitons, phonons, and plasmons with respect to the physical shape of an object is the cause of change in physical and chemical properties of nanosystems. For instance, colored NPs are due to the confinement of electrons and subsequent changes in distinct electronic energy levels.

Isotropic particles are those where the electron confinement is same in all the directions and can be of 0D particles like spheres, while 1, 2, and 3D nanostructures are those in which the electron movements are possible in these many dimensions. Particles of this type are called anisotropic NPs. These are a class of materials in which their properties are direction dependent and more than one parameter is needed to describe them. With their unique and fine-tuned chemical/physical properties, recent focus has been concentrated extensively on the production of diverse-shape NPs for devising new applications.

Widely used relatively green synthesis methods include galvanic replacement reactions, hydro/solvo-thermal route, electrochemical method, template-based technique, seed-mediated and polyol-based method, microwave-based technique, ultrasound-mediated method, and so on (Sajanlal et al., 2011; Millstone et al., 2009). Exploiting the behavior of different biological molecules and harnessing the nature's biomechanisms to produce NPs of various sizes and shapes have been an attractive area of research in recent times. Studying the functionality and using the ability of natural biomolecules satisfy most of the green chemistry principles (Dahl et al., 2007). Numerous intracellular and extracellular biomolecules like proteins and polypeptides and their counterplay dictate the ultimate size and shape of NPs. There is no generalized mechanism till date to describe the formation and growth in these systems, as their sources are abundant and unique from each other. In general, NP growth occurs in either a controlled thermodynamic or kinetic manner. The former often results in the formation of spherical or near-spherical structures, while the latter follows a preferential and directional growth, which results in the anisotropic structure development. The delicate interplay of both thermodynamic and kinetic aspects plays a critical role in influencing the NP shape.

A general mechanism of growth and formation of these morphologies is not established till yet and is under extensive investigation. However, some research groups have proposed a comprehensive mechanism of formation for specific structures like prisms by reviewing their certain preparative methods (Millstone et al., 2009). Here, the authors explained the NP growth on the basis of crystallographic arguments and chemical methods by reviewing different experimental parameters, like pH, temperature, surfactant/capping ligands, and reducing agents.

Moreover, biological processes are much more complex than chemical methods and involve counterplay of many parameters, like proteins, lipids, polysaccharides, peptides, amino acids, and many other metabolites. In biological or biomolecule-assisted NP synthesis, controlled growth is attained by constricted environments and by specific stabilizing/capping agents like proteins. Harnessing the biomolecular machinery and tuning the important experimental process parameters as stated above will result in novel NPs with specific shapes and sizes.

6.1 Microbial based anisotropic NP synthesis

Many biomolecules in microorganisms dictate the shape of materials along with their size: for example, the production of exoskeleton with directed calcium carbonate crystallization by marine organisms using the potential of their intrinsic proteins and polysaccharides as chief biomolecules. Efforts have been made for the genetic transfer of proteins like silicatein (protein responsible for silica deposition) from certain sponges into bacteria hosts for novel NPs like titanium phosphate and to produce anisotropic layered amorphous structures (Dahl et al., 2007).

Endophytic fungus *Aspergillus clavatus* was used to generate mixed Au nanotriangles, nanospheres, and hexagonal plates by biomass alone as well as cell-free fermentative broth (Verma et al., 2011). The authors reported that the percentage of triangular particles (~20–35 nm) is more in extracellular synthesis (in cell filtrate) and not concentrated much on the mechanism of their formation. In a study about the synthesis of Ag NPs using cell-free extract of bacterium *Streptomyces* sp. ERI-3 cells, the authors found self-assembled flower like Ag NPs from 3-month-old spherical NPs (Faghri Zonooz and Salouti, 2011). Although the mechanism of formation of these structures is not yet explained, this study showed the biosynthesis capability to generate anisotropic structures. In a study on fungus, *Sclerotium rolfsii* (Narayanan and Sakthivel, 2011), the authors noticed different shapes of NPs like triangles, hexagons, decahedrons, rods, and few sphericals with a change of ratio of cell-free filtrate to metal ion concentration ($HAuCl_4$). They reported that spherical NPs were formed in the first part of the reaction but became unstable later on due to insufficient capping or protective agent, which resulted in various anisotropic shapes. Production of anisotropic Ag NPs using mycelia-free media by phytopathogenic fungus *Bipolaris nodulosa* was reported (Saha et al., 2010). The influence of metal ion concentration was exclusively noted with the formation of gold nanowire production using cell-free extract of *Rhodopseudomonas capsulate* (He et al., 2008). Here, network-like Au nanowire structures produced at high concentrations of $HAuCl_4$ and bacterial proteins played a key role in the biosynthesis and bioreduction of Au NPs.

Metal-resistant bacterium *Pseudomonas stutzeri* AG259 has the potential to produce various crystalline Ag NPs (~200 nm), such as triangles and hexagons, and can form organic-metal composite structures as NPs, which may be found as deposits between the cell wall and plasma membrane (Klaus et al., 1999). Marine bacteria's potential in NP synthesis was investigated, and a novel strain, *Marinobacter pelagius*, was reported for its Au NP-forming ability with varying shapes (Sharma et al., 2012). Ramanathan et al. studied the synthesis of anisotropic Ag NPs by controlling the growth kinetics of silver-resistant bacteria *Morganella psychrotolerans* (Ramanathan et al., 2010). Variable shape anisotropic NPs were obtained (Fig. 10i) with varying optimum growth conditions for bacteria at different temperatures (25, 20, 15, and 4°C). The authors proposed that kinetics-controlled mechanism (Fig. 10ii) involving slow reduction of metal ions at deviated growth conditions of the bacterium was responsible for the synthesis of anisotropic NPs.

6.2 Plant-based anisotropic NP synthesis

In a study on *Avena sativa* biomass known as oat (Schabes-Retchkiman et al., 2006), the authors reported the influence of pH on the NP production rate and their size. They produced Au NPs of smaller size with a high rate at pH 3 and 4 (~5 to 20 nm) and higher size particles at pH 2 (~25 to 80 nm). They explained that, at higher pH of 5, biomass had a negative charge due to functional groups, and therefore the negatively charged Au (III) ions cannot easily approach the binding sites of inactivated biomass, which leads to lower reduction of the precursor. Various shapes of Au attained in this study are of tetrahedral, decahedral, hexagonal, icosahedral, multitwinned, irregular shape, and rod-shape NPs.

Some plant extracts like the leaf extract of *Hibiscus rosa sinensis* can produce anisotropic Au NPs with less extract to metal ion concentration with SPR peaks ranging from 573 to 548 nm for various shape NPs (Philip, 2010). The author of this study stated that anisotropy resulted due to the sintering effect at low concentration of the leaf extract (Fig. 11A) and tendency towards near-spherical isotropic NPs at higher strength (Fig. 11B). The study indicated that larger quantities of the extract increased the interaction of protective biomolecules and caused size reduction of spherical NPs. This study also reported the formation of Ag NPs from the stated extract with varying pH (from 6.8 to 8.5) and observed isotropic particles (~13 nm) till a pH of 7.5 (with sharp SPR at 399 nm) and anisotropic NPs beyond it with broadening in noted SPR. In another study with lemon grass plant extract, seeded growth process by sintering of spherical NPs resulted in prismatic structures (Shankar et al., 2004).

Leaf extract of *Memecylon edule* produced triangular, circular, and hexagonal shaped Au NPs of size 10 to 45 nm and predominant square-shaped Ag NPs (50 to 90 nm) using the same amount of extract and constant strengths of used metal ions (Elavazhagan and

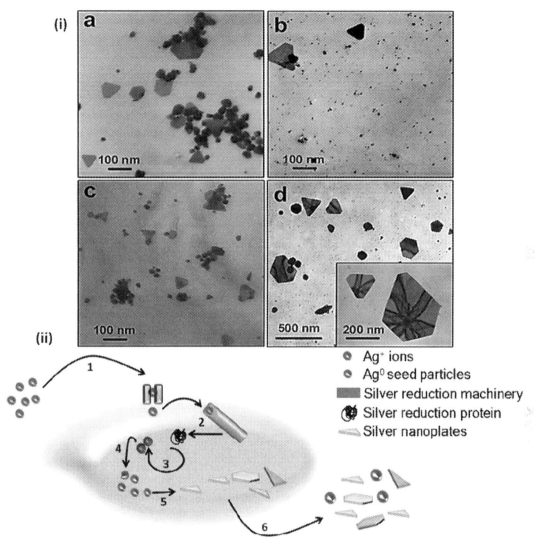

FIG. 10   (i) TEM images of Ag NPs synthesized by *M. psychrotolerans* at (A) 25, (B) 20, (C) 15, and (D) 4°C after 20 h, 24 h, 5 days, and 15 days of reaction, respectively. (ii) Proposed scheme of synthesis mechanism by silver-resistant bacterium (Ramanathan et al., 2010).

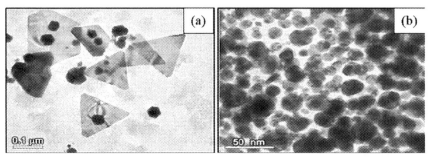

FIG. 11   TEM images of colloidal Au NPs (A) prepared by 10 mL and (B) 30 ml leaf extracts of *Hibiscus rosa-sinensis* with constant metal ion strength (Philip, 2010).

Arunachalam, 2011). The responsible compound identified here for the reduction and capping was a saponin. Although same plant part, same type of metal ions, and almost same methodology were selected for the process, the above studies indicate the effect of plant type and active biomolecules in the attainment of various sizes and shapes of NPs. The influence of process parameters (like temperature, leaf broth, and AgNO$_3$ concentration) on the size of the Ag NPs was shown by Song (2009) on exploring five plant leaf extracts, namely pine, persimmon, ginkgo, magnolia, and platanus extracts. Among all, magnolia leaf broth at 95 °C with 2 mM AgNO$_3$ and 5% leaf broth gave lower size NPs. More than 90% conversion of the added precursor was noted within 11 min of reaction time. The role of pH can be addressed in a study with pine cone extract-mediated Ag NPs in the form of thin, flat, single-crystalline nanohexagonal and nanotriangular particles (Velmurugan et al., 2013). The authors of the study noted higher anisotropic Ag NPs at higher pH 9 and above. *Curcuma longa* tuber extract resulted in more Ag particles than tuber powder at higher temperature and narrow size range particles at alkaline pH, indicating that the additional factors influence apart from the reactants used in the process (Sathishkumar et al., 2010). This phyto-reductive process yielded quasi-spherical, triangular, and small rod-shaped Ag nanoparticles with *C. longa* tuber powder and extract.

The role of hydrolyzable tannins in the formation of anisotropic Au NPs was identified using the aqueous extract of *Terminalia chebula* as the reducing and stabilizing agent (Mohan Kumar et al., 2012). NPs were produced with 6–60 nm size and better antimicrobial activity against gram-positive *S. aureus* compared to gram-negative *E. coli*. Roy et al. have displayed the potential of stem bark extracts of Indian rosewood in the generation of Ag, Au, and Ag-Au bimetallic NPs with interesting shapes (Roy et al., 2012). The exploration was done by preparing two different extracts, namely, aqueous extract and methanol extract, both utilized in lyophilized forms. Different shapes like spheroids, triangles, and hexagons were observed in the study, and the size and shape of the formed NPs were remarkably influenced by the temperature of the process. In summary, Fig. 12 shows the chief factors responsible for anisotropic NP synthesis by biomolecular systems.

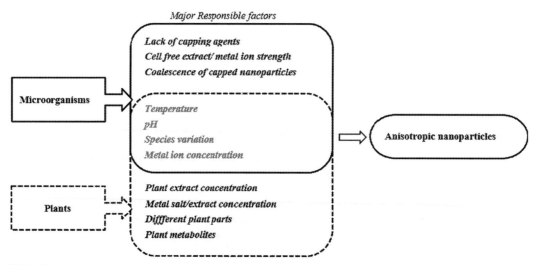

FIG. 12   Key factors affecting the shape control of nanoparticles from major natural biomolecular systems.

6.3 Size and shape control by isolated biomolecules

Morphology of the NPs during synthesis with isolated biomolecules was affected mainly by certain factors, like concentration of reactants, capping agents, and/or seed particles. For example, Sau and Murphy (2004) observed anisometric Au NPs (A of Fig. 13) with ascorbic acid, which is a naturally occurring organic antioxidant compound (Sau and Murphy, 2004). Remarkably, increase in ascorbic acid concentration by keeping all other parameters static resulted in hexagonal and cubic form Au particles (B and C of Fig. 13). On the other hand, triangular particles were formed in more yield instead of cubic particles by increasing the seed Au particles (D of Fig. 13).

Anisotropic Ag NPs in the form of nanoprisms and nanodisks (type 1 and type 2) were produced at ambient conditions by sodium salt of citric acid (trisodium citrate) in two consecutive steps: (1) using PVP and hydrogen peroxide and (2) rapid addition of NaBH$_4$ (Tang et al., 2011). The localized surface plasmon resonance (LSPR) bands of the formed NPs are 718 (nanoprisms), 496 (nanodisk—type 1), and 492 nm (nanodisk—type 2), respectively. The alteration in the shape from nanoprisms to nanodisks (type 1) was attained by varying the concentration of NaBH$_4$ in the process. Nanodisks (type 2) were prepared by halide itching of nanoprism solution being developed by the authors. These nanoprisms and nanodisks were used to functionalize wool fibers and obtain different colored fabrics (Fig. 14) with antimicrobial property.

Soluble starch, a polysaccharide, was employed to generate Ag NPs of size range 10 to 34 nm (Vigneshwaran et al., 2006). The process involves the utilization of linear polymer behavior of amylose content in starch. The aldehyde terminal of soluble starch facilitated silver nitrate reduction. The process gets initiated by autoclaving the aqueous mixture, which

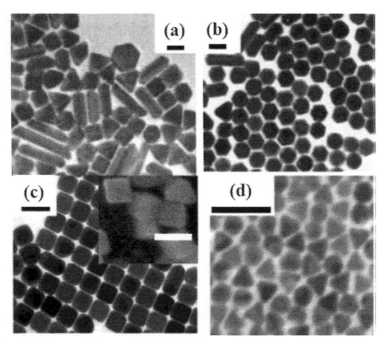

FIG. 13   SEM and TEM (inset of C) images of high-yield solution-phase differently shaped gold NPs (Sau and Murphy, 2004).

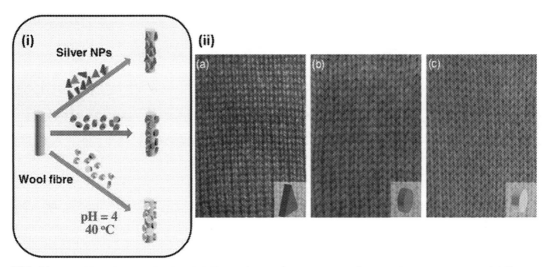

FIG. 14   Wool fiber functionalized with different shapes of Ag NPs at specific pH and temperature (i) and different colors of the fabric when NPs of a particular size and shape are self-assembled (ii). [Shape, color, (length, thickness) in nm] of (a)—nanoprism, *blue* (dark gray in print version) (48.8 ± 3.5, 5.2 ± 0.8), (b)—nanodisk (type 1), *red* (dark gray in print version) (15.7 ± 2.26, 6.1 ± 0.5), (c)—nanodisk (type 2), *yellow* (light gray in print version) (22.1 ± 3.1, 14 ± 1.2) (Tang et al., 2011).

results in the formation of starch-capped Ag NPs. One more study using a polysaccharide biopolymer chitosan (derived from naturally occurring chitin) resulted in producing Ag NPs when allowed to stand for 12 h at 95 °C (Wei et al., 2009).

Ag nanowires (2–20 μm) were synthesized at room temperature by using aqueous tannin solution (Tian et al., 2009). The yield of nanowires was restricted to 5%–10% with a mixture of irregularly shaped NPs. Another study showed spherical Ag NPs of sizes 40 and 70 nm with low and high metal ion starters by hydrolyzable tannin (Bulut and Özacar, 2009). From the mentioned two studies, we can say that NP size and shape are dependent on the process parameters and chemical nature of biomolecules (although they are from the same category). Tannin-capped spherical iron oxide NPs were effectively synthesized using powdered gallic or tannic acid and ferric ammonium sulfate at pH 10 (Herrera-Becerra et al., 2010). Reaction times were 120 and 90 min for gallic and tannic acids with Fe_2O_3 (hematite) obtained by both the sources. Anisotropic or spherical Au nanocrystals were produced using honey by cautious tuning of $HAuCl_4$ and honey concentrations in the aqueous medium (Philip, 2009). The obtained spherical particles were observed to be capped via amine groups, and the size was noted to be ~15 nm. Gum olibanum (*Boswellia serrata*), a renewable natural plant biopolymer, was used to synthesize monodisperse Ag NPs of size 7.5 ± 3.8 nm (Thakor et al., 2011). The process was initiated by autoclaving the aqueous extract with the metal precursor at 121 °C and 103 kPa of pressure for different durations of time. Here, gum concentration and autoclaving time influenced the size of NPs.

7 Extensive use of nanoparticles

Nanomaterials with nanoparticles (NPs) as building blocks have embraced vital sectors, namely engineering, agriculture, medicine, health, and cosmetics. When compared to their larger counterparts, nanomaterials possess unique optical, chemical, photochemical, electronic, and magnetic properties (Mandal et al., 2006; Mody et al., 2010). The attractive physiochemical properties of NPs are due to their high surface-to-volume ratio and dependence on size and shape (Sajanlal et al., 2011). A few vital areas where the extensive use of NPs is on the rise are mentioned below.

Agriculture sector can use NPs as smart delivery vehicles by carrying genes and drugs at the cellular level in plants and animals, which increases the net productivity. NPs as sensors help in detecting soil and food contaminants. Efficient NP photocatalysts were designed to degrade harmful agricultural contaminants like pesticides, and smart nanoscale delivery agents were developed to carry agrochemicals for crop improvement (Baruah and Dutta, 2009).

Biology and medicine have seen significant NP applications as drug carriers, imaging agents, gene-regulating agents, photoresponsive therapeutics, and so on (Giljohann et al., 2010). Functionalized NPs such as colloidal gold were introduced into complex folded structures for diverse applications in therapeutics (Thakor et al., 2011). One example is immunolabeling technique, where NPs were targeted to biomolecules as markers for high-resolution TEM and optical image systems (Salata, 2004). Other NPs which are established for research purposes in cell systems include quantum dots, magnetic NPs, and silver NPs (Nagarajan, 2008). Different stabilizers including surfactants have been found to alter the path of NPs (Europa, 2006; Pal et al., 2007).

Textile manufacturing benefits from carbon NPs in the development of nanofibers and yarns with improved mechanical, chemical, and functional properties. NPs' incorporation generates smart fabrics with antimicrobial and self-cleaning properties. NPs also improve surface properties like wettability, water and oil repellency, and dyeability. Other consumer products which benefit from NPs are flame retardants (formed by coating the foam that has been used in furniture) with carbon nanofibers, lightweight high-strength sporting goods (by silica NPs), vitamin delivery (using nanoemulsions), and UV-blocking materials (like ZnO) in skin care products.

Energy sector uses nanomaterials to generate efficient energy storage devices like photovoltaics at cheaper costs. Novel catalysts for fuel cells, more efficient ultracapacitors and batteries, and improved solar cells are some of the nanomaterial-based products. Carbon in the form of nanotubes and nanoporous aerogels is currently an issue of exploration in this area.

Electronics is the major area with nanomaterials in today's world. Nanomaterials have reduced the power consumption while decreasing the weight and thickness of the products like display screens. Role of carbon nanotubes in designing display panels and integrated circuits is immense.

Environment has been profited with the discovery and development of nanocatalysts and bioactive nanomaterials for organic reactions and contaminant cleanup from polluted sites. Porous manganese oxide with gold NPs, silver nanoclusters, and platinum and iron NPs are some examples of nanocatalysts for soil, water, and air remediation.

Food industry is facing practical problems to implement nanomaterials, although experimental results are satisfactory. Active food packaging is the major research area in this field. NPs are being developed in the form of antimicrobial coatings for storage bins (e.g., nanosilver coatings), UV-blocking and moisture-resistant packing (e.g., ZnO coatings), and nanosensors at food packaging plants to detect bacteria, contaminants, etc.

8 Scope and applications of as-synthesized NPs

The exclusive properties and behavior of NPs triggered an enormous interest in exploring various synthetic pathways, ranging from physical and chemical to biological methods (Thakkar et al., 2010). Irrespective of the methodology employed, monodisperse stable NPs are of great concern in different fields of nanoscience.

Eyeing on the development of clean, nontoxic, and eco-friendly synthesis procedures, recent research has focused on the biological route synthesis of NPs, which employs various plants and microorganisms (Mandal et al., 2006). The key in bio-assisted synthesis is the utilization of bimolecular machineries and processes to aid the required NP production. NPs may be of the same dimensions of some biological molecules, such as proteins and nucleic acids (Salata, 2004). And many of these biomolecules including phytochemicals are in atomic and/or molecular scale. Nanotechnology here is to use these atomic or molecular scale materials for the fabrication of novel NPs with various sizes and shapes. Utilization of as-synthesized NPs with limited or no purification steps will be advantageous, especially for certain biological and environmental applications. Among all the applications, antimicrobial properties of various inorganic NPs are under thorough investigation for novel antimicrobial formulations.

Along with the chemical nature, shape and size of NPs significantly affect the living organisms, especially microbes. The toxicity of NPs has been effectively utilized in various fields like cancer therapies, antibacterial creams and powders, fungicides, biocides, sensors, cleaning of hazardous wastes, and remediation applications. For example, the antimicrobial activity of silver NPs against gram-negative *E. coli* showed shape-dependent interaction along with their size (Pal et al., 2007). Experimentation with nonmetallic NPs like sulfur has also reported antifungal and antimicrobial activities (Choudhury et al., 2011).

8.1 Need of model nanocides

In the early generation, chemicals which are called as antimicrobial substances/antibiotics isolated from natural microorganisms were greatly used to kill or suppress the growth of other harmful microbial species. The later era was followed by synthetic man-made chemicals, including antibiotics, which are very effective in destroying narrow to a very broad range of infectious microbial species. Food crops of farming sectors are profoundly benefited by the application of new chemotherapeutic agents, fungicides, and pesticides for crop improvement. The long-term usage and overdose of antibiotics or pesticides or fungicides resulted in acclimatization, and many harmful microbial species developed resistance towards them. The benign nature of the microorganisms is mainly due to their tunable genetics and development of resistant genes against antimicrobial agents. Development of genetically modified disease-resistant crops by incorporating or modifying the genes seems interesting. However, there are several bottlenecks in this research area as it affects biodiversity including the chance of extinction of the original variety.

Recent research and development focus on the invention of new generation of chemical compounds (biocides) and polymer coatings to combat the microbial infections, majorly by bacteria and fungi. Residual toxicity is the main concern regarding chemical pesticides/antimicrobial agents and can be toxic to human and soil health. The activity of NPs in disturbing the intermolecular forces of well-designed biomolecules may result in valuable or dangerous effect (Europa, 2006; Pal et al., 2007). The model antimicrobial substance should not offer microbial resistance and should have characteristics like eco-friendly with a simple synthesis approach. Further, it should be economical, chemically inert with ease of application, should possess broad range effectiveness towards pathogens and sometimes expected to be multipurpose use. Novel NPs, which satisfy maximum objectives, will be advantageous for instant applicability as it is hard to find all-in-one type of particle.

8.2 Biosynthesized NPs as nanocides

8.2.1 Antimicrobial additives

The biocidal ability of commercial antifungal agents like fluconazole can be enhanced by the use of NPs like silver in the formulation to treat specific fungal species of *Candida albicans*, *Phoma glomerata*, and *Trichoderma* sp. (Gajbhiye et al., 2009). Biologically synthesized Ag NPs (average size of 12.40 nm) from endophytic fungus *Pestalotia* sp. showed increased efficiency in combination of antibiotics like gentamycin and sulfamethizole (Raheman et al., 2011). Ag NPs in combination with gentamycin showed maximum activity (30 mm) against *S. aureus*, followed by sulfamethizole (25 mm). Likewise, similar results were seen in the case of *Salmonella typhi*, where Ag NPs in combination with gentamycin (28 mm) showed more activity than combination of silver NPs and sulfamethizole (24 mm). In another study, Ag NPs produced from *Dioscorea bulbifera* tuber extract exhibited synergistic effect with commercially available antibiotics against various gram-positive and gram-negative bacteria and were also effective towards multidrug-resistant *Acinetobacter baumannii* (Ghosh et al., 2012).

8.2.2 Biocidal agents

Role of starch-stabilized Ag NPs as antimicrobial and antibiofilm agents against human pathogens was effectively demonstrated against *Staphylococcus aureus*, *Pseudomonas aeruginosa*, *Shigella flexneri*, *S. typhi*, and *Mycobacterium smegmatis*, which are gram-positive, gram-negative, and acid-fast bacteria (Mohanty et al., 2012). This study has shown the starch-Ag NPs' augmenting nature towards mouse macrophages in combating the bacterial infection (in vitro) with a promising scope of novel antimicrobial formulations. Ag NPs (∼5–10 nm) by dextran, a polysaccharide, also stood as effective antimicrobials against the bacteria, *Bacillus subtilis*, *Bacillus cereus*, *E. coli*, *S. aureus*, and *P. aeruginosa* (Bankura et al., 2012). Pradhan et al. used in situ Ag NPs as effective antimicrobials, which were synthesized and entrapped in the biofilm of the fungus *Penicillium purpurogenum* (Pradhan et al., 2011). These biomass disks on agar plates showed significant antimicrobial activity against pathogenic gram-negative bacteria like *E. coli* and *P. aeruginosa* and gram-positive bacteria like *S. aureus*.

The effect of size, shape, concentration, and dose of Ag NPs on various multidrug resistance (MDR) bacteria was extensively reviewed by Rai et al., stating various modes of actions responsible for the broad spectrum of bactericidal activity (Fig. 15) (Rai et al., 2012). Antimicrobial properties of various NPs produced by various biological systems are shown in Table 4.

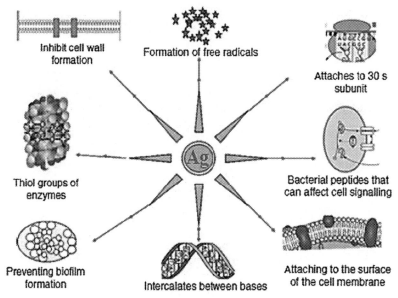

FIG. 15   Different modes of bactericidal action by silver NPs.

TABLE 4   Antimicrobial properties of biosystem-generated NPs (as-synthesized) against harmful microorganisms.

Microbe	Nanoparticle obtained	Size attained	Prep part is effective against	Reference
Bacillus subtilis (FJ460362)	TiO$_2$	15–20/ mostly S	Pond water bacterial biofilm	Dhandapani et al. (2012)
Aeromonas hydrophila	ZnO	57.72/S to oval irregular	*Pseudomonas aeruginosa, Aspergillus flavus* (among all)	Jayaseelan et al. (2012)
Bacillus cereus	Ag	62.8/ irregular	*S. aureus, Klebsiella pneumonia, Salmonella typhi, E. coli*	Silambarasan and Abraham (2012)
P. aeruginosa strain BS-161R	Ag	13/S	Pathogenic gram-positive, gram-negative bacteria and different *Candida* species	Kumar and Mamidyala (2011)
Penicillium purpurogenum	Ag	54.6 (final)/S	Gram-negative: *E. coli* and *P. aeruginosa* Gram-positive: *S. aureus*	Pradhan et al. (2011)

8.2.3 Biocidal nanocomposites

Nanocomposite structures can be effectively used as ready-made antimicrobial agents. Ag-agar composite exhibited its antimicrobial property against *Bacillus pumilus*, and the activity was dose dependent (Shukla et al., 2012). Cu-chitosan composite with iodine stabilization proved to be efficient against *E. coli* causing perforation and irreparable membrane damage (Mallick et al., 2012).

Although there is ample literature about the antimicrobial/antibacterial effects of various NPs, there is a scarcity of reports on "what is happening to the NPs" irrespective of various sizes/shapes, which results in toxicity to microorganisms. Ashkarran et al. discovered the effect of different types (sphere, wire, cube, and triangle) of silver NPs on bacteria (*E. coli*, *B. subtilis*, *S. aureus*) and their antibacterial properties (Ashkarran et al., 2012). The toxicity results showed that each bacterium gave a different response to each shape of NPs of the same amount. The authors claim that the significant differences in bacterial membrane composition are the cause of varied toxicity results. Protein corona composition on NP's surface was studied in detail. Results revealed that proteins have much more affinity to join the sharp edges on NP surface rather than to join the flat surfaces (i.e., more tendencies towards sharp surfaces rather than sphere and wire).

8.3 Other applications of as-synthesized nanoparticles

8.3.1 As biosensors

Au NPs by *Candida albicans* are effectively used to screen liver cancer cells from normal healthy cells by conjugating them with liver cancer cell surface-specific antibodies by binding to the surface antigens of the cancer cells (Chauhan et al., 2011).

Extracellular synthesized metal sulfide (Ag$_2$S) NPs can be used for various applications, such as bioimaging, optical imaging, electronic devices, and solar cell applications, as they are highly biocompatible with no significant toxicity towards tested gram-negative *E. coli*, *Shewanella oneidensis*, and gram-positive *B. subtilis* bacterial species and eukaryotic mouse lung epithelial and macrophage cells (Suresh et al., 2011).

8.3.2 On cancer cell line-targeting cell death

The cytotoxic effect of the as-synthesized Au NPs (from *P. brevicompactum*) was tested with mouse mayo blast cancer C 2C 12 cells (Mishra et al., 2011) in a time-dependent manner. Results of the study showed more controlled cell death with the increase of gold salt concentration and incubation time.

8.3.3 Assisting in heavy metal ion removal

Live biomass reduced selenite salt and generated amorphous selenium NPs. The salt was further used for Hg (II) reduction to Hg (0), which in turn effectively immobilized on the formed selenium NPs (Jiang et al., 2012). This study efficiently removed mercury and selenium with no secondary pollution, such as methylation or volatilization of the formed Hg (0) (which usually occurs on partitioning into air). Thus, the bacterial reduction strategy opened up a wide scope for remediating hazardous metal ions in a safer way.

8.3.4 As catalysts

Biosynthesized Au-doped bio-Pd/Au catalyst was fabricated using the bacterium *S. oneidensis* MR-1 by the reduction of Pd(II) and trivalent Au(III) ions, which can be effectively used to treat pharmaceutical impurities like diclofenac and trichloroethylene in wastewaters (De Corte et al., 2012b). The process incurred is dehalogenation with improved activity when compared to monometallic bio-Pd NPs. Ogi et al. produced bacterial based Pd NPs by the biomass of *S. oneidensis*, and the same was used as an anode catalyst in a polymer electric membrane fuel cell for power production and achieved comparable results with that of commercial palladium catalyst (Ogi et al., 2011).

Extracellular phytosynthesized Ag NPs of size 20–30 nm from the leaf extract of *Coccinia grandis* were effectively used as photocatalysts under UV light for the degradation of Coomassie Brilliant Blue G-250 (Arunachalam et al., 2012). Ag NPs by sodium citrate reduction were used as catalysts in the synthesis of copper NPs from copper nitrate solution with the aid of hydrazine monohydrate as a protectant for copper NPs against oxidation (Grouchko et al., 2007).

9 Summary and future outlook

Inorganic NPs with distinctive properties of metallic, semiconducting, magnetic, and ferroelectric NPs have prompted an escalating interest in these systems. In particular, changes in mechanical, optical, electrical, electro-optical, magnetic, and magneto-optical properties have been explored very well (Sajanlal et al., 2011).

Diverse range of NPs and their prospective significance triggered the exploitation of various cheaper synthesis routes, ranging from chemical to bio-based approaches for large-scale production. Although various chemical routes controlling the physical properties of the particles were reported in the past, most of these are still in the developmental stage.

Green technology for NP synthesis is an important topic receiving momentous attention in recent times, especially when the environmental issues are considered. This technology is driven by 12 principles of green chemistry for efficient, eco-friendly, and safer nanoparticle production methodologies (Dahl et al., 2007). The customary principles include the following: prevent waste, design less hazardous chemical syntheses, use renewable feedstocks, design safer chemicals, reduce derivatives, design for degradation/design for end of life, real-time monitoring and process control, catalysis/use catalysts, not stoichiometric reagents, maximize atom economy, use safer solvents and reaction conditions, design for/increase energy efficiency, and inherently safer chemistry/minimize the potential for accidents.

Among several green routes, microbial and plant-based approaches are the areas which have been explored in a greater extent in recent years for the generation of metallic and/or nonmetallic NPs (Thakkar et al., 2010; Mittal et al., 2013). While comparing microbial and plant extract-based green synthesis routes, it has been found that plant-based methods have some underpinning advantages such as user friendliness, cost-effectiveness, and easy separation and purification processes. Moreover, green synthesis of NPs using plant extracts seems promising for producing a variety of novel NPs (Jha et al., 2009), whereas microbial route suits for some exceptional cases. In addition, careful tuning of the abundant plant and microbial resources can generate different isotropic and anisotropic NPs, which would be advantageous for several novel diverse application fields, such as optics, electronics, medical diagnostics and treatments, sensors, antimicrobial agents, and coatings (Kharissova et al., 2013).

In conclusion, bio-based approaches are enabling novel synthesis of inorganic metal NPs with easy steps. Along with environmentally benign synthetic protocols, they are providing eco-friendly, low-cost, and short reaction conditions. Abundant availability of production methods involving numerous microbial and plant materials for different size and shape control of NPs minimizes the use of harsh and toxic chemicals, maximizes the efficient use of safer materials, and reduces waste products. Novel synthesis strategies with easy steps, mixed hybrid biogenic routes for new type of particles, more than one type of NP synthesis, and transforming the synthesizing products to useable pertinent forms for non-bioapplications are some important challenges of this technology to reach large-scale applicability.

Acknowledgement

We acknowledge funding support from the Department of Biotechnology through the project number PR35666 to the BioInterfaces & Nanomaterials Laboratory facility at Research Park in Vel Tech Rangarajan Dr. Sagunthala R&D Institute of Science and Technology, Avadi, Chennai.

References

Ahmad, A., Mukherjee, P., Mandal, D., Senapati, S., Khan, M.I., Kumar, R., Sastry, M., 2002. Enzyme mediated extracellular synthesis of CdS nanoparticles by the fungus, *Fusarium oxysporum*. J. Am. Chem. Soc. 124, 12108–12109. https://doi.org/10.1021/ja027296o.

Ahmad, A., Mukherjee, P., Senapati, S., Mandal, D., Khan, M.I., Kumar, R., Sastry, M., 2003a. Extracellular biosynthesis of silver nanoparticles using the fungus *Fusarium oxysporum*. Colloids Surf. B: Biointerfaces 28, 313–318. https://doi.org/10.1016/S0927-7765(02)00174-1.

Ahmad, A., Senapati, S., Khan, M.I., Kumar, R., Ramani, R., Srinivas, V., Sastry, M., 2003b. Intracellular synthesis of gold nanoparticles by a novel alkalotolerant actinomycete, Rhodococcus species. Nanotechnology 14, 824–828. https://doi.org/10.1088/0957-4484/14/7/323.

Ahmad, A., Senapati, S., Khan, M.I., Kumar, R., Sastry, M., 2003c. Extracellular biosynthesis of monodisperse gold nanoparticles by a novel extremophilic actinomycete, Thermomonospora sp. Langmuir 19, 3550–3553. https://doi.org/10.1021/la0267721.

Ahmad, A., Senapati, S., Khan, M.I., Kumar, R., Sastry, M., 2005. Extra-/intracellular biosynthesis of gold nanoparticles by an Alkalotolerant fungus, Trichothecium sp. J. Biomed. Nanotechnol. 1, 47–53. https://doi.org/10.1166/jbn.2005.012.

Ahmed, T., Shahid, M., Noman, M., Niazi, M.B.K., Zubair, M., Almatroudi, A., Khurshid, M., Tariq, F., Mumtaz, R., Li, B., 2020. Bioprospecting a native silver-resistant *Bacillus safensis* strain for green synthesis and subsequent antibacterial and anticancer activities of silver nanoparticles. J. Adv. Res. 24, 475–483. https://doi.org/10.1016/j.jare.2020.05.011.

Alani, F., Moo-Young, M., Anderson, W., 2011. Biosynthesis of silver nanoparticles by a new strain of Streptomyces sp. compared with *Aspergillus fumigatus*. World J. Microbiol. Biotechnol. 28, 1081–1086. https://doi.org/10.1007/s11274-011-0906-0.

Anand, P., Isar, J., Saran, S., Saxena, R.K., 2006. Bioaccumulation of copper by *Trichoderma viride*. Bioresour. Technol. 97, 1018–1025. https://doi.org/10.1016/j.biortech.2005.04.046.

Ankamwar, B., Damle, C., Ahmad, A., Sastry, M., 2005. Biosynthesis of gold and silver nanoparticles using *Emblica officinalis* fruit extract, their phase transfer and transmetallation in an organic solution. J. Nanosci. Nanotechnol. 5, 1665–1671. https://doi.org/10.1166/jnn.2005.184.

Armendariz, V., Herrera, I., Peralta-Videa, J.R., Jose-Yacaman, M., Troiani, H., Santiago, P., Gardea-Torresdey, J.L., 2004. Size controlled gold nanoparticle formation by *Avena sativa* biomass: use of plants in nanobiotechnology. J. Nanopart. Res. 6, 377–382. https://doi.org/10.1007/s11051-004-0741-4.

Arunachalam, R., Dhanasingh, S., Kalimuthu, B., Uthirappan, M., Rose, C., Mandal, A.B., 2012. Phytosynthesis of silver nanoparticles using *Coccinia grandis* leaf extract and its application in the photocatalytic degradation. Colloids Surf. B: Biointerfaces 94, 226–230. https://doi.org/10.1016/j.colsurfb.2012.01.040.

Ashkarran, A.A., Ghavami, M., Aghaverdi, H., Stroeve, P., Mahmoudi, M., 2012. Bacterial effects and protein Corona evaluations: crucial ignored factors in the prediction of bio-efficacy of various forms of silver nanoparticles. Chem. Res. Toxicol. 25, 1231–1242. https://doi.org/10.1021/tx300083s.

Bali, R., Razak, N., Lumb, A., Harris, A.T., 2006. The synthesis of metallic nanoparticles inside live plants. In: 2006 International Conference on Nanoscience and Nanotechnology. IEEE, pp. 224–227, https://doi.org/10.1109/ICONN.2006.340592.

Bankura, K.P., Maity, D., Mollick, M.M.R., Mondal, D., Bhowmick, B., Bain, M.K., Chakraborty, A., Sarkar, J., Acharya, K., Chattopadhyay, D., 2012. Synthesis, characterization and antimicrobial activity of dextran stabilized silver nanoparticles in aqueous medium. Carbohydr. Polym. 89, 1159–1165. https://doi.org/10.1016/j.carbpol.2012.03.089.

Bansal, V., Poddar, P., Ahmad, A., Sastry, M., 2006. Room-temperature biosynthesis of ferroelectric barium titanate nanoparticles. J. Am. Chem. Soc. 128, 11958–11963. https://doi.org/10.1021/ja063011m.

Bansal, V., Rautaray, D., Ahmad, A., Sastry, M., 2004. Biosynthesis of zirconia nanoparticles using the fungus *Fusarium oxysporum*. J. Mater. Chem. 14, 3303–3305. https://doi.org/10.1039/B407904C.

Bansal, V., Rautaray, D., Bharde, A., Ahire, K., Sanyal, A., Ahmad, A., Sastry, M., 2005. Fungus-mediated biosynthesis of silica and titania particles. J. Mater. Chem. 15, 2583–2589. https://doi.org/10.1039/B503008K.

Baruah, S., Dutta, J., 2009. Nanotechnology applications in pollution sensing and degradation in agriculture: a review. Environ. Chem. Lett. 7, 191–204. https://doi.org/10.1007/s10311-009-0228-8.

Ben Rogers, J.A., 2008. Nanotechnology: Understanding Small Systems. CRC Press.

Bhainsa, K.C., D'Souza, S.F., 2006. Extracellular biosynthesis of silver nanoparticles using the fungus *Aspergillus fumigatus*. Colloids Surf. B: Biointerfaces 47, 160–164. https://doi.org/10.1016/j.colsurfb.2005.11.026.

Bharde, A., Rautaray, D., Bansal, V., Ahmad, A., Sarkar, I., Yusuf, S.M., Sanyal, M., Sastry, M., 2006. Extracellular biosynthesis of magnetite using fungi. Small 2, 135–141. https://doi.org/10.1002/smll.200500180.

Brayner, R., Yéprémian, C., Djediat, C., Coradin, T., Herbst, F., Livage, J., Fiévet, F., Couté, A., 2009. Photosynthetic microorganism-mediated synthesis of akaganeite (β-FeOOH) nanorods. Langmuir 25, 10062–10067. https://doi.org/10.1021/la9010345.

Bulut, E., Özacar, M., 2009. Rapid, facile synthesis of silver nanostructure using hydrolyzable tannin. Ind. Eng. Chem. Res. 48, 5686–5690. https://doi.org/10.1021/ie801779f.

Busbee, B.D., Obare, S.O., Murphy, C.J., 2003. An improved synthesis of high-aspect-ratio gold nanorods. Adv. Mater. 15, 414–416. https://doi.org/10.1002/adma.200390095.

Castro-Longoria, E., Vilchis-Nestor, A.R., Avalos-Borja, M., 2011. Biosynthesis of silver, gold and bimetallic nanoparticles using the filamentous fungus *Neurospora crassa*. Colloids Surf. B: Biointerfaces 83, 42–48. https://doi.org/10.1016/j.colsurfb.2010.10.035.

Chauhan, A., Zubair, S., Tufail, S., Sherwani, A., Sajid, M., Raman, S.C., Azam, A., Owais, M., 2011. Fungus-mediated biological synthesis of gold nanoparticles: potential in detection of liver cancer. Int. J. Nanomedicine 6, 2305.

Chen, J.C., Lin, Z.H., Ma, X.X., 2003. Evidence of the production of silver nanoparticles via pretreatment of Phoma sp.3.2883 with silver nitrate. Lett. Appl. Microbiol. 37, 105–108. https://doi.org/10.1046/j.1472-765X.2003.01348.x.

Choudhury, S.R., Ghosh, M., Mandal, A., Chakravorty, D., Pal, M., Pradhan, S., Goswami, A., 2011. Surface-modified sulfur nanoparticles: an effective antifungal agent against *Aspergillus niger* and *Fusarium oxysporum*. Appl. Microbiol. Biotechnol. 90, 733–743. https://doi.org/10.1007/s00253-011-3142-5.

Christensen, L., 2011. Biosynthesis of silver nanoparticles using *Murraya koenigii* (curry leaf): an investigation on the effect of broth concentration in reduction mechanism and particle size. Adv. Mater. Lett. 2, 429–434. https://doi.org/10.5185/amlett.2011.4256.

Cittrarasu, V., Kaliannan, D., Dharman, K., Maluventhen, V., Easwaran, M., Liu, W.C., Balasubramanian, B., Arumugam, M., 2021. Green synthesis of selenium nanoparticles mediated from *Ceropegia bulbosa* Roxb extract and its cytotoxicity, antimicrobial, mosquitocidal and photocatalytic activities. Sci. Rep. 11, 1032. https://doi.org/10.1038/s41598-020-80327-9.

Clark, J.H., 1999. Green chemistry: challenges and opportunities. Green Chem. 1, 1–8.

Dahl, J.A., Maddux, B.L.S., Hutchison, J.E., 2007. Toward greener nanosynthesis. Chem. Rev. 107, 2228–2269. https://doi.org/10.1021/cr050943k.

De Corte, S., Hennebel, T., De Gusseme, B., Verstraete, W., Boon, N., 2012a. Bio-palladium: from metal recovery to catalytic applications. Microb. Biotechnol. 5, 5–17. https://doi.org/10.1111/j.1751-7915.2011.00265.x.

De Corte, S., Sabbe, T., Hennebel, T., Vanhaecke, L., De Gusseme, B., Verstraete, W., Boon, N., 2012b. Doping of biogenic Pd catalysts with Au enables dechlorination of diclofenac at environmental conditions. Water Res. 46, 2718–2726. https://doi.org/10.1016/j.watres.2012.02.036.

Deplanche, K., Caldelari, I., Mikheenko, I.P., Sargent, F., Macaskie, L.E., 2010. Involvement of hydrogenases in the formation of highly catalytic Pd(0) nanoparticles by bioreduction of Pd(II) using *Escherichia coli* mutant strains. Microbiology 156, 2630–2640. https://doi.org/10.1099/mic.0.036681-0.

Dhandapani, P., Maruthamuthu, S., Rajagopal, G., 2012. Bio-mediated synthesis of TiO_2 nanoparticles and its photocatalytic effect on aquatic biofilm. J. Photochem. Photobiol. B 110, 43–49. https://doi.org/10.1016/j.jphotobiol.2012.03.003.

Dhanjal, S., Cameotra, S.S., 2010. Aerobic biogenesis of selenium nanospheres by *Bacillus cereus* isolated from coalmine soil. Microb. Cell Factories 9, 52. https://doi.org/10.1186/1475-2859-9-52.

Doyle, M., 2006. Nanotechnology: A Brief Literature Review. Food Research Institute Brief. Internet June.

Durán, N., Marcato, P.D., De Souza, G.I.H., Alves, O.L., Esposito, E., 2007. Antibacterial effect of silver nanoparticles produced by fungal process on textile fabrics and their effluent treatment. J. Biomed. Nanotechnol. 3, 203–208. https://doi.org/10.1166/jbn.2007.022.

Durán, N., Marcato, P.D., Durán, M., Yadav, A., Gade, A., Rai, M., 2011. Mechanistic aspects in the biogenic synthesis of extracellular metal nanoparticles by peptides, bacteria, fungi, and plants. Appl. Microbiol. Biotechnol. 90, 1609–1624. https://doi.org/10.1007/s00253-011-3249-8.

Elavazhagan, T., Arunachalam, K.D., 2011. Memecylon edule leaf extract mediated green synthesis of silver and gold nanoparticles. Int. J. Nanomedicine 6, 1265–1278. https://doi.org/10.2147/IJN.S18347.

El-malek, F.A., Rofeal, M., Farag, A., Omar, S., Khairy, H., 2021. Polyhydroxyalkanoate nanoparticles produced by marine bacteria cultivated on cost effective Mediterranean algal hydrolysate media. J. Biotechnol. 328, 95–105. https://doi.org/10.1016/j.jbiotec.2021.01.008.

Europa, 2006. The Scientific Committee on emerging and newly identified health risks (SCENIHR), Public Health. http://ec.europa.eu/health/opinions2/en/nanotechnologies/l-3/6-health-effects-nanoparticles.htm. (Accessed 31 July 2012).

Faghri Zonooz, N., Salouti, M., 2011. Extracellular biosynthesis of silver nanoparticles using cell filtrate of Streptomyces sp. ERI-3. Sci. Iran. 18, 1631–1635. https://doi.org/10.1016/j.scient.2011.11.029.

Gajbhiye, M., Kesharwani, J., Ingle, A., Gade, A., Rai, M., 2009. Fungus-mediated synthesis of silver nanoparticles and their activity against pathogenic fungi in combination with fluconazole. Nanomed. Nanotechnol. Biol. Med. 5, 382–386. https://doi.org/10.1016/j.nano.2009.06.005.

Gardea-Torresdey, J., Gomez, E., Peralta-Videa, J., 2003. Alfalfa sprouts: a natural source for the synthesis of silver nanoparticles. Langmuir 19, 1357–1361.

Gardea-Torresdey, J.L., Rodriguez, E., Parsons, J.G., Peralta-Videa, J.R., Meitzner, G., Cruz-Jimenez, G., 2005. Use of ICP and XAS to determine the enhancement of gold phytoextraction by *Chilopsis linearis* using thiocyanate as a complexing agent. Anal. Bioanal. Chem. 382, 347–352. https://doi.org/10.1007/s00216-004-2966-6.

Gericke, M., Pinches, A., 2006a. Biological synthesis of metal nanoparticles. Hydrometallurgy 83, 132–140. https://doi.org/10.1016/j.hydromet.2006.03.019.

Gericke, M., Pinches, A., 2006b. Microbial production of gold nanoparticles. Gold Bull. 39, 22–28. https://doi.org/10.1007/BF03215529.

Ghosh, S., Patil, S., Ahire, M., Kitture, R., Kale, S., Pardesi, K., Cameotra, S., Bellare, J., Dhavale, D.D., Jabgunde, A., Chopade, B.A., 2012. Synthesis of silver nanoparticles using *Dioscorea bulbifera* tuber extract and evaluation of its synergistic potential in combination with antimicrobial agents. Int. J. Nanomedicine 7, 483–496.

Giljohann, D.A., Seferos, D.S., Daniel, W.L., Massich, M.D., Patel, P.C., Mirkin, C.A., 2010. Gold nanoparticles for biology and medicine. Angew. Chem. Int. Ed. 49, 3280–3294. https://doi.org/10.1002/anie.200904359.

Grouchko, M., Kamyshny, A., Ben-Ami, K., Magdassi, S., 2007. Synthesis of copper nanoparticles catalyzed by pre-formed silver nanoparticles. J. Nanopart. Res. 11, 713–716. https://doi.org/10.1007/s11051-007-9324-5.

Harris, A.T., Bali, R., 2007. On the formation and extent of uptake of silver nanoparticles by live plants. J. Nanopart. Res. 10, 691–695. https://doi.org/10.1007/s11051-007-9288-5.

Haverkamp, R., Marshall, A., 2009. The mechanism of metal nanoparticle formation in plants: limits on accumulation. J. Nanopart. Res. 11, 1453–1463. https://doi.org/10.1007/s11051-008-9533-6.

He, S., Zhang, Y., Guo, Z., Gu, N., 2008. Biological synthesis of gold nanowires using extract of *Rhodopseudomonas capsulata*. Biotechnol. Prog. 24, 476–480. https://doi.org/10.1021/bp0703174.

Herrera-Becerra, R., Rius, J.L., Zorrilla, C., 2010. Tannin biosynthesis of iron oxide nanoparticles. Appl. Phys. A Mater. Sci. Process. 100, 453–459. https://doi.org/10.1007/s00339-010-5903-x.

Holmes, J.D., Richardson, D.J., Saed, S., Evans-Gowing, R., Russell, D.A., Sodeau, J.R., 1997. Cadmium-specific formation of metal sulfide 'Q-particles' by *Klebsiella pneumoniae*. Microbiology 143, 2521–2530. https://doi.org/10.1099/00221287-143-8-2521.

Hosseinkhani, B., Emtiazi, G., 2011. Synthesis and characterization of a novel extracellular biogenic manganese oxide (bixbyite-like Mn_2O_3) nanoparticle by isolated *Acinetobacter* sp. Curr. Microbiol. 63, 300–305. https://doi.org/10.1007/s00284-011-9971-8.

Hosseinkhani, B., Søbjerg, L.S., Rotaru, A.-E., Emtiazi, G., Skrydstrup, T., Meyer, R.L., 2012. Microbially supported synthesis of catalytically active bimetallic Pd-Au nanoparticles. Biotechnol. Bioeng. 109, 45–52. https://doi.org/10.1002/bit.23293.

Huang, H., Yang, X., 2004. Synthesis of polysaccharide-stabilized gold and silver nanoparticles: a green method. Carbohydr. Res. 339, 2627–2631. https://doi.org/10.1016/j.carres.2004.08.005.

Huang, J., Li, Q., Sun, D., Lu, Y., Su, Y., Yang, X., Wang, H., Wang, Y., Shao, W., He, N., Hong, J., Chen, C., 2007. Biosynthesis of silver and gold nanoparticles by novel sundried *Cinnamomum camphora* leaf. Nanotechnology 18. https://doi.org/10.1088/0957-4484/18/10/105104, 105104.

Ingle, A., Rai, M., Gade, A., Bawaskar, M., 2008. Fusarium solani: a novel biological agent for the extracellular synthesis of silver nanoparticles. J. Nanopart. Res. 11, 2079–2085. https://doi.org/10.1007/s11051-008-9573-y.

Jacob, J.A., Biswas, N., Mukherjee, T., Kapoor, S., 2011. Effect of plant-based phenol derivatives on the formation of Cu and Ag nanoparticles. Colloids Surf. B: Biointerfaces 87, 49–53. https://doi.org/10.1016/j.colsurfb.2011.04.036.

Jagajjanani Rao, K., Paria, S., 2017. Phytochemicals mediated synthesis of multifunctional Ag-Au-TiO$_2$ heterostructure for photocatalytic and antimicrobial applications. J. Clean. Prod. 165, 360–368. https://doi.org/10.1016/j.jclepro.2017.07.147.

Jayaseelan, C., Rahuman, A.A., Kirthi, A.V., Marimuthu, S., Santhoshkumar, T., Bagavan, A., Gaurav, K., Karthik, L., Rao, K.V.B., 2012. Novel microbial route to synthesize ZnO nanoparticles using *Aeromonas hydrophila* and their activity against pathogenic bacteria and fungi. Spectrochim. Acta A Mol. Biomol. Spectrosc. 90, 78–84. https://doi.org/10.1016/j.saa.2012.01.006.

Jebali, A., Ramezani, F., Kazemi, B., 2011. Biosynthesis of silver nanoparticles by Geotricum sp. J. Clust. Sci. 22, 225–232. https://doi.org/10.1007/s10876-011-0375-5.

Jha, A.K., Prasad, K., Kulkarni, A.R., 2009. Synthesis of TiO$_2$ nanoparticles using microorganisms. Colloids Surf. B: Biointerfaces 71, 226–229. https://doi.org/10.1016/j.colsurfb.2009.02.007.

Jiang, S., Ho, C.T., Lee, J.-H., Duong, H.V., Han, S., Hur, H.-G., 2012. Mercury capture into biogenic amorphous selenium nanospheres produced by mercury resistant *Shewanella putrefaciens* 200. Chemosphere 87, 621–624. https://doi.org/10.1016/j.chemosphere.2011.12.083.

Kalathil, S., Lee, J., Cho, M.H., 2011. Electrochemically active biofilm-mediated synthesis of silver nanoparticles in water. Green Chem. 13, 1482. https://doi.org/10.1039/c1gc15309a.

Katepetch, C., Rujiravanit, R., 2011. Synthesis of magnetic nanoparticle into bacterial cellulose matrix by ammonia gas-enhancing in situ co-precipitation method. Carbohydr. Polym. 86, 162–170. https://doi.org/10.1016/j.carbpol.2011.04.024.

Khandel, P., Yadaw, R.K., Soni, D.K., Kanwar, L., Shahi, S.K., 2018. Biogenesis of metal nanoparticles and their pharmacological applications: present status and application prospects. J. Nanostruct. Chem. 8, 217–254. https://doi.org/10.1007/s40097-018-0267-4.

Kharissova, O.V., Dias, H.V.R., Kharisov, B.I., Pérez, B.O., Pérez, V.M.J., 2013. The greener synthesis of nanoparticles. Trends Biotechnol. 31, 240–248. https://doi.org/10.1016/j.tibtech.2013.01.003.

Klaus, T., Joerger, R., Olsson, E., Granqvist, C.G., 1999. Silver-based crystalline nanoparticles, microbially fabricated. Proc. Natl. Acad. Sci. 96, 13611.

Kowshik, M., Ashtaputre, S., Kharrazi, S., Vogel, W., Urban, J., Kulkarni, S.K., Paknikar, K.M., 2003. Extracellular synthesis of silver nanoparticles by a silver-tolerant yeast strain MKY3. Nanotechnology 14, 95–100. https://doi.org/10.1088/0957-4484/14/1/321.

Kowshik, M., Deshmukh, N., Vogel, W., Urban, J., Kulkarni, S.K., Paknikar, K.M., 2002. Microbial synthesis of semiconductor CdS nanoparticles, their characterization, and their use in the fabrication of an ideal diode. Biotechnol. Bioeng. 78, 583–588. https://doi.org/10.1002/bit.10233.

Krumov, N., Oder, S., Perner-Nochta, I., Angelov, A., Posten, C., 2007. Accumulation of CdS nanoparticles by yeasts in a fed-batch bioprocess. J. Biotechnol. 132, 481–486. https://doi.org/10.1016/j.jbiotec.2007.08.016.

Kumar, C.G., Mamidyala, S.K., 2011. Extracellular synthesis of silver nanoparticles using culture supernatant of *Pseudomonas aeruginosa*. Colloids Surf. B: Biointerfaces 84, 462–466. https://doi.org/10.1016/j.colsurfb.2011.01.042.

Kumar, S.A., Abyaneh, M.K., Gosavi, S.W., Kulkarni, S.K., Pasricha, R., Ahmad, A., Khan, M.I., 2007a. Nitrate reductase-mediated synthesis of silver nanoparticles from AgNO$_3$. Biotechnol. Lett. 29, 439–445. https://doi.org/10.1007/s10529-006-9256-7.

Kumar, S.A., Ansary, A.A., Ahmad, A., Khan, M.I., 2007b. Extracellular biosynthesis of CdSe quantum dots by the fungus, fusarium oxysporum. J. Biomed. Nanotechnol. 3, 190–194. https://doi.org/10.1166/jbn.2007.027.

Lakhan, M.N., Chen, R., Shar, A.H., Chand, K., Shah, A.H., Ahmed, M., Ali, I., Ahmed, R., Liu, J., Takahashi, K., Wang, J., 2020. Eco-friendly green synthesis of clove buds extract functionalized silver nanoparticles and evaluation of antibacterial and antidiatom activity. J. Microbiol. Methods 173. https://doi.org/10.1016/j.mimet.2020.105934, 105934.

Lengke, M.F., Fleet, M.E., Southam, G., 2006a. Morphology of gold nanoparticles synthesized by filamentous cyanobacteria from gold(I)-thiosulfate and gold(III)-chloride complexes. Langmuir 22, 2780–2787. https://doi.org/10.1021/la052652c.

Lengke, M.F., Ravel, B., Fleet, M.E., Wanger, G., Gordon, R.A., Southam, G., 2006b. Mechanisms of gold bioaccumulation by filamentous cyanobacteria from gold (III)-chloride complex. Environ. Sci. Technol. 40, 6304–6309.

Li, G., He, D., Qian, Y., Guan, B., Gao, S., Cui, Y., Yokoyama, K., Wang, L., 2012. Fungus-mediated green synthesis of silver nanoparticles using *Aspergillus terreus*. Int. J. Mol. Sci. 13, 466–476. https://doi.org/10.3390/ijms13010466.

Liang, W., Church, T.L., Harris, A.T., 2012. Biogenic synthesis of photocatalytically active Ag/TiO$_2$ and Au/TiO$_2$ composites. Green Chem. 14, 968. https://doi.org/10.1039/c2gc16082j.

Mallick, S., Sharma, S., Banerjee, M., Ghosh, S.S., Chattopadhyay, A., Paul, A., 2012. Iodine-stabilized Cu nanoparticle chitosan composite for antibacterial applications. ACS Appl. Mater. Interfaces 4, 1313–1323. https://doi.org/10.1021/am201586w.

Mandal, D., Bolander, M., Mukhopadhyay, D., Sarkar, G., Mukherjee, P., 2006. The use of microorganisms for the formation of metal nanoparticles and their application. Appl. Microbiol. Biotechnol. 69, 485–492. https://doi.org/10.1007/s00253-005-0179-3.

Manikandan, D.B., Sridhar, A., Krishnasamy Sekar, R., Perumalsamy, B., Veeran, S., Arumugam, M., Karuppaiah, P., Ramasamy, T., 2021. Green fabrication, characterization of silver nanoparticles using aqueous leaf extract of *Ocimum americanum* (hoary basil) and investigation of its in vitro antibacterial, antioxidant, anticancer and photocatalytic reduction. J. Environ. Chem. Eng. 9. https://doi.org/10.1016/j.jece.2020.104845, 104845.

Markus, J., Mathiyalagan, R., Kim, Y.-J., Abbai, R., Singh, P., Ahn, S., Perez, Z.E.J., Hurh, J., Yang, D.C., 2016. Intracellular synthesis of gold nanoparticles with antioxidant activity by probiotic *Lactobacillus kimchicus* DCY51T isolated from Korean kimchi. Enzym. Microb. Technol. 95, 85–93. https://doi.org/10.1016/j.enzmictec.2016.08.018.

Millstone, J.E., Hurst, S.J., Métraux, G.S., Cutler, J.I., Mirkin, C.A., 2009. Colloidal gold and silver triangular nanoprisms. Small 5, 646–664.

Mishra, A., Tripathy, S.K., Wahab, R., Jeong, S.-H., Hwang, I., Yang, Y.-B., Kim, Y.-S., Shin, H.-S., Yun, S.-I., 2011. Microbial synthesis of gold nanoparticles using the fungus *Penicillium brevicompactum* and their cytotoxic effects against mouse mayo blast cancer C2C12 cells. Appl. Microbiol. Biotechnol. 92, 617–630. https://doi.org/10.1007/s00253-011-3556-0.

Mittal, A.K., Chisti, Y., Banerjee, U.C., 2013. Synthesis of metallic nanoparticles using plant extracts. Biotechnol. Adv. 31, 346–356. https://doi.org/10.1016/j.biotechadv.2013.01.003.

Mody, V.V., Siwale, R., Singh, A., Mody, H.R., 2010. Introduction to metallic nanoparticles. J. Pharm. Bioallied Sci. 2, 282–289. https://doi.org/10.4103/0975-7406.72127.

Mohammed Fayaz, A., Girilal, M., Rahman, M., Venkatesan, R., Kalaichelvan, P.T., 2011. Biosynthesis of silver and gold nanoparticles using thermophilic bacterium *Geobacillus stearothermophilus*. Process Biochem. 46, 1958–1962. https://doi.org/10.1016/j.procbio.2011.07.003.

Mohan Kumar, K., Mandal, B.K., Sinha, M., Krishnakumar, V., 2012. Terminalia chebula mediated green and rapid synthesis of gold nanoparticles. Spectrochim. Acta A Mol. Biomol. Spectrosc. 86, 490–494. https://doi.org/10.1016/j.saa.2011.11.001.

Mohanpuria, P., Rana, N.K., Yadav, S.K., 2007. Biosynthesis of nanoparticles: technological concepts and future applications. J. Nanopart. Res. 10, 507–517. https://doi.org/10.1007/s11051-007-9275-x.

Mohanty, S., Mishra, S., Jena, P., Jacob, B., Sarkar, B., Sonawane, A., 2012. An investigation on the antibacterial, cytotoxic, and antibiofilm efficacy of starch-stabilized silver nanoparticles. Nanomed. Nanotechnol. Biol. Med. https://doi.org/10.1016/j.nano.2011.11.007.

Mukherjee, P., Ahmad, A., Mandal, D., Senapati, S., Sainkar, S.R., Khan, M.I., Ramani, R., Parischa, R., Ajayakumar, P.-V., Alam, M., Sastry, M., Kumar, R., 2001. Bioreduction of AuCl$_4$ − ions by the fungus, Verticillium sp. and surface trapping of the gold nanoparticles formed. Angew. Chem. Int. Ed. 40, 3585–3588. https://doi.org/10.1002/1521-3773(20011001)40:19<3585::AID-ANIE3585>3.0.CO;2-K.

Mukherjee, P., Senapati, S., Mandal, D., Ahmad, A., Islam Khan, M., Kumar, R., Sastry, M., 2002. Extracellular synthesis of gold nanoparticles by the fungus *Fusarium oxysporum*. Chembiochem 3, 461–463.

Nagarajan, R., 2008. Nanoparticles: Building Blocks for Nanotechnology. ACS Publications, pp. 2–14.

Nangia, Y., Wangoo, N., Goyal, N., Shekhawat, G., Suri, C.R., 2009. A novel bacterial isolate *Stenotrophomonas maltophilia* as living factory for synthesis of gold nanoparticles. Microb. Cell Factories 8, 39. https://doi.org/10.1186/1475-2859-8-39.

Narayanan, K.B., Sakthivel, N., 2011. Facile green synthesis of gold nanostructures by NADPH-dependent enzyme from the extract of *Sclerotium rolfsii*. Colloids Surf. A Physicochem. Eng. Asp. 380, 156–161. https://doi.org/10.1016/j.colsurfa.2011.02.042.

Nazari, P., Faramarzi, M.A., Sepehrizadeh, Z., Mofid, M.R., Bazaz, R.D., Shahverdi, A.R., 2012. Biosynthesis of bismuth nanoparticles using *serratia marcescens* isolated from the Caspian Sea and their characterisation. NanoBiotechnology 6, 58–62. https://doi.org/10.1049/iet-nbt.2010.0043.

Nguyen, D.T.C., Dang, H.H., Vo, D.-V.N., Bach, L.G., Nguyen, T.D., Tran, T.V., 2021. Biogenic synthesis of MgO nanoparticles from different extracts (flower, bark, leaf) of *Tecoma stans* (L.) and their utilization in selected organic dyes treatment. J. Hazard. Mater. 404. https://doi.org/10.1016/j.jhazmat.2020.124146, 124146.

Noruzi, M., Zare, D., Davoodi, D., 2012. A rapid biosynthesis route for the preparation of gold nanoparticles by aqueous extract of cypress leaves at room temperature. Spectrochim. Acta A Mol. Biomol. Spectrosc. 94, 84–88. https://doi.org/10.1016/j.saa.2012.03.041.

Ogi, T., Honda, R., Tamaoki, K., Saitoh, N., Konishi, Y., 2011. Direct room-temperature synthesis of a highly dispersed Pd nanoparticle catalyst and its electrical properties in a fuel cell. Powder Technol. 205, 143–148. https://doi.org/10.1016/j.powtec.2010.09.004.

Pal, S., Tak, Y.K., Song, J.M., 2007. Does the antibacterial activity of silver nanoparticles depend on the shape of the nanoparticle? A study of the Gram-negative bacterium *Escherichia coli*. Appl. Environ. Microbiol. 73, 1712–1720. https://doi.org/10.1128/AEM.02218-06.

Philip, D., 2009. Honey mediated green synthesis of gold nanoparticles. Spectrochim. Acta A Mol. Biomol. Spectrosc. 73, 650–653. https://doi.org/10.1016/j.saa.2009.03.007.

Philip, D., 2010. Green synthesis of gold and silver nanoparticles using Hibiscus rosa sinensis. Phys. E Low-Dimens. Syst. Nanostruct. 42, 1417–1424. https://doi.org/10.1016/j.physe.2009.11.081.

Ponarulselvam, S., Panneerselvam, C., Murugan, K., Aarthi, N., Kalimuthu, K., Thangamani, S., 2012. Synthesis of silver nanoparticles using leaves of *Catharanthus roseus* Linn. G. Don and their antiplasmodial activities. Asian Pac. J. Trop. Biomed. 2, 574–580. https://doi.org/10.1016/S2221-1691(12)60100-2.

Pradhan, N., Nayak, R.R., Pradhan, A.K., Sukla, L.B., Mishra, B.K., 2011. In situ synthesis of entrapped silver nanoparticles by a fungus-penicillium purpurogenum. Nanosci. Nanotechnol. Lett. 3, 659–665.

Prasad, K.S., Vyas, P., Prajapati, V., Patel, P., Selvaraj, K., 2012. Biomimetic synthesis of selenium nanoparticles using cell-free extract of microbacterium sp. ARB05. Micro Nano Lett. 7, 1–4. https://doi.org/10.1049/mnl.2011.0498.

Qu, J., Luo, C., Cong, Q., Yuan, X., 2012. Carbon nanotubes and Cu–Zn nanoparticles synthesis using hyperaccumulator plants. Environ. Chem. Lett. 10, 153–158. https://doi.org/10.1007/s10311-011-0335-1.

Qu, J., Luo, C., Hou, J., 2011b. Synthesis of ZnO nanoparticles from Zn-hyperaccumulator (*Sedum alfredii* Hance) plants. Micro Nano Lett. 6, 174–176. https://doi.org/10.1049/mnl.2011.0004.

Qu, J., Yuan, X., Wang, X., Shao, P., 2011a. Zinc accumulation and synthesis of ZnO nanoparticles using *Physalis alkekengi* L. Environ. Pollut. 159, 1783–1788. https://doi.org/10.1016/j.envpol.2011.04.016.

Raghunandan, D., Bedre, M.D., Basavaraja, S., Sawle, B., Manjunath, S., Venkataraman, A., 2010. Rapid biosynthesis of irregular shaped gold nanoparticles from macerated aqueous extracellular dried clove buds (*Syzygium aromaticum*) solution. Colloids Surf. B: Biointerfaces 79, 235–240. https://doi.org/10.1016/j.colsurfb.2010.04.003.

Raheman, F., Deshmukh, S., Ingle, A., Gade, A., Rai, M., 2011. Silver nanoparticles: novel antimicrobial agent synthesized from an endophytic fungus Pestalotia sp. Isolated from leaves of *Syzygium cumini* (L). Nano Biomed. Eng. 3, 174–178.

Rai, M.K., Deshmukh, S.D., Ingle, A.P., Gade, A.K., 2012. Silver nanoparticles: the powerful nanoweapon against multidrug-resistant bacteria. J. Appl. Microbiol. 112, 841–852. https://doi.org/10.1111/j.1365-2672.2012.05253.x.

Rajakumar, G., Abdul Rahuman, A., 2012. Acaricidal activity of aqueous extract and synthesized silver nanoparticles from *Manilkara zapota* against Rhipicephalus (Boophilus) microplus. Res. Vet. Sci. 93, 303–309. https://doi.org/10.1016/j.rvsc.2011.08.001.

Rajesh Kumar, R., Poornima Priyadharsani, K., Thamaraiselvi, K., 2012. Mycogenic synthesis of silver nanoparticles by the Japanese environmental isolate *Aspergillus tamarii*. J. Nanopart. Res. 14. https://doi.org/10.1007/s11051-012-0860-2.

Ramanathan, R., O'Mullane, A.P., Parikh, R.Y., Smooker, P.M., Bhargava, S.K., Bansal, V., 2010. Bacterial kinetics-controlled shape-directed biosynthesis of silver nanoplates using *Morganella psychrotolerans*. Langmuir 27, 714–719. https://doi.org/10.1021/la1036162.

Rao, K.J., Paria, S., 2015a. Aegle marmelos leaf extract and plant surfactants mediated green synthesis of Au and Ag nanoparticles by optimizing process parameters using Taguchi method. ACS Sustain. Chem. Eng. https://doi.org/10.1021/acssuschemeng.5b00022.

Rao, K.J., Paria, S., 2015b. Mixed phytochemicals mediated synthesis of multifunctional Ag–Au–Pd nanoparticles for glucose oxidation and antimicrobial applications. ACS Appl. Mater. Interfaces 7, 14018–14025. https://doi.org/10.1021/acsami.5b03089.

Roh, Y., 2001. Microbial synthesis and the characterization of metal-substituted magnetites. Solid State Commun. 118, 529–534. https://doi.org/10.1016/S0038-1098(01)00146-6.

Roy, N., Alam, M.N., Mondal, S., Sk, I., Laskar, R.A., Das, S., Mandal, D., Begum, N.A., 2012. Exploring Indian rosewood as a promising biogenic tool for the synthesis of metal nanoparticles with tailor-made morphologies. Process Biochem. 47, 1371–1380. https://doi.org/10.1016/j.procbio.2012.05.009.

Saha, S., Sarkar, J., Chattopadhyay, D., Patra, S., Chakraborty, A., Acharya, K., 2010. Production of silver nanoparticles by a phytopathogenic fungus bipolaris nodulosa and its antimicrobial activity. Dig. J. Nanomater. Biostruct. 5, 887–895.

Sajanlal, P.R., Sreeprasad, T.S., Samal, A.K., Pradeep, T., 2011. Anisotropic nanomaterials: structure, growth, assembly, and functions. Nano Rev. 2. https://doi.org/10.3402/nano.v2i0.5883.

Salata, O.V., 2004. Applications of nanoparticles in biology and medicine. J. Nanobiotechnol. 2, 3.

Salunkhe, R.B., Patil, S.V., Salunke, B.K., Patil, C.D., Sonawane, A.M., 2011. Studies on silver accumulation and nanoparticle synthesis by *Cochliobolus lunatus*. Appl. Biochem. Biotechnol. 165, 221–234. https://doi.org/10.1007/s12010-011-9245-8.

Sarikaya, M., Tamerler, C., Jen, A.K.-Y., Schulten, K., Baneyx, F., 2003. Molecular biomimetics: nanotechnology through biology. Nat. Mater. 2, 577–585. https://doi.org/10.1038/nmat964.

Sathishkumar, G., Gobinath, C., Karpagam, K., Hemamalini, V., Premkumar, K., Sivaramakrishnan, S., 2012. Phytosynthesis of silver nanoscale particles using *Morinda citrifolia* L. and its inhibitory activity against human pathogens. Colloids Surf. B: Biointerfaces 95, 235–240. https://doi.org/10.1016/j.colsurfb.2012.03.001.

Sathishkumar, M., Sneha, K., Won, S.W., Cho, C.-W., Kim, S., Yun, Y.-S., 2009. Cinnamon zeylanicum bark extract and powder mediated green synthesis of nano-crystalline silver particles and its bactericidal activity. Colloids Surf. B: Biointerfaces 73, 332–338. https://doi.org/10.1016/j.colsurfb.2009.06.005.

Sathishkumar, M., Sneha, K., Yun, Y.-S., 2010. Immobilization of silver nanoparticles synthesized using *Curcuma longa* tuber powder and extract on cotton cloth for bactericidal activity. Bioresour. Technol. 101, 7958–7965. https://doi.org/10.1016/j.biortech.2010.05.051.

Sau, T.K., Murphy, C.J., 2004. Room temperature, high-yield synthesis of multiple shapes of gold nanoparticles in aqueous solution. J. Am. Chem. Soc. 126, 8648–8649. https://doi.org/10.1021/ja047846d.

Schabes-Retchkiman, P.S., Canizal, G., Herrera-Becerra, R., Zorrilla, C., Liu, H.B., Ascencio, J.A., 2006. Biosynthesis and characterization of Ti/Ni bimetallic nanoparticles. Opt. Mater. 29, 95–99. https://doi.org/10.1016/j.optmat.2006.03.014.

Shankar, S.S., Ahmad, A., Sastry, M., 2003. Geranium leaf assisted biosynthesis of silver nanoparticles. Biotechnol. Prog. 19, 1627–1631. https://doi.org/10.1021/bp034070w.

Shankar, S.S., Rai, A., Ankamwar, B., Singh, A., Ahmad, A., Sastry, M., 2004. Biological synthesis of triangular gold nanoprisms. Nat. Mater. 3, 482–488. https://doi.org/10.1038/nmat1152.

Sharma, N., Pinnaka, A.K., Raje, M., Fnu, A., Bhattacharyya, M.S., Choudhury, A.R., 2012. Exploitation of marine bacteria for production of gold nanoparticles. Microb. Cell Factories 11, 86. https://doi.org/10.1186/1475-2859-11-86.

Sharma, V.K., Yngard, R.A., Lin, Y., 2009. Silver nanoparticles: green synthesis and their antimicrobial activities. Adv. Colloid Interf. Sci. 145, 83–96. https://doi.org/10.1016/j.cis.2008.09.002.

Sheny, D.S., Philip, D., Mathew, J., 2012. Rapid green synthesis of palladium nanoparticles using the dried leaf of *Anacardium occidentale*. Spectrochim. Acta. A. Mol. Biomol Spectrosc. 91, 35–38. https://doi.org/10.1016/j.saa.2012.01.063.

Shukla, M.K., Singh, R.P., Reddy, C.R.K., Jha, B., 2012. Synthesis and characterization of agar-based silver nanoparticles and nanocomposite film with antibacterial applications. Bioresour. Technol. 107, 295–300. https://doi.org/10.1016/j.biortech.2011.11.092.

Silambarasan, S., Abraham, J., 2012. Biosynthesis of silver nanoparticles using the bacteria *Bacillus cereus* and their antimicrobial property. Int J Pharm Pharm Sci 4, 536–540.

Sinha, A., Khare, S.K., 2011. Mercury bioaccumulation and simultaneous nanoparticle synthesis by Enterobacter sp. cells. Bioresour. Technol. 102, 4281–4284. https://doi.org/10.1016/j.biortech.2010.12.040.

Song, J., 2009. Rapid biological synthesis of silver nanoparticles using plant leaf extracts. Bioprocess Biosyst. Eng. 32, 79–84. https://doi.org/10.1007/s00449-008-0224-6.

Song, J., Kwon, E.-Y., Kim, B., 2010. Biological synthesis of platinum nanoparticles using *Diopyros kaki* leaf extract. Bioprocess Biosyst. Eng. 33, 159–164. https://doi.org/10.1007/s00449-009-0373-2.

Song, J.Y., Kim, B.S., 2008. Biological synthesis of bimetallic au/ag nanoparticles using persimmon (*Diopyros kaki*) leaf extract. Korean J. Chem. Eng. 25, 808–811. https://doi.org/10.1007/s11814-008-0133-z.

Soundarrajan, C., Sankari, A., Dhandapani, P., Maruthamuthu, S., Ravichandran, S., Sozhan, G., Palaniswamy, N., 2012. Rapid biological synthesis of platinum nanoparticles using *Ocimum sanctum* for water electrolysis applications. Bioprocess Biosyst. Eng. 35, 827–833. https://doi.org/10.1007/s00449-011-0666-0.

Sreekanth, T.V.M., Lee, K.D., 2011. Green synthesis of silver nanoparticles from *Carthamus tinctorius* flower extract and evaluation of their antimicrobial and cytotoxic activities. Curr. Nanosci. 7, 1046–1053.

Srivastava, S., Constanti, M., 2012. Room temperature biogenic synthesis of multiple nanoparticles (Ag, Pd, Fe, Rh, Ni, Ru, Pt, Co, and Li) by *Pseudomonas aeruginosa* SM1. J. Nanopart. Res. 14, 1–10. https://doi.org/10.1007/s11051-012-0831-7.

Suresh, A.K., Doktycz, M.J., Wang, W., Moon, J.-W., Gu, B., Meyer III, H.M., Hensley, D.K., Allison, D.P., Phelps, T.J., Pelletier, D.A., 2011. Monodispersed biocompatible silver sulfide nanoparticles: facile extracellular biosynthesis using the γ-proteobacterium, *Shewanella oneidensis*. Acta Biomater. 7, 4253–4258. https://doi.org/10.1016/j.actbio.2011.07.007.

Tang, B., Wang, J., Xu, S., Afrin, T., Xu, W., Sun, L., Wang, X., 2011. Application of anisotropic silver nanoparticles: multifunctionalization of wool fabric. J. Colloid Interface Sci. 356, 513–518. https://doi.org/10.1016/j.jcis.2011.01.054.

Thakkar, K.N., Mhatre, S.S., Parikh, R.Y., 2010. Biological synthesis of metallic nanoparticles. Nanomed. Nanotechnol. Biol. Med. 6, 257–262.

Thakor, A., Jokerst, J.V., Zavaleta, C.L., Massoud, T.F., Gambhir, S., 2011. Gold nanoparticles a revival in precious metal administration to patients. Nano Lett. 11 (10), 4029–4036.

Tian, X., Li, J., Pan, S., 2009. Facile synthesis of single-crystal silver nanowires through a tannin-reduction process. J. Nanopart. Res. 11, 1839–1844. https://doi.org/10.1007/s11051-009-9700-4.

Velmurugan, P., Lee, S.-M., Iydroose, M., Lee, K.-J., Oh, B.-T., 2013. Pine cone-mediated green synthesis of silver nanoparticles and their antibacterial activity against agricultural pathogens. Appl. Microbiol. Biotechnol., 1–8. https://doi.org/10.1007/s00253-012-3892-8.

Verma, V.C., Singh, S.K., Solanki, R., Prakash, S., 2011. Biofabrication of anisotropic gold nanotriangles using extract of endophytic *Aspergillus clavatus* as a dual functional reductant and stabilizer. Nanoscale Res. Lett. 6, 16. https://doi.org/10.1007/s11671-010-9743-6.

Vigneshwaran, N., Nachane, R.P., Balasubramanya, R.H., Varadarajan, P.V., 2006. A novel one-pot 'green' synthesis of stable silver nanoparticles using soluble starch. Carbohydr. Res. 341, 2012–2018. https://doi.org/10.1016/j.carres.2006.04.042.

Wei, D., Sun, W., Qian, W., Ye, Y., Ma, X., 2009. The synthesis of chitosan-based silver nanoparticles and their antibacterial activity. Carbohydr. Res. 344, 2375–2382. https://doi.org/10.1016/j.carres.2009.09.001.

Yang, Z., Chen, S., Hu, W., Yin, N., Zhang, W., Xiang, C., Wang, H., 2012. Flexible luminescent CdSe/bacterial cellulose nanocomoposite membranes. Carbohydr. Polym. 88, 173–178. https://doi.org/10.1016/j.carbpol.2011.11.080.

Zaki, S., El Kady, M.F., Abd-El-Haleem, D., 2011. Biosynthesis and structural characterization of silver nanoparticles from bacterial isolates. Mater. Res. Bull. 46, 1571–1576. https://doi.org/10.1016/j.materresbull.2011.06.025.

Nanocomposites in food packaging

Debarshi Nath, Rahul Chetri, R. Santhosh, and Preetam Sarkar

Department of Food Process Engineering, National Institute of Technology Rourkela,
Rourkela, India

OUTLINE

1 Introduction

Packaging of food products is done to safeguard them from harmful contaminants and prevent the structural, chemical, and nutritional degradation occurring due to these contaminants' interaction with the product during storage or transit. The packaging materials also protect the packaged food products from different environmental impacts such as shocks, dust, odor, microbe attacks, and vibrations. However, these traditional packaging materials have several shortcomings like poor gas, moisture, and UV barrier properties. Lack of barrier activity results in rapid deterioration of packaged food products and shelf life reduction (de Azeredo, 2009). These limitations of traditional packaging material are countered by the application of nanotechnology.

The application of nanotechnology in food packaging has become one of the most developed areas in food processing and has arisen to be an excellent alternative to traditional packaging methods. These nanocomposites provide several advantages like the extension of shelf life of food product, indicating the quality of products and antimicrobial properties. The nanofillers like nanometal oxide or metal nanoparticles, when incorporated into a polymer matrix, enhance the functional properties of bio-derived and synthetic polymers (Souza and Fernando, 2016). The reinforcement effect occurs primarily due to the large interfacial surface area causing changes in molecular mobility, relaxation behavior, and an increase in interaction between the polymer chains (Ahmed et al., 2017). Recently, several nanometal oxides like ZnO, TiO_2, SiO_2, and metal Ag have gained widespread attention due to their thermal stability, barrier properties, and antimicrobial activity. However, these nanoparticle-incorporated polymers have some drawbacks, like increased opacity and brittleness. Nevertheless, nanocomposites have transpired as a novel food packaging material having new functionalities, which eventually enhance the food quality. Different classes of polymers, which are used to form nanocomposites for food packaging application, are illustrated in Fig. 1.

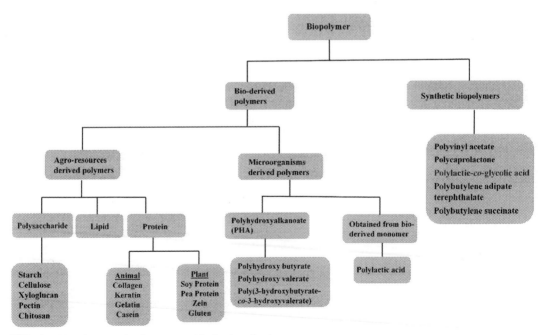

FIG. 1   Classification of polymers used in food packaging.

Essential oils (EOs), an essential source of natural preservatives, are receiving more attention recently and are subjected to extensive studies. Its application as a preservative is mainly due to the broad-spectrum antibacterial activity and several other properties (Ju et al., 2019). Incorporating EOs into the nanocomposite imparts antimicrobial property to the nanocomposites, which improves the overall safety of the packaged product against food-borne pathogens (Burt, 2004). The use of EOs is not limited to antimicrobial benefits, but it also shows benefits relating to human health. Currently, research is done on the various biological properties such as antitumor, analgesics, antidiabetic, antioxidant, and insecticidal (Brahmi et al., 2016; Burt, 2004; Periasamy et al., 2016). It is considered to be safer than synthetic preservatives as they are derived from plant extracts, and the Food and Drug Administration (FDA), United States, has classified EOs as generally recognized as safe (GRAS) and a possible alternative to synthetic compounds (Ribeiro-Santos et al., 2017). EOs are a combination of several complex compounds, such as alkaloids, flavonoids, phenolic acid, aldehydes, and carotenoids. They are almost insoluble in water and are volatile compounds; however, they are susceptible to oxidative decomposition, limiting their packaging industry application (Seow et al., 2014). A proper delivery system suitable for food application is required to release EOs into the food matrix. Nanomaterials can act as excellent carrier materials for the efficient release of EOs. This increases the effectiveness of the EOs, leading to the enhancement of antimicrobial properties of the nanocomposites. Therefore, EO-incorporated nanocomposites are successful in the shelf life extension of highly perishable food products (Buendía et al., 2020; Iamareerat et al., 2018; Sani et al., 2017).

As the use of nanotechnologies increases, the potential consumers and the governments are apprehensive regarding the adverse effects shown by the nanocomposites, like the migration of nanoparticles into the food matrix, causing peril to human health due to their extensive use in the field of food packaging. As proper and concise migration data of nanoparticles into the food matrix is not available due to the unexpected migration properties of nanomaterials compared to bulk material, its use is somewhat restricted. Consumers prefer the products certified as safe by a proper regulatory body by assessing the exposure to the nanoparticles by proper migration testing. Nanoparticles that do not show any migration toward the food matrix or show a minimum level of migrations are approved (Cushen et al., 2014; EFSA, 2012; Magnuson et al., 2011). Therefore, this chapter discusses the different fabrication methods of nanocomposites and the role of nanoparticles in nanocomposite reinforcement. Moreover, the use of EOs for imparting active properties to the nanocomposites, along with the legal aspects regarding the use of nanoparticle in food packaging, is also comprehensively discussed.

2 Fabrication methods of nanocomposites

Nanocomposites are recently used to form the primary packaging material for packaging food items. The morphological and chemical characteristics of film-forming biopolymers, along with the additives, should be adequately known before the fabrication of nanocomposites for distinct applications (Suhag et al., 2020). Furthermore, the incorporation of active agents could be done prior to the film-forming process. There are different ways to produce nanocomposite films and coatings with different polymers and additives, and they are direct casting, extrusion, layer-by-layer assembly, in situ polymerization, etc. (Wang et al., 2018).

2.1 Direct casting

It is the simplest method for film preparation, and hence it is widely used for fabricating nanocomposites. The preparation stages are (1) preparation of film-forming solution by solubilizing the biopolymer into an appropriate solvent and then mixing active bio-compounds or materials like plasticizers and filler; (2) stirring the formulation to obtain homogenous solution; (3) filtration or centrifugation to make the solution free from air bubbles and insoluble particles; (4) pouring on a flat surface or a mold; (5) drying of solution at a certain temperature and humidity for a fixed time; and (6) removing of the casted film from the surface and storing (Shankar and Rhim, 2018). Sometimes, specialized equipment like hot-air oven, vacuum oven, and humidity controller are used for drying. The primary advantages of using the casting method are the ease of fabrication without using any sophisticated equipment and better particle-particle interaction resulting in a more stable film (Siemann, 2005). The prepared films are without any major defects or cracks, which showed its immense promise as a commercial packaging film. However, the nanocomposite film may exhibit structural defects if there is a slight variation in the formula. However, this method is not used commonly for industrial purpose as it is very time consuming and expensive (Suhag et al., 2020). A schematic diagram representing the process of nanocomposite formation by casting method is presented in Fig. 2.

2.2 Extrusion

Extrusion is also being used widely for the production of nanocomposites for different applications. The processes include (1) preparations of different compositions that are formulated from which the film is to be created, (2) mixing and blending of the solution, (3) extrusion of the mix under specific conditions of temperature and moisture content, (4) transforming of the extrudates into pellets using a pelletizer, (5) drying of pellets after some time at a fixed temperature, (6) loading of a twin-screw extruder with the dried pellets and extruding of the pellets into a flat sheet, and (7) using of a blown-film extruder with an annular die to make a blown film (Wang et al., 2018). Extruded films show better mechanical properties and thermal stabilities compared to other methods, although plasticizers are added into the formulation to make the films more flexible (Martínez-Camacho et al., 2013). As this method is a dry method, superior films are produced when the moisture content of the film is minimum. The best films were prepared when the processing temperature was tailored for different nanoparticles (Woranuch and Yoksan, 2013). The main advantages of using the extrusion technique are its low cost, lower energy requirement, and higher transparency for nanocomposites than the casting technique. Other benefits of extrusion include trouble-free handling of polymers with high viscosity, lack of solvents, degree of mixing, etc. (Suhag et al., 2020).

2.3 In situ polymerization

In situ polymerization is a commonly used technique that involves mixing the monomer unit with the nanoparticles in a solvent followed by in situ polymerization. The method includes the dispersion of nanoparticles into the monomer unit, which causes

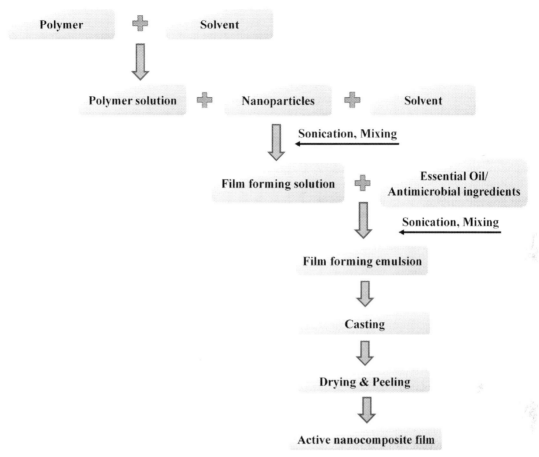

FIG. 2   Fabrication process of nanocomposites by casting method.

the monomer to enter the nanoparticle galleries. Following this, polymerization takes place by adjusting the time and temperature. All these contribute toward stronger interaction between the matrix and the nanoparticles (Shen et al., 2002). Herein, a high level of dispersion of nanoparticles is achieved when compared to other methods. This technique is also referred to as intercalation polymerization. This technique is most notably used for nanocomposite fabrication having thermally unstable or insoluble matrix because such matrices could not be fused or dissolved in solvents. It is believed that this method gives an exfoliated structure to the nanocomposite and provides the freedom of choosing an appropriate polymerization technique and surfactant for better distribution of nanoparticles in the polymer matrix (Rhim and Ng, 2007).

2.4 Layer-by-layer assembly

Layer-by-layer assembly (LBL) is an electrostatic deposition technique, which effectively controls the nanocomposite films' functionality and material properties and can be used to synthesize multicomponent films without using any specialized equipment (Priyadarshi and Rhim, 2020). The change in surface form is mainly due to the deposition of alternating polyelectrolytes with opposite charges on the surface and the mutual attraction between them. Moreover, the films synthesized by this process are not dependent on the substrate shape (Li et al., 2013). The pH is an important parameter for the production of LBL films. The pH governs the electrical charge present in each layer of the LBL film, which results in the variation in the biopolymer accumulation on the previous layers. When pH decreases, the number of layers increases resulting in increased thickness, modulus, and roughness of the nanocomposite (Acevedo-Fani et al., 2015). The combination of the LBL method with other techniques such as immersing technique is often used to prepare functional films in an alternate and easy way. These films are composed of oppositely charged material, and the products are immersed in these charged materials multiple times to form the desired films (Huang et al., 2019). This synergistic method is relatively easy and able to extend shelf life and maintain the food quality effectively.

3 Types of nanoparticles

3.1 Zinc oxide (ZnO)

ZnO is produced by mechanochemical processes and physical vapor synthesis (Casey, 2006). It is produced as a whitish powder and is frequently used in the food industry and sunscreen lotion (Espitia et al., 2012). ZnO can be synthesized by different methods, such as hydrothermal synthesis, thermal decomposition, and precipitation. It can also be prepared using spray pyrolysis and sol-gel method (Espitia et al., 2012). ZnO-incorporated nanocomposites are supposed to act as a cheaper and safer alternative for fabricating active food packaging films due to their antimicrobial action against microbes.

Furthermore, it also has high transparency, excellent barrier properties, thermal stability, and mechanical strength and is cheaper than Ag nanoparticles (Espitia et al., 2012). A study by Kumar et al. (2019) showed that incorporating ZnO resulted in increased mechanical, barrier, and thermal properties. ZnO shows its antimicrobial activity by releasing Zn^{2+} ion and creating reactive oxygen species (ROH), which penetrates the cytoplasm of the microbes, thereby causing the cell to deactivate through cell lysis (Espitia et al., 2012). ZnO nanoparticles display antibacterial activity only when the ZnO comes in contact with the cell, and this also depends on the concentration of ZnO. According to Li et al. (2009), ZnO-incorporated polyvinyl chloride films exhibited bactericidal activity against *Escherichia coli* and *Staphylococcus aureus*, with the latter organism showing higher inhibition. Similarly, according to Pantani et al. (2013), polylactic acid, when incorporated with ZnO, formed biodegradable packaging material with superior UV barrier and moisture barrier properties with antibacterial activity against both gram-positive and gram-negative bacteria.

ZnO-incorporated packaging systems are reported to prolong the shelf life of certain food items significantly. According to Li et al. (2011), PVC-ZnO nanocomposite retarded the growth of fungi and bacteria in a cut apple. Emamifar et al. (2011) reported antibacterial activity against *Lactobacillus plantarum* in orange juice when LDPE is incorporated with ZnO-Ag. Similarly, molds and yeast growths retarded in ZnO-based nanocomposites (Noshirvani et al., 2017b).

3.2 Titanium dioxide (TiO_2)

TiO_2 is a commonly used nanoparticle produced by several methods, among which the sol-gel method is the most preferred one (Ibrahim and Sreekantan, 2011). It forms excellent nanocomposites with enhanced UV barrier properties and absorbs light with a short wavelength, making it photostable. TiO_2 also displays photolytic activities, useful in antimicrobial activity by releasing ROH and self-cleansing under UV rays and black-light illumination. TiO_2 is also economical and has low toxicity, making its use quite common (Carré et al., 2014; Chorianopoulos et al., 2011).

The photocatalytic activity can be useful in preparing active films with antimicrobial properties. Ethylene-vinyl alcohol (EVOH)-TiO_2-incorporated films displayed antimicrobial activity against both gram-negative and gram-positive bacterial strains (Cerrada et al., 2008). Polyvinyl alcohol/amylose starch-based nanocomposite was reinforced by incorporating TiO_2, which improved the water barrier property, tensile strength, and antimicrobial properties (Liu et al., 2015). Black plum peel extracts, when added to chitosan-TiO_2 films, showed ethylene-scavenging activity, along with antioxidant and antimicrobial activity. The mechanical properties and water barrier properties of the film also improved due to the inclusion of TiO_2 (Zhang et al., 2019a). A similar result was obtained for agar-gelatin-TiO_2 films, which delayed the fish oil oxidation (Vejdan et al., 2017). Similar instances of increased food safety and shelf life due to TiO_2 incorporation for products, such as almonds, shrimp, and cheese, are also reported (Gumiero et al., 2013; Luo et al., 2015a; Nasiri et al., 2012). Additionally, the incorporation of TiO_2 can also prevent the accumulation of plastic additives like ethylene glycol on the food product's surface from the packaging material (Farhoodi et al., 2017).

3.3 Silica (SiO_2)

SiO_2 is the most prolific material present on the earth surface, which is found in the form of SiO_2 and silicate (Garcia et al., 2018). SiO_2 nanoparticles are incorporated into polymers to improve their mechanical properties and thermal stability. EVOH, when incorporated with SiO_2, enhanced the gas and water barrier properties, along with the mechanical strength of the nanocomposite (Liu et al., 2010). Similarly, cross-linked alginate-SiO_2 nanocomposite showed improved tensile strength, elongation, thermal stability, and transparency, along with water barrier and UV barrier properties, which would help in the prevention of lipid oxidation (Yang et al., 2016). Whey protein isolate films, when incorporated with SiO_2, also displayed a similar trend and could be used as an environmentally friendly packaging material (Hassannia-Kolaee et al., 2016).

Only a few studies reporting the extension of the shelf life of food products wrapped with SiO$_2$-incorporated nanocomposites have been reported. Nonetheless, in a study by Luo et al. (2015b), LDPE-SiO$_2$ nanocomposite was able to extend the shelf life and maintain the freshness of Pacific white shrimp. Soy protein isolate incorporated with nano-silica was able to extend the shelf life of *Flammulina velutipes* by 10 days (Xu et al., 2019a). Apart from all these properties, LDPE-SiO$_2$-TiO$_2$-Ag blend nanocomposite also showed functional properties, such as regulation of O$_2$ and CO$_2$ level and ethylene-scavenging and ethylene-retarding bacterial growth, thereby prolonging the shelf life of mushrooms (Donglu et al., 2016).

3.4 Silver nanoparticles

Nanosilver (Ag) materials have become a very popular nanomaterial for food packaging due to their inherent reinforcement and antimicrobial property. Ag nanoparticles can be obtained through ex situ fabrication by following regular borohydride reduction of Ag$^+$ ions followed by dispersing it into a polymer solution (Huang et al., 2018). Ag nanoparticles can also be synthesized by the in situ method in a polymer formulation, and it has better dispersibility than the ex situ method. Ag nanomaterial formed through physical reduction has a uniform shape and better dispersibility, while the chemically reduced Ag displays agglomerated structures (de Azeredo, 2013). Properties of Ag include low volatility and imparting thermal stability to the nanocomposite (Cushen et al., 2013).

Besides the enhancements in the mechanical, barrier, and thermal properties, Ag also displays high efficacy against microorganisms and prevents bacteria from spoiling the product. It is effective against both gram-positive and gram-negative bacteria and is successful against multidrug resistance microorganisms (Birla et al., 2009). The Ag nanoparticles get evenly dispersed into the nanocomposite matrix without forming agglomerates, which adds to the antibacterial action and considerably prolongs shelf life. The antimicrobial activity of the nanomaterials is mainly attributed to the direct damage to the cell wall by the Ag$^+$ ions and the reaction of nano-Ag with the dissolved H$^+$ and O$_2$ inducing its antimicrobial action (Chakraborty et al., 2017). Some studies have suggested the formation of ROS by the Ag nanoparticles, which is toxic to the cells. The Ag ions enter the cell wall due to their small size and inhibit DNA replication and protein synthesis, causing cell inhibition (Huang et al., 2015).

According to Arfat et al. (2017b), guar gum/Ag-Cu nanocomposite showed enhanced oxygen and UV-blocking activity, along with increased mechanical strength and thermal stability. The films also exhibited strong antibacterial action against *Salmonella typhimurium* and *Listeria monocytogenes* after incorporating Ag nanoparticles. Similarly, PLA-lignin film, when incorporated with Ag nanoparticles, resulted in nanocomposite displaying superior UV-blocking and water vapor barrier activity, along with antibacterial activity against *E. coli* and *Listeria monocytogenes* (Shankar et al., 2018a). Moreover, Ag-incorporated nanocomposite was reported to successfully increase the shelf life of mangoes (Chi et al., 2019) and chicken sausages (Mathew et al., 2019).

3.5 Halloysite

Nanoclays incorporated with polymers are the earliest nanocomposite that was available on the market as a food packaging material. Nanoclays are widely used in packaging films as nanomaterials and comprise 70% of the nanoparticle-based packaging film market volume (Huang et al., 2015). This is attributed to their low price, abundance, reinforcement property, high stability, and easy processability.

Halloysite (HS) ($Al_2Si_2O_5(OH)_4 \cdot nH_2O$) is a widely used nanoclay and is a subclass of kaolin clay. It is a GRAS packaging material according to the FDA (Abdullayev et al., 2009) and is used extensively in the food packaging industry, owing to its availability, price, environment friendliness, and nonhazardous nature. HS is used to enhance the physical, barrier, and thermal properties of the biocomposite film by forming an exfoliated structure when added to a polymer matrix (Lee et al., 2018). It is observed that the HS could be exfoliated easily due to its shape and composition with a greater aspect ratio, while in kaolinite, it is required to exfoliate for a prolonged period because of its stacked platy shape (Abdullayev et al., 2012). HS consists of a negative silica outer surface and positive alumina inner lumen, which assists in the functionalization of the HS surface.

Nanocomposites can be intercalated and exfoliated depending on the degree of exfoliation of nanomaterials in the polymer matrix. As such, it is recommended that the nanoclay be properly exfoliated as individual platelets and uniformly distributed in the polymeric matrix to use the full potential of the large surface area of the nanoclay (de Abreu et al., 2010). HS, when dispersed adequately into a polymeric matrix, is able to reduce the permeability of the gas through the film surface and provide significant enhancement in barrier properties of the film. HS incorporation also resulted in the improvement of mechanical properties. Moreover, HS-based composites are preferred in active packaging due to their ability to release the desired active characteristics into the food system during long-term food storage (Huang et al., 2015).

4 Essential oils

Essential oils (EOs) are extracted from aromatic plants and are especially known for their antimicrobial and antioxidant activity besides certain reported health benefits, which include anticancer, antidiabetic, antianxiety, antiinflammatory, etc. (Anderson et al., 2016). EOs can be extracted from the plant parts, such as leaves, stem, bark, seeds, and flower. The quality and properties of the EOs are dependent on the plant characteristics, like type of soil used, climate of the region, and development stage while harvesting. However, it relies on the extraction procedure and the solvent used (Hussain et al., 2008; Khajeh et al., 2005).

The application of EOs in packaging and as a food additive is widely being researched due to their ability to replace synthetic additives when incorporated with polymers, as they positively impact the health of the consumer. They are incorporated in a manner in which the EOs release the compound over time into the headspace surrounding the food. Different types of EOs are obtained from various plants, such as *Salvia rosmarinus* (rosemary), *Melaleuca alternifolia* (tea), *Zingiber officinale* (ginger), *Thymus vulgaris* (thyme), *Cinnamomum verum* (cinnamon), *Syzygium aromaticum* (clove), *Brassica nigra* (mustard), *Lavandula angustifolia* (lavender) (Sharma et al., 2020).

EOs are incorporated into nanocomposite because they enhance the functionality of the nanocomposites. The primary function of EOs is to decrease the microbial population on the food surface, along with reducing the process of lipid oxidation, which results in the increased shelf life of food products (Ribeiro-Santos et al., 2017). However, the safety and the potential health benefits are heavily disputed by the FDA, which restricts its sale in the market as the claims of diagnosis, treatment, and disease prevention are yet to be approved by the FDA (Cotton High Tech SL, 2014). Lack of similar actions in their activity by the same EOs is one of the major problems of its usage. Despite multiple variations in the composition of EOs, there may be several qualitative and quantitative changes in the composition of a particular EO, resulting in dissimilar biological effectiveness. Furthermore, the presence of a strong aroma in EOs also limits its use in food applications, particularly meat (Negi, 2012).

5 Effect of the incorporation of nanoparticles and EOs on the properties of the nanocomposite packaging films

The incorporation of nanomaterials and EOs into the polymer matrix greatly affects the properties of nanocomposites. The improvement of the nanocomposite properties helps in the overall acceptability of the nanocomposite. These undesirable effects on the food product can be prevented by developing a food packaging material with high barrier, tensile, and antimicrobial properties. Additionally, nanocomposites incorporated with EOs successfully increase the shelf life of the packaged product due to the presence of active compounds, which inhibits the growth of the microorganism, prevents lipid oxidation, and prevents the premature decaying of food and change in texture, color, taste, and smell. Table 1 summarizes the reinforcements exhibited by the nanocomposites due to nanoparticle incorporation.

TABLE 1   Enhancement in nanocomposite properties due to incorporation of nanoparticles.

Nanoparticles	Load concentrations	Fabricating method	Plasticizer used	Enhanced properties	References
ZnO	0%–4%	Casting method	Glycerol	TS increased while EAB decreased; WVP increased; reduced transparency; thermal stability improved; high antimicrobial activity	Arfat et al. (2016)
	0.5%–1.5%	Casting method	Chloroform	TS and WVP improved significantly	Shankar et al. (2018b)
	0.5%, 1.0% curcumin and 1% carboxymethylcellulose	Casting method	Glycerol	Improved UV and moisture barrier properties; high antioxidant and antimicrobial activity	Roy and Rhim (2020)

TABLE 1 Enhancement in nanocomposite properties due to incorporation of nanoparticles—cont'd

Nanoparticles	Load concentrations	Fabricating method	Plasticizer used	Enhanced properties	References
TiO$_2$	0.5%–3%	Casting method	*Chloroform*	Flexibility of the film increased while Y decreased; UV barrier properties improved; OP and WVP decreased	Baek et al. (2018)
	1%, 3%, 7%	Casting method	Sorbitol	Increased hydrophobicity, melting point, and storage modulus increased; high antimicrobial activity	Salarbashi et al. (2018)
	TiO$_2$: cellulose nanofiber ratio 1:20	Casting method	–	Tortuosity of the film increased but the gas barrier performance is not affected, UV barrier property enhanced	Roilo et al. (2018)
SiO$_2$	0%–0.5%	Casting method	Glycerol	TS increased; gas barrier and moisture barrier properties increased; light transmittance decreased; UV barrier property improved	Zhang et al. (2019b)
	1%–5%	Casting method	Glycerol	TS increased; water absorption increased, solubility improved; water resistance and moisture barrier properties improved; glass transition temperature improved	Hassannia-Kolaee et al. (2016)
	1%–5%	Casting method	Glycerol and sorbitol	TS and Y increased, WVP and OP decreased; light barrier properties improved	Tabatabaei et al. (2018)
Ag	0.1, 1, 1.5 mL	UV irradiation In situ and solvent casting method	Glycerol	High antioxidant and antimicrobial activity	Hajji et al. (2017)
	0%–4%	Casting method	Glycerol	TS and mechanical improved; UV barrier property and transparency improved; thermal stability and glass transition temperature increased	Arfat et al. (2017a)

Continued

TABLE 1   Enhancement in nanocomposite properties due to incorporation of nanoparticles—cont'd

Nanoparticles	Load concentrations	Fabricating method	Plasticizer used	Enhanced properties	References
	0%–0.5%	Casting method	Glycerol	WVP and light transmittance decreased, water solubility increased	Lin et al. (2020)
Halloysite	0%–6%	Casting method	Chloroform	Mechanical, thermal, light, gas, and moisture barrier properties were enhanced	Risyon et al. (2020)
	0%–8%	Casting method	Glycerol	TS and EAB improved; moisture resistance increased; light barrier property improved	Huang et al. (2020)
	0–2 mL	Hot pressing	–	TS and EAB enhanced; light barrier property improved	Yang et al. (2020)

5.1 Mechanical properties

5.1.1 Zinc oxide (ZnO)

Metal oxide nanoparticles (nanoparticles) are commonly incorporated with polymers to improve the mechanical properties of the nanocomposite. It disperses uniformly into the polymer matrix, acting as a filler material, and places itself in the micropores of the polymer, reinforcing the film matrix in the process. The attributes that determine the mechanical properties of the nanocomposite are tensile strength (TS), elongation due to break (EAB), and Young's modulus (Y). In a study by Wang et al. (2019), they synthesized a chitosan (CS)-sodium alginate (SA)-carboxymethyl cellulose (CMC) film with ZnO nanoparticles at different concentrations (0.005%, 0.025%, 0.05%). They reported that the TS of the film increased considerably upon the incorporation of ZnO nanoparticles, while the EAB slightly reduced. CS-SA-CMC-ZnO (0.05%) nanocomposite was found to be the most optimum formulation for increasing the mechanical properties, where the TS increased to 24.38 MPa from 16.25 while the EAB reduced slightly to 25.32% from 28.77% when compared with pure CS films. The increase in TS is attributed to the increase in the interaction and formation of hydrogen bonding between positive CS and negative SA. At the same time, the EAB reduced mainly due to the restriction of the movement of the polymer molecules due to the formation of new bonds between CS and ZnO nanoparticles.

Similarly, Roy and Rhim (2020) prepared a CMC film incorporated with different concentrations of curcumin (0.5%, 1%) and ZnO (1%). They reported that compared to pure CMC films, the TS and Y of the nanocomposite increased to 49.3 MPa and 1.80 GPa, respectively, while the EAB decreased to 12.8% for CMC-Cur (0.5%)-ZnO (1%) nanocomposite. However,

Shankar et al. (2018b) reported that PLA-ZnO (1.5%) nanocomposite exhibited reduced TS, EAB, and EM as compared to PLA-ZnO (1.0%) nanocomposite. This is attributed to the increase in agglomeration of ZnO on the polymer surface creating defect on the film surface. Similar trends were reported by Rojas et al. (2019) for LDPE and Marvizadeh et al. (2017) for tapioca starch/bovine gelatin nanocomposite.

Upon the incorporation of rosemary EO (1%, 2%) along with ZnO nanoparticles (0.5%–2%) into sodium caseinate (SC) polymers, the TS and Y decreased. At the same time, the EAB increased regardless of the presence of ZnO due to the increase in heterogeneity, generation of irregularities in the polymer matrix, and reduction of interactions between the protein's chains. The TS and Y of the film decreased considerably to 2.09 MPa and 16.38 MPa for SC-ZnO (1%)-rosemary EO (2%) nanocomposite compared to control SC films. In comparison, the TS and Y of SC-ZnO (1%) nanocomposite were 4.74 MPa and 17.91%, respectively (Alizadeh-Sani et al., 2020a). A similar result was obtained by Ahmed et al. (2019) for polylactide/clove EOs and Ejaz et al. (2018) for gelatin/clove EO nanocomposite.

5.1.2 Titanium dioxide (TiO_2)

In a study by Siripatrawan and Kaewklin (2018), they synthesized a CS-TiO_2 nanocomposite with different concentrations of TiO_2 (0.25%–2.0%). They observed that the TS of the nanocomposite increased to 16.43 MPa, while the EAB reduced to about 55% for CS-TiO_2 (1%) nanocomposite compared to pure CS films. However, upon further increase in TiO_2 content, both the TS and EAB reduced considerably due to the increase in the agglomeration of excess TiO_2 on the nanocomposite surface. Formation of hydrogen bonds between the —NH and —OH groups of Ti and CS can also lower the EAB of the nanocomposite films. Similarly, Dash et al. (2019) fabricated a potato starch-lemon pectin film incorporated with TiO_2 (0.5%–4%). They reported that the TS and Y of nanocomposite increased with the increase in TiO_2 content, with the highest values reported at 4% TiO_2. However, the EAB of the film was lowest at 4% TiO_2, mainly due to the antiplasticization effect. Similar results were obtained by de Matos Fonseca et al. (2020) for hydroxypropyl methylcellulose/gelatin films and Noori Hashemabad et al. (2017) for LDPE films.

Sago starch-TiO_2 nanocomposite displayed an enhancement in the mechanical properties of the films as compared to pure starch films. However, after incorporating cinnamon EOs (1%–3%), the mechanical properties reduced as the EO concentration increased. The TS and Y of the films increased drastically and were maximum at 5% TiO_2 concentration. However, the incorporation of cinnamon EOs decreased the TS and Y to an extent, with the lowest mechanical properties at 3% cinnamon EO concentration. The opposite trend was observed for EAB, which increased when cinnamon EO was added to the nanocomposite (Arezoo et al., 2020). A similar trend was reported by Alizadeh-Sani et al. (2020b) for cumin EO incorporated into sodium caseinate/guar gum films.

5.1.3 Silicon dioxide (SiO_2)

PVA-xylan films incorporated with nano-SiO_2 (1%–4%) and nano-ZnO (1%–4%) exhibited an enhancement in the TS and EAB values. The TS increased to 22.5 MPa and 20.4 MPa when the SiO_2 and ZnO contents were 3%; EAB of both SiO_2 and ZnO nanocomposite at 3% concentration increased by 210% and 245%, respectively. Further increase in nanoparticle concentration resulted in inferior TS and EAB of the nanocomposite. The TS of the nanocomposite reduced mainly because of particle polymerization, elastic collision between nanoparticles, and phase separation. The EAB reduced due to the formation of a stable network of polymer with the nanoparticles, which eventually reduced the mobility of the polymer chains (Liu et al., 2019). Similarly, agar (AG)/SA nanocomposite films produced by incorporating SiO_2 exhibited a superior TS and EAB. The nanocomposite at 10% nano-SiO_2 displayed a 65.29% and 60.38% increase in TS and EAB, respectively, compared to pure AG/SA film. This was attributed to increased molecular interaction between the polymer and the nanoparticles (Hou et al., 2019). Similar results were reported by Tabatabaei et al. (2018) for gelatin and Wu et al. (2019c) for CS/curcumin films.

5.1.4 Ag nanoparticles

Cellulose, when incorporated with banana peel powder (BPP) as nanofiller and Ag nanoparticles, displayed superior mechanical properties when compared to pure cellulose films. The TS of the nanocomposite increased to 77.41 MPa from 67.46 MPa; the Y increased from 5.58 to 3.83 GPA. However, a drop in EAB was also observed due to nanoparticle incorporation, which decreased to a minimum of 3.68% from 7.92%. The improvement in TS and Y was mainly due to the homogenous dispersion and strong attraction among the components. Moreover, the presence of fibers, carbohydrates, proteins, and polyphenols is suggested to be the cause for the superior attributes of the nanocomposite (Thiagamani et al., 2019). A similar result was obtained by Arfat et al. (2017b) for guar gum and Dehghani et al. (2019) for LDPE films. Furthermore, it was reported by Chu et al. (2017) that incorporation of Ag exhibited better result in enhancing the mechanical properties as compared to ZnO.

In a study by Chi et al. (2018), high pressure (200, 400 MPa) was applied during the fabrication of polylactic acid (PLA)-Ag nanocomposite. The incorporation of nanoparticles improved the mechanical properties of the nanocomposite considerably as compared to pure PLA film, and a further improvement was reported when high pressure was applied. The TS and Y of the nanocomposite were maximum at 3% Ag nanoparticles and 400 MPa pressure, which decreased after the nanoparticle concentration was increased to 5%. Nevertheless, enhancement of TS and Y was noticed when pressure was increased. The reason for this is that high pressure increases the stiffness of nanocomposite by the generation of H-bond. Moreover, high pressure reduced the formation of agglomerates on the film surface due to good dispersibility. Inversely, a reduction of EAB was noticed when both the nanoparticle content and the pressure increased.

According to Ahmed et al. (2018b), the addition of cinnamon EO into LDPE-Ag-Cu polymer improved the EAB by increasing the nanocomposite's flexibility while the mechanical properties slightly reduced. This is attributed to the plasticization effect of the EOs.

5.1.5 Halloysite

Halloysite nanoclays incorporated with PLA exhibited an increase in the tensile strength and Young's modulus at 3% HS concentration, while at the same concentration, the EAB decreased. This was due to the high number of hydrogen bonds developed in the nanocomposites, which restricted the movements of PLA chains (Risyon et al., 2020). Similarly, CS and HS had a synergistic effect on the TS and EAB of agar nanocomposites; 40% CS and 8% HS concentration exhibited the highest TS and EAB. On further increase of CS and HS, the properties declined due to the formation of agglomerations on the surface of the polymer (Huang et al., 2020). Similar results were obtained for Mater-Bi films (Lisuzzo et al., 2020).

In a study by Biddeci et al. (2016), HS nanotubes, when added with peppermint EO and cucurbit in pectin matrix, showed improvement in the TS and Y of the film. Simultaneously, the EAB reduced due to the interaction between pectin and nanofiller, which prevented the sliding of the polymer chains. In another study by Lee et al. (2018), CS/HS films exhibited increased TS and slightly reduced EAB values, but after 10% HS concentration, the TS reduced drastically. The incorporation of clove EO further reduced the TS of the film as the clove EO disrupts the CS-HS matrix. EAB of the nanocomposites increased to 15% HS concentration, after which it declined because the CS matrix disrupted the hydrophobic interaction between clove EO molecules, leading to a reduced tendency of clove EO to agglomerate.

5.2 Moisture barrier properties

5.2.1 Zinc oxide (ZnO)

Moisture barrier properties determine the amount of moisture penetrating the packaging material, and it is measured by the parameter water vapor permeability (WVP) or water vapor transmission rate (WVTR). The reduction in moisture penetration is one reason for the incorporation of nanoparticles. Lower moisture penetration maintains the freshness of the packaged product and increases its shelf life. According to a study by Shahabi-Ghahfarrokhi et al. (2015), kefiran biopolymer, when incorporated with nano-ZnO at different concentrations (1%–3%), displayed a reduction in the WVP of nanocomposite as compared to pure kefiran films. The WVP had a positive effect on the increase in ZnO content, and the lowest WVP of $1.81 \times 10^{-10} \text{ g}^{-1} \text{ s}^{-1} \text{ Pa}^{-1}$ was observed at 3% ZnO content. This was primarily attributed to forming a tortuous pathway, which increases the mean path length for moisture penetration. Moreover, kefiran forms a hydrogen bond with the O atoms of ZnO, which reduces the diffusion of vapor through the films. Similarly, the WVP of ZnO-incorporated fish protein isolate-fish skin gelatin nanocomposite reduced continuously till 3% ZnO to 2.09 and $2.68 \times 10^{-11} \text{ g m}^{-1} \text{ s}^{-1} \text{ Pa}^{-1}$ at pH 11 and 3, respectively. Proper dispersion of ZnO at 3% concentration resulted in fewer paths in the matrix for the vapor to penetrate. However, at 4% ZnO, the WVP increased slightly to 4.19 and $4.17 \times 10^{-11} \text{ g m}^{-1} \text{ s}^{-1} \text{ Pa}^{-1}$ for 11 and 3 pH, which was attributed to the formation of agglomerates on the film surface, creating pores for the vapor to pass through (Arfat et al., 2016). Similar results were reported by Shankar et al. (2018b) for PLA and Jebel and Almasi (2016) for bacterial cellulose films.

PLA-PEG films, when incorporated with cinnamon EOs, had a negative impact on the WVP as it increased to 16.16 g mm^{-1} m^{-2} from 13.75 g mm^{-1} m^{-2}. Moreover, the WVP of PLA-PEG-ZnO-cinnamon EO film increased slightly to 2.97 g mm^{-1} m^{-2} from 1.47 g mm^{-1} m^{-2} of the PLA-PEG-ZnO nanocomposite. This has been concluded that the addition of EOs into the film matrix forms agglomerates on the film surface, increasing the WVP in the process (Ahmed et al., 2019). A similar result was obtained by Alizadeh-Sani et al. (2020a) for sodium caseinate/rosemary EO nanocomposite.

5.2.2 Titanium dioxide (TiO$_2$)

According to Zhang et al. (2019a), chitosan, when incorporated with TiO$_2$ nanoparticles, reduced the WVP of the nanocomposite to 5.82×10^{-11} g m^{-1} s^{-1} Pa^{-1} as compared to pure chitosan film. This is due to the blocking of micro paths in the polymer matrix by the water-insoluble TiO$_2$ nanoparticles. Interestingly, the addition of black plum peel extracts (BPPE) had a better effect than TiO$_2$, where the WVP reduced to 5.31×10^{-11} g m^{-1}s^{-1}Pa^{-1}. Moreover, when TiO$_2$ and BPPE were incorporated together with the chitosan polymer, the WVP further decreased to 5.12×10^{-11} g m^{-1}s^{-1}Pa^{-1}. The presence of heavy aromatic rings in the BPPE structure reportedly obstructs the internal network of the polymer and decreases the affinity toward moisture. Soybean polysaccharide films, when incorporated with nano-TiO$_2$ at different concentrations (1%–5%), exhibited a reduction in WVP with an increase in nanoparticle concentration. The lowest WVP of 4.44×10^{11} g m^{-1} s^{-1} Pa^{-1} was observed at 5% TiO$_2$, and this trend was mainly due to the introduction of tortuosity by incorporating nano-TiO$_2$ (Shaili et al., 2015). A similar result was obtained for fish gelatin-agar films (Vejdan et al., 2016). However, for gelatin-TiO$_2$ nanocomposite, the WVP decreased with an increase in TiO$_2$ concentration. However, at a higher concentration of TiO$_2$, the WVP increased due to uneven dispersion of nanoparticles and formation of agglomerates, which promoted the permeation of moisture (He et al., 2016).

The incorporation of cinnamon EO has a negative impact on the WVP of the sago starch film (Arezoo et al., 2020). The study revealed that the incorporation of TiO$_2$ reduced the WVP of the film significantly. However, after adding cinnamon EO into the matrix, the WVP increased slightly with an increase in cinnamon EO concentration. Though the increase in cinnamon EO content increased the WVP of the nanocomposite film, it was not enough to overcome the positive effects of TiO$_2$ regarding water permeability. A similar result was obtained by Alizadeh-Sani et al. (2020a) for sodium caseinate-cumin EO films.

5.2.3 Silicon dioxide (SiO$_2$)

The WVP of pure soybean polysaccharide films was 7.96×10^{11} g m^{-1}s^{-1}Pa^{-1}, which decreased considerably upon the incorporation of SiO$_2$ nanoparticles. The nanoparticles were incorporated in different concentrations (1%–3%), and it was reported that nanocomposite having 5% SiO$_2$ concentration exhibited the lowest WVP of 4.75×10^{11} g m^{-1} s^{-1} Pa^{-1} and highest tortuosity value of 2.16. This meant that the mean path length covered by the water molecule to penetrate the film would be longer in the case of 5% TiO$_2$, resulting in lower WVP (Ghazihoseini et al., 2015). A similar trend was reported for whey protein isolate-pullulan films (Hassannia-Kolaee et al., 2016) and ethylene vinyl acetate (Li et al., 2016) films. In a study by Xu et al. (2019b), the WVP of pure soy protein and soy protein-nano-silica nanocomposite over 30 days of storage was recorded. The WVP of all the test film of different compositions increased over the storage period. These results may be due to the migration of glycerol and moisture, which disturbs the polymer matrix during storage and increases the stiffness among protein chains. After incorporating nano-SiO$_2$, the WVP of the film decreased considerably, and the increase in WVP over the storage period was also less. This was attributed to the formation of a hydrogen bond between soy protein and SiO$_2$, which resists the diffusion of moisture into the polymer matrix. Also, the formation of the denser matrix due to intermolecular attraction between nano-SiO$_2$ improved the storage stability of the nanocomposite.

5.2.4 Ag nanoparticles

PLA-lignin films, when incorporated with Ag nanoparticles at 0.5% and 1.0% concentration, reduced the WVP of the nanocomposite as compared to the control PLA film. However, the WVP increased slightly at 1% Ag concentration. Furthermore, it was observed that WVP of PLA-lignin film without any Ag (2.36×10^{-11} g m^{-1}s^{-1}Pa^{-1}) showed better results than pure PLA (2.93×10^{-11} g m^{-1}s^{-1}Pa^{-1}) and PLA-lignin-Ag film (2.42×10^{-11} g m^{-1} s^{-1} Pa^{-1} for 0.5% Ag; 2.45×10^{-11} g m^{-1} s^{-1} Pa^{-1} for 1.0% Ag). This was mainly attributed to the hydrophobic nature of lignin, which forms a tortuous pathway (Shankar et al., 2018a). TEMPO-oxidized nanocellulose (TNC), when incorporated with grape seed extracts (GSE), exhibited higher WVP than control TNC due to the hydrophilic GSE, which absorbs more moisture from the atmosphere. The incorporation of Ag into the TNC-GSE nanocomposites decreased the WVP significantly due to the filling of the micropores and defects of the TNC polymer obstructing the diffusion of water vapor (Wu et al., 2019a). A similar result was obtained for CS films incorporated with Ag immobilized in laponite (Wu et al., 2018). Notably, in a study by Chu et al. (2017), it was reported that PLA film incorporated with Ag had lower WVP than PLA-ZnO films. It was suggested that better dispersibility of Ag nanoparticles assisted in lowering the WVP of the nanocomposites.

In a study by Ahmed et al. (2018b), the incorporation of cinnamon EOs had a negative impact on the WVP of LLDPE-Ag-Cu nanocomposites. Ag nanoparticles, when incorporated into LLDPE film at different concentrations of 2% and 4%, lowered the WVP considerably to 0.30 and 0.19 g mm m^{-2} d atm, respectively. Cinnamon EO, when added to LLDPE-4% Ag nanocomposites, increased the WVP to 0.65 mm m^{-2} d atm. This was attributed to the hydrophilic nature of cinnamon EO and the effect on the cohesion forces with the nanocomposite matrix. A similar result was reported by Ahmed et al. (2018a) for sodium caseinate/rosemary EO nanocomposites.

5.2.5 Halloysite

PLA films, when incorporated with HS at different concentrations (1.5%–6%), exhibited a reduction of WVP to 1.1 from 1.4×10^{-5} g mm m^{-2} d Pa till 3% HS, following which the WVP decreased. Higher concentrations of HS formed agglomerates on the surface of the matrix, which hindered the formation of a tortuous pathway (Risyon et al., 2020). Similarly, LDPE, when incorporated with 1% alkaline HS nanotubes, reduced the WVP slightly to 12.10 from 12.20 g m^{-2} day^{-1}, while at 3 and 5% alkaline HS, the WVP increased (Boonsiriwit et al., 2020).

CS biopolymers, when incorporated with clove EO and HS nanotubes, reduced the WVP of the nanocomposite considerably (Lee et al., 2018). They observed that the clove EO-incorporated nanocomposite displayed lower WVP as compared to CS-HS nanocomposite. It is suggested that clove EO incorporation into the nanocomposite resulted in forming H-bonding between the functional groups of CS hydroxyl and amino groups and the phenolic compounds of the clove EO. Similarly, the addition of *Origanum vulgare* EO into the CS/CMX matrix improved the WVP of the nanocomposite (Yousefi et al., 2020).

5.3 Gas barrier properties

5.3.1 Zinc oxide (ZnO)

Gas barrier properties determine the quantity of gas diffusing through the polymer and is quantified by mostly oxygen permeability (OP) and oxygen transmission rate (OTR).

Exposure to oxygen is the main reason for the deteriorations, including enzymatic browning, oil and fat rancidity, and microbial growth of food products, which many packaging systems try to exclude for extending the shelf life of food products. Tapioca starch-bovine gelatin films, when incorporated by nano-ZnO at different concentrations (0.5%–3.5%), positively impacted the oxygen permeability of the nanocomposite. The OP decreased as the ZnO content increased, and the lowest OP of 91.52 cm^3 μm (m^{-2} day^{-1}) was observed at 3.5% ZnO films. This considerable decrease in OP due to ZnO incorporation is because of the formation of a tortuous pathway in the film matrix, making it difficult for the oxygen atoms to penetrate (Marvizadeh et al., 2017). Similar result was obtained for LDPE-polyurethane films (Ahmed et al., 2017). In a study by Marra et al. (2016), they reported a reduction in the OP of PLA film till 3% ZnO content. However, the CO_2 permeability decreased with an increase in ZnO content. The permeability reduction of both O_2 and CO_2 is mainly due to the homogeneous dispersion of ZnO and the formation of tortuous pathway. Similar result was obtained by Ngo et al. (2018) for pectin-alginate films.

According to Ejaz et al. (2018), the OP of the pure gelatin film and films incorporated with ZnO nanorods were adversely affected by the addition of clove EOs. The OP of neat gelatin film was 0.691 cc mm m^{-2} d atm, which increased to 1.13 cc mm m^{-2} d atm on the incorporation of 50% clove EO. Expectedly, the incorporation of ZnO improved the OP of neat gelatin films to 0.468 cc mm m^{-2} d atm; however, addition of 50% clove EO increased it to 0.926 cc mm m^{-2} d atm. The formation of defect and aggregates on the polymer surface is considered as the prime reason for this behavior. A similar result was obtained by Petchwattana et al. (2016) for polybutylene succinate films.

5.3.2 Titanium dioxide (TiO₂)

Soybean polysaccharide, when incorporated with TiO_2, had a positive effect on the OP of the film. The OP of the pure polysaccharide film was 202.14 cm^3 μ mm^{-2} d^{-1} atm^{-1}, which reduced to 98.45 cm^3 μ mm^{-2} d^{-1} atm^{-1} for 5% TiO_2-incorporated nanocomposites (Shaili et al., 2015). Similarly, PLA film incorporated with TiO_2 modified with oleic acid reported decreased OP by 29% compared to pure PLA films (Baek et al., 2018). This occurs mainly due to the increase in tortuosity in the polymer matrix due to the homogenous dispersion of TiO_2 into the polymer matrix. Similar results were reported by Bodaghi et al. (2015) for LDPE films.

The incorporation of cinnamon EO has a negative effect on the OP of the sago starch film (Arezoo et al., 2020). The study reported that the incorporation of TiO_2 reduced the OP of the film significantly. However, addition of cinnamon EO into the matrix increased the OP slightly with the increase in cinnamon EO concentration. The lowest OP was observed for starch-5% TiO_2 nanocomposite. The incorporation of the cinnamon EO increased the OP of the nanocomposite, though it was considerably better than pure sago starch film.

5.3.3 Silicon dioxide (SiO₂)

Gelatin-k-carrageenan films, when incorporated with different concentrations of nano-SiO_2 (1%–5%), reduced the OP of the nanocomposite film. The OP of nanocomposite having 5% SiO_2 displayed the lowest OP of 97.23 cc μm m^{-2} day^{-1} atm^{-1} (Tabatabaei et al., 2018). Similarly, PLA films, when incorporated with nano-silica modified with silane, reported a decrease in O_2 and CO_2 permeability as the quantity of silane increased. The O_2 and CO_2

permeabilities decreased by almost 80% and 50%, respectively, compared to control PLA films (Ortenzi et al., 2015). This has been attributed to the good dispersion of SiO_2 and the reduction of free volume in the polymer matrix. Similar result was reported by Venkatesan and Rajeswari (2019) for PBAT films. In a comparative study by Liu et al. (2019), the effect of nano-SiO_2 and nano-ZnO on the OP of the nanocomposite was calculated, and it was reported that SiO_2 had a better impact on the OP of the film than nano-ZnO due to better dispersion of SiO_2. Additionally, a study by Xu et al. (2019b) showed that the incorporation of nano-silica into soy protein isolate films reduced the OP of the film considerably and lowered the increase of OP during 30 days of storage.

5.3.4 Ag nanoparticles

A study by Arfat et al. (2017b) reported that the addition of Ag-Cu nanoparticles into pure guar gum film resulted in the reduction of OP by 56.1%. The OP of the nanocomposite reduced with the increase in nanoparticle concentration and was lowest at 2% Ag-Cu. Similarly, the CO_2 and O_2 permeabilities of pure PLA films decreased drastically when incorporated with Zn and Cu-Ag nanoparticles. The CO_2 and O_2 permeabilities reduced to 230 and 97 mL m^{-2} day^{-1} from 873 and 1308 mL m^{-2} day^{-1}, respectively, for PLA-ZnO-Cu-Ag (0.5%) nanocomposites (Vasile et al., 2017). Similar result was reported by Wu et al. (2018) for CS films and Wu et al. (2019a) for TNC-GSE films.

For PLA-PEG films, the OP reduced considerably after incorporating Ag-Cu nanoparticles at 2% and 4% nanoparticle concentrations. However, addition of cinnamon EO negatively impacted the OP. The OP values were higher than the control films for PLA-PEG-50% cinnamon EO-4% Ag-Cu nanocomposite. This is attributed to the disruption of the PLA network due to cinnamon EO incorporation, resulting in the higher permeation of O_2 molecules (Ahmed et al., 2018a). Similar result was reported by Ahmed et al. (2018b) for LDPE films incorporated with Ag-Cu and cinnamon EOs.

5.3.5 Halloysite

PLA films, when incorporated with 3% HS nanotubes, showed an improvement of 33% toward OP of the nanocomposite (Risyon et al., 2020). Similarly, OTR and WVTR for LDPE film incorporated with alkaline HS at different concentrations (1%–3%) decreased significantly. The lowest OTR was observed for 1% alkaline HS films, which reduced to 4505.67 from 5380 cm^3 m^{-2} day^{-1} of that of the control film (Boonsiriwit et al., 2020). A similar result was obtained for potato starch films, where the OP decreased with an increase in HS concentration (Sadegh-Hassani and Nafchi, 2014). The enhancement in barrier activity (gas and moisture) in nanocomposites compared to traditional polymers is illustrated in Fig. 3.

5.4 UV barrier properties

5.4.1 Zinc oxide (ZnO)

UV protection is one of the most crucial parameters relating to food packaging as it restricts the penetration of UV light through the packaging material. Exposure to UV light results in lipid oxidation, loss of color, nutrient loss, and food deterioration. In a study by Roy and Rhim (2020), carboxymethylcellulose (CMC) nanocomposite films were produced by incorporating

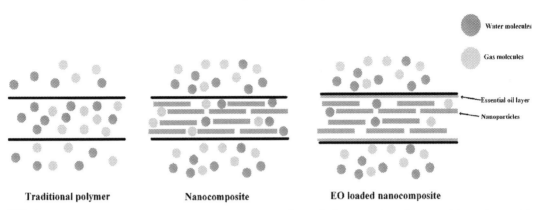

FIG. 3 Formation of tortuosity in the nanocomposite matrix due to the incorporation of nanoparticles and essential oil. *Modified from Duncan, T.V., 2011. Applications of nanotechnology in food packaging and food safety: barrier materials, antimicrobials and sensors. J. Colloid Interface Sci. 363(1), 1–24.*

ZnO and curcumin into the polymer matrix. They reported that as compared to pure CMC film, the addition of curcumin reduced the light penetration through the packaging material. The curcumin extracts were added at different concentrations of 0.5% and 1.0%, and both the nanocomposites exhibited lower UV transmittance at 280 and 660 nm. Furthermore, the addition of ZnO had a synergistic effect on the UV barrier properties, which decreased with an increase in nanofiller concentration. The lowest UV transmittance was observed for CMC-Cur 1%-ZnO 1% at 280 and 660 nm, which were 0.7% and 42.8%, respectively. It is primarily due to the obstruction of light passage by ZnO and absorption of light rays by curcumin and ZnO synergistically. Similar results were obtained by Noshirvani et al. (2017a) for CS-CMC-oleic acid (Valerini et al., 2018) for aluminum-doped ZnO with PLA and Heping Wang et al. (2019) for CS-SA films.

Neem EO, when incorporated into CS-ZnO nanocomposite, increased the transparency of the film to 3.699 from 2.312. However, the addition of ZnO greatly increased transparency as compared to neem EO when incorporated separately. The highest transparency of 5.40 can be observed for CS-ZnO (0.5%) without neem EO incorporation. The high dispersion of ZnO is suggested to be the reason behind this effect, while the neem EO makes the nanocomposite slightly semi-transparent (Sanuja et al., 2015). Similarly, ZnO and rosemary EOs have a synergistic impact on the UV barrier properties of sodium caseinate films, where the % transmission at 600 nm decreased with an increase in ZnO and rosemary EO concentration. The lowest T_{600} of 42.23 was observed for SC-ZnO (2%)-REO (2%) (Alizadeh-Sani et al., 2020a). A similar result was obtained for clove EO-incorporated gelatin films (Ejaz et al., 2018).

5.4.2 Titanium dioxide (TiO₂)

In a study by Ahmadi et al. (2019), light transmittance decreased considerably when CMC polymer was incorporated with TiO_2 and *Miswak* extracts. The T_{280} and T_{600} for pure CMC films were 64.34% and 92.43%, respectively, while the opacity was 0.43. After incorporating 2% TiO_2, the transmittance further reduced, and the opacity increased due to the introduction of crystallinity and homogenous dispersion of TiO_2 into the polymer matrix. The incorporation of *Miswak* extracts synergistically enhanced the UV barrier properties of the nanocomposites. CMC-2% TiO_2 with 450 mg m^{-1} *Miswak* showed 100% and 60% UV blocking at 220 nm and 600 nm, respectively. Similarly, the % transmittance of pure CS films was about 90%, which decreased sharply after TiO_2 incorporation at different concentrations (0.25%–2%). The lowest transmission was reported for CS-2% TiO_2 films due to higher crystallinity and opacity. Low optical clarity is mainly due to the generation of aggregates on the film surface (Siripatrawan and Kaewklin, 2018). Similar results were obtained for gelatin-agar (Vejdan et al., 2016), high-amylose (Liu et al., 2015), and polyvinyl chloride films (Krehula et al., 2017).

Transparency of pure sodium caseinate film was 21.4, which reduced considerably upon incorporating guar gum (GG), TiO_2, and cumin EO. GG and cumin EO reduced the transparency slightly. At the same time, addition of TiO_2 decreased the transparency considerably. The UV transmittance and transparency of sodium caseinate-guar gum film decreased with the increase in the concentration of TiO_2 and cumin EO. The lowest transparency value of 14.3 was observed for SC-GG-2% TiO_2-2% cumin EO films (Alizadeh-Sani et al., 2020a).

5.4.3 Silicon dioxide (SiO_2)

The incorporation of 5% nano-SiO_2 was able to block the UV rays better than pure gelatin-carrageenan. The T_{280} values of pure gelatin-carrageenan films were about 50%, whereas it reduced to about 10% for gelatin-carrageenan films incorporated with 5% nano-SiO_2. Likewise, the T_{660} value of pure gelatin-carrageenan films was 70%, which reduced to around 40% for 5% nano-SiO_2 concentration. This is attributed to the high aspect ratio and aggregated morphology of SiO_2, which promotes the absorption of UV rays (Tabatabaei et al., 2018). Similarly, SiO_2, when added to agar/cellulose film, decreased the light transmittance of the nanocomposite. In the UV region, the transmittance decreased with the increase in SiO_2 and was attributed to the homogenous dispersion of SiO_2 in the polymer matrix. Furthermore, the opacity of the control film was 0.04, which increased with the increase in TiO_2 concentration (2.5%, 5%, 7.5%, 10%), with the highest opacity observed at 10% SiO_2 (Hou et al., 2019). A similar result was obtained for alginate films (Yang et al., 2016). However, Hassannia-Kolaee et al. (2016) and Liu et al. (2019) reported no significant change in transmittance and transparency in SiO_2 films due to the small size of SiO_2 nanoparticles that is smaller than the wavelength of light in the visible range, allowing light to pass through the film surface easily.

5.4.4 Ag nanoparticles

In a study by Lin et al. (2020), fish gelatin (FG)-CS films, when incorporated with TiO_2-Ag nanoparticles at different concentrations (0.1%–0.5%), showed a decrease in light transmission as compared to pure FG-CS films. The %T for pure FG-CS was 87.9%, whereas it decreased drastically with the lowest %T of 54.6 reported at 0.5% Ag nanoparticle concentration. This was attributed to the high refractive index and large specific surface area of TiO_2-Ag, which help in light scattering. Similarly, agar-alginate-collagen composite films, when added with Ag nanoparticles, reduced the % transmission at wavelengths of 280 and 660 nm. The T_{280} and T_{660} values for control film were 39.1% and 87.1%, respectively, which reduced to 12.2% and 30.9% upon Ag incorporation. Furthermore, the addition of grape seed extract decreased the T_{280} to 8.1%. On the other hand, the T_{660} value remains unchanged and would make for excellent transparent packaging material while shielding the product against UV rays (Wang and Rhim, 2015). Agar-lignin film showed around 100% blocking of light rays of 220nm wavelength when incorporated with Ag nanoparticles with different concentrations (0.5%–2%) (Shankar and Rhim, 2017). Similar result was reported for fish skin gelatin nanocomposite films (Arfat et al., 2017a).

Control PLA films displayed inferior UV-blocking properties with %T values of 85.38%, 48.63%, and 70.14% at 600, 280, and 350nm, respectively. After adding Ag-Cu or cinnamon EO into the polymer matrix, the %T value of the nanocomposite reduced considerably. The %T value decreased with the increase in Ag-Cu and cinnamon EO concentration having 0% $T_{200,280,350}$ value at 4% Ag-Cu and 50% cinnamon EO. This was attributed to the light-scattering property of Ag-Cu, and the droplets of cinnamon EO dispersed into the film matrix (Ahmed et al., 2018a). A similar result was obtained for LDPE/cinnamon EO films (Ahmed et al., 2018b).

5.4.5 Halloysite

Agar films, when incorporated with HNT and CS, exhibited a reduction in the UV transmission at 4 and 8% HNT concentration as compared to neat agar film and agar-CS films. This was attributed to the formation of aggregates at higher HNT concentrations (Huang et al., 2020). Similarly, HS, when incorporated with cellulose nanofibers (CNF), displayed superior UV-blocking properties. The light transmission decreased as the nanofiller concentration increased due to the light-scattering properties of the nanofillers. The transmittance of control film at 800nm was 3.4%, which reduced to 1.5%, 0.75%, 0.5%, and 0.3% for 0.5, 1, 1.5, and 2.0mL concentration of HS, respectively (Yang et al., 2020). Similar results were obtained for PVA-starch-HNT films (Abdullah and Dong, 2019).

5.5 Antimicrobial properties

5.5.1 Zinc oxide (ZnO)

Some nanoparticles exhibit strong antimicrobial activity against food-borne and spoilage microorganism, which deteriorates the food quality making it unconsumable. The nanoparticles and EOs, when added as food additives, get absorbed by the microorganism, which penetrates the cell wall and destroys the cytoplasm resulting in cell lysis and inhibition of the microorganism. ZnO shows an antagonistic effect against microorganisms by the generation of ROS, direct contact of ZnO with the microorganism, and creation of hydrogen peroxide, which disintegrates the cell membrane (Marra et al., 2016). They reported that the reduction (%R) of bacterial population is a function of the contact area and quantity of nanoparticles; as expected, pure PLA film did not display any antimicrobial activity, while the PLA-5% ZnO nanocomposite exhibited the highest reduction of 99.99% at 24h for *E. coli*, while the nanocomposite with 1% and 3% concentration showed 99.99% reduction in 5days.

In a study by Liu et al. (2020), soluble soybean polysaccharide (SPSS) films were incorporated with nano-ZnO and myofibril cellulose (MFC), and antimicrobial activities have been reported. Pure SPSS film and SPSS-10% MFC films did not show any antimicrobial activity. Upon addition of nano-ZnO at 10% concentrations, both SPSS and SPSS-MFC (10%) showed antimicrobial activity against both *E. coli* and *Staphylococcus aureus*. The mean values of the zone of inhibition (ZOI) for SPSS-10% ZnO film were 3.1 and 3.5 mm for *Staphylococcus aureus* and *E. coli*, respectively. Similarly, pectin/agar films, when added with different concentrations of ZnO (0.5%, 2.5%, 5%, 25%), exhibited antimicrobial activity against *Aspergillus niger*, *Colletotrichum gloeosporioides*, *E. coli*, and *Saccharomyces cerevisiae*. Pure pectin-alginate films did not possess any antimicrobial activity. The ZOI increased with the increase in ZnO concentration; however, at 25% ZnO, the ZOI did not increase significantly due to a decrease in the diffusion of ZnO into the agar matrix at higher concentrations (Ngo et al., 2018). The mechanism of antimicrobial activity of ZnO is proposed to be the presence of ROS species, penetration of Zn^{2+} ions through the cell wall, and accumulation of nano-ZnO on the surface of the microorganism. Similar results were obtained for carrageenan (Oun and Rhim, 2017) and gelatin (Shankar et al., 2015) films.

Similarly, after loading clove EO into the PLA-ZnO nanocomposite, the antimicrobial activity enhanced as the concentration of EOs increased. The presence of phenolic substance and volatiles released from clove EO is attributed to the high antimicrobial action, which penetrates the cell membrane to rupture the cell (Ahmed et al., 2019). Similar result was reported for gelatin-clove EO films (Ejaz et al., 2018) and CS-*Melissa officinalis* EO (Sani et al., 2019).

5.5.2 Titanium dioxide (TiO$_2$)

TiO$_2$, when added to pure PVA films, showed great antimicrobial activity against *E. coli* and *Staphylococcus aureus*. TiO$_2$ was incorporated with PVA at different concentrations (1%–5%), and the highest ZOI was observed for PVA-TiO$_2$ (4%) nanocomposite, which was 13.34 mm and 11.94 mm for *E. coli* and *Staphylococcus aureus*, respectively. The high antimicrobial activity of TiO$_2$ is attributed to the release of electron from the TiO$_2$ surface, which reacts with O$_2$ to form superoxide ions, resulting in DNA damage and lipid oxidation, ultimately killing the bacteria (Liu et al., 2015). Similarly, for CS films incorporated with black plum peel extracts and TiO$_2$, high antimicrobial activity was observed. Pure CS films reported the lowest antimicrobial activity with the least ZOI, while it improved upon the incorporation of TiO$_2$ due to the generation of ROH, which inhibits bacterial growth by the oxidization of polyunsaturated lipids present in the cell wall. Moreover, CS-BPPE-TiO$_2$ nanocomposite showed the highest activity against microbes due to the presence of several phenolic compounds in the films, along with the synergistic effect of BPPE and TiO$_2$ (Zhang et al., 2019a). Similar results were reported for SPSS (Shaili et al., 2015) and gelatin films (He et al., 2016).

Whey protein isolate-cellulose nanofibers nanocomposite, when incorporated with 1% TiO$_2$ and 2% rosemary EO, reduced the bacterial population of *E. coli*, *Listeria monocytogenes*, and *Staphylococcus aureus* significantly while maintaining the sensory qualities of lamb meat. The shelf life of nanocomposite-packaged lamb meat increased to 15 days, while that of the control meat was only 6 days (Sani et al., 2017). Similar result was obtained for sago starch film incorporated with cinnamon EO, where the ZOI was the highest at 5% TiO$_2$ and 2% cinnamon EO (Arezoo et al., 2020). The high reduction in bacterial population is attributed to the hydrophobic nature of EOs, which aided in destabilizing the cell functioning along with TiO$_2$ leading to cell inhibition.

5.5.3 Silicon dioxide (SiO₂)

Very few studies have been done regarding the antimicrobial property imparted to the nanocomposite due to SiO$_2$ incorporation. In a study by Wu et al. (2019c), CS films were incorporated with curcumin, and nano-silica and their antimicrobial activity against *E. coli* and *Staphylococcus aureus* were studied. They reported that the antimicrobial activity of CS-curcumin films was higher than the CS-curcumin-silica films. This was due to the controlled release of curcumin from the films because of SiO$_2$ incorporation

5.5.4 Ag nanoparticles

The antimicrobial activities of pure fish skin gelatin were enhanced by incorporating Ag-Cu nanoparticles at different concentrations (0.5%–4%). At 0.5% Ag-Cu concentration, the nanocomposite did not show any antimicrobial activity, probably due to low Ag-Cu concentration. The highest log reduction for *Salmonella typhimurium* was observed for FSG-4% Ag-Cu nanocomposite, although for *Listeria monocytogenes*, it was at 3% Ag-Cu. It was attributed to the release of Ag or Cu ions that penetrated the pits and gaps on the cell membrane and reacted with the sulfhydryl group or disulfide group leading to cell inhibition (Arfat et al., 2017a). Similarly, Hajji et al. (2017) studied the antimicrobial activity of CS-PVA films incorporated with Ag nanoparticles on a wide variety of gram-positive and gram-negative bacteria. They reported that nanoparticle-incorporated films showed more bacterial inhibition as compared to pure CS films, while PVA films did not show any bacterial inhibition. However, out of the tested bacterial strains, *Klebsiella pneumoniae* was immune toward the antimicrobial action of the nanocomposite. High antibacterial activity was reported for guar gum (Arfat et al., 2017b) and agar/alginate/collagen films (Wang and Rhim, 2015). Ag nanoparticles, when incorporated with ZnO, showed a synergistic effect against bacterial strains rather than nanocomposite with one type of nanoparticle (Chu et al., 2017).

Similarly, like other nanoparticles, Ag, when added with EOs, reported higher antimicrobial activity. The volatile compounds and phenolic substances present in the EOs assist the Ag nanoparticles in showing synergistic inhibition effect against microorganisms. Ahmed et al. (2018b) found the lowest log$_{10}$ CFU/g values for LDPE films when incorporated with 4% Ag-Cu and 50% cinnamon EOs. A similar result was reported for CS-laurel EO (Wu et al., 2019b) and CS-thyme-clove EO (Cinteza et al., 2018).

5.5.5 Halloysite

Pectin and HS nanocomposites loaded with salicylic acid exhibited antimicrobial properties against *E. coli*, *Pseudomonas aeruginosa*, *Staphylococcus aureus*, and *Salmonella* with ZOI value of 19.0, 16.0, 23.0, and 19.0mm, respectively (Makaremi et al., 2017). *O. vulgare* EO, added to CS-CMX and HS nanocomposite, imparted an antimicrobial property to the nanocomposite due to the presence of carvacrol and thymol. For EO-incorporated films, the highest antibacterial efficacy was approximately 97% and 95% for *E. coli* and *Staphylococcus aureus*, respectively (Yousefi et al., 2020). Similarly, HNT, when loaded with peppermint EOs, exhibited approximately 85% and 38% inhibition of *E. coli* and *Staphylococcus aureus* population, respectively, at 65°C incubation temperature (Biddeci et al., 2016).

A summary of the antimicrobial activity of different nanocomposites is provided in Table 2.

TABLE 2 Antimicrobial properties of nanocomposites.

Nanoparticles	Polymer matrix	Microorganisms	Microbial activity method	Results	References
ZnO	Gelatin Chitosan Nanofiber	*E. coli* *Staphylococcus aureus* *P. aeruginosa*	Disk diffusion method	ZOI was 30.62 mm, 15.06 mm, 10.7 mm for *Staphylococcus aureus, E. coli, P. aeruginosa* ZOI was 33.13 and 25.06 mm, 12.95 mm for *Staphylococcus aureus, E. coli, P. aeruginosa*	Amjadi et al. (2019)
	Fish protein isolate/fish skin gelatin	*Listeria monocytogenes* *P. aeruginosa*	Disc diffusion method	Highest ZOI for *Listeria monocytogenes* and *P. aeruginosa* was 25 and 22 mm, respectively	Arfat et al. (2016)
	Soluble soybean polysaccharide	*E. coli* *Staphylococcus aureus*	Disc diffusion method	ZOI values for *E. coli* and *Staphylococcus aureus* were 2.4 and 1.9 mm	Liu et al. (2020)
	Polylactic acid	*E. coli* *Listeria monocytogenes*	Colony count method	No bacterial colonies were present after 10 days	Shankar et al. (2018b)
TiO$_2$	Chitosan	*E. coli* *Staphylococcus aureus* *P. aeruginosa* *Aspergillus* *Penicillium*	Growth inhibition calculation method	UV-treated films showed greater microbial inactivation; highest growth inhibition was observed for *Staphylococcus aureus* at about 55% for UV-treated films	Siripatrawan and Kaewklin (2018)
	Zein/chitosan	*E. coli* *Salmonella enteritidis* *Staphylococcus aureus*	Disk diffusion method	Samples under UV irradiation showed greater antibacterial inactivity; highest ZOI of *E. coli, Salmonella enteritidis, Staphylococcus aureus* were 32.71, 30.06, 38.02 mm, respectively	Qu et al. (2019)
	Sodium caseinate/ guar gum/ cumin essential oil	*Staphylococcus aureus* *Listeria monocytogenes* *E. coli* *Salmonella enteritidis*	Disk diffusion method	Highest ZOI was observed for essential oil-incorporated films. *Listeria monocytogenes* was the most effected followed by *Staphylococcus aureus. E. coli,* and *Salmonella enteritidis*	Alizadeh-Sani et al. (2020b)

Continued

TABLE 2   Antimicrobial properties of nanocomposites—cont'd

Nanoparticles	Polymer matrix	Microorganisms	Microbial activity method	Results	References
SiO$_2$	Chitosan/ curcumin	E. coli Staphylococcus aureus	Disk diffusion method	Highest ZOI values *for* E. coli and Staphylococcus aureus were about 11 and 8 mm, respectively	Wu et al. (2019c)
Ag	Chitosan-poly (vinyl alcohol)	Staphylococcus aureus Bacillus cereus M. luteus Salmonella enterica E. coli Salmonella typhimurium P. aeruginosa K. pneumoniae	Disk diffusion method	B. cereus and P. aeruginosa Showed the highest ZOI of 16 mm; K. pneumoniae showed no ZOI	Hajji et al. (2017)
	Fish skin gelatin	Salmonella typhimurium Listeria monocytogenes	Log reduction method	Log (CFU ml^{-1}) of Salmonella typhimurium decreased to 0 after 7 days; log (CFU ml^{-1}) of Listeria monocytogenes reduced to 6.5 after 7 days	Arfat et al. (2017a)
	LLDPE/ cinnamon essential oil	Salmonella typhimurium Listeria monocytogenes Campylobacter jejuni	Log reduction method	Log reduction till 0 was observed from 6, 14, and 20 days for Salmonella typhimurium, Listeria monocytogenes, and Campylobacter jejuni, respectively	Ahmed et al. (2018b)
Halloysite	Pectin/ peppermint essential oil	E. coli Staphylococcus aureus	Cell viability method	Percentage of cell viability was highest at 65°C with around 15% and 37% of E. coli and Staphylococcus aureus surviving, respectively	Biddeci et al. (2016)
	Polyethylene/ carvacrol	Aeromonas hydrophila	Plate count and cell viability method	Reduced the cell viability to 20%; the plate count was reduced to 1.5 × 10^4 CFU cm^{-2}	Tas et al. (2019)

6 Regulatory issues

With the increase in the use of nanoparticles in food products, their safety evaluation has become a prime concern. Nanoparticles have been established in several applications due to their high reactivity, organ/cellular infiltration, and superior bioavailability. These distinctive attributes contribute to several advantages in drug delivery, polymer reinforcement, and active packaging areas. However, the smaller particle size (1–100 nm) can easily adhere to human skin, making them susceptible to respiratory and cardiac diseases at irregular usage of nanoparticles. Therefore, appropriate restraint should be implemented to impart nanoparticles in food processing and packaging applications. Regulations, which specify the use of nanoparticles in food packaging, are not enacted yet. However, there are existing laws regulating conventional food products, which are coming in contact with nanoparticles. Only a few regulations refer directly to nanoparticles, and most importantly, they vary among geographical regions.

6.1 European Union

The European Union (EU) has established the most detailed regulations and technical guidance for nanoparticle inclusion in food products. The EU Commission published a report in 2010 on "Towards a Strategic Nanotechnology Action Plan (SNAP) 2010–2015," where the objective was to invite different scientific views on the requirement of nanotechnology in the following 5 years (Bumbudsanpharoke and Ko, 2015). The EU has enacted an all-inclusive regulation concerning the use of nanotechnology in food applications. The European Food Safety Authority (EFSA), in their "Guidance on the risk assessment of the applications of nanoscience and nanotechnologies in the food and feed chain," suggested requirements for a meticulous evaluation of such particles in food packaging applications (EFSA, 2011). Consequently, decision trees were formed to assist the industry in taking appropriate and ethical decisions for manufacturing nanoparticle-based products with ensured safety.

Furthermore, the EFSA requires the migration assessment on substances coming in contact with food products as per the EU regulation No. 1935/2004 (Article 10) (EC, 2004). According to this guideline, a proper sequence of experiments should be undertaken, namely, migration and toxicity studies. The European Commission in 2008 enacted Regulation (EC) No. 1333/2008, which instructed that the food additives produced using nanotechnology should be re-evaluated before promoting (EU, 2008). In 2014, the EFSA published another report listing the application of nanoparticles, which were already in use in the food industry (EFSA, 2015).

The European Parliament in 2009 put forward a comprehensive resolution 2008/2208 (INI) defining the term "nanomaterial" along with various guidelines regarding the safety of nanoparticles and their application in the food industry and various other sectors (European Parliament, 2009). According to them, the nanomaterial is defined as a synthetic or natural material containing particles in the form of free particles or aggregates/agglomerates and where the external diameter of more than 50% particles is in the range of 1–100 nm. The Regulation (EC) No. 450/2009 regulated the use of active and intelligent materials supposed to come into food contact. However, nanoparticles are not specifically mentioned here, but the regulation referred to "substances deliberately engineered to a particle size, which exhibit functional physical and chemical properties that significantly differ from those at a larger scale."

EU Regulation No. 10/2011 stated that in order to use components/plastics in nanometer size range intended to come into food contact, the materials should be listed in Annex I of the regulation (European Commission, 2011). According to the document, the nanoparticles should be assessed on a case-by-case basis. The migration limit for the plastics/additives in cubic packaging of 1 kg food was set at $10 \, mg \, dm^{-2}$, which is equivalent to $60 \, mg \, kg^{-1}$ of food. The EU regulation No. 528/2012 prohibits nanomaterial-based biocidal product without authorization and approval of nanoparticle as safe active substance. Moreover, the exact application and distinct attribute of that particular nanoparticle should also be mentioned (EU, 2012). It was later amended in 2019 to widen the scope of the regulation. Moreover, it was stated that the active substance should be approved for usage initially for 10 years if it meets the criteria ordained in point (b) of Article 19(1).

In 2014, the commission released "European Food Information to Consumers-Regulation (EU) No. 1169/2011 on the provision of prepacked food information to consumers on general food labelling and nutritional labelling" (Bumbudsanpharoke and Ko, 2015). This regulation states that any form of engineered nanomaterial, when added to the food product as ingredients, should be accompanied by the word "nano" in brackets; however, this regulation does not include nanomaterial used in packaging materials.

6.2 North America

In 2011, the US Food and Drug Administration (FDA) drafted a report regarding the use of nanotechnology or containing nanoparticles in FDA-supervised products (Bumbudsanpharoke and Ko, 2015). Although it does not provide regulations for the use of nanoparticles, it acts as a guide to the manufacturer about the potential risks, safety impact, and health concerns resulted from nanoparticle application. In 2011, the National Nanotechnology Initiative (NNI) put forward a strategy for the efficient utilization and application of nanotechnology. This report argues that more research should be undertaken on the physicochemical attributes of nanoparticles to approve their viability and safety (WHO, 2013).

According to the "Guidance for Industry" published in 2014, FDA considers nanoparticles as materials with an average size between 1 mm and 1 μm. Furthermore, it says that having GRAS certification does not guarantee its safety, especially at its nanoscale level (US FDA, 2014a,b). The document also guides different manufacturers of food contact materials and food additives, including color additives and the consumers of such substances. The FDA recommends that the manufacturer should check the toxicity and prepare a detailed toxicity profile of any substance coming in contact with the food before commercializing a certain product. A material that can migrate into the food product through the packaging material or the processing equipment is spared from the regulatory aspect of food additive, provided that the level of migration is low. According to the US FDA (2014b), the material should satisfy specific criteria, such as:

(i) The approximate dietary exposure to the substance
(ii) Its likeness to be a carcinogenic substance

(iii) Absence of any adverse effect on the environment

(iv) Absence of any technical effect on the food system to which it migrates

The FDA approach toward managing nanoparticle usage is product based, i.e., every product is thoroughly checked for its safety on a case-by-case basis. Moreover, if the nanoparticle exhibits antimicrobial activity, they fall under the regulations of Section 409 of the Federal Food, Drug, and Cosmetic Act. It also needs to be registered with the Environmental Protection Agency under the Federal Insecticide, Fungicide, and Rodenticide Act (Misko, 2015).

Canada currently does not have any regulations explicitly referring to nanoparticle application in food. Instead, it relies on existing laws and legislations, including the Canadian Environmental Protection Act, 1999, to regulate the use of nanoparticles in food products without differentiating between nanoform or bulk material (Bumbudsanpharoke and Ko, 2015). However, recently, Canada is also taking measures to develop regulatory guidelines for nanoparticles and hence has established a website related to information concerning nanotechnology programs (Garcia et al., 2018).

6.3 Oceania

The safety concerns of nanoparticles were addressed by the Food Standards Australia New Zealand (FSANZ) in an article entitled "Regulatory Approach to Nanoscale Materials" in the International Food Risk Analysis Journal (WHO, 2013). Furthermore, a report published by FSANZ in 2016 titled "Nanotechnologies in Food Packaging: An Exploratory Appraisal of Safety and Regulations" gave a comprehensive review concerning the migration studies of nanoparticles into food products from packaging materials and inferred that the safety issues associated with nanoparticle migration into food products are unlikely as the migration of nanoparticles from packaging materials into food system is negligible (Drew and Hagen, 2016). Nevertheless, unless further detailed studies are undertaken, the safety status of the nanoparticles remains uncertain.

6.4 Asia

In Japan, emphasis has been placed on nanoparticle research in the Science and Technology Basic Plan for 2006–10. Additionally, the Japanese Government put forward a 6-year plan (2009–14) titled "Research Project on the Potential Hazards, etc. of Nanomaterials," which pointed out the carcinogenic effect of nanoparticles. A survey was undertaken in 2010 regarding the safety of nanoparticle use in the food industry, and it was concluded that specific laws relating to nanoparticles are not yet required (WHO, 2013).

The Republic of Korea in 2011 released a guidance concerning the management of nanoparticle-based products. The main objective of the guidance was to convince the consumers about the beneficial effects and safety of nanotechnology, encourage the establishment of sustainable nanotechnology-based industries, and assure the safety of consumers using nanotechnology-based products. Recently, South Korea released an action plan for nanoparticles, "The First Master Plan on Management of Nanomaterials Safety," during 2012–16. The plan's target was to frame a database for the analysis and assessment of nanoparticles and assess the safety concern of such food products (Hwang et al., 2012).

The South Korean Government has also funded several research projects committing to the safety of the nanoparticles in food packaging and the development of different regulations in controlling its usage (Garcia et al., 2018).

The Food and Drug Administration of Taiwan in 2017 released some regulations regarding the use of nanoparticles used in packaging materials. The guidelines suggested that the nanoparticles should be considered food-contact substances, and thus its incorporation into the packaging material would require thorough safety evaluation to gather market approval. The manufacturer should submit the physicochemical properties and composition of the nanoparticle, production methods, effects imparted by the materials, its application, toxicity profile, migration evaluation data, and international approval status, along with information on the packaging material (ChemicalWatch, 2017).

7　Concluding remarks

The food packaging industry is expanding rapidly, and its necessity has been considered throughout the world. Different techniques are applied to the food packaging materials to make them more acceptable to the consumers. People are becoming more concerned about the safety of the food they consume, which has led to several innovations in packaging materials. Currently, nanotechnology is used commonly to improve the mechanical barrier and antimicrobial properties of the films. In particular, ZnO, TiO$_2$, and Ag incorporation is reported to be the most effective in improving the overall attributes of the films. Some nanocomposites also display antimicrobial activity against gram-positive and gram-negative bacteria and some fungi.

The incorporation of nanocomposites with EOs in the polymer matrix imparts active properties to the nanocomposites. The EO-loaded films compared to EO-excluded films exhibited lower mechanical properties, moisture, and gas permeability, while the UV barrier and antimicrobial properties increased considerably. Therefore, incorporating nanoparticles and EOs into the polymer matrix is an effective strategy for increasing the shelf life of the packaged product. Although the application of nanocomposites in food packaging is limited to selected regions and products only, large-scale production and rapid commercialization of nanocomposites should be undertaken for improved food safety and quality retention.

With the increase in the usage of nanotechnology in food products, there is an urgent need for strict regulations regarding nanoparticles when used in food-contact materials. Even in developed regions like North America, Europe, and Asia, no definite laws have been enacted concerning the use of nanoparticles. Therefore, the production and consumption of food packaged with active nanocomposite need to be regulated strictly. Accordingly, guidelines and legislation should be established to protect the consumers from mishandling and overexposure of nanoparticle in packaging film development.

References

Abdullah, Z.W., Dong, Y., 2019. Biodegradable and water resistant poly (vinyl) alcohol (PVA)/starch (ST)/glycerol (GL)/halloysite nanotube (HNT) nanocomposite films for sustainable food packaging. Front. Mater. 6, 58.

Abdullayev, E., Price, R., Shchukin, D., Lvov, Y., 2009. Halloysite tubes as nanocontainers for anticorrosion coating with benzotriazole. ACS Appl. Mater. Interfaces 1 (7), 1437–1443.

Abdullayev, E., Joshi, A., Wei, W., Zhao, Y., Lvov, Y., 2012. Enlargement of halloysite clay nanotube lumen by selective etching of aluminum oxide. ACS Nano 6 (8), 7216–7226.

Acevedo-Fani, A., Salvia-Trujillo, L., Soliva-Fortuny, R., Martín-Belloso, O., 2015. Modulating biopolymer electrical charge to optimize the assembly of edible multilayer nanofilms by the layer-by-layer technique. Biomacromolecules 16 (9), 2895–2903.

Ahmadi, R., Tanomand, A., Kazeminava, F., Kamounah, F.S., Ayaseh, A., Ganbarov, K., Kafil, H.S., 2019. Fabrication and characterization of a titanium dioxide (TiO2) nanoparticles reinforced bio-nanocomposite containing Miswak (Salvadora persica L.) extract—the antimicrobial, thermo-physical and barrier properties. Int. J. Nanomedicine 14, 3439.

Ahmed, J., Arfat, Y.A., Al-Attar, H., Auras, R., Ejaz, M., 2017. Rheological, structural, ultraviolet protection and oxygen barrier properties of linear low-density polyethylene films reinforced with zinc oxide (ZnO) nanoparticles. Food Packag. Shelf Life 13, 20–26.

Ahmed, J., Arfat, Y.A., Bher, A., Mulla, M., Jacob, H., Auras, R., 2018a. Active chicken meat packaging based on polylactide films and bimetallic Ag–Cu nanoparticles and essential oil. J. Food Sci. 83 (5), 1299–1310.

Ahmed, J., Mulla, M., Arfat, Y.A., Bher, A., Jacob, H., Auras, R., 2018b. Compression molded LLDPE films loaded with bimetallic (Ag-Cu) nanoparticles and cinnamon essential oil for chicken meat packaging applications. LWT 93, 329–338.

Ahmed, J., Mulla, M., Jacob, H., Luciano, G., Bini, T., Almusallam, A., 2019. Polylactide/poly (ε-caprolactone)/zinc oxide/clove essential oil composite antimicrobial films for scrambled egg packaging. Food Packag. Shelf Life 21, 100355.

Alizadeh-Sani, M., Kia, E.M., Ghasempour, Z., Ehsani, A., 2020a. Preparation of active nanocomposite film consisting of sodium caseinate, ZnO nanoparticles and rosemary essential oil for food packaging applications. J. Polym. Environ. 29, 1–11.

Alizadeh-Sani, M., Rhim, J.-W., Azizi-Lalabadi, M., Hemmati-Dinarvand, M., Ehsani, A., 2020b. Preparation and characterization of functional sodium caseinate/guar gum/TiO2/cumin essential oil composite film. Int. J. Biol. Macromol. 145, 835–844.

Amjadi, S., Emaminia, S., Nazari, M., Davudian, S.H., Roufegarinejad, L., Hamishehkar, H., 2019. Application of reinforced ZnO nanoparticle-incorporated gelatin bionanocomposite film with chitosan nanofiber for packaging of chicken fillet and cheese as food models. Food Bioprocess Technol. 12 (7), 1205–1219.

Anderson, R.A., Zhan, Z., Luo, R., Guo, X., Guo, Q., Zhou, J., Stoecker, B.J., 2016. Cinnamon extract lowers glucose, insulin and cholesterol in people with elevated serum glucose. J. Tradit. Complement. Med. 6 (4), 332–336.

Arezoo, E., Mohammadreza, E., Maryam, M., Abdorreza, M.N., 2020. The synergistic effects of cinnamon essential oil and nano TiO2 on antimicrobial and functional properties of sago starch films. Int. J. Biol. Macromol. 157, 743–751.

Arfat, Y.A., Benjakul, S., Prodpran, T., Sumpavapol, P., Songtipya, P., 2016. Physico-mechanical characterization and antimicrobial properties of fish protein isolate/fish skin gelatin-zinc oxide (ZnO) nanocomposite films. Food Bioprocess Technol. 9 (1), 101–112.

Arfat, Y.A., Ahmed, J., Hiremath, N., Auras, R., Joseph, A., 2017a. Thermo-mechanical, rheological, structural and antimicrobial properties of bionanocomposite films based on fish skin gelatin and silver-copper nanoparticles. Food Hydrocoll. 62, 191–202.

Arfat, Y.A., Ejaz, M., Jacob, H., Ahmed, J., 2017b. Deciphering the potential of guar gum/Ag-Cu nanocomposite films as an active food packaging material. Carbohydr. Polym. 157, 65–71.

Baek, N., Kim, Y.T., Marcy, J.E., Duncan, S.E., O'Keefe, S.F., 2018. Physical properties of nanocomposite polylactic acid films prepared with oleic acid modified titanium dioxide. Food Packag. Shelf Life 17, 30–38.

Biddeci, G., Cavallaro, G., Di Blasi, F., Lazzara, G., Massaro, M., Milioto, S., Spinelli, G., 2016. Halloysite nanotubes loaded with peppermint essential oil as filler for functional biopolymer film. Carbohydr. Polym. 152, 548–557.

Birla, S., Tiwari, V., Gade, A., Ingle, A., Yadav, A., Rai, M., 2009. Fabrication of silver nanoparticles by Phoma glomerata and its combined effect against Escherichia coli, Pseudomonas aeruginosa and Staphylococcus aureus. Lett. Appl. Microbiol. 48 (2), 173–179.

Bodaghi, H., Mostofi, Y., Oromiehie, A., Ghanbarzadeh, B., Hagh, Z.G., 2015. Synthesis of clay–TiO2 nanocomposite thin films with barrier and photocatalytic properties for food packaging application. J. Appl. Polym. Sci. 132 (14).

Boonsiriwit, A., Xiao, Y., Joung, J., Kim, M., Singh, S., Lee, Y.S., 2020. Alkaline halloysite nanotubes/low density polyethylene nanocomposite films with increased ethylene absorption capacity: applications in cherry tomato packaging. Food Packag. Shelf Life 25, 100533.

Brahmi, F., Abdenour, A., Bruno, M., Silvia, P., Alessandra, P., Danilo, F., Mohamed, C., 2016. Chemical composition and in vitro antimicrobial, insecticidal and antioxidant activities of the essential oils of *Mentha pulegium* L. and *Mentha rotundifolia* (L.) Huds growing in Algeria. Ind. Crop. Prod. 88, 96–105.

Buendía, L., Sánchez, M.J., Antolinos, V., Ros, M., Navarro, L., Soto, S., López, A., 2020. Active cardboard box with a coating including essential oils entrapped within cyclodextrins and/or halloysite nanotubes. A case study for fresh tomato storage. Food Control 107, 106763.

Bumbudsanpharoke, N., Ko, S., 2015. Nano-food packaging: an overview of market, migration research, and safety regulations. J. Food Sci. 80 (5), R910–R923.

Burt, S., 2004. Essential oils: their antibacterial properties and potential applications in foods—a review. Int. J. Food Microbiol. 94 (3), 223–253.

Carré, G., Hamon, E., Ennahar, S., Estner, M., Lett, M.-C., Horvatovich, P., Andre, P., 2014. TiO2 photocatalysis damages lipids and proteins in *Escherichia coli*. Appl. Environ. Microbiol. 80 (8), 2573–2581.

Casey, P., 2006. Nanoparticle technologies and applications. In: Nanostructure Control of Materials. Elsevier, pp. 1–31.

Cerrada, M.L., Serrano, C., Sánchez-Chaves, M., Fernández-García, M., Fernández-Martín, F., de Andres, A., Fernández-García, M., 2008. Self-sterilized EVOH-TiO2 nanocomposites: interface effects on biocidal properties. Adv. Funct. Mater. 18 (13), 1949–1960.

Chakraborty, S., Chelli, V.R., Das, R.K., Giri, A.S., Golder, A.K., 2017. Bio-mediated silver nanoparticle synthesis: mechanism and microbial inactivation. Toxicol. Environ. Chem. 99 (3), 434–447.

ChemicalWatch, 2017. Taiwan FDA Issues Guidelines on Nanomaterials in Food Packaging. ChemicalWatch. https://chemicalwatch.com/54578/taiwan-fda-issues-guidelines-on-nanomaterials-in-food-packaging.

Chi, H., Xue, J., Zhang, C., Chen, H., Li, L., Qin, Y., 2018. High pressure treatment for improving water vapour barrier properties of poly (lactic acid)/Ag nanocomposite films. Polymers 10 (9), 1011.

Chi, H., Song, S., Luo, M., Zhang, C., Li, W., Li, L., Qin, Y., 2019. Effect of PLA nanocomposite films containing bergamot essential oil, TiO2 nanoparticles, and Ag nanoparticles on shelf life of mangoes. Sci. Hortic. 249, 192–198.

Chorianopoulos, N., Tsoukleris, D., Panagou, E., Falaras, P., Nychas, G.-J., 2011. Use of titanium dioxide (TiO2) photocatalysts as alternative means for *Listeria monocytogenes* biofilm disinfection in food processing. Food Microbiol. 28 (1), 164–170.

Chu, Z., Zhao, T., Li, L., Fan, J., Qin, Y., 2017. Characterization of antimicrobial poly (lactic acid)/nano-composite films with silver and zinc oxide nanoparticles. Materials 10 (6), 659.

Cinteza, L.O., Scomoroscenco, C., Voicu, S.N., Nistor, C.L., Nitu, S.G., Trica, B., Petcu, C., 2018. Chitosan-stabilized Ag nanoparticles with superior biocompatibility and their synergistic antibacterial effect in mixtures with essential oils. Nanomaterials 8 (10), 826.

Cotton High Tech SL, 2014. Inspections, Compliance, Enforcement, and Criminal Investigations. United States Food and Drug Administration.

Cushen, M., Kerry, J., Morris, M., Cruz-Romero, M., Cummins, E., 2013. Migration and exposure assessment of silver from a PVC nanocomposite. Food Chem. 139 (1–4), 389–397.

Cushen, M., Kerry, J., Morris, M., Cruz-Romero, M., Cummins, E., 2014. Evaluation and simulation of silver and copper nanoparticle migration from polyethylene nanocomposites to food and an associated exposure assessment. J. Agric. Food Chem. 62 (6), 1403–1411.

Dash, K.K., Ali, N.A., Das, D., Mohanta, D., 2019. Thorough evaluation of sweet potato starch and lemon-waste pectin based-edible films with nano-titania inclusions for food packaging applications. Int. J. Biol. Macromol. 139, 449–458.

de Abreu, D.A.P., Cruz, J.M., Angulo, I., Losada, P.P., 2010. Mass transport studies of different additives in polyamide and exfoliated nanocomposite polyamide films for food industry. Packag. Technol. Sci. 23 (2), 59–68.

de Azeredo, H.M., 2009. Nanocomposites for food packaging applications. Food Res. Int. 42 (9), 1240–1253.

de Azeredo, H.M., 2013. Antimicrobial nanostructures in food packaging. Trends Food Sci. Technol. 30 (1), 56–69.

de Matos Fonseca, J., Valencia, G.A., Soares, L.S., Dotto, M.E.R., Campos, C.E.M., de Moreira, R.F.P.M., Fritz, A.R.M., 2020. Hydroxypropyl methylcellulose-TiO2 and gelatin-TiO2 nanocomposite films: physicochemical and structural properties. Int. J. Biol. Macromol. 151, 944–956.

Dehghani, S., Peighambardoust, S.H., Peighambardoust, S.J., Hosseini, S.V., Regenstein, J.M., 2019. Improved mechanical and antibacterial properties of active LDPE films prepared with combination of Ag, ZnO and CuO nanoparticles. Food Packag. Shelf Life 22, 100391.

Donglu, F., Wenjian, Y., Kimatu, B.M., Mariga, A.M., Liyan, Z., Xinxin, A., Qiuhui, H., 2016. Effect of nanocomposite-based packaging on storage stability of mushrooms (Flammulina velutipes). Innovative Food Sci. Emerg. Technol. 33, 489–497.

Drew, R., Hagen, T., 2016. Nanotechnologies in food packaging: an exploratory appraisal of safety and regulation. In: A Report Prepared by ToxConsultant Pty Ltd for Food Standards Australia New Zealand. www.foodstandards. gov.au/publications/Documents/Nanotech%20in%20f.ood%20packaging.pdf.

EC, 2004. Regulation (EC) no 1935/2004 of the European parliament and of the council of 27 October 2004 on materials and articles intended to come into contact with food and repealing directives 80/590/EEC and 89/109/EEC. Off. J. Eur. Union 338, 4–17.

EFSA, 2011. Guidance on the risk assessment of the application of nanoscience and nanotechnologies in the food and feed chain. EFSA J. 9 (5), 2140.

EFSA, 2012. Scientific opinion on the safety evaluation of the substance, titanium nitride, nanoparticles, for use in food contact materials. EFSA J. 10 (3), 2641.

EFSA, 2015. Annual Report of the EFSA Scientific Network of Risk Assessment of Nanotechnologies in Food and Feed for 2014. Wiley Online Library.

Ejaz, M., Arfat, Y.A., Mulla, M., Ahmed, J., 2018. Zinc oxide nanorods/clove essential oil incorporated type B gelatin composite films and its applicability for shrimp packaging. Food Packag. Shelf Life 15, 113–121.

Emamifar, A., Kadivar, M., Shahedi, M., Soleimanian-Zad, S., 2011. Effect of nanocomposite packaging containing Ag and ZnO on inactivation of *Lactobacillus plantarum* in orange juice. Food Control 22 (3–4), 408–413.

Espitia, P.J.P., de Soares, N.F.F., dos Reis Coimbra, J.S., de Andrade, N.J., Cruz, R.S., Medeiros, E.A.A., 2012. Zinc oxide nanoparticles: synthesis, antimicrobial activity and food packaging applications. Food Bioprocess Technol. 5 (5), 1447–1464.

EU, 2008. Regulation (EC) No 1333/2008 of the European parliament and of the council of 16 December 2008 on food additive. Off. J. Eur. Union 354, 16–33.

EU, 2012. Regulation (EU) no 528/2012 of the European parliament and of the council of 22 May 2012 concerning the making available on the market and use of biocidal products. Off. J. Eur. Union 167, 1–123.

European Commission, 2011. Commission regulation (EU) no 10/2011 of 14 January 2011 on plastic materials and articles intended to come into contact with food. Off. J. Eur. Union 12, 1–89.

European Parliament, 2009. Regulatory aspects of nanomaterials (2008/2208(INI)). In: Committee on the Environment, Public Health and Food Safety. European Parliament.

Farhoodi, M., Mohammadifar, M.A., Mousavi, M., Sotudeh-Gharebagh, R., Emam-Djomeh, Z., 2017. Migration kinetics of ethylene glycol monomer from PET bottles into acidic food simulant: effects of nanoparticle presence and matrix morphology. J. Food Process Eng. 40 (2), e12383.

Garcia, C.V., Shin, G.H., Kim, J.T., 2018. Metal oxide-based nanocomposites in food packaging: applications, migration, and regulations. Trends Food Sci. Technol. 82, 21–31.

Ghazihoseini, S., Alipoormazandarani, N., Nafchi, A.M., 2015. The effects of nano-SiO2 on mechanical, barrier, and moisture sorption isotherm models of novel soluble soybean polysaccharide films. Int. J. Food Eng. 11 (6), 833–840.

Gumiero, M., Peressini, D., Pizzariello, A., Sensidoni, A., Iacumin, L., Comi, G., Toniolo, R., 2013. Effect of TiO2 photocatalytic activity in a HDPE-based food packaging on the structural and microbiological stability of a short-ripened cheese. Food Chem. 138 (2–3), 1633–1640.

Hajji, S., Salem, R.B.S.-B., Hamdi, M., Jellouli, K., Ayadi, W., Nasri, M., Boufi, S., 2017. Nanocomposite films based on chitosan–poly (vinyl alcohol) and silver nanoparticles with high antibacterial and antioxidant activities. Process. Saf. Environ. Prot. 111, 112–121.

Hassannia-Kolaee, M., Khodaiyan, F., Pourahmad, R., Shahabi-Ghahfarrokhi, I., 2016. Development of ecofriendly bionanocomposite: whey protein isolate/pullulan films with nano-SiO2. Int. J. Biol. Macromol. 86, 139–144.

He, Q., Zhang, Y., Cai, X., Wang, S., 2016. Fabrication of gelatin–TiO2 nanocomposite film and its structural, antibacterial and physical properties. Int. J. Biol. Macromol. 84, 153–160.

Hou, X., Xue, Z., Xia, Y., Qin, Y., Zhang, G., Liu, H., Li, K., 2019. Effect of SiO2 nanoparticle on the physical and chemical properties of eco-friendly agar/sodium alginate nanocomposite film. Int. J. Biol. Macromol. 125, 1289–1298.

Huang, J.-Y., Li, X., Zhou, W., 2015. Safety assessment of nanocomposite for food packaging application. Trends Food Sci. Technol. 45 (2), 187–199.

Huang, Y., Mei, L., Chen, X., Wang, Q., 2018. Recent developments in food packaging based on nanomaterials. Nanomaterials 8 (10), 830.

Huang, J., Cheng, Y., Wu, Y., Shi, X., Du, Y., Deng, H., 2019. Chitosan/tannic acid bilayers layer-by-layer deposited cellulose nanofibrous mats for antibacterial application. Int. J. Biol. Macromol. 139, 191–198.

Huang, D., Zhang, Z., Zheng, Y., Quan, Q., Wang, W., Wang, A., 2020. Synergistic effect of chitosan and halloysite nanotubes on improving agar film properties. Food Hydrocoll. 101, 105471.

Hussain, A.I., Anwar, F., Sherazi, S.T.H., Przybylski, R., 2008. Chemical composition, antioxidant and antimicrobial activities of basil (*Ocimum basilicum*) essential oils depends on seasonal variations. Food Chem. 108 (3), 986–995.

Hwang, M., Lee, E.J., Kweon, S.Y., Park, M.S., Jeong, J.Y., Um, J.H., Yoon, H.J., 2012. Risk assessment principle for engineered nanotechnology in food and drug. Toxicol. Res. 28 (2), 73–79.

Iamareerat, B., Singh, M., Sadiq, M.B., Anal, A.K., 2018. Reinforced cassava starch based edible film incorporated with essential oil and sodium bentonite nanoclay as food packaging material. J. Food Sci. Technol. 55 (5), 1953–1959.

Ibrahim, S.A., Sreekantan, S., 2011. Effect of pH on TiO2 nanoparticles via sol-gel method. In: Paper Presented at the Advanced Materials Research.

Jebel, F.S., Almasi, H., 2016. Morphological, physical, antimicrobial and release properties of ZnO nanoparticles-loaded bacterial cellulose films. Carbohydr. Polym. 149, 8–19.

Ju, J., Xie, Y., Guo, Y., Cheng, Y., Qian, H., Yao, W., 2019. The inhibitory effect of plant essential oils on foodborne pathogenic bacteria in food. Crit. Rev. Food Sci. Nutr. 59 (20), 3281–3292.

Khajeh, M., Yamini, Y., Bahramifar, N., Sefidkon, F., Pirmoradei, M.R., 2005. Comparison of essential oils compositions of *Ferula assa-foetida* obtained by supercritical carbon dioxide extraction and hydrodistillation methods. Food Chem. 91 (4), 639–644.

Krehula, L.K., Papić, A., Krehula, S., Gilja, V., Foglar, L., Hrnjak-Murgić, Z., 2017. Properties of UV protective films of poly (vinyl-chloride)/TiO2 nanocomposites for food packaging. Polym. Bull. 74 (4), 1387–1404.

Kumar, S., Boro, J.C., Ray, D., Mukherjee, A., Dutta, J., 2019. Bionanocomposite films of agar incorporated with ZnO nanoparticles as an active packaging material for shelf life extension of green grape. Heliyon 5 (6), e01867.

Lee, M.H., Kim, S.Y., Park, H.J., 2018. Effect of halloysite nanoclay on the physical, mechanical, and antioxidant properties of chitosan films incorporated with clove essential oil. Food Hydrocoll. 84, 58–67.

Li, X., Xing, Y., Jiang, Y., Ding, Y., Li, W., 2009. Antimicrobial activities of ZnO powder-coated PVC film to inactivate food pathogens. Int. J. Food Sci. Technol. 44 (11), 2161–2168.

Li, X.H., Li, W.L., Xing, Y.G., Jiang, Y.H., Ding, Y.L., Zhang, P.P., 2011. Effects of nano-ZnO power-coated PVC film on the physiological properties and microbiological changes of fresh-cut "Fuji" apple. In: Paper Presented at the Advanced Materials Research.

Li, F., Biagioni, P., Finazzi, M., Tavazzi, S., Piergiovanni, L., 2013. Tunable green oxygen barrier through layer-by-layer self-assembly of chitosan and cellulose nanocrystals. Carbohydr. Polym. 92 (2), 2128–2134.

Li, D., Zhang, J., Xu, W., Fu, Y., 2016. Effect of SiO2/EVA on the mechanical properties, permeability, and residual solvent of polypropylene packaging films. Polym. Compos. 37 (1), 101–107.

Lin, D., Yang, Y., Wang, J., Yan, W., Wu, Z., Chen, H., Tu, Z., 2020. Preparation and characterization of TiO2-Ag loaded fish gelatin-chitosan antibacterial composite film for food packaging. Int. J. Biol. Macromol. 154, 123–133.

Lisuzzo, L., Cavallaro, G., Milioto, S., Lazzara, G., 2020. Effects of halloysite content on the thermo-mechanical performances of composite bioplastics. Appl. Clay Sci. 185, 105416.

Liu, Y., Liu, Y., Wei, S., 2010. Processing Technologies of EVOH/Nano-SiO2 High-Barrier Packaging Composites. Sci Res Publ, Inc-Srp, Irvin.

Liu, C., Xiong, H., Chen, X., Lin, S., Tu, Y., 2015. Effects of nano-TiO2 on the performance of high-amylose starch based antibacterial films. J. Appl. Polym. Sci. 132 (32).

Liu, X., Chen, X., Ren, J., Chang, M., He, B., Zhang, C., 2019. Effects of nano-ZnO and nano-SiO2 particles on properties of PVA/xylan composite films. Int. J. Biol. Macromol. 132, 978–986.

Liu, J., Liu, C., Zheng, X., Chen, M., Tang, K., 2020. Soluble soybean polysaccharide/nano zinc oxide antimicrobial nanocomposite films reinforced with microfibrillated cellulose. Int. J. Biol. Macromol. 159, 793–803.

Luo, Z., Qin, Y., Ye, Q., 2015a. Effect of nano-TiO2-LDPE packaging on microbiological and physicochemical quality of Pacific white shrimp during chilled storage. Int. J. Food Sci. Technol. 50 (7), 1567–1573.

Luo, Z., Xu, Y., Ye, Q., 2015b. Effect of nano-SiO2-LDPE packaging on biochemical, sensory, and microbiological quality of Pacific white shrimp *Penaeus vannamei* during chilled storage. Fish. Sci. 81 (5), 983–993.

Magnuson, B.A., Jonaitis, T.S., Card, J.W., 2011. A brief review of the occurrence, use, and safety of food-related nanomaterials. J. Food Sci. 76 (6), R126–R133.

Makaremi, M., Pasbakhsh, P., Cavallaro, G., Lazzara, G., Aw, Y.K., Lee, S.M., Milioto, S., 2017. Effect of morphology and size of halloysite nanotubes on functional pectin bionanocomposites for food packaging applications. ACS Appl. Mater. Interfaces 9 (20), 17476–17488.

Marra, A., Silvestre, C., Duraccio, D., Cimmino, S., 2016. Polylactic acid/zinc oxide biocomposite films for food packaging application. Int. J. Biol. Macromol. 88, 254–262.

Martínez-Camacho, A., Cortez-Rocha, M., Graciano-Verdugo, A., Rodríguez-Félix, F., Castillo-Ortega, M., Burgos-Hernández, A., Plascencia-Jatomea, M., 2013. Extruded films of blended chitosan, low density polyethylene and ethylene acrylic acid. Carbohydr. Polym. 91 (2), 666–674.

Marvizadeh, M.M., Oladzadabbasabadi, N., Nafchi, A.M., Jokar, M., 2017. Preparation and characterization of bionanocomposite film based on tapioca starch/bovine gelatin/nanorod zinc oxide. Int. J. Biol. Macromol. 99, 1–7.

Mathew, S., Snigdha, S., Mathew, J., Radhakrishnan, E., 2019. Biodegradable and active nanocomposite pouches reinforced with silver nanoparticles for improved packaging of chicken sausages. Food Packag. Shelf Life 19, 155–166.

Misko, G., 2015. EPA Studies Food Packaging with Nanoscale Antimicrobials. Packaging Digest. https://www.packagingdigest.com/food-packaging/epa-studies-food-packaging-nanoscale-antimicrobials.

Nasiri, A., Shariaty-Niasar, M., Akbari, Z., 2012. Synthesis of LDPE/Nano TiO2 nanocomposite for packaging applications. Int. J. Nanosci. Nanotechnol. 8 (3), 165–170.

Negi, P.S., 2012. Plant extracts for the control of bacterial growth: efficacy, stability and safety issues for food application. Int. J. Food Microbiol. 156 (1), 7–17.

Ngo, T.M.P., Dang, T.M.Q., Tran, T.X., Rachtanapun, P., 2018. Effects of zinc oxide nanoparticles on the properties of pectin/alginate edible films. Int. J. Polymer Sci. 2018.

Noori Hashemabad, Z., Shabanpour, B., Azizi, H., Ojagh, S.M., Alishahi, A., 2017. Effect of TiO2 nanoparticles on the antibacterial and physical properties of low-density polyethylene film. Polym.-Plast. Technol. Eng. 56 (14), 1516–1527.

Noshirvani, N., Ghanbarzadeh, B., Mokarram, R.R., Hashemi, M., 2017a. Novel active packaging based on carboxymethyl cellulose-chitosan-ZnO NPs nanocomposite for increasing the shelf life of bread. Food Packag. Shelf Life 11, 106–114.

Noshirvani, N., Ghanbarzadeh, B., Mokarram, R.R., Hashemi, M., Coma, V., 2017b. Preparation and characterization of active emulsified films based on chitosan-carboxymethyl cellulose containing zinc oxide nano particles. Int. J. Biol. Macromol. 99, 530–538.

Ortenzi, M.A., Basilissi, L., Farina, H., Di Silvestro, G., Piergiovanni, L., Mascheroni, E., 2015. Evaluation of crystallinity and gas barrier properties of films obtained from PLA nanocomposites synthesized via "in situ" polymerization of l-lactide with silane-modified nanosilica and montmorillonite. Eur. Polym. J. 66, 478–491.

Oun, A.A., Rhim, J.-W., 2017. Carrageenan-based hydrogels and films: effect of ZnO and CuO nanoparticles on the physical, mechanical, and antimicrobial properties. Food Hydrocoll. 67, 45–53.

Pantani, R., Gorrasi, G., Vigliotta, G., Murariu, M., Dubois, P., 2013. PLA-ZnO nanocomposite films: water vapor barrier properties and specific end-use characteristics. Eur. Polym. J. 49 (11), 3471–3482.

Periasamy, V.S., Athinarayanan, J., Alshatwi, A.A., 2016. Anticancer activity of an ultrasonic nanoemulsion formulation of Nigella sativa L. essential oil on human breast cancer cells. Ultrason. Sonochem. 31, 449–455.

Petchwattana, N., Covavisaruch, S., Wibooranawong, S., Naknaen, P., 2016. Antimicrobial food packaging prepared from poly (butylene succinate) and zinc oxide. Measurement 93, 442–448.

Priyadarshi, R., Rhim, J.-W., 2020. Chitosan-based biodegradable functional films for food packaging applications. Innovative Food Sci. Emerg. Technol. 62, 102346.

Qu, L., Chen, G., Dong, S., Huo, Y., Yin, Z., Li, S., Chen, Y., 2019. Improved mechanical and antimicrobial properties of zein/chitosan films by adding highly dispersed nano-TiO2. Ind. Crop. Prod. 130, 450–458.

Rhim, J.-W., Ng, P.K., 2007. Natural biopolymer-based nanocomposite films for packaging applications. Crit. Rev. Food Sci. Nutr. 47 (4), 411–433.

Ribeiro-Santos, R., Andrade, M., de Melo, N.R., Sanches-Silva, A., 2017. Use of essential oils in active food packaging: recent advances and future trends. Trends Food Sci. Technol. 61, 132–140.

Risyon, N.P., Othman, S.H., Basha, R.K., Talib, R.A., 2020. Characterization of polylactic acid/halloysite nanotubes bionanocomposite films for food packaging. Food Packag. Shelf Life 23, 100450.

Roilo, D., Maestri, C.A., Scarpa, M., Bettotti, P., Checchetto, R., 2018. Gas barrier and optical properties of cellulose nanofiber coatings with dispersed TiO2 nanoparticles. Surf. Coat. Technol. 343, 131–137.

Rojas, K., Canales, D., Amigo, N., Montoille, L., Cament, A., Rivas, L.M., Ribes-Greus, A., 2019. Effective antimicrobial materials based on low-density polyethylene (LDPE) with zinc oxide (ZnO) nanoparticles. Compos. Part B 172, 173–178.

Roy, S., Rhim, J.-W., 2020. Carboxymethyl cellulose-based antioxidant and antimicrobial active packaging film incorporated with curcumin and zinc oxide. Int. J. Biol. Macromol. 148, 666–676.

Sadegh-Hassani, F., Nafchi, A.M., 2014. Preparation and characterization of bionanocomposite films based on potato starch/halloysite nanoclay. Int. J. Biol. Macromol. 67, 458–462.

Salarbashi, D., Tafaghodi, M., Bazzaz, B.S.F., 2018. Soluble soybean polysaccharide/TiO2 bionanocomposite film for food application. Carbohydr. Polym. 186, 384–393.

Sani, M.A., Ehsani, A., Hashemi, M., 2017. Whey protein isolate/cellulose nanofibre/TiO2 nanoparticle/rosemary essential oil nanocomposite film: its effect on microbial and sensory quality of lamb meat and growth of common foodborne pathogenic bacteria during refrigeration. Int. J. Food Microbiol. 251, 8–14.

Sani, I.K., Pirsa, S., Tağı, Ş., 2019. Preparation of chitosan/zinc oxide/*Melissa officinalis* essential oil nano-composite film and evaluation of physical, mechanical and antimicrobial properties by response surface method. Polym. Test. 79, 106004.

Sanuja, S., Agalya, A., Umapathy, M.J., 2015. Synthesis and characterization of zinc oxide–neem oil–chitosan bionanocomposite for food packaging application. Int. J. Biol. Macromol. 74, 76–84.

Seow, Y.X., Yeo, C.R., Chung, H.L., Yuk, H.-G., 2014. Plant essential oils as active antimicrobial agents. Crit. Rev. Food Sci. Nutr. 54 (5), 625–644.

Shahabi-Ghahfarrokhi, I., Khodaiyan, F., Mousavi, M., Yousefi, H., 2015. Preparation of UV-protective kefiran/nano-ZnO nanocomposites: physical and mechanical properties. Int. J. Biol. Macromol. 72, 41–46.

Shaili, T., Abdorreza, M.N., Fariborz, N., 2015. Functional, thermal, and antimicrobial properties of soluble soybean polysaccharide biocomposites reinforced by nano TiO2. Carbohydr. Polym. 134, 726–731.

Shankar, S., Rhim, J.-W., 2017. Preparation and characterization of agar/lignin/silver nanoparticles composite films with ultraviolet light barrier and antibacterial properties. Food Hydrocoll. 71, 76–84.

Shankar, S., Rhim, J.-W., 2018. Preparation of sulfur nanoparticle-incorporated antimicrobial chitosan films. Food Hydrocoll. 82, 116–123.

Shankar, S., Teng, X., Li, G., Rhim, J.-W., 2015. Preparation, characterization, and antimicrobial activity of gelatin/ZnO nanocomposite films. Food Hydrocoll. 45, 264–271.

Shankar, S., Rhim, J.-W., Won, K., 2018a. Preparation of poly (lactide)/lignin/silver nanoparticles composite films with UV light barrier and antibacterial properties. Int. J. Biol. Macromol. 107, 1724–1731.

Shankar, S., Wang, L.-F., Rhim, J.-W., 2018b. Incorporation of zinc oxide nanoparticles improved the mechanical, water vapor barrier, UV-light barrier, and antibacterial properties of PLA-based nanocomposite films. Mater. Sci. Eng. C 93, 289–298.

Sharma, S., Barkauskaite, S., Jaiswal, A.K., Jaiswal, S., 2020. Essential oils as additives in active food packaging. Food Chem., 128403.

Shen, Z., Simon, G.P., Cheng, Y.-B., 2002. Comparison of solution intercalation and melt intercalation of polymer–clay nanocomposites. Polymer 43 (15), 4251–4260.

Siemann, U., 2005. Solvent cast technology—a versatile tool for thin film production. In: Scattering Methods and the Properties of Polymer Materials. Springer, pp. 1–14.

Siripatrawan, U., Kaewklin, P., 2018. Fabrication and characterization of chitosan-titanium dioxide nanocomposite film as ethylene scavenging and antimicrobial active food packaging. Food Hydrocoll. 84, 125–134.

Souza, V.G.L., Fernando, A.L., 2016. Nanoparticles in food packaging: biodegradability and potential migration to food—a review. Food Packag. Shelf Life 8, 63–70.

Suhag, R., Kumar, N., Petkoska, A.T., Upadhyay, A., 2020. Film formation and deposition methods of edible coating on food products: a review. Food Res. Int. 136, 109582.

Tabatabaei, R.H., Jafari, S.M., Mirzaei, H., Nafchi, A.M., Dehnad, D., 2018. Preparation and characterization of nano-SiO2 reinforced gelatin-k-carrageenan biocomposites. Int. J. Biol. Macromol. 111, 1091–1099.

Tas, B.A., Sehit, E., Tas, C.E., Unal, S., Cebeci, F.C., Menceloglu, Y.Z., Unal, H., 2019. Carvacrol loaded halloysite coatings for antimicrobial food packaging applications. Food Packag. Shelf Life 20, 100300.

Thiagamani, S.M.K., Rajini, N., Siengchin, S., Rajulu, A.V., Hariram, N., Ayrilmis, N., 2019. Influence of silver nanoparticles on the mechanical, thermal and antimicrobial properties of cellulose-based hybrid nanocomposites. Compos. Part B 165, 516–525.

US FDA, 2014a. Guidance for Industry Considering whether an FDA-Regulated Product Involves the Application of Nanotechnology. FDA.

US FDA, 2014b. Guidance for Industry: Assessing the Effects of Significant Manufacturing Process Changes, Including Emerging Technologies, on the Safety and Regulatory Status of Food Ingredients and Food Contact Substances, Including Food Ingredients That are Color Additives. pp. 1–29. June.

Valerini, D., Tammaro, L., Di Benedetto, F., Vigliotta, G., Capodieci, L., Terzi, R., Rizzo, A., 2018. Aluminum-doped zinc oxide coatings on polylactic acid films for antimicrobial food packaging. Thin Solid Films 645, 187–192.

Vasile, C., Râpă, M., Ştefan, M., Stan, M., Macavei, S., Darie-Niţă, R., Ştefan, R., 2017. New PLA/ZnO: Cu/Ag bionanocomposites for food packaging. Express Polym Lett 11 (7), 531–544.

Vejdan, A., Ojagh, S.M., Adeli, A., Abdollahi, M., 2016. Effect of TiO2 nanoparticles on the physico-mechanical and ultraviolet light barrier properties of fish gelatin/agar bilayer film. LWT—Food Sci. Technol. 71, 88–95.

Vejdan, A., Ojagh, S.M., Abdollahi, M., 2017. Effect of gelatin/agar bilayer film incorporated with TiO2 nanoparticles as a UV absorbent on fish oil photooxidation. Int. J. Food Sci. Technol. 52 (8), 1862–1868.

Venkatesan, R., Rajeswari, N., 2019. Preparation, mechanical and antimicrobial properties of SiO 2/poly (butylene adipate-co-terephthalate) films for active food packaging. SILICON 11 (5), 2233–2239.

Wang, L.-F., Rhim, J.-W., 2015. Preparation and application of agar/alginate/collagen ternary blend functional food packaging films. Int. J. Biol. Macromol. 80, 460–468.

Wang, H., Qian, J., Ding, F., 2018. Emerging chitosan-based films for food packaging applications. J. Agric. Food Chem. 66 (2), 395–413.

Wang, H., Gong, X., Miao, Y., Guo, X., Liu, C., Fan, Y.-Y., Li, W., 2019. Preparation and characterization of multilayer films composed of chitosan, sodium alginate and carboxymethyl chitosan-ZnO nanoparticles. Food Chem. 283, 397–403.

WHO, 2013. State of the Art on the Initiatives and Activities Relevant to Risk Assessment and Risk Management of Nanotechnologies in the Food and Agriculture Sectors: FAO/WHO Technical Paper. World Health Organization.

Woranuch, S., Yoksan, R., 2013. Eugenol-loaded chitosan nanoparticles: II. Application in bio-based plastics for active packaging. Carbohydr. Polym. 96 (2), 586–592.

Wu, Z., Huang, X., Li, Y.-C., Xiao, H., Wang, X., 2018. Novel chitosan films with laponite immobilized Ag nanoparticles for active food packaging. Carbohydr. Polym. 199, 210–218.

Wu, Z., Deng, W., Luo, J., Deng, D., 2019a. Multifunctional nano-cellulose composite films with grape seed extracts and immobilized silver nanoparticles. Carbohydr. Polym. 205, 447–455.

Wu, Z., Zhou, W., Pang, C., Deng, W., Xu, C., Wang, X., 2019b. Multifunctional chitosan-based coating with liposomes containing laurel essential oils and nanosilver for pork preservation. Food Chem. 295, 16–25.

Wu, C., Zhu, Y., Wu, T., Wang, L., Yuan, Y., Chen, J., Pang, J., 2019c. Enhanced functional properties of biopolymer film incorporated with curcumin-loaded mesoporous silica nanoparticles for food packaging. Food Chem. 288, 139–145.

Xu, L., Cao, W., Li, R., Zhang, H., Xia, N., Li, T., Zhao, X., 2019a. Properties of soy protein isolate/nano-silica films and their applications in the preservation of Flammulina velutipes. J. Food Process. Preserv. 43 (11), e14177.

Xu, L., Liu, Y., Yang, M., Cao, W., Zhang, H., Xia, N., Zhao, X., 2019b. Properties of soy protein isolate/nano-silica bilayer films during storage. J. Food Process Eng. 42 (2), e12984.

Yang, M., Xia, Y., Wang, Y., Zhao, X., Xue, Z., Quan, F., Zhao, Z., 2016. Preparation and property investigation of crosslinked alginate/silicon dioxide nanocomposite films. J. Appl. Polym. Sci. 133 (22).

Yang, X., Zhang, Y., Zheng, D., Yue, J., Liu, M., 2020. Nano-biocomposite films fabricated from cellulose fibers and halloysite nanotubes. Appl. Clay Sci. 190, 105565.

Yousefi, P., Hamedi, S., Garmaroody, E.R., Koosha, M., 2020. Antibacterial nanobiocomposite based on halloysite nanotubes and extracted xylan from bagasse pith. Int. J. Biol. Macromol. 160, 276–287.

Zhang, X., Liu, Y., Yong, H., Qin, Y., Liu, J., Liu, J., 2019a. Development of multifunctional food packaging films based on chitosan, TiO2 nanoparticles and anthocyanin-rich black plum peel extract. Food Hydrocoll. 94, 80–92.

Zhang, R., Wang, X., Wang, J., Cheng, M., 2019b. Synthesis and characterization of konjac glucomannan/carrageenan/nano-silica films for the preservation of postharvest white mushrooms. Polymers 11 (1), 6.

Nano delivery systems for food bioactives

L. Mahalakshmi, K.S. Yoha, J.A. Moses, and C. Anandharamakrishnan

Computational Modeling and Nanoscale Processing Unit, National Institute of Food Technology, Entrepreneurship and Management—Thanjavur, Ministry of Food Processing Industries, Govt. of India, Thanjavur, Tamil Nadu, India

O U T L I N E

1 Introduction

In the current scenario, the development of techniques for the production of healthy functional food products has increased among researchers in the food and pharmaceutical industries due to an increased awareness of unhealthy lifestyles (Bhushani et al., 2017). Bioactives are extra nutritional constituents found in lesser quantities in foods. Those bioactive compounds have several therapeutic effects such as antioxidant, anticarcinogenic, antimutagenic, antiaging, and anti-inflammatory activity. Although the bioactives can be consumed through

Food, Medical, and Environmental Applications of Nanomaterials
https://doi.org/10.1016/B978-0-12-822858-6.00008-X

food, their bioavailability and bioactivity are very low due to their chemical instability, poor solubility, degradation during food processing, and digestion (Aditya et al., 2017). Encapsulation of food bioactives using nano delivery systems are one of the advanced and effective systems that can pave the way to overcome the limitations associated with food bioactives (de Souza Simões et al., 2017). Nano delivery system demonstrated remarkable changes in the delivery of food bioactives because of their higher loading capacity, targeted delivery, and lesser toxicity. Nano delivery systems may also improve food bioactives' stability, solubility, functionality, gastrointestinal stability, cellular uptake, controlled release, and bioavailability (Rani and Yadav, 2018; Borel and Sabliov, 2014; Maria Leena et al., 2020).

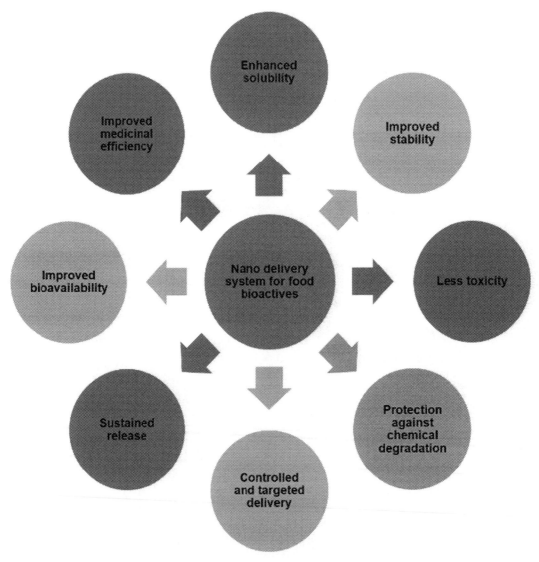

FIG. 1   Various advantages of nano delivery system for food bioactives.

Fig. 1 illustrates the various advantages of nano delivery system for food bioactives. Besides, their small size and specific site delivery provide sufficient bioavailability of the bioactives, reduce regular dosage levels, and enhance the blood circulation duration, and cellular internalization (Rani and Yadav, 2018). Characteristics such as the large surface-area-to-volume ratio and the effect of physical and chemical interactions between materials occurring at nanoscale structures have a substantial effect on the overall functionalities of those systems. Carrier materials such as inorganic compounds, organic compounds, synthetic and biopolymer compounds have been used for fabricating the nano delivery system. Though the synthetic polymers offer higher encapsulation efficiency and better entanglement, their health constraints extend researchers to use organic and bio-based carriers obtained from animal or plant sources for the development of a nano delivery system (Turasan and Kokini, 2020a).

Due to the distinct properties of food bioactives and the advantages of nano delivery systems, several food scientists explore ways to design customized nano delivery systems for the particular bioactive compound. Various nano delivery systems such as nanoemulsions, nanoliposomes, nanostructured lipid carriers, nanohydrogels, layer-by-layer nanoparticles, nanofibers, and biopolymer-based delivery systems have been widely used for the delivery of food bioactives (Bhushani et al., 2017; Ezhilarasi et al., 2013). In this chapter, nano delivery systems developed from various approaches will be condensed by studying the current studies.

2 Requirement of nano delivery system

The most important challenges associated with food bioactives delivery are physical and chemical instability, degradation during food processing, storage, transportation, and digestion.

This section describes the requirement of nano delivery systems for food bioactives at different phases involved in food product development and marketing.

2.1 Poor oral bioavailability

Poor oral bioavailability is a major drawback associated with various food bioactives. The bioavailability of hydrophobic molecules is very poor due to their very low aqueous solubility (Rezaei et al., 2019). In many cases, hydrophilic bioactives with high molecular weight and also the irreversible binding of bioactives with proteins in gastrointestinal conditions lead to very low bioavailability of hydrophilic food bioactives. Degradation of food bioactives before delivering to their active adsorption site due to various environmental conditions such as pH, ionic strength, temperature, enzyme, oxygen, etc., limits their bioavailability. So hydrophobic and certain hydrophilic nutraceuticals (e.g., quercetin, catechin) are required delivery systems to overcome the problems related to their bioavailability (Aditya et al., 2017).

2.2 Development of functional food

Many foods and nutraceutical industries facing problems in the development of functional food due to their chemical and physical instability during product development. This causes product destabilization. Also, the food system is composed of various ingredients with diverse physicochemical properties. Food bioactives undergo interaction with other ingredients such as oxidizing and reducing agents, transition metals, and hydrogen ions which induce chemical degradation in the food system. Various bioactive compounds are very sensitive to pH, temperature, oxygen, and light. The development of food products involves various extreme conditions that food ingredient undergoes like thermal sterilization, high-pressure mixing, blending, etc. leads to rapid oxidation, hydrolysis, and reduction of bioactives results in degradation or very less active ingredient in product (Aditya et al., 2015). Hydrophilic and hydrophobic bioactives have a problem in the incorporation of lipid enriched and aqueous food products respectively, due to their solubility. It affects food product stability and sensory attributes due to the mass transfer of bioactive compounds from one site to another site (McClements, 2015; Yao et al., 2014; Patel and Velikov, 2011).

2.3 Consumer and regulatory acceptance

Consumer acceptance and regulations are a serious concern in the nano delivery system. A serious safety concern arises due to the possibility of smaller size nanoparticle transportation to cells and tissues higher than that of normally accepted concentration. Type of bioactive compound, their concentration, physical and chemical properties, and their interaction with living cells are needed to be concentrated and should follow proper regulatory aspects for designing nano delivery system. It is important to find the proper clearance of such a delivery system after gastrointestinal digestion and its probable toxicity effects (Borel and Sabliov, 2014; Sampathkumar et al., 2020).

3 Properties of the delivery system

The delivery system has various functional and technical properties suitable for different applications in food industries. Functional properties such as loading capacity, loading efficiency, delivery efficiency, delivery mechanism, protection against degradation, and bioavailability.

- **Loading capacity** is defined as the ratio of the mass of the core material to the mass of the carrier material. The delivery system should have a higher loading capacity.
- **Loading efficiency** is the capacity of the delivery system to hold the core material over time. Loading efficiency is preferred to be higher throughout the processing, storage, transport, and delivery.
- **Delivery efficiency** is the capacity of the delivery system to deliver the core material at the required site of action. It should be higher at the targeted site of its activity.
- **The delivery mechanism** is depending on the design of the delivery system that carries the core material and delivers it to the targeted site. The release of core material may have to be at a controlled rate or it may be triggered by environmental factors such as pH, ionic strength, enzyme activity, and temperature.
- **Protection against degradation**: Various factors such as heat, light, pH, oxygen or chemicals promote degradation reaction during processing, storage, and transport.

- **Bioavailability** is the amount of the active material that is available in systemic circulation for its bioactivity in the targeted site. The delivery system should enhance the bioavailability of the bioactive compound.

Technical properties of the delivery system are food bioactives should be made of food-grade ingredients (GRAS: Generally Recognized as Safe), compatible with food matrix, compatible with the processing system, and economically effective. The delivery system should be compatible with the food matrix that it should not affect the texture, flavor, color, stability, and nature of the food product. Also, the delivery system should be compatible with food processing conditions such as thermal processing, drying, freezing, chilling, pH, high ionic strengths, and mechanical stresses. The delivery system should be developed with inexpensive ingredients. Benefits obtained from delivery systems should compensate for the costs utilized for developing delivery systems (McClements and Li, 2010).

4 Nano delivery system

The demand for the development of a suitable delivery system for bioactive compounds rises to control, localize, and improve the delivery of food bioactives. Various delivery systems have been developed based on lipid, surfactant, biopolymer, and layer-by-layer assembly.

4.1 Lipid-based delivery system

The lipid-based delivery system consists of lipids, surfactants, and hydrophilic co-solvents. A lipid-based delivery system reduces the poor and incomplete solubility of poorly soluble bioactives. It also facilitates the formation of solubilized structures after digestion from which absorption occurs (Cerpnjak et al., 2013). Some of the recent lipid-based nano delivery systems are listed in Table 1.

4.1.1 Nanoemulsion

Nanoemulsion is the development of a colloidal delivery system to encapsulate bioactive ingredients. It is a nonhomogeneous system in which oil droplets are dispersed in an aqueous medium and stabilized by surfactants. Fig. 2B shows the nanoemulsion formation. Nanoemulsion has high colloidal stability, optical transparency, and a high interfacial area to volume ratio due to the presence of nanometric size droplets in the range of 10–200 nm (Donsì, 2018; McClements, 2011, 2012). Surfactants are the surface-active compounds that play a major role in assisting the stability of the nanoemulsion by adsorbing at the oil-water interface which reduces the interfacial tension between the oil and water phase (Gasa-Falcon et al., 2020). The formation, stability, and functionality of the nanoemulsion are depended on the properties of core, wall, surfactant, carrier material, and processing condition (Silva et al., 2012). Two different methods are commonly used for the production of nanoemulsion namely the high-energy method and the low-energy method. A high-energy method such as high-speed/high-pressure homogenization, ultrasonication, membrane emulsification, and microfluidization applies mechanical shear force and disruptive force which break up

TABLE 1   Recent lipid-based nano delivery system for food bioactives.

Nano delivery system	Bioactive compound	Production method	Main findings	References
Nanoemulsion	Curcumin	High-pressure homogenization	Improved stability with medium-chain triglyceride oil and Tween 80 Higher oil concentration showed increased curcumin content and less stability	Ma et al. (2017)
	Betacarotene	High-speed homogenizer	Improved stability for long-term storage	Moreira et al. (2019)
	Resveratrol	Ultrasonication	Production against degradation and good antioxidant activity	Kumar et al. (2017)
	Docosahexaenoic acid	Micro fluidization	Increased stability and extended lipid digestion	Karthik and Anandharamakrishnan (2016)
	Vitamin E	Dual-channel microfluidizer	Improved bioaccessibility	Lv et al. (2018)
SLN	Betacarotene	High shear homogenization	Effective protection in gastric and intestinal condition; better stability during storage	Selvakumar et al. (2019)
	Curcumin	Maillard conjugate	Showed homogenous spherical particles; enhanced entrapment efficiency; improved stability against pH, ionic strength, and gastrointestinal condition; and highly controlled release of curcumin	Huang et al. (2020)
	Curcumin and resveratrol	Probe sonication	Improved antioxidant and bioavailability	Gumireddy et al. (2019)
	Hydroxycitric acid	Hot homogenization	Excellent storage stability; controlled release; higher bioavailability	Ezhilarasi et al. (2016)
	Curcumin and capsaicin	High shear homogenization	Better encapsulation efficiency; high storage stability	Nishihira et al. (2019)
	Fish oil	High-speed blending and spray drying	Improved physical stability of emulsions	Azizi et al. (2019)

TABLE 1 Recent lipid-based nano delivery system for food bioactives—cont'd

Nano delivery system	Bioactive compound	Production method	Main findings	References
	Curcumin	High shear homogenization and ultrasonication	Higher bioaccessibility (>91%) of curcumin	Ban et al. (2020)
NLC	β-Carotene	Hot high shear homogenization	97.7% encapsulation efficiency was achieved and remained stable up to14 days at 25°C	Pezeshki et al. (2019)
	Vitamin A Palmitate	Hot homogenization method	98.5% encapsulation was observed; good storage stability	Pezeshki et al. (2014)
	Quercetin and linseed oil	High-pressure homogenization technique	NLC showed good stability for more than 3 months	Huang et al. (2017)
	Coenzyme Q10	Hot high-pressure homogenization technique	98.4% encapsulation efficiency was achieved	Wang et al. (2012)
	Cinnamon essential oil	Homogenization	82% encapsulation efficiency was obtained	Bashiri et al. (2020)
Nanoliposomes	Resveratrol	Film dispersion method	Improved physical stability and retention of resveratrol	Shao et al. (2018)
	Curcumin	Thin-film evaporation and dynamic high-pressure microfluidization	The addition of pluronic enhanced the pH and thermal stability of liposomes	Li et al. (2018)

the droplets into a smaller size. Lower energy methods such as self-emulsification and phase inversion use the internal chemical energy of the system without or with gentle stirring which permits the development of small droplets (Gasa-Falcon et al., 2020; Jiang et al., 2020).

The incorporation of food bioactives in the lipid phase of the nanoemulsion facilitates (1) improved dispersibility of food bioactives in the aqueous phase, (2) minimizes the phase separation, (3) protecting the food bioactives from chemical interaction with food ingredients, retaining its functional properties, and prevents food deterioration due to lipid oxidation, (4) minimize the influence on organoleptic properties of food, (5) improves the bioaccessibility, absorption, and bioavailability due to its small size which enhances the transport mechanism across the cell membrane. In a nanoemulsion system, oil droplets behave as Brownian motion due to their nano-size and do not interact with each other thereby

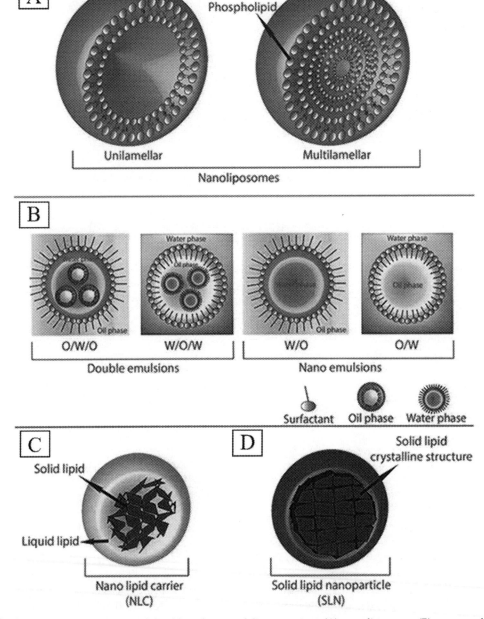

FIG. 2 Schematic illustration of lipid-based nano delivery system (A) nanoliposome, (B) nanoemulsion, (C) nanostructured lipid carrier, and (D) solid lipid nanoparticles. *Reproduced with permission from Akhavan, S., Assadpour, E., Katouzian, I., Jafari, S.M., 2018. Lipid nano scale cargos for the protection and delivery of food bioactive ingredients and nutraceuticals. Trends Food Sci. Technol. 74 (February), 132–146. https://doi.org/10.1016/j.tifs.2018.02.001.*

promoting better stability against sedimentation, gravitational separation, aggregation, creaming, flocculation coalescence, coagulation, and precipitation (Donsì et al., 2011). Nanoemulsion is the easiest and convenient delivery system for the development of aqueous-based functional food systems (such as salad dressing, dips, beverages, sauces, desserts, confectionery, and dairy products such as ice cream, yogurt, milk, and table spreads) (Dasgupta et al., 2019). Various food bioactives such as curcumin (Artiga-Artigas et al., 2018), betacarotene (Moreira et al., 2019), resveratrol (Kumar et al., 2017), omega 3 fatty acid (Karthik and Anandharamakrishnan, 2016), green tea catechins (Anu Bhushani and Anandharamakrishnan, 2014), etc. have been encapsulated using nanoemulsion technique with improved stability and bioavailability.

4.1.2 Solid lipid nanoparticles

Solid lipid nanoparticles (SLN) is a colloidal nano delivery system composed of a solid lipid core containing food bioactives in the lipid matrix with monolayer phospholipid as shell material (Sivakamasundari et al., 2020). The SLN is stabilized by a single or mixture of various surfactants (Cruz et al., 2015). The structure of the SLN is shown in Fig. 2D. The composition of the carrier vehicle plays a major role in controlling the structure and characteristics of the SLN. Materials (solid lipids, phospholipids, and emulsifiers) used for the preparation of SLN are generally recognized as safe, biocompatible, and nontoxic. Lipid matrix made up of mono and di-glycerides enhances the solubility of the poor soluble food bioactives. Lipids used for SLN should be extremely stable against degradation due to oxidation (Aditya and Ko, 2015). Most commonly used lipids for SLN formation include triglycerides, waxes, fatty acids, steroids, and partial glycerides. A low degree of polymorphism in long-chain triglycerides is highly suitable for the formation of SLN compared to short-chain triglycerides. The stability and release of bioactives encapsulated in SLN can be controlled effectively by controlling the physical state of the lipid matrix (da Silva Santos et al., 2019). SLN is formed by heating the solid lipid above its melting point and the surfactant is added to hot liquid lipid for stabilization. Then it remains solid particles at room temperature (Sivakamasundari et al., 2020). SLN can be produced possibly from the oil in water nanoemulsion or water in oil nanoemulsion or multilayer nanoemulsion. SLN size depends on the emulsion droplets' size. Various approaches such as high-pressure homogenization, inclusion complexation, ultrasonication, microfluidization, nanoprecipitation, solvent injection, spray drying, freeze drying, spray freeze drying, coacervation, and supercritical fluid technique can be used for the production of SLN with a wide range of particle size from 10 to 1000nm (Lin et al., 2017). SLN can encapsulate the food bioactives with enhanced stability against degradation, entrapment efficiency, improved bioavailability, and controlled delivery of the food bioactives on the target site (Weiss et al., 2008). SLN has some limitations such as less loading capacity due to its highly ordered crystalline structure, the low ability for encapsulation of water-soluble food bioactives, the possibility of the expulsion of bioactives from SLN due to polymorphic transformation of crystalline lipid matrix during storage (Sivakamasundari et al., 2020).

4.1.3 Nanoliposomes

Nanoliposomes are amphiphilic lipid-based delivery systems formed from phospholipids in the nanoscale range (10 to 1000 nm) shown in Fig. 2A. Polar heads of phospholipids are faced in the aqueous part of the internal and external media. Nonpolar tails are connected in a bilayer manner. In nanoliposomes, hydrophilic bioactives can be incorporated within an aqueous cavity in the central region whereas lipophilic bioactives can be encapsulated in a nonpolar layer at the peripheral region of the resultant capsules (Dutta et al., 2018). Nanoliposomes are formed by the interaction between the aqueous part and amphiphilic lipid. The existence of hydrophilic and lipophilic parts enables the efficient entrapment and delivery of hydrophilic, lipophilic, and amphiphilic bioactives (Akhavan et al., 2018). Lecithin is the most commonly used phospholipid in the liposomal delivery system.

Four fundamental steps (Ghorbanzade et al., 2017) involved in the preparation of nanoliposomes are as follows:

- Analyzing the generated nanoliposomes
- Scattering the lipid in an aqueous media via high-energy means
- Drying down lipids from an organic solvent
- Purifying the ultimate yield

Thin layer hydration techniques and noncholesterol techniques are the most commonly used traditional techniques for the development of nanoliposomes. The thin layer hydration technique uses cholesterol whereas the noncholesterol method uses solvent (e.g., Glycerol) instead of cholesterol and heating it in a water bath. The size of the liposomes was further reduced to nano-size using high-energy methods such as ultrasonication. These traditional methods are not efficient for the development of complex nanoliposomes. Nowadays, high-energy techniques such as ultrasonication, extrusion, microfluidization, and freeze-thawing have been used for nanoliposomes development with higher stability and minuscule and monolithic structures (Akhavan et al., 2018).

The main drawbacks associated with nanoliposomes (Mortazavi et al., 2007) are as follows:

- Residual of utilized organic solvents
- Use of synthetic solvents and detergents for desolvation
- Lack of scalability and affordability
- Physical stability and chemical lability

4.1.4 Nanostructured lipid carriers

Nanostructured lipid carriers (NLC) are formed by mixing solid and liquid lipids at room temperature causes melting point depression due to the formation of highly disordered solid matrices (Nobari Azar et al., 2020). In NLC, crystallization of lipid droplets occurs partially with an amorphous structure (Park et al., 2017) results in increased loading capacity and release of bioactives compared to the SLN matrix. Surfactants are used for the stabilization of lipid dispersion by lowering the interfacial tension between the lipid and water phases. The structure of NLC is illustrated in Fig. 2C. Unlike SLN, NLC is produced from more liquid lipids preferably in the range of the ratio of 70:30 of liquid lipid to solid lipid 99.9:0.1 ratio. The physical, chemical, and colloidal stability of the NLC are strongly affected by lipids. The selection of suitable solid and liquid lipid is more important for the formulation of

NLC (Schäfer-Korting et al., 2007; Jenning et al., 2000). Lipids should have more solubility for bioactives, more miscibility, and compatibility with solid lipid (Han et al., 2012; Joshi et al., 2008). Solid and liquid lipid should be added in proper proportion for the formation of NLC to avoid instability or phase separation at temperatures below the melting point (Tamjidi et al., 2013). Liquid lipid should not involve in solid lipid crystallization and also solid lipid crystals should not dissolve in an oil phase. The most commonly used liquid lipid for NLC productions is medium-chain triglycerides (MCT) oil and oleic acid. MCT oil digests faster and has more stability against oxidative degradation. Both MCT oil and oleic acid are odorless. Different methods such as high-pressure homogenization, ultrasonication, microemulsion, solvent evaporation, phase inversion, solvent diffusion, solvent injection or solvent displacement, and membrane contractor have been used for the production of NLC (Nobari Azar et al., 2020). Release of bioactives from the lipid matrix occurred by diffusion mechanism and degradation of the lipid occurs concurrently in the body. The rate of release and degradation depends on the type and properties of the lipids. Shorter the fatty acids chains of the triglycerides fasten the rate of degradation (Olbrich et al., 2002).

The release of bioactives from the matrix is influenced by the below-listed factors (Jaiswal et al., 2016).

- Higher surface-area-to-volume ratio of nanoparticles
- High movement of the bioactive compound
- The low partition coefficient of the bioactives
- Highly disordered dispersion of the bioactive compound in the lipid matrix
- Poor crystallinity of the lipid carrier

NLC can accommodate higher bioactive compounds due to structural imperfection during solidification. Also, NLC has higher encapsulation efficiency due to the higher solubility of bioactives in liquid lipid than solid lipid matrix. Further, NLC facilitates higher stability during storage due to its stronger inclusion of food bioactives. NLC can encapsulate both hydrophilic and hydrophobic food bioactives (Poonia et al., 2016).

4.1.5 Lyotropic liquid crystalline nanostructures

Lyotropic liquid crystalline (LLC) nanostructures composed of amphiphilic lipids with the polar head group and a nonpolar tail group capable of self-assembly into different liquid crystalline structures when exposed to the aqueous solution (Aditya et al., 2017). These structures are called mesophases or liquid crystals. Three different structures such as cubic, hexagonal, and lamellar can be produced by the LLC system (Karami and Hamidi, 2016). The biphasic structures of LLC are thermodynamically stable nano delivery systems and occur in an equilibrium state with higher aqueous media. The temperature used for the development of LLC and water content has a great impact on the type of structure formation and alignment in aqueous media (Mezzenga et al., 2019). In cubic LLC, it is in form of the bi-continuous domain of the aqueous and lipid phases. Therefore, it is possible to load both hydrophilic and hydrophobic compounds in cubic LLC. Whereas in hexagonal structure LLC, lipids are formed in for of column which separates the water from the inner and outer side of the structure. Thus, this hexagonal structured LLC is suitable for the delivery of hydrophilic bioactives because of its aqueous core-site. Fig. 3 illustrates the cubic (Fig. 3A) and hexagonal (Fig. 3B) LLC nanostructures. In the case of lamellar structures, an aqueous compartment segregates the

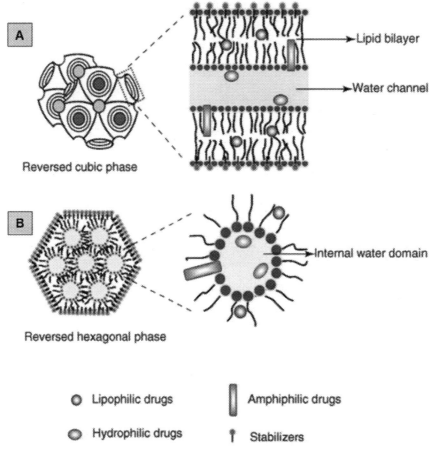

FIG. 3   Schematic illustration of lyotropic liquid crystalline nanostructures (A) cubic LLC and (B) hexagonal LLC. *Reproduced with permission from Guo, C., Wang, J., Cao, F., Lee, R.J., Zhai, G., 2010. Lyotropic liquid crystal systems in drug delivery. Drug Discov. Today 15(23–24), 1032–1040. https://doi.org/:10.1016/j.drudis.2010.09.006.*

planar lipid bilayers. It has unique morphological characteristics, hydrophilic and hydrophobic domains, the possibility of a wide range of order-order transition depends on small changes in the external environment (Guo et al., 2010).

The most commonly used lipid information of the LLC delivery system is glycerol monooleate by both top-down and bottom-up approaches. In the case of the top-down method, high-energy techniques such as lipid mixed with surfactant high-pressure homogenization, sonication, microfluidization for dispersing the lipid mixed with surfactant in the water phase (Esposito et al., 2003). In the bottom-up method, hydrotropes are responsible for lipid solubilization to avoid a more viscous lipid phase and then mixed with surfactant. Then, this mixture is dispersed in an aqueous phase with the application of minimal energy inputs (Spicer et al., 2001). The bottom approach has more advantages such as low-energy input, no possibility of thermal degradation, and is suitable for encapsulation of heat-sensitive food bioactives. It forms uniform particles with higher stability (Karami and Hamidi, 2016).

4.2 Surfactant-based delivery system

The lipid-based delivery system involves high-cost ingredients, health issues, and instability in presence of the acidic condition, and bile salt has demanded the design of a surfactant-based delivery system which are inexpensive and stable during the formation of complex food product and also in the gastrointestinal system.

4.2.1 Reverse micelle

A reverse micelle nano delivery system involves the aggregation of surfactants containing hydrophilic molecules as inner core material within bulk nonpolar solvent in nanometric size (Senske et al., 2018) as shown in Fig. 4. Generally, a reverse micelle is formed by dissolving the surfactant completely in a nonpolar solvent, and then the aqueous buffer is added to the surfactant solution. Reverse micelle is thermodynamically stable, transparent, clear, homogenous in size, and sequester the bioactive molecule in an aqueous core (Orellano et al., 2017). Morphology, internal structure, aggregation number, and viscosity are the major characteristics that need to be studied in a reverse micelle delivery system (Sun and Bandara, 2019). Various surfactants such as anionic, cationic, nonionic, zwitterionic, and mixed surfactants can be used for the formation of a reverse micelle delivery system (Lépori et al., 2016). The most commonly used surfactant for the formation of the reverse micelle is sulphosuccinic acid bis (2-ethylhexyl) ester sodium salt since it does not require any co-surfactant and can incorporate a large number of active molecules (Fuglestad et al., 2016). Aqueous content in reverse micelle has a direct impact on the size of the reverse micelle and it is commonly expressed by W_o (mol [water]/mol [surfactant]). The values of the Wo are depending on the temperature and type of polar and nonpolar solvent used for the development of reverse micelle formation. The encapsulated water molecules are highly immobilized due to the interaction between water and surfactant head groups (counterions and dipole) which is different from the bulk form of water. Mobility of the water increases with an increase in water content. When W_o is more than 16, the characteristics of AOT containing reverse micelle are the same as that of bulk water form. Advantages of the reverse micelle are excellent potential in scaling-

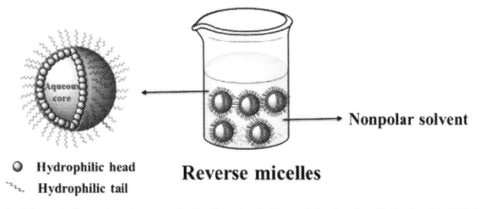

FIG. 4 Schematic illustration of reverse micelles. *Reproduced with permission from Sun, X., Bandara, N., 2019. Applications of reverse micelles technique in food science: a comprehensive review. Trends Food Sci. Technol. 91(April), 106–115. https://doi.org/10.1016/j.tifs.2019.07.001.*

up, low cost, more convenience, and protection of bioactives from degradation. Recently, a reverse micelle system was developed for encapsulation of nisin with vegetable oil, distilled monoglycerides, ethanol, and water. In brief, the nonpolar phase was prepared by mixing ethanol and vegetable oil. Then distilled monoglycerides were added to the nonpolar phase. Then water was added slowly to develop the reverse micelle delivery system (Sun and Bandara, 2019).

4.2.2 Niosomes

Niosomes are the vesicular delivery system developed from self-assembly of nonionic surfactants in aqueous solution led to the formation of closed bilayer structure depending on its amphiphilicity (Matos et al., 2019). Niosomes can be developed with a size ranging from 10 nm to 3 μm (Moghassemi and Hadjizadeh, 2014). Niosomes are formed due to the interaction between surfactant and water molecules. Niosomes can be used to encapsulate hydrophilic, hydrophobic, and amphiphilic bioactive molecules (Mahale et al., 2012). Lipophilic bioactive molecules can be loaded in the lipidic membrane whereas hydrophilic molecules can be loaded in the inner aqueous section. Fig. 5 illustrates the scheme of niosome structure. Niosomes have the potential for better skin permeation, sustained release of bioactive molecules, higher stability, and a low-cost delivery system. Niosomes are osmotically active, have high stability against chemical degradation, and have long-term storage stability. Surface modification on niosomes is simple due to the presence of a functional group on the hydrophilic head. Niosomes are biocompatible and nonimmunogenic. Variables used for the niosomal delivery system can be easily controllable and possess low toxicity due to their nonionic behavior (Matos et al., 2019).

The vesicular delivery system is divided into two types namely unilamellar and multilamellar vesicles shown in Fig. 6. Unilamellar vesicles are developed from a unique bilayer containing aqueous media which can be small (< 100 nm) or large (100 to 1000 nm) or giant (>1000 nm) unilamellar vesicles. Multilayer lamellar vesicles are formed by bilayer containing aqueous media which have a greater number of double layers. Generally, the multilayer lamellar vesicle size is greater than 1000 nm (Matos et al., 2019).

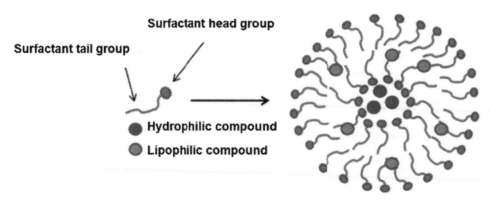

FIG. 5   Schematic illustration of noisome. *Reproduced with permission from Matos, M., Pando, D., Gutiérrez, G., 2019. Elsevier. https://doi.org/10.1016/b978-0-12-815673-5.00011-8.*

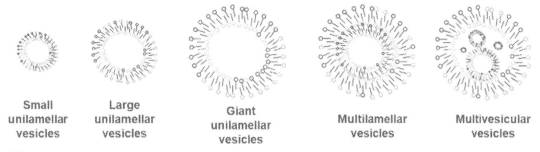

Small unilamellar vesicles   Large unilamellar vesicles   Giant unilamellar vesicles   Multilamellar vesicles   Multivesicular vesicles

FIG. 6 Schematic illustration of the classification of vesicle structure. *Reproduced with permission from Matos, M., Pando, D., Gutiérrez, G., 2019. Elsevier. https://doi.org/10.1016/b978-0-12-815673-5.00011-8.*

Key factor influenced the formation of niosomes

- Type, concentration, and structure of nonionic surfactant (Ghanbarzadeh et al., 2015).
- Additives such as membrane additive (e.g., cholesterol) (Ghanbarzadeh et al., 2015), steric additive (e.g., polyethylene glycol), and surface additives (Ghanbarzadeh et al., 2015) (e.g., diacetyl phosphate).
- Temperature: hydrating temperature used for the preparation of niosomes should be greater than the gel-liquid phase transition temperature for higher entrapment efficiency (Ghanbarzadeh et al., 2015).

Various approaches used for the development of niosomes are as follows (Moghassemi and Hadjizadeh, 2014; Mokhtari et al., 2017):

- Agitation-sonication method
- Thin-film hydration method
- Dehydration-rehydration vesicle method
- Reverse phase evaporation method
- Ether injection method
- Freeze and thaw method
- Injection of melted surfactants
- Injection of solid surfactants
- Enzymatic method
- Transformation of liquid lamellar crystals
- Extrusion method
- Single-pass technique
- Microfluidization
- Microfluidic flow-focusing method
- Membrane contractors
- Mozafari method
- Bubbling nitrogen method
- Supercritical method
- Proniosome method
- Transmembrane pH gradient method

Techniques used for the purification of niosomes are as follow:

- Gel filtration
- Dialysis
- Centrifugation and ultracentrifugation

4.2.3 Bilosomes

Bilosomes are the modified version of niosomes. Bilosomes are formed by surfactants along with bile salts which form globular concentric bilayer structures. Fig. 7 shows the schematic illustration of the structure and composition of bilosomes. Bilosomes are highly stable in the gastrointestinal system. The most commonly used bile salt for the development of bilosomes is sodium cholate, taurocholate, deoxycholic acid, deoxycholate, etc. Bile salts are easily available, low cost, and act as penetration enhancers for various drugs and vaccines. The application of bilosomes in nutraceutical delivery is not still explored (Matos et al., 2019).

4.3 Biopolymer-based nano delivery system

Protein and polysaccharides are the food-grade biopolymers most commonly used for the development of nano delivery systems suitable encapsulation and delivery of food bioactives. The use of biopolymers attracted interest among the research groups in the food and pharmaceutical industries due to their biodegradability, excellent biomimetic activity, ease of design, and structural variations (Faridi Esfanjani and Jafari, 2016). Also, various nanostructured delivery systems such as lipid-based nanoencapsulation, nanoemulsion,

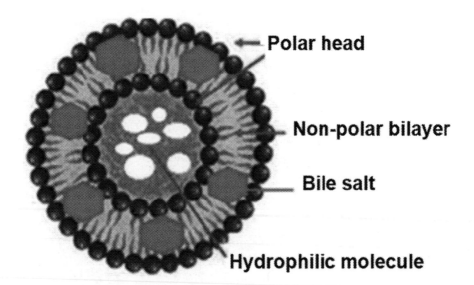

FIG. 7 Schematic illustration of structure and composition of bilosomes. *Reproduced with permission from Shukla, A., Mishra, V., Kesharwani, P., 2016. Bilosomes in the context of oral immunization: development, challenges and opportunities. Drug Discov. Today 21(6), 888–899. https://doi.org/10.1016/j.drudis.2016.03.013.*

complex nanoparticles, nanohydrogels, polymeric nanoparticles, nanotubes, and nanofibers can be developed using food biopolymers (proteins, carbohydrates) along with other food-grade ingredients such as fats, surfactants, and copolymers to deliver a wide of range of food bioactives. These structures can be fabricated using various techniques including nanoprecipitation, electrospinning, coacervation, injection, layer-by-layer self-assembly, and gelation method. A biopolymer colloidal nano delivery system can be developed from protein and polysaccharides by different top-down and bottom-up approaches such as biopolymer aggregation, segregation, and disruption. The inherent variability and complexity associated with biopolymers is a challenge to develop a delivery system with well-defined physicochemical and functional characteristics (Matalanis et al., 2011).

4.3.1 Complex nanoparticles

Various biopolymer molecules have the binding capacity with bioactive compounds forming the complex system. The bioactive molecule can be bound with individual biopolymer molecules or within the cluster of single biopolymer or multiple biopolymers (Matalanis et al., 2011). Bioactive molecules bind with one or more active sites of individual biopolymer molecules through specific or nonspecific interaction with another molecule. Globular proteins (b-lactoglobulin and bovine serum albumin) tend to bind with bioactives such as resveratrol, docosahexaenoic acid, and linoleic acid, and other surfactants in the hydrophobic sites on its surface (Kelly et al., 2003). Caseinate is a kind of flexible protein bound with food bioactives and forms a complex molecular structure that remains dispersed in aqueous media (Semo et al., 2007). The binding ability of bioactives and biopolymers is due to either hydrophilic or hydrophobic or electrostatic interaction depending on the nature of the bioactives and biopolymers. The nonpolar site of bioactives and surfactants binds with nonpolar sites of the biopolymers through hydrophobic attraction. The protein binds with oppositely charged ionic surfactant through electrostatic attraction (Bao et al., 2008). The stability and structure of the delivery system are depending on the type and concentration of surfactant, external environmental conditions, and interaction between ingredients. Surface-active bioactives bind with biopolymers and are formed moreover as monomer or micelle-based clusters depending on concentration, type, and several surfactants, biopolymers, and interactions (Semo et al., 2007; Portnaya et al., 2006). There are binary and ternary complex nano delivery systems with two or three biopolymers for encapsulation of food bioactives (Xie et al., 2019; Hu et al., 2020). Protein such as whey protein, zein, soy protein, bovine serum albumin, wheat protein, barley protein, etc., and polysaccharides and hydrocolloids are used for the development of complex nanoparticles. Fig. 8 shows the complex nanoparticles.

4.3.2 Nanohydrogels

Nanohydrogels are composed of 3D cross-linked polymer network at 10 to 1000 nm in diameter. These hydrogels are highly hydrophilic and superabsorbent material (Peppas et al., 2000). It has a large surface area and interior network for bioactive entrapment and also high swelling and water holding capacity. Cross-linking polymer provides structural stability to the delivery system and prevents the dissolution of polymer in aqueous media. It possesses high mechanical properties, is renewable, has high flexibility, is versatile, has biodegradability, biocompatibility, and is cheaper. The characteristics of nano hydrogel including size,

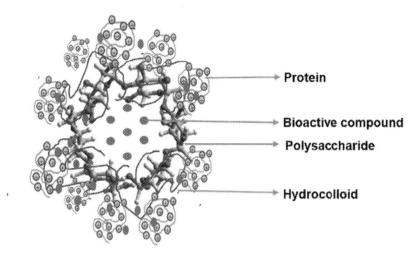

FIG. 8   Schematic illustration of complex nanoparticle.

porosity, chemical functions, degradation, swelling, and softness can be controlled by altering the composition of ingredients (Schexnailder and Schmidt, 2009).

Nanohydrogels are classified into different types based on the source of origin (natural and synthetic), type of cross-linking (physical and chemical), and response due to external stimulus (biochemical, physical, or chemical). In chemical cross-linking, linkage of the network is made by covalent bonding. The type of chemical linkage and functional groups responsible for the formation of a linkage network defines the nature and behavior of hydrogel. Photon-induced, amine, and di-sulfide-based chemical cross-linking forms hydrogels. Physical cross-linking involves various physical processes such as electrostatic interactions, complexation or crystallization, association, aggregation, crystallization, Van der Waal's forces, and hydrogen bonding. Physical cross-linking is weaker in characteristics compared to chemical cross-linking due to covalent bonding (Setia and Ahuja, 2018).

Nanogels can be formed by using polymer precursors and heterogeneous polymerization of monomers. Polymer precursors consist of active sites modified with functional groups that develop interaction between polymers either by physical or chemical cross-linking. The second method involves two steps that are polymerization and then the formation of nanogels (Raemdonck et al., 2009).

Techniques used for nanohydrogels formation are as follows (Setia and Ahuja, 2018):

- Emulsion polymerization technique
- Water in oil heterogeneous emulsion methodologies
- Photolithographic techniques
- Micro molding method
- Preparation of copolymers and biopolymers
- Heterogeneous controlled/living-radical polymerization
- Physical self-assembly of interactive polymers
- Transfer of macroscopic gels into nanogels
- Associating polymer-based nanogels

- Chemical cross-linking
- Heterogeneous free-radical polymerization

The swelling and shrinkage behavior of nanohydrogels are very sensitive to environmental changes such as pH, temperature, ionic strength, and other factors that have to be examined in detail because of their great potential.

The mechanism involved in the release of food bioactives from the nanohydrogels including (Setia and Ahuja, 2018)

➤ pH-responsive
➤ Diffusion
➤ Degradation
➤ Photochemical internalization and photoisomerization
➤ Thermal-sensitive and volume transition mechanism

4.3.3 Nanopolymerosomes

Nanopolymerosomes are a form of bilayer vesicles developed from the amphiphilic copolymers in which food bioactives can be entrapped in the cavity at nanoscale size as same as nanoliposomes (Bouwmeester et al., 2009). The main difference between nanoliposome and nanopolymerosomes is amphiphilic copolymer is used for the development of nanopolymerosomes instead of phospholipids which are used for liposomes. Polymersomes can be used to encapsulate both hydrophilic and lipophilic bioactive molecules. Polymersomes increase stability, versatility, and provide controlled release. Polymersomes can be made with tri-block copolymers (polymer A-polymer B-polymer A) and synthesized using various methods including atomization spraying, salting out, solvent displacement, nanoprecipitation, desolvation, emulsion evaporation, pH cycling, thermal treatment, and the use of supercritical fluids (Rastogi et al., 2009).

4.4 Layer-by-layer self-assembly

Layer-by-layer (LBL) self-assembly delivery system can be developed from the interaction between oppositely charged electrolytes through electrostatic attraction. Nanoscale LBL self-assembled system can be developed using solid charged supporting material as the base plate and it can be dissolved after the development of LBL nanoparticles (Turasan and Kokini, 2020b). The first layer of LBL is formed by placing the oppositely charged polyelectrolytes with respect to the charge of the solid supporting template. Then the oppositely charged second polyelectrolyte with respect to the first polyelectrolyte is introduced over the first electrolyte forms the first bilayer structure. Then, the multiple bilayers can be prepared by repeating the deposition of polyelectrolytes depending on the required thickness of the system. Both spherical and tubular shapes can be obtained by placing spherical and cylinder-shaped supporting solid templates. Bioactive compounds can be incorporated in the inner core of the particle or in between the layers. In the case of a cylinder-shaped structure, bioactives are incorporated in the inner portion of the cylinder-shaped tubular structure (Bastarrachea et al., 2015).

4.4.1 Spherical-shaped nanoparticle through LBL self-assembly

Encapsulation of curcumin in spherical nanoparticles was developed using sorghum and kafirin protein to improve the bioavailability of curcumin. The desolvation method was used for the formation of LBL assembly with ethanol and water as solvent and nonsolvent. Sodium carbonate was used as a basic supporting template. Curcumin mixed kafirin solution was added over the sodium carbonate. Then sodium carbonate was removed by leaving in water which forms a hollow kafirin layer. Then the first layer was formed by deposition of negatively charged dextran sulfate over the kafirin layer. Then positively charged chitosan was placed over dextran sulfate which forms the first bilayer structure. Then, the deposition of dextran sulfate and chitosan was repeated to obtain the double bilayer structure as shown in Fig. 9 (Li et al., 2019).

4.4.2 Nanotube through LBL assembly

Nano tubular structure was developed through layer-by-layer self-assembly using bovine serum albumin (negative charge) and poly-D-lysine (positive charge) for encapsulation of curcumin. pH significantly influences the zeta potential of the polyelectrolytes. The highest potential difference between bovine serum albumin and poly-D-lysine was observed at pH 7.4 and a nanotube was developed at that pH. A hollow cylinder-shaped polycarbonate track-etched membrane was used as a solid template. Poly-D-lysine has deposited over the solid template forms the first layer. Then, bovine serum albumin was deposited over the poly-D-lysine and forms the second layer. This was repeated to form multiple bilayers. Fig. 10 shows the LBL nanotubes with bovine serum albumin and sodium alginate using polycarbonate as a template with various pore sizes. Then, the solid template was dissolved in N, N-dimethylformamide solution and the developed LBL nanotubes were freeze dried (Turasan and Kokini, 2020b).

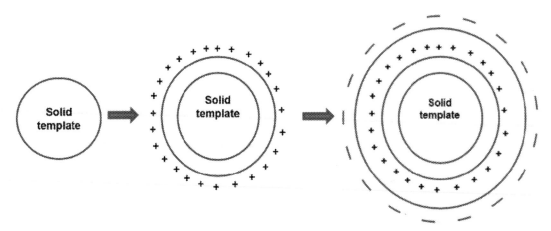

FIG. 9   LBL self-assembly of polyelectrolytes on a supporting solid template.

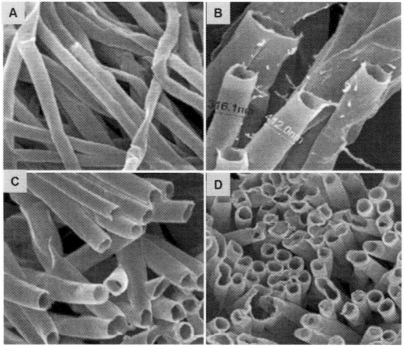

FIG. 10   LBL nanotubes with bovine serum albumin and sodium alginate were developed using a polycarbonate template with a different pore size (200 nm (A), 400 nm (B), 600 nm (C), and 800 nm (D)). *Reproduced with permission from Turasan, H., Kokini, J.L., 2020. Enhancing the Bioavailability of Nutrients by Nanodelivery Systems. INC. https://doi.org/10.1016/b978-0-12-815866-1.00009-1.*

5 Conclusion and future perspective

The major challenges associated with the delivery of food bioactives are stability during processing, storage, and transportation. Also, the physicochemical properties of the food bioactives, their chemical instability against the digestive environment, and poor intestinal permeability cause very low oral bioavailability. In this context, the delivery of food bioactives in the nanoscale system has shown effective improvement in its bioavailability. Nano delivery system protects the bioactives from degradation during processing, storage, transportation, and digestion. Further, the nano delivery system provides controlled smart and sustained delivery of bioactives in the targeted site, thus improving bioavailability. With this background, this chapter summarized the challenges and recent studies on nano delivery systems for food bioactives. Also, the chapter described the detailed mechanism and advantages of lipid, biopolymer, surfactant, and layer-by-layer self-assembly-based nano delivery systems. However, the characteristics of different nano delivery systems have different from one other. Therefore, the clearance from the body and toxicity effect has to be studied individually. With proper guidelines and detailed knowledge of the physicochemical properties of nano delivery systems, food bioactives can be delivered effectively.

References

Aditya, N.P., Ko, S., 2015. Solid lipid nanoparticles (SLNs): delivery vehicles for food bioactives. RSC Adv. 5 (39), 30902–30911. https://doi.org/10.1039/c4ra17127f.

Aditya, N.P., Aditya, S., Yang, H., Kim, H.W., Park, S.O., Ko, S., 2015. Co-delivery of hydrophobic curcumin and hydrophilic catechin by a water-in-oil-in-water double emulsion. Food Chem. 173, 7–13. https://doi.org/10.1016/j.foodchem.2014.09.131.

Aditya, N.P., Espinosa, Y.G., Norton, I.T., 2017. Encapsulation systems for the delivery of hydrophilic nutraceuticals: food application. Biotechnol. Adv. 35 (4), 450–457. https://doi.org/10.1016/j.biotechadv.2017.03.012.

Akhavan, S., Assadpour, E., Katouzian, I., Jafari, S.M., 2018. Lipid nano scale cargos for the protection and delivery of food bioactive ingredients and nutraceuticals. Trends Food Sci. Technol. 74, 132–146. https://doi.org/10.1016/j.tifs.2018.02.001.

Anu Bhushani, J., Anandharamakrishnan, C., 2014. Electrospinning and electrospraying techniques: potential food based applications. Trends Food Sci. Technol. 38 (1), 21–33. https://doi.org/10.1016/j.tifs.2014.03.004.

Artiga-Artigas, M., Lanjari-Pérez, Y., Martín-Belloso, O., 2018. Curcumin-loaded nanoemulsions stability as affected by the nature and concentration of surfactant. Food Chem. 266, 466–474. https://doi.org/10.1016/j.foodchem.2018.06.043.

Azizi, M., Li, Y., Kaul, N., Abbaspourrad, A., 2019. Study of the physicochemical properties of fish oil solid lipid nanoparticle in the presence of palmitic acid and quercetin. J. Agric. Food Chem. 67 (2), 671–679. https://doi.org/10.1021/acs.jafc.8b02246.

Ban, C., Jo, M., Park, Y.H., Kim, J.H., Han, J.Y., Lee, K.W., Kweon, D.-H., Choi, Y.J., 2020. Enhancing the oral bioavailability of curcumin using solid lipid nanoparticles. Food Chem. 302, 125328. https://doi.org/10.1016/j.foodchem.2019.125328.

Bao, H., Li, L., Gan, L.H., Zhang, H., 2008. Interactions between ionic surfactants and polysaccharides in aqueous solutions. Macromolecules 41 (23), 9406–9412. https://doi.org/10.1021/ma801957v.

Bashiri, S., Ghanbarzadeh, B., Ayaseh, A., Dehghannya, J., Ehsani, A., 2020. Preparation and characterization of chitosan-coated nanostructured lipid carriers (CH-NLC) containing cinnamon essential oil for enriching milk and anti-oxidant activity. LWT 119. https://doi.org/10.1016/j.lwt.2019.108836, 108836.

Bastarrachea, L.J., Denis-Rohr, A., Goddard, J.M., 2015. Antimicrobial food equipment coatings: applications and challenges. Annu. Rev. Food Sci. Technol. 6 (1), 97–118. https://doi.org/10.1146/annurev-food-022814-015453.

Bhushani, A., Harish, U., Anandharamakrishnan, C., 2017. Nanodelivery of Nutrients for Improved Bioavailability. Elsevier, https://doi.org/10.1016/b978-0-12-804304-2.00010-x.

Borel, T., Sabliov, C.M., 2014. Nanodelivery of bioactive components for food applications: types of delivery systems, properties, and their effect on ADME profiles and toxicity of nanoparticles. Annu. Rev. Food Sci. Technol. 5 (1), 197–213. https://doi.org/10.1146/annurev-food-030713-092354.

Bouwmeester, H., Dekkers, S., Noordam, M.Y., Hagens, W.I., Bulder, A.S., de Heer, C., ten Voorde, S.E.C.G., Wijnhoven, S.W.P., Marvin, H.J.P., Sips, A.J.A.M., 2009. Review of health safety aspects of nanotechnologies in food production. Regul. Toxicol. Pharmacol. 53 (1), 52–62. https://doi.org/10.1016/j.yrtph.2008.10.008.

Cerpnjak, K., Zvonar, A., Gašperlin, M., Vrečer, F., 2013. Lipid-based systems as a promising approach for enhancing the bioavailability of poorly water-soluble drugs. Acta Pharm. (Zagreb, Croatia) 63 (4), 427–445. https://doi.org/10.2478/acph-2013-0040.

Cruz, Z., García-Estrada, C., Olabarrieta, I., Rainieri, S., 2015. Lipid nanoparticles: delivery system for bioactive food compounds. In: Sagis, L.M.C. (Ed.), Microencapsulation and Microspheres for Food Applications. Academic Press, pp. 313–331.

da Silva Santos, V., Badan Ribeiro, A.P., Andrade Santana, M.H., 2019. Solid lipid nanoparticles as carriers for lipophilic compounds for applications in foods. Food Res. Int. 122, 610–626. https://doi.org/10.1016/j.foodres.2019.01.032.

Dasgupta, N., Ranjan, S., Gandhi, M., 2019. Nanoemulsions in food: market demand. Environ. Chem. Lett. 17, 1003–1009. https://doi.org/10.1007/s10311-019-00856-2.

de Souza Simões, L., Madalena, D.A., Pinheiro, A.C., Teixeira, J.A., Vicente, A.A., Ramos, Ó.L., 2017. Micro- and nano bio-based delivery systems for food applications: in vitro behavior. Adv. Colloid Interface Sci. 243, 23–45. https://doi.org/10.1016/j.cis.2017.02.010.

Donsì, F., 2018. Applications of nanoemulsions in foods. In: Jafari, S.M., McClements, D.J. (Eds.), Nanoemulsions—Formulation, Applications, and Characterization. Academic Press, pp. 349–377, https://doi.org/10.1016/B978-0-12-811838-2.00011-4.

Donsì, F., Sessa, M., Mediouni, H., Mgaidi, A., Ferrari, G., 2011. Encapsulation of bioactive compounds in nanoemulsion-based delivery systems. Procedia Food Sci. 1, 1666–1671. https://doi.org/10.1016/j.profoo.2011.09.246.

Dutta, S., Moses, J.A., Anandharamakrishnan, C., 2018. Encapsulation of nutraceutical ingredients in liposomes and their potential for cancer treatment. Nutr. Cancer 70 (8), 1184–1198. https://doi.org/10.1080/01635581.2018.1557212.

Esposito, E., Eblovi, N., Rasi, S., Drechsler, M., Di Gregorio, G.M., Menegatti, E., Cortesi, R., 2003. Lipid-based supramolecular systems for topical application: a preformulatory study. AAPS PharmSci 5 (4), E30. https://doi.org/10.1208/ps050430.

Ezhilarasi, P.N., Karthik, P., Chhanwal, N., Chinnaswamy, A., 2013. Nanoencapsulation techniques for food bioactive components: a review. Food Bioproc. Tech. 6, 628–647. https://doi.org/10.1007/s11947-012-0944-0.

Ezhilarasi, P.N., Muthukumar, S.P., Anandharamakrishnan, C., 2016. Solid lipid nanoparticle enhances bioavailability of hydroxycitric acid compared to a microparticle delivery system. RSC Adv. 6 (59), 53784–53793. https://doi.org/10.1039/C6RA04312G.

Faridi Esfanjani, A., Jafari, S.M., 2016. Biopolymer nano-particles and natural nano-carriers for nano-encapsulation of phenolic compounds. Colloids Surf. B: Biointerfaces 146, 532–543. https://doi.org/10.1016/j.colsurfb.2016.06.053.

Fuglestad, B., Gupta, K., Wand, A.J., Sharp, K.A., 2016. Characterization of cetyltrimethylammonium bromide/hexanol reverse micelles by experimentally benchmarked molecular dynamics simulations. Langmuir 32 (7), 1674–1684. https://doi.org/10.1021/acs.langmuir.5b03981.

Gasa-Falcon, A., Odriozola-Serrano, I., Oms-Oliu, G., Martín-Belloso, O., 2020. Nanostructured lipid-based delivery systems as a strategy to increase functionality of bioactive compounds. Foods 9 (3). https://doi.org/10.3390/foods9030325.

Ghanbarzadeh, S., Khorrami, A., Arami, S., 2015. Nonionic surfactant-based vesicular system for transdermal drug delivery. Drug Deliv. 22 (8), 1071–1077. https://doi.org/10.3109/10717544.2013.873837.

Ghorbanzade, T., Jafari, S.M., Akhavan, S., Hadavi, R., 2017. Nano-encapsulation of fish oil in nano-liposomes and its application in fortification of yogurt. Food Chem. 216, 146–152. https://doi.org/10.1016/j.foodchem.2016.08.022.

Gumireddy, A., Christman, R., Kumari, D., Tiwari, A., North, E.J., Chauhan, H., 2019. Preparation, characterization, and in vitro evaluation of curcumin- and resveratrol-loaded solid lipid nanoparticles. AAPS PharmSciTech 20 (4), 145. https://doi.org/10.1208/s12249-019-1349-4.

Guo, C., Wang, J., Cao, F., Lee, R.J., Zhai, G., 2010. Lyotropic liquid crystal systems in drug delivery. Drug Discov. Today 15 (23–24), 1032–1040. https://doi.org/10.1016/j.drudis.2010.09.006.

Han, F., Yin, R., Che, X., Yuan, J., Cui, Y., Yin, H., Li, S., 2012. Nanostructured lipid carriers (NLC) based topical gel of flurbiprofen: design, characterization and in vivo evaluation. Int. J. Pharm. 439 (1), 349–357. https://doi.org/10.1016/j.ijpharm.2012.08.040.

Hu, Q., Hu, S., Fleming, E., Lee, J.Y., Luo, Y., 2020. Chitosan-caseinate-dextran ternary complex nanoparticles for potential oral delivery of astaxanthin with significantly improved bioactivity. Int. J. Biol. Macromol. 151, 747–756. https://doi.org/10.1016/j.ijbiomac.2020.02.170.

Huang, J., Wang, Q., Li, T., Xia, N., Xia, Q., 2017. Nanostructured lipid carrier (NLC) as a strategy for encapsulation of quercetin and linseed oil: preparation and in vitro characterization studies. J. Food Eng. 215, 1–12. https://doi.org/10.1016/j.jfoodeng.2017.07.002.

Huang, S., He, J., Cao, L., Lin, H., Zhang, W., Zhong, Q., 2020. Improved physicochemical properties of curcumin-loaded solid lipid nanoparticles stabilized by sodium caseinate-lactose Maillard conjugate. J. Agric. Food Chem. 68 (26), 7072–7081. https://doi.org/10.1021/acs.jafc.0c01171.

Jaiswal, P., Gidwani, B., Vyas, A., 2016. Nanostructured lipid carriers and their current application in targeted drug delivery. Artif. Cells Nanomed. Biotechnol. 44 (1), 27–40. https://doi.org/10.3109/21691401.2014.909822.

Jenning, V., Gysler, A., Schäfer-Korting, M., Gohla, S., 2000. Vitamin A loaded solid lipid nanoparticles for topical use: occlusive properties and drug targeting to the upper skin. Eur. J. Pharm. Biopharm. 49 (3), 211–218. https://doi.org/10.1016/S0939-6411(99)00075-2.

Jiang, T., Liao, W., Charcosset, C., 2020. Recent advances in encapsulation of curcumin in nanoemulsions: a review of encapsulation technologies, bioaccessibility and applications. Food Res. Int. 132, 109035. https://doi.org/10.1016/j.foodres.2020.109035.

Joshi, M., Pathak, S., Sharma, S., Patravale, V., 2008. Design and in vivo pharmacodynamic evaluation of nanostructured lipid carriers for parenteral delivery of artemether: nanoject. Int. J. Pharm. 364 (1), 119–126. https://doi.org/10.1016/j.ijpharm.2008.07.032.

Karami, Z., Hamidi, M., 2016. Cubosomes: remarkable drug delivery potential. Drug Discov. Today 21 (5), 789–801. https://doi.org/10.1016/j.drudis.2016.01.004.

Karthik, P., Anandharamakrishnan, C., 2016. Enhancing omega-3 fatty acids nanoemulsion stability and in-vitro digestibility through emulsifiers. J. Food Eng. 187, 92–105. https://doi.org/10.1016/j.jfoodeng.2016.05.003.

Kelly, K.L., Coronado, E., Zhao, L.L., Schatz, G.C., 2003. The optical properties of metal nanoparticles: the influence of size, shape, and dielectric environment. J. Phys. Chem. B 107 (3), 668–677. https://doi.org/10.1021/jp026731y.

Kumar, R., Kaur, K., Uppal, S., Mehta, S.K., 2017. Ultrasound processed nanoemulsion: a comparative approach between resveratrol and resveratrol cyclodextrin inclusion complex to study its binding interactions, antioxidant activity and UV light stability. Ultrason. Sonochem. 37, 478–489. https://doi.org/10.1016/j.ultsonch.2017.02.004.

Lépori, C.M.O., Correa, N.M., Silber, J.J., Falcone, R.D., 2016. How the cation 1-butyl-3-methylimidazolium impacts the interaction between the entrapped water and the reverse micelle interface created with an ionic liquid-like surfactant. Soft Matter 12 (3), 830–844. https://doi.org/10.1039/C5SM02421H.

Li, Z.l., Peng, S.f., Chen, X., Zhu, Y.q., Zou, L.q., Liu, W., Liu, C.m., 2018. Pluronics modified liposomes for curcumin encapsulation: sustained release, stability and bioaccessibility. Food Res. Int. 108, 246–253. https://doi.org/10.1016/j.foodres.2018.03.048.

Li, X., Maldonado, L., Malmr, M., Rouf, T.B., Hua, Y., Kokini, J., 2019. Development of hollow kafirin-based nanoparticles fabricated through layer-by-layer assembly as delivery vehicles for curcumin. Food Hydrocoll. 96, 93–101. https://doi.org/10.1016/j.foodhyd.2019.04.042.

Lin, C.H., Chen, C.H., Lin, Z.C., Fang, J.Y., 2017. Recent advances in oral delivery of drugs and bioactive natural products using solid lipid nanoparticles as the carriers. J. Food Drug Anal. 25 (2), 219–234. https://doi.org/10.1016/j.jfda.2017.02.001.

Lv, S., Gu, J., Zhang, R., Zhang, Y., Tan, H., McClements, D.J., 2018. Vitamin E encapsulation in plant-based nanoemulsions fabricated using dual-channel microfluidization: formation, stability, and bioaccessibility. J. Agric. Food Chem. 66 (40), 10532–10542. https://doi.org/10.1021/acs.jafc.8b03077.

Ma, P., Zeng, Q., Tai, K., He, X., Yao, Y., Hong, X., Yuan, F., 2017. Preparation of curcumin-loaded emulsion using high pressure homogenization: impact of oil phase and concentration on physicochemical stability. LWT 84, 34–46. https://doi.org/10.1016/j.lwt.2017.04.074.

Mahale, N., Thakkar, P., Mali, R.G., Walunj, D.R., Chaudhari, S., 2012. Niosomes: novel sustained release nonionic stable vesicular systems—an overview. Adv. Colloid Interface Sci. 183–184, 46–54. https://doi.org/10.1016/j.cis.2012.08.002.

Maria Leena, M., Mahalakshmi, L., Moses, J.A., Anandharamakrishnan, C., 2020. Nanoencapsulation of Nutraceutical Ingredients. Elsevier, https://doi.org/10.1016/b978-0-12-816897-4.00014-x.

Matalanis, A., Grif, O., Mcclements, D.J., 2011. Structured biopolymer-based delivery systems for encapsulation, protection, and release of lipophilic compounds. Food Hydrocoll. 25 (8), 1865–1880. https://doi.org/10.1016/j.foodhyd.2011.04.014.

Matos, M., Pando, D., Gutiérrez, G., 2019. Nanoencapsulation of food ingredients by niosomes. In: Jafari, S.M. (Ed.), Nanoencapsulation in the Food Industry, Lipid-Based Nanostructures for Food Encapsulation Purposes. 2. Elsevier, pp. 447–481.

McClements, D., 2011. Edible nanoemulsions: fabrication, properties, and functional performance. Soft Matter 7, 2297–2316. https://doi.org/10.1039/C0SM00549E.

McClements, D.J., 2012. Nanoemulsions versus microemulsions: terminology, differences, and similarities. Soft Matter 8 (6), 1719–1729. https://doi.org/10.1039/c2sm06903b.

McClements, D.J., 2015. Encapsulation, protection, and release of hydrophilic active components: potential and limitations of colloidal delivery systems. Adv. Colloid Interface Sci. 219, 27–53. https://doi.org/10.1016/j.cis.2015.02.002.

McClements, D.J., Li, Y., 2010. Structured emulsion-based delivery systems: controlling the digestion and release of lipophilic food components. Adv. Colloid Interface Sci. 159 (2), 213–228. https://doi.org/10.1016/j.cis.2010.06.010.

Mezzenga, R., Seddon, J.M., Drummond, C.J., Boyd, B.J., Schröder-Turk, G.E., Sagalowicz, L., 2019. Nature-inspired design and application of lipidic lyotropic liquid crystals. Adv. Mater. 31 (35), 1–19. https://doi.org/10.1002/adma.201900818.

Moghassemi, S., Hadjizadeh, A., 2014. Nano-niosomes as nanoscale drug delivery systems: an illustrated review. J. Control. Release 185, 22–36. https://doi.org/10.1016/j.jconrel.2014.04.015.

Mokhtari, S., Jafari, S.M., Khomeiri, M., Maghsoudlou, Y., Ghorbani, M., 2017. The cell wall compound of *Saccharomyces cerevisiae* as a novel wall material for encapsulation of probiotics. Food Res. Int. 96, 19–26. https://doi.org/10.1016/j.foodres.2017.03.014.

Moreira, J.B., Goularte, P.G., de Morais, M.G., Costa, J.A.V., 2019. Preparation of beta-carotene nanoemulsion and evaluation of stability at a long storage period. Food Sci. Technol. 39, 599–604.

Mortazavi, S.M., Mohammadabadi, M.R., Khosravi-Darani, K., Mozafari, M.R., 2007. Preparation of liposomal gene therapy vectors by a scalable method without using volatile solvents or detergents. J. Biotechnol. 129 (4), 604–613. https://doi.org/10.1016/j.jbiotec.2007.02.005.

Nishihira, V.S.K., Rubim, A.M., Brondani, M., Dos Santos, J.T., Pohl, A.R., Friedrich, J.F., de Lara, J.D., Nunes, C.M., Feksa, L.R., Simão, E., et al., 2019. In vitro and in silico protein corona formation evaluation of curcumin and capsaicin loaded-solid lipid nanoparticles. Toxicol. In Vitro 61, 104598. https://doi.org/10.1016/j.tiv.2019.104598.

Nobari Azar, F.A., Pezeshki, A., Ghanbarzadeh, B., Hamishehkar, H., Mohammadi, M., 2020. Nanostructured lipid carriers: promising delivery systems for encapsulation of food ingredients. J. Agric. Food Res. 2, 100084. https://doi.org/10.1016/j.jafr.2020.100084.

Olbrich, C., Kayser, O., Müller, R.H., 2002. Lipase degradation of Dynasan 114 and 116 solid lipid nanoparticles (SLN)—effect of surfactants, storage time and crystallinity. Int. J. Pharm. 237 (1), 119–128. https://doi.org/10.1016/S0378-5173(02)00035-2.

Orellano, M.S., Porporatto, C., Silber, J.J., Falcone, R.D., Correa, N.M., 2017. AOT reverse micelles as versatile reaction media for chitosan nanoparticles synthesis. Carbohydr. Polym. 171, 85–93. https://doi.org/10.1016/j.carbpol.2017.04.074.

Park, S.J., Garcia, C.V., Shin, G.H., Kim, J.T., 2017. Development of nanostructured lipid carriers for the encapsulation and controlled release of vitamin D3. Food Chem. 225, 213–219. https://doi.org/10.1016/j.foodchem.2017.01.015.

Patel, A.R., Velikov, K.P., 2011. Colloidal delivery systems in foods: a general comparison with oral drug delivery. LWT Food Sci. Technol. 44 (9), 1958–1964. https://doi.org/10.1016/j.lwt.2011.04.005.

Peppas, N.A., Bures, P., Leobandung, W., Ichikawa, H., 2000. Hydrogels in pharmaceutical formulations. Eur. J. Pharm. Biopharm. 50 (1), 27–46. https://doi.org/10.1016/S0939-6411(00)00090-4.

Pezeshki, A., Ghanbarzadeh, B., Mohammadi, M., Fathollahi, I., Hamishehkar, H., 2014. Encapsulation of vitamin A palmitate in nanostructured lipid carrier (NLC)—effect of surfactant concentration on the formulation properties. Adv. Pharm. Bull. 4 (Suppl. 2), 563–568. https://doi.org/10.5681/apb.2014.083.

Pezeshki, A., Hamishehkar, H., Ghanbarzadeh, B., Fathollahy, I., Keivani Nahr, F., Khakbaz Heshmati, M., Mohammadi, M., 2019. Nanostructured lipid carriers as a favorable delivery system for β-carotene. Food Biosci. 27, 11–17. https://doi.org/10.1016/j.fbio.2018.11.004.

Poonia, N., Kharb, R., Lather, V., Pandita, D., 2016. Nanostructured lipid carriers: versatile oral delivery vehicle. Future Sci. OA 2 (3), FSO135. https://doi.org/10.4155/fsoa-2016-0030.

Portnaya, I., Cogan, U., Livney, Y.D., Ramon, O., Shimoni, K., Rosenberg, M., Danino, D., 2006. Micellization of bovine β-casein studied by isothermal titration microcalorimetry and cryogenic transmission electron microscopy. J. Agric. Food Chem. 54 (15), 5555–5561. https://doi.org/10.1021/jf060119c.

Raemdonck, K., Demeester, J., De Smedt, S., 2009. Advanced nanogel engineering for drug delivery. Soft Matter 5, 707–715. https://doi.org/10.1039/b811923f.

Rani, V., Yadav, U.C.S., 2018. Functional Food and Human Health. Springer, pp. 1–694, https://doi.org/10.1007/978-981-13-1123-9.

Rastogi, R., Anand, S., Koul, V., 2009. Flexible polymerosomes—an alternative vehicle for topical delivery. Colloids Surf. B: Biointerfaces 72 (1), 161–166. https://doi.org/10.1016/j.colsurfb.2009.03.022.

Rezaei, A., Fathi, M., Jafari, S.M., 2019. Nanoencapsulation of hydrophobic and low-soluble food bioactive compounds within different nanocarriers. Food Hydrocoll. 88, 146–162. https://doi.org/10.1016/j.foodhyd.2018.10.003.

Sampathkumar, K., Tan, K.X., Loo, S.C.J., 2020. Developing nano-delivery systems for agriculture and food applications with nature-derived polymers. iScience 23 (5). https://doi.org/10.1016/j.isci.2020.101055, 101055.

Schäfer-Korting, M., Mehnert, W., Korting, H.-C., 2007. Lipid nanoparticles for improved topical application of drugs for skin diseases. Adv. Drug Deliv. Rev. 59 (6), 427–443. https://doi.org/10.1016/j.addr.2007.04.006.

Schexnailder, P., Schmidt, G., 2009. Nanocomposite polymer hydrogels. Colloid Polym. Sci. 287 (1), 1–11. https://doi.org/10.1007/s00396-008-1949-0.

Selvakumar, S., Janakiraman, A., Michael, M., Arthur, M., Chinnaswamy, A., 2019. Formulation and characterization of B-carotene loaded solid lipid nanoparticles. J. Food Process. Preserv. 43. https://doi.org/10.1111/jfpp.14212.

Semo, E., Kesselman, E., Danino, D., Livney, Y.D., 2007. Casein micelle as a natural nano-capsular vehicle for nutraceuticals. Food Hydrocoll. 21 (5), 936–942. https://doi.org/10.1016/j.foodhyd.2006.09.006.

Senske, M., Xu, Y., Bäumer, A., Schäfer, S., Wirtz, H., Savolainen, J., Weingärtner, H., Havenith, M., 2018. Local chemistry of the surfactant's head groups determines protein stability in reverse micelles. Phys. Chem. Chem. Phys. 20 (13), 8515–8522. https://doi.org/10.1039/C8CP00407B.

Setia, A., Ahuja, P., 2018. Nanohydrogels. Elsevier, https://doi.org/10.1016/B978-0-12-813663-8.00008-7.

Shao, P., Wang, P., Niu, B., Kang, J., 2018. Environmental stress stability of pectin-stabilized resveratrol liposomes with different degree of esterification. Int. J. Biol. Macromol. 119, 53–59. https://doi.org/10.1016/j.ijbiomac.2018.07.139.

Silva, A.C., Santos, D., Ferreira, D., Lopes, C.M., 2012. Lipid-based nanocarriers as an alternative for oral delivery of poorly water-soluble drugs: peroral and mucosal routes. Curr. Med. Chem. 19 (26), 4495–4510. https://doi.org/10.2174/092986712803251584.

Sivakamasundari, S.K., Leena, M., Moses, J.A., Chinnaswamy, A., 2020. Solid lipid nanoparticles: formulation and applications in food bioactive delivery. In: Knoerzer, K., Muthukumarappan, K. (Eds.), Innovative Food Processing Technologies—A Comprehensive Review, Reference Module in Food Science. Elsevier, pp. 580–604, https://doi.org/10.1016/B978-0-08-100596-5.23028-X.

Spicer, P.T., Hayden, K.L., Lynch, M.L., Ofori-Boateng, A., Burns, J.L., 2001. Novel process for producing cubic liquid crystalline nanoparticles (cubosomes). Langmuir 17 (19), 5748–5756. https://doi.org/10.1021/la010161w.

Sun, X., Bandara, N., 2019. Applications of reverse micelles technique in food science: a comprehensive review. Trends Food Sci. Technol. 91, 106–115. https://doi.org/10.1016/j.tifs.2019.07.001.

Tamjidi, F., Shahedi, M., Varshosaz, J., Nasirpour, A., 2013. Nanostructured lipid carriers (NLC): a potential delivery system for bioactive food molecules. Innov. Food Sci. Emerg. Technol. 19, 29–43. https://doi.org/10.1016/j.ifset.2013.03.002.

Turasan, H., Kokini, J.L., 2020a. Delivery of bioactives using biocompatible nanodelivery technologies. In: Hussain, C.M. (Ed.), Handbook of Functionalized Nanomaterials for Industrial Applications—Micro and Nano Technologies. Elsevier, pp. 133–166, https://doi.org/10.1016/b978-0-12-816787-8.00006-5.

Turasan, H., Kokini, J.L., 2020b. Enhancing the bioavailability of nutrients by nanodelivery systems. In: Jafari, S.M. (Ed.), Handbook of Food Nanotechnology—Applications and Approaches. Academic Press, pp. 345–375, https://doi.org/10.1016/b978-0-12-815866-1.00009-1.

Wang, J., Wang, H., Zhou, X., Tang, Z., Liu, G., Liu, G., Xia, Q., 2012. Physicochemical characterization, photo-stability and cytotoxicity of coenzyme Q10-loading nanostructured lipid carrier. J. Nanosci. Nanotechnol. 12 (3), 2136–2148. https://doi.org/10.1166/jnn.2012.5790.

Weiss, J., Decker, E.A., Mcclements, D., Kristbergsson, K., Helgason, T., Awad, T., 2008. Solid lipid nanoparticles as delivery systems for bioactive food components. Food Biophys. 3, 146–154. https://doi.org/10.1007/s11483-008-9065-8.

Xie, H., Xiang, C., Li, Y., Wang, L., Zhang, Y., Song, Z., Ma, X., Lu, X., Lei, Q., Fang, W., 2019. Fabrication of ovalbumin/κ-carrageenan complex nanoparticles as a novel carrier for curcumin delivery. Food Hydrocoll. 89, 111–121. https://doi.org/10.1016/j.foodhyd.2018.10.027.

Yao, M., Xiao, H., McClements, D.J., 2014. Delivery of lipophilic bioactives: assembly, disassembly, and reassembly of lipid nanoparticles. Annu. Rev. Food Sci. Technol. 5 (1), 53–81. https://doi.org/10.1146/annurev-food-072913-100350.

Nanostructures for improving food structure and functionality

Sophia Devi Nongmaithem and Nishant Rachayya Swami Hulle

Department of Food Engineering and Technology, Tezpur University, Tezpur, Assam, India

1 Introduction

Nanotechnology is an integrative field of studies and innovation involving the production, development, processing, and application of substances or tools in the nanometer range having one or more dimensions. It is an emerging branch of science that has influenced several production sectors like pharmaceuticals, cosmetics, food processing (Mozafari et al., 2008). In 1974, the term nanotechnology was first used by Norio Taniguchi to trace controlled thin film deposition. It is also generally described as the design, development, and application of structure or systems to monitor by regulating the size and shape of the material at meter-scale dimensions. Nanotechnology has many applications in the food industry, including nano-based food material processing, nanoencapsulation, food safety biosensors, improved delivery mechanism for nutrients, and it could be the most effective technique for smart food packaging. It provides significant prospects toward the development of innovative products and applied in various industrial sectors which include main agricultural production, livestock feed, novel food processing, nutritional supplements, and food packaging (Peters et al., 2016).

The physicochemical and the biochemical properties in nanoscale structures and systems vary considerably from those of macroscale systems due to interrelationship between atoms and molecules thereby offering unique and innovative practical applications. The surface-to-volume proportion generally increases when the particle size decreases to the nanoscale leading to an increase in the rate of reactions and also affecting the electrical, mechanical, and optical characteristics of particles (Neethirajan and Jayas, 2011).

The nanostructured materials have shown great potential application in functional food systems. Nanocapsules (liposomes, micelles), microemulsions, nanoemulsions, and polymeric nanoparticles are the most researched nanostructured materials and are currently in use in industry. The nanostructures help in improving solubility, enhancing bioavailability, controlled delivery, prevent micronutrients and bioactive components during processing and storage (Pathakoti et al., 2017). Nanostructures improve the protection of encapsulated materials, enhance solubilization and dispensability in water-based systems of lipid-soluble substances. They also regulate the compound release and increase the absorption rate of nutraceuticals, and showing decent improvement than bigger particle size systems (Acevedo-Fani et al., 2017). The present chapter discusses an overview of present technologies for nanostructure formulations and their applications in food structure and functionality.

2 Overview of methods for nanostructure formations

The methods for the formation of nanostructures include encapsulation, homogenization, electrospraying, supercritical fluids, ultrasound, and others using various ingredients like proteins, carbohydrates, lipids, and other bioactives or chemical compounds.

Nanostructured components are materials with modified physical and chemical characteristics to serve the desired functions and applications. It can be prepared in the form of layered films, wire structures, and atomic clusters. The distinctive characteristics of such nanostructures or nanomaterials, particularly physicochemical and biochemical properties vary significantly from their bulk form in changing the understanding of biochemical and physical occurrence in systems of food (Pathakoti et al., 2017). The functionality of nanostructures in food characterizes its range of applicability. Based on its functions, food nanomaterials affect bioavailability and nutritional properties (He and Hwang, 2016).

Naturally, various nano-sized structures are contained in the food system includes higher-order self-assembled structures like carbohydrates and fats are different from synthetic nanostructures. Therefore, such ingredients are used in making nanoemulsions, nanoencapsulation, and food-based polymers (Bajpai et al., 2018). Some food protein particles are globular in shapes that are varying in size from 10 to 100 nm. Polysaccharides and lipids are mostly polymers that have a chain that is less than 1 nm. Coagulation or emulsification is associated with the formation of a reticular or two- and three-dimension nanostructure. Naturally, milk products contain nanomaterials such as milk proteins where the size of fat globules is about 100 nm (Greßler et al., 2010).

2.1 Encapsulation

The encapsulation technique is used for protecting bioactive components, micronutrients, nutraceuticals, and other nutrients by insulating them and preventing unfavorable environmental changes. It also helps in controlled release at the desired active site. Nanoencapsulation in food systems is a technique where the small particles of substances are packed and encapsulated then applied in the food system (Ezhilarasi et al., 2013). During processing and storage, functional foods (vitamins, antioxidants, and flavors) are quite sensitive to the external environment. The encapsulation approach primarily aims to resolve these undesirable conditions by masking flavor, preventing oxidation and external damage caused by heat, moisture, oxygen, and light. It also eliminates vaporization losses, enhances stability and safety of the bioactive constituents, particle size, shape, solubility in oil/water, and color (Bratovcic and Suljagic, 2019).

The nanoencapsulation process involves the development of encapsulation in the form of films, layers, coatings, or simply microdispersion on the nanoscale level. As the encapsulation layer is a nanometer-scale that forms a protective layer on the molecules or ingredients in the food or taste (Paredes et al., 2016). The advancements of nanoencapsulation offer the opportunity to overcome the uncertainties of the food processing industry regarding the effective production of functional health ingredients and targeted delivery of different flavors. Using nanocomposites, nano emulsification, and nanostructuring, nanoencapsulation holds miniature materials and offers the characteristic product. Nanoencapsulation takes a role as a vehicle and helps in transporting active ingredients to the anticipated site of action. It tends to prevent the beneficial active ingredient from biochemical degradation while processing and storage. It can regulate the release of functional ingredients. Eventually, the delivery system is therefore consistent with the physicochemical and functional properties of the finished product. Generally, nanocarrier structures may be carbohydrate, protein, or lipid-based (Bratovcic and Suljagic, 2019).

Several forms of proteins including albumin, casein, lutein, zein, lectin, polysaccharides such as cellulose, starch, dextrin, chitosan, alginate, and lipids including phospholipids, triglycerides, cholesterol derived materials have been extensively used in encapsulation with some characteristics of water solubility, biocompatibility, and biodegradability (Khandelwal et al., 2016). Generally, coacervation emulsification, ionic gelation, and spray drying are the quite commonly used encapsulation methods among the various nanoencapsulation innovations to enhance the preservative efficiency of essential oils and their bioactive components. Different types of encapsulated essential oils or bioactive components such as nanoemulsion, nanofiber, nanogels, nanoliposome, etc. are generated based on encapsulation methods (Tiwari et al., 2020).

Lipid-based nanoparticles and lipid-polymers are the two suitable techniques used in the nanoencapsulation of essential oils. The lipid-based method is mainly applicable in the food preservation process compared to the polymer-based as essential oils are lipophilic (Prakash et al., 2018). Some common examples include nanoliposomes, nanoemulsions, solid lipid nanoparticles (Pandit et al., 2016). A study by Hasheminejad et al. (2019) has shown that nano-encapsulated essential oil can be utilized as a fungicide to extend the storage stability of fresh fruits and vegetables. They observed that emulsions loaded with clove essential oil into chitosan nanoparticles increased the antifungal effect of oils against *Aspergillus niger*. It was mainly because under in vitro conditions due to the controlled release of the volatile oil from chitosan nanoparticles during the study, which resulted in an increased inhibitory effect and even the inhibitory influence of the chitosan nanoparticles themselves. Findings by Bratovcic and Suljagic (2019) also mentioned that essential oils after encapsulation can be effectively used in meat products preservation for bioavailability enhancement, increased stability, and stimulating targeted sites benefiting preserving the sensory properties.

A preliminary report of Medeiros et al. (2019) also reveals the addition of yogurt incorporated with nanocapsule cantaloupe melon carotenoid in porcine gelatin could remain relatively stable for 60 days more so than crude extract. The utilization of gelatin results in enhanced water solubility and increases the tendency of melon carotenoids to be applied as natural dyes in food. Numerous studies reported nanoencapsulation of thyme and oregano essential oils for in vitro antiinflammatory, antioxidant function, and predominant constituents (thymol and carvacrol). Nano-encapsulated thyme essential oil was effectively encapsulated in chitosan having an average size of about 9.1 nm reported the effect against the pathogenic strains studied. The highest inhibitory effect against *B. cereus* (halo inhibition 1.9 cm) is reported as 40 μL of MIV (Sotelo-Boyás et al., 2017a,b).

Moghimi et al. (2016) observed the antibacterial activity of *Thymus daenensis* essential oils in control and nanoemulsion forms. Thymus essential oils showed an antibacterial effect against *E. coli* and this effect was higher against *E. coli* when it transformed to nanoemulsion. The addition of nano-essential oils in food packaging is a much more significant way of introducing nano-encapsulated essential oils and their bioactive components in food systems. Moreover, edible coatings films containing functional compounds like essential oil may be an effective approach for food packaging. Edible coatings derived from alginate-based nanoemulsion packed with thymus essential oil showed significant antibacterial effects against *E. coli* decreased the number of these bacteria by 4.71 logs within 12 h (Boskovic et al., 2019).

2.2 High-pressure homogenization

High-pressure homogenization (HPH) is the most commonly used method in the prevention of agglomerates of nanoparticles and delays the separation of compounds by gravitation compared to conventional emulsions. The extremely high shear pressure such as hydraulic shear, intense turbulence, and cavitations are applied together to a mixture and gets pumped through a tight valve and causes very fine emulsion droplets (Nethaji and Parambil, 2017) (Fig. 1). The preparation process runs until the product achieves the targeted droplet size and the polydispersity index (PDI) is used to specify the uniformity of the size of the droplet (Jaiswal et al., 2015). If the droplet size uniformity is low, higher value in the PDI. PDI value is less than 0.08 in monodisperse samples, narrow size distribution states PDI between 0.08 and 0.3, while PDI higher than 0.3 indicates a wide size distribution (Zhang, 2011).

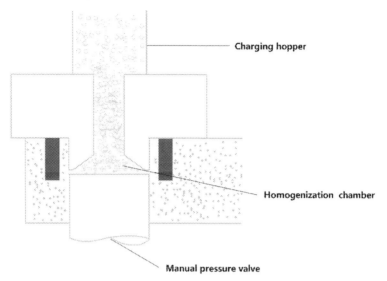

FIG. 1 Internal structure of high-pressure homogenizer. *Adapted from Huang, J., Wang, Q., Li, T., Xia, N., Xia, Q., 2017. Nanostructured lipid carrier (NLC) as a strategy for encapsulation of quercetin and linseed oil: preparation and in vitro characterization studies. J. Food Eng. 215, 1–12.*

High-energy mixers and rotor-stator systems use high-pressure methods for the preparation of nanoemulsion. By increasing the mixing speed of these instruments, the emulsion droplet size of the emulsion phase can be reduced. However, it is difficult to produce emulsions with an average droplet size of less than 200–300 nm (Koroleva and Yurtov, 2012). Ali et al. (2020) reported the effect of nanoencapsulation using HPH on the extraction, the antioxidant potential of *Origanum glandulosum* Desf hydro-distilled oil. They encapsulated essential oil by high-speed homogenization into nanocapsules and by using high-pressure homogenization into nanoemulsions. Application of high-pressure homogenization in nanoemulsion adversely affected the active ingredients in hydro-distilled oil mainly in carvacrol and thymol, whereas the application of high-speed homogenization resulted in major variations in the volatile composition in hydro-distilled oil and nanocapsules, however, produced the matching profile. The hydro-distilled essential oil showed higher antioxidant capacity than nanocapsules and nanoemulsion (IC50 78.50 mg/mL), correlated with variation in phenolics, flavonoids, and volatiles identified in hydro-distilled and nanoparticles, while nanocapsules exhibited the strong cytotoxic effect on the Hep-G2 liver cancer cell line (54.93 μg/mL) compared to hydro-distilled oil (78.50 mg/mL).

2.3 Supercritical fluids

Supercritical fluid (SCF) is fluid at its critical temperature, pressure, and density; at these parameters, the fluid is present in a single phase. Supercritical fluids show both gas-like and liquid-like thermophysical properties and these can be easily altered by varying the pressure and temperature of SCF (Cansell and Aymonier, 2009). Several studies have reported using

SCF for nanostructure formations and synthesis (Akbari-Alavijeh et al., 2020). However, synthesizing nanostructures using SCF requires information about the influence of various parameters during the interaction of the substrate with SCF. The interaction between nanomaterial and solvent medium used in the SCF process have an important role in nanostructure formation. The very low viscosity of SCF and wide contact area with nanomaterials are beneficial for the effective formation of nanostructures. However, the understanding of phase behavior, kinetics, and interaction of nanomaterials with solvent and transport properties are essential.

The process of using supercritical fluid for deposition of metal nanoparticles on porous solid surfaces or inside polymers was described by Zhang and Erkey (2006) as represented in Fig. 2. The process consisted of dissolving the metallic precursor in SCF and interacting the substrate with the solvent. After the interaction of the substrate with the precursor, the metallic precursor can be transformed to its metal form by methods such as chemical reduction in the SCF using a reduction agent, temperature-induced reduction, or chemical conversion with different gases after lowering the pressure to atmospheric conditions.

Controlling the interaction between the materials and the medium is imperative in the formation of structures. This is particularly true for nanomaterials because their interaction energy is very large. Regulating the interaction between nanomaterials and the medium is crucial, together with the surface design. Supercritical fluids' physical properties and phase behavior can be easily altered, thus changing the interaction between nanomaterials which helps in various controlled applications.

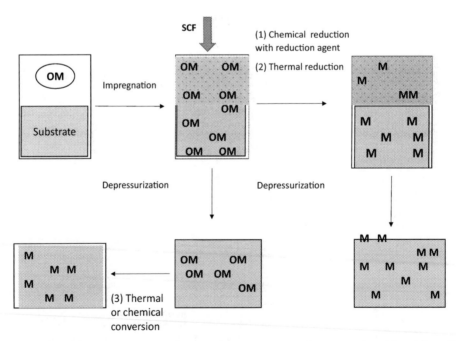

FIG. 2   The process of using supercritical fluid to synthesize supported nanoparticles. *Adapted from Zhang, Y., Erkey, C., 2006. Preparation of supported metallic nanoparticles using supercritical fluids: a review. J. Supercrit. Fluids 38(2), 252–267.*

2.4 Electrospraying

The electrospray technique is one of the latest methods for the preparation of biopolymers nanoparticles. By applying, a high voltage to the polymer solution that produces the jet stream across a nozzle where liquid droplets are formed. Electrospraying is one of the methods for producing stable nanostructures (Ghorani and Tucker, 2015). Presently, electro-dynamic atomization has been anticipated as an effective process for the production of micro or nanoscale particles by the fortification of bioactives, with several applications compared to conventional techniques (Arya et al., 2009; Jaworek and Sobczyk, 2008).

Electrodynamic atomization can be carried out from an aqueous solution, and at low or ambient temperatures. The principles of the electrospraying technique are based on the ability of an electric field to act on a liquid drop. Here, the electrostatic force created by high voltage atomizes the liquid into fine droplets and the solvent evaporation takes place while the flight of droplets toward the ground electrode (Alehosseini et al., 2018). A typical setup for electrospraying with components is shown in Fig. 3. Vaze et al. (2018) reported Engineered Water Nanostructures (EWNS) synthesized using electrospray and ionization of water. They reported that increased reactive oxygen species in EWNS led to an increase in antimicrobial ability which inactivated *E. coli* (up to 4 log reduction).

In the production of protein nanoparticles, electrospray is also utilized. Gliadin and elastin peptide nanoparticles are often produced by using the approach. A high voltage is given to a protein solution via an emitter, which then emits a liquid jet resulting in an aerosolized liquid containing medication and nucleic acid (Verma et al., 2018).

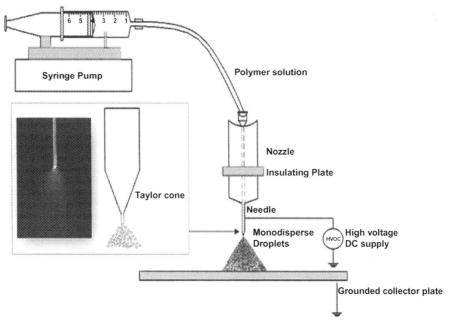

FIG. 3 A typical electrospraying setup. *Adapted from Bhushani, J.A., Anandharamakrishnan, C., 2014. Electrospinning and electrospraying techniques: potential food based applications.* Trends Food Sci. Technol. 38(1), 21-33.

2.5 Ultrasound

Emulsion droplets are produced through cavitation of immiscible liquids subjected to high-frequency sound waves in the presence of a surfactant, which produces extreme shock waves in the adjacent fluid medium. Thus resulting in the production of high-speed liquid jets, essential for emulsion droplets formation (Nethaji and Parambil, 2017). Ultrasonic emulsification is one of the effective methods for reducing droplets size. The energy is produced through sonotrodes by the sonicator probe and comprises of piezoelectric quartz crystal with a mechanism to expand and contract in response to an alternating electric voltage. It stimulates mechanical vibration and when the tip of the sonicator reaches the liquid, cavitation occurs. It can be possible to be used in laboratories in the production of emulsion droplet sizes as small as 0.2 μm (Jaiswal et al., 2015).

Several studies have also reported using ultrasonic techniques in food application. In the study of Walia et al. (2017), vitamin D molecule, which is lipophilic, has been encapsulated in fish oil. The oil-in-water nanoemulsion was prepared with droplet size ranging 300–450 nm and more than 90 days shelf life using ultrasonic technology. Encapsulation effectiveness of the prepared nanoemulsion was found in the range of 95.7%–98.2%. In addition, nanoemulsion through the virtual gastrointestinal tract showed an improved bioavailability relative to nonencapsulated vitamins. For different lipophilic materials, the mixture should also be used as a drug delivery tool (Ramalingam, 2017).

2.6 Other approaches

Nanostructures have a vital role in improving the performance of electrochemical affinity biosensors, because of their small size, quantum size, and interface effects particularly in terms of sensitivity, selectivity, and stability (Zhang and Wei, 2016). They are synthesized with low-cost wet-chemical synthesis methods, earth-abundant material, and nanostructures employed in the fabrication of electrochemical affinity biosensors (Nunes et al., 2019). Gold nanoparticles are a common nanomaterial used to make nanostructures and have wide applications in electrochemical affinity biosensing as an electrode modifier, signal elements carrier catalytic label, and electron transfer regulator (Campuzano et al., 2019).

Biochemical gas sensors composed of metal oxide nanoparticles were being utilized to evaluate and identify the standard and quality of various foods to determine the shelf life, freshness, and maturity of fruits, vegetables, and grains, as well as to evaluate real-time process monitoring, such as harvesting, manufacturing, and storage (Galstyan et al., 2018).

Recently, Ma et al. (2018), designed a method for detecting food contamination or spoiling that was both easy and low-cost and was based on a nanostructured conductive polymer wireless sensor. The researchers developed a gas sensor with a sensitivity of 46% and 17% for 5 ppm putrescine and cadaverine, respectively, as well as a high sensitivity of $R/R0 = 225\%$. This gas sensor serves as a switch in the NFC tag's circuit, allowing a smartphone to detect meat deterioration as the concentration of target gas is high. Moreover, biosensing platforms based on nanomaterials could improve the performance of analytical properties like sensitivity, selectivity, and stability (Campuzano et al., 2020). In other approaches, nanostructure in food offers a lot of potential for the processing and preservation of food. The application of suitable nanotechnology can contribute to the development of high, nutritional food products and ingredients like food additives, nutritional supplements, and functional food components have. Nanocochleates, which are based on a soya bean-derived phosphatidylserine carrier, are widely deemed safe nanomicelle-based carriers for nutraceuticals and nutritional supplements. Calcium ions are added to tiny phosphatidylserine vesicles to produce them and the nanocochleate system can protect the micronutrients and antioxidants in the food (Singh et al., 2017).

3 Sources of biopolymers for nanostructure development

3.1 Protein-based nanostructures

Proteins are essential nutrients in the human diet and they have specific functional properties for building functional, textural, and sensory properties in foods (Abaee et al., 2017; Martins et al., 2018). In addition to excellent nutrient content, antitoxicity, strong biocompatibility, and the potential to develop new textures, they have a tendency to form gel particles from large to the nanoscale. Its functional properties have a strong effect on the organoleptic (color, taste, and smell), kinesthetic (mouthfeel, stiffness, and smoothness), and textural (elasticity, cohesion, chewiness, and adhesion) properties of the final food products.

Proteins are selected for nanostructures formation based on their extrinsic factors (temperature, chemical environment, pH, etc.) and the intrinsic properties (molecular structure, composition, solubility, etc.) (Oliver et al., 2006).

The molecular, structural, electrical, and other configurations are described by the composition of proteins and the number of amino acid sequences. Proteins can function as an excellent carrier of bioactive components in the prepared food system due to their ion-binding properties and nutritional value. The protein-based nanostructures are reportedly used for the encapsulation of various forms of functional compounds with hydrophilic or lipophilic compounds of varying solubility or molecular weight (Abaee et al., 2017). The different sources of proteins and their application reported in nanostructure formation are represented in Table 1. The high surface area of these nano-systems enhances the chances of uptake of active compounds and enhances their absorption and bioavailability (Tapia-Hernández et al., 2017). It is important to minimize the matrix size for nonsolid and semisolid food so that it can be added without reducing the sensory properties of the food (Augustin and Oliver, 2014). Moreover, by minimizing the structure size from micro to nanoscale, it is indeed possible to modify protein structures with better delivery properties.

The methodologies commonly used for the fabrication of protein-based nanostructures are gelation, segregative separation, interactive polymer, water-in-oil heterogeneous gelation, micro-molding, photolithography, and microfluidic preparation.

Gelation usually involves a motive to release the structure of the protein, followed by an aggregation process to provide a 3D system. Several structures formed are generated by protein-protein interactions and the formation of protein aggregates due to different processing interventions. Thermally-induced gelation is applied in the gel formation of globular protein in the presence of heat. Temperature influences the thermal properties of whey proteins by affecting the efficiency of molecular collisions and improving interreaction

TABLE 1   Sources of proteins for nanostructure formation and their applications.

Protein source	Applications	References
Soy proteins	Controlled release of bioactive compounds	Teng et al. (2012)
	Promising nanocarriers to facilitate the oral delivery of vitamin B12	Zhang et al. (2015)
	Using stable and biocompatible nanocarriers for bioactives and drugs	Cheng et al. (2017)
	Controlled release of bioactive substances	Jin et al. (2016)
	Manufacturing of food products for vegetarians	Ansarifar et al. (2017)
	Controlled release nanocarrier for local drug delivery to the inner ear	Yu et al. (2014)
	Delivery of anticancer drugs	Wang et al. (2016)
	Entrapment/encapsulation of bioactive molecules	Sadeghi et al. (2013)
Whey proteins	Applicable as an enriching agent in clear and nonclear food beverages	Abbasi et al. (2014)
	Useful in cancer treatment	Jain et al. (2018)
	Enriching of acidic beverages	Mohammadian et al. (2019)
	Delivery of bioactive compounds	Sullivan et al. (2014)
Bovine serum albumin	Could be employed as carriers to improve the controlled release of bioactives	Sadeghi et al. (2014)
Egg albumin	Delivery of bioactive molecules	Delfiya et al. (2016)
Ovalbumin	Delivery systems for curcumin to fortify functional foods, drinks, and beverages	Feng et al. (2016)
	Production of microcapsules for encapsulation purposes	Humblet-Hua et al. (2011)
Lysozyme	Controlled release of anticancer drugs	Zhu et al. (2013)
	Anticancer drug delivery	Lin et al. (2015)
Zein	Could be considered as a prospective drug delivery system for cancer chemotherapy	Dong et al. (2016)
	Enhancing the chemical stability and biological activity of liable nutraceuticals	Feng et al. (2018)
	Enriching of food and pharmaceutical formulations	Li et al. (2019)
	Carrying of payloads for intracellular drug delivery	Xu et al. (2011)
	Controlled delivery of anticancer therapeutics	Xu et al. (2015)
	Encapsulation and oral delivery of lipophilic bioactives	Hu et al. (2016)
	Stabilization of light-sensitive bioactives	Fernandez et al. (2009)
	Oral delivery of aceclofenac with reduced side effects	Karthikeyan et al. (2012)

TABLE 1 Sources of proteins for nanostructure formation and their applications—cont'd

Protein source	Applications	References
Lactoferrin-glycomacropeptide	Controlled release of bioactive compounds	Bourbon et al. (2016)
β-Lactoglobulin	Encapsulation of lipophilic bioactive ingredients	Serfert et al. (2014)
	Internalization of nanoparticles into living cells with enhanced transport properties	Bolisetty et al. (2014)
	Iron fortification of solid and liquid food products	Shen et al. (2017)
α-Lactalbumin	Applicable in food and cosmetic formulations	Fuciños et al. (2017)
	Could be served as gelling agents and the carriers of natural colorants in different food formulations	Tarhan and Harsa (2014)
Gelatin	Controlled drug delivery	Yang et al. (2007)
	Antibacterial wound dressing	Li et al. (2016)
	Applicable in food products where the maximum protection of bioactives is needed	Tavassoli-Kafrani et al. (2018)

Adapted from Mohammadian, M., Salami, M., Emam-Djomeh, Z., Momen, S., Moosavi-Movahedi, A.A., 2018. Gelation of oil-in-water emulsions stabilized by heat-denatured and nanofibrillated whey proteins through ion bridging or citric acid-mediated cross-linking. Int. J. Biol. Macromol. 120, 2247–2258.

between hydrophobic particles, which is a critical stage in the production of functional protein aggregates (Creamer and MacGibbon, 1996).

Acid-induced gelation occurs at low pH, increasing the net load of protein molecules and the interactions between protein and solvent. Repulsive forces decrease as the protein has a zero net charge, and aggregation occurs. The fractal aggregation theory may describe the processes of acid-induced gel-forming (Lucey and Singh, 1997). As the salt is mixed with a protein solution, the ionic strength increases and protects the solution from the electrostatic interaction on aggregates. Thereby repulsive forces are reduced between the molecules followed by gel formation (Thomä-Worringer et al., 2006). The enzymes are used in artificial covalent cross-linking of food proteins, which can only regulate the aggregation state of proteins that are complimentary to heat-induced gelling. Microbial transglutaminase is an example that catalyzes cross-linkages involving lysine and glutamine residues (Saricay et al., 2013). Milk caseins are excellent proteins for transglutaminase catalyzed cross-linking. Similarly, whey proteins are polymerized using transglutaminase when partially unfolded (Otte et al., 1999).

In segregative separation, a repulsive force acts between two polymers whenever the polymers are uncharged leading to phase separation. Nanoemulsion including water and oil was been developed by segregative separation and the process was considered kinetically balanced and can lead to innovative microstructures for the food industry with new functional and rheological properties (Feng et al., 2018). An interactive polymer mechanism is also known as a physical self-assembly. This mechanism approach with charged polymers produced protein-based structures, whereby, interactions in between biopolymers occur as in noncovalent interactions.

Heterogeneous water-in-oil gelation comprises the emulsifying process of aqueous protein droplets (gelling agents) which are constantly spread in a continuous organic process using oil-soluble surfactants (Mohammadian et al., 2018). Inverse emulsion, reverse micellar phase and membrane emulsification are the common methods for water-in-oil heterogeneous gelation (Atkinson et al., 1995). Micro-molding, photolithography, and microfluidic preparation techniques are often used for the development of biopolymer structures, including nanoparticle gels. A biopolymer solution is added to the mold in the micro-molding process and is produced by controlling variables such as temperature or gelling agents to form a gel. The microfluidic technique allows the formation of multiple biopolymer-based particles based on the arrangement of various microfluidic channels. Such methods permit the formulation of specific delivery systems for the 1 µm particle size along with different shapes. The production of unique molds, photolithographic molds, and microfluidic devices are important for the preparation of particles using these methods (Shewan and Stokes, 2013). Spray drying is a well-known method for producing active substances in the food and pharmaceutical industries. By altering process parameters or customizing the spray dryer configuration, it is possible to manage numerous product attributes such as particle size, bulk density, and flow qualities (Bourbon et al., 2016).

3.2 Polysaccharide-based nanostructures

Polysaccharides are abundantly present food ingredients having wide applications in food formulations. Polysaccharides are desirable biomaterials in both food and pharmaceutical formulations. They are environmentally friendly in terms of biodegradability, hence are promising raw materials for nanostructures formation (Bilal et al., 2021). Polysaccharides are biologically compatible due to their hydrophilicity and they contain different functional groups like hydroxyl, amino and carboxylic acids, which further assist in conjugation with other target molecules (Myrick et al., 2014). The surface charge on the polysaccharides (cationic or anionic) can target the interaction with different molecules and assess the bioavailability of the conjugates. There have been many polysaccharides used in the formation of nanoparticle conjugates with food and drug molecules like cyclodextrin, carboxymethyl-hexanoyl chitosan, and chitosan. Pectin found abundantly in fruit processing waste is also a well-explored source for food processing applications. Apart from the product formulations in products like jams, jellies preserve it has applications in micro and nanostructure formation using different methodologies like emulsions, hydrogel formations (Massironi et al., 2020).

Starch is a promising ingredient for nanostructure formations because of its wide availability and compatibility in food systems. It has many applications in food, drug delivery, food packaging, and so on. Starch is an enzyme-responsive carrier digested by digestive enzymes and converted into monosaccharides (Yu et al., 2021). The modification of the amorphous domain of starch granule makes the starch nanocrystal whereas, gelatinized starch is used for the formation of starch nanoparticles. Nanoprecipitation is another promising method for making controlled-size nanoparticles and is a simple method to produce the nanoparticles. In this method, the adjustment of the concentration of ingredients, solvent concentration, and mixing conditions controls the resulting particle size (Joye and McClements, 2013; Qin et al., 2016).

Biopolymeric nanocarriers are submicron-sized biopolymeric particles that can enclose bioactive nanoparticles. Individual biopolymers, such as lactoferrin nanoparticles, starch nanoparticles, alginate nanohydrogels, and lactalbumin nanotubes, as well as complexation/conjugation of two unique biopolymers, such as protein-polysaccharides, can be used to construct biopolymeric nanocarriers (Assadpour and Jafari, 2019).

4 Application on nanostructures in food systems

4.1 Nanoemulsions

Emulsions with droplet sizes ranging from 5 to 200 nm are referred to as nanoemulsions, translucent or transparent emulsions, ultrafine emulsions, submicron emulsions, and mini emulsions (Caldero et al., 2011). Nanoemulsions are the dispersion of two immiscible liquid phases as oil-in-water or water-in-oil in such a way that the first liquid (dispersion phase) is dissolved into a second liquid as a droplet (continuous phase). The mean diameter of the oil-in-water droplet varies from 20 to 200 nm (Acevedo-Fani et al., 2017). According to the droplet's size and stability nature, emulsions are classified as coarse emulsions, microemulsions, and nanoemulsions (Komaiko and McClements, 2016). The solubilization potential is higher in nanoemulsion than the micellar dispersion and higher kinetic stability reported in nanoemulsion than the coarse emulsion. Nanoemulsions are used in various fields including food processing, cosmetics, and pharmaceutical manufacturing sectors. It comprises of many dosage formulations such as liquids, gels aerosol, creams, sprays, foams of aerosols, and these can be delivered similarly through different routes such as topical, oral, intravenous, intranasal, pulmonary, and ocular (Singh et al., 2017).

In the current scenario, nanoemulsion is widely used in the food processing sectors for the production of pharmaceuticals, coloring, and food additives, and antimicrobial agents. For the development of biodegradable coating and packaging films, nanoemulsion formulations of active compounds help in improving the quality of food, functional characteristics, nutritional value, and extent of the shelf life of food products (Aswathanarayan and Vittal, 2019).

Generally, due to facilitating passive transport through a biological membrane in nanoemulsion resulting increased in the bioavailability of bioactive compounds. Nanoemulsion helps in maintaining organoleptic stability as it protects the functional substances in food when reacting with the food matrix (Aboalnaja et al., 2016). Sari et al. (2015) reported that nanoencapsulation enhanced solubility and also increase the absorption rate of bioactive compounds at the gastrointestinal level. Besides, the application of nanoencapsulation together with nanoemulsions improves protection, lowers its metabolism, and reduces efflux transporter activity. There may have several advantages of nanoemulsions as compared to conventional emulsions relative to the size of the droplets they produce. Some of the advantages of nanoemulsion over conventional methods include high optical clarity and physical stability; increased bioavailability of encapsulated substances that make them in the food application.

The optical characteristic of nanoemulsion is a very important criterion for their application in the food industry. In nanoemulsions, particles are optically transparent or faintly turbid, varying on droplet size, its opacity is expressed in terms of turbidity (τ) and it is defined by measurements of transmission (Aswathanarayan and Vittal, 2019). In the beverage processing sector, indeed a high level of transparency and purity is considered as the most acceptable and favorable attribute of nanoemulsions as the small droplet size enables weak light scattering of the byproducts. So, the incorporation of functional components encapsulated emulsion will not change the appearance visibly (de Oca-Ávalos et al., 2017).

4.1.1 Materials require in nanoemulsion production

Based on the surfactant applied, various approaches to promoting stability can be obtained. One of the important steps is the stabilization of emulsions to provide good stability. At the interface, surfactant adsorption is the common way to stabilize (Devarajan and Ravichandran, 2011). Solvents like *n*-hexane, *n*-decane, acetone, etc. are commonly used for the preparation of nanoemulsion. Active functional compounds like lycopene, lidocaine, curcumin are the required compounds to be added to the food system and again they can be converted into antioxidants, phytosterols, carotenoids, and fatty substances (Nethaji and Parambil, 2017). Several factors should be considered while nanoemulsion production. The surfactant should be carefully selected so that ultra-low interfacial stress can be established, which is the main requirement for the production of nanoemulsion. The concentration of surfactants must be enough to maintain microdroplets. The surfactant must be versatile to facilitate the development of nanoemulsion factors (Jaiswal et al., 2015).

4.1.2 Preparation of nanoemulsion

Nanoemulsion is often used to develop active foods as this technology enables the delivery of active nutrients and anti with low water solubility. Aqueous lipid droplet solution (size 100 nm approx.) is prepared in nanoemulsion production using two different techniques; high-energy and low-energy methods. Usually, nanoemulsion is produced with a high-energy process by applying mechanical energy input into high-pressure homogenizers, ultrasound generators, and high shear stirring (Sole et al., 2012). The input energy density in high-energy methods is around-W (Gupta et al., 2016).

The low-energy method is also used in the preparation of self-nano-emulsifying drug delivery systems (SNEDDS). The energy needed to interfere with the oil droplets varies on the interfacial tension, and that in turn depends upon the type of emulsifier. Low-energy techniques are based on the involuntary production of small droplets of oil so that the composition or temperature of the system is manipulated in a unique direction (Çınar, 2017).

There are some advantages and limitations of both the methodologies for the production of nanoemulsion. High-energy methods are being the appropriate techniques for nanoemulsion preparation from emulsions and oils. The nanoemulsion of capsaicin-loaded nanoemulsion of droplet size ranging between 50 and 70 nm is shown in Fig. 4 as reported by Akbas et al. (2018). They prepared these nanoemulsions using a high-pressure homogenizer by passing the mixture of distilled water, surfactant, and oil. Low-energy method needs lower surfactant-to-oil ratios, while high-energy requires quite expensive homogenizers to acquire and retain. On the contrarily, low-energy methods do not require any advanced equipment to be installed and are easy and affordable to introduce. As usual, it requires surfactant-to-oil ratios and is the most effective tiny synthetic molecule surfactant. Therefore, based upon the bioactive component which will be utilized, a suitable method should be applied for the nanoemulsion formulation (Martínez-Ballesta et al., 2018). High-pressure

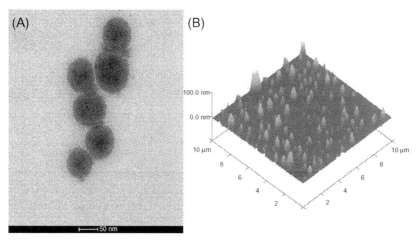

FIG. 4 Transmission electron microscopy (A) and atomic force microscopy image (B) of capsaicin-loaded nanoemulsion particles prepared with high-pressure homogenization. *Adapted from Akbas, E., Soyler, B., Oztop, M.H., 2018. Formation of capsaicin loaded nanoemulsions with high pressure homogenization and ultrasonication. LWT 96, 266–273.*

homogenization, ultra-speed, and ultrasound are mechanical operations that have been categorized into different categories based on the methods used to produce nanoemulsions high-energy emulsion formation is controlled by the type of energy quality and composition to be used.

4.2 Food packaging

Integrating nanotechnology into food packaging processes has led to significant advancements in recent years. A number of studies on increasing barrier characteristics, improving antimicrobial qualities, integrating sensors, refining biodegradability, and compatibility are available (Cerqueira et al., 2018). To fulfill industrial and consumer expectations, nanostructures of various materials, dimensions, forms, and functionalities are being utilized in the introduction of unique food packaging solutions.

Nanostructures improve food packaging by enhancing the barrier properties by altering the water vapor and gas permeability, and by improving the material characteristics like tensile strength, elastic modulus, and thermal stability resulting in enhanced functionality to food packaging systems (Mei and Wang, 2020). Antimicrobial activity of silver-based nanostructures are reportedly been studied extensively against over 650 species (Siddiqui et al., 2018). The effectiveness and applications of silver-based antimicrobial agents (AgNPs, AgNCs, etc.) vary depending on their nanostructure. AgNPs (silver nanoparticles) in food packaging materials range in size from 10 to 100 nm and are made by reducing $AgNO_3$ with sodium borohydrite or other reducing agents. Silver nano-clusters (AgNCs) are produced by groups of silver atoms and are around 2–4 nm.

Polyethylene, ethylene vinyl alcohol, polyvinyl chloride, polyvinylpyrrolidone, are the commonly used polymers used to accommodate silver-based nanoparticles for packaging development. Low-density polyethylene (LDPE) has been extensively researched as a matrix for Ag-based nanostructures and is commonly utilized as a covering film for fresh food goods. Low-temperature plasma-treated LDPE films had even more hydrophilic groups coated on the film surface, according to Sadeghnejad et al. (2014), which increased antibacterial activity and enhanced the hydrophilicity and reactivity of the films. According to Zhao et al. (2007), the top layer of free energy of metal-polymer composites changes based on the metal and polymer ratios. Because of their poor binding affinity, surfaces with less energy have less microbial adhesion and are more sensitive to cleaning agents.

Nanoclays are aluminum silicates and have a sheet-like structure that occurs naturally and is about 1-nm thick. The negatively charged layers are connected by interlayer cations or Van der Waals forces. The high surface area (700–800 m^2/g) and huge aspect ratio (50–1000) of the MMT plate make it an efficient reinforcing filler for increasing the gas barrier performance of films (Majeed et al., 2013). Because of this structure, MMT plates can function as physical barriers with a more challenging route and lower gas permeability.

The two most prevalent cellulose-based nanostructures are cellulose nanofibers (CNFs) and cellulose nanocrystals (CNCs), which are made using different methods (Vilarinho et al., 2018). The addition of CNCs and CNFs to polymers could improve the structural and barrier properties of the latter. CNCs offer good structural qualities because their molecular structure includes highly cryogenic components. This may be due to the presence of highly crystalline parts in their chemical structure and the ability to form a complex network. Jiang et al. (2019) reported enhancement of tensile strength from 1.3 to 3.15 MPa by incorporation of CNCs to whey protein concentrate–cast film and reduction WVP from 4.91×10^{12} (Pa s m^2) to 3.01×10^{12} (Pa s m^2) with better mechanical properties.

4.3 Nanosensors

In this current scenario in the area of agriculture and food processing, nanosensors become one of the most innovative devices and provide a drastic improvement in selectivity, sensitivity, and speed compared to conventional biochemical methods. It is also used in the detection of microorganisms, contaminants, and freshness in food (Joyner and Kumar, 2015).

The configuration of nanosensors is the same as conventional sensors, however, their output is at the nanoscale and has sensing dimensions up to 100 nm. Consequently, it is defined as an incredibly small tool that can attach to anything to be sensed and give back a signal. Nanosensor is much more applicable in packaging and transport in the food production sector. It can easily able to detect the defection of the packaging material of the food through their chemical and electro-optical properties. Nanosensors make sure that the customers afford cheap and nutritious foods and reduce the risk of food contamination that promotes food safety (Abdel-Karim et al., 2020).

Nanosensors are used in the agriculture and food industries to identify hazards in the incidence of suspected food poisoning, or as a nano tracer in packaging to reveal the origin of the food product and to verify whether it is of standard acceptable quality at any given moment (Omanović-Miklićanina and Maksimović, 2016). In food packaging, the sensor is used to detect microorganism growth and color change when the threshold is achieved. Nanosensors are being used in online process control as they are capable of tracking storage conditions to avoid food poisoning (Augustin and Sanguansri, 2009). Researchers have produced and coated gold nanoparticles with molecules that can stick to compounds such as pesticides. Farmers sprayed these nanoparticles on their fields to detect a chemical like a pesticide. Also, nanosensors using Raman spectroscopy are considered suitable for food forensics. Food forensics is a study of food sources, adulteration, and pollution. The use of nanosensors contributes to the reliability of the methodology and enables the impact of various analytes that may be analyzed.

4.4 Other applications

The implementation of nanomaterials in the agricultural food industry was first discussed in a roadmap published in 2003 by the Department of Agriculture of the United States (USDA) (Rashidi and Khosravi-Darani, 2011). Nanotechnology recently opened up a new significant revolution of food industries in both developed and developing nations. It also provides more prospects in different fields such as agriculture, food, and medicine with the production of new food system properties and materials structure application from production, processing to packaging, nanotechnology provides excellent food solutions. Various associations, researchers, and manufacturing companies have been implementing novel methodology or techniques and products that directly apply nanotechnology to food science (Singh et al., 2017). Food processing methods involving nanomaterials which include the production of nutraceuticals, gelation and thickening agent, nutrient distribution, fortification of minerals and vitamins, and flavor nanoencapsulation (Pradhan et al., 2015).

5 Conclusion

The advancements in nanotechnology have enabled the development of novel nanomaterials for food processing applications. Techniques like electrospray, high-pressure homogenization, supercritical fluid processing have enabled promising alternative methods for the formation and development of nanostructures. This certainly has enabled the formation of novel functional ingredients for the development of stable functional food products. Still, there are further studies required to assess the efficacy of nanostructure, its effect on human health, and meeting regulatory requirements. The availability of advanced characterization tools in spectroscopy and chromatography with computational methods has enabled more understanding of the interaction of nanostructures and their functionality in food systems. It is hoped that this will enable the wider application of nanostructures in food systems with more acceptability by the food industry.

References

Abaee, A., Mohammadian, M., Jafari, S.M., 2017. Whey and soy protein-based hydrogels and nano-hydrogels as bioactive delivery systems. Trends Food Sci. Technol. 70, 69–81.
Abbasi, A., Emam-Djomeh, Z., Mousavi, M.A.E., Davoodi, D., 2014. Stability of vitamin D3 encapsulated in nanoparticles of whey protein isolate. Food Chem. 143, 379–383.

Abdel-Karim, R., Reda, Y., Abdel-Fattah, A., 2020. Nanostructured materials-based nanosensors. J. Electrochem. Soc. 167 (3), 037554.

Aboalnaja, K.O., Yaghmoor, S., Kumosani, T.A., McClements, D.J., 2016. Utilization of nanoemulsions to enhance bio-activity of pharmaceuticals, supplements, and nutraceuticals: nanoemulsion delivery systems and nanoemulsion excipient systems. Exp. Opin. Drug Deliv. 13 (9), 1327–1336.

Acevedo-Fani, A., Soliva-Fortuny, R., Martín-Belloso, O., 2017. Nanostructured emulsions and nanolaminates for delivery of active ingredients: improving food safety and functionality. Trends Food Sci. Technol. 60 (c), 12–22. https://doi.org/10.1016/j.tifs.2016.10.027. Elsevier Ltd.

Akbari-Alavijeh, S., Shaddel, R., Jafari, S.M., 2020. Encapsulation of food bioactives and nutraceuticals by various chitosan-based nanocarriers. Food Hydrocoll. 105, 105774.

Akbas, E., Soyler, B., Oztop, M.H., 2018. Formation of capsaicin loaded nanoemulsions with high pressure homogenization and ultrasonication. LWT 96, 266–273.

Alehosseini, A., Ghorani, B., Sarabi-Jamab, M., Tucker, N., 2018. Principles of electrospraying: a new approach in protection of bioactive compounds in foods. Crit. Rev. Food Sci. Nutr. 58 (14), 2346–2363.

Ali, H., Al-Khalifa, A.R., Aouf, A., Boukhebti, H., Farouk, A., 2020. Effect of nanoencapsulation on volatile constituents, and antioxidant and anticancer activities of Algerian Origanum glandulosum Desf. essential oil. Sci. Rep. 10 (1), 1–9.

Ansarifar, E., Mohebbi, M., Shahidi, F., Koocheki, A., Ramezanian, N., 2017. Novel multilayer microcapsules based on soy protein isolate fibrils and high methoxyl pectin: production, characterization and release modeling. Int. J. Biol. Macromol. 97, 761–769.

Arya, N., Chakraborty, S., Dube, N., Katti, D.S., 2009. Electrospraying: a facile technique for synthesis of chitosan-based micro/nanospheres for drug delivery applications. J. Biomed. Mater. Res. B Appl. Biomater. 88 (1), 17–31.

Assadpour, E., Jafari, S.M., 2019. An overview of biopolymer nanostructures for encapsulation of food ingredients. In: Biopolymer Nanostructures for Food Encapsulation Purposes. Elsevier, pp. 1–35.

Aswathanarayan, J.B., Vittal, R.R., 2019. Nanoemulsions and their potential applications in food industry. Front. Sustain. Food Syst. 3, 95.

Atkinson, P.J., Robinson, B.H., Howe, A.M., Pitt, A.R., 1995. Characterisation of water-in-oil microemulsions and organo-gels based on sulphonate surfactants. Colloids Surf. A Physicochem. Eng. Asp. 94 (2–3), 231–242.

Augustin, M.A., Oliver, C.M., 2014. Use of milk proteins for encapsulation of food ingredients. In: Microencapsulation in the Food Industry. Elsevier, pp. 211–226.

Augustin, M.A., Sanguansri, P., 2009. Nanostructured materials in the food industry. Adv. Food Nutr. Res. 58, 183–213.

Bajpai, V.K., et al., 2018. Prospects of using nanotechnology for food preservation, safety, and security. J. Food Drug Anal. 26 (4), 1201–1214. https://doi.org/10.1016/j.jfda.2018.06.011. Elsevier Ltd.

Bilal, M., Gul, I., Basharat, A., Qamar, S.A., 2021. Polysaccharides-based bio-nanostructures and their potential food applications. Int. J. Biol. Macromol. 176, 540–557.

Bolisetty, S., Boddupalli, C.S., Handschin, S., Chaitanya, K., Adamcik, J., Saito, Y., Manz, M.G., Mezzenga, R., 2014. Amyloid fibrils enhance transport of metal nanoparticles in living cells and induced cytotoxicity. Biomacromolecules 15 (7), 2793–2799.

Bourbon, A.I., Cerqueira, M.A., Vicente, A.A., 2016. Encapsulation and controlled release of bioactive compounds in lactoferrin-glycomacropeptide nanohydrogels: curcumin and caffeine as model compounds. J. Food Eng. 180, 110–119.

Bratovcic, A., Suljagic, J., 2019. Micro- and nano-encapsulation in food industry. Croat. J. Food Sci. Technol. 11 (1), 113–121.

Caldero, G., Maria, J.G.C., Solans, C., 2011. Formation of polymeric nano-emulsions by a low-energy method and their use for nanoparticle preparation. J. Colloid Interface Sci. 353, 406–411.

Campuzano, S., Gamella, M., Serafín, V., Pedrero, M., Yáñez-Sedeño, P., Pingarrón, J.M., 2019. Biosensing and delivery of nucleic acids involving selected well-known and rising star functional nanomaterials. Nanomaterials 9 (11), 1614.

Campuzano, S., Yáñez-Sedeño, P., Pingarrón, J.M., 2020. Electrochemical affinity biosensors based on selected nanostructures for food and environmental monitoring. Sensors 20 (18), 5125.

Cansell, F., Aymonier, C., 2009. Design of functional nanostructured materials using supercritical fluids. J. Supercrit. Fluids 47 (3), 508–516.

Cerqueira, M.A., Vicente, A.A., Pastrana, L.M., 2018. Nanotechnology in food packaging: opportunities and challenges. In: Nanomaterials for Food Packaging. Elsevier, pp. 1–11.

Cheng, X., Wang, X., Cao, Z., Yao, W., Wang, J., Tang, R., 2017. Folic acid-modified soy protein nanoparticles for enhanced targeting and inhibitory. Mater. Sci. Eng. C 71, 298–307.

Çınar, K., 2017. A review on nanoemulsions: preparation methods and stability. Trakya Univ. J. Eng. Sci. 18 (1), 73–83. Department of Food Engineering, Trakya University, Edirne, Turkey.

Creamer, L.K., MacGibbon, A.K., 1996. Some recent advances in the basic chemistry of milk proteins and lipids. Int. Dairy J. 6 (6), 539–568.

de Oca-Ávalos, J.M.M., Candal, R.J., Herrera, M.L., 2017. Nanoemulsions: stability and physical properties. Curr. Opin. Food Sci. 16, 1–6.

Delfiya, D.A., Thangavel, K., Amirtham, D., 2016. Preparation of curcumin loaded egg albumin nanoparticles using acetone and optimization of desolvation process. Protein J. 35 (2), 124–135.

Devarajan, V., Ravichandran, V., 2011. Nanoemulsions: as modified drug delivery tool. Int. J. Compr. Pharm. 2 (4), 1–6.

Dong, F., Dong, X., Zhou, L., Xiao, H., Ho, P.Y., Wong, M.S., Wang, Y., 2016. Doxorubicin-loaded biodegradable self-assembly zein nanoparticle and its anti-cancer effect: preparation, in vitro evaluation, and cellular uptake. Colloids Surf. B: Biointerfaces 140, 324–331.

Ezhilarasi, P.N., Karthik, P., Chhanwal, N., Anandharamakrishnan, C., 2013. Nanoencapsulation techniques for food bioactive components: a review. Food Bioprocess Technol. 6 (3), 628–647.

Feng, J., Wu, S., Wang, H., Liu, S., 2016. Improved bioavailability of curcumin in ovalbumin-dextran nanogels prepared by Maillard reaction. J. Funct. Foods 27, 55–68.

Feng, W., Yue, C., Ni, Y., Liang, L., 2018. Preparation and characterization of emulsion-filled gel beads for the encapsulation and protection of resveratrol and α-tocopherol. Food Res. Int. 108, 161–171.

Fernandez, A., Torres-Giner, S., Lagaron, J.M., 2009. Novel route to stabilization of bioactive antioxidants by encapsulation in electrospun fibers of zein prolamine. Food Hydrocoll. 23 (5), 1427–1432.

Fuciños, C., Míguez, M., Fuciños, P., Pastrana, L.M., Rúa, M.L., Vicente, A.A., 2017. Creating functional nanostructures: encapsulation of caffeine into α-lactalbumin nanotubes. Innovative Food Sci. Emerg. Technol. 40, 10–17.

Galstyan, V., Bhandari, M.P., Sberveglieri, V., Sberveglieri, G., Comini, E., 2018. Metal oxide nanostructures in food applications: quality control and packaging. Chemosensors 6 (2), 16.

Ghorani, B., Tucker, N., 2015. Fundamentals of electrospinning as a novel delivery vehicle for bioactive compounds in food nanotechnology. Food Hydrocoll. 51, 227–240.

Greßler, S., Gazsó, A., Simkó, M., Nentwich, M., Fiedeler, U., 2010. Nanoparticles and nanostructured materials in the food industry (NanoTrust Dossier No. 004en–December 2010).

Gupta, A., Eral, H.B., Hatton, T.A., Doyle, P.S., 2016. Nanoemulsions: formation, properties and applications. Soft Matter 12, 2826–2841.

Hasheminejad, N., Khodaiyan, F., Safari, M., 2019. Improving the antifungal activity of clove essential oil encapsulated by chitosan nanoparticles. Food Chem. 275, 113–122.

He, X., Hwang, H.M., 2016. Nanotechnology in food science: functionality, applicability, and safety assessment. J. Food Drug Anal. 24 (4), 671–681. https://doi.org/10.1016/j.jfda.2016.06.001. Elsevier Ltd.

Hu, S., Wang, T., Fernandez, M.L., Luo, Y., 2016. Development of tannic acid cross-linked hollow zein nanoparticles as potential oral delivery vehicles for curcumin. Food Hydrocoll. 61, 821–831.

Humblet-Hua, K.N.P., Scheltens, G., Van Der Linden, E., Sagis, L.M.C., 2011. Encapsulation systems based on ovalbumin fibrils and high methoxyl pectin. Food Hydrocoll. 25 (4), 569–576.

Jain, A., Sharma, G., Ghoshal, G., Kesharwani, P., Singh, B., Shivhare, U.S., Katare, O.P., 2018. Lycopene loaded whey protein isolate nanoparticles: an innovative endeavor for enhanced bioavailability of lycopene and anti-cancer activity. Int. J. Pharm. 546 (1–2), 97–105.

Jaiswal, M., Dudhe, R., Sharma, P.K., 2015. Nanoemulsion: an advanced mode of drug delivery system. 3 Biotech 5 (2), 123–127.

Jaworek, A.T.S.A., Sobczyk, A.T., 2008. Electrospraying route to nanotechnology: an overview. J. Electrostat. 66 (3–4), 197–219.

Jiang, S.J., Zhang, T., Song, Y., Qian, F., Tuo, Y., Mu, G., 2019. Mechanical properties of whey protein concentrate based film improved by the coexistence of nanocrystalline cellulose and transglutaminase. Int. J. Biol. Macromol. 126, 1266–1272.

Jin, B., Zhou, X., Li, X., Lin, W., Chen, G., Qiu, R., 2016. Self-assembled modified soy protein/dextran nanogel induced by ultrasonication as a delivery vehicle for riboflavin. Molecules 21 (3), 282.

Joye, I.J., McClements, D.J., 2013. Production of nanoparticles by anti-solvent precipitation for use in food systems. Trends Food Sci. Technol. 34 (2), 109–123.

Joyner, J.J., Kumar, D.V., 2015. Nanosensors and their applications in food analysis: a review. Int. J. Sci. Technol. 3 (4), 80.

Karthikeyan, K., Guhathakarta, S., Rajaram, R., Korrapati, P.S., 2012. Electrospun zein/eudragit nanofibers based dual drug delivery system for the simultaneous delivery of aceclofenac and pantoprazole. Int. J. Pharm. 438 (1–2), 117–122.

Khandelwal, N., Barbole, R.S., Banerjee, S.S., Chate, G.P., Biradar, A.V., Khandare, J.J., Giri, A.P., 2016. Budding trends in integrated pest management using advanced micro and nano-materials: challenges and perspectives. J. Environ. Manage. 184, 157–169.

Komaiko, J.S., McClements, D.J., 2016. Formation of food-grade nanoemulsions using low-energy preparation methods: a review of available methods. Compr. Rev. Food Sci. Food Saf. 15 (2), 331–352.

Koroleva, M.Y., Yurtov, E.V., 2012. Nanoemulsions: the properties, methods of preparation and promising applications. Russ. Chem. Rev. 81 (1), 21.

Li, H., Wang, M., Williams, G.R., Wu, J., Sun, X., Lv, Y., Zhu, L.M., 2016. Electrospun gelatin nanofibers loaded with vitamins A and E as antibacterial wound dressing materials. RSC Adv. 6 (55), 50267–50277.

Li, H., Wang, D., Liu, C., Zhu, J., Fan, M., Sun, X., Wang, T., Xu, Y., Cao, Y., 2019. Fabrication of stable zein nanoparticles coated with soluble soybean polysaccharide for encapsulation of quercetin. Food Hydrocoll. 87, 342–351.

Lin, L., Xu, W., Liang, H., He, L., Liu, S., Li, Y., Li, B., Chen, Y., 2015. Construction of pH-sensitive lysozyme/pectin nanogel for tumor methotrexate delivery. Colloids Surf. B: Biointerfaces 126, 459–466.

Lucey, J.A., Singh, H., 1997. Formation and physical properties of acid milk gels: a review. Food Res. Int. 30 (7), 529–542.

Ma, Z., Chen, P., Cheng, W., Yan, K., Pan, L., Shi, Y., Yu, G., 2018. Highly sensitive, printable nanostructured conductive polymer wireless sensor for food spoilage detection. Nano Lett. 18 (7), 4570–4575.

Majeed, K., Jawaid, M., Hassan, A.A.B.A.A., Bakar, A.A., Khalil, H.A., Salema, A.A., Inuwa, I., 2013. Potential materials for food packaging from nanoclay/natural fibres filled hybrid composites. Mater. Des. 46, 391–410.

Martínez-Ballesta, M., Gil-Izquierdo, Á., García-Viguera, C., Domínguez-Perles, R., 2018. Nanoparticles and controlled delivery for bioactive compounds: outlining challenges for new "smart-foods" for health. Foods 7 (5), 72.

Martins, J.T., Bourbon, A.I., Pinheiro, A.C., Fasolin, L.H., Vicente, A.A., 2018. Protein-based structures for food applications: from macro to nanoscale. Front. Sustain. Food Syst. 2, 77.

Massironi, A., Morelli, A., Puppi, D., Chiellini, F., 2020. Renewable polysaccharides micro/nanostructures for food and cosmetic applications. Molecules 25 (21), 4886.

Medeiros, A.K.D.O.C., de Carvalho Gomes, C., de Araújo Amaral, M.L.Q., de Medeiros, L.D.G., Medeiros, I., Porto, D.L., Aragão, C.F.S., Maciel, B.L.L., de Araújo Morais, A.H., Passos, T.S., 2019. Nanoencapsulation improved water solubility and color stability of carotenoids extracted from Cantaloupe melon (Cucumis melo L.). Food Chem. 270, 562–572.

Mei, L., Wang, Q., 2020. Advances in using nanotechnology structuring approaches for improving food packaging. Annu. Rev. Food Sci. Technol. 11, 339–364.

Moghimi, R., Ghaderi, L., Rafati, H., Aliahmadi, A., McClements, D.J., 2016. Superior antibacterial activity of nanoemulsion of Thymus daenensis essential oil against E. coli. Food Chem. 194, 410–415.

Mohammadian, M., Salami, M., Emam-Djomeh, Z., Momen, S., Moosavi-Movahedi, A.A., 2018. Gelation of oil-in-water emulsions stabilized by heat-denatured and nanofibrillated whey proteins through ion bridging or citric acid-mediated cross-linking. Int. J. Biol. Macromol. 120, 2247–2258.

Mohammadian, M., Salami, M., Momen, S., Alavi, F., Emam-Djomeh, Z., Moosavi-Movahedi, A.A., 2019. Enhancing the aqueous solubility of curcumin at acidic condition through the complexation with whey protein nanofibrils. Food Hydrocoll. 87, 902–914.

Mozafari, M.R., et al., 2008. Nanoliposomes and their applications in food nanotechnology. J. Liposome Res. 18 (4), 309–327. https://doi.org/10.1080/08982100802465941.

Myrick, J.M., Vendra, V.K., Krishnan, S., 2014. Self-assembled polysaccharide nanostructures for controlled-release applications. Nanotechnol. Rev. 3 (4), 319–346.

Neethirajan, S., Jayas, D.S., 2011. Nanotechnology for the food and bioprocessing industries. Food Bioproc. Tech. 4 (1), 39–47. https://doi.org/10.1007/s11947-010-0328-2.

Nethaji, D.K., Parambil, K.A., 2017. Development and applications of nano emulsion in food technology. Int. J. Sci. Eng. Manage. 2 (12), 60–61.

Nunes, D., Pimentel, A., Gonçalves, A., Pereira, S., Branquinho, R., Barquinha, P., Fortunato, E., Martins, R., 2019. Metal oxide nanostructures for sensor applications. Semicond. Sci. Technol. 34 (4), 043001.

Oliver, C.M., Melton, L.D., Stanley, R.A., 2006. Creating proteins with novel functionality via the Maillard reaction: a review. Crit. Rev. Food Sci. Nutr. 46 (4), 337–350.

Omanović-Mikličanina, E., Maksimović, M., 2016. Nanosensors applications in agriculture and food industry. Bull. Chem. Technol. Bosnia Herzegovina 47, 59–70.

Otte, J., Schumacher, E., Ipsen, R., Ju, Z.Y., Qvist, K.B., 1999. Protease-induced gelation of unheated and heated whey proteins: effects of pH, temperature, and concentrations of protein, enzyme and salts. Int. Dairy J. 9 (11), 801–812.

Pandit, J., Aqil, M., Sultana, Y., 2016. Nanoencapsulation technology to control release and enhance bioactivity of essential oils. In: Encapsulations. Academic Press, Amsterdam, pp. 597–640.

Paredes, A.J., Asensio, C.M., Llabot, J.M., Allemandi, D.A., Palma, S.D., 2016. Nanoencapsulation in the food industry: manufacture, applications and characterization. J. Food Bioeng. Nanoprocess. 1 (1), 56–79.

Pathakoti, K., Manubolu, M., Hwang, H.M., 2017. Nanostructures: current uses and future applications in food science. J. Food Drug Anal. 25 (2), 245–253. https://doi.org/10.1016/j.jfda.2017.02.004. Elsevier Ltd.

Peters, R.J., Bouwmeester, H., Gottardo, S., Amenta, V., Arena, M., Brandhoff, P., Marvin, H.J., Mech, A., Moniz, F.B., Pesudo, L.Q., Rauscher, H., 2016. Nanomaterials for products and application in agriculture, feed and food. Trends Food Sci. Technol. 54, 155–164.

Pradhan, N., et al., 2015. Facets of nanotechnology as seen in food processing, packaging, and preservation industry. Biomed. Res. Int. 2015. https://doi.org/10.1155/2015/365672.

Prakash, B., Kujur, A., Yadav, A., Kumar, A., Singh, P.P., Dubey, N.K., 2018. Nanoencapsulation: an efficient technology to boost the antimicrobial potential of plant essential oils in food system. Food Control 89, 1–11.

Qin, Y., Liu, C., Jiang, S., Xiong, L., Sun, Q., 2016. Characterization of starch nanoparticles prepared by nanoprecipitation: influence of amylose content and starch type. Ind. Crop Prod. 87, 182–190.

Rashidi, L., Khosravi-Darani, K., 2011. The applications of nanotechnology in food industry. Crit. Rev. Food Sci. Nutr. 51 (8), 723–730. https://doi.org/10.1080/10408391003785417.

Sadeghi, R., Kalbasi, A., Emam-Jomeh, Z., Razavi, S.H., Kokini, J., Moosavi-Movahedi, A.A., 2013. Biocompatible nanotubes as potential carrier for curcumin as a model bioactive compound. J. Nanopart. Res. 15 (11), 1–11.

Sadeghi, R., Moosavi-Movahedi, A.A., Emam-Jomeh, Z., Kalbasi, A., Razavi, S.H., Karimi, M., Kokini, J., 2014. The effect of different desolvating agents on BSA nanoparticle properties and encapsulation of curcumin. J. Nanopart. Res. 16 (9), 1–14.

Sadeghnejad, A., Aroujalian, A., Raisi, A., Fazel, S., 2014. Antibacterial nano silver coating on the surface of polyethylene films using corona discharge. Surf. Coat. Technol. 245, 1–8.

Sari, T.P., Mann, B., Kumar, R., Singh, R.R.B., Sharma, R., Bhardwaj, M., Athira, S., 2015. Preparation and characterization of nanoemulsion encapsulating curcumin. Food Hydrocoll. 43, 540–546.

Saricay, Y., Wierenga, P., de Vries, R., 2013. Nanostructure development during peroxidase catalysed cross-linking of α-lactalbumin. Food Hydrocoll. 33 (2), 280–288.

Serfert, Y., Lamprecht, C., Tan, C.P., Keppler, J.K., Appel, E., Rossier-Miranda, F.J., Schroen, K., Boom, R.M., Gorb, S., Selhuber-Unkel, C., Drusch, S., 2014. Characterisation and use of β-lactoglobulin fibrils for microencapsulation of lipophilic ingredients and oxidative stability thereof. J. Food Eng. 143, 53–61.

Shen, Y., Posavec, L., Bolisetty, S., Hilty, F.M., Nyström, G., Kohlbrecher, J., Hilbe, M., Rossi, A., Baumgartner, J., Zimmermann, M.B., Mezzenga, R., 2017. Amyloid fibril systems reduce, stabilize and deliver bioavailable nanosized iron. Nat. Nanotechnol. 12 (7), 642–647.

Shewan, H.M., Stokes, J.R., 2013. Review of techniques to manufacture micro-hydrogel particles for the food industry and their applications. J. Food Eng. 119 (4), 781–792.

Siddiqui, N., Bhardwaj, A., Hada, R., Yadav, V.S., Goyal, D., 2018. Synthesis, characterization and antimicrobial study of poly (methyl methacrylate)/Ag nanocomposites. Vacuum 153, 6–11.

Singh, Y., Meher, J.G., Raval, K., Khan, F.A., Chaurasia, M., Jain, N.K., Chourasia, M.K., 2017. Nanoemulsion: concepts, development and applications in drug delivery. J. Control. Release 252, 28–49. https://doi.org/10.1016/j.jconrel.2017.03.008.

Sole, I., Solans, C., Maestro, A., Gonzalez, C., Gutierrez, J.M., 2012. Study of nano-emulsion formation by dilution of microemulsions. J. Colloid Interface Sci. 376, 133–139.

Sotelo-Boyás, M., Correa-Pacheco, Z., Bautista-Baños, S., y Gómez, Y.G., 2017a. Release study and inhibitory activity of thyme essential oil-loaded chitosan nanoparticles and nanocapsules against foodborne bacteria. Int. J. Biol. Macromol. 103, 409–414.

Sotelo-Boyás, M.E., Correa-Pacheco, Z.N., Bautista-Baños, S., Corona-Rangel, M.L., 2017b. Physicochemical characterization of chitosan nanoparticles and nanocapsules incorporated with lime essential oil and their antibacterial activity against food-borne pathogens. LWT 77, 15–20.

Sullivan, S.T., Tang, C., Kennedy, A., Talwar, S., Khan, S.A., 2014. Electrospinning and heat treatment of whey protein nanofibers. Food Hydrocoll. 35, 36–50.

Tapia-Hernández, J.A., Rodríguez-Félix, F., Katouzian, I., 2017. Nanocapsule formation by electrospraying. In: Nanoencapsulation Technologies for the Food and Nutraceutical Industries. Academic Press, pp. 320–345.

Tarhan, O., Harsa, S., 2014. Nanotubular structures developed from whey-based α-lactalbumin fractions for food applications. Biotechnol. Prog. 30, 1301–1310.

Tavassoli-Kafrani, E., Goli, S.A.H., Fathi, M., 2018. Encapsulation of orange essential oil using cross-linked electrospun gelatin nanofibers. Food Bioprocess Technol. 11 (2), 427–434.

Teng, Z., Luo, Y., Wang, Q., 2012. Nanoparticles synthesized from soy protein: preparation, characterization, and application for nutraceutical encapsulation. J. Agric. Food Chem. 60 (10), 2712–2720.

Thomä-Worringer, C., Sørensen, J., López-Fandiño, R., 2006. Health effects and technological features of caseinomacropeptide. Int. Dairy J. 16 (11), 1324–1333.

Tiwari, S., Singh, B.K., Dubey, N.K., 2020. Encapsulation of essential oils-a booster to enhance their bio-efficacy as botanical preservatives. J. Sci. Res. 64 (1). https://doi.org/10.37398/JSR.2020.640125.

Vaze, N., Jiang, Y., Mena, L., Zhang, Y., Bello, D., Leonard, S.S., Morris, A.M., Eleftheriadou, M., Pyrgiotakis, G., Demokritou, P., 2018. An integrated electrolysis–electrospray–ionization antimicrobial platform using Engineered Water Nanostructures (EWNS) for food safety applications. Food Control 85, 151–160.

Verma, D., Gulati, N., Kaul, S., Mukherjee, S., Nagaich, U., 2018. Protein based nanostructures for drug delivery. J. Pharm. 2018, 9285854.

Vilarinho, F., Sanches Silva, A., Vaz, M.F., Farinha, J.P., 2018. Nanocellulose in green food packaging. Crit. Rev. Food Sci. Nutr. 58 (9), 1526–1537.

Walia, N., Dasgupta, N., Ranjan, S., Chen, L., Ramalingam, C., 2017. Fish oil based vitamin D nanoencapsulation by ultrasonication and bioaccessibility analysis in simulated gastro-intestinal tract. Ultrason. Sonochem. 39, 623–635.

Wang, K., Zhang, Y., Wang, J., Yuan, A., Sun, M., Wu, J., Hu, Y., 2016. Self-assembled IR780-loaded transferrin nanoparticles as an imaging, targeting and PDT/PTT agent for cancer therapy. Sci. Rep. 6 (1), 1–11.

Xu, H., Jiang, Q., Reddy, N., Yang, Y., 2011. Hollow nanoparticles from zein for potential medical applications. J. Mater. Chem. 21 (45), 18227–18235.

Xu, H., Shen, L., Xu, L., Yang, Y., 2015. Controlled delivery of hollow corn protein nanoparticles via non-toxic crosslinking: in vivo and drug loading study. Biomed. Microdevices 17 (1), 1–8.

Yang, D., Li, Y., Nie, J., 2007. Preparation of gelatin/PVA nanofibers and their potential application in controlled release of drugs. Carbohydr. Polym. 69 (3), 538–543.

Yu, Z., Yu, M., Zhang, Z., Hong, G., Xiong, Q., 2014. Bovine serum albumin nanoparticles as controlled release carrier for local drug delivery to the inner ear. Nanoscale Res. Lett. 9 (1), 1–7.

Yu, M., Ji, N., Wang, Y., Dai, L., Xiong, L., Sun, Q., 2021. Starch-based nanoparticles: stimuli responsiveness, toxicity, and interactions with food components. Compr. Rev. Food Sci. Food Saf. 20 (1), 1075–1100.

Zhang, J., 2011. Novel emulsion-based delivery systems (Dissertation). University of Minnesota.

Zhang, Y., Erkey, C., 2006. Preparation of supported metallic nanoparticles using supercritical fluids: a review. J. Supercrit. Fluids 38 (2), 252–267.

Zhang, Y., Wei, Q., 2016. The role of nanomaterials in electroanalytical biosensors: a mini review. J. Electroanal. Chem. 781, 401–409.

Zhang, J., Field, C.J., Vine, D., Chen, L., 2015. Intestinal uptake and transport of vitamin B 12-loaded soy protein nanoparticles. Pharm. Res. 32 (4), 1288–1303.

Zhao, Q., Wang, C., Liu, Y., Wang, S., 2007. Bacterial adhesion on the metal-polymer composite coatings. Int. J. Adhes. Adhes. 27 (2), 85–91.

Zhu, K., Ye, T., Liu, J., Peng, Z., Xu, S., Lei, J., Deng, H., Li, B., 2013. Nanogels fabricated by lysozyme and sodium carboxymethyl cellulose for 5-fluorouracil controlled release. Int. J. Pharm. 441 (1–2), 721–727.

CHAPTER

10

Nanotechnology in microbial food safety

Abhinandan Pal and Kanishka Bhunia

Agricultural and Food Engineering Department, Indian Institute of Technology Kharagpur, Kharagpur, WB, India

OUTLINE

1 Introduction

Nanotechnology is an advanced arena, which comprises of the development, fabrication, and application of matters in various technical fields (including food technology) by maintaining the dimension in the range of nanometer-scale (1–100 nm) (Di Giulio et al., 2020; McClements and Xiao, 2017). Nanoparticles and nanomaterials are the two main nanocomponents used in this ground, which can constitute several modified nanoforms like nanotubes, nanorods, nanosheets, nanocoating, nanocapsule, and nanofiber (Handford et al., 2014). Due to high surface reactivity and mass transfer rate, nanoparticles exhibit advanced quantum and catalytic behavior, which result in innovative chemical, physical, biological, and enzymatic reactivity (Sahoo et al., 2021). In recent days, the application of nanotechnology in the food industry has brought a huge potential to improve food preservation and safety. Microbial decontamination, food packaging and preservation, surface coating, shelf-life enhancement, nutraceutical and bioactive components transportation are the most renowned applications of nanotechnology in the food industry (Nile et al., 2020; Singh et al., 2017). Nanomaterials or nanoparticles have been applied during food production to both detect and inhibit or prevent or delay the growth of food pathogens (Hoseinnejad et al., 2018; Mustafa and Andreescu, 2020). Direct application of nanoparticles in food safety includes nanocoating on food, nanomaterials based active food packaging, food contact surfaces, hurdle technology, etc., in which spoilage bacteria come in contact with nanoparticles and are killed or inhibited due to oxidative stress, ROS production, photocatalysis, ion release, pH change, physical damages and other various mechanisms (Egodage et al., 2017; Hoseinnejad et al., 2018; Torres Dominguez et al., 2019). In the indirect applications, nanostructures are used to detect different food contaminants (e.g., adulterants, heavy metals, toxins, antibiotics, and pesticides residuals), and foodborne pathogens like *Salmonella*, *Escherichia coli* (*E. coli*), and *Listeria monocytogenes* (*L. monocytogenes)* by fabrication of nanosensors and nanomaterial-based assays to ensure microbial food safety (Fig. 1) (Mustafa and Andreescu, 2020; Singh et al., 2017).

Depending upon the applications, the choice of nanomaterials or nanoparticles and suitable technologies are important considerations, because technologies based on a variety of nanostructures exhibit different functionalities. For example, different nanoparticles (nano-engineered) are used to enhance antimicrobial activity and bioavailability; nanoencapsulation (delivery-based technology) can impart taste, flavor, and protect substituents from external environments; active and smart nano packaging (packaging technology) have the ability to improve durability, flexibility, and mechanical strength of the packaging materials; nanofilter and membrane (nanofiltration) can generate high-quality products; and fluorescent nanoparticles (nanosensor) are used to detect the food pathogens, etc. (Neethirajan and Jayas, 2011). Also, the aggregation state, size, charges, and shapes of the food-grade nanoparticles are the crucial factors (McClements et al., 2016) (Fig. 2).

The nanomaterials used in microbial food safety are generally engineered nanomaterials (ENMs), which are intentionally manufactured with specific properties or compositions (FAO/WHO, 2010). Based on the application in food processing, The ENMs can be divided into three main categories, i.e., inorganic, organic, and functionalized nanomaterials. Silver (Ag), gold (Au), zinc oxide (ZnO), titanium oxide (TiO_2), silica (SiO_2), and copper oxides (nCuO) are the most used food-grade inorganic nanomaterials (INMs) (FAO/WHO, 2010;

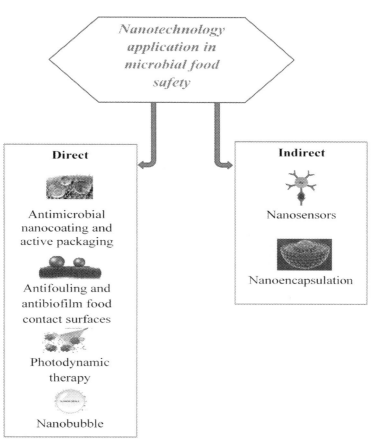

FIG. 1   Application of nanotechnology in microbial food safety.

McClements and Xiao, 2017). The major areas of application of the ENMs in food applications include antimicrobial and antifouling agents, anticaking agents, active and smart food packaging, and sensing of unwanted substances (e.g., food pathogens) present in food materials. The organic nanomaterials include carbohydrate, peptide, and lipid-based nano substances (e.g., nanoemulsion, solid lipid nanoparticles, nanocellulose, micelles, etc.), and different food pigments (e.g., lycopene, carotenoids), which are used to carry bioactive compounds and nutraceuticals, formulate antimicrobials and control the release of encapsulated antimicrobials (FAO/WHO, 2010; Sahoo et al., 2021). The antimicrobial activity of NPs (both inorganic and organic) and nanostructures (nanofiber, graphene oxide, quantum dots, and nanowire) for different microorganisms is summarized in Tables 1 and 2.

This chapter is mainly focused on the applications of different nanotechnologies for microbial food safety. It also summarizes the role of different nanomaterials and nanostructures in microbial food safety for preventing food quality deterioration and thereby increasing the shelf life of foods.

FIG. 2 Schematic diagram of important physicochemical factors of food-grade nanoparticles.

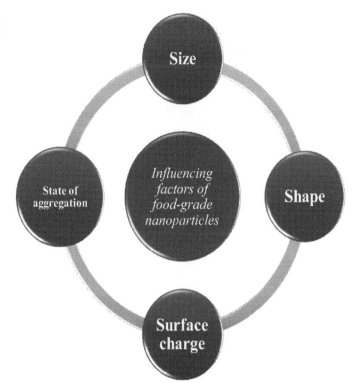

2 Interaction between nanoparticles and microbes

Different NPs respond in different ways after coming in contact with the microbes depending on the types of microbes as well as nanostructures used (Slavin et al., 2017). The nanoparticles with smaller particle sizes (<10 nm) are considered to be more detrimental compared to larger ones and can significantly damage the essential biomolecules, including DNA, proteins, and lipids of the targeted microbes because of their relatively larger surface area to volume ratio (Slavin et al., 2017). In this section, the antibacterial, antifungal, and antiviral activity of the nanoparticles have been demonstrated.

2.1 Bactericidal mechanism of the nanoparticles

The two major groups of bacteria identified in food products and food-based substances are Gram-positive and Gram-negative bacteria. The Gram-positive bacteria own a thick outer layer of peptidoglycan (20–80 nm) on their cell wall, whereas, the Gram-negative bacteria contain a thin (<10 nm) layer of peptidoglycan in their cell wall including an additional porous layer made up of lipopolysaccharide (Mai-Prochnow et al., 2016; Slavin et al., 2017). That is why the Gram-negative bacteria (e.g., *Escherichia coli*) are more fragile against the

TABLE 1 Different types of nanoparticles and their mode of actions against different pathogens.

Type of nanoparticles	Name of the nanoparticles (NPs)	Mode of action	The main factor affecting the antimicrobial activity	Mechanism of action	Pathogens	Reference
Metal and metal-oxide nanoparticles	Silver (Ag)	Ion release, ROS generation	Particle size and shape	• Adsorption of silver ions on the microbial cell wall or membrane due to electrostatic interaction or affinity towards sulfur protein, and results in alteration of permeability and disruption of microbial cell • Due to the presence of Ag^0 core, AgNP generates ROS, which interacts with the cells and produces oxidative stress	*E. coli, S. aureus, C. albicans, B. subtilis, B. cereus, A. niger, A. flavus, L. monocytogene, P. aeruginosa, Campylobacter*	Yin et al. (2020), Le Ouay and Stellacci (2015), Anees Ahmad et al. (2020), Duffy et al. (2018)
	Gold (Au)	Vesicle formation after interacting with carbohydrate, protein and lipid. ROS generation. Inhibit transcription.	Particle size	• Pore generation due to adsorption of AuNPs on the lipid surfaces by coulomb potential. Adsorption of AuNPs results in positional change of phosphocholine from inclined to the vertical position by dipole-charge. Then by electrostatic interaction between lipid molecules and AuNPs, the pore is generated. • Due to affinity towards the thiol group, AuNPs binds with membrane protein, which with different receptors allow AuNPs inside the bacteria, resulting in cell death. • Binds with DNA and disturbs transcription. • AuNP increases intracellular ROS production.	*E. coli, S. aureus, K. pneumonia, B. subtilis.* (Although this NP was found more effective against Gram-negative bacteria compared to Gram-positive one)	Ortiz-Benitez et al. (2019), Shamaila et al. (2016)
	ZnO	Photocatalysis. ROS generation. Zn^{2+} ion release.	• Facet dependent • Concentration	• Upon UV-illumination loosely bound oxygen is desorbed, which further by reduction produces superoxide radicals ($°O_2$). After the reaction between $°O_2$ and electron-hole pairs, hydroxyl radicals ($°OH$) are formed. Both these $°O_2$ and $°OH$ damage carbohydrates, lipid, and nucleic acid, causing death.	*S. aureus, E. coli, C. jejuni* (a common foodborne pathogen), *S. enterica, S. enteritidis, P. aeruginosa,* etc.	Sirelkhatim et al. (2015), Navarro-López et al. (2021)

Continued

TABLE 1 Different types of nanoparticles and their mode of actions against different pathogens—cont'd

Type of nanoparticles	Name of the nanoparticles (NPs)	Mode of action	The main factor affecting the antimicrobial activity	Mechanism of action	Pathogens	Reference
				• Released Zn^{2+} ions generate electrostatic force after reacting with negatively charged microbial cells and penetrate inside. Which further react with -SH groups of the surface proteins and cause cell death		
	TiO_2	Photooxidation. ROS production.	• Crystal structure • UV-light intensity • Type of artificial light • Shape and size	• Dissolution of outer membranes of microorganisms by generating ROS (mainly hydroxyl radicals (-OH)), resulting in phospholipid peroxidation and cell death. • Bacterial inactivation by photolytic oxidation reaction under UV-light exposer.	Gram-positive and Gram-negative bacteria.	Azizi-Lalabadi et al. (2019), Subhapriya and Gomathipriya (2018)
	CuO	ROS production. Ion release. Direct interaction with protein and fumarase enzyme.	• Size • Concentration • Morphology • Dissolution of copper ions in differed medium	• After direct interaction with the cell components, Cu(II) is reduced to Cu(I) by sulfhydryl, which is responsible for deactivating microorganisms by generating oxidative stress through Cu(I) mediated ROS.	*E.coli, P. aeruginosa, Campylobacter, B. subtilis* and so on	Meghana et al. (2015), Bezza et al. (2020), Duffy et al. (2018)
Non-metal nanoparticles	Carbon nanotubes (CNTs) e.g. Single-walled carbon nanotubes (SWCNT)	Physical membrane damage. ROS production. Altering metabolic pathways.	• Length • Residual catalyst • Diameter • Surface chemistry • Surface functional groups	• During wrapping the SWCNT interacts with microbes by hydrogen bonding and electrostatic absorption, causing disintegration of the cell wall and cell death. • ROS generation on the SWCNT surface could damage the protein, lipid, and nucleic acid and further generate oxidative stress to the microbes • Direct bacterium-SWCNT contact facilitates DNA damage, protein disruption, and cell block	Several Gram-positive and Gram-negative bacteria like *E. coli*, *S. aureus*, etc.	Liu et al. (2011), Chen et al. (2013), Teixeira-Santos et al. (2021)

Natural nanomaterial	Chitosan	pH change. Ionic interaction.	• Polycationic structure • Chitosan concentration • The molecular weight of chitosan • Positive charge density • Hydrophilic and hydrophobic properties • Chelating capacity • Ionic strength	• Antimicrobial activity by surface to surface or local interaction, rather than throughout contact mode. • Interaction with the negatively charged cell wall of the bacteria, resulting disrupting the cell wall, alteration of membrane permeability, inhibition of DNA replication of the microorganisms	*E. coli, K. pneumonia, S. aureus,* and fungi	Yılmaz Atay (2020), Goy et al. (2016)
	Nanoclay	Electrostatic interactions (to attract microbes) and hydrophobic interaction between ammonium surfactant and lipophilic components.	• Cationic surface • pH	• With the presence of ammonium surfactants, nanoclay alters the membrane permeability of the microbial cells, which allows intracellular ions and metabolites to diffuse out, causing cell death.	*S. aureus, E. coli*	Yahiaoui et al. (2015)

TABLE 2 Antimicrobial activity of different nanostructure conjugates against food pathogens.

Type of nanostructures	Materials used for various nanostructures	Mode of actions	Targeted food microorganisms	Major findings	Source
Nanofibers	Tea-polyphenol (TP)-polylactic acid (PLA) nanofiber		E. coli and S. aureus	Inhibition of E. coli and S. aureus up to 92.26% and 94.58% respectively at the ratio of 3:1 (PLA: TP)	Liu et al. (2018), Ge et al. (2012)
	Poly(butylene adipate-co-terephthalate) incorporated chitosan Nanofiber (CS-NF)		**Gram-positive** S. aureus and B. subtilis. **Gram-negative** Salmonella enteritidis and E. coli bacteria.	Concentration-dependent antimicrobial activity of CS-NF (wt%), where, at the highest concentration (12 wt %), the final growth of Gram-negative bacteria was 3.7 log CFU.	Díez-Pascual and Díez-Vicente (2015)
	Glucose oxidase immobilized Poly (vinyl alcohol) / Chitosan/tea extract nanofibrous membranes		E. coli and S. aureus	97% reduction of E. coli after the reaction time of 90 min. Almost 100% reduction of S. aureus after 40 min.	Ge et al. (2012)
	Liquid smoke loaded chitosan nanofibers (LSCN) and chitosan (CN) nanofiber		TMABc (total mesophilic aerobic bacteria count), TPBc (total psychrophilic bacteria count), and TYMc (total yeast and mold count) in Bream fish fillet	40%–50% inhibition of microbial growth. Antimicrobial activity of CN ≥ LSCN during cold storage	Ceylan et al. (2017)
	Multi- and Double-walled carbon nanotube incorporated thermoplastic polyurethane (TPU) nanofibers		S. aureus, P. aeruginosa, K. pneumoniae and C. albicans	Concentration-dependent antimicrobial activity. At 100 µg/mL concentration, killing was maximum. Functionalized MWCNT exhibited more antibacterial activity than DWCNT.	Saleemi et al. (2020)

Material	Mechanism	Test organisms	Findings	References
Silver-cellulose nanofiber	Oxidative-stress mediated antimicrobial activity	*S. aureus* and *E. coli*	Formation of higher halo width with CMC-Ag nanofiber samples for both *Staphylococcus aureus* and *Escherichia coli* than AgNPs/CMC nanofiber samples. ~48% of silver release from CMC-Ag compared to ~27% from AgNPs/CMC, resulting in higher antimicrobial activity	Gopiraman et al. (2016)
Ag NP incorporated PVA-lignin nanofiber		*Bacillus circulans* and *E. coli.*	The size of the inhibition zone against Gram-negative and Gram-positive *bacteria* was 1.1 cm and 1.3 cm, respectively.	Aadil et al. (2018)
FeNP incorporated Polyacrylonitrile nanofiber		*S. aureus* and *E. coli*	Better antibacterial activity against *S. aureus* ($R > 1$) compared to *E. coli* ($R < 1$)	Kalwar et al. (2021)
PVA based ginger nanofiber biocomposite film		**Gram-positive bacteria** *S. aureus, B. subtilis* **Gram-negative bacteria** *E. coli, P. aeruginosa* **Fungi** *C. albicans*	Good antibacterial activity but antifungal activity was not found because of the low concentration of ginger fiber. Sensitive to Gram-positive compared to Gram-negative bacteria	Abral et al. (2020)
Silver NP/nylon-6 nanofiber incorporated polypropylene food packaging film		*E. coli and S. aureus*	The inhibition percentage (PI) against *E.coli* and *S. aureus* was 100% and 79% after electrospinning for 30 min. PI of 100% was obtained against *S. aureus* after 60 min of electrospinning. More effective against *E.coli* than *S. aureus*.	Cheng et al. (2018)

Continued

TABLE 2 Antimicrobial activity of different nanostructure conjugates against food pathogens—cont'd

Type of nanostructures	Materials used for various nanostructures	Mode of actions	Targeted food microorganisms	Major findings	Source
	Pleurocidin induced PVA (Ple-PVA) electrospun fibers		S. aureus and E. coli O157:H7	All the cells of Both bacterial were inhibited within 5 h after treating with 0.25% Pleurocidin/PVA fiber ma Antimicrobial activity of released Pleurocidin was more than free Pleurocidin against E. coli O157:H7	Wang et al. (2015)
	Silver nanoparticles integrated poly(vinyl alcohol) (PVA) nanofiber		P. aeruginosa, B. cereus, S. aureus, E. coli, and P. aeruginosa	All the tested bacteria were significantly inhibited after the treatment. However, S. aureus was affected maximum with an average halos diameter of 21.47 mm	Kowsalya et al. (2019)
Graphene oxide (GO)	Graphene oxide nanomaterials (GO-NMs) against microbial activity	• Physical mode of action ("Nanoknives", "Wrapping" and "photothermal effect") • Chemical mode of action (ROS depended or independent)	**Antimicrobial activity** **Bacteria** S. aureus, E. coli, B. subtilis. **Virus** Rotavirus **Fungi** A. niger, A. flavus	• Around 69.3% reduction of E. coli • 80% decline of colony number of S. aureus and B. subtilis after treating with PLLA-GO film • Significant reduction of virus and fungi	Liu et al. (2011), Kumar et al. (2019a), Fan et al. (2014), Nguyen et al. (2019), Xia et al. (2019)
	Graphene oxide nanomaterials (GO-NMs) against biofilm activity	• Hydrogen bond interaction between the extracellular polymeric substances (EPS) of biofilm and GO. • π–π interaction between DNA/RNA and GO	**Antibiofilm activity** S. aureus and P. aeruginosa	GO reduces the (i) new biofilm growth of S. aureus and P. aeruginosa by 55.05% and 44.18% (ii) mature biofilm growth of S. aureus and P. aeruginosa by 70.24% and 63.68% .	Di Giulio et al. (2020)

Material	Nanomaterial	Mechanism	Target bacteria	Finding	References
Quantum dots (QDs)	Graphene oxide quantum dots (GOQDs)	• phototoxicity against microbes	*S. epidermitis*	60% loss of cell viability after treating with GOQD	Smith et al. (2019), Nichols and Chen (2020)
	CdSeQDs and ZnSQDs	• Photo-oxidation • Generation of free radicals	*E. coli, A. baumanni* and *Bacillus subtilis*	CdSe QDs showed the more significant antibacterial activity compared to ZnS QDs due to the smaller size and larger surface area	Kumari et al. (2020), Dumas et al. (2009), Oetiker et al. (2020), Rojas-Andrade et al. (2020)
Nanowire	Ag (silver), and	• Ag^+ ions release	*S. aureus,* and *E. coli*	1.2–1.4 log reduction of *S. aureus* and 1.5–3.7 log reduction of *E. coli*	Kim et al. (2020)
	TiO_2	• Photocatalytic activity • Positive charge	Gram-negative bacteria (e.g., *E. coli, K. pneumoniae* etc.) and Gram-positive bacteria (e.g., *S. aureus*)	TiO_2 nanowire has more antimicrobial efficacy than its subsequent nanoparticle form.	Munisparan et al. (2018), Nataraj et al. (2014)
	Silicon (Si)	• Generation of heat and ROS after exposing to sunlight.	*E. coli*	100% inhibition of *E. coli* within 4 min after treating with Au/Ag NPs induced silicon nanowire in sunlight.	Wang et al. (2019a)
	CuO	—	*S. aureus* and *E. coli*	Antimicrobial activity depends on the annealing time.	Mahmoodi et al. (2018)

nanoparticles compared to the Gram-positive bacteria (e.g., *staphylococcus aureus*). Another probability behind the susceptibility of Gram-negative bacteria towards the NPs is due to the presence of negatively charged lipopolysaccharide in their cell wall. Most of the NPs release ions into the medium are positively charged. After coming in contact with the bacteria (Gram-negative), these ions affiliate towards its lipopolysaccharide, which leads to the disintegration of the cell wall structure, change the viscosity of the cell membrane, and imbalance the transport mechanism. As a result, the death of the bacteria occurs (Behera et al., 2019; Hou et al., 2018; Slavin et al., 2017).

Although the precise mechanisms of the bactericidal effect of the NPs are still not much distinguishable, it has been hypothesized that the inhibition of the microbial growth occurs after the direct contact between the nanoparticles and the bacterial cell wall by electrostatic interaction, Van der Waals forces, receptor-ligand, and hydrophobic interaction (Wang et al., 2017).

There are some identified mechanisms by which the nanoparticles inhibit the bacteria (Fig. 3), such as,

• *Electrostatic interaction between sulfur proteins and the ions*: The ions released by the nanoparticles are adsorbed on the microbial cells due to electrostatic interaction between

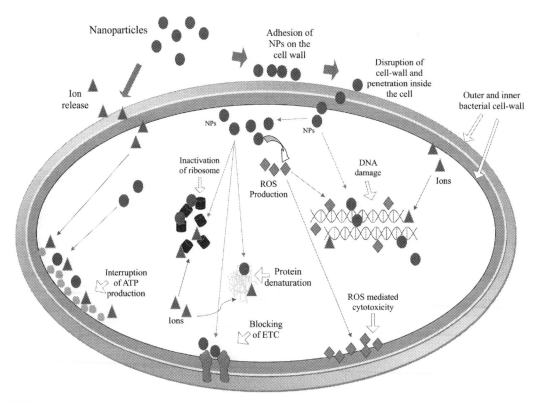

FIG. 3   Schematic diagram of antibacterial activity of nanoparticles in food applications.

sulfur proteins and the ions, which result in inflation of the cytoplasmic membrane permeability, followed by disruption of the bacterial envelope. It further leads to the internalization of the adsorbed ions. After taking up into the cells, free ions deactivate the respiratory enzymes, denature ribosomes, degrades intracellular ATP, and interfere with essential metabolic pathways (Singh et al., 2019; Yin et al., 2020).

- *Genotoxicity:* In this process, NP causes toxicity against the genetic materials causing oxidative damage to the DNA molecules and leading to the disintegration of DNA strands (both single- and double-stranded) by opening the deoxyribose ring, which facilitates bacterial death (Nile et al., 2020).

- *Generation of Reactive Oxygen species (ROS):* Among them, ROS-mediated antimicrobial activity is the most decisive one. The NPs generate reactive oxygen species (ROS) on the surface of the bacteria (*Bacillus subtilis*, *E. coli*, etc.), resulting in oxidative stress creation on the bacterial membrane causing damage to the membrane. After diffusing into the cell, it damages the lipid, protein, DNA, and other intracellular materials. Without treating with the NPs, ROS can also be generated in bacteria but those naturally produced ROS remain under the stress condition, which can be balanced by the different scavenging enzymes (e.g., SOD) present in the bacteria (Behera et al., 2019; Wang et al., 2017). Excessive production of oxidative stress by the NPs alter the permeability of cell membrane, and thereby cause cytotoxicity, lipid peroxidation, protein denaturation, and damaging of the DNA structure, which leads to cell death (Slavin et al., 2017). Among different ROS, the $^{\cdot}$OH is the most effective for bacterial inhibition. Different types of nanoparticles can generate different kinds of ROS by reducing the oxygen molecules, for example, calcium and magnesium oxide produce superoxide anion radicals, zinc oxide generates both hydroxyl radical ($^{\cdot}$OH) and hydrogen peroxide (H_2O_2), and copper oxide NPs can produce all the oxidants depending on the reaction properties (Wang et al., 2017).

2.2 Antiviral activity of nanoparticles

The most common foodborne viruses are Norovirus (NoV), Human Rotavirus (HRV), Hepatitis A (HAV), and E (HEV) virus, which can be easily transmitted by foods and cause several chronic health problems in human (FAO/WHO, 2008). The application of nanoparticles can be a better option for their eliminations. For example, different biopolymers or polymers incorporated silver (AgNPs) and copper (CuNPs) nanoparticles are effective against the NoV (Anvar et al., 2021; Randazzo et al., 2018). It has also been established that nanoparticles induced food packaging (AgNPs, CuNPs, and ZnNPs) can resist the SARS-CoV-2 (commonly "COVID-19") (Anvar et al., 2021). The nanoparticles can show antiviral activity inside or outside the host cells by the following mechanisms of actions:

- After interacting with the viruses the nanoparticles release their respective ions, which then adhere to the outer surface of the viruses and form the pits, through which intercellular materials of the targeted viruses come out, leading to death (Ishida, 2019).
- Another identified mechanism to inhibit viral infection is the transduction-inhibition effect, where, the nanoparticles bind with the viral genome or the viral particles and interrupt the communication between the virus and host cell, that is necessary for infection.

As a result, the virus gets inactivated before entry to the host cell (de Souza et al., 2016; Maduray and Parboosing, 2020).

- The nanoparticles interact with the glycoprotein structure of the virus envelope (e.g., gp 120 for HIV, which is used by the virus to enter the host cell) and compete with the virus for the binding sites of the host cell, followed by inferencing with viral attachment. Therefore, NPs block the viral-host binding sites, resulting in the prevention of the penetration or attachment of the virus to the host cell (Maduray and Parboosing, 2020).
- The nanoparticles generate ROS, which can inhibit viral infection (Li et al., 2016). It has been assumed that ROS leads to the peroxidation of the envelope membrane, alters the structure of the virus surface protein, and suppresses the synthesis of negative-strand RNA, and inhibit the virus from infection (Gurunathan et al., 2020; Seo et al., 2020).

2.3 Antifungal activity of nanoparticles

The nanoparticles show effective antifungal activity against *Aspergillus* spp., *Penicillium* spp., and *Candida albicans* (*C. albicans*) depending on the size, shape, and capping protein attached to the NPs. After binding with the NPs, pores are generated on the surface of the fungal cells which destroy cellular structures due to depolarization of the membrane. The formation of pores also lead to the extraction of intracellular materials from the cells and alter metabolism pathways (Singh et al., 2019). Additionally, after interacting with the nanoparticles, lipid and protein molecules of the fungal cells become oxidized and the DNA becomes disabled for replication due to the oxidative stress, resulting in damaging of ribosomal proteins expression (Farias et al., 2021; Yassin et al., 2021; Yilmaz et al., 2020a).

3 Antimicrobial nanocoating

Food surfaces are the primary site for microbial contamination. Microbes can attach themselves to the food surfaces and become able to proliferate. Therefore, an appropriate coating of food components can prevent the growth of pathogenic and spoilage microorganisms at the initial stages and prolong the shelf life (Kazemeini et al., 2021).

In nanocoating, either the second phase particles distributed into the matrix are nanostructured or the thickness of the coating is <100 nm. The nanocoating technique in food production is constituted with various antimicrobial and, antifouling properties. Compare to conventional coating, nano-coating contributes a superior distribution and homogeneity on the surface of food components due to their large surface area, consistency, flexibility, and adhesiveness between the coating layer and the substrate. It also provides a better appearance to the food materials. The food components are treated with this coating mainly by the application of spraying, immersion, or rubbing technique (Dhital et al., 2017; Joshi and Adak, 2019).

Different biopolymers, such as polysaccharides (chitosan, starch, and pectins), resin, protein (whey, gelatin, gluten, etc.), poly-ε-caprolactone (PCL), polylactic acid (PLA), poly-D,L-lactide-*co*-glycolide (PLGA), cellulose acetate phthalate (CAP), ethylcellulose and alginate are used as the backbone of the nanocoating to cover up the foods (Zambrano-Zaragoza et al., 2018). Nanocomposites, containing nano-ZnO, nano-Ag, nano-TiO$_2$, and nano-SiO$_2$ also provide a suitable structural matrix during nanocoating and can be used to coat uncut fruits and vegetables (Shi et al., 2018; Zambrano-Zaragoza et al., 2018). Antimicrobial nanocoating can be made of different metallic NPs, nano clay, chitosan, photocatalytic nanoparticles, etc. (Montazer and Harifi, 2017).

The antimicrobial nanocoating is grouped into two major categories such as microbiocidal coating and microbial-resistant coating (Cloutier et al., 2015; Kaur and Liu, 2016; Montazer and Harifi, 2017).

3.1 Different types of antimicrobial nanocoating for food safety

In this technique, foodstuffs are coated with a coating containing antimicrobial compounds to inhibit microbial growth. This approach is further subdivided into two groups:

3.1.1 Antimicrobial leachable nanocoating

This nanocoating inhibits microbial growth by leaching the antimicrobial metal ions or nanoparticles over time, which interact with the microbes by either diffusion, releasing the free radicals, or hydrolysis of covalent bond, and causes cell death of the microorganisms. However, during the real-time application of nanocoating, the leached components could get into the food and interfere with its quality, that's why this approach is quite questionable (Cloutier et al., 2015).

3.1.2 Contact killing based nanocoating

In this approach, the antimicrobial compounds are immobilized to the surface of coating materials by a covalent bond, which becomes active upon microbial contact. This method is more sustainable than the leaching technique, because, the biocidal components are not allowed to move freely into the matrix due to immobilization, as a result, food quality is not affected. Moreover, in this method, the compounds either produce oxidative stress or cause physical damage to the cell wall instead of targeting specific sites of the microbes, which leads to more microbial inhibition (Cloutier et al., 2015; Kaur and Liu, 2016).

3.1.3 Microbial resistant nano-coating

In this technique, foods are coated with nano-surfaces to resist the microbes from colonization instead of killing due to the superhydrophobicity and surface roughness of the coated material. These two physical modifications during coating can dramatically reduce the interaction between microorganisms and the foodstuffs (Cloutier et al., 2015).

3.2 Antimicrobial nanocoating on different solid food materials

3.2.1 Fruits and vegetables

- Attempts to maintain the quality and extend the shelf life of the fruits are the top priority of every producer. Studies have shown that the shelf life of the apples can be prolonged (up to 8 weeks in refrigeration whereas, 4 weeks in marketing conditions) by candelilla wax/ tarbush bioactive-based nanocoating (De León-Zapata et al., 2018).

- Strawberries are highly perishable and prone to fungal (*B. cinerea*) infections during the preservation. To prevent this, curcumin and D-limonene incorporated methyl cellulose-based nanocoating was found to be highly effective against this fungal contamination. The coating can also increase the shelf life (5–14 days) of the strawberries significantly (Dhital et al., 2017).
- Nanocoating of papaya with protein-based organo-clay nanocomposite (montmorillonite) has the high ability to resist the contamination of *Salmonella* spp., *E. coli*, molds, and yeasts in papaya and can extend the shelf life up to 12 days (Cortez-Vega et al., 2014).
- Silver-montmorillonite nanoparticles loaded calcium-alginate nanocoating protects the raw fresh carrot against *Pseudomonas* spp., *Enterobacteriaceae*, yeast, and psychrotrophic bacteria (Costa et al., 2012).
- Coating of oregano oil incorporated nano-emulsion on the fresh lettuce can reduce the proliferation of *Listeria monocytogenes* (*L. monocytogenes*), *Salmonella typhimurium* (*S. typhimurium*), and *Escherichia coli* (*E. coli*) O157:H7 up to 3.57, 3.26, and 3.35 log CFU/g reductions, respectively (Bhargava et al., 2015).
- Nanoemulsion of essential oils incorporated Chitosan-based nanocoating on the green beans is effective against the contamination of *Escherichia* (*E. coli*) O157:H7 (1.7 log CFU/g after 7 days; initially 2.45 log CFU/g) and *Salmonella typhimurium* (*S. typhimurium*) (1 log reduction after 7 days). The efficiency can further be tuned by the interference of the Gamma-ray and modified atmospheric packaging (MAP) (Severino et al., 2015).
- During cold storage, the microbial contamination (*Escherichia coli*, *Pseudomonas*, yeast, and molds) of shiitake mushroom (*Lentinus edodes*) can be restrained up to 12 days by alginate/ nano-Ag coating (Jiang et al., 2013).

3.2.2 Meat and fish

- Cress (*Lepidium sativum*) seed gum (CSG) based nano-coating of chitosan (CN) extends the shelf life of the beef at cold storage and can inhibit both the *Pseudomonas aeruginosa* (*P. aeruginosa*) and *Staphylococcus aureus* (*S. aureus*) growth significantly (Esmaeili et al., 2021).
- In the refrigerated condition (4 °C), nanocoating with rosemary essential oil activated chitosan/clay nanocomposites can reduce the total viable and psychrotrophic count of the silver carp (*Hypophthalmichthys molitrix*) fillet and can extend the shelf life up to 13 days (Abdollahi et al., 2014).
- Eugenol exhibits effective antimicrobial and antifungal properties, Eugenol-*Aloe vera*-based nanocoating can significantly inhibit microbial contamination in deep fat-fried shrimp during the refrigerating condition (Sharifimehr et al., 2019).

4 Anti-fouling surface

Fouling refers to unwanted adsorption and deposition of residue components on the surface or in the pores of machinery during processing. The deposited matrices, also called foulant, can be particulate components, organic components, microorganisms (biofouling), or salts (Liu et al., 2019a). Fouling is one of the major problems in the food industry, occurring during the processing of liquid and semiliquid food materials. Deposition of food foulant on the surfaces of processing equipments such as heat exchanger, filter membrane, conveyer belt, etc., result in the reduction of operating efficiency, decreasing productivity, and influencing biofilm formation. It can also manipulate heat transfer rate, pressure fluctuation, food quality alteration, and product cost increment (Merian and Goddard, 2012).

In the food industry, membrane separation operation and heat transfer operation are mainly affected by fouling. Fouling in membrane separation is occurred due to the accumulation of residues at the membrane surface. After prolonged usage, suspended solids and colloidal particles of the juices, raw milk, etc., deposit on the equipment surfaces and block the membrane pores, resulting in decreased permeability of the pores and membrane flux, which therefore reduces filtration efficiency (Wilson, 2018). This reduction of permeability can be calculated by fouling index (F.I),

$$F.I = \left(1 - \frac{P_1}{P_0}\right) \times 100 \tag{1}$$

where, P_1 is the permeability of permeates at constant volume through the active area of the membrane after fouling and P_0 is the initial permeability (Avram et al., 2017). On the other hand, the formation of fouling layer during heat transfer operation causes the rise of local temperature, reduction of thermal conductivity, solidification, and protein rearrangement, which would further result in increased capital, operational, and maintenance costs (Nakao et al., 2017; Wilson, 2018). In both the cases (membrane and heat exchanger), adsorption of foulant significantly affects fluid flowrate, machine efficiency and productivity of the process also.

4.1 Application of nanotechnology in developing anti-fouling surfaces

Nanocoating methods such as Plasma modification, ion implantation, chemical cross-linking, electroless nickel-polytetrafluoroethylene (EN-PTFE) deposition, and redox grafting have been established to provide antifouling properties to the food equipment surfaces (Barish and Goddard, 2013; Lu et al., 2021). Among them, EN-PTEE coating has effective anti-fouling characteristics against biofouling, mineral, and protein deposition.

Various food contact surfaces are widely used in the food processing industry, which frequently comes in contact with food materials and becomes prone to fouling. For example, due to hydrophilicity and high surface energy, stainless steel is sensitive to protein, minerals, and biofouling. Huang and Goddard (2015) and Barish and Goddard (2013) modified stainless steel surface of heat-exchanger with nickel -polytetrafluoroethylene (Ni-PTFE) (fluorinated nanoparticle modified steel) by the electroless nickel (EN) plating process to study the fouling behavior during pasteurization of raw milk and different liquid dairy products. They found that fouling due to raw whole milk (about 60% reduction) and liquid dairy products (40%–50% reduction) on the modified stainless steel surface was significantly reduced compared to the native stainless steel. This reduction could be better explained in terms of the grain boundaries and surface energy. In native steel, due to the presence of grain boundaries and high surface energy, the fouling was suitable for well-anchoring on its surface, whereas, both the grain boundary and surface energy were absent on the modified steel, which caused the lesser accumulation of fouling on the surface. The reduction of fouling was lesser in the case of dairy products (e.g., chocolate milk, chocolate milk with carrageenan, etc.) due to the increasing size of casein protein aggregates than that of pure milk.

Polytetrafluoroethylene (PTFE) nanoparticles induced Nickel (Ni) coating on stainless steel (SS) can be considered as an effective anti-biofouling surface also. It can reduce the microbial adhesion and subsequent growth of biofilm of pathogenic microbes (e.g., *Bacillus cereus*) on the food contact surfaces and therefore, enhances food safety. This reduction of biofilm formation on modified SS surface is due to lower surface energy compared to the native SS surface, which shrinks the effective surface area between microbial film and the stainless steel substratum (Huang et al., 2016).

The polymeric coating also exhibits excellent antifouling and antimicrobial activities. Styrene maleic anhydride (SMA) copolymer-Polyethylenimine (PEI) coated polypropylene (PP) can terminate the initiation of biofilm formation due to the presence of a surface cationic amine. After direct contact with the microbes (e.g., *P. aeruginosa*), the PP-PEI-SMA-PEI surface disrupts the cell membrane of the microbes and decreases the total cell mass of the biofilm on the surface (about 90%–99% reduction in viable cells during 24–48 h). Besides, it also alters the structure of protein A, which is responsible for extracellular polymeric surfaces (EPS) formation in biofilms. As a result the initial cell adhesion can be reduced by about 45% in modified polypropylene compared to native polypropylene. This enhanced anti-biofouling activity of PP-PEI-SMA-PEI surface is also due to cross-linking between antimicrobial agent and styrene-maleic anhydride (SMA) after interaction, which can reduce effective surface energy for cell adhering (Werner et al., 2019).

An antifouling surface can also be formulated by coating with a zwitterionic polymer. These polymeric ions contain an equal amount of cationic and anionic charges in their structures. Zwitterionic polymers are hydrophilic due to the presence of a rich source of ions and a strong hydration layer (Zheng et al., 2017). [2-(methacryloyloxy)ethyl]dimethyl-(3-sulfopropyl)ammonium hydroxide (SBMA) is a zwitterionic component, which has a potential antifouling activity in food packaging. Poly (vinyl alcohol-co-ethylene) (EVOH) nanofibrous membrane, a food packaging material, can be modified into an antimicrobial and antifouling film by grafting with biocidal compound N-halamine precursor and antifouling compound SBMA. This modified nanofibrous structure shows two useful features: (i) zwitterionic moieties can both reduce the attachment of microorganisms (e.g., *L. innocua* and *E. coli*) and remove bacterial debris significantly due to their hydrophilicity, and (ii) - N-halamine has ability to inhibit the attached microbial growth (Ma et al., 2019). N-halamine can be replaced by 3,3′,4,4′-benzophenonetetracarboxylic dianhydride (BPTCD) photosensitizer. The modified surface membrane, therefore, can act as photoactive antimicrobials and antifoulants both. The SBMA weakens the interaction between microbes and the membrane, making them difficult to be adsorbed. However, further if any bacteria adheres, BPTCD immediately kills them by generating ROS after photoexcitation. Moreover, due to superhydrophilicity of SBMA, the microbial debris can be easily removed from the nanofibrous antifouling surface (Ma et al., 2020).

5 Antimicrobial nanomaterials for biofilm

"Biofilm" is defined as the three-dimensional community of living bacteria, enclosed in a self-produced hydrated polymeric matrix, which increases the virulence of the pathogenic bacteria, ensures protection from the adverse environment, and provides a barrier against antibiotics and host defenses. The biofilm is a multicellular behavior of microorganisms, which creates an interaction and communication channel between multiple bacterial species (Flemming and Wingender, 2010).

In the food industry environment, biofilm-related infections are quite common. Among them, some of the biofilm-forming species are human pathogens and can be developed on both the hard (e.g., food processing equipment, transport medium, storage components, etc.) and biological surfaces (e.g., fish, meat, vegetables, etc.). Biofilm formation in food industrial equipment in static or dynamic conditions results in a huge loss of capital and production efficiency. Also, it is a possible threat to food safety due to the involvement of several foodborne pathogens including *Bacillus cereus* (*B. cereus*), *Escherichia coli* (*E. coli*), *Listeria monocytogenes* (*L. monocytogenes*), *Salmonella enterica* (*S. enterica*), and *Staphylococcus aureus* (*S. aureus*), which are responsible for different foodborne diseases (Galie et al., 2018; Qayyum and Khan, 2016).

5.1 Application of nanotechnology against biofilm

The interaction between the nanoparticles (NPs) and biofilm largely depends on their physicochemical properties. The NPs can react with the biofilm by three steps (i) NPs are transferred to targeted biofilms, (ii) attachment to the biofilm and occurrence of electrostatic interaction, hydrophobic interaction, and steric repulsion between the attached NPs and the biofilm, (iii) migration through the film by diffusion. The electrostatic interaction between nanoparticles and the biofilm depends on the zeta potential of the NPs and the charge distribution of the matrix. Mostly, the bacteria composes negatively charged film matrices, which can interact with the positively charged metal ion NPs through electrostatic interaction, resulting in diffusion of NPs through the matrix (Fulaz et al., 2019; Shkodenko et al., 2020). Generation of ROS after diffusion alters the gene expression, destroys cell membrane, and causes leaching of intracellular materials, which results in microbial death as well as the destruction of biofilm (Qayyum and Khan, 2016). There are certain parameters such as, size, surface, shape, and interior of the nanoparticles of the nanostructures, which significantly influence the interaction of NPs with biofilms (Liu et al., 2019b).

As per the antibiofilm activity, nanoparticles are divided into the following categories:

I. **Metallic and metal oxide nanoparticles** including AgNP, AuNP, CuNP, IONP, ZnO, TiO_2, etc., can eliminate biofilm by releasing metal ions, generating ROS, and biofilm matrix disruption. They are effective against pathogens such as *Bacillus subtilis* (*B. subtilis*), *Escherichia coli* (*E. coli*), *Salmonella typhimurium* (*S. typhimurium*), *Candida albicans* (*C. albicans*), and *Pseudomonas aeruginosa* (*P. aeruginosa*) like pathogenic microbes. These metallic NPs are more effective than other type of NPs due to their non-specificity towards the microbes, although sometimes it is considered a major demerit because of the negative effect of metallic NPs on symbiotic microbes. (LewisOscar et al., 2015; Sanyasi et al., 2016; Yu et al., 2016).

II. **Carbon-based NPs** also have the high potential to battle against biofilm. Graphene-based NPs are effective against different types of biofilms, e.g., *Pseudomonas putida* (*P. putida*), *Escherichia coli* (*E. coli*), *Staphylococcus aureus* (*S. aureus*), *Pseudomonas aeruginosa* (*P. aeruginosa*), etc. Their antibiofilm activity depends on different factors, like concentration of NPs, maturity of biofilms, etc. (Di Giulio et al., 2020; Fallatah et al., 2019; Sun et al., 2014).

III. The activity of biofilm can be inhibited by **polymer-based NPs** also. Among the polymer NPs, the most effective anti-biofilm activity is reported by chitosan. Due to functional amino groups (NH_2) of *N*-acetylglucosamine units of the polycationic chitosan, it attributes significant antibiofilm activity. After interacting with the biofilm positively charged chitosan electrostatically reacts with the negatively charged EPS (mainly structural protein and lipids) and inhibits biofilm formation of *Staphylococcus aureus*, *Listeria, monocytogenes, Bacillus cereus, Pseudomonas aeruginosa*, and *Escherichia coli* (Khan et al., 2020).

6 Nanoencapsulation

Encapsulation is a technique, by which small particles of bioactive food components (probiotics, antimicrobial agent, antioxidants, food colorants, flavors, essential fatty acids, vitamins, pigments, enzymes, bioactive peptides, and minerals) are packed inside a shell material to form a capsule (Jafari, 2017). Newly emerged nanoencapsulation (1–1000 nm-sized capsulation) has the advantage over microencapsulation (1–5 μm), as it increases the bioavailability, stability, and controlled and targeted release of the trapped bioactive components. Due to these advanced characteristics, nanoencapsulation is preferred in the application of food products (Chopde et al., 2020).

In nanoencapsulation, the bioactive nanoconstituents of food components are encased (known as the core) by a polymeric wall (functioning as an envelope), called shell; together which is called the "nanocapsule". The "core" part usually contains the sensitive compound such as essential oils (EOs), fish, and linseed oil containing omega-3-fatty acids, etc., whereas, shell components are made up of different solid nanoparticles, liposomes, emulsion, polysaccharides, proteins (albumin, lecithin, gelatin, and legumin), lipids (phospholipids, niosomes, cubosomes, hexosomes, nanoemulsion, etc.), and biopolymers. (Jafari, 2017; Rostamabadi et al., 2019).

6.1 Antimicrobial nanoencapsulation

Natural antimicrobials exhibit strong efficacy against the microbial populations involved in food-borne illness. But, their high volatility, sensitivity, rapid degradability, unacceptable taste (in some cases like essential oils), and low water solubility can reduce antimicrobial activity as well as can affect the quality and appearance of food. To overcome this problem the antimicrobials are nanoencapsulated.

Different types of bioactive compounds nanoencapsulated with shell components, are significantly efficient to inhibit microbial growth in foodstuff. For instance, enzymes, essential oils, phenolic compounds, and probiotics play major roles.

6.1.1 Enzymes and peptides nanoencapsulation

Proteolytic enzymes show significant antimicrobial activity against different food pathogens like *E. coli*, *Listeria monocytogenes* (*L. monocytogenes*), *Bacillus subtilis* (*B. subtilis*), and *Enterobacter cloacae* (*E. cloacae*). Papain and bromelain, nanoencapsulated with chitosan and alginates, are significantly effective against *Alicyclobacillus* (Anjos et al., 2018). Antimicrobial peptide Nisin and lysozyme, encapsulated with both chitosan and nanoliposome, exhibit excellent inhibitory mechanism against both Gram-positive (e.g., *B. subtilis*) and Gram-negative foodborne pathogens (e.g., *E.coli*) by cleaving the peptidoglycan component of the bacterial cell wall, causing cell death (Matouskova et al., 2016). However, liposome-encapsulated lysozyme is effective against *L. monocytogenes* but, not against *Salmonella enteritidis* (*S. enteritidis*), whereas, liposome-encapsulated mixture of nisin/lysozyme can inhibit both the Gram-positive and Gram-negative bacteria (Lopes et al., 2019).

6.1.2 Bioactive oils nanoencapsulation

Bioactive oils (also called "Essential oils") are aromatic, concentrated natural plants' extracts, which were proved to be good sources of secondary metabolites with significant antimicrobial properties (Ferreira and Nunes, 2019). Several essential oils, obtained from different Indian plant species (*Cymbopogon flexuosus*, *Pogostemon cablin*, *Curcuma caesia*, *Psidium guajava*, *Ocimum sanctum*, and so on) show effective antimicrobial activity against various microorganisms, including Gram-positive and Gram-negative bacteria (e.g., *S. aureus*, *B. cereus*, *B. subtilis*, *S. typhimurium*, and *E. coli*) and various fungi (e.g., *Aspergillus niger*, *Aspergillus fumigatus*, *Saccharomyces cerevisiae* and *Candida albicans*) (Munda et al., 2019). But one of the key complications with these bioactive oils is the loss of essential volatile components during storage. This happens due to the presence of PUFA and other constituents (carotenoids, xanthophylls, sterols, flavonols, monoterpenes, etc.), which are sensitive to different environmental factors like moisture, oxygen, heat, and light (Ferreira and Nunes, 2019). Nanoencapsulation of the oils can overcome this limitation and increase their stability, efficacy, and antimicrobial activity. Nanoencapsulation of essential oils, such as cinnamon, rosemary, lavender, tea tree, peppermint, eucalyptus, lemongrass, clove, thyme, and lemon oils, within liposome and sodium alginates shells have an effective role to inhibit the growth of both fungi (e.g., *C. albicans*) and bacteria (e.g., *E. coli*) (Haggag et al., 2021; Liakos et al., 2014). Chitosan nanostructure-based encapsulation of EO also has great effectiveness against microbial contamination of *Staphilococcus aureus* (*S. aureus*), *Bacillus cereus* (*B. cereus*), *Escherichia coli* (*E. coli*), *Aspergillus niger* (*A. niger*), *Listeria monocytogenes* (*L. monocytogenes*) (Detsi et al., 2020).

6.1.3 Probiotics nanoencapsulation

Probiotics are the beneficial living microflora in the gastrointestinal tract, responsible for the treatment of various acute and chronic diseases. The health-promoting effects of probiotics are attributed to their capability of providing essential nutrients, metabolization, and inhibiting pathogens. As some probiotics are sensitive to several negative circumstances, they can be nanoencapsulated to retain their stability and activity (Yilmaz et al., 2020b). A recent trend shows that the survival percentage of various probiotics, like, *Lactobacillus paracasei* KS-199 (*L. paracasei*), *Lactobacillus plantarum* (*L. plantarum*), *E. coli* Nissle 1917

(*E. coli*), *Pediococcus pentosaceus* Li05 (*P. pentosaceus*), and *Lactobacillus brevis* ST-69 (*L. brevis*) can be increased considerably, after nanoencapsulation with electrospun alginate nanofibers, nanocellulose, chitosan/alginate, MgO NPs induced alginate-gelatin microbeads and starch NCs/alginate composites respectively (Razavi et al., 2021; Yilmaz et al., 2020b).

6.1.4 Polyphenols encapsulation

Polyphenols, a secondary metabolite of plant species, are capable of inhibition of several foodborne pathogens, such as *Listeria monocytogenes* (*L. monocytogenes*), *Bacillus subtilis* (*B. subtilis*), *Salmonella enteritidis* (*S. enteritidis*), *Staphilococcus aureus* (*S. aureus*), *Escherichia coli* (*E. coli*), and *Pseudomonas aeruginosa* (*P. aeruginosa*) (Bouarab-Chibane et al., 2019). But, due to low bioavailability, polyphenols are nanoencapsulated with different polymeric, vesicular, or inorganic nanocarriers to enhance their biocidal activities (Rambaran, 2020).

The controlled release of antimicrobial agents is also a factor of concern, which is improved after nanoencapsulation, resulting in a longer reaction time with microbes and more lethality. This releasing is the summation of the three mechanisms (i) diffusion, (ii) erosion, and (iii) swelling. Where, the last two deal with the hydrophilic nanoencapsulants (protein and carbohydrate) (Mircioiu et al., 2019).

7 Nanophotosensitizer

Photosensitization is a concept, where a required molecule, which is unable to activate itself for a specific photochemical reaction, gets excited by the involvement of a donor molecule, which is capable of absorbing light. This donor molecule is called "Photosensitizer (PS)". Upon absorption of light, photosensitizers get activated and generate toxic components, that facilitate the inactivation of pathogenic microorganisms (Ghorbani et al., 2018; Spagnul et al., 2015).

The triplet state of photosensitizer (generated after photoirradiation) reacts with the substrate by two different photosynthesized reactions: electron transfer (Type I) and energy transfer (Type II) (dos Santos et al., 2021) (Fig. 4). The first one involves the electron transfer between triplet PS (PS(T)) and the biomolecules within the cells, resulting in the initiation of radical chain reactions. These reactions lead to the formation of superoxide free radicals, which further produce H_2O_2; by reduction of which, hydroxyl radicals (HO•) are generated. Whereas, in Type II, the conversion of the excitation energy is occurred directly from the triplet PS to molecular oxygen ($^3O_2$), leading to the generation of the ROS like singlet oxygen ($^1O_2$), which can kill foodborne pathogens after oxidative reaction (Huang et al., 2012; Spagnul et al., 2015). This inactivation of microbes is largely associated with the rate of ROS generation, which further depends on the level of molecular oxygen. Both Gram-positive and Gram-negative bacteria, such as *E. coli*, *S. aureus*, *Streptococcus mutans* (*S. mutans*), *Porphyromonas gingivalis* (*P. gingivalis*), and, *P. aeruginosa* are susceptible to photosensitization under appropriate conditions. Among them, Gram-positive bacteria are more susceptible to photodynamic reaction than Gram-negative bacteria, because they possess an outer wall, which allows photosensitizer to diffuse inside the cell (Gualdesi et al., 2019; Spagnul et al., 2015).

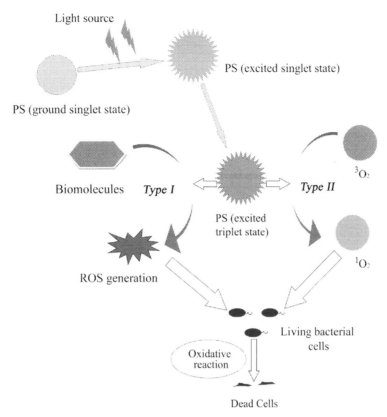

FIG. 4 Schematic representation of the photosensitizing mechanism against foodborne pathogens.

Cationic photosensitizers are significantly effective against a wide range of microbes, including Gram-negative bacteria. Also, they are advantageous because of providing the selectivity for uptake into the microbes' cells, which reduces the chances of damaging the useful cells during antimicrobial treatment (Rout et al., 2017). However, certain physicochemical and biological characteristics need to be present in an ideal photosensitizer (Vázquez-Hernández et al., 2018):

   I. Purity, photo-stability, and structural stability for an extended photoirradiation time.
  II. Amphiphilic (consists of both hydrophilicity and lipophilicity) property of the photosensitizer.
 III. High quantum yield.
 IV. The near-infrared wavelength absorption capability.
  V. Rapid accumulation in target cells.
 VI. No mutagenic properties.
VII. Negligible dark-cytotoxicity.

7.1 Nanoparticles in photosensitization

To overcome the limitations of the traditional PS, in recent days, nanoparticles have emerged in the field of photosensitization, where photosensitizers are conjugated or encapsulated by nanoparticles, and depending on their physicochemical characteristics and mode of attachment, the amount of deliverable PS to the targeted cells is increased due to the large effective surface area. Most of the PSs lack amphiphilic property. Therefore, NPs impart significant amphiphilicity to the PSs after conjugation and can also improve the controlled release, which prevents accumulation of the PSs in normal cells. During antimicrobial action, PSs can diffuse inside the microbial cell with the help of the NPs by altering the permeability of the cells, leading to cell death. Additionally, after conjugating with the NPs, the PSs can be turned into multifunctional agents, which can enhance their efficacy (Lucky et al., 2015).

Nanoparticles used in photosensitization, are of two categories: (a) biodegradable nanoparticles (e.g., lipid nanoparticles) and (b) non-biodegradable nanoparticles (e.g., metal oxide nanoparticles).

7.1.1 Lipid nanoparticles

The use of lipid nanoparticles can improve some of the desirable properties, such as viscosity, stability, control release, etc. They are spherical vesicles, evolved to carry the bioactive molecules. Lipid NPs encapsulated PSs exhibit significant antimicrobial effects against pathogenic microorganisms. When the microbial cell is treated with the PS loaded lipid NPs, the lipid NPs are taken up by the cells by endocytosis, followed by the opening of endosomal escape at low pH, which allows the PSs to release in the cytoplasm (Let's talk about lipid nanoparticles, 2021). Moreover, the lipid nanoparticles impose careful selectivity, slightly negative zeta potential, and allow a much lower average particle size. These factors combined result in effective cellular uptake and penetration of the PS inside the microbial cells and initiate cytotoxicity. For example, Toluidine Blue O (TBO), a cationic photosensitizer, encapsulated inside lipid NPs (prepared by eucalyptus oil, glycerol, and surfactant), can prevent the growth of both *P. aeruginosa* (almost total inhibition), *E. coli* (~4 logs) and *S. aureus* (~3 logs) significantly by generating ROS after photooxidation (Rout et al., 2017).

As the ROS generation is directly proportional to the level of molecular oxygen, therefore, a very fast rate of generation of singlet oxygen will deplete the level of molecular oxygen rapidly, and the production of ROS may stop after a certain time. Lipid NPs enhance the singlet oxygen production in photosensitizer but at a slow rate, which is beneficial. Therefore, nanoencapsulation of PS with lipid nanoparticles can enhance antimicrobial efficacy against pathogens (Rout et al., 2017).

7.1.2 Metal-oxide nanoparticles

Certain metal oxide nanoparticles have been widely used in photosensitization because of their multifunctionality and biocompatibility. The major metal oxides, used for antimicrobial photodynamic therapy are:

7.1.3 Iron oxide nanoparticles (IONPs)

IONPs have owned great attention in photosensitization due to their super magnetism and high saturation magnetization. Moreover, they consist of a high surface area to volume ratio, which helps in the functionalization of photosensitizers on their surface. Iron oxide NPs based PSs have the high potential to kill the microorganisms by introducing oxidative stress, which is generated by highly reactive oxygen species (ROS), such as singlet oxygen ($^1O_2$), hydroxyl radical, hydrogen peroxide, and anion superoxide. Although this killing is dependent on the timing of the PSs release from NPs and the level of the molecular oxygen. Functionalized IONPs increase the efficacy of the treatment by avoiding the premature release of the photosensitizers (Toledo et al., 2020). For example, Methylene blue (MB) is a cost-effective photosensitizer, which has a significant antimicrobial effect. But, after introducing IONPs in MB, it becomes more effective to eliminate pathogens like *E.coli* (Toledo et al., 2020). Additionally, porphyrin photosensitizer functionalized with SPION (Superparamagnetic ION), is proved to be beneficial to inactivate the Gram-positive bacteria, such as *Streptococcus mutans* (*S. mutans*) and *Staphylococcus aureus* (*S. aureus*) (Thandu et al., 2017).

TiO$_2$ nanoparticles

TiO$_2$ is the naturally occurring oxide of Titanium, also known as "Titania". In photosensitization, Titania plays an important role due to their photo-induced superhydrophilicity, photocatalytic activity, low toxicity, and biocompatibility (Ochiai and Fujishima, 2012). They can exhibit their photocatalytic activity either by their own or as a nanocarrier of other photosensitizers. In the first process, it generates photoactive electron-band pairs after irradiating TiO$_2$ with UV light. These excited electrons and holes can react with surrounding O$_2$ and H$_2$O to generate ROS, which is detrimental to microbes, like, *L. monocytogenes*, *S. aureus*, *E. coli*, and *S. typhimurium* (Yemmireddy and Hung, 2017). While the second one exhibits a photocatalytic effect and increased ROS production after conjugating with other photosensitizers. As an example, when TiO$_2$ NPs are coated with the ruthenium(II) photosensitizer, the transferring of electrons occurs from ruthenium to TiO$_2$, leading to an increase in ROS production (Gilson et al., 2017).

ZnO nanoparticles

ZnO nanoparticle, a well-known photocatalyst, is the suitable carrier to transport different photosensitizers, such as 5-aminolevulinic acid (ALA), photofrin, meso-tetra(o-amino phenyl) porphyrin (MTAP), etc. due to their some exceptional properties (e.g., large surface area to volume ratio, absorption of UV light, etc.). Furthermore, conventional photosensitizers cannot be excited by UV radiation. This limitation can be overcome by using of ZnO nanocarrier, which is excited by UV light and generates white light, which is subsequently absorbed by the PSs and become excited. Moreover, the ROS generation from both ZnO NPs and PSs leads to severe damage to the targeted cells (e.g., Microbes) as well (Yi et al., 2020).

8 Application of nanotechnology in microbial food safety

The application of different nanomaterials in food production largely depends on their characteristics. In different sectors of food industries, nanomaterials are widely applied, for example in packaging, food contact surfaces, filtrations, hurdle technology, and so on. In this section, some of their applications have been discussed.

8.1 Active packaging in microbial food safety

The involvement of nanotechnology in active packaging not only retains the food quality but can also solve multidirectional problems during food packaging, such as, inhibition of microbial activity, enhancement of mechanical strength, and gas barrier properties of the food packagings (Basavegowda et al., 2020; Bhunia et al., 2012; Lim, 2015; Paidari and Ibrahim, 2021; Sharma et al., 2017).

In the following section applications of different categories of nanoparticles and nanomaterials in active packaging are discussed to extend microbial food safety:

8.1.1 Inorganic nanoparticles in active food packaging

Different inorganic nanomaterials (metallic, metal-oxides, and nonmetal) have indispensable roles in diverse applications of active packaging. Such as,

Silver nanoparticles

Appropriate use of silver NPs in food packaging enhances their stability and ability against large variants of food pathogenic microorganisms such as *Staphylococcus aureus* (*S. aureus*), *Bacillus subtilis* (*B. subtilis*), *Escherichia coli* (*E. coli*), *Pseudomonas aeruginosa* (*P. aeruginosa*), and many more (Zorraquin-Pena et al., 2020). For example, it was observed that wrapping of litchi fruits with Ag-NPs and laponite introduced chitosan film could enhance the shelf life of litchi from 4 days to one week (Omerovic et al., 2021).

Gold nanoparticles

The suitability of gold nanoparticles in food packaging is due to their inert and nontoxicity, therapeutic properties, and oxidative catalytic activity. Moreover, their high surface reactivity can inhibit various food pathogens, like, *S. aureus*, *B. subtilis*, *E. coli*, *Klebsiella pneumoniae* (*K. pneuminiae*), and *P. aeruginosa* (Gu et al., 2021). For example, cross-linking of gold nanoparticles as a filler into poly(vinyl) alcohol (PVA) composite films (PVA-glyoxal-AuNPs) can enhance the antimicrobial activity (target pathogen: *E. coli*) of the films as well as the shelf life of the perishable foods, like bananas up to 5 days (Chowdhury et al., 2020). The nanocomposite is also used in food packaging applications for enhanced antimicrobial activity, for example, incorporation of Ag and Au NPs on cellulose film to reduce *E. coli* JM109 proliferation (Tsai et al., 2017).

Sulfur nanoparticles

Sulfur nanoparticles themselves or as composites with other nanoconstituents play a significant role to inhibit pathogenic microorganisms, i.e., *L. monocytogenes*, *E. coli* and, *S. aureus* without harming human health (Shankar et al., 2018). Priyadarshi et al. (2021) investigated the antimicrobial effect of sulfur nanoparticles induced calcium-linked alginate films against *E. coli* and L. *monocytogenes*. They reported that nanocomposites film had great efficacy against tested pathogenic microorganisms, whereas, no biocidal activity was found for alginate film itself. However, Gram-positive bacteria (*L. monocytogenes*) are more sensitive (0 log CFU/mL) to the sulfur nanoparticle-based films than the Gram-negative ones (*E. coli*) ($\sim$ 3 log CFU/mL) after twelve hours of exposure. This packaging film was also suitable for frozen food wrapping. The possible antimicrobial mechanisms of sulfur nanoparticles (SNPs) include (a) interaction of SNPs with the microbial cell wall, leading to dissimilation metabolic pathways; (b) generation of a large number of sulfur ions and H_2S after the interaction, causing defunctionalization of cellular materials; (c) blocking of essential sulfhydryl (SH)-enzymes by SNPs inside bacterial cells, resulting in cell death (Priyadarshi et al., 2021).

TiO_2 nanoparticles

For being a nontoxic nanoparticle, the American Food and Drug Administration (FDA) has permitted TiO_2 nanoparticles as safe for application (e.g., as colorants) in the food industry (Noori Hashemabad et al., 2017). Moreover, their excellent ethylene scavenging activity makes them suitable for active packaging applications (Zhang et al., 2019). The antimicrobial activity of TiO_2-based active packaging films such as LDPE-TiO_2 nanocomposite film, chitosan-TiO_2-black plum peel extract film, TiO_2-coated oriented-polypropylene (OPP) film, and Cellulose- PVA-TiO_2 nanocomposites is well documented against several Gram-positive (*Bacillus cereus*, *S. aureus* and *L. monocytogenes*) and Gram-negative (*E. coli*, *Salmonella*, *Pseudomonas* spp.) food pathogenic bacteria and yeast cells (*Rhodotorula mucilaginosa*) under the exposer of the UV radiation (Bodaghi et al., 2013; Chawengkijwanich and Hayata, 2008; Dhungana and Bhandari, 2021; Noori Hashemabad et al., 2017; Ramesh et al., 2018; Zhang et al., 2019). These TiO_2 NPs-coated packaging films can shrink the microbial infection on the surface of solid food products as well as prevent the pathogens development on fresh-cut produce.

ZnO nanoparticles

ZnO is considered a suitable ingredient in food packaging applications, as it is enlisted as "generally recognized as safe" (GRAS) by the US Food and Drug Administration (FDA). With good chemical stability and biocompatibility, semiconductor-type ZnO nanoparticles have effective antimicrobial, antioxidant, and photocatalytic properties also, which show excellent inhibition against a wide range of food pathogenic microorganisms. Moreover, in active packaging, ZnO nanoparticles provide excellent mechanical strength and gas barrier properties also, to protect packaging film from unfavorable conditions (Rojas et al., 2019; Yadav et al., 2021). Several researchers have reported the applications of ZnO nanoparticles in food packaging film aiming to increase the shelf life of food commodities with enhanced antimicrobial properties. Yadav et al. (2021) studied the properties of Chitosan (Ch) induced zinc oxide nanoparticles loaded gallic-acid films (Ch-ZnO@gal) and found that incorporation of nanostructures increased the mechanical strength as well as barrier properties, whereas, reduced solubility and swelling properties of the film. Additionally, significant concentration-dependent antimicrobial activity of Ch-ZnO@gal film against *Bacillus subtilis* (*B. subtilis*) and *E. coli* was also observed. Other ZnO NPs based packaging films like poly (vinyl alcohol)/ Chitosan/ZnO-SiO$_2$ nanocomposites, hydroxyethylcellulose/citric acid/ZnO films, polylactic acid/ZnO membrane, and LDPE/ZnO film can introduce improved biocidal properties into active packaging against *Escherichia coli* (*E. coli*) and, *Staphylococcus aureus* (*S. aureus*) food pathogens (Al-Tayyar et al., 2020; El Fawal et al., 2020; Rojas et al., 2019; Zhang et al., 2021).

8.1.2 Organic nanoparticles

Nanocellulose

Nanocellulose or cellulose nano-whisker, a novel bio-nanomaterial with a diameter less than 100 nm, is extracted from natural cellulose sources. It is composed of alternating crystalline and amorphous sequences, where, crystalline regions consist of strong hydrogen bonds, resulting in enhanced stability of the nanostructured cellulose. It is fit for application in food packaging due to its biodegradability, biocompatibility, enhanced mechanical and gas barrier properties (Yang et al., 2020). It also shows improved axial stiffness, which is the most important requirement for filler material in the packaging film (Chaudhary et al., 2020; Yang et al., 2020). To prevent microbial activity in food, nanocellulose can also be used in active packaging. For example, sugarcane bagasse-nanocellulose/nisin nanohybrid films have significant inhibition activity against *Listeria monocytogenes* (*L. monocytogenes*) (inhibited up to 7 days at 4° C storage), where the hydrogen bond between the hydroxyl group of nano-cellulose and an amino group of nisin improves the controlled release of the antimicrobial agents (Yang et al., 2020). The chitosan-nanocellulose film also exhibits a lethal effect after coming in contact with Gram-positive (*S. aureus*) as well as Gram-negative (*E. coli* and *S. enteritidis*) bacteria (Dehnad et al., 2014).

Nanostarch

Nanostarch (a novel form of modified starch) is a natural, biodegradable, and nontoxic polysaccharide, which is applied in food production, packaging materials, cosmetics, and medicinal sectors in a wide range. Different antimicrobial compounds (benzoic acid, chitosan, nisin, and ZnO) embedded with nanostarch influence its biocidal activity against a large number of food pathogens (Guleria et al., 2020). Abreu et al. (2015) constructed nanoclay (C30B) and silver (Ag) nanoparticle incorporated nanostarch-based antimicrobial food packaging films, which were significantly effective against both the bacteria (*E. coli* and *S. aureus*) and the yeast (*C. albicans*). Although no antimicrobial activity was found when *C. albicans* was treated with C30B-nanostarch based films. The antibacterial activity of C30B-nanostarch film against the *E. coli* and *S. aureus* was due to the release of ammonium salts from the nanoclay, which altered the bacterial cell membrane. Although, nanostarch, as a filler, make the packaging film extensively reactive and also enhances the physical and barrier properties of the film significantly due to their hydrophilic characteristics (Kim et al., 2015).

8.2 Nanotechnology in smart food packaging

Smart food packaging (SFP) is an advanced approach in packaging technology, where the quality of the packed food commodities in the supply chain can be constantly monitored by modification of the packaging film with different quality indicators (e.g., RFID, TTI, etc.) and sensors (e.g., gas sensors, etc.) (Mohammadian et al., 2020). The emergence of nanotechnology plays a significant role during the modification of conventional food packaging to smart packaging. It contributes in many aspects related to smart food packaging, for example, (i) pathogen detector, (ii) time-temperature indicator (TTI), (iii) O_2 and CO_2 indicator, (iv) moisture indicator, and (v) food contaminating toxin indicator, which has been represented in Table 3.

TABLE 3 Application of different nanostructures in smart food packaging.

Nanostructures	Indicator/sensor types	Mechanism of action	Food products	References
Gold, silver, CNTs, and magnetofluorescent nanoparticles	Pathogen indicator (e.g., E. coli 0157:H7, Salmonella typhimurium, and S. aureus majorly)	Immunosensing, Raman-based detection, fluorescence microscopy.	Milk and milk powder, pork, and chicken carcass.	Kumar et al. (2020), Mocan et al. (2017), Chen et al. (2015), Chen et al. (2021), Stephen Inbaraj and Chen (2016)
Gold nanoparticle-induced silver NP based core-shell nanorods, chitosan/AuNPs nanocomposites and, Anthocyanin induced chitosan/PVA film	Time-temperature indicator (TTI)	Colorimetric	Milk quality in the cold chain, frozen meat, fish, and pasteurized milk.	Wang et al. (2018), Gao et al. (2021), Pereira et al. (2015)
Titanium dioxide (TiO$_2$), Platinum (Pt), iron NPs (IONPs), carbon nanotubes (CNT)	Gas absorber (O$_2$, CO$_2$, and ethylene)	Photocatalysis, electrochemical, redox reaction, electrochemical	Fruits in the cold chain (ethylene detection by TiO$_2$), packed perishable foods (O$_2$ detection by IONP/silicon film), fresh meat (CO$_2$ detection by carbon nanotube)	Hussain et al. (2011), Satter et al. (2018), Foltynowicz et al. (2017), Mustafa and Andreescu (2020), Lin et al. (2015)
Gold, iron oxide, cerium oxide, zinc oxide nanoparticles. SWNT and MWCNT	Food contaminating toxins (Botulinum neurotoxin type B, aflatoxin M1, Ochratoxin A, Erythromycin and palytoxin)	Electrochemical, immunoassay, electrochemiluminescence	Milk, red wine, honey, beer, and meat	Kumar et al. (2017), Manoj et al. (2021)
Silicon dioxide, Zinc oxide nanoparticles	Volatile compounds due to microbial spoilage (milk volatiles, ethanol)	Electrical, colorimetric	Milk	Ziyaina et al. (2019), Wang et al. (2019b)

In recent years, applications of two-dimensional (2-D) materials have gained attention in the field of smart and active packaging due to their advanced characteristics, including enhanced conductivity and catalytic activity. These unique materials have a thickness of approximately one to a few molecular atoms with lateral dimensions varying from nanometers to micrometers. Newly developed 2D materials like nanosheets of graphene, graphene oxide, and reduced graphene oxide can be successfully induced in smart packaging applications, as they have the potential to detect leakage, food spoilage biomarkers, ethanol and also to use in TTI (Yu et al., 2021).

Therefore, it has been determined that nanotechnology contributes a significant role in the field of smart food packaging, which turns smart packaging from conventional to future food packaging. Although, the proper safety and guidelines must be maintained during the addition of active and intelligent materials in the packaging films, and must not change the quality and safety parameters of the foodstuffs also. In this regard, European Parliament (2004b) stated in Regulation (EC) No 1935/2004 of the European Parliament and the Council, *"Active and intelligent food contact materials and articles should not change the composition or the organoleptic properties of food or give information about the condition of the food that could mislead consumers. For example, active food contact materials and articles should not release or absorb substances such as aldehydes or amines in order to mask an incipient spoilage of the food. Such changes which could manipulate signs of spoilage could mislead the consumer and they should therefore not be allowed. Similarly, active food contact materials and articles which produce colour changes to the food that give the wrong information concerning the condition of the food could mislead the consumer and therefore should not be allowed either"*.

8.3 Application in food contact surfaces (FCSs)

Food contact surfaces (FCSs) comprise all the surfaces in food processing units, in which food materials come in contact during manufacturing, processing, packaging, storage, and transportations. According to the EU regulation No. 852/2004, the food contact surfaces should be the *"surfaces (including surfaces of equipment) in areas where foods are handled and in particular, those in contact with food are to be maintained in a sound condition and be easy to clean and, where necessary, to disinfect. This will require the use of smooth, washable corrosion-resistant and non-toxic materials unless food business operators can satisfy the competent authority that other materials used are appropriate"* (European Parliament, 2004a).

The FCSs used in various food sectors like milk, fish, and meat processing are susceptible to different harmful microorganisms like *E. coli*, *L. monocytogenes*, *Salmonella*, Coliforms, and *S. aureus*, which after cross-contamination in food matrices can result in food quality deterioration and several health issues (Torres Dominguez et al., 2019).

The novel nanomaterial-based technology, such as nanocoating on the food contact surfaces as well as on food itself not only protect them from the adverse conditions and enhance their durability but plays a significant role against fouling formation and pathogenic proliferations (Al-Naamani et al., 2018; Kumar et al., 2019b; Mohan et al., 2020; Park et al., 2017). The general purposes of antimicrobial coating on the FCSs include packaging quality improvement, microbial decontamination, and incorporation of antimicrobial properties (Torres Dominguez et al., 2019).

Yemmireddy et al. (2015) and Yemmireddy and Hung (2015) applied TiO_2 nanocoatings on the stainless steel (mostly used food contact surface) by using different proportions of organic and inorganic binding agents, e.g., PVA, PEG, polyurethane (PU), shellac resin, polycrylic, sodium and potassium silicates, to study antimicrobial activity. They reported that, due to the unique photocatalytic activities of the TiO_2, all binders with TiO_2 nanocoating exposed biocidal activities against *E. coli*. However, repeated use of TiO_2 nanocoated stainless steel could result in loss of their stability and biocidal activity.

Several other food contact surfaces, where the application of nanotechnology is widely found, are discussed in Table 4.

8.4 Photodynamic (PD) packaging

Photodynamic packaging is a photo-oxidative-stress mediated antimicrobial packaging technique where photosensitizers are used to inhibit microbial growth. Treatment of packaging film by non-thermal photodynamic therapy (PDT) is a promising way to photo-inactivate microorganisms, as they (microorganisms) are absent of any resistance against PDT (Luksiene et al., 2009; Luksiene and Paskeviciute, 2011). In the PDT, local cytotoxicity is generated after irradiation of the photosensitizers with visible light. There are three main components of the PDT, e.g., light, oxygen, and photosensitizers. The selection of those could be vital during the treatment because the efficacy of the inactivation significantly depends on these three factors (Ghate et al., 2019).

Photosensitizers can inactivate food pathogens in active packaging after illuminating under Light-emitting diodes (LEDs), operating in the visible spectra (Pérez et al., 2021). A study showed that riboflavin (RB) photosensitizer-induced chitosan (CS) food packaging films had a significant ability to inactivate food pathogenic microorganisms (*L. monocytogenes*, *Vibrio parahaemolyticus*, and *Shewanella baltica*) under blue LED irradiation for two hours due to the generation of singlet oxygen. Although, the inactivation of the bacteria was largely associated with both the radiation dosages ($3.42 J/cm^2$ to $13.68 J/cm^2$) and the concentration of the photosensitizers (1% to 5% w/w). Additionally a good ultraviolet (UV)-barrier property was found within this RB-CS food packaging film, which is an important property to prevent UV-light transmission (as UV-light can promote oxidation of lipid, alteration of color, odor, and flavor of the foods) (Su et al., 2021). Chlorophyll-mediated photodynamic therapy is also a sound approach for antimicrobial food packaging, as chlorophyll and its derivatives are known as effective photosensitizers, which can prevent the microbes of the inside environment of the packaging (Luksiene and Paskeviciute, 2011). Spores forming *Bombylius aureus* and non-spore-forming *L. monocytogenes* biofilms are the most common contaminants in foods, mostly coming from food packaging and food processing equipment. Chlorophyllin-based photosensitizers incorporated polyolefine packaging film can significantly prevent these biofilms under the LED light at a wavelength of 405 nm and energy density $12 mW/cm^2$. Moreover, this PDT treatment is advantageous because of being non-thermal and environment-friendly nature (Luksiene and Paskeviciute, 2011). 5-aminolevulinic acid (ALA) photosensitizer also exhibits good photodynamic microbial inactivation on the polyolefine packaging film with LED at 400 nm wavelength with the energy density of $20 mW/cm^2$. At a 7.5 mM concentration of ALA, the PDT can prevent the adhesion of *L. monocytogenes* vegetative cells on the packaging film by 3.7 log cycle whereas, biofilm by 3.1 log cycle (Buchovec et al., 2010). This limitation of inactivation of less biofilm compared to

TABLE 4   Application of nanostructures into the food contact surfaces.

Types of food contact surfaces	Examples	Nanomaterial used for coating	purpose	References
Plastic	LDPE	Silver nanoparticles	Antimicrobial milk packaging film (against *S. aureus* and *E. coli*)	Bayani Bandpey et al. (2017), Sadeghnejad et al. (2014)
	PVC and PE	Silver nanoparticle doped hybrid coating	Antimicrobial activity against (*S. aureus* and *E. coli*)	Marini et al. (2007)
	HDPE	Titanium dioxide	Antimicrobial activity	Hashim and Salih (2016)
	LDPE	Titanium dioxide, nanoclay	Barrier properties and antimicrobial activity	Moghaddam et al. (2014)
	PET	Silver nanoparticle	Antimicrobial activity against *P. aeruginosa*, *S. aureus*, and *C. albicans*	Deng et al. (2015)
	PP	Silver nanoparticle	Antimicrobial activity against *E. coli*	Egodage et al. (2017)
Stainless steel	316 L stainless steel	Electroless nickel with embedded polytetrafluoroethylene nanoparticles	Antifouling surface of plate heat exchanger	Barish and Goddard (2013)
	Stainless steel	Silver nanoparticles	Antimicrobial activity against *E. coli*	Chen et al. (2010)
	AISI 304 Stainless steel	nanoliposomes and the polymeric nanocapsules encapsulated carvacrol surface sanitizer	Antimicrobial activities against *S. aureus*, *E. coli*, and *Salmonella*	Ayres Cacciatore et al. (2020)
	AISI 304 L stainless steel alloy	silver nanocluster/silica nanocomposite based coating	Antimicrobial activity after high thermal treatment (450 °C) and ability to withstand cleaning procedure	Ferraris et al. (2017)
	AISI 304 L Stainless steel coupons	TiO_2 with binders	Antimicrobial activity against *E. coli*	Yemmireddy and Hung (2015)
Wood	Wood-based food packaging material	Silver, ZnO, TiO_2, SiO_2, CeO_2, and cellulose nanofibers	Antimicrobial activity	Jasmani et al. (2020)

LDPE, *low density polyethylene*; HDPE, *high density polyethylene*; PVC, *polyvinyl chloride*; PE, *polyethylene*; PP, *polypropylene*; PET, *polyethylene terephthalate.*

vegetative cells adhered to the film surface was due to the presence of polysaccharide matrix in biofilms (Buchovec et al., 2010).

Different types of nanomaterials (e.g., inorganic nanoparticles, polymers, liposomes, dendrimers, and micelles) are applied in PDT in the field of medicine and food packaging film to control the release of the PSs and to increase the singlet oxygen generation (Park et al., 2021). In the packaging film, gold nanostructures can be incorporated to enhance photosensitization during PDT (Krajczewski et al., 2019). Different studies reported about the development of modified polymeric (polyurethane) films by using gold nanorods and gold nanoparticles embedded with crystal violet (CV) dye, which exhibited significant antimicrobial activity against both Gram-positive and Gram-negative bacteria under both light and dark conditions. However, a limited antimicrobial activity was found in dark conditions. Polymeric matrices containing only gold nanoparticles can also be used in PDT against microorganisms. The inhibition of bacteria was caused by a significant amount of ROS generation after photo-irradiation of CV dye, which led to producing oxidative stress to the bacteria and resulted in cell death (Macdonald et al., 2016; Rossi et al., 2020; Rossi et al., 2019). However, these photosensitizer-polymeric films are less effective against Gram-negative bacteria (*E. coli*) compared to Gram-positive (*S. aureus*), which can be attributed to the differences in their cell wall structures (Macdonald et al., 2016). Moreover, the Gram-negative bacteria can produce superoxidase dismutase (SOD) enzymes, which lead to the conversion of the ROS and peroxides to their deactivated forms such as, water and molecular oxygen (Rossi et al., 2020).

Carbon quantum dots (CQDs) are also widely used as photosensitizers for photodynamic treatment. Several types of modified CQDs and CQD-NPs composites (N-CQDs, Spd-CQDs, CdTe-QDs, CQDs-ZnO, CQDs-TiO$_2$, hCQDs) exposed excellent antimicrobial activities against *E. coli, S. aureus, B. cereus, P. aeruginosa, K. pneumoniae,* and *L. monocytogenes* like harmful food pathogens. Besides, When CQDs are introduced into polymeric films, several physicochemical properties are enhanced after modification. For example, polyurethane/N-CQDs composites can be used as oxygen sensors, chitosan/CQDs nanocomposite hydrogel films are efficient for the prevention of light emission, PVA-CQDs films are appropriate for intelligent packaging and CQDs induced PMMA polymer film can be used as UV-radiation blocker in food packaging industries (Kováčová et al., 2020).

TiO$_2$ NPs can be used as photosensitizers in photodynamic therapy due to their ability to generate ROS and singlet oxygen under UV light. The generation of singlet oxygen can also be achieved in visible and NIR spectra by functionalization of TiO$_2$ with catechol, salicylic acid, and gold nanoparticle or other semiconductors like CDS (Buchalska et al., 2013; Etacheri et al., 2015; Glass et al., 2018; Saito and Nosaka, 2014; Zhu et al., 2018). TiO$_2$ functionalization in different packaging films like LDPE, HDPE, and chitosan matrix has effective photocatalytic activity against *Pseudomonas* spp., *R. mucilaginosa, E. coli, B. subtilis,* lactic acid bacteria (LAB), and coliforms (Bodaghi et al., 2013; Chawengkijwanich and Hayata, 2008; Zhu et al., 2018).

Fullerenes can also be used in photodynamic therapy (PDT) as photosensitizers after increasing their solubility by functionalizing the hydrophilic group because pristine fullerenes are highly hydrophobic (Maleki Dizaj et al., 2015). Due to extended *π-conjugation of the fullerenes*, visible light can be adsorbed and subsequently ROS (singlet oxygen and superoxide anion radicals) will be generated due to high triplet yield (Hamblin, 2018). Fullerenes are advantageous due to their versatility, ability to run *Type 1, Type 2, and electron transfer photochemical reactions,* more cytotoxicity towards microbial cells than mammalian cells, oxygen-independent photoinactivation, and resistance to photobleaching (Hamblin, 2018). It has been found that it has 99.9% lethality against bacteria and fungi (Hamblin, 2018; Maleki Dizaj et al., 2015). Although, more studies are needed before application in food production due to certain limitations, e.g., hydrophobicity, etc.

In recent days, curcumin (a yellow-colored polyphenol of *Curcuma longa*) based PDT is also a promising approach to inhibit the growth of different food pathogens (Mirzahosseinipour et al., 2020). It has been found that, under a blue LED source (465 nm), curcumin-silica nanoparticle-based photosensitizer can significantly reduce the number of both *P. aeruginosa* (1 log reduction), and *S. aureus* (1.2 log reduction) in planktonic condition as well as showed effective anti-biofilm activity against those two bacterial biofilms also (Mirzahosseinipour et al., 2020).

8.5 Photodynamic therapy in hurdle technology

The hurdle is well-defined as *"the deliberate combination of existing and novel preservation techniques to establish a series of preservative factors (hurdles) that any microorganisms present should not be able to overcome"* (Oh et al., 2019). Hurdle technology is the combination of the different food preservation strategies, which ensures improved food quality, shelf life, and safety. There exist more than 60 hurdles, which can be divided into physical, chemical, microbiological, and miscellaneous hurdles. Among those, refrigeration, thermal treatment, pressure, water activity, redox potential, acidity, and modified atmosphere are considered as the major hurdles (Pilizota, 2014). Although, all the hurdles applied to inhibit pathogens must be above the critical limits, if not, then the portions of the microorganisms will survive and respond to further contaminations. There are four mechanisms by which the hurdle technology affects microbial propagation (Aaliya et al., 2021; Singh and Shalini, 2016). They are described in the following:

- **Homeostasis**: Homeostasis is a condition by which the internal status of the microbes becomes stabilized and uniform. The key target of the hurdle technology is to imbalance the homeostatic conditions of the microorganisms by the application of more than one hurdles or techniques, which impart nonstop physicochemical and environmental stresses to the microbes, causing the death of the microorganisms (Aaliya et al., 2021).
- **Stress reactions**: Due to inequity of internal status, some pathogens may become more lethal or resistant to applied hurdle or generate stresses by generating protective stress shock proteins (SSP). Multiple hurdles could hamper the synthesis of these proteins by disturbing the gene activation mechanism required for synthesis, thus microbes become metabolically weak (Tsironi et al., 2020).
- **Metabolic exhaustion:** The microbes in multiple-hurdles-treated stable foods become metabolically exhausted to recreate their homeostasis by utilizing their existing energy, which results in "auto-sterilization" of the food matrix. Hence, the microbial stable foods become safe to store at ambient temperature (Aaliya et al., 2021).
- **Multi-target preservation:** This phenomenon is directly associated with the deprivation of the synthesizing of microbial shock proteins (MSPs). An excellent antimicrobial activity is achieved, when the hurdles in foods attack the multi-targets in microbial cells (cell membrane, DNA, and enzymes) and cause the imbalance of the homeostasis, resulting in cell death (Tsironi et al., 2020).

Nowadays, different non-thermal processing technologies are being combined with conventional thermal processing to achieve excellent decontamination of microbes without compromising food quality safety (Aaliya et al., 2021).

PDT, based on neutral and anionic photosensitizers sometime shows limited antimicrobial activity against Gram-negative bacteria (Buchovec et al., 2017). Therefore, a hurdle approach using PDT and high-power pulsed light (HPPL) may provide a significant effect. HPPL has been approved as a safe food surface decontamination technique by FDA in 2000 (Kairyte et al., 2012). A study reported that only hypericin (Hyp)-based PSs reduced the population of *Listeria* and *Salmonella* by seven log cycles and one log cycle, respectively, whereas, HPPL and PDT (Hyp based PS) combined reduced both the *Listeria* and *Salmonella* population by 6.7 to 7 log cycles (Kairyte et al., 2012). Authors suggested that photosensitizers alone cannot exhibit phototoxicity against Gram-negative bacteria due to their cell-wall structures. They (PSs) can pass only through the outer membrane but not the inner membrane of gram-negative bacteria. That's why ROS generated on the outer surface could not impart any detrimental effect to intracellular components. However, PSs along with HPPL can enhance the antimicrobial activity significantly against those bacteria (Kairyte et al., 2012). The combined lethality of PDT and high-power pulsed light (HPPL) has also been proven very effective against the *S. enterica* viability (7.5 log reduction) without any recovery of the microbial population (Irina Buchovec et al., 2017). Thus this innovative hurdle technology is effective and safe for use in antimicrobial food safety.

8.6 Nanobubble

Nanobubbles (NBs) are ultrafine nanoscopic gaseous cavities (<200nm) in the aqueous medium. The effectiveness of the NBs largely depends on the gas present inside the bubbles, characteristics of the surfactants, electrolyte concentration, temperature, and applied pressure (Thi Phan et al., 2020). In food applications, air, nitrogen, CO_2, oxygen and their combinations are mainly chosen for the nanobubble applications. These gases are considered safe for human and animal consumptions, whereas, application of other gases can arise food safety issues (Amamcharla et al., 2017). Nanobubbles have two excellent features: (1) highly effective surface area, and (2) excessive internal pressure. Both the internal pressure and area promote the mass transfer between the liquid medium and gas which facilitates different chemical reactions, physical adsorptions, ROS production, and dissolution of oxygen (O_2) and ozone (O_3) (S. Liu et al., 2016; Phan et al., 2021).

The potential antimicrobial activity of the NBs can be explained in terms of ROS generation and shock wave generation. Nanobubbles contain high internal pressure compared to the outer environment. Thus, when the NBs burst out, an excessive amount of surface energy is released, which results in the conversion of molecular oxygen to ROS, causing microbial cell death (Thi Phan et al., 2020). For a long period of time, ozone has been considered as an antioxidant with promising antimicrobial activity. Ozone mainly targets the lipids of the membrane and oxidizes them. Further, it breaks down the surface layer of the pathogens. But due to their instability, they need special treatment for antimicrobial activity. It has been found that the stability and the antimicrobial activity of ozone can be increased after storing them in liposomes as a nanobubble form (Alkan et al., 2021). At 1000ppm and 1500ppm concentrations of Ozone nanobubbles, the significant inhibition of microbial populations can be achieved against *Escherichia coli* (*E. coli*) and *Staphylococcus aureus* (*S. aureus*) respectively. It has been found that the stability and the effectiveness of the ozone nanobubbles can be retained over a year (Alkan et al., 2021).

The NBs expose strong intention to prevent microbial biomass and alter the compositions of biofilms in food contact surfaces due to high-temperature raises, high pressure, and the generation of hydroxyl radicals during the outburst of the nanobubbles. Furthermore, nanobubbles have the ability to reduce the deposition of mineral contents (e.g., carbonate, silicate, phosphate, and quartz) on the food contact surfaces (Xiao et al., 2020). In the future food production, as a novel and emerging technology, nanobubbles have high potential. Although, more researches need to be carried out to gain more informations about the nanobubbles.

8.7 Nanosensor

Nanosensor is an advanced sensing method, used to determine food safety with the help of nanomaterials (NMs). The main purpose of the application of NMs in food sensing is for improved and rapid target (microbes, adulterants, VOCs) identification, better output signals, advanced selectivity, and sensitivity. Quantum dots, nanowire, nanorod, metal nanoparticles, and carbon nanotubes are the main NMs used in the nanosensors. These sensors can be used to detect several food pathogens like *E. coli*, *Salmonella*, and *S. aureus*, as well as several toxins (e.g., Aflatoxin B1, and Ochratoxin A) produced by the microbes, which are responsible for food spoilage and contaminations (Malik et al., 2013; Yang et al., 2016).

9 Risk assessment

The burgeoning applications of nanotechnology in the food sectors also pose a public safety concern. The risk associated with the use of different nanomaterials in foods should equally be evaluated with the benefits of using nanotechnology. Nanotechnology-enabled food products have been found to be risky to human health and the environment in some aspects due to their exceptional dynamic, kinetic, and catalytic properties, exposures, and widespread uses (Amenta et al., 2015; Nile et al., 2020). Thus, the assessment of nano-related risk factors, the establishment of regulations, and public awareness should be taken into consideration.

Due to the nano-scale size, nanoparticles contain high surface reactivity, which leads to easy penetration through biological barriers and can cause circulatory, cardiovascular, and pulmonary diseases (Tarhan, 2020). The exposure of nanoparticles can be impregnated into the human body in three ways including,

I. **Dermal exposure:** The nanomaterials have the ability to penetrate through the human skin due to their ultrafine structures and after that can translocate in some parts of the body, e.g., lymph nodes. Although this path is quite unfamiliar with the food incorporated nanomaterials (De Matteis, 2017; Tarhan, 2020).

II. Ingestion: Ingestion is the most important route for nanostructure to be established inside the body, occurring by nanomaterials-induced food ingredients, supplements, and additives. Through direct ingestion of nanofoods, the nanoparticles are absorbed in blood stream, and microvilli of the intestinal surface. Therefore, in intestinal cells, the nanoparticles can trigger oxidative stress, DNA damage, and inflammation (De Matteis, 2017).

III. Inhalation: Another important exposure way is inhalation of NPs during breathing, where, the smaller nanoparticles can reach the alveolar region of the respiratory systems and impart nanotoxicity. Therefore, smaller nanoparticles are more dangerous compared to larger ones (De Matteis, 2017).

Both organic and inorganic nanomaterials are widely used in food processing due to their multidirectional functionalities. However, with the several benefits, various types of potential nanotoxicity also have been observed.

- **Organic nanomaterials**, used in food applications are derived from carbohydrates, lipids, and proteins. Although the applications of these types of nanoparticles are not risky, but they may cause some allergenic responses to some extent (Tarhan, 2020).
- The **inorganic nanomaterials** may cause acute or chronic toxicity due to long-term ingestion of nanofoods, which depends on the ability of nanomaterials to harm human cells or organs (McClements and Xiao, 2017). The risk behind using nanoparticles also depends on their shape, size, dosage, ion releases, and particle dissolutions. Widely used Ag NPs, TiO_2 NPs, ZnO NPs, IONPs, and SiO_2 NPs can be deposited in various vital organs and cause GIT, kidney, and hepatic injuries as well as can result lung damage also. Moreover, due to the production of ROS, RNS (reactive nitrogen species) and oxidative stress, several inflammatory reactions, cytotoxicity, and genotoxicity can occur, which further may form carcinogenic cells (McClements and Xiao, 2017; Nile et al., 2020).

Besides, in recent days, conventional synthesis of nanoparticles is categorized into two approaches; "top-down" and "bottom-up", which are further sub-grouped into two methods, such as physical and, chemical methods. (Jadoun et al., 2021). These processes require chemical reductants, stabilizers, high thermal load, and high radiation for NPs synthesis, which are harmful to the environment as well as to human health also. Moreover, the generation of hazardous intermediates from these methods also refers to a threat to both the environment and our ecosystems (Marouzi et al., 2021). To overcome these problems, nanoparticles can be synthesized by the green method, which is considered as safe, sustainable, and biocompatible (Jadoun et al., 2021). Besides, the nanoparticles synthesized greenly, are with diverse nature, greater stability, and uniform distribution (Chand et al., 2020).

Due to the above reasons, different legal bodies from countries have proposed some regulatory frameworks regarding nanomaterials based applications in foods with proper safety.

10 Regulatory and legislative aspects

Applications of nanotechnology include nano-encapsulation of bioactive compounds, antimicrobial nanocoating, antifouling surfaces, active and intelligent food packaging which commercially exist in the market. However, they also pose a threat to human health due to their shape, size, and unique properties. Efforts are being made to regulate the production of nanomaterials and their safe usage in foods or food contact materials with adequate legislation and recommendations. The regulations adopted by the USA, EC, and India are mainly emphasized here.

In the USA, the Food and Drug Administration (FDA) is the major body for providing guidance and regulatory aspects to ensure food safety. To date, any specification or regulation for nanotechnology-based products has not yet been developed by FDA. Nevertheless, FDA has published food safety guidelines for the industry, naming "Considering Whether an FDA-Regulated Product Involves the Application of Nanotechnology". In this guideline, FDA mentioned about the regulatory status, safety procedures, and health impact of the use of nanotechnology in food products (USFDA, 2014). The FDA also published another guideline, about nanotechnology in animal food, called "Draft Guidance for Industry on Use of Nanomaterials in Food for Animals", which is applicable when the food consists of a nanomaterial as a whole or nanomaterial as a component or include nanotechnology (USFDA, 2015).

In Europe, nanomaterials are considered as hazards during labeling or packaging if they exhibit any property mentioned in the Regulation (EC) No 1272/2008 of the European Parliament and the Council (Amenta et al., 2015). The European Food Safety Authority (EFSA) provided a guideline on the risk assessment of the application of nanoscience and nanotechnologies in the food and feed chain about the potential risk of the application of nanotechnology in different food components and feed additives, and recommended certain parameters that must be followed during incorporation of nanotechnology in food materials (EFSA, 2018).

In India, there is a lack of legislation to prove nanomaterials as hazards, although they incorporate high risk due to uncontrolled emission from various sectors into the environment, resulting in health issues. The Food Safety and Standards Authority of India (FSSAI) is the principal regulatory body on food safety and regulations and it is necessary to generate strict regulations about the application of nanomaterials in foods to control their usage. Although, the Department of Science and Technology (DST), India, has issued guidelines about the safe handling of nanomaterials in research, industries, and laboratories (Amenta et al., 2015; Mishra et al., 2019). In 2020, The Energy and Resources Institute (TERI) with the support of the Department of Biotechnology (DBT), Government of India published the "GUIDELINES FOR EVALUATION OF NANO-BASED AGRI-INPUT AND FOOD PRODUCTS IN INDIA", where various characteristics and safety assessments of the manufactured nano-agri-input products (NAIPs), nano-agri products (NAPs), nano-composites and sensors made from NMs are provided. These guidelines are also applied for the nanomaterials which directly come in contact with crops, food, and feed (PIB, 2020).

In other countries like China, Japan, Brazil, Australia, and New Zealand are also aware of producing nanomaterial-based regulations.

11 Final remarks

Nanotechnology in food industry is a novel and advanced approach, which has great importance to overcome several food-related difficulties; microbial food safety is one of them.

Compared to other conventional antimicrobial approaches, nanotechnology can eliminate, prevent or delay the growth of a wide range of different food pathogen species without compromising the food quality and appearance. Different nanostructured-based technologies have emerged in this field, that have great efficiencies against microbial contaminations in foods, but more researches need to be conducted for further commercial applications. However, with the multiple benefits, different nanostructures and nanomaterials pose potential risk to human. Systematic studies regarding potential health-related concerns, environmental safety, and consumers' acceptance towards nanotechnology in food must be carried out. Along with this, to harness the benefits of nanotechnology, appropriate legislations need to be formulated and strictly followed during food applications. Nevertheless, this technology has the great possibility to exhibit microbial food safety.

References

Aadil, K.R., Mussatto, S.I., Jha, H., 2018. Synthesis and characterization of silver nanoparticles loaded poly(vinyl alcohol)-lignin electrospun nanofibers and their antimicrobial activity. Int. J. Biol. Macromol. 120, 763–767. https://doi.org/10.1016/j.ijbiomac.2018.08.109.

Aaliya, B., Valiyapeediyekkal Sunooj, K., Navaf, M., Parambil Akhila, P., Sudheesh, C., Ahmed Mir, S., et al., 2021. Recent trends in bacterial decontamination of food products by hurdle technology: a synergistic approach using thermal and non-thermal processing techniques. Food Res. Int. https://doi.org/10.1016/j.foodres.2021.110514, 110514.

Abdollahi, M., Rezaei, M., Farzi, G., 2014. Influence of chitosan/clay functional bionanocomposite activated with rosemary essential oil on the shelf life of fresh silver carp. Int. J. Food Sci. Technol. 49 (3), 811–818. https://doi.org/10.1111/ijfs.12369.

Abral, H., Ariksa, J., Mahardika, M., Handayani, D., Aminah, I., Sandrawati, N., et al., 2020. Highly transparent and antimicrobial PVA based bionanocomposites reinforced by ginger nanofiber. Polym. Test. 81. https://doi.org/10.1016/j.polymertesting.2019.106186, 106186.

Abreu, A.S., Oliveira, M., de Sá, A., Rodrigues, R.M., Cerqueira, M.A., Vicente, A.A., Machado, A.V., 2015. Antimicrobial nanostructured starch based films for packaging. Carbohydr. Polym. 129, 127–134. https://doi.org/10.1016/j.carbpol.2015.04.021.

Alkan, P.E., Güneş, M.E., Özakin, C., Sabanci, A.Ü., 2021. New antibacterial agent: nanobubble ozone stored in liposomes: the antibacterial activity of nanobubble ozone in liposomes and their thymol solutions. Ozone Sci. Eng., 1–5. https://doi.org/10.1080/01919512.2021.1904205.

Al-Naamani, L., Dutta, J., Dobretsov, S., 2018. Nanocomposite zinc oxide-chitosan coatings on polyethylene films for extending storage life of okra (Abelmoschus esculentus). Nanomaterials 8 (7), 479. https://doi.org/10.3390/nano8070479.

Al-Tayyar, N.A., Youssef, A.M., Al-Hindi, R.R., 2020. Antimicrobial packaging efficiency of ZnO-SiO_2 nanocomposites infused into PVA/CS film for enhancing the shelf life of food products. Food Packag. Shelf Life 25. https://doi.org/10.1016/j.fpsl.2020.100523, 100523.

Amamcharla, J., Li, B., Liu, Z., 2017. United States Patent: use of micro and nanobubbles in liquid processing. WO 2017/127636 AI. PCT/US2017/014272.

Amenta, V., Aschberger, K., Arena, M., Bouwmeester, H., Botelho Moniz, F., Brandhoff, P., et al., 2015. Regulatory aspects of nanotechnology in the agri/feed/food sector in EU and non-EU countries. Regul. Toxicol. Pharmacol. 73 (1), 463–476. https://doi.org/10.1016/j.yrtph.2015.06.016.

Anees Ahmad, S., Sachi Das, S., Khatoon, A., Tahir Ansari, M., Afzal, M., Saquib Hasnain, M., Kumar Nayak, A., 2020. Bactericidal activity of silver nanoparticles: a mechanistic review. Mater. Sci. Energy Technol. 3, 756–769. https://doi.org/10.1016/j.mset.2020.09.002.

Anjos, M.M., Endo, E.H., Leimann, F.V., Gonçalves, O.H., Dias-Filho, B.P., Abreu Filho, B.A.D., 2018. Preservation of the antibacterial activity of enzymes against Alicyclobacillus spp. through microencapsulation. LWT Food Sci. Technol. 88, 18–25. https://doi.org/10.1016/j.lwt.2017.09.039.

Anvar, A.A., Ahari, H., Ataee, M., 2021. Antimicrobial properties of food nanopackaging: a new focus on foodborne pathogens. Front. Microbiol. 12, 1945. Retrieved from https://www.frontiersin.org/article/10.3389/fmicb.2021.690706.

Avram, A.M., Morin, P., Brownmiller, C., Howard, L.R., Sengupta, A., Wickramasinghe, S.R., 2017. Concentrations of polyphenols from blueberry pomace extract using nanofiltration. Food Bioprod. Process. 106, 91–101. https://doi.org/10.1016/j.fbp.2017.07.006.

Ayres Cacciatore, F., Dalmás, M., Maders, C., Ataíde Isaía, H., Brandelli, A., da Silva Malheiros, P., 2020. Carvacrol encapsulation into nanostructures: characterization and antimicrobial activity against foodborne pathogens adhered to stainless steel. Food Res. Int. 133. https://doi.org/10.1016/j.foodres.2020.109143, 109143.

Azizi-Lalabadi, M., Ehsani, A., Divband, B., Alizadeh-Sani, M., 2019. Antimicrobial activity of titanium dioxide and zinc oxide nanoparticles supported in 4A zeolite and evaluation the morphological characteristic. Sci. Rep. 9 (1), 17439. https://doi.org/10.1038/s41598-019-54025-0.

Barish, J.A., Goddard, J.M., 2013. Anti-fouling surface modified stainless steel for food processing. Food Bioprod. Process. 91 (4), 352–361. https://doi.org/10.1016/j.fbp.2013.01.003.

Basavegowda, N., Mandal, T.K., Baek, K.-H., 2020. Bimetallic and Trimetallic nanoparticles for active food packaging applications: a review. Food Bioprocess Technol. 13 (1), 30–44. https://doi.org/10.1007/s11947-019-02370-3.

Bayani Bandpey, N., Aroujalian, A., Raisi, A., Fazel, S., 2017. Surface coating of silver nanoparticles on polyethylene for fabrication of antimicrobial milk packaging films. Int. J. Dairy Technol. 70 (2), 204–211. https://doi.org/10.1111/1471-0307.12320.

Behera, N., Arakha, M., Priyadarshinee, M., Pattanayak, B.S., Soren, S., Jha, S., Mallick, B.C., 2019. Oxidative stress generated at nickel oxide nanoparticle interface results in bacterial membrane damage leading to cell death. RSC Adv. 9 (43), 24888–24894. https://doi.org/10.1039/C9RA02082A.

Bezza, F.A., Tichapondwa, S.M., Chirwa, E.M.N., 2020. Fabrication of monodispersed copper oxide nanoparticles with potential application as antimicrobial agents. Sci. Rep. 10 (1), 16680. https://doi.org/10.1038/s41598-020-73497-z.

Bhargava, K., Conti, D.S., da Rocha, S.R.P., Zhang, Y., 2015. Application of an oregano oil nanoemulsion to the control of foodborne bacteria on fresh lettuce. Food Microbiol. 47, 69–73. https://doi.org/10.1016/j.fm.2014.11.007.

Bhunia, K., Dhawan, S., Sablani, S.S., 2012. Modeling the oxygen diffusion of nanocomposite-based food packaging films. J. Food Sci. 77 (7), N29–N38. https://doi.org/10.1111/j.1750-3841.2012.02768.x.

Bodaghi, H., Mostofi, Y., Oromiehie, A., Zamani, Z., Ghanbarzadeh, B., Costa, C., et al., 2013. Evaluation of the photocatalytic antimicrobial effects of a TiO_2 nanocomposite food packaging film by in vitro and in vivo tests. LWT Food Sci. Technol. 50 (2), 702–706. https://doi.org/10.1016/j.lwt.2012.07.027.

Bouarab-Chibane, L., Forquet, V., Lantéri, P., Clément, Y., Léonard-Akkari, L., Oulahal, N., et al., 2019. Antibacterial properties of polyphenols: characterization and QSAR (quantitative structure–activity relationship) models. Front. Microbiol. 10, 829. Retrieved from https://www.frontiersin.org/article/10.3389/fmicb.2019.00829.

Buchalska, M., Łabuz, P., Bujak, Ł., Szewczyk, G., Sarna, T., Maćkowski, S., Macyk, W., 2013. New insight into singlet oxygen generation at surface modified nanocrystalline TiO_2 – the effect of near-infrared irradiation. Dalton Trans. 42 (26), 9468–9475. https://doi.org/10.1039/C3DT50399B.

Buchovec, I., Lukseviciūtė, V., Kokstaite, R., Labeikyte, D., Kaziukonyte, L., Luksiene, Z., 2017. Inactivation of gram (−) bacteria salmonella enterica by chlorophyllin-based photosensitization: mechanism of action and new strategies to enhance the inactivation efficiency. J. Photochem. Photobiol. B Biol. 172, 1–10. https://doi.org/10.1016/j.jphotobiol.2017.05.008.

Buchovec, I., Paskeviciute, E., Luksiene, Z., 2010. Photosensitization-based inactivation of food pathogen listeria monocytogenes in vitro and on the surface of packaging material. J. Photochem. Photobiol. B 99 (1), 9–14. https://doi.org/10.1016/j.jphotobiol.2010.01.007.

Ceylan, Z., Unal Sengor, G.F., Sağdıç, O., Yilmaz, M.T., 2017. A novel approach to extend microbiological stability of sea bass (Dicentrarchus labrax) fillets coated with electrospun chitosan nanofibers. LWT Food Sci. Technol. 79, 367–375. https://doi.org/10.1016/j.lwt.2017.01.062.

Chand, K., Cao, D., Eldin Fouad, D., Hussain Shah, A., Qadeer Dayo, A., Zhu, K., et al., 2020. Green synthesis, characterization and photocatalytic application of silver nanoparticles synthesized by various plant extracts. Arab. J. Chem. 13 (11), 8248–8261. https://doi.org/10.1016/j.arabjc.2020.01.009.

Chaudhary, P., Fatima, F., Kumar, A., 2020. Relevance of nanomaterials in food packaging and its advanced future prospects. J. Inorg. Organomet. Polym. Mater., 1–13. https://doi.org/10.1007/s10904-020-01674-8.

Chawengkijwanich, C., Hayata, Y., 2008. Development of TiO$_2$ powder-coated food packaging film and its ability to inactivate *Escherichia coli* in vitro and in actual tests. Int. J. Food Microbiol. 123 (3), 288–292. https://doi.org/10.1016/j.ijfoodmicro.2007.12.017.

Chen, H., Wang, B., Gao, D., Guan, M., Zheng, L., Ouyang, H., et al., 2013. Broad-spectrum antibacterial activity of carbon nanotubes to human gut bacteria. Small 9 (16), 2735–2746. https://doi.org/10.1002/smll.201202792.

Chen, L., Zheng, L., Lv, Y., Liu, H., Wang, G., Ren, N., et al., 2010. Chemical assembly of silver nanoparticles on stainless steel for antimicrobial applications. Surf. Coat. Technol. 204 (23), 3871–3875. https://doi.org/10.1016/j.surfcoat.2010.05.003.

Chen, M., Pan, L., Tu, K., 2021. A fluorescence biosensor for *Salmonella typhimurium* detection in food based on the nano-self-assembly of alendronic acid modified upconversion and gold nanoparticles. Anal. Methods 13 (21), 2415–2423. https://doi.org/10.1039/D1AY00493J.

Chen, X., Wu, X., Gan, M., Xu, F., He, L., Yang, D., et al., 2015. Rapid detection of *Staphylococcus aureus* in dairy and meat foods by combination of capture with silica-coated magnetic nanoparticles and thermophilic helicase-dependent isothermal amplification. J. Dairy Sci. 98 (3), 1563–1570. https://doi.org/10.3168/jds.2014-8828.

Cheng, T.-H., Lin, S.-B., Chen, L.-C., Chen, H.-H., 2018. Studies of the antimicrobial ability and silver ions migration from silver nitrate-incorporated electrospun nylon nanofibers. Food Packag. Shelf Life 16, 129–137. https://doi.org/10.1016/j.fpsl.2018.03.003.

Chopde, S., Datir, R., Deshmukh, G., Dhotre, A., Patil, M., 2020. Nanoparticle formation by nanospray drying & its application in nanoencapsulation of food bioactive ingredients. J. Agric. Food Res. 2. https://doi.org/10.1016/j.jafr.2020.100085.

Chowdhury, S., Teoh, Y.L., Ong, K.M., Rafflisman Zaidi, N.S., Mah, S.-K., 2020. Poly(vinyl) alcohol crosslinked composite packaging film containing gold nanoparticles on shelf life extension of banana. Food Packag. Shelf Life 24. https://doi.org/10.1016/j.fpsl.2020.100463, 100463.

Cloutier, M., Mantovani, D., Rosei, F., 2015. Antibacterial coatings: challenges, perspectives, and opportunities. Trends Biotechnol. 33 (11), 637–652. https://doi.org/10.1016/j.tibtech.2015.09.002.

Cortez-Vega, W.R., Pizato, S., de Souza, J.T.A., Prentice, C., 2014. Using edible coatings from Whitemouth croaker (*Micropogonias furnieri*) protein isolate and organo-clay nanocomposite for improve the conservation properties of fresh-cut 'Formosa' papaya. Innovative Food Sci. Emerg. Technol. 22, 197–202. https://doi.org/10.1016/j.ifset.2013.12.007.

Costa, C., Conte, A., Buonocore, G.G., Lavorgna, M., Del Nobile, M.A., 2012. Calcium-alginate coating loaded with silver-montmorillonite nanoparticles to prolong the shelf-life of fresh-cut carrots. Food Res. Int. 48 (1), 164–169. https://doi.org/10.1016/j.foodres.2012.03.001.

De León-Zapata, M.A., Ventura-Sobrevilla, J.M., Salinas-Jasso, T.A., Flores-Gallegos, A.C., Rodríguez-Herrera, R., Pastrana-Castro, L., et al., 2018. Changes of the shelf life of candelilla wax/tarbush bioactive based-nanocoated apples at industrial level conditions. Sci. Hortic. 231, 43–48. https://doi.org/10.1016/j.scienta.2017.12.005.

De Matteis, V., 2017. Exposure to inorganic nanoparticles: routes of entry, immune response, biodistribution and in vitro/in vivo toxicity evaluation. Toxics 5 (4), 29. https://doi.org/10.3390/toxics5040029.

de Souza, E.S.J.M., Hanchuk, T.D., Santos, M.I., Kobarg, J., Bajgelman, M.C., Cardoso, M.B., 2016. Viral inhibition mechanism mediated by surface-modified silica nanoparticles. ACS Appl. Mater. Interfaces 8 (26), 16564–16572. https://doi.org/10.1021/acsami.6b03342.

Dehnad, D., Mirzaei, H., Emam-Djomeh, Z., Jafari, S.M., Dadashi, S., 2014. Thermal and antimicrobial properties of chitosan-nanocellulose films for extending shelf life of ground meat. Carbohydr. Polym. 109, 148–154. https://doi.org/10.1016/j.carbpol.2014.03.063.

Deng, X., Yu Nikiforov, A., Coenye, T., Cools, P., Aziz, G., Morent, R., et al., 2015. Antimicrobial nano-silver nonwoven polyethylene terephthalate fabric via an atmospheric pressure plasma deposition process. Sci. Rep. 5 (1), 10138. https://doi.org/10.1038/srep10138.

Detsi, A., Kavetsou, E., Kostopoulou, I., Pitterou, I., Pontillo, A.R.N., Tzani, A., et al., 2020. Nanosystems for the encapsulation of natural products: the case of chitosan biopolymer as a matrix. Pharmaceutics 12 (7), 669. https://doi.org/10.3390/pharmaceutics12070669.

Dhital, R., Joshi, P., Becerra-Mora, N., Umagiliyage, A., Chai, T., Kohli, P., Choudhary, R., 2017. Integrity of edible nano-coatings and its effects on quality of strawberries subjected to simulated in-transit vibrations. LWT Food Sci. Technol. 80, 257–264. https://doi.org/10.1016/j.lwt.2017.02.033.

Dhungana, P., Bhandari, B., 2021. Development of a continuous membrane nanobubble generation method applicable in liquid food processing. Int. J. Food Sci. Technol. https://doi.org/10.1111/ijfs.15182.

Di Giulio, M., Di Lodovico, S., Fontana, A., Traini, T., Di Campli, E., Pilato, S., et al., 2020. Graphene oxide affects Staphylococcus aureus and *Pseudomonas aeruginosa* dual species biofilm in Lubbock chronic wound biofilm model. Sci. Rep. 10 (1), 18525. https://doi.org/10.1038/s41598-020-75086-6.

Díez-Pascual, A.M., Díez-Vicente, A.L., 2015. Antimicrobial and sustainable food packaging based on poly(butylene adipate-co-terephthalate) and electrospun chitosan nanofibers. RSC Adv. 5 (113), 93095–93107. https://doi.org/10.1039/C5RA14359D.

dos Santos, A.F., Arini, G.S., de Almeida, D.R.Q., Labriola, L., 2021. Nanophotosensitizers for cancer therapy: a promising technology? J. Phys. Mater. 4 (3). https://doi.org/10.1088/2515-7639/abf7dd, 032006.

Duffy, L.L., Osmond-McLeod, M.J., Judy, J., King, T., 2018. Investigation into the antibacterial activity of silver, zinc oxide and copper oxide nanoparticles against poultry-relevant isolates of Salmonella and Campylobacter. Food Control 92, 293–300. https://doi.org/10.1016/j.foodcont.2018.05.008.

Dumas, E.M., Ozenne, V., Mielke, R.E., Nadeau, J.L., 2009. Toxicity of CdTe quantum dots in bacterial strains. IEEE Trans. Nanobioscience 8 (1), 58–64. https://doi.org/10.1109/TNB.2009.2017313.

EFSA, 2018. Guidance on risk assessment of the application of nanoscience and nanotechnologies in the food and feed chain: part 1, human and animal health. EFSA J. 16 (7). https://doi.org/10.2903/j.efsa.2018.5327, e05327.

Egodage, D.P., Jayalath, H.T.S., Samarasekara, A.M.P.B., Amarasinghe, D.A.S., Madushani, S.P.A., Senerath, S.M.N.S., 2017. Novel antimicrobial nano coated polypropylene based materials for food packaging systems. In: Paper Presented at the 2017 Moratuwa Engineering Research Conference (MERCon), 29–31 May 2017.

El Fawal, G., Hong, H., Song, X., Wu, J., Sun, M., He, C., et al., 2020. Fabrication of antimicrobial films based on hydroxyethylcellulose and ZnO for food packaging application. Food Packag. Shelf Life 23. https://doi.org/10.1016/j.fpsl.2020.100462, 100462.

Esmaeili, M., Ariaii, P., Nasiraie, L.R., Pour, M.Y., 2021. Comparison of coating and nano-coating of chitosan-*Lepidium sativum* seed gum composites on quality and shelf life of beef. J. Food Meas. Charact. 15 (1), 341–352. https://doi.org/10.1007/s11694-020-00643-6.

Etacheri, V., Di Valentin, C., Schneider, J., Bahnemann, D., Pillai, S.C., 2015. Visible-light activation of TiO$_2$ photocatalysts: advances in theory and experiments. J. Photochem. Photobiol. C: Photochem. Rev. 25, 1–29. https://doi.org/10.1016/j.jphotochemrev.2015.08.003.

European Parliament, C. o. t. E. U, 2004a. Regulation (Ec) No 852/2004 of the European Parliament and of the council of 29 April 2004 on the hygiene of foodstuffs. Off. J. Eur. Communities 47, 1–54.

European Parliament, C. o. t. E. U, 2004b. Regulation (Ec) no 1935/2004 of the European Parliament and of the council of 27 October 2004 on materials and articles intended to come into contact with food and repealing directives 80/590/EEC and 89/109/EEC. Off. J. Eur. Union 47, 4–17.

Fallatah, H., Elhaneid, M., Ali-Boucetta, H., Overton, T.W., El Kadri, H., Gkatzionis, K., 2019. Antibacterial effect of graphene oxide (GO) nano-particles against *Pseudomonas putida* biofilm of variable age. Environ. Sci. Pollut. Res. Int. 26 (24), 25057–25070. https://doi.org/10.1007/s11356-019-05688-9.

Fan, Z., Yust, B., Nellore, B.P., Sinha, S.S., Kanchanapally, R., Crouch, R.A., et al., 2014. Accurate identification and selective removal of rotavirus using a plasmonic-magnetic 3D graphene oxide architecture. J. Phys. Chem. Lett. 5 (18), 3216–3221. https://doi.org/10.1021/jz501402b.

FAO/WHO, 2008. Viruses in Food: Scientific Advice to Support Risk Management Activities. Retrieved from Rome: https://www.who.int/foodsafety/publications/micro/Viruses_in_food_MRA.pdf.

FAO/WHO, 2010. FAO/WHO Expert Meeting on the Application of Nanotechnologies in the Food and Agriculture Sectors: Potential Food Safety Implications. Meeting report. Available from: http://www.fao.org/docrep/012/i1434e/i1434e00.pdf.

Farias, I.A.P., Santos, C.C.L., Xavier, A.L., Batista, T.M., Nascimento, Y.M., Nunes, J.M.F.F., et al., 2021. Synthesis, physicochemical characterization, antifungal activity and toxicological features of cerium oxide nanoparticles. Arab. J. Chem. 14 (1). https://doi.org/10.1016/j.arabjc.2020.10.035.

Ferraris, M., Perero, S., Ferraris, S., Miola, M., Vernè, E., Skoglund, S., et al., 2017. Antibacterial silver nanocluster/silica composite coatings on stainless steel. Appl. Surf. Sci. 396, 1546–1555. https://doi.org/10.1016/j.apsusc.2016.11.207.

Ferreira, C.D., Nunes, I.L., 2019. Oil nanoencapsulation: development, application, and incorporation into the food market. Nanoscale Res. Lett. 14 (1), 9. https://doi.org/10.1186/s11671-018-2829-2.

Flemming, H.C., Wingender, J., 2010. The biofilm matrix. Nat. Rev. Microbiol. 8 (9), 623–633. https://doi.org/10.1038/nrmicro2415.

Foltynowicz, Z., Bardenshtein, A., Sängerlaub, S., Antvorskov, H., Kozak, W., 2017. Nanoscale, zero valent iron particles for application as oxygen scavenger in food packaging. Food Packag. Shelf Life 11, 74–83. https://doi.org/10.1016/j.fpsl.2017.01.003.

Fulaz, S., Vitale, S., Quinn, L., Casey, E., 2019. Nanoparticle-biofilm interactions: the role of the EPS matrix. Trends Microbiol. 27 (11), 915–926. https://doi.org/10.1016/j.tim.2019.07.004.

Galie, S., Garcia-Gutierrez, C., Miguelez, E.M., Villar, C.J., Lombo, F., 2018. Biofilms in the food industry: health aspects and control methods. Front. Microbiol. 9, 898. https://doi.org/10.3389/fmicb.2018.00898.

Gao, T., Sun, D.-W., Tian, Y., Zhu, Z., 2021. Gold–silver core-shell nanorods based time-temperature indicator for quality monitoring of pasteurized milk in the cold chain. J. Food Eng. 306. https://doi.org/10.1016/j.jfoodeng.2021.110624, 110624.

Ge, L., Zhao, Y.-S., Mo, T., Li, J.-R., Li, P., 2012. Immobilization of glucose oxidase in electrospun nanofibrous membranes for food preservation. Food Control 26 (1), 188–193. https://doi.org/10.1016/j.foodcont.2012.01.022.

Ghate, V.S., Zhou, W., Yuk, H.-G., 2019. Perspectives and trends in the application of photodynamic inactivation for microbiological food safety. Compr. Rev. Food Sci. Food Saf. 18 (2), 402–424. https://doi.org/10.1111/1541-4337.12418.

Ghorbani, J., Rahban, D., Aghamiri, S., Teymouri, A., Bahador, A., 2018. Photosensitizers in antibacterial photodynamic therapy: an overview. Laser Ther. 27 (4), 293–302. https://doi.org/10.5978/islsm.27_18-RA-01.

Gilson, R.C., Black, K.C.L., Lane, D.D., Achilefu, S., 2017. Hybrid TiO_2-ruthenium nano-photosensitizer synergistically produces reactive oxygen species in both hypoxic and normoxic conditions. Angew. Chem. Int. Ed. Eng. 56 (36), 10717–10720. https://doi.org/10.1002/anie.201704458.

Glass, S., Trinklein, B., Abel, B., Schulze, A., 2018. TiO_2 as photosensitizer and photoinitiator for synthesis of photoactive TiO_2-PEGDA hydrogel without organic photoinitiator. Front. Chem. 6, 340. https://doi.org/10.3389/fchem.2018.00340.

Gopiraman, M., Jatoi, A.W., Hiromichi, S., Yamaguchi, K., Jeon, H.Y., Chung, I.M., Ick Soo, K., 2016. Silver coated anionic cellulose nanofiber composites for an efficient antimicrobial activity. Carbohydr. Polym. 149, 51–59. https://doi.org/10.1016/j.carbpol.2016.04.084.

Goy, R.C., Morais, S.T.B., Assis, O.B.G., 2016. Evaluation of the antimicrobial activity of chitosan and its quaternized derivative on *E. coli* and *S. aureus* growth. Rev. Bras. Farm. 26 (1), 122–127. https://doi.org/10.1016/j.bjp.2015.09.010.

Gu, X., Xu, Z., Gu, L., Xu, H., Han, F., Chen, B., Pan, X., 2021. Preparation and antibacterial properties of gold nanoparticles: a review. Environ. Chem. Lett. 19 (1), 167–187. https://doi.org/10.1007/s10311-020-01071-0.

Gualdesi, M.S., Aiassa, V., Vara, J., Alvarez Igarzabal, C.I., Ortiz, C.S., 2019. Development and evaluation of novel nanophotosensitizers as photoantimicrobial agents against *Staphylococcus aureus*. Mater. Sci. Eng. C Mater. Biol. Appl. 94, 303–309. https://doi.org/10.1016/j.msec.2018.09.040.

Guleria, S., Kumar, M., Kumari, S., Kumar, A., Kumari, S., 2020. Nano-starch films as effective antimicrobial packaging materials. In: Nanotechnological Approaches in Food Microbiology, first ed. CRC Press, Boca Raton, p. 494.

Gurunathan, S., Qasim, M., Choi, Y., Do, J.T., Park, C., Hong, K., et al., 2020. Antiviral potential of nanoparticles-can nanoparticles fight against coronaviruses? Nanomaterials (Basel) 10 (9). https://doi.org/10.3390/nano10091645.

Haggag, M.G., Shafaa, M.W., Kareem, H.S., El-Gamil, A.M., El-Hendawy, H.H., 2021. Screening and enhancement of the antimicrobial activity of some plant oils using liposomes as nanoscale carrier. Bull. Natl. Res. Cent. 45 (1), 38. https://doi.org/10.1186/s42269-021-00497-y.

Hamblin, M.R., 2018. Fullerenes as photosensitizers in photodynamic therapy: pros and cons. Photochem. Photobiol. Sci. 17 (11), 1515–1533. https://doi.org/10.1039/c8pp00195b.

Handford, C.E., Dean, M., Henchion, M., Spence, M., Elliott, C.T., Campbell, K., 2014. Implications of nanotechnology for the agri-food industry: opportunities, benefits and risks. Trends Food Sci. Technol. 40 (2), 226–241. https://doi.org/10.1016/j.tifs.2014.09.007.

Hashim, A., Salih, W., 2016. Anti-bacterial high density polyethylene/nano titanium dioxide composite synthesis and characterization. Int. J. Sci. Eng. Res. 7 (2), 81–85.

Hoseinnejad, M., Jafari, S.M., Katouzian, I., 2018. Inorganic and metal nanoparticles and their antimicrobial activity in food packaging applications. Crit. Rev. Microbiol. 44 (2), 161–181. https://doi.org/10.1080/1040841X.2017.1332001.

Hou, J., Wu, Y., Li, X., Wei, B., Li, S., Wang, X., 2018. Toxic effects of different types of zinc oxide nanoparticles on algae, plants, invertebrates, vertebrates and microorganisms. Chemosphere 193, 852–860. https://doi.org/10.1016/j.chemosphere.2017.11.077.

Huang, K., Goddard, J.M., 2015. Influence of fluid milk product composition on fouling and cleaning of Ni–PTFE modified stainless steel heat exchanger surfaces. J. Food Eng. 158, 22–29. https://doi.org/10.1016/j.jfoodeng.2015.02.026.

Huang, K., McLandsborough, L.A., Goddard, J.M., 2016. Adhesion and removal kinetics of *Bacillus cereus* biofilms on Ni-PTFE modified stainless steel. Biofouling 32 (5), 523–533. https://doi.org/10.1080/08927014.2016.1160284.

Huang, L., Xuan, Y., Koide, Y., Zhiyentayev, T., Tanaka, M., Hamblin, M.R., 2012. Type I and type II mechanisms of antimicrobial photodynamic therapy: an in vitro study on gram-negative and gram-positive bacteria. Lasers Surg. Med. 44 (6), 490–499. https://doi.org/10.1002/lsm.22045.

Hussain, M., Bensaid, S., Geobaldo, F., Saracco, G., Russo, N., 2011. Photocatalytic degradation of ethylene emitted by fruits with TiO_2 nanoparticles. Ind. Eng. Chem. Res. 50 (5), 2536–2543. https://doi.org/10.1021/ie1005756.

Ishida, T., 2019. Review on the role of Zn^{2+} ions in viral pathogenesis and the effect of Zn^{2+} ions for host cell-virus growth inhibition. Am. J. Biomed. Sci. Res. 2 (1), 28–37. https://doi.org/10.34297/ajbsr.2019.02.000566.

Jadoun, S., Arif, R., Jangid, N.K., Meena, R.K., 2021. Green synthesis of nanoparticles using plant extracts: a review. Environ. Chem. Lett. 19 (1), 355–374. https://doi.org/10.1007/s10311-020-01074-x.

Jafari, S.M., 2017. Chapter 1: An overview of nanoencapsulation techniques and their classification. In: Jafari, S.M. (Ed.), Nanoencapsulation Technologies for the Food and Nutraceutical Industries. Academic Press, pp. 1–34.

Jasmani, L., Rusli, R., Khadiran, T., Jalil, R., Adnan, S., 2020. Application of nanotechnology in wood-based products industry: a review. Nanoscale Res. Lett. 15 (1), 207. https://doi.org/10.1186/s11671-020-03438-2.

Jiang, T., Feng, L., Wang, Y., 2013. Effect of alginate/nano-Ag coating on microbial and physicochemical characteristics of shiitake mushroom (*Lentinus edodes*) during cold storage. Food Chem. 141 (2), 954–960. https://doi.org/10.1016/j.foodchem.2013.03.093.

Joshi, M., Adak, B., 2019. Advances in nanotechnology based functional, smart and intelligent textiles: a review. In: Comprehensive Nanoscience and Nanotechnology. Elsevier.

Kairyte, K., Lapinskas, S., Gudelis, V., Luksiene, Z., 2012. Effective inactivation of food pathogens listeria monocytogenes and salmonella enterica by combined treatment of hypericin-based photosensitization and high power pulsed light. J. Appl. Microbiol. 112 (6), 1144–1151. https://doi.org/10.1111/j.1365-2672.2012.05296.x.

Kalwar, K., Xi, J., Dandan, L., Gao, L., 2021. Fabrication of PAN/FeNPs electrospun nanofibers: nanozyme and an efficient antimicrobial agent. Mater. Today Commun. 26. https://doi.org/10.1016/j.mtcomm.2021.102168, 102168.

Kaur, R., Liu, S., 2016. Antibacterial surface design – contact kill. Prog. Surf. Sci. 91 (3), 136–153. https://doi.org/10.1016/j.progsurf.2016.09.001.

Kazemeini, H., Azizian, A., Adib, H., 2021. Inhibition of *Listeria monocytogenes* growth in Turkey fillets by alginate edible coating with Trachyspermum ammi essential oil nano-emulsion. Int. J. Food Microbiol. 344. https://doi.org/10.1016/j.ijfoodmicro.2021.109104, 109104.

Khan, F., Pham, D.T.N., Oloketuyi, S.F., Manivasagan, P., Oh, J., Kim, Y.-M., 2020. Chitosan and their derivatives: antibiofilm drugs against pathogenic bacteria. Colloids Surf. B: Biointerfaces 185. https://doi.org/10.1016/j.colsurfb.2019.110627, 110627.

Kim, H.-Y., Park, S.S., Lim, S.-T., 2015. Preparation, characterization and utilization of starch nanoparticles. Colloids Surf. B: Biointerfaces 126, 607–620. https://doi.org/10.1016/j.colsurfb.2014.11.011.

Kim, J.-H., Ma, J., Jo, S., Lee, S., Kim, C.S., 2020. Enhancement of antibacterial properties of a silver nanowire film via electron beam irradiation. ACS Appl. Bio Mater. 3 (4), 2117–2124. https://doi.org/10.1021/acsabm.0c00003.

Kováčová, M., Špitalská, E., Markovic, Z., Špitálský, Z., 2020. Carbon quantum dots as antibacterial photosensitizers and their polymer nanocomposite applications. Part. Part. Syst. Charact. 37 (1), 1900348. https://doi.org/10.1002/ppsc.201900348.

Kowsalya, E., MosaChristas, K., Balashanmugam, P., Tamil Selvi, A., Chinna Rani I, J., 2019. Biocompatible silver nanoparticles/poly(vinyl alcohol) electrospun nanofibers for potential antimicrobial food packaging applications. Food Packag. Shelf Life 21. https://doi.org/10.1016/j.fpsl.2019.100379, 100379.

Krajczewski, J., Rucińska, K., Townley, H.E., Kudelski, A., 2019. Role of various nanoparticles in photodynamic therapy and detection methods of singlet oxygen. Photodiagn. Photodyn. Ther. 26, 162–178. https://doi.org/10.1016/j.pdpdt.2019.03.016.

Kumar, H., Kuča, K., Bhatia, S.K., Saini, K., Kaushal, A., Verma, R., et al., 2020. Applications of nanotechnology in sensor-based detection of foodborne pathogens. Sensors 20 (7), 1966. https://doi.org/10.3390/s20071966.

Kumar, P., Huo, P., Zhang, R., Liu, B., 2019a. Antibacterial properties of graphene-based nanomaterials. Nanomaterials 9 (5), 737. https://doi.org/10.3390/nano9050737.

Kumar, S., Ye, F., Dobretsov, S., Dutta, J., 2019b. Chitosan nanocomposite coatings for food, paints, and water treatment applications. Appl. Sci. 9 (12). https://doi.org/10.3390/app9122409.

Kumar, V., Guleria, P., Mehta, S.K., 2017. Nanosensors for food quality and safety assessment. Environ. Chem. Lett. 15 (2), 165–177. https://doi.org/10.1007/s10311-017-0616-4.

Kumari, A., Thakur, N., Vashishtt, J., Singh, R.R., 2020. Structural, luminescent and antimicrobial properties of ZnS and CdSe/ZnS quantum dot structures originated by precursors. Spectrochim. Acta A Mol. Biomol. Spectrosc. 229. https://doi.org/10.1016/j.saa.2019.117962, 117962.

Le Ouay, B., Stellacci, F., 2015. Antibacterial activity of silver nanoparticles: a surface science insight. Nano Today 10 (3), 339–354. https://doi.org/10.1016/j.nantod.2015.04.002.

Let's talk about lipid nanoparticles, 2021. Nat. Rev. Mater. 6 (2), 99. https://doi.org/10.1038/s41578-021-00281-4.

LewisOscar, F., MubarakAli, D., Nithya, C., Priyanka, R., Gopinath, V., Alharbi, N.S., Thajuddin, N., 2015. One pot synthesis and anti-biofilm potential of copper nanoparticles (CuNPs) against clinical strains of *Pseudomonas aeruginosa*. Biofouling 31 (4), 379–391. https://doi.org/10.1080/08927014.2015.1048686.

Li, Y., Lin, Z., Zhao, M., Xu, T., Wang, C., Hua, L., et al., 2016. Silver nanoparticle based codelivery of oseltamivir to inhibit the activity of the H1N1 influenza virus through ROS-mediated signaling pathways. ACS Appl. Mater. Interfaces 8 (37), 24385–24393. https://doi.org/10.1021/acsami.6b06613.

Liakos, I., Rizzello, L., Scurr, D.J., Pompa, P.P., Bayer, I.S., Athanassiou, A., 2014. All-natural composite wound dressing films of essential oils encapsulated in sodium alginate with antimicrobial properties. Int. J. Pharm. 463 (2), 137–145. https://doi.org/10.1016/j.ijpharm.2013.10.046.

Lim, L.T., 2015. Chapter 8: Enzymes for food-packaging applications. In: Yada, R.Y. (Ed.), Improving and Tailoring Enzymes for Food Quality and Functionality. Woodhead Publishing, pp. 161–178.

Lin, Z., Young, S., Chang, S., 2015. CO_2 gas sensors based on carbon nanotube thin films using a simple transfer method on flexible substrate. IEEE Sensors J. 15 (12), 7017–7020. https://doi.org/10.1109/JSEN.2015.2472968.

Liu, L., Luo, X.-B., Ding, L., Luo, S.-L., 2019a. Chapter 4: Application of nanotechnology in the removal of heavy metal from water. In: Luo, X., Deng, F. (Eds.), Nanomaterials for the Removal of Pollutants and Resource Reutilization. Elsevier, pp. 83–147.

Liu, S., Oshita, S., Kawabata, S., Makino, Y., Yoshimoto, T., 2016. Identification of ROS produced by nanobubbles and their positive and negative effects on vegetable seed germination. Langmuir 32 (43), 11295–11302. https://doi.org/10.1021/acs.langmuir.6b01621.

Liu, S., Zeng, T.H., Hofmann, M., Burcombe, E., Wei, J., Jiang, R., et al., 2011. Antibacterial activity of graphite, graphite oxide, graphene oxide, and reduced graphene oxide: membrane and oxidative stress. ACS Nano 5 (9), 6971–6980. https://doi.org/10.1021/nn202451x.

Liu, Y., Liang, X., Wang, S., Qin, W., Zhang, Q., 2018. Electrospun antimicrobial polylactic acid/tea polyphenol nanofibers for food-packaging applications. Polymers (Basel) 10 (5). https://doi.org/10.3390/polym10050561.

Liu, Y., Shi, L., Su, L., van der Mei, H.C., Jutte, P.C., Ren, Y., Busscher, H.J., 2019b. Nanotechnology-based antimicrobials and delivery systems for biofilm-infection control. Chem. Soc. Rev. 48 (2), 428–446. https://doi.org/10.1039/c7cs00807d.

Lopes, N.A., Barreto Pinilla, C.M., Brandelli, A., 2019. Antimicrobial activity of lysozyme-nisin co-encapsulated in liposomes coated with polysaccharides. Food Hydrocoll. 93, 1–9. https://doi.org/10.1016/j.foodhyd.2019.02.009.

Lu, C., Bao, Y., Huang, J.-Y., 2021. Fouling in membrane filtration for juice processing. Curr. Opin. Food Sci. 42, 76–85. https://doi.org/10.1016/j.cofs.2021.05.004.

Lucky, S.S., Soo, K.C., Zhang, Y., 2015. Nanoparticles in photodynamic therapy. Chem. Rev. 115 (4), 1990–2042. https://doi.org/10.1021/cr5004198.

Luksiene, Z., Buchovec, I., Paskeviciute, E., 2009. Inactivation of food pathogen *Bacillus cereus* by photosensitization in vitro and on the surface of packaging material. J. Appl. Microbiol. 107 (6), 2037–2046. https://doi.org/10.1111/j.1365-2672.2009.04383.x.

Luksiene, Z., Paskeviciute, E., 2011. Novel approach to decontaminate food-packaging from pathogens in non-thermal and not chemical way: chlorophyllin-based photosensitization. J. Food Eng. 106 (2), 152–158. https://doi.org/10.1016/j.jfoodeng.2011.04.024.

Ma, Y., Li, J., Si, Y., Huang, K., Nitin, N., Sun, G., 2019. Rechargeable antibacterial N-halamine films with antifouling function for food packaging applications. ACS Appl. Mater. Interfaces 11 (19), 17814–17822. https://doi.org/10.1021/acsami.9b03464.

Ma, Y., Zhang, Z., Nitin, N., Sun, G., 2020. Integration of photo-induced biocidal and hydrophilic antifouling functions on nanofibrous membranes with demonstrated reduction of biofilm formation. J. Colloid Interface Sci. 578, 779–787. https://doi.org/10.1016/j.jcis.2020.06.037.

Macdonald, T.J., Wu, K., Sehmi, S.K., Noimark, S., Peveler, W.J., du Toit, H., et al., 2016. Thiol-capped gold nanoparticles swell-encapsulated into polyurethane as powerful antibacterial surfaces under dark and light conditions. Sci. Rep. 6 (1). https://doi.org/10.1038/srep39272.

Maduray, K., Parboosing, R., 2020. Metal nanoparticles: a promising treatment for viral and arboviral infections. Biol. Trace Elem. Res. 199, 3159–3176. https://doi.org/10.1007/s12011-020-02414-2.

Mahmoodi, A., Solaymani, S., Amini, M., Nezafat, N.B., Ghoranneviss, M., 2018. Structural, morphological and antibacterial characterization of CuO nanowires. SILICON 10 (4), 1427–1431. https://doi.org/10.1007/s12633-017-9621-2.

Mai-Prochnow, A., Clauson, M., Hong, J., Murphy, A.B., 2016. Gram positive and gram negative bacteria differ in their sensitivity to cold plasma. Sci. Rep. 6 (1), 38610. https://doi.org/10.1038/srep38610.

Maleki Dizaj, S., Mennati, A., Jafari, S., Khezri, K., Adibkia, K., 2015. Antimicrobial activity of carbon-based nanoparticles. Adv. Pharm. Bull. 5 (1), 19–23. https://doi.org/10.5681/apb.2015.003.

Malik, P., Katyal, V., Malik, V., Asatkar, A., Inwati, G., Mukherjee, T.K., 2013. Nanobiosensors: concepts and variations. ISRN Nanomater. 2013. https://doi.org/10.1155/2013/327435, 327435.

Manoj, D., Shanmugasundaram, S., Anandharamakrishnan, C., 2021. Nanosensing and nanobiosensing: concepts, methods, and applications for quality evaluation of liquid foods. Food Control 126. https://doi.org/10.1016/j.foodcont.2021.108017, 108017.

Marini, M., De Niederhausern, S., Iseppi, R., Bondi, M., Sabia, C., Toselli, M., Pilati, F., 2007. Antibacterial activity of plastics coated with silver-doped organic–inorganic hybrid coatings prepared by sol–gel processes. Biomacromolecules 8 (4), 1246–1254. https://doi.org/10.1021/bm060721b.

Marouzi, S., Sabouri, Z., Darroudi, M., 2021. Greener synthesis and medical applications of metal oxide nanoparticles. Ceram. Int. 47 (14), 19632–19650. https://doi.org/10.1016/j.ceramint.2021.03.301.

Matouskova, P., Marova, I., Bokrova, J., Benesova, P., 2016. Effect of encapsulation on antimicrobial activity of herbal extracts with lysozyme. Food Technol. Biotechnol. 54 (3), 304–316. https://doi.org/10.17113/ftb.54.03.16.4413.

McClements, D.J., DeLoid, G., Pyrgiotakis, G., Shatkin, J.A., Xiao, H., Demokritou, P., 2016. The role of the food matrix and gastrointestinal tract in the assessment of biological properties of ingested engineered nanomaterials (iENMs): state of the science and knowledge gaps. NanoImpact 3-4, 47–57. https://doi.org/10.1016/j.impact.2016.10.002.

McClements, D.J., Xiao, H., 2017. Is nano safe in foods? Establishing the factors impacting the gastrointestinal fate and toxicity of organic and inorganic food-grade nanoparticles. npj Sci. Food 1 (1), 6. https://doi.org/10.1038/s41538-017-0005-1.

Meghana, S., Kabra, P., Chakraborty, S., Padmavathy, N., 2015. Understanding the pathway of antibacterial activity of copper oxide nanoparticles. RSC Adv. 5 (16), 12293–12299. https://doi.org/10.1039/C4RA12163E.

Merian, T., Goddard, J.M., 2012. Advances in nonfouling materials: perspectives for the food industry. J. Agric. Food Chem. 60 (12), 2943–2957. https://doi.org/10.1021/jf204741p.

Mircioiu, C., Voicu, V., Anuta, V., Tudose, A., Celia, C., Paolino, D., et al., 2019. Mathematical modeling of release kinetics from supramolecular drug delivery systems. Pharmaceutics 11 (3). https://doi.org/10.3390/pharmaceutics11030140.

Mirzahosseinipour, M., Khorsandi, K., Hosseinzadeh, R., Ghazaeian, M., Shahidi, F.K., 2020. Antimicrobial photodynamic and wound healing activity of curcumin encapsulated in silica nanoparticles. Photodiagn. Photodyn. Ther. 29. https://doi.org/10.1016/j.pdpdt.2019.101639, 101639.

Mishra, M., Dashora, K., Srivastava, A., Fasake, V.D., Nag, R.H., 2019. Prospects, challenges and need for regulation of nanotechnology with special reference to India. Ecotoxicol. Environ. Saf. 171, 677–682. https://doi.org/10.1016/j.ecoenv.2018.12.085.

Mocan, T., Matea, C.T., Pop, T., Mosteanu, O., Buzoianu, A.D., Puia, C., et al., 2017. Development of nanoparticle-based optical sensors for pathogenic bacterial detection. J. Nanobiotechnol. 15 (1), 25. https://doi.org/10.1186/s12951-017-0260-y.

Moghaddam, H.M., Khoshtaghaza, M.H., Salimi, A., Barzegar, M., 2014. The TiO$_2$–clay-LDPE nanocomposite packaging films: investigation on the structure and physicomechanical properties. Polym.-Plast. Technol. Eng. 53 (17), 1759–1767. https://doi.org/10.1080/03602559.2014.919647.

Mohammadian, E., Alizadeh-Sani, M., Jafari, S.M., 2020. Smart monitoring of gas/temperature changes within food packaging based on natural colorants. Compr. Rev. Food Sci. Food Saf. 19 (6), 2885–2931. https://doi.org/10.1111/1541-4337.12635.

Mohan, S., Nair, S.S., Ajay, A.V., Senthil Saravanan, M.S., Vishnu, B.R., Sivapirakasam, S.P., Surianarayanan, M., 2020. Corrosion behaviour of ZrO$_2$-TiO$_2$ nano composite coating on stainless steel under simulated marine environment. Mater. Today Proc. 27, 2492–2497. https://doi.org/10.1016/j.matpr.2019.09.224.

Montazer, M., Harifi, T., 2017. Chapter 16: New approaches and future aspects of antibacterial food packaging: from nanoparticles coating to nanofibers and nanocomposites, with foresight to address the regulatory uncertainty. In: Grumezescu, A.M. (Ed.), Food Packaging. Academic Press, pp. 533–565.

Munda, S., Dutta, S., Pandey, S.K., Sarma, N., Lal, M., 2019. Antimicrobial activity of essential oils of medicinal and aromatic plants of the north East India: a biodiversity hot spot. J. Essent. Oil Bear. Plants 22 (1), 105–119. https://doi.org/10.1080/0972060x.2019.1601032.

Munisparan, T., Yang, E.C.Y., Paramasivam, R., Dahlan, N.A., Pushpamalar, J., 2018. Optimisation of preparation conditions for Ti nanowires and suitability as an antibacterial material. IET Nanobiotechnol. 12 (4), 429–435. https://doi.org/10.1049/iet-nbt.2017.0186.

Mustafa, F., Andreescu, S., 2020. Nanotechnology-based approaches for food sensing and packaging applications. RSC Adv. 10 (33), 19309–19336. https://doi.org/10.1039/d0ra01084g.

Nakao, A., Valdman, A., Costa, A.L.H., Bagajewicz, M.J., Queiroz, E.M., 2017. Incorporating fouling modeling into shell-and-tube heat exchanger design. Ind. Eng. Chem. Res. 56 (15), 4377–4385. https://doi.org/10.1021/acs.iecr.6b03564.

Nataraj, N., Anjusree, G.S., Madhavan, A.A., Priyanka, P., Sankar, D., Nisha, N., et al., 2014. Synthesis and anti-staphylococcal activity of TiO$_2$ nanoparticles and nanowires in ex vivo porcine skin model. J. Biomed. Nanotechnol. 10 (5), 864–870. https://doi.org/10.1166/jbn.2014.1756.

Navarro-López, D.E., Garcia-Varela, R., Ceballos-Sanchez, O., Sanchez-Martinez, A., Sanchez-Ante, G., Corona-Romero, K., et al., 2021. Effective antimicrobial activity of ZnO and Yb-doped ZnO nanoparticles against *Staphylococcus aureus* and *Escherichia coli*. Mater. Sci. Eng. C 123. https://doi.org/10.1016/j.msec.2021.112004, 112004.

Neethirajan, S., Jayas, D.S., 2011. Nanotechnology for the food and bioprocessing industries. Food Bioprocess Technol. 4 (1), 39–47. https://doi.org/10.1007/s11947-010-0328-2.

Nguyen, H.N., Chaves-Lopez, C., Oliveira, R.C., Paparella, A., Rodrigues, D.F., 2019. Cellular and metabolic approaches to investigate the effects of graphene and graphene oxide in the fungi *Aspergillus flavus* and *Aspergillus niger*. Carbon 143, 419–429. https://doi.org/10.1016/j.carbon.2018.10.099.

Nichols, F., Chen, S., 2020. Graphene oxide quantum dot-based functional nanomaterials for effective antimicrobial applications. Chem. Rec. 20 (12), 1505–1515. https://doi.org/10.1002/tcr.202000090.

Nile, S.H., Baskar, V., Selvaraj, D., Nile, A., Xiao, J., Kai, G., 2020. Nanotechnologies in food science: applications, recent trends, and future perspectives. Nano-Micro Lett. 12 (1), 45. https://doi.org/10.1007/s40820-020-0383-9.

Noori Hashemabad, Z., Shabanpour, B., Azizi, H., Ojagh, S.M., Alishahi, A., 2017. Effect of TiO$_2$ nanoparticles on the antibacterial and physical properties of low-density polyethylene film. Polym.-Plast. Technol. Eng. 56 (14), 1516–1527. https://doi.org/10.1080/03602559.2016.1278022.

Ochiai, T., Fujishima, A., 2012. Photoelectrochemical properties of TiO$_2$ photocatalyst and its applications for environmental purification. J. Photochem. Photobiol. C: Photochem. Rev. 13 (4), 247–262. https://doi.org/10.1016/j.jphotochemrev.2012.07.001.

Oetiker, N., Munoz-Villagran, C., Vasquez, C.C., Bravo, D., Perez-Donoso, J.M., 2020. Bacterial phototoxicity of biomimetic CdTe-GSH quantum dots. J. Appl. Microbiol. https://doi.org/10.1111/jam.14957.

Oh, D.-H., Khan, I., Tango, C.N., 2019. Hurdle enhancement of electrolyzed water with other techniques. In: Ding, T., Oh, D.-H., Liu, D. (Eds.), Electrolyzed Water in Food: Fundamentals and Applications. Springer Singapore, Singapore, pp. 231–260.

Omerovic, N., Djisalov, M., Zivojevic, K., Mladenovic, M., Vunduk, J., Milenkovic, I., et al., 2021. Antimicrobial nanoparticles and biodegradable polymer composites for active food packaging applications. Compr. Rev. Food Sci. Food Saf. 20 (3), 2428–2454. https://doi.org/10.1111/1541-4337.12727.

Ortiz-Benítez, E.A., Velázquez-Guadarrama, N., Durán Figueroa, N.V., Quezada, H., Olivares-Trejo, J.D.J., 2019. Antibacterial mechanism of gold nanoparticles on *Streptococcus pneumoniae*. Metallomics 11 (7), 1265–1276. https://doi.org/10.1039/C9MT00084D.

Paidari, S., Ibrahim, S.A., 2021. Potential application of gold nanoparticles in food packaging: a mini review. Gold Bull. 54 (1), 31–36. https://doi.org/10.1007/s13404-021-00290-9.

Park, J., Lee, Y.-K., Park, I.-K., Hwang, S.R., 2021. Current limitations and recent progress in nanomedicine for clinically available photodynamic therapy. Biomedicine 9 (1), 85. https://doi.org/10.3390/biomedicines9010085.

Park, J.H., Choi, S., Moon, H.C., Seo, H., Kim, J.Y., Hong, S.P., et al., 2017. Antimicrobial spray nanocoating of supramolecular Fe(III)-tannic acid metal-organic coordination complex: applications to shoe insoles and fruits. Sci. Rep. 7 (1), 6980. https://doi.org/10.1038/s41598-017-07257-x.

Pereira, V.A., de Arruda, I.N.Q., Stefani, R., 2015. Active chitosan/PVA films with anthocyanins from brassica oleraceae (red cabbage) as time–temperature Indicators for application in intelligent food packaging. Food Hydrocoll. 43, 180–188. https://doi.org/10.1016/j.foodhyd.2014.05.014.

Pérez, C., Zúñiga, T., Palavecino, C.E., 2021. Photodynamic therapy for treatment of *Staphylococcus aureus* infections. Photodiagn. Photodyn. Ther. 34. https://doi.org/10.1016/j.pdpdt.2021.102285, 102285.

Phan, K.K.T., Truong, T., Wang, Y., Bhandari, B., 2021. Formation and stability of carbon dioxide nanobubbles for potential applications in food processing. Food Eng. Rev. 13 (1), 3–14. https://doi.org/10.1007/s12393-020-09233-0.

PIB, 2020. Guidelines for Evaluation of Nano-Based Agri-Input and Food Products in India. Department of Biotechnology, New Delhi. Available from https://pib.gov.in/PressReleasePage.aspx?PRID=1637011.

Piližota, V., 2014. Chapter 9: Fruits and vegetables (including herbs). In: Motarjemi, Y., Lelieveld, H. (Eds.), Food Safety Management. Academic Press, San Diego, pp. 213–249.

Priyadarshi, R., Kim, H.-J., Rhim, J.-W., 2021. Effect of sulfur nanoparticles on properties of alginate-based films for active food packaging applications. Food Hydrocoll. 110. https://doi.org/10.1016/j.foodhyd.2020.106155, 106155.

Qayyum, S., Khan, A.U., 2016. Nanoparticles vs. biofilms: a battle against another paradigm of antibiotic resistance. Med. Chem. Commun. 7 (8), 1479–1498. https://doi.org/10.1039/C6MD00124F.

Rambaran, T.F., 2020. Nanopolyphenols: a review of their encapsulation and anti-diabetic effects. SN Appl. Sci. 2 (8), 1335. https://doi.org/10.1007/s42452-020-3110-8.

Ramesh, S., Kim, H.S., Kim, J.-H., 2018. Cellulose–polyvinyl alcohol–nano-TiO_2 hybrid nanocomposite: thermal, optical, and antimicrobial properties against pathogenic bacteria. Polym.-Plast. Technol. Eng. 57 (7), 669–681. https://doi.org/10.1080/03602559.2017.1344851.

Randazzo, W., Fabra, M.J., Falcó, I., López-Rubio, A., Sánchez, G., 2018. Polymers and biopolymers with antiviral activity: potential applications for improving food safety. Compr. Rev. Food Sci. Food Saf. 17 (3), 754–768. https://doi.org/10.1111/1541-4337.12349.

Razavi, S., Janfaza, S., Tasnim, N., Gibson, D.L., Hoorfar, M., 2021. Nanomaterial-based encapsulation for controlled gastrointestinal delivery of viable probiotic bacteria. Nanoscale Adv. 3 (10), 2699–2709. https://doi.org/10.1039/D0NA00952K.

Rojas, K., Canales, D., Amigo, N., Montoille, L., Cament, A., Rivas, L.M., et al., 2019. Effective antimicrobial materials based on low-density polyethylene (LDPE) with zinc oxide (ZnO) nanoparticles. Compos. Part B 172, 173–178. https://doi.org/10.1016/j.compositesb.2019.05.054.

Rojas-Andrade, M.D., Nguyen, T.A., Mistler, W.P., Armas, J., Lu, J.E., Roseman, G., et al., 2020. Antimicrobial activity of graphene oxide quantum dots: impacts of chemical reduction. Nanoscale Adv. 2 (3), 1074–1083. https://doi.org/10.1039/C9NA00698B.

Rossi, F., Khoo, E.H., Su, X., Thanh, N.T.K., 2020. Study of the effect of anisotropic gold nanoparticles on plasmonic coupling with a photosensitizer for antimicrobial film. ACS Appl. Bio Mater. 3 (1), 315–326. https://doi.org/10.1021/acsabm.9b00838.

Rossi, F., Thanh, N.T.K., Su, X.D., 2019. Gold nanorods embedded in polymeric film for killing bacteria by generating reactive oxygen species with light. ACS Appl. Bio Mater. 2 (7), 3059–3067. https://doi.org/10.1021/acsabm.9b00343.

Rostamabadi, H., Falsafi, S.R., Jafari, S.M., 2019. Nanoencapsulation of carotenoids within lipid-based nanocarriers. J. Control. Release 298, 38–67. https://doi.org/10.1016/j.jconrel.2019.02.005.

Rout, B., Liu, C.H., Wu, W.C., 2017. Photosensitizer in lipid nanoparticle: a nano-scaled approach to antibacterial function. Sci. Rep. 7 (1), 7892. https://doi.org/10.1038/s41598-017-07444-w.

Sadeghnejad, A., Aroujalian, A., Raisi, A., Fazel, S., 2014. Antibacterial nano silver coating on the surface of polyethylene films using corona discharge. Surf. Coat. Technol. 245, 1–8. https://doi.org/10.1016/j.surfcoat.2014.02.023.

Sahoo, M., Vishwakarma, S., Panigrahi, C., Kumar, J., 2021. Nanotechnology: current applications and future scope in food. Food Front. 2 (1), 3–22. https://doi.org/10.1002/fft2.58.

Saito, H., Nosaka, Y., 2014. Mechanism of singlet oxygen generation in visible-light-induced photocatalysis of gold-nanoparticle-deposited titanium dioxide. J. Phys. Chem. C 118 (29), 15656–15663. https://doi.org/10.1021/jp502440f.

Saleemi, M.A., Yong, P.V.C., Wong, E.H., 2020. Investigation of antimicrobial activity and cytotoxicity of synthesized surfactant-modified carbon nanotubes/polyurethane electrospun nanofibers. Nano-Struct. Nano-Objects 24. https://doi.org/10.1016/j.nanoso.2020.100612, 100612.

Sanyasi, S., Majhi, R.K., Kumar, S., Mishra, M., Ghosh, A., Suar, M., et al., 2016. Polysaccharide-capped silver nanoparticles inhibit biofilm formation and eliminate multi-drug-resistant bacteria by disrupting bacterial cytoskeleton with reduced cytotoxicity towards mammalian cells. Sci. Rep. 6, 24929. https://doi.org/10.1038/srep24929.

Satter, S.S., Yokoya, T., Hirayama, J., Nakajima, K., Fukuoka, A., 2018. Oxidation of trace ethylene at 0°C over platinum nanoparticles supported on silica. ACS Sustain. Chem. Eng. 6 (9), 11480–11486. https://doi.org/10.1021/acssuschemeng.8b01543.

Seo, Y., Park, K., Hong, Y., Lee, E.S., Kim, S.S., Jung, Y.T., et al., 2020. Reactive-oxygen-species-mediated mechanism for photoinduced antibacterial and antiviral activities of Ag_3PO_4. J. Anal. Sci. Technol. 11 (1), 21. https://doi.org/10.1186/s40543-020-00220-y.

Severino, R., Ferrari, G., Vu, K.D., Donsì, F., Salmieri, S., Lacroix, M., 2015. Antimicrobial effects of modified chitosan based coating containing nanoemulsion of essential oils, modified atmosphere packaging and gamma irradiation against Escherichia coli O157:H7 and salmonella typhimurium on green beans. Food Control 50, 215–222. https://doi.org/10.1016/j.foodcont.2014.08.029.

Shamaila, S., Zafar, N., Riaz, S., Sharif, R., Nazir, J., Naseem, S., 2016. Gold nanoparticles: an efficient antimicrobial agent against enteric bacterial human pathogen. Nanomaterials 6 (4), 71. https://doi.org/10.3390/nano6040071.

Shankar, S., Pangeni, R., Park, J.W., Rhim, J.-W., 2018. Preparation of sulfur nanoparticles and their antibacterial activity and cytotoxic effect. Mater. Sci. Eng. C 92, 508–517. https://doi.org/10.1016/j.msec.2018.07.015.

Sharifimehr, S., Soltanizadeh, N., Goli, S.A.H., 2019. Physicochemical properties of fried shrimp coated with bio-nano-coating containing eugenol and Aloe vera. LWT Food Sci. Technol. 109, 33–39. https://doi.org/10.1016/j.lwt.2019.03.084.

Sharma, C., Dhiman, R., Rokana, N., Panwar, H., 2017. Nanotechnology: an untapped resource for food packaging. Front. Microbiol. 8, 1735. https://doi.org/10.3389/fmicb.2017.01735.

Shi, C., Wu, Y., Fang, D., Pei, F., Mariga, A.M., Yang, W., Hu, Q., 2018. Effect of nanocomposite packaging on postharvest senescence of flammulina velutipes. Food Chem. 246, 414–421. https://doi.org/10.1016/j.foodchem.2017.10.103.

Shkodenko, L., Kassirov, I., Koshel, E., 2020. Metal oxide nanoparticles against bacterial biofilms: perspectives and limitations. Microorganisms 8 (10), 1545. https://doi.org/10.3390/microorganisms8101545.

Singh, J., Vishwakarma, K., Ramawat, N., Rai, P., Singh, V.K., Mishra, R.K., et al., 2019. Nanomaterials and microbes' interactions: a contemporary overview. 3 Biotech 9 (3), 68. https://doi.org/10.1007/s13205-019-1576-0.

Singh, S., Shalini, R., 2016. Effect of hurdle technology in food preservation: a review. Crit. Rev. Food Sci. Nutr. 56 (4), 641–649. https://doi.org/10.1080/10408398.2012.761594.

Singh, T., Shukla, S., Kumar, P., Wahla, V., Bajpai, V.K., 2017. Application of nanotechnology in food science: perception and overview. Front. Microbiol. 8, 1501. https://doi.org/10.3389/fmicb.2017.01501.

Sirelkhatim, A., Mahmud, S., Seeni, A., Kaus, N.H.M., Ann, L.C., Bakhori, S.K.M., et al., 2015. Review on zinc oxide nanoparticles: antibacterial activity and toxicity mechanism. Nano-Micro Lett. 7 (3), 219–242. https://doi.org/10.1007/s40820-015-0040-x.

Slavin, Y.N., Asnis, J., Hafeli, U.O., Bach, H., 2017. Metal nanoparticles: understanding the mechanisms behind antibacterial activity. J. Nanobiotechnol. 15 (1), 65. https://doi.org/10.1186/s12951-017-0308-z.

Smith, A.T., LaChance, A.M., Zeng, S., Liu, B., Sun, L., 2019. Synthesis, properties, and applications of graphene oxide/reduced graphene oxide and their nanocomposites. Nano Mater. Sci. 1 (1), 31–47. https://doi.org/10.1016/j.nanoms.2019.02.004.

Spagnul, C., Turner, L.C., Boyle, R.W., 2015. Immobilized photosensitizers for antimicrobial applications. J. Photochem. Photobiol. B Biol. 150, 11–30. https://doi.org/10.1016/j.jphotobiol.2015.04.021.

Stephen Inbaraj, B., Chen, B.H., 2016. Nanomaterial-based sensors for detection of foodborne bacterial pathogens and toxins as well as pork adulteration in meat products. J. Food Drug Anal. 24 (1), 15–28. https://doi.org/10.1016/j.jfda.2015.05.001.

Su, L., Huang, J., Li, H., Pan, Y., Zhu, B., Zhao, Y., Liu, H., 2021. Chitosan-riboflavin composite film based on photodynamic inactivation technology for antibacterial food packaging. Int. J. Biol. Macromol. 172, 231–240. https://doi.org/10.1016/j.ijbiomac.2021.01.056.

Subhapriya, S., Gomathipriya, P., 2018. Green synthesis of titanium dioxide (TiO$_2$) nanoparticles by *Trigonella foenumgraecum* extract and its antimicrobial properties. Microb. Pathog. 116, 215–220. https://doi.org/10.1016/j.micpath.2018.01.027.

Sun, H., Gao, N., Dong, K., Ren, J., Qu, X., 2014. Graphene quantum dots-band-aids used for wound disinfection. ACS Nano 8 (6), 6202–6210. https://doi.org/10.1021/nn501640q.

Tarhan, Ö., 2020. Chapter 16: Safety and regulatory issues of nanomaterials in foods. In: Jafari, S.M. (Ed.), Handbook of Food Nanotechnology. Academic Press, pp. 655–703.

Teixeira-Santos, R., Gomes, M., Gomes, L.C., Mergulhão, F.J., 2021. Antimicrobial and anti-adhesive properties of carbon nanotube-based surfaces for medical applications: a systematic review. iScience 24 (1). https://doi.org/10.1016/j.isci.2020.102001, 102001.

Thandu, M.M., Cavalli, S., Rossi, G., Rizzardini, C.B., Goi, D., Comuzzi, C., 2017. Biological evaluation of a porphyrin-SPION nanoconjugate as an antimicrobial magnetic photosensitizer. J. Porphyrins Phthalocyanines 21 (09), 581–588. https://doi.org/10.1142/S1088424617500560.

Thi Phan, K.K., Truong, T., Wang, Y., Bhandari, B., 2020. Nanobubbles: fundamental characteristics and applications in food processing. Trends Food Sci. Technol. 95, 118–130. https://doi.org/10.1016/j.tifs.2019.11.019.

Toledo, V.H., Yoshimura, T.M., Pereira, S.T., Castro, C.E., Ferreira, F.F., Ribeiro, M.S., Haddad, P.S., 2020. Methylene blue-covered superparamagnetic iron oxide nanoparticles combined with red light as a novel platform to fight non-local bacterial infections: a proof of concept study against Escherichia coli. J. Photochem. Photobiol. B Biol. 209. https://doi.org/10.1016/j.jphotobiol.2020.111956, 111956.

Torres Dominguez, E., Nguyen, P.H., Hunt, H.K., Mustapha, A., 2019. Antimicrobial coatings for food contact surfaces: legal framework, mechanical properties, and potential applications. Compr. Rev. Food Sci. Food Saf. 18 (6), 1825–1858. https://doi.org/10.1111/1541-4337.12502.

Tsai, T.T., Huang, T.H., Chang, C.J., Yi-Ju Ho, N., Tseng, Y.T., Chen, C.F., 2017. Antibacterial cellulose paper made with silver-coated gold nanoparticles. Sci. Rep. 7 (1), 3155. https://doi.org/10.1038/s41598-017-03357-w.

Tsironi, T., Houhoula, D., Taoukis, P., 2020. Hurdle technology for fish preservation. Aquacult. Fish. 5 (2), 65–71. https://doi.org/10.1016/j.aaf.2020.02.001.

USFDA, 2014. Guidance for industry: Assessing the effects of significant manufacturing process changes, including emerging technologies, on the safety and regulatory status of food ingredients and food contact substances, including food ingredients that are color additives. Available from https://www.fda.gov/regulatory-information/search-fda-guidance-documents/considering-whether-fda-regulated-product-involves-application-nanotechnology.

USFDA, 2015. Guidence document: CVM GFI #220 Use of Nanomaterials in Food for Animals. Available from https://www.fda.gov/regulatory-information/search-fda-guidance-documents/cvm-gfi-220-use-nanomaterials-food-animals.

Vázquez-Hernández, F., Granada-Ramírez, D.A., Arias-Cerón, J.S., Rodriguez-Fragoso, P., Mendoza-Álvarez, J.G., Ramón-Gallegos, E., et al., 2018. Chapter 20: Use of nanostructured materials in drug delivery. In: Narayan, R. (Ed.), Nanobiomaterials. Woodhead Publishing, pp. 503–549.

Wang, C., Wang, Z.-G., Xi, R., Zhang, L., Zhang, S.-H., Wang, L.-J., Pan, G.-B., 2019b. In situ synthesis of flower-like ZnO on GaN using electrodeposition and its application as ethanol gas sensor at room temperature. Sensors Actuators B Chem. 292, 270–276. https://doi.org/10.1016/j.snb.2019.04.140.

Wang, L., Hu, C., Shao, L., 2017. The antimicrobial activity of nanoparticles: present situation and prospects for the future. Int. J. Nanomedicine 12, 1227–1249. https://doi.org/10.2147/IJN.S121956.

Wang, X., Yue, T., Lee, T.-C., 2015. Development of pleurocidin-poly(vinyl alcohol) electrospun antimicrobial nanofibers to retain antimicrobial activity in food system application. Food Control 54, 150–157. https://doi.org/10.1016/j.foodcont.2015.02.001.

Wang, Y.-C., Mohan, C.O., Guan, J., Ravishankar, C.N., Gunasekaran, S., 2018. Chitosan and gold nanoparticles-based thermal history indicators and frozen indicators for perishable and temperature-sensitive products. Food Control 85, 186–193. https://doi.org/10.1016/j.foodcont.2017.09.031.

Wang, Z., Huang, X., Jin, S., Wang, H., Yuan, L., Brash, J.L., 2019a. Rapid antibacterial effect of sunlight-exposed silicon nanowire arrays modified with Au/Ag alloy nanoparticles. J. Mater. Chem. B 7 (40), 6202–6209. https://doi.org/10.1039/c9tb01472a.

Werner, B.G., Wu, J.Y., Goddard, J.M., 2019. Antimicrobial and antifouling polymeric coating mitigates persistence of *Pseudomonas aeruginosa* biofilm. Biofouling 35 (7), 785–795. https://doi.org/10.1080/08927014.2019.1660774.

Wilson, D.I., 2018. Fouling during food processing – progress in tackling this inconvenient truth. Curr. Opin. Food Sci. 23, 105–112. https://doi.org/10.1016/j.cofs.2018.10.002.

Xia, M.Y., Xie, Y., Yu, C.H., Chen, G.Y., Li, Y.H., Zhang, T., Peng, Q., 2019. Graphene-based nanomaterials: the promising active agents for antibiotics-independent antibacterial applications. J. Control. Release 307, 16–31. https://doi.org/10.1016/j.jconrel.2019.06.011.

Xiao, Y., Jiang, S.C., Wang, X., Muhammad, T., Song, P., Zhou, B., et al., 2020. Mitigation of biofouling in agricultural water distribution systems with nanobubbles. Environ. Int. 141. https://doi.org/10.1016/j.envint.2020.105787, 105787.

Yadav, S., Mehrotra, G.K., Dutta, P.K., 2021. Chitosan based ZnO nanoparticles loaded gallic-acid films for active food packaging. Food Chem. 334. https://doi.org/10.1016/j.foodchem.2020.127605, 127605.

Yahiaoui, F., Benhacine, F., Ferfera-Harrar, H., Habi, A., Hadj-Hamou, A.S., Grohens, Y., 2015. Development of antimicrobial PCL/nanoclay nanocomposite films with enhanced mechanical and water vapor barrier properties for packaging applications. Polym. Bull. 72 (2), 235–254. https://doi.org/10.1007/s00289-014-1269-0.

Yang, T., Huang, H., Zhu, F., Lin, Q., Zhang, L., Liu, J., 2016. Recent progresses in nanobiosensing for food safety analysis. Sensors 16 (7), 1118. https://doi.org/10.3390/s16071118.

Yang, Y., Liu, H., Wu, M., Ma, J., Lu, P., 2020. Bio-based antimicrobial packaging from sugarcane bagasse nanocellulose/nisin hybrid films. Int. J. Biol. Macromol. 161, 627–635. https://doi.org/10.1016/j.ijbiomac.2020.06.081.

Yassin, M.A., Elgorban, A.M., El-Samawaty, A., Almunqedhi, B.M.A., 2021. Biosynthesis of silver nanoparticles using *Penicillium verrucosum* and analysis of their antifungal activity. Saudi J. Biol. Sci. 28 (4), 2123–2127. https://doi.org/10.1016/j.sjbs.2021.01.063.

Yemmireddy, V.K., Farrell, G.D., Hung, Y.C., 2015. Development of titanium dioxide (TiO$_2$) nanocoatings on food contact surfaces and method to evaluate their durability and photocatalytic bactericidal property. J. Food Sci. 80 (8), N1903–N1911. https://doi.org/10.1111/1750-3841.12962.

Yemmireddy, V.K., Hung, Y.-C., 2015. Effect of binder on the physical stability and bactericidal property of titanium dioxide (TiO$_2$) nanocoatings on food contact surfaces. Food Control 57, 82–88. https://doi.org/10.1016/j.foodcont.2015.04.009.

Yemmireddy, V.K., Hung, Y.-C., 2017. Using photocatalyst metal oxides as antimicrobial surface coatings to ensure food safety—opportunities and challenges. Compr. Rev. Food Sci. Food Saf. 16 (4), 617–631. https://doi.org/10.1111/1541-4337.12267.

Yi, C., Yu, Z., Ren, Q., Liu, X., Wang, Y., Sun, X., et al., 2020. Nanoscale ZnO-based photosensitizers for photodynamic therapy. Photodiagn. Photodyn. Ther. 30. https://doi.org/10.1016/j.pdpdt.2020.101694, 101694.

Yilmaz Atay, H., 2020. Antibacterial activity of chitosan-based systems. In: Functional Chitosan: Drug Delivery and Biomedical Applications. Springer, pp. 457–489, https://doi.org/10.1007/978-981-15-0263-7_15.

Yilmaz, M.T., İspirli, H., Taylan, O., Dertli, E., 2020a. Synthesis and characterisation of alternan-stabilised silver nanoparticles and determination of their antibacterial and antifungal activities against foodborne pathogens and fungi. LWT 128. https://doi.org/10.1016/j.lwt.2020.109497.

Yilmaz, M.T., Taylan, O., Karakas, C.Y., Dertli, E., 2020b. An alternative way to encapsulate probiotics within electrospun alginate nanofibers as monitored under simulated gastrointestinal conditions and in kefir. Carbohydr. Polym. 244. https://doi.org/10.1016/j.carbpol.2020.116447, 116447.

Yin, I.X., Zhang, J., Zhao, I.S., Mei, M.L., Li, Q., Chu, C.H., 2020. The antibacterial mechanism of silver nanoparticles and its application in dentistry. Int. J. Nanomedicine 15, 2555–2562. https://doi.org/10.2147/IJN.S246764.

Yu, Q., Li, J., Zhang, Y., Wang, Y., Liu, L., Li, M., 2016. Inhibition of gold nanoparticles (AuNPs) on pathogenic biofilm formation and invasion to host cells. Sci. Rep. 6, 26667. https://doi.org/10.1038/srep26667.

Yu, Y., Zheng, J., Li, J., Lu, L., Yan, J., Zhang, L., Wang, L., 2021. Applications of two-dimensional materials in food packaging. Trends Food Sci. Technol. 110, 443–457. https://doi.org/10.1016/j.tifs.2021.02.021.

Zambrano-Zaragoza, M.L., Gonzalez-Reza, R., Mendoza-Munoz, N., Miranda-Linares, V., Bernal-Couoh, T.F., Mendoza-Elvira, S., Quintanar-Guerrero, D., 2018. Nanosystems in edible coatings: a novel strategy for food preservation. Int. J. Mol. Sci. 19 (3), 705. https://doi.org/10.3390/ijms19030705.

Zhang, R., Lan, W., Ji, T., Sameen, D.E., Ahmed, S., Qin, W., Liu, Y., 2021. Development of polylactic acid/ZnO composite membranes prepared by ultrasonication and electrospinning for food packaging. LWT Food Sci. Technol. 135. https://doi.org/10.1016/j.lwt.2020.110072, 110072.

Zhang, X., Liu, Y., Yong, H., Qin, Y., Liu, J., Liu, J., 2019. Development of multifunctional food packaging films based on chitosan, TiO$_2$ nanoparticles and anthocyanin-rich black plum peel extract. Food Hydrocoll. 94, 80–92. https://doi.org/10.1016/j.foodhyd.2019.03.009.

Zheng, L., Sundaram, H.S., Wei, Z., Li, C., Yuan, Z., 2017. Applications of zwitterionic polymers. React. Funct. Polym. 118, 51–61. https://doi.org/10.1016/j.reactfunctpolym.2017.07.006.

Zhu, Z., Cai, H., Sun, D.-W., 2018. Titanium dioxide (TiO$_2$) photocatalysis technology for nonthermal inactivation of microorganisms in foods. Trends Food Sci. Technol. 75, 23–35. https://doi.org/10.1016/j.tifs.2018.02.018.

Ziyaina, M., Rasco, B., Coffey, T., Ünlü, G., Sablani, S.S., 2019. Colorimetric detection of volatile organic compounds for shelf-life monitoring of milk. Food Control 100, 220–226. https://doi.org/10.1016/j.foodcont.2019.01.018.

Zorraquin-Pena, I., Cueva, C., Bartolome, B., Moreno-Arribas, M.V., 2020. Silver nanoparticles against foodborne bacteria. Effects at intestinal level and health limitations. Microorganisms 8 (1). https://doi.org/10.3390/microorganisms8010132.

Electroconductive nanofibrillar biocomposite platforms for cardiac tissue engineering

Tarun Agarwal[a,*], *Sheri-Ann Tan*[b,*], *Lei Nie*[c,*],
Ensieh Zahmatkesh[d,*], *Aafreen Ansari*[e], *Niloofar Khoshdel*
Rad[d], *Ibrahim Zarkesh*[d], *Tapas Kumar Maiti*[a],
and Massoud Vosough[d,f]

[a]Department of Biotechnology, Indian Institute of Technology, Kharagpur, West Bengal, India
[b]Department of Bioscience, Faculty of Applied Sciences, Tunku Abdul Rahman University
College, Kuala Lumpur, Malaysia [c]College of Life Sciences, Xinyang Normal University, Xinyang,
China [d]Department of Stem Cells and Developmental Biology, Cell Science Research Center,
Royan Institute for Stem Cell Biology and Technology, ACECR, Tehran, Iran [e]Department of
Biotechnology and Medical Engineering, National Institute of Technology Rourkela, Rourkela,
Orissa, India [f]Department of Regenerative Medicine, Cell Science Research Centre, Royan
Institute for Stem Cell Biology and Technology, ACECR, Tehran, Iran

*Equal contribution.

Abbreviations

3D	three-dimensional
Anf	atrial natriuretic factor
ATP2a2	sarco/endoplasmic reticulum
AuNP	gold nanoparticle
BSA	bovine serum albumin
Ca2+	ATPase (SERCA) isoform SERCA2
Ca^{2+}	calcium ions
CD	carbon dots
CM	cardiomyocytes
CND	carbon nanodots
CNT	carbon nanotubes
CQD	carbon quantum dots
CS	chitosan
cTnC	cardiac troponin C
cTnI	cardiac troponin I
cTnT	cardiac troponin T
CVD	cardiovascular disease
Cx43	connexin 43
ECM	extracellular matrix
ESC	embryonic stem cell
FDA	Food and Drug Administration
GATA4	GATA binding protein 4
GelMA	gelatin methacryloyl
GO	graphene oxide
GQD	graphene quantum dots
HA	hyaluronic acid
hCPC	human cardiac progenitor cells
hiPSC	human induced pluripotent stem cells
HPLys	hyperbranched poly-L-lysine
HUVEC	human umbilical vein endothelial cells
Hz	hertz
MHC	myosin heavy chain
MI	myocardial infarction
MoS_2	molybdenum disulfide
MSC	mesenchymal stem cells
MWCNT	multiwalled carbon nanotubes
Nkx2.5	NK2 homeobox 5
PANi	polyaniline
PCL	polycaprolactone

PD	polymer dots
PEDOT	poly (3,4-ethylenedioxythiophene)
PEO	poly (ethylene oxide)
PGS	poly (glycerol sebacate)
PLA	poly L-lactic acid
PLGA	poly (lactic-*co*-glycolic acid)
PPy	polypyrrole
PSCs	pluripotent stem cells
PSS	polystyrene sulfonate
PT	polythiophene
PU	polyurethane
PVA	polyvinyl alcohol
rGO	reduced graphene oxide
SF	silk fibroin
SWCNT	single-walled carbon nanotubes
TBX-18	T-box transcription factor 18
TMDC	transition metal dichalcogenide

1 Introduction

The adult human heart is a complex physiological pump that delivers blood through the circulatory system (Venugopal et al., 2012). Such constant pumping activity is attributed to the tightly coordinated conduction of impulses, leading to the generation of action potential and contraction in the cardiac muscle cells, cardiomyocytes (CMs). Any acute or chronic impairment of the heart muscle is considered a prognosis of heart failure.

Myocardial infarction (MI) is a type of cardiac-related disease; wherein a specific part of the cardiac tissue undergoes irreversible necrosis, due to the prolonged lack of oxygen supply (Ferrini et al., 2019; Wang and Guan Jianjun, 2010). Owing to the negligible proliferative capability of the CMs, the damaged cardiac tissue is not repaired; instead, it becomes populated by fibroblast-like cells that deposit collagen, thereby forming a scar (Jiang and Lian, 2020; Wang and Guan Jianjun, 2010). Such scared tissue imparts stiffness to the organ, reduces transmission of electrical pulses, and alteration in muscle contractile behavior, leading to a decrease in the overall organ functionality (Monteiro et al., 2017). The unavailability of readily accessible transplants and the incapability of existing clinical treatments to reestablish functional myocardial tissue have become potent challenges for designing a practical therapeutic strategy.

Two of the potential approaches for restoration of damaged myocardium are direct injection of fresh cells to the site of injury and the replacement of the impaired myocardium with tissue-engineered implants (Domenech et al., 2016; Mazzola and Di Pasquale, 2020; Monteiro et al., 2017). Several research groups have tried injecting fresh cells to the site of damage, but the success rate of this strategy is limited by cell death, poor localization of cells at the site of damage, and high rejection rate by recipient heart tissue (Chaudhuri et al., 2017; Oh et al., 2016). In the scope of addressing these complications, tissue engineering approaches are being explored widely.

Tissue engineering incorporates biological and medical concepts into engineering ideas. Rather than general medication, it utilizes biomaterials for facilitating the regenerative abilities of the organ (Fleischer and Dvir, 2013; Schwach and Passier, 2019; Vunjak-Novakovic et al., 2010). In order to stimulate the regenerative capacity of the heart, bioengineered scaffolds are designed. These scaffolds support and protect cells and provide native-like, three-dimensional (3D) geometry and biophysicochemical cues that modulate their phenotypic characteristics (Ghafar-Zadeh et al., 2011). To date, various natural and synthetic polymers like collagen, gelatin, alginate, polycaprolactone (PCL), poly(lactic-*co*-glycolic acid) (PLGA), chitosan, fibrin, polyurethane (PU), and decellularized cardiac-specific extracellular matrix (ECM) have been explored to engineer cardiac tissue constructs (Domenech et al., 2016; Venugopal et al., 2012). Integrating biophysical cues, including nanoscale topology, anisotropy, stiffness, and electrical stimulus, individually and in combination, have been shown to further aid in improving the regenerative efficiencies of the outcomes (Chan et al., 2013; Chen et al., 2019a; Hernández et al., 2016; Stoppel et al., 2016; Thavandiran et al., 2013).

In particular, in the last 10 years, a huge wave of enthusiasm for developing electroactive nanofibrillar tissue constructs for cardiac regenerative therapy has been observed. With this perspective, in the current chapter, we have reviewed the current advances in the domain.

2 Nanotopologies and electrical stimulation—Intrinsic biophysical determinant of CMs

2.1 Nanotopology

Cell-material interactions are one of the key criteria for designing an improved scaffold and its integration with the in vivo organs. The majority of the biological reactions occur on the surface interface, making the constructs' surface properties a vital feature (Salmasi, 2015). The cell adhesion, its binding strength, spreading ability, shape, and fate are governed by various biomaterial characteristics such as surface topology, chemistry, and mechanical properties (Ermis et al., 2018). Among them, the micro/nano-surface architecture of biological scaffolds is considered an essential tissue engineering criterion and has been explored widely (Phong et al., 2010; Tay et al., 2011).

As regards to the myocardial tissue, it is composed of complex and highly aligned cellular and extracellular matrix organization, consisting of various nanofibrous elements (like elastin and collagen), which significantly regulate the cellular activity (Sharma et al., 2019; Thavandiran et al., 2013). Besides, the integration of nanoscale components in 3D cardiac constructs allows the action potentials to propagate through the CMs quickly and enables actively coordinated contractions in the engineered cardiac constructs. Furthermore, these structures regulate the growth and differentiation of the CMs during the regeneration process (Fleischer and Dvir, 2013; Kankala et al., 2018). Besides, the recent research trend has been directed toward developing anisotropic fibrous scaffolds over hydrogels that can better imitate the micro and nanoscale fibrous structure of the myocardium (Lei et al., 2019; Wang et al., 2013).

2.2 Electrical stimulation

The surface potential generated by the heart is far greater than skeletal or nerve-muscle, making it the largest bioelectrical source (Hart and Gandhi, 1998). The electrically simulative aspect of the heart has widely been explored by researchers for investigating defibrillation, pacing, arrhythmia, and cardiac development (Tandon et al., 2009). The myocardium depends on the complex coordination of Ca^{2+} signal transmission and action potential for the synchronous heartbeat and constant blood circulation. The rhythmic contraction of the heart is achieved by the synergistic effect of intracellular junctions, cardiac pacemaker conduction system, and membrane potential depolarization (Anderson et al., 2009; Solazzo et al., 2019). The electroconductive value of the native cardiac tissue falls in the range of 0.005–0.16 S/m (Mostafavi et al., 2020). Moreover, electrical signals play a major role in the development of the embryonic heart; pulsatile signals aid in the cardiac syncytium development; whereas the direct current signals are associated with cell relocation during cardiac left-right asymmetry and primitive streak formation (Levin, 2003; Nuccitelli, 1992; Ypey et al., 1979).

Ischemic insult of myocardial muscle results in the replacement of the functional myocardium by a fibrotic scar which is comparatively nonelectroconductive and isolates the distant CMs, disrupting the communication of the tissue (Thygesen et al., 2007). Such disruption of cellular communication results in organ failure. Implantation of electroactive cardiac patches in vivo, in this regard, has been shown to improve regenerative and reparative outcomes.

Besides, exogenous electrical stimulation of cardiac cell culture has been found to be necessary for CMs maturation. Such stimulation can help the cells to achieve a relatively native structure and phenotype (Lasher et al., 2012; Radisic et al., 2004; Tandon et al., 2009). It has been reported that a low frequency of symmetric biphasic square pulse leads to a higher cell density, increased expression of connexin 43 (Cx43), and decreased excitation threshold, as compared to nonstimulated cells and monophasic pulses (Chiu et al., 2011). Also, an ordered increase in the electrical stimulation frequency over 1 week (1–3 and 1–6 Hz) leads to better maturation of human pluripotent stem cells (PSCs)-derived CMs (Sun and Nunes, 2016).

3 Strategies for fabricating electroactive nanofibrous platforms

Electrospinning is considered a straightforward strategy for developing nanofibrous scaffolds and has witnessed huge applicability in the realm of tissue engineering and regenerative medicines (Murugan and Ramakrishna, 2007). This technique, principally, employs the polymer solution to make ultrafine fibers ranging from several micrometers to several hundreds of nanometers using a high-voltage power supply, and the volume feed rate can be controlled by a capillary pump (Park, 2011). The technique is compatible with a wide range of biopolymers, including polyvinyl alcohol (PVA), chitosan (CS), silk fibroin (SF), gelatin, and others; however, it is still unsuitable while considering the addition of living cells during the fabrication process. This is due to the use of toxic organic solvents in the process.

As regard to imparting electroconductive properties, electroactive materials are generally incorporated into the scaffolds. Intrinsically conductive polymers such as polyaniline (PANi), polypyrrole (PPy), poly(3,4-ethylenedioxythiophene) (PEDOT), and polythiophene (PT) are often used for the development of electroconductive cardiac scaffolds. Moreover, conductive nanomaterials such as graphene, graphene oxide, carbon nanotubes, metal nanoparticles, transition metal dichalcogenides (TMDCs) could also be combined with nonconductive biomaterials to give extrinsically conductive materials (Ashtari et al., 2019; Solazzo et al., 2019).

For the fabrication of nanofibrous constructs with electroactive properties, three potential strategies could be opted—(i) direct blending (Li et al., 2006; Tondnevis et al., 2020), (ii) chemical conjugation with electrospun scaffolds (He et al., 2018), and (iii) physical coating over electrospun constructs (Zhao et al., 2018).

Besides electrospinning, recently developed solution-based and melt-based electrohydrodynamic printing strategies have also been shown to develop nanofibrous electroactive scaffolds. Principally, this strategy relies on the electrohydrodynamically induced flow of materials to fabricate user-specific constructs with micro/nanoscale features or patterns (Lei et al., 2019; Zhang et al., 2016).

4 Recent developments in electroconductive nanofibrillar platforms for CTE

As highlighted in the above sections, nanotopology and electroconductivity are imperative features for efficient cardiac regeneration. This section would discuss the current developments in the realm of electroconductive nanofibrillar platforms, highlighting cellular responses both in vitro and in vivo. The discussion is subclassified based on the type of electroactive material employed and summarized in Table 1.

4.1 Carbon nanofibers and carbon nanotubes

Carbon nanofibers (CNFs) and carbon nanotubes (CNT) are both allotrope forms of carbon. CNFs have average diameters ranging from 125 to 150 nm, and lengths ranging from 50 to 100 μm (Mombini et al., 2019). CNFs have a smaller nanofiber in diameter than that of conventional carbon fibers. CNFs have excellent electrical, chemical, and physical properties, including chemical stability and inertness in a physiological environment, biocompatible, electrically robust, and conductive for signal detection, high surface-to-volume ratio, and high spatial resolution (Zhang et al., 2012). On the other hand, CNT is also efficient in endowing electrical transport potential to low-conductivity matter (Lekawa-Raus et al., 2014). CNT can be either single-walled (SWCNTs) or multiwalled (MWCNTs); each of these materials has slightly varied properties depending on the available surfaced number in CNTs (Veetil and Ye, 2009).

PLGA/CNF composite scaffolds, containing varied PLGA: CNF ratio (100:0, 75:25, 50:50, 25:75, and 0:100 wt%) were fabricated and tested for CTE (Stout et al., 2011). The results suggested that the incorporation of CNF improved the electrical properties of the scaffolds. CMs cultured on all the conductive scaffolds showed higher cell adhesion density and proliferative potential than nonconductive bare PLGA scaffolds. However, proliferative behavior was highly dependent on the diameter of the CNF used. As an extension of this study, the authors upgraded the scaffolds by modulating the alignment of CNFs in PLGA (Webster et al., 2014). This modification significantly enhanced the mechanical strength and electrical conductivity of the fabricated composite. Furthermore, adhesion and proliferation of CMs were remarkably improved on scaffolds with aligned CNF than those with random CNF orientation.

TABLE 1 Electroconductive nanofibrous scaffolds for cardiac tissue engineering.

Electroconductive material	Biopolymeric formulation	Fabrication method	Size of fibrils	Cells used	In vitro responses on conductive scaffolds	In vivo of conductive scaffolds	Response to an external field (if applied)	References
CNF	Gelatin	Electrospinning (polymeric blend)	–	Rat CMs	Upregulation of the cardiac genes (Actin4, and Cx43)	Subcutaneous implantation—improved angiogenesis and cell infiltration	–	Mehrabi et al. (2020)
CNT	PGS and gelatin	Electrospinning (polymeric blend)	167 ± 82 nm	Rat CMs	Good cell alignment, spontaneous and synchronous contractile function, improved expression of Cx43 expression	–	–	Kharaziha et al. (2014)
Single wall CNT	PU and gelatin	Electrospinning (polymeric blend)	140 nm	Rat CMs and HUVEC	Improved proliferation and spreading	–	–	Tondnevis et al. (2020)
Multiwall CNT	PU	Electrospinning (PU) and simultaneous electrospraying (CNT)	94–113 nm	H9C2 cells and HUVEC	Improved cell adhesion and proliferation	–	–	Shokraei et al. (2019)
CNT (suspended in gum Arabic)	SF and PEO	Electrospinning (polymeric blend)	~400 nm	Neonatal rat CMs	Enhanced expression of cardiac markers (α-actinin, cTnI, Cx43)	–	–	Zhao et al. (2020)
Single wall CNT	Chitosan and PVA	Electrospinning (polymeric blend)	292–307 nm	MSCs	Cells expressed cardiac markers (Nkx2.5, cTnI, and MHC-β)	–	Enhanced expression of cardiac markers upon stimulation	Mombini et al. (2019)

Continued

TABLE 1 Electroconductive nanofibrous scaffolds for cardiac tissue engineering—cont'd

Electroconductive material	Biopolymeric formulation	Fabrication method	Size of fibrils	Cells used	In vitro responses on conductive scaffolds	In vivo of conductive scaffolds	Response to an external field (if applied)	References
Multiwall CNT	PLGA	Electrospinning (polymeric blend)	–	Neonatal rat CMs	Enhanced cell elongation, improved expression of cardiac markers (α-actinin and cTnI)	–	–	Liu et al. (2016)
Graphene	Gelatin and PCL	Electrospinning (polymeric blend)	551–595 nm	Neonatal rat CMs	Improved expression of cardiac markers (Cx43 and cTnT)	No toxicity observed	–	Chen et al. (2019b)
Graphene	PCL	Electrospinning (polymeric blend)	690 nm	Mouse ESCs-derived CMs	Enhanced expression of cardiac markers (MHC, cTnT, Cx43) and improved calcium handling	–	–	Hitscherich et al. (2018)
rGO	SF/PEO	Electrospinning (SF/PEO) and coating-reduction (rGO)	–	Neonatal rat CMs	Improved expression of cardiac markers (α-actinin, cTnI, and Cx43)	–	Enhanced expression of cardiac marker and beating characteristics upon electrical stimulation	Zhao et al. (2018)
GO	PET	Electrospinning (polymeric blend)	147 nm (solid) and 253 nm (core-shell)	H9C2 and HUVEC	Enhanced cell adhesion and spreading	–	–	Ghasemi et al. (2019)
MoS2 or rGO	SF/PEO	Electrospinning (polymeric blend)	450 nm	Human iPSCs (TBX-18 transfected)	Upregulation of cardiac functional genes (GATA4, cTnT, and MHC-α)	–	–	Nazari et al. (2020)

rGO-Ag	PU	Electrospinning (polymeric blend)	284–369 nm	Human cardiac progenitor cells (hCPCs)	Upregulation of cardiac functional genes (TBX-18, cTnT, and MHC-α)	—	Nazari et al. (2019b)
rGO	PEA or PEA-chitosan	Electrospinning (polymeric blend)	50–700 nm	iPSC-derived MSCs	Enhanced GATA-4 and Nkx2.5 expression upon differentiation	—	Stone et al. (2019)
rGO	PU	Electrospinning (polymeric blend)	246 nm (aligned)—383 nm (random)	Mouse muscle satellite cells	Improved cell anchoring and enhanced cTnI expression	—	Azizi et al. (2019)
CQD (p-phenylenediamine functionalized)	SF and PLA	Electrospinning (polymeric blend)	—	Rat CMs	Improved cardiac marker genes (cTnC, cTnT, Cx43, Anf, and ATP2a2)	—	Yan et al. (2020)
PPy	Gelatin and PCL	Electrospinning (polymeric blend)	191–239 nm	Rabbit CMs	Enhanced cell proliferation and Cx43 gene expression	—	Kai et al. (2011)
PPy (GelMA-PPy NPs)	GelMA and PCL	Electrospinning (GelMA-PCL blend) and conjugation of NPs	948 nm	Rat CMs, HUVEC	Improved expression of cardiac markers (Cx-43 and α-actinin), spontaneous beating patterns, improved angiogenesis phenotype	Enhanced recovery and cardiac functionality post-MI, revascularization observed	He et al. (2018)

Continued

TABLE 1 Electroconductive nanofibrous scaffolds for cardiac tissue engineering—cont'd

Electroconductive material	Biopolymeric formulation	Fabrication method	Size of fibrils	Cells used	In vitro responses on conductive scaffolds	In vivo of conductive scaffolds	Response to an external field (if applied)	References
PANi	Gelatin	Electrospinning (polymeric blend)	61–600 nm	H9C2 cells	Supported cell proliferation	—	—	Li et al. (2006)
PANi	PLA	Electrospinning (polymeric blend)	~500 nm	H9C2 cells and Neonatal rat CMs	Better myotube formation and cell elongation upregulated gene expression (α-actinin, Cx43), higher and more synchronous beating	—	—	Wang et al. (2017)
PANi	PLGA	Electrospinning (polymeric blend)	59–185 nm	Neonatal rat CMs	Expression of cardiac markers (cTnI and Cx43) observed	—	Improved beating pattern under electrical stimulation	Hsiao et al. (2013)
PANi	Collagen and HA	Electrospinning (polymeric blend)	120 nm	Neonatal rat CMs	Observable longer contraction time, higher contractile amplitude, and lower beating rates, upregulation of Cx43 expression	—	—	Roshanbinfar et al. (2020)
PANi	PES	Electrospinning (polymeric blend)	Aligned nanofiber = 294 ± 66 nm; random nanofiber = 267 ± 67 nm	iPSCs (CVD patient-derived)	Upregulation of cardiac markers (Nkx2.5, GATA-4, Anf, and cTnT)	—	Electrical stimulation improved the differentiation efficiency	Mohammadi Amirabad et al. (2017)

PEDOT:PSS-PEO	PCL	Solution-based (PEDOT:PSS-PEO) and melt-based (PCL) EHD printing	470 nm (PEDOT:PSS-PEO conductive fiber)	Neonatal rat CMs	Upregulated cardiac gene (α-actinin and Cx43) expression, increased beating frequencies	–	Lei et al. (2019)
PEDOT:PSS	Chitosan and PVA	Electrospinning (polymeric blend)	84–117 nm	Rat bone marrow MSC	Improved cell proliferation	–	Abedi et al. (2019)
AuNP	PCL, SF, vitamin B12, *Aloe vera*	Electrospinning (polymeric blend)	229 nm	Rabbit CMs and MSCs	Trans-differentiation of MSCs into the cardiac lineage, upregulation of cardiac marker (actinin and cTnT)	–	Sridhar et al. (2015)
AuNP	PVA and BSA	Electrospinning (polymeric blend)	273–278 nm	Human MSCs	Enhanced cardiomyogenic differentiation with expression of genes (α-actinin, cTnT, and Cx43)	–	Ravichandran et al. (2014)
AuNP	PCL	Electrospinning (PCL) and e-beam evaporation (Au)	–	Neonatal rat CMs	Aligned and elongated cellular morphology, increased actinin striation, higher contraction rate	–	Fleischer et al. (2014)

Continued

TABLE 1 Electroconductive nanofibrous scaffolds for cardiac tissue engineering—cont'd

Electroconductive material	Biopolymeric formulation	Fabrication method	Size of fibrils	Cells used	In vitro responses on conductive scaffolds	In vivo of conductive scaffolds	Response to an external field (if applied)	References
Melanin	PLCL and gelatin	Electrospinning (polymeric blend)	153–366 nm	Human CMs	Enhanced expression of cardiac genes (α-actinin and Cx43)	–	Electrical stimulation improved early proliferation	Kai et al. (2013)
MoS$_2$	Nylon6	Electrospinning (polymeric blend)	161 nm	Mouse embryonic cardiac cells	Upregulation of cardiac markers (GATA-4, cTnT, Nkx2.5, and MHC-α)	–	–	Nazari et al. (2019a)

An electrospun gelatin/CNF cardiac patch, with randomly oriented nanofibers, were prepared using electrospinning (Mehrabi et al., 2020). These conductive scaffolds exhibited good biocompatibility and supported a higher differentiated phenotype of CMs, as revealed by higher expression of cardiac-specific genes. Interestingly, subcutaneous implantation of this patch in mice showed higher vascularization than pure gelatin scaffolds.

A cardiac electrospun scaffold made from poly(glycerol sebacate) (PGS), gelatin, and CNTs (coated with methacrylated gelatin (GelMA)) were shown to display good strength, flexibility, and electroconductivity (Kharaziha et al., 2014). Rat CMs cultured on these scaffolds were able to proliferate rapidly without demonstrating any cytotoxic effects. Moreover, cell alignment, upregulated cardiac markers, and spontaneous-synchronous contractile activities were evident features of the study. Notably, CNT concentration greatly affected cellular activities.

A combined electrospinning and electrospraying method was developed to fabricate PU/MWCNT scaffolds (Shokraei et al., 2019). This scaffold had randomly arranged fibrillar morphology and exhibited electroactive properties while conferring high cellular adhesion, viability, and proliferation of H9C2 and HUVEC cells. Another composite scaffold was tested in the same year, consisting of PVA, chitosan, and CNT (at three different concentrations: 1, 3, and 5%) (Mombini et al., 2019). The results suggested that the scaffold with the lowest CNT concentration (1%) was nontoxic to the mesenchymal stem cells (MSCs) but a higher concentration of CNTs showed cytotoxic effects. Besides, these conductive scaffolds promoted cardiomyogenic differentiation of MSCs. External electrical stimulation was further shown to augment the differentiation potency of MSCs.

Due to the intrinsic hydrophobic nature of the CNTs, they are often dispersed in organic solvents or surfactants, which have cytotoxic effects. In order to overcome this limitation, recently, FDA-approved gum Arabic solution was used to disperse CNTs. The CNT dispersed solution was then added to SF/poly (ethylene oxide) (PEO) solution and further electrospun to prepare mats with either random or aligned fibrillar arrangement (Zhao et al., 2020). The fabricated scaffold demonstrated improved mechanical and electroactive properties. Rat CMs seeded on the CNT/SF material exhibited good biocompatibility and improved cardiac marker expression. Moreover, CMs on the aligned scaffolds had augmented cell alignment and gene expression profiles.

4.2 Graphene and its oxides

Graphene, graphene oxide (GO), and reduced graphene oxide (rGO) belong to the family of two-dimensional nanomaterials. Graphene is an atomically-thin, 2D sheet of sp^2 carbon atoms in a honeycomb structure (Chen et al., 2019b; Hitscherich et al., 2018). Graphene has widely been used to prepare polymer-based nanocomposites due to its high mechanical strength, electrical conductivity, and molecular barrier abilities; however, the use of graphene is still challenging because of its poor solubility, leading to agglomeration in solution (Cheng et al., 2018). In this regard, its oxide forms, GO and rGO, have gained popularity lately. The structure of GO is similar to graphene but it has multiple oxygen-based functional groups, such as hydroxyl (—OH), alkoxy (C—O—C), carbonyl (C=O), and carboxylic acid (—COOH) (Wright et al., 2019). The possibility of surface functionalization of GO has made this material more promising to be incorporated into nanocomposites compared to plain graphene. However, GO is shown to have high toxic effects in its pristine form (Makvandi et al., 2020; Vuppaladadium et al., 2020). In contrast, in its reduced form, rGO is less toxic but has limited solubility in the aqueous medium due to its hydrophobic character (Makvandi et al., 2020; Zhao et al., 2018). Nevertheless, all these materials have been shown to possess the potential to regulate cellular behavior. Owing to these aspects, they have found broad applicability in the biomedical sector (Shang et al., 2019).

Recently, graphene, at different percentages (0.3%, 0.5%, 0.8% and 1.0%), was blended into the gelatin/PCL formulation and electrospun to develop scaffolds for CTE (Chen et al., 2019b). The addition of graphene improved scaffolds' conductive properties in a concentration-dependent manner; however, significant cytotoxic responses were observed in neonatal rat ventricular CMs, when cultured on substrates with higher graphene concentration (0.8%–1%). The cells cultured on scaffolds with 0.5% graphene in vitro, retained the expression of cardiac-specific markers. When implanted in rats in vivo, no toxicity was detected even after 12 weeks. Infiltration of the surrounding cells was also evident in gelatin/PCL/graphene hybrid scaffolds.

A nanocomposite conductive scaffold produced by electrospinning PCL and graphene demonstrated promising results in CTE (Hitscherich et al., 2018). The fabricated scaffolds showed no adverse effects on mouse ESC-CMs and improved the expression of cardiac-specific markers with retention of their intrinsic contractile nature. Fiber alignments in the scaffolds critically influenced cellular behavior.

In another study, rGO was uniformly coated on electrospun SF using the unique vacuum filtration method (Zhao et al., 2018). The reduction process was carried out using ascorbic acid treatment. The developed nanocomposite was found to be nontoxic toward neonatal rat CMs. The results indicated that nanocomposite scaffolds with rGO density between 0.005 and 0.01 mg/cm^2 significantly improved gene expression profiles and beating patterns compared to nonconductive ones. The study further highlighted the augmentative effects of fiber alignment and electrical stimulation on cell behavior. Notably, electrical stimulation in the aligned scaffolds improved cells' functional phenotype, while the randomly aligned constructs were associated with decreased functionalities compared to nonstimulated controls.

4.3 Carbon quantum nanodots

Carbon quantum dots (CQDs), also named carbon dots (CDs), are novel zero-dimensional carbon-based nanomaterials with strong fluorescence characteristics (Wang et al., 2019). CQDs include graphene quantum dots (GQDs), carbon nanodots (CNDs), and polymer dots (PDs) (Wang et al., 2019). CQDs possess a large number of oxygen-containing groups on the surface. They are featured with good cytocompatibility, good water solubility, chemical stability, photo-bleaching resistance, and ease in surface functionalization (Wang et al., 2019). Compared to graphene, CQDs are much more hydrophilic and have higher electrocatalytic activity.

In a recent study, *p*-phenylenediamine surface-functionalized CQDs were loaded into SF/PLA formulation and then electrospun to form conductive nanofibrous scaffolds for CTE (Fig. 1; Yan et al., 2020). The CQD integrated construct's young modulus increased by threefold compared to the bare SF/PLA scaffold. H9C2 rat CMs were able to attach and proliferate on the nanocomposite scaffolds. Moreover, enhanced cardiac-specific gene profiles were observed on the CQD@SF/PLA material.

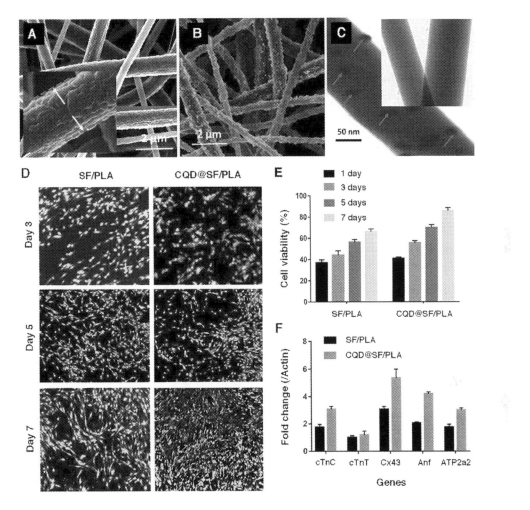

FIG. 1   Transmission electron microscopy (TEM) images of (A) SF/PLA and (B, C) CQD@SF/PLA nanofibrous scaffolds. *Red* (*light gray* in print version) *arrows* show the presence of CQDs in the nanofibers. (D) Representative fluorescent images of rat CMs at day 3, 5, and 7 of culture on both the substrates. (E) Quantification of cell viability of rat CMs at different days by MTT assay on both the substrates. (F) Gene expression profiling of rat CMs cultured on both the scaffolds. Gene expression was normalized to the actin. *Reproduced with permission from Yan, C., Ren, Y., Sun, X., Jin, L., Liu, X., Chen, H., Wang, K., Yu, M., Zhao, Y., 2020. Photoluminescent functionalized carbon quantum dots loaded electroactive silk fibroin/PLA nanofibrous bioactive scaffolds for cardiac tissue engineering. J. Photochem. Photobiol. B Biol. 202, 111680. https://doi.org/10.1016/j.jphotobiol.2019.111680, Elsevier B.V.*

4.4   Electroactive polymers

4.4.1   PANi

PANi belongs to a semiflexible polymer family that has captured much attention in building electroconductive nanofibrillar platforms. This is mainly due to its interesting electronic, optical, and electro-optical properties (Kaur et al., 2015). PANi has gained much popularity in biosensors advantaged by impressive signal amplification and elimination of electrode fouling (Dhand et al., 2011). Owing to their excellent electrical properties and biocompatibility, they have also been employed for CTE.

PANi could be blended with (PLA) to develop conductive nanofibrous sheets as CMs-based 3D bioactuators in both tubular and folded shapes for CTE (Wang et al., 2017). With an increase of PANi content from 0 to 3 wt%, PLA/PANi nanofibrous scaffolds' conductivity also increased. The fabricated PLA/PANi nanofibrous sheets also promoted the differentiation of H9C2 cells, cell-cell interaction, maturation, and spontaneous beating.

In another study, an aligned electrospun nanofibrous mesh of PANI and PLGA, doped with HCl, was fabricated (Fig. 2; Hsiao et al., 2013). Doping using HCl enhanced the conductivity of PANI and transformed the neutral scaffold into a positive charge. Rat CMs adhered and oriented along the fiber length and expressed cardiac-specific markers. Furthermore, external electrical stimulation was able to synchronize the beating of all isolated cellular clusters mimicking the native cardiac tissue.

Homogenous electrospun PANI nanofibers grafted with hyperbranched poly-L-lysine dendrimers (HPLys) were also tested (Fernandes et al., 2010). The diameter size of the electrospun fibers (69 to 80 nm) at PANI concentration of 1.5% w/w matched with the size of collagen fiber (50 to 500 nm), which was one of the major constituents of the ECM. The conductive scaffold was shown to be nontoxic to rat CMs and promoted their proliferation.

Recently, an electrospun mat consisting of PANi, collagen, and hyaluronic acid (HA) was developed (Roshanbinfar et al., 2020). These composite scaffolds possessed average mechanical strength and were biocompatible with neonatal rat CMs and human-iPSC-derived CMs. The cells were able to attach, proliferate, and contract on the PANI/collagen/HA nanofibrous mat. The hiPSC-derived CMs exhibited a faster beating pace but at lower amplitude when grown on these electroactive scaffolds than those cultured on the collagen/HA scaffolds.

4.4.2 PPy

Another conducting polymer experimented for engineering cardiac patches is PPy. It has wide applications due to its simple synthetic process, good electrical conductivity under physiological conditions, environmental stability, well-characterized biocompatibility in vitro and in vivo. Besides, the electrical and biological activities of this polymer could be adjusted easily through chemical modifications (He et al., 2018; Kai et al., 2011).

In a study, PPy was mixed with PCL and gelatin solution and electrospun to develop scaffolds for CTE (Kai et al., 2011). The addition of PPy was shown to reduce the fiber diameter while improving the scaffolds' mechanical and electroconductive properties. Culture of rabbit CMs on these composite scaffolds was associated with better adhesion, spreading, proliferation, and expression of cardiac-specific markers.

In the same milieu, conductive scaffolds were fabricated by chemical functionalization of GelMA/PPy nanoparticles (at various concentrations 10, 20, and 50 mg/mL) onto GelMA/PCL electrospun membranes using a mussel-inspired dopamine-$N'N'$-methylene-bisacrylamide (dopamine-MBA) crosslinker (Fig. 3; He et al., 2018). This crosslinking strategy enabled a uniform distribution of the nanoparticles on the scaffolds. As regards to the biological evaluation, the developed scaffolds showed excellent biocompatibility (regardless of the nanoparticle concentration) and were able to promote the growth and functionality of rat CMs and HUVECs in vitro. Moreover, implantation of these conductive scaffolds in MI rat

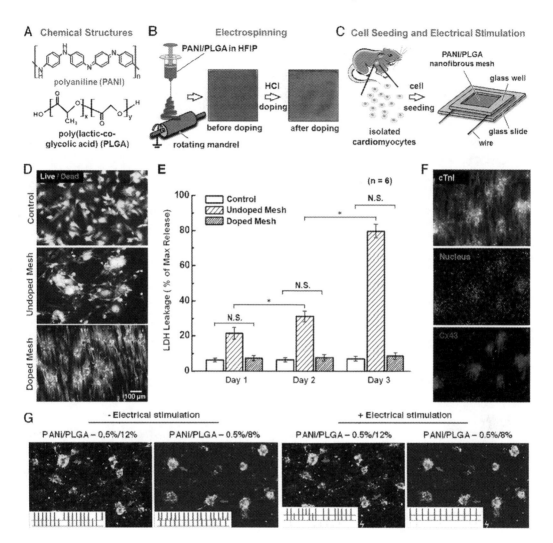

FIG. 2 (A) Chemical structure of PLGA and PANi. (B) Schematics for the preparation of PANi/PLGA scaffolds using electrospinning and HCl doping. (C) Schematics for seeding CMs on PANi/PLGA nanofibrous mesh and stimulating them with electrical impulses. (D) Live/dead images of neonatal rat CMs on undoped mesh, doped meshes, and tissue culture plates. (E) Viability assessment of neonatal rat CMs quantified by LDH assay. (F) Fluorescence micrographs of neonatal rat CMs on doped mesh, immunostained with cTnI and Cx43. Nucleus was stained with DAPI. (G) Beating frequencies of 2 CMs clusters [red (*light gray* in print version) and *blue* (*dark gray* in print version)] on low and high conductivity doped meshes with and without electrical stimulation. *Reproduced with permission from Hsiao, C.-W., Bai, M.-Y., Chang, Y., Chung, M.-F., Lee, T.-Y., Wu, C.-T., Maiti, B., Liao, Z.-X., Li, R.-K., Sung, H.-W., 2013. Electrical coupling of isolated cardiomyocyte clusters grown on aligned conductive nanofibrous meshes for their synchronized beating. Biomaterials 34(4), 1063–1072. https://doi.org/10.1016/j.biomaterials.2012.10.065, Elsevier B.V.*

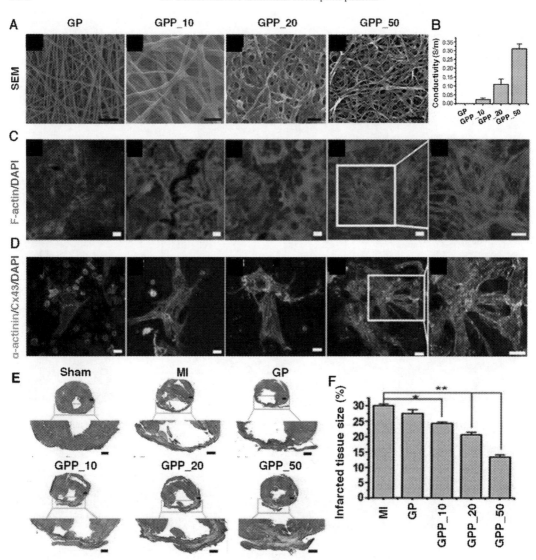

FIG. 3   (A) Scanning electron micrographs of GelMA/PCL (GP) and those containing GelMA/PPy nanoparticles at 10 (GPP_10), 20 (GPP_20), and 50 (GPP_50) mg/mL concentration. (B) Quantification of conductivity of the fabricated scaffolds. (C) Representative fluorescent micrographs of CMs cultured on nanofibrous scaffolds; stained for F-actin and nucleus. (D) Representative micrographs of CMs cultured on different scaffolds; immunostained for α-actinin and Cx43. Nucleus was stained with DAPI. (E) Masson trichrome staining of cardiac tissue [red (*light gray* in print version*)* and fibrous tissue (*blue* (*dark gray* in print version)] in different groups, namely sham, MI, and MI treated with different scaffolds. (F) Quantitative representation of percentage infarcted tissue size in different animal groups. *Reproduced from He, Y., Ye, G., Song, C., Li, C., Xiong, W., Yu, L., Qiu, X., Wang, L., 2018. Mussel-inspired conductive nanofibrous membranes repair myocardial infarction by enhancing cardiac function and revascularization. Theranostics 8(18), 5159–5177. https://doi.org/10.7150/thno.27760.*

models was associated with faster structural and functional restoration of infarcted cardiac tissue along with higher neovascularization compared to models implanted with nonconductive GelMA/PCL scaffolds.

4.4.3 PEDOT:PSS

PEDOT: polystyrene sulfonate (PSS) is another important and successfully employed conducting polymer, which possesses outstanding advantages in terms of thermal stability, tunability, and improved conductivity achieved by secondary doping (Shi et al., 2015). Furthermore, the conductivity of PEDOT: PSS could be remarkably enhanced by thermal and light treatments as well as chemical modifications using organic solvents, salts, ionic liquids, and zwitterions (Shi et al., 2015).

Chitosan/PVA scaffolds containing different PEDOT:PSS concentrations (0, 0.3, 0.6, and 1%) were fabricated through the electrospinning process (Abedi et al., 2019). Significant reduction in the fiber diameter with improved mechanical and electroconductive properties was evident with an increase in the PEDOT:PSS concentration. The conductive scaffolds supported better adhesion, spreading, and proliferation of rat bone marrow MSCs than nonconductive ones.

In order to mimic the hierarchical microarchitectures of the native myocardium, the microfibrous scaffold was prepared by combining solution-based and melt-based electrohydrodynamic printing approaches (Lei et al., 2019). The scaffold consisted of multitiered PCL layers with submicro scale PEDOT: PSS fibers inserted in between each layer of different orientations. Both H9C2 and primary CMs, on these conductive scaffolds, were able to attach and align themselves on the anisotropic structure. Moreover, cells on these conductive scaffolds were associated with enhanced expression of cardiac-specific markers and better beating frequencies and synchronous contractions than the scaffold without the PEDOT:PSS, indicating the importance of electroactive polymers for CTE.

4.5 Metal-based nanomaterials

Both noble metal- and TMDC-based nanomaterials are now being integrated into the development of electroconductive platforms for CTE application.

Gold nanoparticles (AuNPs), noble metal-based nanomaterials, can easily be synthesized using chemical, thermal, and electrochemical methods. AuNPs with sizes between 1 and 500 nm are inert, biocompatible, and can easily be modified for a wide range of applications. AuNPs are successfully used in CTE to produce a unique cellular milieu coupled with tunable conductivity. In a study, AuNPs-laden PVA/BSA electrospun nanofibrous mats with electroconductive properties were prepared (Ravichandran et al., 2014). The composite scaffolds were shown to efficiently support adhesion, growth, and cardiomyogenic differentiation of human MSCs as compared to bare PVA/BSA scaffolds. Based on another work, electroconductive PCL scaffolds were constructed and integrated with AuNPs using electron-beam evaporation (Fig. 4; Fleischer et al., 2014). Aligned & elongated cellular morphology, increased actinin striation, and higher contraction rate of neonatal rat CMs were observed on the conductive scaffolds.

As regards to TMDCs, molybdenum disulfide (MoS_2), consisting of S-Mo-S tri-layered structure, has attracted tremendous interest due to its properties such as high surface

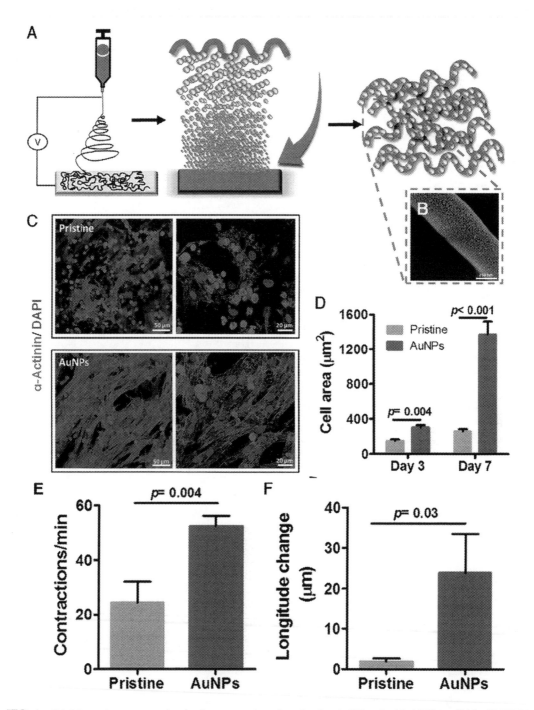

FIG. 4 (A) Schematic representation for the preparation of conductive AuNP-embedded PCL scaffolds. (B) ESEM images of the AuNP-embedded PCL fiber. (C) Representative micrographs of CMs cultured on pristine or AuNP-embedded scaffolds; immunostained for α-actinin. Nucleus was stained with DAPI. (D) Quantification of CMs area on pristine and AuNP-embedded scaffolds at day 3 and 7. (E) Contraction rate and (F) longitude change of the constructs on day 7. *Reproduced with permission from Fleischer, S., Shevach, M., Feiner, R., Dvir, T., 2014. Coiled fiber scaffolds embedded with gold nanoparticles improve the performance of engineered cardiac tissues. Nanoscale 6(16), 9410–9414. https:// doi.org/10.1039/C4NR00300D, Royal Society of Chemistry.*

area/mass ratio, good electrical conductivity, and acceptable biocompatibility (Yadav et al., 2019). MoS_2 nanosheets were mixed with nylon polymeric formulation and electrospun to produce nanofibers with strong mechanical and electroactive properties (Nazari et al., 2019a). The composite's hydrophobicity was reduced significantly using plasma treatment postfabrication, enabling the culture of the mouse embryonic cardiac cells. The cells on nylon/MoS_2 material attached, elongated, and aligned along the nanofibers better than those seeded on bare nylon nanofibers, without any evidence of cytotoxicity. Moreover, conductive scaffolds were associated with improved differentiation and maturation of cells, as observed via gene expression analysis and immunofluorescence studies.

5 Conclusion and outlook

As highlighted in the current chapter, nano-topological cues and electroconductive aspects could synergistically impact the cardiac cells' physiological and functional phenotype. Moreover, they have an augmentative effect on stem cells' cardiomyogenic differentiation. These material platforms could also support efficient electromechanical integration with the host cardiac system posttransplantation, thereby enhancing the reparative outcomes in vivo.

Despite significant developments in this domain, the technology is still far-reaching from the actual clinical translation. This is mainly due to multiple factors. First, the scaffold designs are still primitive and lack essential biophysical and biochemical cues that are necessary to maintain cardiac cell functionality. Moreover, our limited understanding of the cardiac microenvironment and electrophysiological aspects further restrain developments in this realm. Second, most of the research in this domain is concentrated on the use of CMs; however, other cardiac cells are also crucial and should be considered to improve constructs' therapeutic efficacy. Third, the safety and stability of the materials, posttransplantation; should also be considered for clinical applicability. Moreover, to further progress in this domain, consistent efforts from the various interdisciplinary research groups, clinicians, and stockholders are necessary.

Acknowledgment

The authors would like to acknowledge the INSPIRE scheme, Department of Science and Technology, Government of India, for providing the fellowship to TA.

Conflict of interest

The authors declare no conflicts of interest.

References

Abedi, A., Hasanzadeh, M., Tayebi, L., 2019. Conductive nanofibrous chitosan/PEDOT:PSS tissue engineering scaffolds. Mater. Chem. Phys. 237, 121882. https://doi.org/10.1016/j.matchemphys.2019.121882.

Anderson, R.H., Yanni, J., Boyett, M.R., Chandler, N.J., Dobrzynski, H., 2009. The anatomy of the cardiac conduction system. Clin. Anat. 22 (1), 99–113. https://doi.org/10.1002/ca.20700.

Ashtari, K., Nazari, H., Ko, H., Tebon, P., Akhshik, M., Akbari, M., Alhosseini, S.N., Mozafari, M., Mehravi, B., Soleimani, M., Ardehali, R., Ebrahimi Warkiani, M., Ahadian, S., Khademhosseini, A., 2019. Electrically conductive nanomaterials for cardiac tissue engineering. Adv. Drug Deliv. Rev. 144, 162–179. https://doi.org/10.1016/j.addr.2019.06.001.

Azizi, M., Navidbakhsh, M., Hosseinzadeh, S., Sajjadi, M., 2019. Cardiac cell differentiation of muscle satellite cells on aligned composite electrospun polyurethane with reduced graphene oxide. J. Polym. Res. 26 (11), 258. https://doi.org/10.1007/s10965-019-1936-9.

Chan, Y.-C., Ting, S., Lee, Y.-K., Ng, K.-M., Zhang, J., Chen, Z., Siu, C.-W., Oh, S.K.W., Tse, H.-F., 2013. Electrical stimulation promotes maturation of cardiomyocytes derived from human embryonic stem cells. J. Cardiovasc. Transl. Res. 6 (6), 989–999. https://doi.org/10.1007/s12265-013-9510-z.

Chaudhuri, R., Ramachandran, M., Moharil, P., Harumalani, M., Jaiswal, A.K., 2017. Biomaterials and cells for cardiac tissue engineering: current choices. Mater. Sci. Eng. C 79, 950–957. https://doi.org/10.1016/j.msec.2017.05.121.

Chen, C., Bai, X., Ding, Y., Lee, I.-S., 2019a. Electrical stimulation as a novel tool for regulating cell behavior in tissue engineering. Biomater. Res. 23 (1), 25. https://doi.org/10.1186/s40824-019-0176-8.

Chen, X., Feng, B., Zhu, D.-Q., Chen, Y.-W., Ji, W., Ji, T.-J., Li, F., 2019b. Characteristics and toxicity assessment of electrospun gelatin/PCL nanofibrous scaffold loaded with graphene in vitro and in vivo. Int. J. Nanomedicine 14, 3669–3678. https://doi.org/10.2147/IJN.S204971.

Cheng, X., Wan, Q., Pei, X., 2018. Graphene family materials in bone tissue regeneration: perspectives and challenges. Nanoscale Res. Lett. 13 (1), 289. https://doi.org/10.1186/s11671-018-2694-z.

Chiu, L.L.Y., Iyer, R.K., King, J.-P., Radisic, M., 2011. Biphasic electrical field stimulation aids in tissue engineering of multicell-type cardiac organoids. Tissue Eng. A 17 (11–12), 1465–1477. https://doi.org/10.1089/ten.tea.2007.0244.

Dhand, C., Das, M., Datta, M., Malhotra, B.D., 2011. Recent advances in polyaniline based biosensors. Biosens. Bioelectron. 26 (6), 2811–2821. https://doi.org/10.1016/j.bios.2010.10.017.

Domenech, M., Polo-Corrales, L., Ramirez-Vick, J.E., Freytes, D.O., 2016. Tissue engineering strategies for myocardial regeneration: acellular versus cellular scaffolds? Tissue Eng. B Rev. 22 (6), 438–458. https://doi.org/10.1089/ten.teb.2015.0523.

Ermis, M., Antmen, E., Hasirci, V., 2018. Micro and nanofabrication methods to control cell-substrate interactions and cell behavior: a review from the tissue engineering perspective. Bioact. Mater. 3 (3), 355–369. https://doi.org/10.1016/j.bioactmat.2018.05.005.

Fernandes, E.G.R., Zucolotto, V., De Queiroz, A.A.A., 2010. Electrospinning of hyperbranched poly-L-lysine/polyaniline nanofibers for application in cardiac tissue engineering. J. Macromol. Sci. A 47 (12), 1203–1207. https://doi.org/10.1080/10601325.2010.518847.

Ferrini, A., Stevens, M.M., Sattler, S., Rosenthal, N., 2019. Toward regeneration of the heart: bioengineering strategies for immunomodulation. Front. Cardiovasc. Med. 6. https://doi.org/10.3389/fcvm.2019.00026.

Fleischer, S., Dvir, T., 2013. Tissue engineering on the nanoscale: lessons from the heart. Curr. Opin. Biotechnol. 24 (4), 664–671. https://doi.org/10.1016/j.copbio.2012.10.016.

Fleischer, S., Shevach, M., Feiner, R., Dvir, T., 2014. Coiled fiber scaffolds embedded with gold nanoparticles improve the performance of engineered cardiac tissues. Nanoscale 6 (16), 9410–9414. https://doi.org/10.1039/C4NR00300D.

Ghafar-Zadeh, E., Waldeisen, J.R., Lee, L.P., 2011. Engineered approaches to the stem cell microenvironment for cardiac tissue regeneration. Lab Chip 11 (18), 3031. https://doi.org/10.1039/c1lc20284g.

Ghasemi, A., Imani, R., Yousefzadeh, M., Bonakdar, S., Solouk, A., Fakhrzadeh, H., 2019. Studying the potential application of electrospun polyethylene terephthalate/graphene oxide nanofibers as electroconductive cardiac patch. Macromol. Mater. Eng. 304 (8), 1900187. https://doi.org/10.1002/mame.201900187.

Hart, R.A., Gandhi, O.P., 1998. Comparison of cardiac-induced endogenous fields and power frequency induced exogenous fields in an anatomical model of the human body. Phys. Med. Biol. 43 (10), 3083–3099. https://doi.org/10.1088/0031-9155/43/10/027.

He, Y., Ye, G., Song, C., Li, C., Xiong, W., Yu, L., Qiu, X., Wang, L., 2018. Mussel-inspired conductive nanofibrous membranes repair myocardial infarction by enhancing cardiac function and revascularization. Theranostics 8 (18), 5159–5177. https://doi.org/10.7150/thno.27760.

Hernández, D., Millard, R., Sivakumaran, P., Wong, R.C.B., Crombie, D.E., Hewitt, A.W., Liang, H., Hung, S.S.C., Pébay, A., Shepherd, R.K., Dusting, G.J., Lim, S.Y., 2016. Electrical stimulation promotes cardiac differentiation of human induced pluripotent stem cells. Stem Cells Int. 2016, 1–12. https://doi.org/10.1155/2016/1718041.

Hitscherich, P., Aphale, A., Gordan, R., Whitaker, R., Singh, P., Xie, L., Patra, P., Lee, E.J., 2018. Electroactive graphene composite scaffolds for cardiac tissue engineering. J. Biomed. Mater. Res. A 106 (11), 2923–2933. https://doi.org/10.1002/jbm.a.36481.

Hsiao, C.-W., Bai, M.-Y., Chang, Y., Chung, M.-F., Lee, T.-Y., Wu, C.-T., Maiti, B., Liao, Z.-X., Li, R.-K., Sung, H.-W., 2013. Electrical coupling of isolated cardiomyocyte clusters grown on aligned conductive nanofibrous meshes for their synchronized beating. Biomaterials 34 (4), 1063–1072. https://doi.org/10.1016/j.biomaterials.2012.10.065.

Jiang, Y., Lian, X.L., 2020. Heart regeneration with human pluripotent stem cells: prospects and challenges. Bioact. Mater. 5 (1), 74–81. https://doi.org/10.1016/j.bioactmat.2020.01.003.

Kai, D., Prabhakaran, M.P., Jin, G., Ramakrishna, S., 2011. Polypyrrole-contained electrospun conductive nanofibrous membranes for cardiac tissue engineering. J. Biomed. Mater. Res. A 99A (3), 376–385. https://doi.org/10.1002/jbm.a.33200.

Kai, D., Prabhakaran, M.P., Jin, G., Ramakrishna, S., 2013. Biocompatibility evaluation of electrically conductive nanofibrous scaffolds for cardiac tissue engineering. J. Mater. Chem. B 1 (17), 2305. https://doi.org/10.1039/c3tb00151b.

Kankala, R.K., Zhu, K., Sun, X.-N., Liu, C.-G., Wang, S.-B., Chen, A.-Z., 2018. Cardiac tissue engineering on the nanoscale. ACS Biomater. Sci. Eng. 4 (3), 800–818. https://doi.org/10.1021/acsbiomaterials.7b00913.

Kaur, G., Adhikari, R., Cass, P., Bown, M., Gunatillake, P., 2015. Electrically conductive polymers and composites for biomedical applications. RSC Adv. 5 (47), 37553–37567. https://doi.org/10.1039/C5RA01851J.

Kharaziha, M., Shin, S.R., Nikkhah, M., Topkaya, S.N., Masoumi, N., Annabi, N., Dokmeci, M.R., Khademhosseini, A., 2014. Tough and flexible CNT–polymeric hybrid scaffolds for engineering cardiac constructs. Biomaterials 35 (26), 7346–7354. https://doi.org/10.1016/j.biomaterials.2014.05.014.

Lasher, R.A., Pahnke, A.Q., Johnson, J.M., Sachse, F.B., Hitchcock, R.W., 2012. Electrical stimulation directs engineered cardiac tissue to an age-matched native phenotype. J. Tissue Eng. 3 (1). https://doi.org/10.1177/2041731412455354. 204173141245535.

Lei, Q., He, J., Li, D., 2019. Electrohydrodynamic 3D printing of layer-specifically oriented, multiscale conductive scaffolds for cardiac tissue engineering. Nanoscale 11 (32), 15195–15205. https://doi.org/10.1039/C9NR04989D.

Lekawa-Raus, A., Patmore, J., Kurzepa, L., Bulmer, J., Koziol, K., 2014. Electrical properties of carbon nanotube based fibers and their future use in electrical wiring. Adv. Funct. Mater. 24 (24), 3661–3682. https://doi.org/10.1002/adfm.201303716.

Levin, M., 2003. Motor protein control of ion flux is an early step in embryonic left-right asymmetry. BioEssays 25 (10), 1002–1010. https://doi.org/10.1002/bies.10339.

Li, M., Guo, Y., Wei, Y., Macdiarmid, A., Lelkes, P., 2006. Electrospinning polyaniline-contained gelatin nanofibers for tissue engineering applications. Biomaterials 27 (13), 2705–2715. https://doi.org/10.1016/j.biomaterials.2005.11.037.

Liu, Y., Liang, X., Wang, S., Hu, K., 2016. Electrospun poly(lactic-co-glycolic acid)/multiwalled carbon nanotube nanofibers for cardiac tissue engineering. J. Biomater. Tissue Eng. 6 (9), 719–728. https://doi.org/10.1166/jbt.2016.1496.

Makvandi, P., Ghomi, M., Ashrafizadeh, M., Tafazoli, A., Agarwal, T., Delfi, M., Akhtari, J., Zare, E.N., Padil, V.V., Zarrabi, A., Pourreza, N., Miltyk, W., Maiti, T.K., 2020. A review on advances in graphene-derivative/polysaccharide bionanocomposites: therapeutics, pharmacogenomics and toxicity. Carbohydr. Polym. 250, 116952. https://doi.org/10.1016/j.carbpol.2020.116952.

Mazzola, M., Di Pasquale, E., 2020. Toward cardiac regeneration: combination of pluripotent stem cell-based therapies and bioengineering strategies. Front. Bioeng. Biotechnol. 8. https://doi.org/10.3389/fbioe.2020.00455.

Mehrabi, A., Baheiraei, N., Adabi, M., Amirkhani, Z., 2020. Development of a novel electroactive cardiac patch based on carbon nanofibers and gelatin encouraging vascularization. Appl. Biochem. Biotechnol. 190 (3), 931–948. https://doi.org/10.1007/s12010-019-03135-6.

Mohammadi Amirabad, L., Massumi, M., Shamsara, M., Shabani, I., Amari, A., Mossahebi Mohammadi, M., Hosseinzadeh, S., Vakilian, S., Steinbach, S.K., Khorramizadeh, M.R., Soleimani, M., Barzin, J., 2017. Enhanced cardiac differentiation of human cardiovascular disease patient-specific induced pluripotent stem cells by applying unidirectional electrical pulses using aligned electroactive Nanofibrous scaffolds. ACS Appl. Mater. Interfaces 9 (8), 6849–6864. https://doi.org/10.1021/acsami.6b15271.

Mombini, S., Mohammadnejad, J., Bakhshandeh, B., Narmani, A., Nourmohammadi, J., Vahdat, S., Zirak, S., 2019. Chitosan-PVA-CNT nanofibers as electrically conductive scaffolds for cardiovascular tissue engineering. Int. J. Biol. Macromol. 140, 278–287. https://doi.org/10.1016/j.ijbiomac.2019.08.046.

Monteiro, L.M., Vasques-Nóvoa, F., Ferreira, L., Pinto-do-Ó, P., Nascimento, D.S., 2017. Restoring heart function and electrical integrity: closing the circuit. Npj Regener. Med. 2 (1), 9. https://doi.org/10.1038/s41536-017-0015-2.

Mostafavi, E., Medina-Cruz, D., Kalantari, K., Taymoori, A., Soltantabar, P., Webster, T.J., 2020. Electroconductive nanobiomaterials for tissue engineering and regenerative medicine. Bioelectricity 2 (2), 120–149. https://doi.org/10.1089/bioe.2020.0021.

Murugan, R., Ramakrishna, S., 2007. Design strategies of tissue engineering scaffolds with controlled fiber orientation. Tissue Eng. 13 (8), 1845–1866. https://doi.org/10.1089/ten.2006.0078.

Nazari, H., Heirani-Tabasi, A., Alavijeh, M.S., Jeshvaghani, Z.S., Esmaeili, E., Hosseinzadeh, S., Mohabatpour, F., Taheri, B., Tafti, S.H.A., Soleimani, M., 2019a. Nanofibrous composites reinforced by MoS2 nanosheets as a conductive scaffold for cardiac tissue engineering. ChemistrySelect 4 (39), 11557–11563. https://doi.org/10.1002/slct.201901357.

Nazari, H., Azadi, S., Hatamie, S., Zomorrod, M.S., Ashtari, K., Soleimani, M., Hosseinzadeh, S., 2019b. Fabrication of graphene-silver/polyurethane nanofibrous scaffolds for cardiac tissue engineering. Polym. Adv. Technol. 30 (8), 2086–2099. https://doi.org/10.1002/pat.4641.

Nazari, H., Heirani Tabasi, A., Hajiabbas, M., Khalili, M., Shahsavari Alavijeh, M., Hatamie, S., Mahdavi Gorabi, A., Esmaeili, E., Ahmadi Tafti, S.H., 2020. Incorporation of two-dimensional nanomaterials into silk fibroin nanofibers for cardiac tissue engineering. Polym. Adv. Technol. 31 (2), 248–259. https://doi.org/10.1002/pat.4765.

Nuccitelli, R., 1992. Endogenous ionic currents and DC electric fields in multicellular animal tissues. Bioelectromagnetics 13 (S1), 147–157. https://doi.org/10.1002/bem.2250130714.

Oh, H., Ito, H., Sano, S., 2016. Challenges to success in heart failure: cardiac cell therapies in patients with heart diseases. J. Cardiol. 68 (5), 361–367. https://doi.org/10.1016/j.jjcc.2016.04.010.

Park, J.-S., 2011. Electrospinning and its applications. Adv. Nat. Sci. Nanosci. Nanotechnol. 1 (4), 043002. https://doi.org/10.1088/2043-6262/1/4/043002.

Phong, H.Q., Wang, S.-L., Wang, M.-J., 2010. Cell behaviors on micro-patterned porous thin films. Mater. Sci. Eng. B 169 (1–3), 94–100. https://doi.org/10.1016/j.mseb.2010.01.009.

Radisic, M., Park, H., Shing, H., Consi, T., Schoen, F.J., Langer, R., Freed, L.E., Vunjak-Novakovic, G., 2004. Functional assembly of engineered myocardium by electrical stimulation of cardiac myocytes cultured on scaffolds. Proc. Natl. Acad. Sci. 101 (52), 18129–18134. https://doi.org/10.1073/pnas.0407817101.

Ravichandran, R., Sridhar, R., Venugopal, J.R., Sundarrajan, S., Mukherjee, S., Ramakrishna, S., 2014. Gold nanoparticle loaded hybrid nanofibers for cardiogenic differentiation of stem cells for infarcted myocardium regeneration. Macromol. Biosci. 14 (4), 515–525. https://doi.org/10.1002/mabi.201300407.

Roshanbinfar, K., Vogt, L., Ruther, F., Roether, J.A., Boccaccini, A.R., Engel, F.B., 2020. Nanofibrous composite with tailorable electrical and mechanical properties for cardiac tissue engineering. Adv. Funct. Mater. 30 (7), 1908612. https://doi.org/10.1002/adfm.201908612.

Salmasi, S., 2015. Role of nanotopography in the development of tissue engineered 3D organs and tissues using mesenchymal stem cells. World J. Stem Cells 7 (2), 266. https://doi.org/10.4252/wjsc.v7.i2.266.

Schwach, V., Passier, R., 2019. Native cardiac environment and its impact on engineering cardiac tissue. Biomater. Sci. 7 (9), 3566–3580. https://doi.org/10.1039/C8BM01348A.

Shang, L., Qi, Y., Lu, H., Pei, H., Li, Y., Qu, L., Wu, Z., Zhang, W., 2019. Graphene and graphene oxide for tissue engineering and regeneration. In: Theranostic Bionanomaterials. Elsevier, pp. 165–185, https://doi.org/10.1016/B978-0-12-815341-3.00007-9.

Sharma, D., Ferguson, M., Kamp, T.J., Zhao, F., 2019. Constructing biomimetic cardiac tissues: a review of scaffold materials for engineering cardiac patches. Emergent Mater. 2 (2), 181–191. https://doi.org/10.1007/s42247-019-00046-4.

Shi, H., Liu, C., Jiang, Q., Xu, J., 2015. Effective approaches to improve the electrical conductivity of PEDOT:PSS: a review. Adv. Electron. Mater. 1 (4), 1500017. https://doi.org/10.1002/aelm.201500017.

Shokraei, N., Asadpour, S., Shokraei, S., Nasrollahzadeh Sabet, M., Faridi-Majidi, R., Ghanbari, H., 2019. Development of electrically conductive hybrid nanofibers based on CNT-polyurethane nanocomposite for cardiac tissue engineering. Microsc. Res. Tech. 82 (8), 1316–1325. https://doi.org/10.1002/jemt.23282.

Solazzo, M., O'Brien, F.J., Nicolosi, V., Monaghan, M.G., 2019. The rationale and emergence of electroconductive biomaterial scaffolds in cardiac tissue engineering. APL Bioeng. 3 (4), 041501. https://doi.org/10.1063/1.5116579.

Sridhar, S., Venugopal, J.R., Sridhar, R., Ramakrishna, S., 2015. Cardiogenic differentiation of mesenchymal stem cells with gold nanoparticle loaded functionalized nanofibers. Colloids Surf. B: Biointerfaces 134, 346–354. https://doi.org/10.1016/j.colsurfb.2015.07.019.

Stone, H., Lin, S., Mequanint, K., 2019. Preparation and characterization of electrospun rGO-poly(ester amide) conductive scaffolds. Mater. Sci. Eng. C. https://doi.org/10.1016/j.msec.2018.12.122.

Stoppel, W.L., Kaplan, D.L., Black, L.D., 2016. Electrical and mechanical stimulation of cardiac cells and tissue constructs. Adv. Drug Deliv. Rev. 96 (96), 135–155. https://doi.org/10.1016/j.addr.2015.07.009.

Stout, D.A., Basu, B., Webster, T.J., 2011. Poly(lactic–co-glycolic acid): carbon nanofiber composites for myocardial tissue engineering applications. Acta Biomater. 7 (8), 3101–3112. https://doi.org/10.1016/j.actbio.2011.04.028.

Sun, X., Nunes, S.S., 2016. Biowire platform for maturation of human pluripotent stem cell-derived cardiomyocytes. Methods 101, 21–26. https://doi.org/10.1016/j.ymeth.2015.11.005.

Tandon, N., Cannizzaro, C., Chao, P.-H.G., Maidhof, R., Marsano, A., Au, H.T.H., Radisic, M., Vunjak-Novakovic, G., 2009. Electrical stimulation systems for cardiac tissue engineering. Nat. Protoc. 4 (2), 155–173. https://doi.org/10.1038/nprot.2008.183.

Tay, C.Y., Irvine, S.A., Boey, F.Y.C., Tan, L.P., Venkatraman, S., 2011. Micro-/nano-engineered cellular responses for soft tissue engineering and biomedical applications. Small 7 (10), 1361–1378. https://doi.org/10.1002/smll.201100046.

Thavandiran, N., Nunes, S.S., Xiao, Y., Radisic, M., 2013. Topological and electrical control of cardiac differentiation and assembly. Stem Cell Res Ther 4 (1), 14. https://doi.org/10.1186/scrt162.

Thygesen, K., Alpert, J.S., White, H.D., 2007. Universal definition of myocardial infarction. Circulation 116 (22), 2634–2653. https://doi.org/10.1161/CIRCULATIONAHA.107.187397.

Tondnevis, F., Keshvari, H., Mohandesi, J.A., 2020. Fabrication, characterization, and in vitro evaluation of electrospun polyurethane-gelatin-carbon nanotube scaffolds for cardiovascular tissue engineering applications. J Biomed Mater Res B Appl Biomater 108 (5), 2276–2293. https://doi.org/10.1002/jbm.b.34564.

Veetil, J.V., Ye, K., 2009. Tailored carbon nanotubes for tissue engineering applications. Biotechnol. Prog. 25 (3), 709–721. https://doi.org/10.1002/btpr.165.

Venugopal, J.R., Prabhakaran, M.P., Mukherjee, S., Ravichandran, R., Dan, K., Ramakrishna, S., 2012. Biomaterial strategies for alleviation of myocardial infarction. J. R. Soc. Interface 9 (66), 1–19. https://doi.org/10.1098/rsif.2011.0301.

Vunjak-Novakovic, G., Tandon, N., Godier, A., Maidhof, R., Marsano, A., Martens, T.P., Radisic, M., 2010. Challenges in cardiac tissue engineering. Tissue Eng. B Rev. 16 (2), 169–187. https://doi.org/10.1089/ten.teb.2009.0352.

Vuppaladadium, S.S.R., Agarwal, T., Kulanthaivel, S., Mohanty, B., Barik, C.S., Maiti, T.K., Pal, S., Pal, K., Banerjee, I., 2020. Silanization improves biocompatibility of graphene oxide. Mater. Sci. Eng. C 110, 110647. https://doi.org/10.1016/j.msec.2020.110647.

Wang, F., Guan Jianjun, J., 2010. Cellular cardiomyoplasty and cardiac tissue engineering for myocardial therapy. Adv. Drug Deliv. Rev. 62 (7–8), 784–797. https://doi.org/10.1016/j.addr.2010.03.001.

Wang, B., Wang, G., To, F., Butler, J.R., Claude, A., McLaughlin, R.M., Williams, L.N., de Jongh Curry, A.L., Liao, J., 2013. Myocardial scaffold-based cardiac tissue engineering: application of coordinated mechanical and electrical stimulations. Langmuir 29 (35), 11109–11117. https://doi.org/10.1021/la401702w.

Wang, L., Wu, Y., Hu, T., Guo, B., Ma, P.X., 2017. Electrospun conductive nanofibrous scaffolds for engineering cardiac tissue and 3D bioactuators. Acta Biomater. 59 (59), 68–81. https://doi.org/10.1016/j.actbio.2017.06.036.

Wang, X., Feng, Y., Dong, P., Huang, J., 2019. A mini review on carbon quantum dots: preparation, properties, and electrocatalytic application. Front. Chem. 7. https://doi.org/10.3389/fchem.2019.00671.

Webster, T.J., Asiri, A., Marwani, H., Khan, S.B., 2014. Greater cardiomyocyte density on aligned compared with random carbon nanofibers in polymer composites. Int. J. Nanomedicine 5533. https://doi.org/10.2147/IJN.S71587.

Wright, Z.M., Arnold, A.M., Holt, B.D., Eckhart, K.E., Sydlik, S.A., 2019. Functional graphenic materials, graphene oxide, and graphene as scaffolds for bone regeneration. Regener. Eng. Transl. Med. 5 (2), 190–209. https://doi.org/10.1007/s40883-018-0081-z.

Yadav, V., Roy, S., Singh, P., Khan, Z., Jaiswal, A., 2019. 2D MoS2-based nanomaterials for therapeutic, bioimaging, and biosensing applications. Small 15 (1), 1803706. https://doi.org/10.1002/smll.201803706.

Yan, C., Ren, Y., Sun, X., Jin, L., Liu, X., Chen, H., Wang, K., Yu, M., Zhao, Y., 2020. Photoluminescent functionalized carbon quantum dots loaded electroactive silk fibroin/PLA nanofibrous bioactive scaffolds for cardiac tissue engineering. J. Photochem. Photobiol. B Biol. 202, 111680. https://doi.org/10.1016/j.jphotobiol.2019.111680.

Ypey, D.L., Clapham, D.E., DeHaan, R.L., 1979. Development of electrical coupling and action potential synchrony between paired aggregates of embryonic heart cells. J. Membr. Biol. 51 (1), 75–96. https://doi.org/10.1007/BF01869344.

Zhang, H., Yu, M., Xie, L., Jin, L., Yu, Z., 2012. Carbon-nanofibers-based micro-/nanodevices for neural-electrical and neural-chemical interfaces. J. Nanomater. 2012, 1–6. https://doi.org/10.1155/2012/280902.

Zhang, B., He, J., Li, X., Xu, F., Li, D., 2016. Micro/nanoscale electrohydrodynamic printing: from 2D to 3D. Nanoscale 8 (34), 15376–15388. https://doi.org/10.1039/C6NR04106J.

Zhao, G., Qing, H., Huang, G., Genin, G.M., Lu, T.J., Luo, Z., Xu, F., Zhang, X., 2018. Reduced graphene oxide functionalized nanofibrous silk fibroin matrices for engineering excitable tissues. NPG Asia Mater. 10 (10), 982–994. https://doi.org/10.1038/s41427-018-0092-8.

Zhao, G., Zhang, X., Li, B., Huang, G., Xu, F., Zhang, X., 2020. Solvent-free fabrication of carbon nanotube/silk fibroin electrospun matrices for enhancing cardiomyocyte functionalities. ACS Biomater. Sci. Eng. 6 (3), 1630–1640. https://doi.org/10.1021/acsbiomaterials.9b01682.

Impacts of nanotechnology in tissue engineering

Mh Busra Fauzi, Jia Xian Law, Min Hwei Ng, Yogeswaran Lokanathan, Nadiah Sulaiman, and Atiqah Salleh

Centre for Tissue Engineering and Regenerative Medicine, Faculty of Medicine, National University of Malaysia, Kuala Lumpur, Malaysia

1 Nanomaterials for skin repair and regeneration

1.1 Introduction

Skin is the largest and the most vulnerable human organ owning to undertaking many protective functions against external triggers. Skin damage or injuries caused by several circumstances, such as burns, traumas, and other injuries, has been for decades one of the most significant health issues for millions of patients around the globe. Due to the multilayered nature of the skin and the fact that several variables are involved in the regeneration process, regular wound healing activities last for a long time, especially in large, acute, or chronic wounds. Standard wound dressings offer limited involvement in the wound healing process more than shielding the wound from external contaminants with a frequent need for replacement, which makes it rather painful for patients.

A wide range of natural and synthetic polymeric biomaterials seeded human dermal fibroblast (HDFs), and human epidermal keratinocytes (HEKs) have been introduced as constructs and scaffolds for skin repair and wound healing. Some of these mimicry biomaterials were proven to promote cell growth, proliferation, migration, and differentiation in vitro, and accelerate wound healing in the preclinical trials. However, these polymeric biomaterials (mainly hydrogels) offer low mechanical properties, fast biodegradation, and lack antibacterial activity, which can develop into serious immune rejection and tissue regeneration failure postimplantation.

Among many recent wound dressing advances, nanomaterials were reported to have a high potential to enhance the performance of skin scaffolds by offering excellent cell-to-material interaction. A variety of shapes, types, and sizes of nanoparticles were fabricated and used in wound healing products as drag nanocarriers, matrix reinforcement nanofibers and nanorods, and antibacterial agent nanospheres. Noteworthy that appropriate particle size, dose, and concentration should be considered once using nanoparticles in biomedical therapeutics to avoid toxicity, especially with metallic nanoparticles.

When the skin is inflicted with an acute wound such as a scratch, the superficial or epidermal layer of the skin will replace itself as the cells on the upper layer of the skin proliferate regularly. For a chronic wound, the loss in the dermal is permanent because the body does not have the ability to regenerate the dermal tissue naturally, especially in large wounds. Hence, scientists are still searching for permanent treatments that replace currently available therapies. Nanomaterials, however, have a tremendous opportunity in increasing the efficacy of these emerging treatments. In this section, we will discuss nanomaterial's contribution to skin dermal and epidermal regeneration in the wound healing process.

1.2 Nanomaterials for epidermal regeneration

The outermost thin layer of the skin is the epidermal layer which gives the skin its color, protects the body from UV radiations, regulates water release and body temperature. In addition to melanocytes and inflammatory cells, the epidermal is mainly lined up of keratinocytes (represent 90% of the total epidermal size) which proliferate and differentiate rapidly in a continuous skin renewal process. As a response to wounding, keratinocytes surrounding the wound borders migrate across the wound and cover it with a white epithelial layer in a process called re-epithelialization. Nevertheless, the speed of re-epithelialization depends on the severity of skin damage which can cause complicated and slow keratinocytes migration. Epidermal injuries can be treated with several wound dressings that can improve and accelerate re-epithelialization (Reyes-Ortega et al., 2015).

To enhance the antibacterial activity of epidermal and promote cell proliferation postinjury, Chereddy et al. (2014) investigated poly (lactic-*co*-glycolic acid) nanoparticles loaded antimicrobial peptide LL37. PLGA-LL37 NPs scaffold seeded human epidermal keratinocytes were applied on the dorsal of mouse models in vivo. They found that the treatment with PLGA-LL37 nanoparticles significantly accelerated wound healing compared to PLGA-LL37 alone. PLGA-LL37 NPs-treated wounds demonstrated improved angiogenesis, better tissue granulation, re-epithelization, and neovascular formation, with enhanced cell migration and proliferation. Additionally, PLGA-LL37 NPs showed good antimicrobial activity against *Escherichia coli*. In another study, bacterial cellulose nanofilms combined poly(2-hydroxyethyl methacrylate) (PHEMA) were seeded in human adipose-derived mesenchymal stem cells to evaluate their wound healing potential. BC-PHEMA nanocomposites showed no cytotoxicity, high cell growth, and proliferation, making it a promising dry wound dressing with an excellent regenerative capacity (Figueiredo et al., 2013).

Nanomaterials are also used as nanocarriers which provide more efficient drug delivery. Nanocarriers can induce wound healing by proffering better angiogenic response, cutting off inflammation and antimicrobial activity, and they are used widely as antioxidants, antiinflammatory, antibacterial, and antifungal agents (Bernal-Chávez et al., 2019). Dave et al. formulated lipid polymer hybrid nanoparticles (LPNs) as drug nanocarriers of norfloxacin to treat bacterial infections in burns. Norfloxacin-loaded nanoparticles were found to exhibit desirable 24 h drug release with an extended-lasting property to avoid frequent application. It also showed high antimicrobial efficacy against *Staphylococcus aureus* and *Pseudomonas aeruginosa* (Dave et al., 2017).

Wounds usually develop into scars on the epidermal layer, that can last for some time after wound healing. Scar contracture developed after traumatic injury and burns can be permanent, causing physiological and psychological problems to patients. Nanomaterials have also been used as antiscarring agents. Xiao et al. (2019) studied the contribution of cuprous oxide nanoparticles (CONPs) in minimizing hypertrophic scar formation. Following complete re-epithelialization at day 14 of a rabbit ear scar model, CONPs were introduced. On day 35, they observe improved scar appearance with better collagen alignment and decreased scar elevation index (SEI). By regulating the inhibition of human dermal fibroblast proliferation and activating their apoptosis, CONPs can be considered as excellent therapeutic potential in the treatment of hypertrophic scars.

1.3 Nanomaterials for dermal regeneration

The inner layer is the dermal layer, made up of interconnective tissue, which function as the mechanical shield against external forces. The dermis matrix is mainly lined up of long collagen fibrils, elastin fibers, and fibroblasts. However, this layer is difficult to be repaired naturally due to the severe loss of the extracellular matrix cells require to migrate and proliferate, which necessitates skin grafting therapeutics. Many biomaterials have been developed to encapsulate skin cells and used as wound coverage.

Nanoparticles have developed unique characteristics as they tend to be more effective in comparison with their bulk structures and are biocompatible due to their nano-size. Nanoparticles have two major strategies in enhancing the wound healing therapies: (1) act as a drug delivery system for controlled release of medication and (2) act as an antimicrobial agent to prevent infection in the wound site. The application of nanoparticles in drug delivery or nano-drug delivery systems function to carry the bioactive substances to the targeted region, control the release of the drug, and thus maximize the therapeutics effect of the drugs (Wang et al., 2019).

Polymeric nanoparticles have been intensively used as drug carriers due to their nontoxic nature. Polymeric nanoparticles also can control the drug release and improve the mechanical strength of the drug, which lowers the biodegradation rate of the drug. Bairagi et al. (2018) were able to utilize poly-lactic-co-glycolic acid (PLGA) nanoparticles to carry ferulic acid used to treat diabetic wounds. The in vivo result shows the treated group has faster epithelization compared with the nontreated group. Liposomes are nanoparticles that can mimic the phospholipids by creating bilayer vesicles from amphiphilic molecules. Besides, providing sustained protection of drugs, liposomes that have inner water cavities, able to develop a moist microenvironment at the wound site, which is ideal for the wound healing process.

Apart from being a drug carrier, nanoparticles can also act as an antimicrobial agent to prevent prolonged infection in the wound site. Infections have been a significant threat in dermal regeneration as these harmful bacteria start to colonize and form biofilm in the wound area making it harder for the antibodies to remove them, thus prolonging the inflammation stage of wound healing. The ability of nanoparticles as drug nanocarriers that help in the regeneration of cells in the wound site can also be used to carry antibiotics such as penicillin to the wound site (Garcia-Orue et al., 2016). Nanoparticles can also combine with currently available treatments such as hydrogel to enhance the antibacterial properties of wound dressings.

Metallic nanoparticles such as silver and gold nanoparticles exhibit strong antibacterial properties due to their unique characteristic, thus widening their mechanism of action. The bacteria have a hard time developing resistance to the metallic nanoparticles due to the wide range of antibacterial activity in the nanoparticles. Embedment of the metallic nanoparticles in wound dressing can be lower the toxicity of these nanoparticles and prevent the infection on the wound site. Alipour et al. (2019) have incorporated the silver nanoparticles in the electrospun nanofiber and found useful antibacterial properties as well as promoting wound healing capabilities.

2 Nanomaterial technology for eye regeneration

2.1 Introduction

Eye injury caused by mechanical trauma, chemicals, and radiation is common and may lead to very serious consequences including permanent loss of vision and blindness. Loss of vision has a huge impact on the patient life. It is a daunting task to restore vision as the eye has poor regeneration capacity. Currently available pharmacological treatments and surgical managements show limited success in restoring the structure and function of ocular tissue. In addition, there is a significant shortage in the number of donors for ocular tissue such as cornea internationally.

Ocular tissue engineering has been introduced to support eye regeneration. Tissue engineering has been tested in many preclinical studies to replace and regenerate the damaged or loss cornea, lens, and retina tissue. There are two main approaches for ocular tissue engineering, i.e., using the cell-based strategy or the scaffold-based strategy, to promote eye regeneration. Many types of cells have been tested for the cell-based approach. Similarly for the scaffold-based approach whereby many types of biomaterials have been tested using the animal eye injury models.

More recently, nanomaterials have been reported to be useful for ocular tissue engineering as they have the capability to acquire the shape of the defect and fill the gap, serving as a template to promote tissue regeneration. Various types of nanomaterials, including nanofiber scaffolds, self-assembling peptides, 3D self-assembled nanomatrix, and nanostructured surfaces have been used to promote ocular regeneration. The nanomaterials have tailored physical, chemical, and biological properties that mimic the native ocular tissue which render them ideal for ocular regeneration. In addition, the nanoscaffolds also have high porosity to support nutrient and oxygen transport as well as waste removal in order to maintain cell survival in vivo. Nanomaterials can be loaded with biologics such as cells, growth factors, and genes as well as drugs and serve as delivery vehicles to modulate the wound environment and create a regenerative environment that enhances eye regeneration. However, due to its small sizes, ranging from 1 to 100 nm, it is also very challenging when using nanomaterials. In this section, we will discuss the nanomaterials used for cornea, lens, and retina repair and regeneration.

2.2 Nanomaterials for cornea regeneration

The cornea is a highly specialized, transparent, and avascular connective tissue that covers the front part of the eye. The cornea serves as a barrier and refracts the light entering the eye. The corneal comprises three cell layers, from the outermost epithelial cell layer to the middle stroma layer that mainly consists of collagen fibers, and the innermost endothelial cell layer (Chaurasia et al., 2015). Upon injury, the limbal epithelial stem cells residing at the limbus will proliferate and migrate to the corneal basal layer to replace the loss of corneal epithelial cells. However, the endothelial cell layer cannot regenerate after injury (Sahle et al., 2019). Injured cornea loses its transparency and thus leads to visual dysfunction. Cornea injury can be treated with drugs applied as eye drops. However, poor patient compliance and low drug penetration limited the success of this technique. Replacement of damaged cornea with donor corneal tissue is hindered by donor insufficiency.

Ma et al. (2013) seeded the adipose tissue-derived stem cells (ASCs) on the polylactic-*co*-glycolic acid (PLGA) scaffold to repair corneal defects. The engineered tissue was transplanted in vivo and was found to promote cornea regeneration without promoting tissue angiogenesis. The transplanted cells were reported to differentiate into functional keratocytes which produced the stroma matrices. In a different study, Duan and Sheardown (2006) reported the fabrication of dendrimer crosslinked collagen which supported human corneal epithelial cell proliferation and attachment. Generally, most of the studies combined natural and synthetic biomaterials to prepare nanoscaffold with excellent biocompatibility and appropriate mechanical property for corneal tissue regeneration. In addition, the nanofiber alignment also affects corneal regeneration. Wray and Orwin (2009) found that aligned collagen nanofibers reduced the expression of alpha-smooth muscle actin (α-sma) expression of rabbit corneal fibroblasts compared to those expanded on random nanofibers.

Apart from nanofibers, nanomaterials also have been prepared in nanoparticle form to enhance corneal regeneration. Recently, Nagai et al. (2019) reported that a combination of magnesium hydroxide nanoparticles and sericin can hasten corneal wound healing in vivo. Nagai et al. (2017) fabricated nanoparticles based on zirconia beads containing dexamethasone to improve the drug permeability through the cornea to reduce tissue inflammation.

Self-assembling peptide is an attractive nanomaterial for corneal regeneration as it can be injected and formed a nanofibrous structure in situ. Uzunalli et al. (2014) injected laminin-mimetic YIGSR-peptide amphiphile molecules that self-assembled into nanofibers to rabbit corneal and found that it promoted keratocyte migration and stroma regeneration.

2.3 Nanomaterials for lens regeneration

Cataract is a clouding of the lens in the eye which may lead to blindness. Cataract is the most common lens disease. The treatment of cataracts involves the surgical removal of the affected lens and replacing it with an artificial lens. However, certain eye conditions may prohibit the implantation of artificial lenses.

Endogenous lens epithelial stem cells have been investigated as a potential source of stem cells to regenerate the lens. Nanomaterials are used to facilitate lens regeneration by stem cells by providing a nanostructure that facilitates cell survival and proliferation. Nibourg et al. (2016) applied nanofiber-based hydrogels consisting of fibers of a low molecular weight hydrogelator as a matrix to support lens epithelial cell growth in ex vivo cultured porcine lens. It was found that a lens with nanofiber-based hydrogels has a lower formation of capsular opacification, lower expression α-sma, and less lens epithelial cell transformation. Xi et al. (2013) cultured human lens epithelial cells on polyurethane nanofibers and found that the nanofibers with the anisotropic-latitude arrangement are more potent in enhancing lens epithelial cell migration compared to the nanofibers with anisotropic-longitude and random arrangement.

2.4 Nanomaterials for retina regeneration

The retina is the thin light-sensitive layer of tissue lining the inner surface at the back of the eye. Retina senses light from the lens converts it to electrical impulses and transmits the impulses to the brain. The retina is made of 10 distinct layers containing multiple types of cells, including the photoreceptors, bipolar cells, retinal ganglion cells, horizontal cells, amacrine cells, and retinal pigment epithelial cells (RPE) (Willermain et al., 2014; Wilson, 2001). Retinal disease is a leading cause of blindness. Macular degeneration, diabetic eye disease, retinitis pigmentosa, retinal detachment, and floaters are the common retinal diseases.

Multiple types of nanomaterials have been investigated to support retina regeneration. Chan et al. (2017) produced electrospun pectin-polyhydroxybutyrate nanofibers and found that it supports the attachment and proliferation of human RPE cells. Warnke et al. fabricated a nanofibrous scaffold using PLGA and collagen type I and seeded human RPE cells on it. The authors found that the RPE cells grown on the scaffold and the cells showed great similarity to the native human RPE cells by having a polygonal cell shape and with the presence of microvilli on the apical surface (Warnke et al., 2013).

Pritchard et al. (2010) coated the poly(glycerolco-sebacic acid) membrane with electrospun laminin-poly(caprolactone) (PCL) nanofibers and found that nanofiber coating enhances the attachment of the photoreceptor layer isolated from embryonic retina tissue. Shahmoradi et al. modified the surface PCL nanofibrous scaffold using the alkaline hydrolysis method to increase the hydrophilicity. The authors found that surface modification enhanced the proliferation of human RPE cells and supported cell survival up to 45 days in vitro (Shahmoradi et al., 2017). Importantly, the RPE cells retained its polygonal morphology and the expression of cell-specific markers.

2.5 Summary

Nanomaterials show great promise in ocular regeneration, with the aim of preventing or treating vision loss and blindness. Nanomaterials can serve as a drug carrier and also as a supportive structure for cells to regenerate the lost tissue. An effective nanomaterial for tissue engineering should be able to facilitate cell transplantation, adhesion, proliferation, migration, and differentiation in order to restore tissue function. Currently, most of the studies are still in the in vitro and preclinical phase. More effects are needed to identify the suitable nanomaterials that are suitable to be used clinically, which can proceed to clinical study.

3 Nanostructured biomaterial used in bone regeneration

3.1 Introduction

Bone, which serves as the structural framework for the body, constantly undergoes self-renewal and remodeling (Florencio-Silva et al., 2015; Han et al., 2018). The bone tissue comprises 40% dry weight of organic (90% are collagen type 1 fibers) and 60% inorganic compound (calcium and phosphate in the form of hydroxyapatite). The collagen fibers provide bone tissue with flexural resilience and toughness while the mineral crystallites HA the mechanical strength and stiffness. Within the bone are four types of bone cells, i.e., osteoblast, osteoclast, bone lining cell, and osteocyte that play an intricate and dynamic role in maintaining its function. Bone serves various functions in the body including providing support, locomotion, protection of soft tissue, production of blood cells (marrow), a reservoir for minerals (calcium and phosphorus) and energy (fat), and playing endocrine regulation roles (Robling et al., 2006; Datta et al., 2008).

In 1952, Robinson et al. has reported that HA nanoparticles of approximately 50 nm long, 25 nm wide, and 2–5 nm, thick constitute approximately 70% of the native bone (Robinson, 1952). Recent understanding of the complex interactions of its various cellular components and their environment (Florencio-Silva et al., 2015; Han et al., 2018) including the revelation of the hierarchical organization of the bone extracellular matrix at the nanoscale (ref) and the interaction of cells with nanostructures around them (Reznikov et al., 2018), provides further insights into the required complexity of future endeavors in bone regeneration.

In this section, a review of three important aspects of nanotechnology being applied in bone regeneration will be covered, i.e., (i) nanoscale biomaterials for improved bone assimilation and mechanical strength of implants, (ii) nanostructures at the material-cell interphase to facilitate bone cell behavior, and (iii) nanoparticles as a carrier for drugs and biologics that can trigger bone regeneration.

3.2 Nanomaterials as bone fillers and improved biomaterial mechanical strength

As collagen type I and hydroxyapatite are a basic part of the bone, nanosized hydroxyapatite and collagen particles have been delivered on its own or incorporated with other implant materials to be assimilated into host bone tissues (Eliaz and Metoki, 2017) Calcium phosphates-based material, especially hydroxyapatites [HA, $Ca_{10}(PO_4)_6(OH)_2$] has already been successfully used as bone graft due to its similarity with the mineral crystal of bone. Such material will degrade over time whereby the calcium and phosphate ions, will be reassimilated into the bone during bone mineralization (Ng et al., 2014). As such, other bone minerals could be explored such as magnesium, potassium and fluoride, and silicon.

Nano-ceramics primarily nano-hydroxyapatite has been widely used as commercial products with approval by the US Food and Drug Administration (FDA) through the 510(k) process. The first is nannOss (Angstrom Medica Inc., MA, United States), followed by Ostim (Osartis GmbH & Co. KG, Dieburg, Germany) and FortrOss (Pioneer Surgical Technology, Inc). Healos (DePuy Spine, MA, United States), a nanocomposite of type I collagen fibers and nonsintered calcium phosphate mineral nanoparticles also have been granted marketing approval from the FDA.

Furthermore, it has been widely reported that the addition of nanoparticles such as nano-HA to polymeric scaffolds improves their mechanical stability, biocompatibility, and biological activity (Wang et al., 2016; Molino et al., 2020; Montalbano et al., 2020; Hill et al., 2019). Besides collagen and hydroxyapatite, nanostructured carbon nanotube (Stocco et al., 2019) and graphene (Elkhenany et al., 2017), and metals such as gold (Vial et al., 2017), silver (Zhang et al., 2015), and titanium oxide (TiO_2) (Rasoulianboroujeni et al., 2019) have demonstrated interesting and unique properties of their own to be incorporated as fillers in bone tissue engineering scaffolds.

3.3 Nanostructures on implant surfaces to facilitate cell behavior

The cellular response to the nanoscale features has drawn tremendous attention. This is not surprising as cells through their filopodia and lamellipodia are structures on the nanoscale that interact with the ECM proteins in vivo, which are characterized by nanometric collagen fibrils (Anselme et al., 2010). In our observation using aligned electrospun nanofibers, cells were found to adhere and grow along the aligned fibers within minutes of cell seeding. Nanostructures can be introduced on implant surfaces to improve implant osteointegration via interaction with host cells (Dalby et al., 2007).

The effect of surface topography on cell behavior has been extensively reviewed by Hayes and Richards (2010) for metal surfaces, whereby macro-topography (features >100 μm) affects primary fixation and the surface micro-topography (between 1 and 100 μm) influences cell recruitment, adhesion, orientation, and morphology and even gene expression while the submicron and nano topography influence the focal contacts and cytoskeletal arrangement, the orientation of the cells and their communication, as well as adsorption and conformation of the proteins and biomolecules on the surface.

Webster et al. (2000) have shown in a series of studies comparing nano-HA (<100 nm) with microHA, the enhanced osteoblast adhesion in vitro was due to the higher protein adsorption on the nano-HA. Liao et al. indicated that nano-HA adsorbed vitronectin which facilitates osteoblast adhesion. It has been shown that a diameter of 20 nm exhibits the highest osteogenic effect for primary osteoblasts, while 30–50 nm particles for the human adipose-derived stem cells performed the best (Liao et al., 2004). It was further demonstrated that the enhanced protein adsorption is due to the presence of functional groups such as oxygen-containing group (OH) and amino groups (Dalby et al., 2007; Neděla et al., 2017).

3.4 Nanoparticles as a carrier for biological or synthetic molecules for bone regeneration

As nanoarchitectures provide a high surface area to volume ratio, hence provide better interaction with host cells. Nanoparticles theoretically can diffuse within the nanoscale extracellular matrix of bone cells such as lacunar-canaliculi space for its assimilation. A wide variety of nanoparticles as delivery carriers of drugs, growth factors, and genetic material-promoting bone tissue regeneration have been widely investigated (De Witte et al., 2018). Collagen and ceramic-based nanoparticles have been widely used as carriers due to their natural presence in bone tissues (Chatterjee et al., 2009). The biological or synthetic molecules can be coupled to the nanomaterials via adsorption, covalent linking, or encapsulation into a resorbable carrier.

Local delivery of osteogenic factors (mainly bone morphogenic protein-2 (BMP2), FGF) and angiogenic growth factors (VEGF) in nanomaterials have shown a promising strategy for overall enhanced osteogenesis (Kuttappan et al., 2018). Specifically, BMP2 enhances osteogenic differentiation of osteoprogenitor cells (Huang et al., 2005), VEGF promotes angiogenesis (Lü et al., 2018) and bFGF promotes cell proliferation and migration of endothelial cells (Lee et al., 2015). Sequential/temporal release of various growth factors using the unique properties of different nanomaterials has been shown to accelerate vascularized bone formation (Yilgor et al., 2010).

For a more sustained delivery of a biomolecule, researchers can resort to a gene (plasmid DNA vector) delivery system. Krebs et al. reported that BMP-2 encoding plasmid loaded onto calcium phosphate nanoparticles could be released in a sustainable manner in vitro. These plasmid-loaded nanoparticles were then incorporated in an injectable alginate hydrogel, together with MC3T3-E1 pre-osteoblast cells, and injected subcutaneously in the back of mice. Ectopic bone formation was generated within the hydrogel after 2.5 weeks demonstrating the feasibility of this approach (Krebs et al., 2010).

Another approach involves the intracellular delivery of microRNAs (miRNA), small molecules that regulate posttranscriptional events, to control the osteogenic differentiation (Mencía Castaño et al., 2015). A plethora of miRNAs has been identified to be associated with osteogenesis such as miR-21, miR-148b, etc. (Leng et al., 2020). Qureshi et al. (2015) elegantly designed a photo-responsive system that delivers miR-148b in a temporally controlled manner to induce osteogenic differentiation in human adipose-derived stem cells (ASCs). The miR-148b was tethered to silver nanoparticles via a photo-cleavable linker, which breaks when photo-activated by an external light source. This allowed a spatial/temporal-controlled miR-148b release which subsequently affected ASCs differentiation in vitro and resident host stem cells in vivo.

Besides biologics, drugs such as bisphosphonate used for treating osteoporosis have also been incorporated in nanoparticles of bone implants. As bone in our body undergoes a continuous process of bone remodeling whereby osteoclast mediates bone resorption and osteoblast mediates bone formation in a coordinated fashion, it is equally important to mediate the bone resorption process especially in osteoporotic patients (Ikeda and Takeshita, 2014). Wang et al. (2018) developed mesoporous silicate nanoparticle (MSN)-based electrospun polycaprolacton/gelatin nanofibers for the dual delivery of alendronate and silicate ions. The strategy was based on the premise that alendronate could inhibit the bone-resorbing process via inhibiting guanosine triphosphate-related protein expression, while silicate ions could enhance the bone-forming process via promoting angiogenesis and bone mineralization. Implantation of the MSN-loaded nanofibers in a rat critical-sized cranial defect model revealed an accelerated bone healing time (4 weeks postimplantation), which was three times faster in comparison with the nanofibers carrying only ALN or MSN.

3.5 Conclusion and future prospects

Nanotechnology has been widely adopted and applied in bone tissue engineering and its regeneration. Appreciation of the intricate interplay between bone cells and its niche has opened a new area for the creation of bone biomimetics. The future lies in amalgamating knowledge in biology, engineering, and data analytics in the creation of next-generation bone substitutes. The fabrication of bone substitutes using 3D printing technology is currently limited by its resolution and range of bioinks. Customization of bone substitutes using computed tomography scan data from a patient's bone defect has already made it to the clinic. In due course, precision medicine using autologous MSCs or iPSCs for the generation of a multitude of cell types including blood vessels and nerves will become mainstream. The next advancement is expected with the development of picotechnology.

4 Nanomaterials in management of chronic respiratory diseases and mucosal injury

4.1 Introduction

Chronic respiratory diseases (CRDs) such as asthma, chronic obstructive pulmonary disease (COPD), pulmonary fibrosis, pneumonia, and lung cancer are the third highest cause of mortality worldwide, which caused 7% of all global deaths (Khaltaev and Axelrod, 2019). Besides CRDs, the mucosal layer of the airway also can be damaged by infections, pollution, smoking, and surgical interventions such as intubation and sinus surgery (Cao et al., 2020; Jahshan et al., 2020; Robinet et al., 2005). Furthermore, allergies, infections, pollution, and occupational hazards are also worsening the condition of patients suffering from CRDs (Yang et al., 2017). As the efficacious treatment modalities for respiratory diseases and mucosal injuries are inadequate, more efficient pharmacological approaches or bioengineering-based treatment modalities are essential to managing the CRDs and mucosal injury (Kumar et al., 2017; Selvarajah et al., 2020).

Inhalation is one of the important ways of drug delivery to the airways including the lung and also to the systemic route. The advantage of drug delivery thru inhalation or pulmonary delivery includes a relatively lower dose of the drug is required, lower incidence of systemic side effects, rapid uptake into a systemic route, avoid various limitation associated with drug intake through the gastrointestinal (GI) route (oral administration) such as poor absorption in GI route, avoid invasive administration by the intravenous route, and rapid onset of action for some drugs. However, the pulmonary delivery comes with its own challenges which include the need for high adherence or deposition of the aerosol particle to the airways, the removal and degradation of drug by mucociliary action, mechanical barriers such as narrow airways, and particle engulfment by alveolar macrophages (Newman, 2017). Various nanoparticles are being developed and tested in alleviating CRDs and also repairing damaged respiratory mucosa.

4.2 Asthma

Asthma is CRD that is characterized by inflamed airways that swell and have increased secretion of mucus. The narrowed airway tract causes shortness of breath and wheezing. Salbutamol is the common bronchodilator that is used to relieve and prevent asthma. A study that produced nano-salbutamol sulfate (SBS) dry powder inhalers found that the drug had sustained release for more than 8h after inhalation, higher total lung deposition, and even distribution within different lung regions (Bhavna et al., 2009). Nasr et al. (2014) used polyamidoamine (PAMAM) dendrimers as nanocarriers for pulmonary delivery of poorly soluble antiasthma drug beclometasone dipropionate (BDP). This study found increased solubilization of BDP and sustained release of BDP for more than 8h. More recent approaches include the delivery of small molecules, proteins, and nucleic acids using nanocarriers to deliver these drugs to targeted cells or tissues with increased efficiency. As Bcl-2 protein inhibition is known to reduce the number of inflammatory cells in allergen-induced airway inflammation, a Bcl-2 inhibitor, ABT-199, was nanoformulated (Nf-ABT-199) to deliver ABT-199 to inflammatory cells in a mouse model of allergic asthma (Tian et al., 2019). The study found that Nf-ABT-199 was able to accumulate in inflammatory cells of the lungs and efficiently reduced allergic asthma without any toxic effect to the other cells and without affecting the airway epithelial barrier and liver function. The nanoformulated Bcl-2 inhibitor Nf-ABT-199 accumulates in the mitochondria of inflammatory cells and efficiently alleviates allergic asthma. Furthermore, poly(lactic-co-glycolic acid) (PLGA) nanoparticle-based vaccine against dust mite allergies also have been developed and this vaccine was shown to prevent dust mite allergies successfully in the mice model (Joshi et al., 2014; Salem, 2014).

4.3 Chronic obstructive pulmonary disease

COPD is a chronic inflammatory lung condition that causes breathing difficulties and cough with sputum production. COPD is commonly caused by long-term exposure to inhaled toxic particles or fumes, such as tobacco smoke, dust, air pollution, and also due to untreated or prolonged asthma that damages the airway walls (Eapen et al., 2017; Passi et al., 2020). Nanotechnology enables the sustained release of a drug or a combination of drugs in therapeutic concentrations for a prolonged time period. Furthermore, direct delivery of drugs to the lungs enables the drug's activity to be retained for a longer time period, enable specific cell targeting and reduce systemic toxicity (Passi et al., 2020). The advantage of nanocarrier in COPD treatment is that the smaller nanoparticle can readily cross the thick and highly viscous mucous of COPD patients compared to the larger nanoparticles (Alp and Aydogan, 2020; Liu et al., 2020). For example, cysteamine is currently used in tablet form as a systemic treatment of COPD. However, nanoparticle (PLGA-PEG- or dendrimer) mediated cysteamine delivery is able to produce a more sustained release of cysteamine and higher bioavailability compared in the tablet form because the nanoformulated cysteamine is able to penetrate the thick mucus and inflammatory barrier of COPD patient's airways. Furthermore, the nanoformulated cysteamine also showed increased antibacterial activity and also was able to restore the impaired autophagy function in COPD and cystic fibrosis (Vij, 2017, 2020).

4.4 Lung cancer

Lung cancer, which was uncommon cancer in the 20th century, has become the second common cancer and the top mortality-causing cancer type (Deng et al., 2020). Tobacco smoking is the major risk factor of lung cancer and the other risk factors are including environmental and occupational exposures to toxic gases and particles, CRDs, lung infections, and other lifestyle/behavioral factors (Bade and Dela Cruz, 2020). Although chemotherapy is the main treatment modality of lung cancer, it causes various adverse effects. Thus, the scientist is desperately searching for lung cancer drugs and delivery methods that are more effective but cause less side effects. Various nanoparticle-based inhalable drug delivery systems have been designed for more targeted delivery of chemotherapeutic drugs to tumor tissues and hence reduce the side effects caused by systemic delivery of those drugs (Abdelaziz et al., 2018). Furthermore, these inhalable nanoparticle carriers were found to efficiently carry the chemotherapeutic drugs, able to be made into aerosols, tolerate nebulization forces, and target the cancer tissues in the lung. The nanocarriers also were able to avoid mucociliary clearance and lung phagocytic mechanisms while reaching the lung tissues, thus enhancing the residence of the therapeutic agent and its release in a sustained manner (Abdelaziz et al., 2018).

Various type of nanoparticles, which includes polymeric, lipid, hybrid lipid-polymeric, and inorganic nanocarriers, are being studied to carry the chemotherapeutic drugs, genes, or their combinations to the cancerous tissue at lung. The polymeric nanoparticles made of either albumin, gelatin, hyaluronan, poly(ethylene glycol)-poly(L-lysine) (PLL), or poly(isobutyl cyanoacrylate, BIPCA) have been used to carry chemotherapeutic drugs such as cisplatin, doxorubicin, in animal models. These studies showed enhanced efficacy in chemotherapeutics delivery, sustained release, and mode of action. In term of gene therapy for cancer, viral vectors that were initially proposed to deliver the therapeutic genes was a cause of concern. This concern was alleviated with polymeric nanocarriers that are proven to be less toxic, nonimmunogenic, highly stable during storage, easily producible in large-scale production, most importantly it would not cause any viral vectors integration into the human genome. Polyethyleneimine (PEI), PLGA, chitosan, poly(amidoamine) (PAMAM), poly(propylene imine) (PPI) and combination of spermine (SPE) and glycerol propoxylate-triacrylate (GPT) have been used as nanocarriers for transfection of genes into tissues (Abdelaziz et al., 2018; Amreddy et al., 2017).

Besides the polymeric nanocarriers, liposomes have been widely studied for delivery of lipid-based drugs as these bilayered phospholipid vesicles are biocompatible, nonimmunogenic, able to increase solubility and permeability of chemotherapeutic agents, and have the capacity to incorporate both the hydrophilic and hydrophobic drugs in their cores and lipophilic bilayers (Abdelaziz et al., 2018). A study by Tagami et al., using a hybrid lipid-polymeric nanocarrier composed of phospholipid (DPPC) and PEGylated block-copolymer (Poloxamer 188) found that the hybrid nanocarrier can rapidly release the encapsulated hydrophilic drug, doxorubicin in the presence of phospholipase A_2 (PLA$_2$), which is an enzyme expressed by various cancerous tissue and have been identified as a target for cancer therapy including lung cancer (Lu and Dong, 2017; Tagami et al., 2017). Besides that, gold, various types of quantum dots (QDs), mesoporous silica, and superparamagnetic iron oxide NPs (SPIONs) are some examples of inorganic materials that have been studied to produce inhalable nanocarriers for lung cancer treatment. The inorganic compounds' superior physico-chemical properties and the ability to modify and control the parameters such as size, shape, surface charge, and concentration to increase the drug delivery efficiency make these inorganic compounds potential inhalable nanocarriers for lung cancer drug delivery (Kreyling et al., 2018; Mukherjee et al., 2019).

5 Biomaterials in cardiovascular tissue engineering and regenerative medicine

5.1 Introduction

Cardiovascular diseases (CVD) are the number one cause of death worldwide. CVD involves the heart and blood vessels, therefore covering coronary heart disease, cerebrovascular disease, peripheral arterial disease, rheumatic heart disease, congenital heart disease, deep vein thrombosis, and pulmonary embolism.

Current clinical interventions such as coronary artery bypass, balloon angioplasty, valve repair and replacement, heart transplantation, and artificial heart operations could initially solve the problem and slow down the progression of the diseases. Unfortunately, patients undergoing these interventions are predisposed to future re-intervention procedures or other clinical complications due to nonadequate remodeling of the implanted graft or the use of prolonged immunosuppressant.

The emergence of tissue engineering and regenerative medicine (TERM) research focusing on cardiovascular diseases are developed based on the tissue engineering triad, which employs the cells, scaffold, and regulatory signals in devising a product or approach to combat CVDs. These triad approaches enable researchers to look into different perspectives that influence the applicability of engineered tissue. The notable need for several tissues for cardiovascular repair such as small-diameter vascular grafts, valves, and cardiac patches drives researchers to develop such tissue via TERM approach. In this subchapter, the biomaterials, fabrication methods, and currently available products in the clinical trials are discussed in detail.

5.2 Choices of biomaterials

Extracellular matrices are the base materials in building tissue. It is produced by cells and is where cells reside on. In considering materials for cardiovascular TERM, researchers usually either go for either synthetic, biological, or a combination of both making a hybrid biomaterial. Each class of biomaterials has its advantages and disadvantages.

Synthetic materials such as polytetrafluoroethylene (ePTFE or Teflon) and polyethylene-terephthalate (PET or Dacron) are used successfully in a large diameter (>6 mm) vascular graft and prosthetic heart valve (Daenens et al., 2009; Steinvil et al., 2017). Both ePTFE and PET have very robust mechanical properties that could withstand the high-pressure environment of the heart. Unfortunately, both Teflon and Dacron are prone to thrombosis and intimal hyperplasia thus additional research is done to overcome the apparent problem (Martin et al., 1999; Ao et al., 2000; Caro et al., 2005). Once Teflon and Dacron are implanted, they remain in the body until death as they are not biodegradable. Researchers are interested in producing a graft that could biodegrade and exploits the body's own ability to regenerate and heal tissue. Therefore, the use of polylactic acid (PLL), polyglycolic acid (PGA), poly(ε-caprolactone) (PCL), and polyglycerol sebacate (PGS) was researched as potential materials in fabricating cardiovascular graft and were proven to be of possible good candidate (Mrówczyński et al., 2014; Sugiura et al., 2017). These synthetic materials have to be modified to improve its biocompatibility with cells that are intended to repopulate the acellular grafts (Mrówczyński et al., 2014; Serrano et al., 2005; Huang et al., 2010; Pektok et al., 2008).

Biological or natural biomaterials on the other hand have better biocompatibility and are another choice of material in fabricating cardiovascular grafts. Gelatin, collagen, elastin, and decellularized extracellular matrix (dECM) are well characterized, naturally occurring biopolymers. These biomaterials are superior that its synthetic counterparts as they are readily available compounds of the body. In the tissue engineering triad, biomaterials act as the scaffold, but a scaffold that does not favor cell attachment, growth, and proliferation would not produce a good cardiovascular graft. Collagen and elastin promote cells attachment and proliferation (Robinet et al., 2005; Park et al., 2019; Hodge and Quint, 2020). The only disadvantages of natural biomaterials are its mechanical strength that is not as robust as synthetic biomaterials or at par with the native tissue. Research in overcoming these hurdles is done by acclimatizing the biomaterials in an environment that stimulates the native environment. In vitro and ex vivo studies on collagen and elastin shows mechanical stimulation improves graft strength (Engbers-Buijtenhuijs et al., 2006; Berglund et al., 2003).

Decellularized natural biomaterials bypass several fabrication steps in producing cardiovascular graft. dECM possess both the mechanical strength and biocompatibility that is crucial in translational use of the graft in vivo and in clinical settings. dECM could be either autograft, allograft, and xenograft. The first dECM for vascular graft is the bovine carotid artery that was used as an arteriovenous fistula (AVF) and was observed to not be significantly better than ePTFE that are commonly used (Butler et al., 1977; Hurt et al., 1983). dECM for vascular graft has advanced further since then, Amensag et al. (2017) reported a patent graft after 4 weeks of implantation of human amnion membrane on rabbit carotid artery with no significant decrease in diameter and blood flow.

Hybrid biomaterials exploit the use of both synthetic and biological material together to balances both material advantages and disadvantages. Atala and colleagues combined both PLA and implanted it in the rabbit at the aortoiliac position and reported that the graft was patent after 1 month in vivo with little cells invasion and thrombosis observed (Tillman et al., 2009). Wise et al. (2011) report similar success with the combination of PCL and elastin that was patent after 1 month implantation on rabbit carotid artery.

5.3 Fabrication of biomaterials

There are various fabrication methods to produce a cardiovascular graft. The most use methods in the fabrication of cardiovascular grafts are conventional casting, electrospinning, and 3D printing or bioprinting.

Conventional casting or molding has been used in biomaterials fabrication in vitro. Desired shape and form could be very easily achieved with a mold. The complex shape of a heart valve, for example, is fabricated by using PLA and PGA in a heart valve cast before being seeded with cells hence creating neotissue (Fallahiarezoudar et al., 2015; Mol et al., 2009). The cardiac patch was also developed by casting, whereby collagen type 1 was molded in culture plates to get a disk spherical sheet before evaluation in both murine and porcine models (Serpooshan et al., 2013, 2014).

Electrospinning is done by loading biomaterials in a syringe before being pumped through the needles to form strands of nano- to micrometer size fibers. These fibers are a resemblance of ECM, e.g., collagen strands and allow arrangement of these strands as in native tissues. Alignment of fibers could be manipulated to allow maximum cells attachment and adhesion onto the matrix (Kim and Cho, 2016). Dacron vascular graft fabricated via electrospinning allows endothelialization and vascularization through the pores of the alignment of the fibers (Blakeney et al., 2011; Schmidt et al., 1984).

3D printing or bioprinting is the most complex fabrication method that allows minute arrangements of multiple types of biomaterials with precision. In contrast to electrospinning where the biomaterials are fabricated into its desired shape before cells were introduced in the complex, 3D bioprinting allows both biomaterials and cells to be printed out simultaneously (Guillemette et al., 2010). Thus, permitting cells to be placed in crevices and locations that are previously hard to reach (Badie and Bursac, 2009). 3D bioprinting of endothelial cells (EC) onto polyethylene glycol (PEG) functionalized PLL created a vascular network that resembles a capillary network in nature (Pacharra et al., 2019).

5.4 Clinical applications

Cardiovascular TERM has come a long way from the first use of decellularized bovine carotid artery as AVF shunt in 1977 (Butler et al., 1977). To date, promising clinical applications include vascular graft and heart valve.

Tissue-engineered vascular graft (TEVG) is a significantly advanced niche with the clinical trial application. Shinoka and colleagues have worked on TEVG from in vitro and in vivo until they successfully transplanted a segment of the pulmonary artery in a 4-year-old female patient (Matsuzaki et al., 2019; Sugiura et al., 2018). They implanted more TEVG in patients with extracardiac total cavopulmonary connection (TCPC) surgery. Implanted TEVG was patent with no rupture, aneurysm, or calcification and no graft-related complication mortalities. Commercially, CorMatrix is a graft derived from decellularized porcine small intestinal submucosa ECM that is marketed by CorMatrix Cardiovascular Inc. CorMatrix is FDA approved product derived from the work of Badylack and colleagues (Lantz et al., 1993). The application of CorMatrix in a clinical setting is positive albeit most studies highlighted the need for long-term follow-up (Nelson et al., 2016; Mosala Nezhad et al., 2016; Woo et al., 2016; Rosario-Quinones et al., 2015).

Another promising product of cardiovascular TERM is the tissue engineering heart valve (TEHV). Clinical application of TEHV starts fairly rough with the initial success of decellularized porcine valve, Synergraft in adult humans (Elkins et al., 2001). More implantation of Synergraft into pediatric patients was done but mortality due to graft complications was then reported (Simon et al., 2003; Bechtel et al., 2008). Therefore, future applications of TEHV were done with caution especially with xenograft, where complete removal of xenogeneic antigens remains challenging.

5.5 Summary

Cardiovascular TERM is a rapidly developing field alongside the development of biomaterials and the advancement of fabrication methods. These interdisciplinary field warrants technological advancement for translational application. Clinical applications of cardiovascular TERM products need to be done very stringently to avoid any unnecessary mortality due to negligence in product development. Quality control and quality assurance measure must be taken into consideration in any translational application. A guideline or criteria in assuring the safety of products need to be addressed. Criteria to be met in the decellularization of tissues and the desired mechanical properties are crucial in determining the safety and efficacy of future clinical products. With the collaborative partnership of interdisciplinary research community, safe translational product in cardiovascular TERM is attainable.

References

Abdelaziz, H.M., Gaber, M., Abd-Elwakil, M.M., Mabrouk, M.T., Elgohary, M.M., Kamel, N.M., et al., 2018. Inhalable particulate drug delivery systems for lung cancer therapy: nanoparticles, microparticles, nanocomposites and nanoaggregates. J. Control. Release 269, 374–392. https://doi.org/10.1016/j.jconrel.2017.11.036. Available from: https://www.sciencedirect.com/science/article/pii/S0168365917310325.

Alipour, R., Khorshidi, A., Shojaei, A.F., Mashayekhi, F., Moghaddam, M.J.M., 2019. Skin wound healing acceleration by Ag nanoparticles embedded in PVA/PVP/Pectin/Mafenide acetate composite nanofibers. Polym. Test. 79 (July). https://doi.org/10.1016/j.polymertesting.2019.106022, 106022.

Alp, G., Aydogan, N., 2020. Lipid-based mucus penetrating nanoparticles and their biophysical interactions with pulmonary mucus layer. Eur. J. Pharm. Biopharm. 149, 45–57. https://doi.org/10.1016/j.ejpb.2020.01.017.

Amensag, S., Goldberg, L.A., O'Malley, K.A., Rush, D.S., Berceli, S.A., McFetridge, P.S., 2017. Pilot assessment of a human extracellular matrix-based vascular graft in a rabbit model. J. Vasc. Surg. 65 (3), 839–847. https://doi.org/10.1016/j.jvs.2016.02.046.

Amreddy, N., Babu, A., Muralidharan, R., Munshi, A., Ramesh, R., 2017. Polymeric nanoparticle-mediated gene delivery for lung cancer treatment. Top. Curr. Chem. 375 (2), 35. https://doi.org/10.1007/s41061-017-0128-5.

Anselme, K., Davidson, P., Popa, A.M., Giazzon, M., Liley, M., Ploux, L., 2010. The interaction of cells and bacteria with surfaces structured at the nanometre scale. Acta Biomater. 6 (10), 3824–3846. https://doi.org/10.1016/j.actbio.2010.04.001.

Ao, P., Hawthorne, W., Vicaretti, M., Fletcher, J., 2000. Development of intimal hyperplasia in six different vascular prostheses. Eur. J. Vasc. Endovasc. Surg. 20 (3), 241–249. https://doi.org/10.1053/ejvs.2000.1177.

Bade, B.C., Dela Cruz, C.S., 2020. Lung cancer 2020: epidemiology, etiology, and prevention. Clin. Chest Med. 41 (1), 1–24. https://doi.org/10.1016/j.ccm.2019.10.001.

Badie, N., Bursac, N., 2009. Novel micropatterned cardiac cell cultures with realistic ventricular microstructure. Biophys. J. 96 (9), 3873–3885. https://doi.org/10.1016/j.bpj.2009.02.019. Available from: https://linkinghub.elsevier.com/retrieve/pii/S0006349509005724.

Bairagi, U., Mittal, P., Singh, J., Mishra, B., 2018. Preparation, characterization, and in vivo evaluation of nano formulations of ferulic acid in diabetic wound healing. Drug Dev. Ind. Pharm. 44 (11), 1783–1796. https://doi.org/10.1080/03639045.2018.1496448.

Bechtel, J.F.M., Stierle, U., Sievers, H.-H., 2008. Fifty-two months' mean follow up of decellularized SynerGraft-treated pulmonary valve allografts. J. Heart Valve Dis. 17 (1), 98–104. Available from: http://www.ncbi.nlm.nih.gov/pubmed/18365576.

Berglund, J.D., Mohseni, M.M., Nerem, R.M., Sambanis, A., 2003. A biological hybrid model for collagen-based tissue engineered vascular constructs. Biomaterials 24 (7), 1241–1254. https://doi.org/10.1016/s0142-9612(02)00506-9. Available from: https://linkinghub.elsevier.com/retrieve/pii/S0142961202005069.

Bernal-Chávez, S., Nava-Arzaluz, M.G., Quiroz-Segoviano, R.I.Y., Ganem-Rondero, A., 2019. Nanocarrier-based systems for wound healing. Drug Dev. Ind. Pharm. 45 (9), 1389–1402. https://doi.org/10.1080/03639045.2019.1620270. Available from: https://www.tandfonline.com/doi/full/10.1080/03639045.2019.1620270.

Bhavna, A.F.J., Mittal, G., Jain, G.K., Malhotra, G., Khar, R.K., et al., 2009. Nano-salbutamol dry powder inhalation: a new approach for treating broncho-constrictive conditions. Eur. J. Pharm. Biopharm. 71 (2), 282–291. https://doi.org/10.1016/j.ejpb.2008.09.018.

Blakeney, B.A., Tambralli, A., Anderson, J.M., Andukuri, A., Lim, D.-J., Dean, D.R., et al., 2011. Cell infiltration and growth in a low density, uncompressed three-dimensional electrospun nanofibrous scaffold. Biomaterials 32 (6), 1583–1590. https://doi.org/10.1016/j.biomaterials.2010.10.056. Available from: https://linkinghub.elsevier.com/retrieve/pii/S0142961210013840.

Butler, H.G., Baker, L.D., Johnson, J.M., 1977. Vascular access for chronic hemodialysis: polytetrafluoroethylene (PTFE) versus bovine heterograft. Am. J. Surg. 134 (6), 791–793. https://doi.org/10.1016/0002-9610(77)90326-9. Available from: https://linkinghub.elsevier.com/retrieve/pii/0002961077903269.

Cao, Y., Chen, M., Dong, D., Xie, S., Liu, M., 2020. Environmental pollutants damage airway epithelial cell cilia: implications for the prevention of obstructive lung diseases. Thorac. Cancer 11 (3), 505–510. https://doi.org/10.1111/1759-7714.13323.

Caro, C.G., Cheshire, N.J., Watkins, N., 2005. Preliminary comparative study of small amplitude helical and conventional ePTFE arteriovenous shunts in pigs. J. R. Soc. Interface 2 (3), 261–266. https://doi.org/10.1098/rsif.2005.0044.

Chan, S.Y., Chan, B.Q.Y., Liu, Z., Parikh, B.H., Zhang, K., Lin, Q., et al., 2017. Electrospun pectin-polyhydroxybutyrate nanofibers for retinal tissue engineering. ACS Omega 2 (12), 8959–8968. https://doi.org/10.1021/acsomega.7b01604.

Chatterjee, U., Jewrajka, S.K., Guha, S., 2009. Dispersion of functionalized silver nanoparticles in polymer matrices: stability, characterization, and physical properties. Polym. Compos. 30 (6), 827–834. https://doi.org/10.1002/pc.20655.

Chaurasia, S., Lim, R., Lakshminarayanan, R., Mohan, R., 2015. Nanomedicine approaches for corneal diseases. J. Funct. Biomater. 6 (2), 277–298. https://doi.org/10.3390/jfb6020277. Available from: http://www.mdpi.com/2079-4983/6/2/277.

Chereddy, K.K., Her, C.-H., Comune, M., Moia, C., Lopes, A., Porporato, P.E., et al., 2014. PLGA nanoparticles loaded with host defense peptide LL37 promote wound healing. J. Control. Release 194 (2018), 138–147. https://doi.org/10.1016/j.jconrel.2014.08.016. Available from: https://linkinghub.elsevier.com/retrieve/pii/S0168365914005987.

Daenens, K., Schepers, S., Fourneau, I., Houthoofd, S., Nevelsteen, A., 2009. Heparin-bonded ePTFE grafts compared with vein grafts in femoropopliteal and femorocrural bypasses: 1- and 2-year results. J. Vasc. Surg. 49 (5), 1210–1216. https://doi.org/10.1016/j.jvs.2008.12.009.

Dalby, M.J., Gadegaard, N., Tare, R., Andar, A., Riehle, M.O., Herzyk, P., et al., 2007. The control of human mesenchymal cell differentiation using nanoscale symmetry and disorder. Nat. Mater. 6 (12), 997–1003. https://doi.org/10.1038/nmat2013.

Datta, H.K., Ng, W.F., Walker, J.A., Tuck, S.P., Varanasi, S.S., 2008. The cell biology of bone metabolism. J. Clin. Pathol. 61 (5), 577–587. https://doi.org/10.1136/jcp.2007.048868.

Dave, V., Kushwaha, K., Yadav, R.B., Agrawal, U., 2017. Hybrid nanoparticles for the topical delivery of norfloxacin for the effective treatment of bacterial infection produced after burn. J. Microencapsul. 34 (4), 351–365. https://doi.org/10.1080/02652048.2017.1337249. Available from: https://www.tandfonline.com/doi/full/10.1080/02652048.2017.1337249.

De Witte, T.-M., Fratila-Apachitei, L.E., Zadpoor, A.A., Peppas, N.A., 2018. Bone tissue engineering via growth factor delivery: from scaffolds to complex matrices. Regen. Biomater. 5 (4), 197–211. https://doi.org/10.1093/rb/rby013.

Deng, Y., Zhao, P., Zhou, L., Xiang, D., Hu, J., Liu, Y., et al., 2020. Epidemiological trends of tracheal, bronchus, and lung cancer at the global, regional, and national levels: a population-based study. J. Hematol. Oncol. 13 (1), 98–114. https://doi.org/10.1186/s13045-020-00915-0.

Duan, X., Sheardown, H., 2006. Dendrimer crosslinked collagen as a corneal tissue engineering scaffold: mechanical properties and corneal epithelial cell interactions. Biomaterials 27 (26), 4608–4617. https://doi.org/10.1016/j.biomaterials.2006.04.022.

Eapen, M.S., Myers, S., Walters, E.H., Sohal, S.S., 2017. Airway inflammation in chronic obstructive pulmonary disease (COPD): a true paradox. Expert Rev. Respir. Med. 11 (10), 827–839. https://doi.org/10.1080/17476348.2017.1360769.

Eliaz, N., Metoki, N., 2017. Calcium phosphate bioceramics: a review of their history, structure, properties, coating technologies and biomedical applications. Mater (Basel, Switz.) 10 (4), 334–438. https://doi.org/10.3390/ma10040334.

Elkhenany, H., Bourdo, S., Hecht, S., Donnell, R., Gerard, D., Abdelwahed, R., et al., 2017. Graphene nanoparticles as osteoinductive and osteoconductive platform for stem cell and bone regeneration. Nanomedicine 13 (7), 2117–2126. https://doi.org/10.1016/j.nano.2017.05.009.

Elkins, R.C., Dawson, P.E., Goldstein, S., Walsh, S.P., Black, K.S., 2001. Decellularized human valve allografts. Ann. Thorac. Surg. 71 (5 Suppl), S428–S432. https://doi.org/10.1016/s0003-4975(01)02503-6. Available from: https://linkinghub.elsevier.com/retrieve/pii/S0003497501025036.

Engbers-Buijtenhuijs, P., Buttafoco, L., Poot, A.A., Dijkstra, P.J., de Vos, R.A.I., Sterk, L.M.T., et al., 2006. Biological characterisation of vascular grafts cultured in a bioreactor. Biomaterials 27 (11), 2390–2397. https://doi.org/10.1016/j.biomaterials.2005.10.016. Available from: https://linkinghub.elsevier.com/retrieve/pii/S0142961205009208.

Fallahiarezoudar, E., Ahmadipourroudposht, M., Idris, A., Mohd, Y.N., 2015. A review of: application of synthetic scaffold in tissue engineering heart valves. Mater. Sci. Eng. C Mater. Biol. Appl. 48, 556–565. https://doi.org/10.1016/j.msec.2014.12.016. Available from: https://linkinghub.elsevier.com/retrieve/pii/S0928493114008145.

Figueiredo, A.G.P.R., Figueiredo, A.R.P., Alonso-Varona, A., Fernandes, S.C.M., Palomares, T., Rubio-Azpeitia, E., et al., 2013. Biocompatible bacterial cellulose-poly(2-hydroxyethyl methacrylate) nanocomposite films. Biomed. Res. Int. 2013, 1–14. https://doi.org/10.1155/2013/698141. Available from: http://www.hindawi.com/journals/bmri/2013/698141/.

Florencio-Silva, R., Sasso, G.R.D.S., Sasso-Cerri, E., Simões, M.J., Cerri, P.S., 2015. Biology of bone tissue: structure, function, and factors that influence bone cells. Biomed. Res. Int. 2015, 421746. https://doi.org/10.1155/2015/421746.

Garcia-Orue, I., Gainza, G., Villullas, S., Pedraz, J.L., Hernandez, R.M., Igartua, M., 2016. Nanotechnology approaches for skin wound regeneration using drug-delivery systems. In: Nanobiomaterials in Soft Tissue Engineering: Applications of Nanobiomaterials. Elsevier Inc, pp. 31–55, https://doi.org/10.1016/B978-0-323-42865-1.00002-7.

Guillemette, M.D., Park, H., Hsiao, J.C., Jain, S.R., Larson, B.L., Langer, R., et al., 2010. Combined technologies for microfabricating elastomeric cardiac tissue engineering scaffolds. Macromol. Biosci. 10 (11), 1330–1337. https://doi.org/10.1002/mabi.201000165.

Han, Y., You, X., Xing, W., Zhang, Z., Zou, W., 2018. Paracrine and endocrine actions of bone-the functions of secretory proteins from osteoblasts, osteocytes, and osteoclasts. Bone Res. 6, 16–27. https://doi.org/10.1038/s41413-018-0019-6.

Hayes, J.S., Richards, R.G., 2010. Surfaces to control tissue adhesion for osteosynthesis with metal implants: in vitro and in vivo studies to bring solutions to the patient. Expert Rev. Med. Devices 7 (1), 131–142. https://doi.org/10.1586/erd.09.55.

Hill, M.J., Qi, B., Bayaniahangar, R., Araban, V., Bakhtiary, Z., Doschak, M.R., et al., 2019. Nanomaterials for bone tissue regeneration: updates and future perspectives. Nanomedicine (Lond.) 14 (22), 2987–3006. https://doi.org/10.2217/nnm-2018-0445.

Hodge, J., Quint, C., 2020. Tissue engineered vessel from a biodegradable electrospun scaffold stimulated with mechanical stretch. Biomed. Mater. 15 (5), 055006. https://doi.org/10.1088/1748-605X/ab8e98. Available from: https://iopscience.iop.org/article/10.1088/1748-605X/ab8e98.

Huang, Y.-C., Kaigler, D., Rice, K.G., Krebsbach, P.H., Mooney, D.J., 2005. Combined angiogenic and osteogenic factor delivery enhances bone marrow stromal cell-driven bone regeneration. J. Bone Miner. Res. Off. J. Am. Soc. Bone Miner. Res. 20 (5), 848–857. https://doi.org/10.1359/JBMR.041226.

Huang, X., Zauscher, S., Klitzman, B., Truskey, G.A., Reichert, W.M., Kenan, D.J., et al., 2010. Peptide interfacial biomaterials improve endothelial cell adhesion and spreading on synthetic polyglycolic acid materials. Ann. Biomed. Eng. 38 (6), 1965–1976. https://doi.org/10.1007/s10439-010-9986-5. Available from: http://link.springer.com/10.1007/s10439-010-9986-5.

Hurt, A.V., Batello-Cruz, M., Skipper, B.J., Teaf, S.R., Sterling, W.A., 1983. Bovine carotid artery heterografts versus polytetrafluoroethylene grafts. A prospective, randomized study. Am. J. Surg. 146 (6), 844–847. https://doi.org/10.1016/0002-9610(83)90356-2.

Ikeda, K., Takeshita, S., 2014. Factors and mechanisms involved in the coupling from bone resorption to formation: how osteoclasts talk to osteoblasts. J. Bone Metab. 21 (3), 163–167. https://doi.org/10.11005/jbm.2014.21.3.163.

Jahshan, F., Ertracht, O., Eisenbach, N., Daoud, A., Sela, E., Atar, S., et al., 2020. A novel rat model for tracheal mucosal damage assessment of following long term intubation. Int. J. Pediatr. Otorhinolaryngol. 128, 109738. https://doi.org/10.1016/j.ijporl.2019.109738. 109738. doi:31698244.

Joshi, V.B., Adamcakova-Dodd, A., Jing, X., Wongrakpanich, A., Gibson-Corley, K.N., Thorne, P.S., et al., 2014. Development of a poly (lactic-co-glycolic acid) particle vaccine to protect against house dust mite induced allergy. AAPS J. 16 (5), 975–985. https://doi.org/10.1208/s12248-014-9624-5.

Khaltaev, N., Axelrod, S., 2019. Chronic respiratory diseases global mortality trends, treatment guidelines, life style modifications, and air pollution: preliminary analysis. J. Thorac. Dis. 11, 2643–2655. https://doi.org/10.21037/jtd.2019.06.08.

Kim, P.-H., Cho, J.-Y., 2016. Myocardial tissue engineering using electrospun nanofiber composites. BMB Rep. 49 (1), 26–36. https://doi.org/10.5483/BMBRep.2016.49.1.165. Available from: http://koreascience.or.kr/journal/view.jsp?kj=E1MBB7&py=2016&vnc=v49n1&sp=26.

Krebs, M.D., Salter, E., Chen, E., Sutter, K.A., Alsberg, E., 2010. Calcium phosphate-DNA nanoparticle gene delivery from alginate hydrogels induces in vivo osteogenesis. J. Biomed. Mater. Res. A 92 (3), 1131–1138. https://doi.org/10.1002/jbm.a.32441.

Kreyling, W.G., Möller, W., Holzwarth, U., Hirn, S., Wenk, A., Schleh, C., et al., 2018. Age-dependent rat lung deposition patterns of inhaled 20 nanometer gold nanoparticles and their quantitative biokinetics in adult rats. ACS Nano 12 (8), 7771–7790. https://doi.org/10.1021/acsnano.8b01826.

Kumar, P., Vrana, N.E., Ghaemmaghami, A., 2017. Prospects and challenges in engineering functional respiratory epithelium for in vitro and in vivo applications. Microphysiol. Syst. 1 (2), 1–20. https://doi.org/10.21037/mps.2017.09.01.

Kuttappan, S., Mathew, D., Jo, J.-I., Tanaka, R., Menon, D., Ishimoto, T., et al., 2018. Dual release of growth factor from nanocomposite fibrous scaffold promotes vascularisation and bone regeneration in rat critical sized calvarial defect. Acta Biomater. 78, 36–47. https://doi.org/10.1016/j.actbio.2018.07.050.

Lantz, G.C., Badylak, S.F., Hiles, M.C., Coffey, A.C., Geddes, L.A., Kokini, K., et al., 1993. Small intestinal submucosa as a vascular graft: a review. J. Investig. Surg. 6 (3), 297–310. https://doi.org/10.3109/08941939309141619. Available from: http://www.tandfonline.com/doi/full/10.3109/08941939309141619.

Lee, J., Lee, Y.J., Cho, H., Kim, D.W., Shin, H., 2015. The incorporation of bFGF mediated by heparin into PCL/gelatin composite fiber meshes for guided bone regeneration. Drug Deliv. Transl. Res. 5 (2), 146–159. https://doi.org/10.1007/s13346-013-0154-y.

Leng, Q., Chen, L., Lv, Y., 2020. RNA-based scaffolds for bone regeneration: application and mechanisms of mRNA, miRNA and siRNA. Theranostics 10 (7), 3190–3205. https://doi.org/10.7150/thno.42640.

Liao, S., Cui, F.-Z., Zhu, Y., 2004. Osteoblasts adherence and migration through three-dimensional porous mineralized collagen based composite: nHAC/PLA. J. Bioact. Compat. Polym. 19, 117–130. https://doi.org/10.1177/0883911504042643.

Liu, Q., Guan, J., Qin, L., Zhang, X., Mao, S., 2020. Physicochemical properties affecting the fate of nanoparticles in pulmonary drug delivery. Drug Discov. Today 25 (1), 150–159. https://doi.org/10.1016/j.drudis.2019.09.023.

Lu, S., Dong, Z., 2017. Overexpression of secretory phospholipase A2-IIa supports cancer stem cell phenotype via HER/ERBB-elicited signaling in lung and prostate cancer cells. Int. J. Oncol. 50 (6), 2113–2122. https://doi.org/10.3892/ijo.2017.3964.

Lü, L., Deegan, A., Musa, F., Xu, T., Yang, Y., 2018. The effects of biomimetically conjugated VEGF on osteogenesis and angiogenesis of MSCs (human and rat) and HUVECs co-culture models. Colloids Surf. B: Biointerfaces 167, 550–559. https://doi.org/10.1016/j.colsurfb.2018.04.060.

Ma, X.-Y., Bao, H.-J., Cui, L., Zou, J., 2013. The graft of autologous adipose-derived stem cells in the corneal stromal after mechanic damage. PLoS One 8 (10), e76103. https://doi.org/10.1371/journal.pone.0076103. Available from: https://doi.org/10.1371/journal.pone.0076103.

Martin, L.G., MacDonald, M.J., Kikeri, D., Cotsonis, G.A., Harker, L.A., Lumsden, A.B., 1999. Prophylactic angioplasty reduces thrombosis in virgin ePTFE arteriovenous dialysis grafts with greater than 50% stenosis: subset analysis of a prospectively randomized study. J. Vasc. Interv. Radiol. 10 (4), 389–396. https://doi.org/10.1016/s1051-0443(99)70054-0. Available from: https://linkinghub.elsevier.com/retrieve/pii/S1051044399700540.

Matsuzaki, Y., John, K., Shoji, T., Shinoka, T., 2019. The evolution of tissue engineered vascular graft technologies: from preclinical trials to advancing patient care. Appl. Sci. (Switz.) 9 (7), 1274–1295. https://doi.org/10.3390/app9071274.

Mencía Castaño, I., Curtin, C.M., Shaw, G., Murphy, J.M., Duffy, G.P., O'Brien, F.J., 2015. A novel collagen-nanohydroxyapatite microRNA-activated scaffold for tissue engineering applications capable of efficient delivery of both miR-mimics and antagomiRs to human mesenchymal stem cells. J. Control. Release 200, 42–51. https://doi.org/10.1016/j.jconrel.2014.12.034.

Mol, A., Smith, A.I.P.M., Bouten, C.V.C., Baaijens, F.P.T., 2009. Tissue engineering of heart valves: advances and current challenges. Expert Rev. Med. Devices 6 (3), 259–275. https://doi.org/10.1586/erd.09.12.

Molino, G., Palmieri, M.C., Montalbano, G., Fiorilli, S., Vitale-Brovarone, C., 2020. Biomimetic and mesoporous nano-hydroxyapatite for bone tissue application: a short review. Biomed. Mater. 15 (2), 22001. https://doi.org/10.1088/1748-605X/ab5f1a.

Montalbano, G., Molino, G., Fiorilli, S., Vitale-Brovarone, C., 2020. Synthesis and incorporation of rod-like nano-hydroxyapatite into type I collagen matrix: a hybrid formulation for 3D printing of bone scaffolds. J. Eur. Ceram. Soc. 40 (11), 3689–3697. https://doi.org/10.1016/j.jeurceramsoc.2020.02.018. Available from: https://www.sciencedirect.com/science/article/pii/S0955221920301114.

Mosala Nezhad, Z., Poncelet, A., De Kerchove, L., Gianello, P., Fervaille, C., El Khoury, G., 2016. Small intestinal submucosa extracellular matrix (CorMatrix®) in cardiovascular surgery: a systematic review. Interact. Cardiovasc. Thorac. Surg. 22 (6), 839–850. https://doi.org/10.1093/icvts/ivw020.

Mrówczyński, W., Mugnai, D., De Valence, S., Tille, J.C., Khabiri, E., Cikirikcioglu, M., et al., 2014. Porcine carotid artery replacement with biodegradable electrospun poly-e-caprolactone vascular prosthesis. J. Vasc. Surg. 59 (1), 210–219. https://doi.org/10.1016/j.jvs.2013.03.004.

Mukherjee, A., Paul, M., Mukherjee, S., 2019. Recent progress in the theranostics application of nanomedicine in lung cancer. Cancers (Basel) 11 (5), 597–615. https://doi.org/10.3390/cancers11050597.

Nagai, N., Nakazawa, Y., Ito, Y., Kanai, K., Okamoto, N., Shimomura, Y., 2017. A nanoparticle-based ophthalmic formulation of dexamethasone enhances corneal permeability of the drug and prolongs its corneal residence time. Biol. Pharm. Bull. 40 (7), 1055–1062. https://doi.org/10.1248/bpb.b17-00137.

Nagai, N., Iwai, Y., Deguchi, S., Otake, H., Kanai, K., Okamoto, N., et al., 2019. Therapeutic potential of a combination of magnesium hydroxide nanoparticles and sericin for epithelial corneal wound healing. Nanomaterials 9 (5), 768. https://doi.org/10.3390/nano9050768. Available from: https://www.mdpi.com/2079-4991/9/5/768.

Nasr, M., Najlah, M., D'Emanuele, A., Elhissi, A., 2014. PAMAM dendrimers as aerosol drug nanocarriers for pulmonary delivery via nebulization. Int. J. Pharm. 461 (1–2), 242–250. https://doi.org/10.1016/j.ijpharm.2013.11.023.

Neděla, O., Slepička, P., Švorčík, V., 2017. Surface modification of polymer substrates for biomedical applications. Mater (Basel, Switz.) 10 (10), 1115–1137. https://doi.org/10.3390/ma10101115.

Nelson, J.S., Heider, A., Si, M.-S., Ohye, R.G., 2016. Evaluation of explanted CorMatrix intracardiac patches in children with congenital heart disease. Ann. Thorac. Surg. 102 (4), 1329–1335. https://doi.org/10.1016/j.athoracsur.2016.03.086. Available from: https://linkinghub.elsevier.com/retrieve/pii/S0003497516302211.

Newman, S.P., 2017. Drug delivery to the lungs: challenges and opportunities. Ther. Deliv. 8 (8), 647–661. https://doi.org/10.4155/tde-2017-0037.

Ng, M.H., Duski, S., Tan, K.K., Yusof, M.R., Low, K.C., Rose, I.M., et al., 2014. Repair of segmental load-bearing bone defect by autologous mesenchymal stem cells and plasma-derived fibrin impregnated ceramic block results in early recovery of limb function. Biomed. Res. Int. 2014, 345910. https://doi.org/10.1155/2014/345910.345910.

Nibourg, L.M., Gelens, E., de Jong, M.R., Kuijer, R., van Kooten, T.G., Koopmans, S.A., 2016. Nanofiber-based hydrogels with extracellular matrix-based synthetic peptides for the prevention of capsular opacification. Exp. Eye Res. 143, 60–67. https://doi.org/10.1016/j.exer.2015.10.001.

Pacharra, S., Ortiz, R., McMahon, S., Wang, W., Viebahn, R., Salber, J., et al., 2019. Surface patterning of a novel PEG-functionalized poly-l-lactide polymer to improve its biocompatibility: applications to bioresorbable vascular stents. J. Biomed. Mater. Res. B Appl. Biomater. 107 (3), 624–634. https://doi.org/10.1002/jbm.b.34155.

Park, S., Kim, J., Lee, M.-K., Park, C., Jung, H.-D., Kim, H.-E., et al., 2019. Fabrication of strong, bioactive vascular grafts with PCL/collagen and PCL/silica bilayers for small-diameter vascular applications. Mater. Des. 181, 108079. https://doi.org/10.1016/j.matdes.2019.108079. Available from: https://linkinghub.elsevier.com/retrieve/pii/S0264127519305179.

Passi, M., Shahid, S., Chockalingam, S., Sundar, I.K., Packirisamy, G., 2020. Conventional and nanotechnology based approaches to combat chronic obstructive pulmonary disease: implications for chronic airway diseases. Int. J. Nanomedicine 15, 3803–3826. https://doi.org/10.2147/IJN.S242516.

Pektok, E., Nottelet, B., Tille, J.C., Gurny, R., Kalangos, A., Moeller, M., et al., 2008. Degradation and healing characteristics of small-diameter poly(ε-caprolactone) vascular grafts in the rat systemic arterial circulation. Circulation 118 (24), 2563–2570. https://doi.org/10.1161/CIRCULATIONAHA.108.795732.

Pritchard, C.D., Arnér, K.M., Neal, R.A., Neeley, W.L., Bojo, P., Bachelder, E., et al., 2010. The use of surface modified poly(glycerol-co-sebacic acid) in retinal transplantation. Biomaterials 31 (8), 2153–2162. https://doi.org/10.1016/j.biomaterials.2009.11.074.

Qureshi, A.T., Doyle, A., Chen, C., Coulon, D., Dasa, V., Del Piero, F., et al., 2015. Photoactivated miR-148b-nanoparticle conjugates improve closure of critical size mouse calvarial defects. Acta Biomater. 12, 166–173. https://doi.org/10.1016/j.actbio.2014.10.010.

Rasoulianboroujeni, M., Fahimipour, F., Shah, P., Khoshroo, K., Tahriri, M., Eslami, H., et al., 2019. Development of 3D-printed PLGA/TiO(2) nanocomposite scaffolds for bone tissue engineering applications. Mater. Sci. Eng. C Mater. Biol. Appl. 96, 105–113. https://doi.org/10.1016/j.msec.2018.10.077.

Reyes-Ortega, F., Cifuentes, A., Rodríguez, G., Aguilar, M.R., González-Gómez, Á., Solis, R., et al., 2015. Bioactive bilayered dressing for compromised epidermal tissue regeneration with sequential activity of complementary agents. Acta Biomater. 23, 103–115. https://doi.org/10.1016/j.actbio.2015.05.012. Available from: https://linkinghub.elsevier.com/retrieve/pii/S1742706115002299.

Reznikov, N., Bilton, M., Lari, L., Stevens, M.M., Kröger, R., 2018. Fractal-like hierarchical organization of bone begins at the nanoscale. Science 360 (6388), 2189–2218. https://doi.org/10.1126/science.aao2189.

Robinet, A., Fahem, A., Cauchard, J.-H., Huet, E., Vincent, L., Lorimier, S., et al., 2005. Elastin-derived peptides enhance angiogenesis by promoting endothelial cell migration and tubulogenesis through upregulation of MT1-MMP. J. Cell Sci. 118 (Pt 2), 343–356. https://doi.org/10.1242/jcs.01613. Available from: http://jcs.biologists.org/cgi/doi/10.1242/jcs.01613.

Robinson, R.A., 1952. An electron-microscopic study of the crystalline inorganic component of bone and its relationship to the organic matrix. J. Bone Joint Surg. Am. 34-A (2), 389–435.

Robling, A.G., Castillo, A.B., Turner, C.H., 2006. Biomechanical and molecular regulation of bone remodeling. Annu. Rev. Biomed. Eng. 8, 455–498. https://doi.org/10.1146/annurev.bioeng.8.061505.095721.

Rosario-Quinones, F., Magid, M.S., Yau, J., Pawale, A., Nguyen, K., 2015. Tissue reaction to porcine intestinal submucosa (CorMatrix) implants in pediatric cardiac patients: a single-center experience. Ann. Thorac. Surg. 99 (4), 1373–1377. https://doi.org/10.1016/j.athoracsur.2014.11.064. Available from: https://linkinghub.elsevier.com/retrieve/pii/S0003497514023297.

Sahle, F.F., Kim, S., Niloy, K.K., Tahia, F., Fili, C.V., Cooper, E., et al., 2019. Nanotechnology in regenerative ophthalmology. Adv. Drug Deliv. Rev. 148, 290–307. https://doi.org/10.1016/j.addr.2019.10.006. Available from: https://linkinghub.elsevier.com/retrieve/pii/S0169409X19301917.

Salem, A.K., 2014. A promising CpG adjuvant-loaded nanoparticle-based vaccine for treatment of dust mite allergies. Immunotherapy 6, 1161–1163. https://doi.org/10.2217/imt.14.97.

Schmidt, S.P., Hunter, T.J., Sharp, W.V., Malindzak, G.S., Evancho, M.M., 1984. Endothelial cell–seeded four-millimeter Dacron vascular grafts: effects of blood flow manipulation through the grafts. J. Vasc. Surg. 01 (3), 434–441. avs00104346237211.

Selvarajah, J., Bin, S.A., Bt Hj Idrus, R., Lokanathan, Y., 2020. Current and alternative therapies for nasal mucosa injury: a review. Int. J. Mol. Sci. 21 (2), 480–495. https://doi.org/10.3390/ijms21020480.

Serpooshan, V., Zhao, M., Metzler, S.A., Wei, K., Shah, P.B., Wang, A., et al., 2013. The effect of bioengineered acellular collagen patch on cardiac remodeling and ventricular function post myocardial infarction. Biomaterials 34 (36), 9048–9055. https://doi.org/10.1016/j.biomaterials.2013.08.017.

Serpooshan, V., Zhao, M., Metzler, S.A., Wei, K., Shah, P.B., Wang, A., et al., 2014. Use of bio-mimetic three-dimensional technology in therapeutics for heart disease. Bioengineered 5 (3), 193–197. https://doi.org/10.4161/bioe.27751. Available from: http://www.tandfonline.com/doi/abs/10.4161/bioe.27751.

Serrano, M.C., Portolés, M.T., Vallet-Regí, M., Izquierdo, I., Galletti, L., Comas, J.V., et al., 2005. Vascular endothelial and smooth muscle cell culture on NaOH-treated poly(ε-caprolactone) films: a preliminary study for vascular graft development. Macromol. Biosci. 5 (5), 415–423. https://doi.org/10.1002/mabi.200400214.

Shahmoradi, S., Yazdian, F., Tabandeh, F., Soheili, Z.-S., Hatamian Zarami, A.S., Navaei-Nigjeh, M., 2017. Controlled surface morphology and hydrophilicity of polycaprolactone toward human retinal pigment epithelium cells. Mater. Sci. Eng. C Mater. Biol. Appl. 73, 300–309. https://doi.org/10.1016/j.msec.2016.11.076. Available from: https://linkinghub.elsevier.com/retrieve/pii/S0928493116314631.

Simon, P., Kasimir, M.T., Seebacher, G., Weigel, G., Ullrich, R., Salzer-Muhar, U., et al., 2003. Early failure of the tissue engineered porcine heart valve SYNERGRAFT in pediatric patients. Eur. J. Cardiothorac. Surg. 23 (6), 1002–1006. discussion 1006. https://doi.org/10.1016/s1010-7940(03)00094-0. Available from: https://academic.oup.com/ejcts/article-lookup/doi/10.1016/S1010-7940(03)00094-0.

Steinvil, A., Bernardo, N., Rogers, T., Koifman, E., Buchanan, K., Alraies, M.C., et al., 2017. Use of an ePTFE-covered nitinol self-expanding stent graft for the treatment off pre-closure device failure during transcatheter aortic valve replacement. Cardiovasc. Revasc. Med. 18 (2), 128–132. https://doi.org/10.1016/j.carrev.2016.10.006. Available from: https://linkinghub.elsevier.com/retrieve/pii/S1553838916302871.

Stocco, T.D., Antonioli, E., Elias, C.d.M.V., Rodrigues, B.V.M., Siqueira, I.A.W.d.B., Ferretti, M., et al., 2019. Cell viability of porous poly(d,l-lactic acid)/vertically aligned carbon nanotubes/nanohydroxyapatite scaffolds for osteochondral tissue engineering. Mater (Basel, Switz.) 12 (6), 849. https://doi.org/10.3390/ma12060849.

Sugiura, T., Tara, S., Nakayama, H., Yi, T., Lee, Y.-U., Shoji, T., et al., 2017. Fast-degrading bioresorbable arterial vascular graft with high cellular infiltration inhibits calcification of the graft. J. Vasc. Surg. 66 (1), 243–250. https://doi.org/10.1016/j.jvs.2016.05.096. Available from: https://linkinghub.elsevier.com/retrieve/pii/S0741521416308230.

Sugiura, T., Matsumura, G., Miyamoto, S., Miyachi, H., Breuer, C.K., Shinoka, T., 2018. Tissue-engineered vascular grafts in children with congenital heart disease: intermediate term follow-up. Semin. Thorac. Cardiovasc. Surg. 30 (2), 175–179. https://doi.org/10.1053/j.semtcvs.2018.02.002.

Tagami, T., Ando, Y., Ozeki, T., 2017. Fabrication of liposomal doxorubicin exhibiting ultrasensitivity against phospholipase A(2) for efficient pulmonary drug delivery to lung cancers. Int. J. Pharm. 517 (1–2), 35–41. https://doi.org/10.1016/j.ijpharm.2016.11.039.

Tian, B.-P., Li, F., Li, R., Hu, X., Lai, T.-W., Lu, J., et al., 2019. Nanoformulated ABT-199 to effectively target Bcl-2 at mitochondrial membrane alleviates airway inflammation by inducing apoptosis. Biomaterials 192, 429–439. https://doi.org/10.1016/j.biomaterials.2018.06.020.

Tillman, B.W., Yazdani, S.K., Lee, S.J., Geary, R.L., Atala, A., Yoo, J.J., 2009. The in vivo stability of electrospun polycaprolactone-collagen scaffolds in vascular reconstruction. Biomaterials 30 (4), 583–588. https://doi.org/10.1016/j.biomaterials.2008.10.006. Available from: https://linkinghub.elsevier.com/retrieve/pii/S0142961208007813.

Uzunalli, G., Soran, Z., Erkal, T.S., Dagdas, Y.S., Dinc, E., Hondur, A.M., et al., 2014. Bioactive self-assembled peptide nanofibers for corneal stroma regeneration. Acta Biomater. 10 (3), 1156–1166. https://doi.org/10.1016/j.actbio.2013.12.002.

Vial, S., Reis, R.L., Oliveira, J.M., 2017. Recent advances using gold nanoparticles as a promising multimodal tool for tissue engineering and regenerative medicine. Curr. Opin. Solid State Mater. Sci. 21 (2), 92–112. https://doi.org/10.1016/j.cossms.2016.03.006. Available from: https://www.sciencedirect.com/science/article/pii/S1359028616300201.

Vij, N., 2017. Nano-based rescue of dysfunctional autophagy in chronic obstructive lung diseases. Expert Opin. Drug Deliv. 14 (4), 483–489. https://doi.org/10.1080/17425247.2016.1223040.

Vij, N., 2020. Synthesis and evaluation of dendrimers for autophagy augmentation and alleviation of obstructive lung diseases. Methods Mol. Biol. 2118, 155–164. https://doi.org/10.1007/978-1-0716-0319-2_12.

Wang, Q., Yan, J., Yang, J., Li, B., 2016. Nanomaterials promise better bone repair. Mater. Today 19 (8), 451–463. https://doi.org/10.1016/j.mattod.2015.12.003. Available from: https://www.sciencedirect.com/science/article/pii/S1369702115004046.

Wang, Y., Cui, W., Zhao, X., Wen, S., Sun, Y., Han, J., et al., 2018. Bone remodeling-inspired dual delivery electrospun nanofibers for promoting bone regeneration. Nanoscale 11 (1), 60–71. https://doi.org/10.1039/c8nr07329e.

Wang, W., Lu, K.J., Yu, C.H., Huang, Q.L., Du, Y.Z., 2019. Nano-drug delivery systems in wound treatment and skin regeneration. J. Nanobiotechnol. 17 (1), 1–15. https://doi.org/10.1186/s12951-019-0514-y.

Warnke, P.H., Alamein, M., Skabo, S., Stephens, S., Bourke, R., Heiner, P., et al., 2013. Primordium of an artificial Bruch's membrane made of nanofibers for engineering of retinal pigment epithelium cell monolayers. Acta Biomater. 9 (12), 9414–9422. https://doi.org/10.1016/j.actbio.2013.07.029.

Webster, T.J., Ergun, C., Doremus, R.H., Siegel, R.W., Bizios, R., 2000. Enhanced functions of osteoblasts on nanophase ceramics. Biomaterials 21 (17), 1803–1810. https://doi.org/10.1016/s0142-9612(00)00075-2.

Willermain, F., Libert, S., Motulsky, E., Salik, D., Caspers, L., Perret, J., et al., 2014. Origins and consequences of hyperosmolar stress in retinal pigmented epithelial cells. Front. Physiol. 5, 199. https://doi.org/10.3389/fphys.2014.00199 (Frontiers Media S.A.).

Wilson, H.R., 2001. Vision, low-level theory of. In: Smelser, N.J., Baltes PBBT-IE of the S& BS (Eds.), International Encyclopedia of the Social & Behavioral Sciences. Elsevier, Oxford, pp. 16232–16237. Available from: https://linkinghub.elsevier.com/retrieve/pii/B0080430767006719.

Wise, S.G., Byrom, M.J., Waterhouse, A., Bannon, P.G., Ng, M.K.C., Weiss, A.S., 2011. A multilayered synthetic human elastin/polycaprolactone hybrid vascular graft with tailored mechanical properties. Acta Biomater. 7 (1), 295–303. https://doi.org/10.1016/j.actbio.2010.07.022.

Woo, J.S., Fishbein, M.C., Reemtsen, B., 2016. Histologic examination of decellularized porcine intestinal submucosa extracellular matrix (CorMatrix) in pediatric congenital heart surgery. Cardiovasc. Pathol. 25 (1), 12–17. https://doi.org/10.1016/j.carpath.2015.08.007. Available from: https://linkinghub.elsevier.com/retrieve/pii/S1054880715001076.

Wray, L.S., Orwin, E.J., 2009. Recreating the microenvironment of the native cornea for tissue engineering applications. Tissue Eng. Part A 15 (7), 1463–1472. https://doi.org/10.1089/ten.tea.2008.0239.

Xi, Y., Dong, H., Sun, K., Liu, H., Liu, R., Qin, Y., et al., 2013. Scab-inspired cytophilic membrane of anisotropic nanofibers for rapid wound healing. ACS Appl. Mater. Interfaces 5 (11), 4821–4826. https://doi.org/10.1021/am4004683.

Xiao, Y., Xu, D., Song, H., Shu, F., Wei, P., Yang, X., et al., 2019. Cuprous oxide nanoparticles reduces hypertrophic scarring by inducing fibroblast apoptosis. Int. J. Nanomedicine 14, 5989–6000. https://doi.org/10.2147/IJN.S196794. Available from: https://www.dovepress.com/cuprous-oxide-nanoparticles-reduces-hypertrophic-scarring-by-inducing--peer-reviewed-article-IJN.

Yang, Y., Mao, J., Ye, Z., Li, J., Zhao, H., Liu, Y., 2017. Risk factors of chronic obstructive pulmonary disease among adults in Chinese mainland: a systematic review and meta-analysis. Respir. Med. 131, 158–165. https://doi.org/10.1016/j.rmed.2017.08.018.

Yilgor, P., Hasirci, N., Hasirci, V., 2010. Sequential BMP-2/BMP-7 delivery from polyester nanocapsules. J. Biomed. Mater. Res. A 93 (2), 528–536. https://doi.org/10.1002/jbm.a.32520.

Zhang, R., Lee, P., Lui, V.C.H., Chen, Y., Liu, X., Lok, C.N., et al., 2015. Silver nanoparticles promote osteogenesis of mesenchymal stem cells and improve bone fracture healing in osteogenesis mechanism mouse model. Nanomedicine 11 (8), 1949–1959. https://doi.org/10.1016/j.nano.2015.07.016.

CHAPTER

13

Piezoelectric nanomaterials for biomedical applications

Akash Roy[a], Dipanjan Dwari[a], Mukesh Kumar Ram[a], and Pallab Datta[b]

[a]Centre for Healthcare Science and Technology, Indian Institute of Engineering Science and Technology (IIEST), Howrah, West Bengal, India [b]Department of Pharmaceutics, National Institute of Pharmaceutical Education and Research (NIPER), Kolkata, West Bengal, India

OUTLINE

1 Introduction and origin of piezoelectricity

Piezoelectric materials are materials that produces a voltage when force or stress is applied. This effect also applies in the reverse manner, that is a voltage across the sample produces stress within the sample. The term originated from the Greek vocabulary "piezo" which denotes "to press." In other words, piezoelectricity implies that electricity can be obtained due to application of pressure and application of electricity can also cause deformation in the material ade that bend, expanded or contracted when a voltage is applied or vice-versa.

Among the all-smart materials discussed nowadays, the piezoelectric material is the oldest one which was involved in a study long back. The effect was pioneered in 1880 by Pierre and Jacques Curie, who demonstrated that certain crystalline material produces charges on its surfaces under applied stress. Pierre Curie got Nobel prize in 1903 for this. After that it was subsequently demonstrated the converse effect when electric field is applied, crystal change shape and size, which is known as reverse piezoelectricity and for that Gabriel Lippmann got Noble prize in 1908. At that time there were actually two sources of electricity—one is contact electricity or static electricity which is generated from friction; second one is pyroelectricity which is generated from crystal due to heating, and then the third one was discovered which is the piezoelectricity generating due to pressure. This piezoelectric effect was due to the electric dipoles present in the material spontaneously aligning to the electric field. Because of this alignment there is a clear change in the shape and size according to the intensity of the diploe moment features. The first application of the piezoelectric material (actually a composite of steel plate and quartz) was in sonar transducer by Langvein in 1917. Presently, piezoelectric materials may belong to one of the following categories: natural biomaterials, ceramic crystals occurring in nature, synthetic ceramics, piezoelectric polymers, or composites of the above materials. The ceramic crystals can be monocrystalline or polycrystalline (Mason, 1981; Cross and Heywang, 2008; Uchino, 2010; Manbachi and Cobbold, 2011; Yu et al., 2020). Fig. 1 depicts the structures of commonly occurring piezoelectric materials.

Crystals are represented by their organized and repeating structure of atoms in their dimensional space called unit cells. In case of piezoelectric crystals, the atoms of the unit cell are polarized in nature. But due to their orientation the effect of the charges cancels out leaving no net charge, i.e., the net dipole moment is zero, on application of stress the structure of the unit cell undergoes deformation. Due to the deformation some positively charged atoms come toward one surface while negatively charged atoms approach another surface. This creates a separation of charges and a net dipole moment. The surface charging tendency is altered if the mechanical force is applied tension instead of compression (Bogdanov, 2002; Kholkin et al., 2008; Malgrange et al., 2014; Trolier-McKinstry, 2008). For piezoelectricity, the compound should possess a noncentrosymmetric atomic structure, though some centrosymmetric materials in nanometer dimensions or nonequilibrium conditions may exhibit piezoelectric (Lang et al., 2013). For example, in lead zirconia titanate (PZT), titanium ion or in case of aluminum nitride, nitrogen atom in the unit cell changes its position when a load is applied, resulting in charge development. This is a reversable process. In biopiezoelectric materials of biological origin like collagen or keratin, the dipole reorientation may occur due to rearrangement of hydrogen bonds while several other components of biological tissue also exhibit piezoelectricity (Guerin et al., 2019; Wojnar, 2012; Kar et al., 2021). In organic polymers, piezoelectricity develops due to reorientation of polymers under high electrical field or mechanical drawing. The most commonly used piezopolymer, polyvinylidene difluoride, can exist in different crystal phases (α, β, γ, δ, ε). Among them, all molecular dipoles are

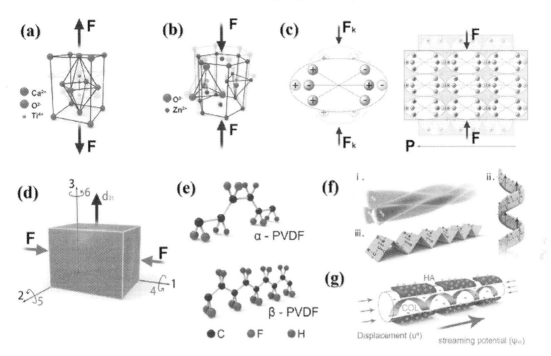

FIG. 1 Representative piezoelectric materials and their structure: (A) ceramics; (B) perovskites; (C) wurtzites; (D) anisotropy of piezoelectricity; (E) synthetic polymers; (F) naturally-occurring polymers (i) triple helix (ii) collagen; (iii) silk; (G) bone tissue. *Reproduced from Kapat, K., Shubhra, Q.T.H., Zhou, M. and Leeuwenburgh, S., 2020. Piezoelectric nano-biomaterials for biomedicine and tissue regeneration. Adv. Funct. Mater. 1909045.*

aligned parallelly in the β phase and thus this phase possess highest piezoelectricity. One interesting property of PVDF over PZT is that it possesses negative piezoelectric constant, thus it undergoes compression under applied electric field (You et al., 2019; Li et al., 2011; Mokhtari et al., 2019). In case of poly-L-lactic acid, the thermodynamically stable α-form consists of randomly oriented C=O dipoles whereas upon transformation to the β-form, dipoles get aligned along the direction of loading (Smith et al., 2017; Sultana et al., 2017). The general mechanism of piezoelectricity is illustrated in Fig. 2. The naturally occurring phenomenon of piezoelectricity is put into many practical life appliances like many medical devices (Zaszczyńska et al., 2020), mobile phones, various sensors used in cars, industries and everywhere.

Piezoelectric materials tend to produce large force when their natural expansion is constrained determined by the stiffness of the material. As the electric field across the material varies the stiffness also varies accordingly. On the other hand, the generated charge due to the straining of the material can be measured by connecting electrodes across the material. So, the piezoelectric material is very much comfortable to be used as both sensor and actuators as per one's requirement. Piezoelectric materials can be both naturally occurring and synthetic. Monocrystalline materials like berlinite, quartz, rochelle salt, topaz and tourmaline comprises the former class whereas synthetic polycrystalline materials like Barium titanate, Lead Zirconate Titanate (PZT), etc., are the classical examples of the synthetic category. However, the concerns over the toxicity of lead later gave more emphasis on lead-free piezoelectric ceramics (Ichinose and Kimura, 1991; Bhalla et al., 2000; Panda and Sahoo, 2015; Wei et al., 2018).

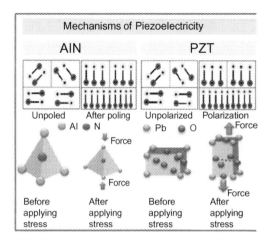
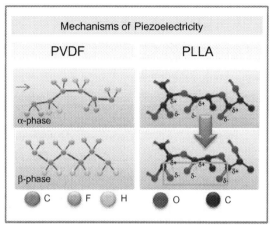

FIG. 2 Mechanisms of piezoelectricity in different inorganic and organics materials. *Reproduced from Zaszczyńska, A., Gradys, A., Sajkiewicz, P., 2020. Progress in the applications of smart piezoelectric materials for medical devices. Polymers (Basel) 12 2754.*

However, the piezoelectric effect of most naturally occurring piezoelectric materials is very low in magnitude for most practical applications. Synthetic piezoelectric materials mainly PZT offers significantly higher amount of piezoelectric effect providing larger mechanical forces for relatively small applied voltage and vice versa. The PZT ceramics can be manufactured easily and they have strong coupling between electrical and mechanical domain. That is why PZT patches have excellent application as piezoelectric material in various fields now.

Piezoelectric sensors are very much mechanically durable. They exhibit excellent linearity property for wide range of application. The piezoelectric effect is insensitive to the different environmental factors thus functioning under harsh conditions. Materials like tourmaline or gallium-phosphate have excellent thermal resilience and made to work under 1000°C. The main disadvantages of the piezoelectric sensors are, however their inability to measure static forces as they yield a static number of charges on the piezoelectric material.

2 Preparation of piezoelectric materials

Conventionally, polycrystalline ceramics are used to fabricate piezoelectric devices. The devices are fabricated in a two-step process—starting with preparation of ceramic powder, followed by sintering to obtain the desired shape and structure. Single crystals are more efficient in performance than polycrystalline forms but they are costlier. To achieve performance equivalent to single crystal, temperature grain growth techniques are used. Detailed account of the preparation methods may be found in excellent literature focusing on the topic (Uchino, 2009, 2017; Miclea, 2012). A brief overview is presented here.

2.1 Preparation of ceramic powder by mixed oxide technology

In order to maintain reproducibility of piezoelectricity the key factors needed to be controlled are particle size distribution or shape and compositional homogeneity to obtain reproducible performance. Oxide-mixing technique is the usual method to get the ceramic powder, in this process the calcination of oxide powder is followed by crushing into fine particles. The mixing of oxide technique can cause microstructural nonuniformities, and is not efficient to obtain particles less than 1 μm. In addition, the milling media also causes contamination of the samples. The wet chemical method has become more popular to overcome the above problem. On the other hand, if the materials are milled for long duration such that size reduction increases their chemical reactivity, the process is called as mechanochemical synthesis and is an effective method to obtain piezoelectric nanomaterials.

2.2 Coprecipitation

In this method, a precipitant is added to the solution of metals leading to more homogeneous precipitation. Homogeneous powders are obtained from the precipitate by using thermal dissolution. Followed by co-precipitation, chemical reactions can be carried out in an autoclave, i.e., controlled temperature and pressure conditions in a technique called as hydrothermal synthesis.

2.3 Alkoxide hydrolysis

Metal alkoxides are mixed with alcohol and water, initiating hydrolytic reaction yielding metal oxides or hydrates. This is also known as the sol-gel method and can produce very pure ceramics. The purification can be achieved using conventional distillation. The hydrolysis and condensation reaction are schematically shown as (Uchino, 2009):

Hydrolysis

$$H — O + M — OR \rightarrow H — O — M + ROH$$
$$|$$
$$H$$

Alkoxylation

$$M — O + M — OR \rightarrow M — O — M + ROH$$
$$|$$
$$H$$

Oxolation

$$M — O + M — OH \rightarrow M — O — M + OH2$$
$$|$$
$$H$$

Under appropriate pH, crystalline stoichiometric powder of very fine size, 99.98% purity can be obtained by this process.

2.4 Sintering

The agglomerated powder is molded into desired shape and then fired at very high temperatures. Due to the surface energy accelerated diffusion of atoms causes bonding of crystals at the interfacial contacts of adjacent powders and provides mechanical consolidation of the ceramic preserving the shape of the mold. In this process, the pores are eliminated and the density is increased.

2.5 Single crystal

Single crystal piezoelectrics are getting their attention to drastic improvements in actuators, sensors, and transducers technology. The most attractive properties of these materials include a high piezoelectric coefficient, low hysteresis, high strain levels, and ultra-high electromechanical coupling. These materials have promising bandwidth and sensitivity. The piezoelectric single crystals have variety of application in medical transducers such as high frequency transducers, ultrasound harmonic imaging, etc. However, small size, high cost, and low yields limit the manufacturing of the single crystal piezoelectric material. The Bridgman method is a suitable method to fabricate, in which a molten charge is poured in a crucible and moves along stationary temperature gradient leading to solidification in a single direction. In this technique the limiting factor is the segregation coefficient.

2.6 Templated grain growth

Crystallographically textured piezoelectric ceramic offers high strain actuation. In single crystal there is crystallographic orientation, due to that high actuation results but it is very expensive. Highly textured ceramic can give another pathway to aligned crystal but with lower cost. In textured ceramics the crystallites are crystallographically oriented. Into a fine precursor, large anisotropic grains are added. To minimize the limitations of sintering the template content is kept below 10 volume %. Due to additional heat treatment the template grows by the oriented matrix. That results in a strong crystallographic texture.

Among the polymeric piezoelectric materials, PVDF is generally synthesized from the monomer 1,1-difluoroethylene using a suspension or emulsion-free radical polymerization process whereas for PLA synthesis different methods like direct polycondesation or ring opening polymerization techniques have been used (Pholharn et al., 2017; Sharma et al., 2017). The piezoelectric ceramics or polymers can then be processed by several of nano and microfabrication techniques like lithography, thin film deposition, electrospinning, spin coating, etc., depending upon the application.

The charge displacement in piezoelectric takes place either by redistribution of electronic clouds, separation of ions, reorientation of dipoles, or by polarization at the interfaces. The charge-stress relationships can be expressed by various linear functions (Kapat et al., 2020;

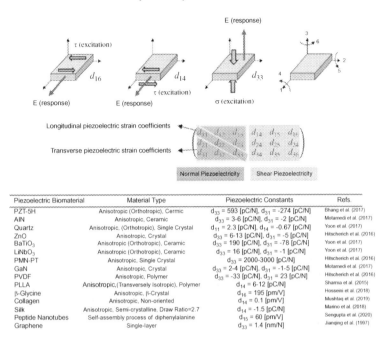

<table>
<tr><th>Piezoelectric Biomaterial</th><th>Material Type</th><th>Piezoelectric Constants</th><th>Refs.</th></tr>
</table>

Piezoelectric Biomaterial	Material Type	Piezoelectric Constants	Refs.
PZT-5H	Anisotropic (Orthotropic), Cermic	d_{33} = 593 [pC/N], d_{31} = -274 [pC/N]	Bhang et al. (2017)
AlN	Anisotropic, Ceramic	d_{33} = 3-6 [pC/N], d_{31} = -2 [pC/N]	Motamedi et al. (2017)
Quartz	Anisotropic, (Orthotropic), Single Crystal	d_{11} = 2.3 [pC/N], d_{14} = -0.67 [pC/N]	Yoon et al. (2017)
ZnO	Anisotropic, Crystal	d_{33} = 6-13 [pC/N], d_{31} = -5 [pC/N]	Hitscherich et al. (2016)
BaTiO$_3$	Anisotropic (Orthotropic), Ceramic	d_{33} = 190 [pC/N], d_{31} = -78 [pC/N]	Yoon et al. (2017)
LiNbO$_3$	Anisotropic (Orthotropic), Ceramic	d_{33} = 16 [pC/N], d_{31} = -1 [pC/N]	Yoon et al. (2017)
PMN-PT	Anisotropic, Single Crystal	d_{33} = 2000-3000 [pC/N]	Hitscherich et al. (2016)
GaN	Anisotropic, Crystal	d_{33} = 2-4 [pC/N], d_{31} = -1-5 [pC/N]	Motamedi et al. (2017)
PVDF	Anisotropic, Polymer	d_{33} = -33 [pC/N], d_{31} = 23 [pC/N]	Hitscherich et al. (2016)
PLLA	Anisotropic,(Transversely Isotropic), Polymer	d_{14} = 6-12 [pC/N]	Sharma et al. (2015)
β-Glycine	Anisotropic, β-Crystal	d_{16} = 195 [pm/V]	Hosseini et al. (2018)
Collagen	Anisotropic, Non-oriented	d_{14} = 0.1 [pm/V]	Mushtaq et al. (2019)
Silk	Anisotropic, Semi-crystalline, Draw Ratio=2.7	d_{14} = -1.5 [pC/N]	Marino et al. (2018)
Peptide Nanotubes	Self-assembly process of diphenylalanine	d_{15} = 60 [pm/V]	Sengupta et al. (2020)
Graphene	Single-layer	d_{33} = 1.4 [nm/N]	Jianqing et al. (1997)

FIG. 3 Development of electric fields due to applied stresses, and glimpse of piezoelectric coefficients of different materials used for biomedical applications. *Reproduced with permission from Chorsi, M.T., Curry, E.J., Chorsi, H.T., Das, R., Baroody, J., Purohit, P.K., Ilies, H., Nguyen, T.D., 2019. Piezoelectric biomaterials for sensors and actuators. Adv. Mater. 31. Copyright Wiley.*

Zaszczyńska et al., 2020; Chorsi et al., 2019; Sezer and Koç, 2021; Trolier-McKinstry et al., 2018; Singh, 2014). Such polarization can be achieved or enhanced by electrical or thermal poling processes. Piezoelectric potential of the materials is characterized by using parameters such as electromechanical coupling coefficients (k), piezoelectric charge constant (dij) or strain coefficients (d), and voltage coefficients. Fig. 3 represents the piezoelectric coefficients of some commonly employed piezoelectric materials in biomedical applications. The ratio of accumulated mechanical energy in the material to magnitude of electric strength applied gives the electromechanical coupling coefficient. The strain induced in the material per unit electric field or electrical polarization developed per unit of stress applied characterizes the piezoelectric charge constant and is expressed in units of Coulomb/Newton. This quantity is anisotropic and is therefore indicated with direction-dependent subscripts. The piezoelectric voltage constant is quantified by measuring generated charge per unit of applied shear forces. There exists a linear coupling between the applied field and strains generated. Both the direct and reverse piezoelectric effects are thermodynamically equivalent. The piezoelectric coefficients indicate the magnitude of effect, whereas electromechanical coupling factor indicate the work that can be done from the material based upon the phenomenon. Therefore, the measurement of observed piezoelectricity can be further expressed as shown in Eq. (1):

$$D = d \times T + \varepsilon \times E \qquad (1)$$

where, E denotes field strength, T is stress applied, D is the observed displacement, and d and ε denote the piezoelectric coefficient and material permittivity of the material.

In an extension, reverse piezoelectricity is calculated using Eq. (2) as

$$X = s \times T + d \times E \tag{2}$$

With s denoting the compliance. Thus, a material with greater piezoelectric coefficient will mean higher output voltage for the same magnitude of deformation.

The work done from the piezoelectric material is characterized by Eq. (3):

$$k = \sqrt{\frac{\text{energy converted}}{\text{input energy}}} \tag{3}$$

and for any given geometry, k can be expressed as Eq. (4):

$$k^2 = \frac{d^2}{\varepsilon s} \tag{4}$$

Nanomaterials showed a marked difference in piezoelectricity than bulk materials because of size-scale effect (Zhang and Meguid, 2016; Wang and Liu, 2012; Fridkin and Ducharme, 2014a,b; Zhang et al., 2014). Apart from size effects, temperature, crystal orientation, geometry influence the piezoelectric properties of several nanomaterials to a greater extent.

3 Biomedical applications of piezoelectric nanomaterials

3.1 Flexible, organic piezoelectric membranes for implantable applications

Modern development in mechanics, material science and manufacturing engineering allow the production of piezoelectric devices in fine dimensions in a malleable or mechanically flexible form. These flexible electronics can be attached to any type of surfaces maintaining the same operational characteristics as that of rigid devices. Electrical energy can be harvested from mechanical forces from human body movements, operating machinery, and other sources. These flexible piezoelectric materials have growing application in implantable and wearable biomedical devices. The operating principle and design guideline are different for the flexible devices than the rigid technologies. As an example, if we consider a recent study on nanoscale ribbons (nearly 200 nm thickness) of PZT created on a silicon wafer, the electromechanical activities can be explained via an analytical model not by conventional model (Dagdeviren et al., 2014; Salim et al., 2018).

PVDF and its copolymer, PVDF-TrFE, are organic polymers commonly used for piezoelectric membrane applications. These materials are manufactured into aligned fibers form by electrospinning. The β-phase of PVDF is the solid form with high piezoelectric coefficient due to dipoles being oriented parallelly because of surface charge separation is enhanced due to electrospinning resulting in high piezoelectric response. They have high piezoelectric constant (nearly −57.6 pm/V), detect pressure changes down to 0.1 Pa which is very applicable for high sensitivity touch interfaces. As a device application of flow sensing the implantable biomedical device the PVDF-TrFE thin film can be developed as pressure sensors for pressure values in 0–300 mmHg with quick response time of around 0.2 s (Sharma et al., 2013).

Polyamide 66 yarns coated with Ag are connected with the PVDF monofilaments yielding a β-phase content 80% using melt spinning approach (Anand et al., 2016). Textile-based piezoelectric which is more flexible than conventional fabrics can be manufactured by a knitting technique integrating PVDF monofilaments. Other type of technique relies on gold-coated polyester matrices as the upper and lower electrodes incorporated with polyethylene and ZnO nanowire. The Ag-coated textile and PE film interact and generate electrostatic energy. The mechanical deformation occurs due to piezoelectricity. The interaction between two phenomena causes enhancement in overall efficient system for wearable applications particularly the output voltage.

Inorganic-organic composites provide great advantages by controlling mechanical and piezoelectric properties by controlling the ratio of the composites. For example, incorporation of PZT fibers in polymer matrices provides materials with well-balanced mechanical and electrical properties. Compared to flexible platforms, stretchable materials have better application in biomedical devices particularly to covering large area of internal organs.

Recently portable and wireless micro and nano-scaled based devices are being widely used in many fields like medical devices, environment monitoring, defense technology, industry safety, etc. In all micro- and nano-scaled devices, the major problem is the battery, though modern researchers have developed battery with very small dimensions and very high battery life, still it requires replacement after certain year. The flexible piezoelectric materials use nanogenerators to provide very long-life power supply to the micro and nano-scaled devices.

A very interesting example of flexible piezoelectric is energy harvesting from the natural movement of the heart by using PZT-based ultra-flexible device without any unnecessary load or harm to heart. In vivo test was done on a swine animal model for different surgical condition. From this experiment it was demonstrated that one can get peak-to-peak 3 V when ultra-flexible device is connected between the apex of left ventricle to right ventricle (Lu et al., 2015). This demonstrated flexible coverage of system to have sustainable energy source for implantable devices from motion of heart movements.

3.2 Applications as energy harvesting materials

The piezoelectric effect can also transform kinetic energy like shocks or vibrations into electrical energy (charge). Energy harvesting is the technique, which offers a reliable and robust solution by transforming wasted shock or vibration energy of the environment to usable electrical energy with the help of piezoelectric generators or energy harvesters in domain of biomedical applications. Energy harvesting technologies aim to increase the power delivered to portable devices, implantable biosensors, wireless electronics, and flexible (stretchable) electronics, with minimum device size.

From a materials perspective, piezoelectric voltage is constant for piezoelectric nanogenerator. However, for practical applications, the voltage and power needs further improvement compared to the present state of the art is needed.

The working of the nanogenerator may be affected due to limitations in the fabrication of semiconductor piezoelectric nanomaterials. Development of a crystal nanowire of diameters less than 100 nm and longer than 50 μm are difficult. The output voltage of the nanogenerator is still controlled because of the small length of nanowires.

To efficiently convert mechanical energy into electrical form, higher piezoelectric voltage, and dielectric constants and are optimal attributes of active materials. One of the most common such used material is PZT. One fabrication process of the nanogenerator device starts with the electrospinning of PZT nanofiber. Arrays of platinum fine wire interdigitated electrodes of radius 25 μm are then deposited. Electrodes of platinum fine wire arrays are assembled on a silicon substrate. Before hand, PZT can be synthesized by sol-gel technique in which the diameter of PZT is controlled at 60 μm. Two adjacent electrodes are separated by 500 μm.

Annealing at 650°C for 25 min, yields a pure perovskite phase. Soft PDMS polymer is applied on the top of the PZT nanofibers. The fine platinum electrodes are connected to extraction electrodes for transferring the electrons harvested to the outer circuit. Nanofibers of PZT are polled in electric field of 4 V/μm around electrodes maintained more than 140°C temperature for 24 h (approx).

When one applies pressure on the upper surface of the nano-generator, through the PDMS matrix pressure is transferred to the PZT nanofibers and results in charge production due to the combination of bending and tensile stresses on PZT nanofibers. Due to this separation of charge potential difference is induced between both neighboring electrodes. The output voltage of 1.63 V and power of 0.03 μW under periodic stress are achieved. Advantages of PZT are more efficient nano-generator having low-frequency vibration is obtained, which can be made into composites or woven into fabrics (Lu et al., 2015).

3.3 Cellulose-based nanostructures for solar energy harvesting

Such devices give a renewable green resource that is bio-compatible, sustainable, cost-effective, and, biodegradable. Cellulose is one of the amplest natural polymers on our planet. Solar energy-harvester needs good charge transport with high surface area for absorption and conversion of photons into electrical energy effectively. Cellulose essentially is an insulator. The transparent substrate is the most popular use of cellulose nanofibrils in the development of the solar cell. For maximum photon absorption, both high amount of light scattered diffusely by transparent surface and high optical transparency is appropriate for solar cell development.

Paper of nanocellulose with a thickness of 40 μm (approx) can be fabricated by compression of pulp of carboxymethylated cellulose at 70°C, 1 bar pressure into a sheet format. Apply nano-cellulose paper as a substrate of poly(3,4-ethylene dioxythiophene): poly(styrene sulfonate), (3-hexylthiophene)(P3HT), and [6,6]-phenyl-C61-butyric acid methyl ester.

The performance achieved with these systems are represented as follows:

$$j_{sc} \text{ (short-circuit current)} \approx 2.4 \, \text{mAmp cm}^{-2}, \text{PCE of } 0.2\% \text{ and open-circuit voltage } (V_{oc}) \approx 0.4 \, \text{V}.$$

Cellulose nanocrystal CNC, the cellulose nanocrystal substrates offer similar optical properties to cellulose nanofibers. Highly conductive Ag electrode or a polyethyleneimine-modified Ag electrode can be further employed to increase the efficiency of CNC substrates to upto 4.0%. These materials can be used as supercapacitors, solar cells, photoelectrochemical electrodes, and nanogenerators (Hu et al., 2013; Wang et al., 2017).

3.4 Applications as bio actuators

Properties of piezoelectric actuators are high resolution up to subnanometer values, i.e., small changes in supply voltage can be detected and converted into linear motion, ultra-precise positioning in dynamic or static situations (with μm accuracy more than 10.000 N loads can be positioned), rapid response time (<1 ms), generation and handling of high pressures or forces, higher efficiency, lower driving voltage, and no friction. Movements produced by the actuators are parallel to the, applied electric field. A major limitation of piezoelectric actuators are the rate of independent hysteresis between input voltage and the output displacement, which minimizes positioning accuracy, increases undesirable oscillations, and trigger system instability.

As nanoparticles, piezoelectric materials are attractive for actuating and energy harvesting applications because of their reduced size, excellent electromechanical coupling, wave propagation vibration. Due to these properties, piezoelectric nanomaterials are used to develop biodevices, for sensing biological forces, analyze or diagnosing medical problems, electrifying (stimulating) tissue growth, and healing.

The hybrid nanowires can be fabricated using a CNT coating and zinc oxide (ZnO) crystalline layer. Due to the electrically conducting CNT layer, synergy of stiff, and the ZnO layer with semiconducting and piezoelectric properties, this nanomaterial is an attractive actuator. Both components of hybrid nanowire combine with one other by "Van der Waals" force. The components of hybrid nanowires are cast as "Euler beams." While the interphase van der Waals interaction leads to an intensifying piezoelectric effect. The lower frequency vibration of the hybrid nanowire lowers the stiffness of the structure. Functioning of hybrid nanowire-based materials/nanodevices leans on the electromechanical responses of the hybrid nanowire (Xiao et al., 2016).

The CNT-based piezoelectric bioactuators can be used for cell culture studies using acoustic waves. By controlled mechanical stimuli, the study of cellular response and adhesive forces is possible by using these systems. Such studies have further potential for cancer diagnostic, as the mechanical behavior of cell change with their proliferate conditions (Strobl et al., 2004; Fu et al., 2017).

For cochlear stimulation, Mota et al. employed piezoelectric nanocomposites consisting of polyvinylidene fluoride (PVDF) fibers and barium titanate (BT) nanoparticles. PVDF/BT nano-particles composite fabricated by electrospinning nanofibers were electrospun have also been employed as electromechanical transducers for restoring function in cochlea (Strobl et al., 2004; Fu et al., 2017). PVDF nanofibers were fabricated using electrospinning and then pressed sputtered with gold (Au) electrode on each surface poled with an electric field inside silicon oil bath for enhancing the piezoelectric property of the sensors. The fabricated actuators have also been used for energy harvesting. The PCL magnetic nanofilms with PVDF-trifluoromethyl microfibers electrospun PVDF-TrFE are collected in a spin-coated solution of PCL. PCL spin-coated over PVDF-trifluoromethyl (TrFE) electrospun nanofibers have been employed as magnetic nanofilms for the culture of cardiac cells (Lv and Cheng, 2021; Arumugam et al., 2019; Gouveia et al., 2017). A methodology for the fabrication of such microfibers is shown in Fig. 4. Similarly, PLLA and hydroxyapatite granules (gHA) membranes obtained by electrospinning have been employed for bone regeneration application (Santos et al., 2017).

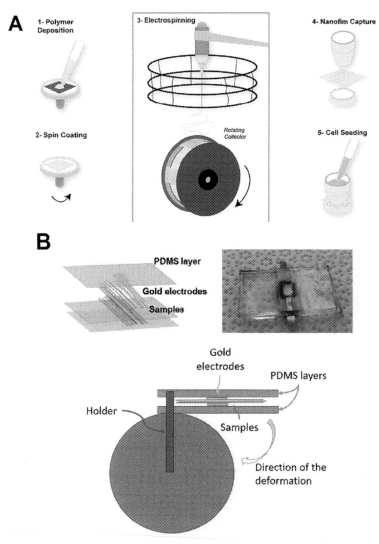

FIG. 4   A methodology for the (A) preparation and (B) assessment of the piezoelectricity in microfibers. *Reproduced from Gouveia, P.J., Rosa, S., Ricotti, L., Abecasis, B., Almeida, H.V., Monteiro, L., Nunes, J., Carvalho, F.S., Serra, M., Luchkin, S., Kholkin, A.L., Alves, P.M., Oliveira, P.J., Carvalho, R., Menciassi, A., das Neves, R.P., Ferreira, L.S., 2017. Flexible nanofilms coated with aligned piezoelectric microfibers preserve the contractility of cardiomyocytes. Biomaterials 139, 213–28.*

3.5 Applications as sensors

With the recent advancements in material science and various manufacturing techniques for different microelectromechanical systems (MEMS), several sensors and actuators are going through huge innovations for biomedical applications. Bio-devices based on biocompatible piezoelectric materials are now being designed to be safely integrated with different physiological systems for diagnosing various diseases and discomfort by sensing different biological forces as well as are used for stimulating tissue growth and healing. In terms of functionality, wearable and implantable systems with flexible electronic components are increasingly becoming popular for personalized health monitoring and point of care diagnostic system.

Wearable electronics and human-machine interactive systems are going through a period of rapid development to upgrade the level of existing individual-centered diagnostic and healthcare monitoring system. Flexible and stretchable piezoelectric nanomaterials are extensively explored for wide range of applications as wearable electronics varying form pulse sensors based on nanogenerators to stretchable and flexible electronic skins.

Electronic skin or e-skin is a good candidate as biomimetic skin which can regenerate tactile sensing function of cells and also can enable the perception of touch in prosthetics and robotics devices. Inspired by human skin which possesses a sophisticated sensing network of pressure, strain and vibration sensing mechanism to delicately perceive a wide range of physical sensation, an emerging concept of wearable e-skin has attracted the scientific community for different biomedical applications.

To perform different dynamic and sophisticated functions of wearable devices like stretching, twisting, and bending, e-skin need to be conformably connected to surface of skin. Fabrication of such a surface with the required shape adaptability and at the same time maintaining high sensitivity is a prevailing challenge. Considerable efforts have been devoted to develop a shape-adaptive flexible electronic skin to fulfill the specification of comfort as well as efficiency for its desired application.

The coaxial fiber-based electronic skins are now proposed which can be conformably mounted onto various curved as well as moving surfaces like joints to quantify the movement of these joints associated with different activities in our daily life. For example, Zhu et al. have fabricated coaxial structures of intrinsically flexible and stretchable material which have ability to absorb high mechanical deformations shows improved piezoelectric sensitivity boosted up to 10.89 mV kPa^{-1} for pressure value of 80–230 kPa. These power efficient shape-adaptive e-skin with deformable coaxial structure possesses enormous potential for biomedical applications based on motion sensing systems (Zhu et al., 2020).

Ultrathin, flexible piezoelectric sensors are also being explored for real-time monitoring of the movement of skin surfaces for making assessment of different physiological conditions. For example, real-time monitoring of eyelid motion for assessing the fatigue state of eyes is one such possible application. Eye fatigue has now become an important public health issue due to the increased time spent by of people on electronic devices. Focusing eyes on such devices for continuous prolonged duration may induce various ophthalmic diseases which can be monitored by using wearable devices made up of piezoelectric nanoribbons array patterned on a conformal flexible substrate. Similarly Lü et al. have developed a wearable sensor incorporated with an electromechanical model for detection of strain in eyes by monitoring the blinking of eyes (see Fig. 5) (Lü et al., 2018).

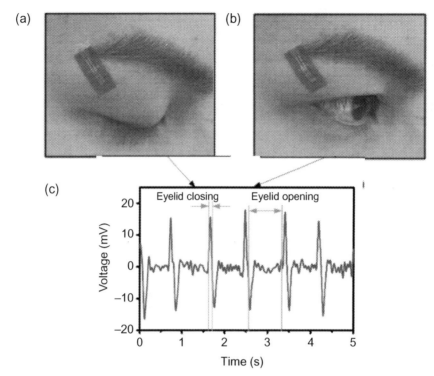

FIG. 5 In vivo application of piezo-device for eyelid motion sensing (A) open and (B) closed positions and (C) corresponding voltage fluctuations. *Reproduced with permission from Lü, C., Wu, S., Lu, B., Zhang, Y., Du, Y., Feng, X., 2018. Ultrathin flexible piezoelectric sensors for monitoring eye fatigue. J. Micromech. Microeng. 28, 25010.*

Self-powered nanogenerators (SANG) based pulse sensors for measuring blood pressure are also being explored due to their high sensitivity and improved signal-to-noise ratio. A flexible, self-arched structure of elastomers like ecoflex silicone and piezoelectric polymer like polydimethylsiloxane (PDMS), can effectively convert tiny mechanical signals into electric signal. These self-powered nanogenerators work on the combined effect of piezoelectricity and triboelectricity by taking advantage of the stress mismatch which forms a natural curvature at the interface of the two films. In another work, Zoua et al. demonstrated a SANG with enhanced sensitivity and stability with ability to monitor pulse waveforms in real time like that of radial artery for noninvasive cardiovascular disease monitoring and prediction (Zou et al., 2020).

Along with progress in conformable wearable electronics, hybrid self-powered piezoelectric sensors combined with triboelectric nanogenerators have improved energy conversion efficiency as they can harvest a variety of small mechanical energies. Fang et al. reported a triboelectric nanogenerators (TENG) based on electrostatically induced charges and contact friction possessing advantages of high efficiency, in addition to lightweight and robust construction. Research has been going on to develop these kind of hybrid nanogenerators to generate energies from different human movements like blinking or movement of facial muscles to subsequently harvest this energy for detecting biomechanical motions triggered on the skin surface (Kar et al., 2021; Zou et al., 2020).

Researches on piezoelectric force touch panels based on piezoelectric sensors and machine learning models are going on to improve the man-machine interactive system. Advancements in electronics and information technologies have fostered the development of these devices with economic power consumption, simplicity in design of panels as well as high sensitivity. However, various touch orientations destabilize the stress-voltage response of such tactile panels, which constrain the applications of interactive displays. To overcome the constraints, different machine learning models have been proposed, in which touch-induced capacitive pattern of the users are used to train artificial neural networks (ANN). As an example, Gao et al. (2020) have used a piezoelectric touch panel of polyethylene terephthalate (PET) and polyvinylidene fluoride (PVDF) along with triaxial gyroscope, accelerometer, and magnetometer to fetch the user data and trained a machine learning model to achieve an accuracy of 95.7% of in terms of distinct touch-angle classification and an accuracy of detection of 90% for force detection of different tactile orientation.

Portable biosensors for label-free detection of different biomarkers have also received much attention due to their promising applications in point of care (POC) systems for early diagnosis. Electroacoustic resonator-based piezoelectric sensors work on different acoustic wave modes, such as film bulk acoustic resonators (FBAR), quartz crystal microbalances (QCM) and are being explored for different immune-sensing application.

Quartz crystal microbalance (QCM) is an electroacoustic resonator, which can convert very small changes of mass occurring during electrochemical process to an electrical signal. In this type of piezoelectric immunosensor, chemical or physical interactions of the receptors, immobilized on resonator layer, with that of the target molecules from test analyte, creates perturbations on the surface, and changes the resonance states of the crystal. Perturbations cause changes in resonance, amplitude, frequency and phase differences to indicate presence of biomarkers. Performances of these highly sensitive portable immunosensors are comparable to that of standard ELISA technique and thus these sensors can be used to detect different biomarkers noninvasively without using any bulky detection systems.

Exploring the applications of the sensors in cell biology and toxicology, Han et al. (2018) investigated the barrier function of a monolayer of endothelial cells treated with inflammatory agents using a piezoelectric immunosensor composed of biocompatible conductive polymer membranes of poly-pyrrole, poly-glutamate, and poly-lysine. Liu et al. (2020) also demonstrated the fabrication of a portable sensor for assessment of cardiac biomarker troponins I leveraging thickness shear mode (TSM) electroacoustic resonance. The micro-fabricated device consisted of a label-free immunologic sensitive element along with a film bulk acoustic resonator (FBAR) fabricated employing ZnO film. Response of immunoreactions between the antibodies and the target antigens were converted to a quantitative frequency signal and the changes in the frequency spectrum were analyzed in real-time for measurement of troponins I (Liu et al., 2020).

3.6 Implantable in vivo physiological force sensing piezoelectric sensors

Commonly used piezoelectric materials are either toxic or nonbiodegradable in nature, which raises a significant concern in terms of safety issue, thus restricting their in vivo applications. Biodegradable and biocompatible piezoelectric materials such as glycine or PLA nanofibers are now proposed as piezoelectric materials for implantable devices. The ability to produce electrical energy under biomechanical stimulation provides great prospect for employing these piezoelectric materials for noninvasive therapeutics like, regeneration of nerves which can be an effective treatment in case of injury to the peripheral or central nervous system. With highly stable, efficient and controllable piezoelectric performances, these biodegradable piezoelectric materials can be employed for monitoring vital physiological pressures and can be incorporated in controlled drug delivery as a biodegradable ultrasonic transducer.

Exposure to optimum mechanical pressures is known to accelerate wound healing, though with the risk of hindering the natural healing process as excess pressure can tear small tissues and disrupt the vessels. Thus, the healing process can be significantly affected by monitoring the applied stress in the vicinity of wound area and the effectiveness of resultant strain, which can be achieved by integrating piezoelectric stress sensor of flexible, biocompatible materials such as glycine with wound dressing materials. A significant amount of research also has been done to improve the efficacy of the treatment of nonunions of femur through the use of high-frequency, low-magnitude vibrations.

As wearable electronic systems are getting popular for personalized health monitoring, newer technologies and materials are being explored for self-powered biodegradable sensors to display different physiological states and broaden the application of these wearable electronics. The ability to generate electrical power under mechanical stimulus paves the way for various physiological force and pressure sensing applications. Piezoelectric materials like $BaTiO_3$, PZT, etc., possess high piezoelectric coefficients but these are toxic, nonrenewable, and nonbiodegradable in nature. In addition, fabrication of flexible membrane using these materials is difficult as they require a high temperature and a strong electric field which restricts application of these materials for wearable piezoelectric sensors.

New piezoelectric materials based on biomolecules like amino acids, cellulose, peptide nanotubes and collagen have attracted the scientific community due to their biocompatibility and biodegradable nature. But most of these biomolecules have weak piezoelectricity, and are also difficult to fabricate into a unidirectional, uniform-polarized structure. Amino acids such as glycine are explored as piezoelectric material to enhance the performance of these sensors. They form self-assembled unidirectional polar crystals having uniform structure which makes them appropriate for large scale fabrication of pressure sensors based on biomolecules.

Hosseini et al. (2020) reported a highly stable and uniform β glycine membrane formed by allowing glycine crystals to grow in an elastomeric dielectric polymer matrix of chitosan. Under ambient conditions, Glycine can be crystallized into three polymorphs (α, β, and γ), among which β and γ phases have piezoelectric properties with noncentrosymmetric polar configuration. The bioresorbable chitosan polymers have improved the orientation of collagen in the collagen-chitosan components with enhanced piezoelectric responses (Hosseini et al., 2020).

Apart from these new bio-based piezoelectric materials, a lot of efforts have been put for improving the flexibility and shape adaptability of the existing piezoelectric materials. A highly shape adaptive e-skin with real-time tactile mapping is being proposed for human movement detection. Zhu et al. reported a coaxial fiber e-skin containing resilient elastic polyurethane (PU) film supporting the substrate. Flexible piezoelectric membranes can be prepared from electrospinning of inorganic $BaTiO_3$ nanoparticles as the core, surrounded by graphene-oxide-doped-PVDF (GO) (Zhu et al., 2020).

However, performance of pressure sensors based on conventional organic or inorganic piezoelectric composite materials faces the bottlenecks of cracks and defects due to poor dispersion. Yang et al. in their work modified the surface of inorganic piezoelectric materials like that of barium titanate (BaTiO$_3$) with a surface modifying agent such as polydopamine (PDA) and combined it with the piezoelectric polymer matrix of PVDF to yield a homogeneous composite. PDA functionalization strategy reduced occurrence of cracks and interfacial hole defects between the two materials and improved overall nanoparticle dispersion in PVDF phase (Yang et al., 2020).

Electroactive piezoelectric materials are explored to improve the therapeutic efficiency of bone regeneration as they can reintegrate the local osteogenic electrical microenvironment to induce stem cells. Inspired by piezoelectric properties of bone and the structural and functional properties of periosteum, Zhao et al. (2020) reported a periosteum mimicking scaffolds for critical-sized bone defect.

Biomimetic nonfibrous bioactive glass surface significantly improved the osteogenesis differentiation, adhesion, and proliferation of bone marrow stem cells. This remarkably enhanced the critical-sized bone regeneration and the formation of periosteum-like tissue at center of the defect. A compression bandage integrated with a biodegradable pressure sensor like this has the potential to accelerate the healing process. Hosseini et al. (2018) reported a flexible stress device based on piezoelectric biodegradable γ-glycine crystals embedded in PDMS layer to apply controlled mechanical pressure at the wound site. This enhanced the tissue regeneration with cellular proliferation and accelerated the healing process (Hosseini et al., 2018).

Piezoelectric scaffolds of poly(3-hydroxybutyrate) (PHB) are also explored for skin, nerve and bone regeneration. It is biodegradable and biocompatible piezoelectric material with asymmetric crystal cell structure, and has been used as a synthetic polymer in tissue engineering. Zviagin et al. (2019) have enhanced the piezoelectric properties of this PHB-based sensors by electrospinning the fibrous PHB scaffolds with ZnO. The rod-like nanostructure of ZnO improved the wettability of the scaffolds as well as improved the piezoelectric coefficient from $2.9 \pm 0.1\,\mathrm{pC\,N^{-1}}$ for pristine PHB scaffolds to $13.7 \pm 1.6\,\mathrm{pC\,N^{-1}}$ (Zviagin et al., 2019).

Piezoelectric sensors have also been known to be used along with catheters inside the body to measure blood, bladder, or intracranial pressures. Curry et al. (2020) proposed a ultrasonic transducer based on biocompatible and biodegradable polymer PLLA [poly(L-lactic acid)], which is found to exhibit piezoelectricity, thereby offering an excellent platform for opening the blood-brain barrier, thus avoiding complicated implant removal surgeries.

3.7 Drug delivery applications

Piezoelectric materials can be exploited for targeted drug delivery by fabricating them into biomimetic structure called smart nanorobots. Nanorobots can perform precise navigation to target sites and release drug cargo at the site powered by an external stimulus temperature or pH. Smart nanorobots use magnetic, electric, or acoustic energy to propel in the biological fluids like blood. For example, magnetic nanorobots can be designed in a such a way that rotating fields can be leveraged for forward locomotion and alternating fields for drug delivery.

For example, doxorubicin-loaded a nanorobots made of P(VDF-TrFE) tail and nickel (Ni) ring-polypyrrole (PPy) nanowire head demonstrated wavy motion under magnetic actuation and further pulsatile release of the drug under magnetic field stimulation of different frequency (Mushtaq et al., 2019). Piezoelectric properties of magnetostrictive FeGA and PVDF-TrFE core/shell nanowire has been used for paclitaxel delivery (Chen et al., 2017). Doxorubicin has also been complexed with barium titanate nanoparticles for delivery to neuronal cancer cells (Ciofani et al., 2010) whereas the same piezoelectric nanoparticles loaded with the drug temozolomide and functionalized with antitransferrin were shown to improve delivery across the blood-brain barrier (Marino et al., 2019). The drug was released after ultrasonic stimulation and enhanced the therapeutic efficacy of the anticancer drug. In another application, functionalization of barium titanate nanoparticles with anti-HER2 antibodies by biotin-streptovidin conjugation method and subsequent chronic piezoelectric stimulation (ultrasound-driven) showed significant cell cycle arrest and antiproliferative potential as observed by the upregulated an inward rectifier potassium channel, disturbed Ca^{2+} homeostasis, and disarranged mitotic spindles during cell division (Marino et al., 2018). Further, Jariwala et al. leveraged the inherent negative zeta potential of P(VDF-TrFE) nanofibers to deliver cationic drugs molecules (Jariwala et al., 2021). External mechanical stimuli changed the surface potential of nanofibers causing controlled release of the molecules. Similarly, Miar et al. delivered biotinylated growth factors from polypyrrole coated PVDF nanofibers (Miar et al., 2021). In fact, PVDF can also be combined with other conducting polymers to enhance piezoelectric performances (Sengupta et al., 2020, 2021).

3.8 Tissue engineering and regenerative medicine

Several cells of the human body show altered phenotype or genotypes or epigenetics (Acharya et al., 2021) in response to biomaterial stimulation resulting in enhanced proliferation or differentiation. Moreover, in compression, negative potentials are generated in bone tissue while positive potentials are generated in tension, contributing to bone regeneration. Such effects are mediated by piezoelectric properties of collagen and hydroxyapatite nanocrystals. Thus, piezoelectric scaffolds have been employed for bone tissue engineering as exemplified by piezoelectric bismuth ferrite nanofilms on strontium titanate forming an electrical field of +75 mV and interacting with electrical potential (−52 to −87 mV) of tissue (Liu et al., 2017). After implantation and under external mechanical stimulation, piezoelectric scaffolds undergo dipole reorientation and form negative charges, triggering cell signaling and Ca^{2+} channels modulation. To provide such exogenous stimulation, a PVDF actuators driven by Li-battery was used to provide electrical stimulation resulting in bone formation at injured site (Reis et al., 2012). Similarly, piezoelectric potential of PLLA nanofibers mediated protein adsorption, and callus formation (Gao et al., 2019). Much earlier, bone formation under barium titanate-hydroxyapatite scaffolds was observed in doh jawbones (Jianqing et al., 1997), which also revealed the anisotropic nature of piezoelectric scaffold-driven regeneration. Frias et al. have used a PVDF film sandwiched between printed silver electrodes to strain osteoblasts directly and used the assembly to study static and dynamic effects on cell behavior (Frias et al., 2010). Damaraju et al. have also shown the output-voltage-dependent effect of piezoelectric nanomaterials on chondrogenic or osteogenic differentiation as well as observed the superior effect of electromechanical stimulation over only-mechanical stimulation on differentiation behavior (Damaraju et al., 2017).

Exogenous electric fields can also enhance neuronal regeneration. In contrast to some wound healing where AC currents of 150–1200 mV are involved, nerve growth involves DC currents of 70–250 mV mm^{-1} because of the transmembrane potentials (Rajabi et al., 2015; Genchi et al., 2016; Bhang et al., 2017). Such properties have been exploited by using PVDF-TrFE scaffolds after positive poling for formation of myelinated axons, ultrasound-powered PZT for wireless intervention in causing muscle-twitches, as well as barium titanate ZnO nanowires and barium nitride nanotubes as transducers (Kapat et al., 2020). PVDF nanocomposites with barium titanate have also exhibited favorable response in restoring hair cell function (Motamedi et al., 2017). In the skin tissue also involves transepithelial potentials, with potentials to the tune of 500 mV mm^{-1}, being recorded for wound healing (Kapat et al., 2020). The piezoelectric principle has also been applied for mechano-electrical transduction for in vitro myogenic differentiation of mesenchymal stem cells using a poly(N-isopropylacrylamide) (pNIPAAm)-grafted PDMS and ZnO nanorods as dispersed phase (Yoon et al., 2017). With respect to cardiac tissue engineering, role of transmembrane piezo proteins, in governing cellular alignment and mechanotransduction indicate the possible role of piezoelectric materials. In some initial works, PVDF and PVDF-TrFE electrospun nanofiber membranes have been employed to culture endothelial cells and cardiomyocytes (Hitscherich et al., 2016).

Conversely, piezoelectric materials can also be used for measuring several mechanical properties of the cells. Such application has been demonstrated by the use of PZT nanoribbons for determination of viscoelastic properties of soft tissues as well as by for measurement of cellular deformations. Such PZT nanostructures are potentially less invasive for such measurements. PVDF-TrFE core-shell nanofibers with a conducting polymer (poly(3,4-ethylenedioxythiophene) (PEDOT)) can also be used for functionalizing catheters to enable live flow measurements (Sharma et al., 2015). Tajitsu et al. have demonstrated PLLA biodegradable tweezer can be used for removal of thrombotic plaques (Tajitsu, 2006). The PLLA tweezers can be inserted inside body by catheter and stimulated by AC fields applied externally. Apart from the effects on eukaryotic cells, piezoelectric nanomaterials can also influence bacterial growth and behavior after piezoelectric stimulation, which may find important applications in minimizing biomaterial-associated infections (Carvalho et al., 2019; Vatlin et al., 2020; Swain et al., 2020).

4 Conclusions

In conclusion, it can be stated that a wide range of nanomaterials with piezoelectric properties are being used for wide range of biomedical applications like sensors, actuators, drug delivery, wearables as well as implantables and tissue engineering. Future materials and fabrication innovations are going to drive design of several new devices and hitherto unknown applications may also be explored. However, biocompatibility evaluation of the materials have not kept equal pace and will require more rigorous investigations. Therefore, apart from new materials, new devices with biocompatibility, stability, and reliability of the devices should be accorded more attention.

References

Acharya, V., Ghosh, A., Chowdhury, A.R., Datta, P., 2021. Tannic acid-crosslinked chitosan matrices enhance osteogenic differentiation and modulate epigenetic status of cultured cells over glutaraldehyde crosslinking. Soft Mater. https://doi.org/10.1080/1539445X.2021.1933032.

Anand, S., Soin, N., Shah, T.H., Siores, E., 2016. Energy harvesting "3-D knitted spacer" based piezoelectric textiles. IOP Conf. Ser. Mater. Sci. Eng. 141, 12001.

Arumugam, R., Srinadhu, E.S., Subramanian, B., Nallani, S., 2019. β-PVDF based electrospun nanofibers—a promising material for developing cardiac patches. Med. Hypotheses 122, 31–34.

Bhalla, A.S., Guo, R., Roy, R., 2000. The perovskite structure—a review of its role in ceramic science and technology. Mater. Res. Innov. 4, 3–26.

Bhang, S.H., Jang, W.S., Han, J., Yoon, J.-K., La, W.-G., Lee, E., Kim, Y.S., Shin, J.-Y., Lee, T.-J., Baik, H.K., Kim, B.-S., 2017. Zinc oxide Nanorod-based piezoelectric dermal patch for wound healing. Adv. Funct. Mater. 27, 1603497.

Bogdanov, S.V., 2002. The origin of the piezoelectric effect in pyroelectric crystals. IEEE Trans. Ultrason. Ferroelectr. Freq. Control 49, 1469–1473.

Carvalho, E.O., Fernandes, M.M., Padrao, J., Nicolau, A., Marqués-Marchán, J., Asenjo, A., Gama, F.M., Ribeiro, C., Lanceros-Mendez, S., 2019. Tailoring bacteria response by piezoelectric stimulation. ACS Appl. Mater. Interfaces 11, 27297–27305.

Chen, X.-Z., Hoop, M., Shamsudhin, N., Huang, T., Özkale, B., Li, Q., Siringil, E., Mushtaq, F., Di Tizio, L., Nelson, B.J., Pané, S., 2017. Hybrid magnetoelectric nanowires for nanorobotic applications: fabrication, magnetoelectric coupling, and magnetically assisted in vitro targeted drug delivery. Adv. Mater. 29, 1605458.

Chorsi, M.T., Curry, E.J., Chorsi, H.T., Das, R., Baroody, J., Purohit, P.K., Ilies, H., Nguyen, T.D., 2019. Piezoelectric biomaterials for sensors and actuators. Adv. Mater. 31.

Ciofani, G., Danti, S., D'Alessandro, D., Moscato, S., Petrini, M., Menciassi, A., 2010. Barium Titanate nanoparticles: highly cytocompatible dispersions in glycol-chitosan and doxorubicin complexes for cancer therapy. Nanoscale Res. Lett. 5, 1093.

Cross, L.E., Heywang, W., 2008. Introduction. In: Heywang, W., Lubitz, K., Wersing, W. (Eds.), Piezoelectricity: Evolution and Future of a Technology. Springer Berlin Heidelberg, Berlin, Heidelberg, pp. 1–5.

Curry, E.J., Le, T.T., Das, R., Ke, K., Santorella, E.M., Paul, D., Chorsi, M.T., Tran, K.T.M., Baroody, J., Borges, E.R., Ko, B., Golabchi, A., Xin, X., Rowe, D., Yue, L., Feng, J., Daniela Morales-Acosta, M., Wu, Q., Chen, I.P., Tracy Cui, X., Pachter, J., Nguyen, T.D., 2020. Biodegradable nanofiber-based piezoelectric transducer. Proc. Natl. Acad. Sci. U. S. A. 117, 214–220.

Dagdeviren, C., Yang, B.D., Su, Y., Tran, P.L., Joe, P., Anderson, E., Xia, J., Doraiswamy, V., Dehdashti, B., Feng, X., Lu, B., Poston, R., Khalpey, Z., Ghaffari, R., Huang, Y., Slepian, M.J., Rogers, J.A., 2014. Conformal piezoelectric energy harvesting and storage from motions of the heart, lung, and diaphragm. Proc. Natl. Acad. Sci. 111, 1927–1932.

Damaraju, S.M., Shen, Y., Elele, E., Khusid, B., Eshghinejad, A., Li, J., Jaffe, M., Arinzeh, T.L., 2017. Three-dimensional piezoelectric fibrous scaffolds selectively promote mesenchymal stem cell differentiation. Biomaterials 149, 51–62.

Frias, C., Reis, J., Capela E Silva, F., Potes, J., Simões, J., Marques, A.T., 2010. Polymeric piezoelectric actuator substrate for osteoblast mechanical stimulation. J. Biomech. 43, 1061–1066.

Fridkin, V., Ducharme, S., 2014a. Critical size in ferroelectricity. In: Fridkin, V., Ducharme, S. (Eds.), Ferroelectricity at the Nanoscale: Basics and Applications. Springer Berlin Heidelberg, Berlin, Heidelberg, pp. 17–27.

Fridkin, V., Ducharme, S., 2014b. Switching kinetics at the nanoscale. In: Fridkin, V., Ducharme, S. (Eds.), Ferroelectricity at the Nanoscale: Basics and Applications. Springer Berlin Heidelberg, Berlin, Heidelberg, pp. 87–120.

Fu, Y.Q., Luo, J.K., Nguyen, N.T., Walton, A.J., Flewitt, A.J., Zu, X.T., Li, Y., McHale, G., Matthews, A., Iborra, E., Du, H., Milne, W.I., 2017. Advances in piezoelectric thin films for acoustic biosensors, acoustofluidics and lab-on-chip applications. Prog. Mater. Sci. 89, 31–91.

Gao, E., Zhang, C., Wang, J., 2019. Effects of budesonide combined with noninvasive ventilation on PCT, sTREM-1, chest lung compliance, humoral immune function and quality of life in patients with AECOPD complicated with type II respiratory failure. Open Med. 14, 271–278.

Gao, S., Guo, R., Shao, M., Xu, L., 2020. A touch orientation classification-based force-voltage responsivity stabilization method for piezoelectric force sensing in interactive displays. IEEE Sensors J. 20, 8147–8154.

Genchi, G.G., Marino, A., Rocca, A., Mattoli, V., Ciofani, G., 2016. Barium titanate nanoparticles: promising multitasking vectors in nanomedicine. Nanotechnology. 27, 232001.

Gouveia, P.J., Rosa, S., Ricotti, L., Abecasis, B., Almeida, H.V., Monteiro, L., Nunes, J., Carvalho, F.S., Serra, M., Luchkin, S., Kholkin, A.L., Alves, P.M., Oliveira, P.J., Carvalho, R., Menciassi, A., das Neves, R.P., Ferreira, L.S., 2017. Flexible nanofilms coated with aligned piezoelectric microfibers preserve the contractility of cardiomyocytes. Biomaterials 139, 213–228.

Guerin, S., Tofail, S.A.M., Thompson, D., 2019. Organic piezoelectric materials: milestones and potential. NPG Asia Mater. 11, 10.

Han, J., Tong, F., Chen, P., Zeng, X., Duan, Z., 2018. Study of inflammatory factors' effect on the endothelial barrier using piezoelectric biosensor. Biosens. Bioelectron. 109, 43–49.

Hitscherich, P., Wu, S., Gordan, R., Xie, L.H., Arinzeh, T., Lee, E.J., 2016. The effect of PVDF-TrFE scaffolds on stem cell derived cardiovascular cells. Biotechnol. Bioeng. 113, 1577–1585.

Hosseini, E.S., Manjakkal, L., Dahiya, R., 2018. Bio-organic glycine based flexible piezoelectric stress sensor for wound monitoring. In: 2018 IEEE SENSORS Proceedingspp. 1–4.

Hosseini, E.S., Manjakkal, L., Shakthivel, D., Dahiya, R., 2020. Glycine-chitosan-based flexible biodegradable piezo-electric pressure sensor. ACS Appl. Mater. Interfaces 12, 9008–9016.

Hu, L., Zheng, G., Yao, J., Liu, N., Weil, B., Eskilsson, M., Karabulut, E., Ruan, Z., Fan, S., Bloking, J.T., McGehee, M.D., Wågberg, L., Cui, Y., 2013. Transparent and conductive paper from nanocellulose fibers. Energy Environ. Sci. 6, 513–518.

Ichinose, N., Kimura, M., 1991. Preparation and properties of Lead zirconate-titanate piezoelectric ceramics using ultrafine particles. Jpn. J. Appl. Phys. 30, 2220–2223.

Jariwala, T., Ico, G., Tai, Y., Park, H., Myung, N.V., Nam, J., 2021. Mechano-responsive piezoelectric nanofiber as an on-demand drug delivery vehicle. ACS Appl. Bio Mater. 4, 3706–3715.

Jianqing, F., Huipin, Y., Xingdong, Z., 1997. Promotion of osteogenesis by a piezoelectric biological ceramic. Biomaterials 18, 1531–1534.

Kapat, K., Shubhra, Q.T.H., Zhou, M., Leeuwenburgh, S., 2020. Piezoelectric nano-biomaterials for biomedicine and tissue regeneration. Adv. Funct. Mater. 1909045.

Kar, E., Barman, M., Das, S., Das, A., Datta, P., Mukherjee, S., Tavakoli, M., Mukherjee, N., Bose, N., 2021. Chicken feather fiber-based bio-piezoelectric energy harvester: an efficient green energy source for flexible electronics. Sustain. Energy Fuels 5, 1857–1866.

Kholkin, A.L., Pertsev, N.A., Goltsev, A.V., 2008. Piezoelectricity and crystal symmetry. In: Safari, A., Akdoğan, E.K. (Eds.), Piezoelectric and Acoustic Materials for Transducer Applications. Springer US, Boston, MA, pp. 17–38.

Lang, S.B., Tofail, S.A.M., Kholkin, A.L., Wojtaś, M., Gregor, M., Gandhi, A.A., Wang, Y., Bauer, S., Krause, M., Plecenik, A., 2013. Ferroelectric polarization in nanocrystalline hydroxyapatite thin films on silicon. Sci. Rep. 3, 2215.

Li, J., Meng, Q., Li, W., Zhang, Z., 2011. Influence of crystalline properties on the dielectric and energy storage properties of poly(vinylidene fluoride). J. Appl. Polym. Sci. 122, 1659–1668.

Liu, Y., Zhang, X., Cao, C., Zhang, Y., Wei, J., Jun, L.Y., Liang, W., Hu, Z., Zhang, J., Wei, Y., Deng, X., 2017. Built-in electric fields dramatically induce enhancement of Osseointegration. Adv. Funct. Mater. 27, 1703771.

Liu, J., Chen, D., Wang, P., Song, G., Zhang, X., Li, Z., Wang, Y., Wang, J., Yang, J., 2020. A microfabricated thickness shear mode electroacoustic resonator for the label-free detection of cardiac troponin in serum. Talanta. 215, 120890.

Lu, B., Chen, Y., Ou, D., Chen, H., Diao, L., Zhang, W., Zheng, J., Ma, W., Sun, L., Feng, X., 2015. Ultra-flexible piezoelectric devices integrated with heart to harvest the biomechanical energy. Sci. Rep. 5, 16065.

Lü, C., Wu, S., Lu, B., Zhang, Y., Du, Y., Feng, X., 2018. Ultrathin flexible piezoelectric sensors for monitoring eye fatigue. J. Micromech. Microeng. 28, 25010.

Lv, J., Cheng, Y., 2021. Fluoropolymers in biomedical applications: state-of-the-art and future perspectives. Chem. Soc. Rev. 50, 5435–5467.

Malgrange, C., Ricolleau, C., Schlenker, M., 2014. Crystal thermodynamics. In: Malgrange, C., Ricolleau, C., Schlenker, M. (Eds.), Piezoelectricity – Symmetry and Physical Properties of Crystals. Springer Netherlands, Dordrecht, pp. 311–339.

Manbachi, A., Cobbold, R.S.C., 2011. Development and application of piezoelectric materials for ultrasound generation and detection. Ultrasound 19, 187–196.

Marino, A., Battaglini, M., De Pasquale, D., Degl'Innocenti, A., Ciofani, G., 2018. Ultrasound-activated piezoelectric nanoparticles inhibit proliferation of breast cancer cells. Sci. Rep. 8, 6257.

Marino, A., Almici, E., Migliorin, S., Tapeinos, C., Battaglini, M., Cappello, V., Marchetti, M., de Vito, G., Cicchi, R., Pavone, F.S., Ciofani, G., 2019. Piezoelectric barium titanate nanostimulators for the treatment of glioblastoma multiforme. J. Colloid Interface Sci. 538, 449–461.

Mason, W.P., 1981. Piezoelectricity, its history and applications. J. Acoust. Soc. Am. 70, 1561–1566.

Miar, S., Ong, J.L., Bizios, R., Guda, T., 2021. Electrically stimulated tunable drug delivery from polypyrrole-coated polyvinylidene fluoride. Front. Chem. 9, 4.

Miclea, C., 2012. Preparation of piezoelectric nanoparticles. In: Ciofani, G., Menciassi, A. (Eds.), Piezoelectric Nanomaterials for Biomedical Applications. Springer Berlin Heidelberg, Berlin, Heidelberg, pp. 29–61.

Mokhtari, F., Foroughi, J., Latifi, M., 2019. Enhancing β crystal phase content in electrospun PVDF nanofibers. In: Energy Harvesting Properties of Electrospun Nanofibers. IOP Publishing Ltd, pp. 5–28.

Motamedi, A.S., Mirzadeh, H., Hajiesmaeilbaigi, F., Bagheri-Khoulenjani, S., Shokrgozar, M.A., 2017. Piezoelectric electrospun nanocomposite comprising Au NPs/PVDF for nerve tissue engineering. J. Biomed. Mater. Res. Part A 105, 1984–1993.

Mushtaq, F., Torlakcik, H., Hoop, M., Jang, B., Carlson, F., Grunow, T., Läubli, N., Ferreira, A., Chen, X.-Z., Nelson, B.J., Pané, S., 2019. Motile piezoelectric Nanoeels for targeted drug delivery. Adv. Funct. Mater. 29, 1808135.

Panda, P.K., Sahoo, B., 2015. PZT to lead free piezo ceramics: a review. Ferroelectrics 474, 128–143.

Pholharn, D., Srithep, Y., Morris, J., 2017. Effect of initiators on synthesis of poly(L-lactide) by ring opening polymerization. IOP Conf. Ser. Mater. Sci. Eng. 213, 12022.

Rajabi, A.H., Jaffe, M., Arinzeh, T.L., 2015. Piezoelectric materials for tissue regeneration: a review. Acta Biomater. 24, 12–23.

Reis, J., Frias, C., Canto E Castro, C., Botelho, M.L., Marques, A.T., Simões, J.A.O., Capela E Silva, F., Potes, J., 2012. A new piezoelectric actuator induces bone formation *in vivo*: a preliminary study. (Ed. C Wu)J. Biomed. Biotechnol. 2012, 613403.

Salim, M., Salim, D., Chandran, D., Aljibori, H.S., Kherbeet, A.S., 2018. Review of nano piezoelectric devices in biomedicine applications. J. Intell. Mater. Syst. Struct. 29, 2105–2121.

Santos, D., Silva, D.M., Gomes, P.S., Fernandes, M.H., Santos, J.D., Sencadas, V., 2017. Multifunctional PLLA-ceramic fiber membranes for bone regeneration applications. J. Colloid Interface Sci. 504, 101–110.

Sengupta, A., Das, S., Dasgupta, S., Sengupta, P., Datta, P., 2021. Flexible Nanogenerator from electrospun PVDF–polycarbazole nanofiber membranes for human motion energy-harvesting device applications. ACS Biomater. Sci. Eng. 7, 1673–1685.

Sengupta, P., Ghosh, A., Bose, N., Mukherjee, S., Roy Chowdhury, A., Datta, P.A., 2020. Comparative assessment of poly(vinylidene fluoride)/conducting polymer electrospun nanofiber membranes for biomedical applications. J. Appl. Polym. Sci. 49115.

Sezer, N., Koç, M., 2021. A comprehensive review on the state-of-the-art of piezoelectric energy harvesting. Nano Energy. 80, 105567.

Sharma, T., Aroom, K., Naik, S., Gill, B., Zhang, J.X.J., 2013. Flexible thin-film PVDF-TrFE based pressure sensor for smart catheter applications. Ann. Biomed. Eng. 41, 744–751.

Sharma, T., Naik, S., Langevine, J., Gill, B., Zhang, J.X.J., 2015. Aligned PVDF-TrFE nanofibers with high-density PVDF nanofibers and PVDF core–shell structures for endovascular pressure sensing. IEEE Trans. Biomed. Eng. 62, 188–195.

Sharma, P.P., Gahlot, S., Kulshrestha, V., 2017. One pot synthesis of PVDF based copolymer proton conducting membrane by free radical polymerization for electro-chemical energy applications. Colloids Surf. A Physicochem. Eng. Asp. 520, 239–245.

Singh, A., 2014. Preparation and characterization of piezoelectric materials. In: International Conference for Convergence for Technology-2014pp. 1–5.

Smith, M., Calahorra, Y., Jing, Q., Kar-Narayan, S., 2017. Direct observation of shear piezoelectricity in poly-l-lactic acid nanowires. APL Mater. 5, 74105.

Strobl, C.J., Schäflein, C., Beierlein, U., Ebbecke, J., Wixforth, A., 2004. Carbon nanotube alignment by surface acoustic waves. Appl. Phys. Lett. 85, 1427–1429.

Sultana, A., Ghosh, S.K., Sencadas, V., Zheng, T., Higgins, M.J., Middya, T.R., Mandal, D., 2017. Human skin interactive self-powered wearable piezoelectric bio-e-skin by electrospun poly-l-lactic acid nanofibers for non-invasive physiological signal monitoring. J. Mater. Chem. B 5, 7352–7359.

Swain, S., Padhy, R.N., Rautray, T.R., 2020. Polarized piezoelectric bioceramic composites exhibit antibacterial activity. Mater. Chem. Phys.. 239, 122002.

Tajitsu, Y., 2006. Development of electric control catheter and tweezers for thrombosis sample in blood vessels using piezoelectric polymeric fibers. Polym. Adv. Technol. 17, 907–913.

Trolier-McKinstry, S., 2008. Crystal chemistry of piezoelectric materials. In: Safari, A., Akdoğan, E.K. (Eds.), Piezo-electric and Acoustic Materials for Transducer Applications. Springer US, Boston, MA, pp. 39–56.

Trolier-McKinstry, S., Zhang, S., Bell, A.J., Tan, X., 2018. High-performance piezoelectric crystals, ceramics, and films. Annu. Rev. Mater. Res. 48, 191–217.

Uchino, K., 2009. Designing with materials and devices and fabrication processes. In: Ferroelectric Devices. CRC Press.

Uchino, K., 2010. 1—The development of piezoelectric materials and the new perspective. In: Uchino, K. (Ed.), Woodhead Publishing Series in Electronic and Optical Materials. Woodhead Publishing, pp. 1–85.

Uchino, K., 2017. Manufacturing methods for piezoelectric ceramic materials. In: Uchino, K. (Ed.), Woodhead Publishing in Materials. second ed. Woodhead Publishing, pp. 385–421 (Chapter 10).

Vatlin, I.S., Chernozem, R.V., Timin, A.S., Chernova, A.P., Plotnikov, E.V., Mukhortova, Y.R., Surmeneva, M.A., Surmenev, R.A., 2020. Bacteriostatic effect of piezoelectric Poly-3-hydroxybutyrate and polyvinylidene fluoride polymer films under ultrasound treatment. Polymers (Basel) 12, 240.

Wang, Z.L., Liu, Y., 2012. Piezoelectric effect at nanoscale. In: Bhushan, B. (Ed.), Encyclopedia of Nanotechnology. Springer Netherlands, Dordrecht, pp. 2085–2099.

Wang, X., Yao, C., Wang, F., Li, Z., 2017. Cellulose-based nanomaterials for energy applications. Small 13, 1702240.

Wei, H., Wang, H., Xia, Y., Cui, D., Shi, Y., Dong, M., Liu, C., Ding, T., Zhang, J., Ma, Y., Wang, N., Wang, Z., Sun, Y., Wei, R., Guo, Z., 2018. An overview of lead-free piezoelectric materials and devices. J. Mater. Chem. C 6, 12446–12467.

Wojnar, R., 2012. Piezoelectric phenomena in biological tissues. In: Ciofani, G., Menciassi, A. (Eds.), Piezoelectric Nanomaterials for Biomedical Applications. Springer Berlin Heidelberg, Berlin, Heidelberg, pp. 173–185.

Xiao, Y., Wang, C., Feng, Y., 2016. Vibration of Piezoelectric ZnO-SWCNT Nanowires. Nanomaterial (Basel, Switz.) 6, 242.

Yang, Y., Pan, H., Xie, G., Jiang, Y., Chen, C., Su, Y., Wang, Y., Tai, H., 2020. Flexible piezoelectric pressure sensor based on polydopamine-modified BaTiO3/PVDF composite film for human motion monitoring. Sensors Actuators A Phys.. 301, 111789.

Yoon, J.-K., Misra, M., Yu, S.J., Kim, H.Y., Bhang, S.H., Song, S.Y., Lee, J.-R., Ryu, S., Choo, Y.W., Jeong, G.-J., Kwon, S.P., Im, S.G., Il, L.T., Kim, B.-S., 2017. Thermosensitive, stretchable, and piezoelectric substrate for generation of myogenic cell sheet fragments from human mesenchymal stem cells for skeletal muscle regeneration. Adv. Funct. Mater. 27, 1703853.

You, L., Zhang, Y., Zhou, S., Chaturvedi, A., Morris, S.A., Liu, F., Chang, L., Ichinose, D., Funakubo, H., Hu, W., Wu, T., Liu, Z., Dong, S., Wang, J., 2019. Origin of giant negative piezoelectricity in a layered van der Waals ferroelectric. Sci. Adv.. 5, eaav3780.

Yu, J., Loh, X.J., Luo, Y., Ge, S., Fan, X., Ruan, J., 2020. Insights into the epigenetic effects of nanomaterials on cells. Biomater. Sci. 8, 763–775.

Zaszczyńska, A., Gradys, A., Sajkiewicz, P., 2020. Progress in the applications of smart piezoelectric materials for medical devices. Polymers (Basel) 12, 2754.

Zhang, J., Meguid, S.A., 2016. Piezoelectric response at nanoscale. In: Meguid, S.A. (Ed.), Advances in Nanocomposites: Modeling, Characterization and Applications. Springer International Publishing, Cham, pp. 41–76.

Zhang, J., Wang, C., Bowen, C., 2014. Piezoelectric effects and electromechanical theories at the nanoscale. Nanoscale 6, 13314–13327.

Zhao, F., Zhang, C., Liu, J., Liu, L., Cao, X., Chen, X., Lei, B., Shao, L., 2020. Periosteum structure/function-mimicking bioactive scaffolds with piezoelectric/chem/nano signals for critical-sized bone regeneration. Chem. Eng. J.. 402, 126203.

Zhu, M., Lou, M., Abdalla, I., Yu, J., Li, Z., Ding, B., 2020. Highly shape adaptive fiber based electronic skin for sensitive joint motion monitoring and tactile sensing. Nano Energy. 69, 104429.

Zou, Y., Liao, J., Ouyang, H., Jiang, D., Zhao, C., Li, Z., Qu, X., Liu, Z., Fan, Y., Shi, B., Zheng, L., Li, Z., 2020. A flexible self-arched biosensor based on combination of piezoelectric and triboelectric effects. Appl. Mater. Today 20.

Zviagin, A.S., Chernozem, R.V., Surmeneva, M.A., Pyeon, M., Frank, M., Ludwig, T., Tutacz, P., Ivanov, Y.F., Mathur, S., Surmenev, R.A., 2019. Enhanced piezoelectric response of hybrid biodegradable 3D poly(3-hydroxybutyrate) scaffolds coated with hydrothermally deposited ZnO for biomedical applications. Eur. Polym. J. 117, 272–279.

CHAPTER

14

Nanotechnology-based interventions for interactions with the immune system

Sayandeep Saha, Shalini Dasgupta, and Ananya Barui

Centre for Healthcare Science and Technology, Indian Institute of Engineering Science and Technology, Shibpur, Howrah, West Bengal, India

OUTLINE

1 Introduction

In the 21st century, the requirements for active components to strengthen our immune system to tackle various physiological disorders has improved manifold. With the scarcity of newer pharmacological lead molecules being found for active usage against diseases, it is essential to focus on innovative methods that would entail in improving the drug efficiency and delivery. The increase in growth of various antibiotic-resistant microorganisms and other pathogens has led to the demand for newer and advanced treatment strategies. Nanotechnology is a field which has taken quite a leap in this decade, with myriad applications when it comes to providing carriers for existing drugs, improving drug efficacy and immunotherapy. Ideally, the definition of immunotherapy is referred to as the improvement of the resistant immune system of the body with the help of various medical treatments and techniques to combat diseases and to improve overall immunity of the host.

Nano-sized particles are specially designed to encourage positive immune response through methodologies that have diverse biological advantages as well as through the attachment to specific molecules for activation. Moreover, nanoscale level of point-of-care diagnostics offers compact and effective treatments that can direct the utilization of nanoscale immunotherapies (Fig. 1).

From being significant carriers to improving the potential of the immunotherapies by latent or stepwise activation techniques, nanotechnology has the potential to revolutionize the field of pharmacology for incredibly tricky diseases. Many potential molecules which are designed to supplement the immune system often fail to activate to its true potential. These can be appropriately activated by specific nanoparticles. The adaptive nature of nanoparticles in terms of drug conveyance and target-specific action makes it very effective. By attaching hormones, glycomolecules, proteins and any other type of medication on nanoparticles, we can utilize this method to cure many diseases. In this chapter, we will examine the different nanotechnology-based therapies which can be utilized to strengthen our immune framework.

2 Emerging clinical needs of human immune physiology

The defence barriers of our body have numerous layers of security which are significant for managing different infections. The resistant immune system has been developed to battle an expansive scope of pathogens and other microorganisms. However, it is challenging, and current medical physiology is yet to be utilized to its maximum capacity for eradicating new and advancing illnesses. Most pharmacological innovations and revelations are set up with the plan to give extra support to the immune system. Indeed, antibodies are perhaps the best discovery in clinical history due to the utility of being able to eliminate persistent pathogens. A contributor to the issue has been the inaccessibility of the target location due to passage through various other physiological systems which take away the potential efficacy. There has been a wide range of advancements and strategies to expand the proficiency of the immune system. Designing T cells (Huye and Dotti, 2010) and numerous different other procedures are, as of now going through different preliminaries in many research labs.

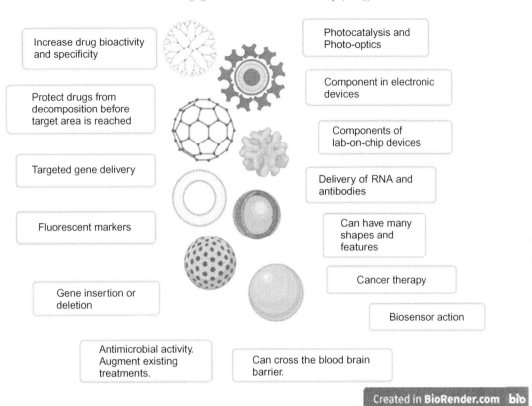

FIG. 1 Various functions and treatment methods utilized for the immune system with the help of nanoparticles (created by biorender.com).

Adding to this field is the extent of the utilization of nanobiotechnology. Upcoming improvement for advancing the aspects of nano-pharmaceutical delivery can activate the natural defences. This guarantees the enhancement of the effectively settled components of prophylactic immunizations and active immunotherapy.

Different pathogens all through the body collectively attempt to infiltrate inside the checkpoints and endeavor to cause unalterable harm. Pathogens can cause damage by uncontrolled expression of harmful proteins, causing immune system issues or overexpression of any oncogenic gene. There are numerous dangers our physiology faces at a consistent schedule. Pathogenesis includes some connection inside the variations of regular gene expression, as genetic aberrations assume an etiologic function in roughly 20% of disease cases around the world. Although antimicrobial therapies have helped in decreasing the disease rates and the occurrence of related malignancies, treatments considered for tumors stay restricted. Immunological issues and malignant growths are connected as they happen all the more frequently among immunosuppressed patients as contrasted to healthier people.

Along these lines, there is a dire requirement for remedial techniques that can evoke a healthy immune response and help in antitumorigenic response as well. Moreover, the investigation of microorganism defence when counteracted by the host immune system may give knowledge into improved treatments for malignancy and other incurable diseases. There have been many techniques involving the improvement and amplification of the T cells and B cells for treating such diseases. The up-and-coming procedures have shown positive implications to the incorporation of genetically modified T cells and usage of adoptive T-cell transfer therapy.

2.1 Engineered T cells and CAR T cells

Nowadays, normal thymus cells are simply not enough to fight against infections or remove tumors. This is because either the T cells tend to have off-target focus, not having had enough exposure, or they are weak against the pathogen. These issues can be solved by simply engineering genetically the T cells to carry specific genes or to be activated previously so that efficacy is increased. The basic procedure to prepare CAR T cells are as follows. First, they must be obtained from the various patients. After this, the specific genetic code important to activate the action against cancer is inserted in the T cells. After insertion of DNA, various chimeric antigen receptors begin to grow on the surface. Then they are cultured and grown in sufficient quantities for the patient (Fig. 2). The development for designed

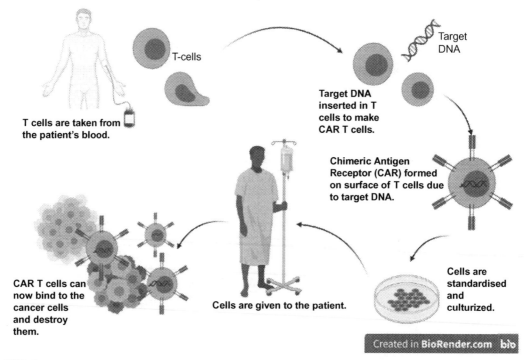

FIG. 2   Process of formation and usage of CAR T cells (created in biorender.com).

thymus lymphocytes as a type of malignancy treatment denotes the start of the age of advancement in medication, giving a ground-breaking method to battle complex illnesses. While ready to revolutionize disease treatment, the hopefulness about T-cell malignant growth treatments stays tempered by worries about wellbeing and off-target effects (Huye and Dotti, 2010).

Designed T cells are essential for a lot more extensive action for immunology; however, the factor that is required for the success of these treatments is the idea of utilizing any live immune cell as a helpful vector (Huye and Dotti, 2010). Here are the examples of engineering T cells which are being considered for therapy in the recent days in Table 1.

Simultaneously, these cells are tested for control and production. This method combines the formation of cells with a more directed sense of action and response. Chimeric antigen receptors (CARs) are engineered cells that permit more adaptable retargeting of T cells. Vehicles defeat a few impediments of the TCR, for example, the requirement for MHC articulation, MHC character, and co-stimulation (Bertoletti and Tan, 2020). Eshaar et al., had showed different engineered receptor particles empowered MHC-free processing by T cells. One constraint of current CAR, various progenitor thymus cell systems function in a way that they must be activated by various specific physiological processes before they are directed to attack tumors (Lim and June, 2017). Scientific research has assisted with characterizing the significant difficulties which should be fulfilled to prepare designed lymphocytes as a reliable, safe, and powerful medium that can be sent against a broad scope of tumors. Development of various scientific studies for cell building is giving us an extensively extended arrangement of apparatuses for programming insusceptible cells.

CAR lymphocytes focusing on CD19 as a lead molecule has risen as the lead research for built T-cell treatments in malignant growth and delineate the collaborations of joining genetics with immunology (Davila and Brentjens, 2016). A few highlights should be kept in mind that have added to the accomplishment in the clinical investigations focusing on CD19 (Park et al., 2016). CD19 was picked for the primary vehicle for its continuous and significant level articulation in memory cell tumors. CD19 is an important marker to activate memory cells. In any case, the loss of B cells has numerous disadvantages as well. Patients who have been effectively given the treatment of CD19 CARs show significant B cell deterioration, and loss of

TABLE 1 Types of engineered T cells and their functions.

Types of cells	Function	References
Tumor infiltrating lymphocytes	Cells are obtained from the tumorigenic tissue	Rosenberg et al. (2008)
Endogenous T-cell therapy	The specific effector cells are prepared from the peripheral blood	Ruffner et al. (2017)
CAR thymus cells	When the targeted extrinsic activating gene is inserted within the target T cell in order to make a specific targeted response	(Yin et al., 2011)
TCR transduced cells	When the genes of the T-cell receptors are inserted into the other peripheral cells in order to attack the tumor cells	Sakai et al. (2020)

B cell aplasia frequently predetermines the phase for a potential remission (Chu et al., 2017). These outcomes show the effect these engineered lymphocytes could have on track off-tumor impacts as well, but can be moderated by utilizing on-off gated CAR lymphocytes (Boyiadzis et al., 2018). However, considering an idea of gated engineered lymphocytes still theoretical and generally calculated and has not indicated any enormous scope substantial outcomes starting at yet. Real outcomes with different CARs demonstrate the off-targeting of lymphocytes' responses will not be a particular model yet might be commonly seen with other genealogy subordinate targets (Maude et al., 2015).

2.2 Challenges associated with CAR T cells

Notwithstanding their prosperity, a few issues stay when it comes to CAR lymphocytes. For situations of intense lymphoblastic leukemia, as more than 70% of patients enter reduction, the main objective is to diminish effectiveness. Notwithstanding, numerous different approaches are being examined for checking the chance of fortifying the resistant immune system reaction. In such situations, the CAR T-cell strategy is generally applied for viral contaminations and related tumors starting at now.

Viral diseases stay a significant reason for issues in patients with immunodeficiency, for example, beneficiaries of hemopoietic progenitor stem cell transplantation (Ruffner et al., 2017). Pathogen exposure for explicit cytotoxic T lymphocytes is a methodology to reestablish infection insusceptibility to forestall or treat viral ailments. It has been tried in the clinical setting for over 20 years. A few research groups have utilized extended infection direct T-cell items explicit for one or various infections to both reconstitute antiviral resistance after transplantation and to treat dynamic viral diseases (Yin et al., 2011).

2.3 Adaptive thymus cell transfer treatment in cancer and viral infections

There have been many different methods of developing immunity for the body. One of these methods includes adaptive thymus cell transfer treatment. This technique utilizes activation of tumor invading lymphocytes by IL-2 and the subsequent growth and release back to the affected body. They are extremely sensitized to a specific tumor and help in mounting a tumoricidal response in that region. The presence of CD4 with CD8 targeted thymus cells in chronic infections are represented by the degradation due to the presence of T cells, because of the overexpression of inhibitory molecules like PD-1 and TIM-3. Thus the adoptive transfer of independent T cells with specific TCRs attuned for the viral markers such as protein coats, in along with the anti-PD1 and the anti-TIM3 effectors can be useful for the perfect methodology for specific therapy of virally induced malignancies. Adaptive cellular immunotherapy includes an assortment as well as development of activated as well as modified T cells consisting of same type of cells once more into the human body, where redirection of these cells can be conducted to the tumor and advance focusing on tumor degradation (Perica et al., 2015). Tumor-explicit antigens introduced on MHC class I particles give an incredible objective to separating harmful cells from typical cells. For instance, T cells focusing on melanoma-related antigen perceived by the specific antigen known as MART-1 were demonstrated as extremely effective (Redeker and Arens, 2016). Lymphomas which occur to ENV have been treated with such methods as before. In particular, the phenomenon of posttransplant lympho-proliferative disease (PTLD) emerges for the overload of hormones which keep the immune system in check. This can lead to the activation of the latent EBV that was present in the body. The progression of PTLDs can lead to many various transitions of carcinoma from polyclonal hyperplasia to the formation of even Hodgkin's lymphoma (Visser et al., 2012). The declaration of EBV-explicit antigens on harmful cells gives a case of how tumor-explicit antigens can make such diseases especially appropriate for focused cell treatments.

This immunotherapeutic methodology additionally was utilized for various patients suffering from for example, diseases such as carcinoma of the nasopharyngeal tract (NPC) as well as Hodgkin lymphoma (Chou et al., 2008). The presence of EBV contributing to the above types of carcinoma and lymphoma was only counteracted by antibodies which focused on specific proteins from the EBNA3 family. For example, antibodies which were exclusively based on LMP2 protein showed greater efficacy when it came to diseases like Hodgekin's Lymphoma (Bhatia et al., 2011). Lamentably, these outcomes were regularly brief in all likelihood due to the unpredictability of the disease and erratic results which were observed in vivo. Receptive Thymus progenitor cell systems were likewise being explored to see if any viable treatment can be found for those suffering from melanoma. Hence, melanoma has exceptionally specific tumor-explicit antigens for T-cell treatment.

When the T-cells have been exposed and engineered to provide a specific response against the specific antigens against any tumor or other disease, it is referred to adaptive T-cell transfer therapy (Perica et al., 2015). Tumor-explicit antigens introduced on MHC class I particles give a specific objective to separating harmful from typical cells (Redeker and Arens, 2016). Adaptive cell transfer therapy is also referred to as ACT.

2.4 Challenges associated with receptive T-cell transfer treatment

While this treatment had a lot of assurance, there are unmistakable troubles that stonewalled the headway with its productive and effective usage. The major problems with T-cell transfer therapy comes with the complications associated with oncoviral infections. Issues like T-cell exhaustion, deficiency, and overall saturation are always a major problem with this kind of treatment. Lympho-deletion remains a very pertinent issue. Mixing the T cells and activating them with the requisite hormones such as IL-2 might help out in this regard (Powell et al., 2007). With the lympho depletion, the native lymphocytes which are not CD45-specific get destroyed and instead the remaining genetically modified ones are utilized with the help of other interleukin cytokines for activation such as IL-5 (Romano et al., 2019). However, as a process where the inherent T cells must be destroyed for the engineered T cells to grow, the risks that might occur have to be considered (Bertoletti and Tan, 2020). Additionally, lympho-depletion may be liable for a large group of other potential issues that may happen caused by the destruction of the host thymus cells.

Thus, for treatments of carcinoma it should be remembered as the helper thymus cells have a proper well-defined target. However, the tumor cells often obfuscate the targeted location by decrease in expression of HLA-1. This leads to scattering of the thymus cells that are designed to attack based on their specific expression. The reduction of expression can also be counteracted with the help of interferon gamma, or by radiation therapy, which in turn can be utilized for recalibrating a target for T cells to attack oncogenic cells.

These are not the only challenges faced by the adaptive T-cell receptor therapy. One more issue remains and that is the T-cell exhaustion. Thymus cells might lose their own targeted objective to attack harmful pathogens due to constant exposure to only one kind of antigen. WBC cell exhaustion is best described in LCMV affected mice (Srivastava and Riddell, 2018). Various different molecules detecting T-cell exhaustion have been researched. Two checkpoint blockade inhibitors, such as CTL4 and PD-1 have shown some agonist/antagonist reactions with the phenomenon in the sense. CTLA-4 works to retain homeostasis in the body and helps in activation of the regulatory section of thymus cells, whereas PD-1 helps to activate the T cells into a specific defence against the presence of tumors.

Finally, it has been understood that careful clinical measures of dosages of the inhibitors of CTLA4 and PD1 will ultimately provide a respite from the T-cell exhaustion and help in controlling the carcinogenic tissue growing within the human body (Rotte, 2019). A lot of research is required in this particular area for the T-cell technology to overcome this challenge.

Disease cells here and there discover approaches to utilize these checkpoints to abstain from being discovered by the immune system. However, drugs that focus on these checkpoints hold a great deal of guarantee as malignant growth inhibitors. These medications are called checkpoint inhibitors. By restraining these checkpoints, specific T-cells may likewise be initiated and used against the harmful cells. Methodologies to release T cells against tumors are incredibly convincing, as the action of these cells presents significant highlights that are invaluable over other malignant growth treatments (Iwasaki et al., 2017).

2.5 Immune check point blockade

Safe checkpoints are molecules that can increment or lessen the signs of the immune system, and they are viewed as fundamental factors in treating contaminations, malignant growths, and immune system sicknesses. Presently, invulnerable checkpoint treatment is viewed as a mainstay of malignant growth treatment. Among the specific checkpoint treatments, those including PD-1 and CTLA-4 might be the best (Rotte, 2019). These two receptors have a significant function as immune checkpoints and might hold the key in being viewed as an expected objective for disease treatment.

These checkpoints help shield insusceptible reactions from being excessively substantial and once in a while can shield T cells from executing malignant growth cells. At the point when these checkpoints are hindered, T cells can slaughter disease cells better. CTLA-4 and PDL-1 are two different checkpoints which help in handling the immune response of T cells (Intlekofer and Thompson, 2013). Preclinical examinations have exhibited that focusing on various other T-cell surface epitopes, including both positive and negative resistant controllers, additionally has effective antitumor action.

2.6 Challenges associated with immune system blockade

The immune barricade eliminates inhibitory signs of T-cell activation, which empowers tumor-receptive T cells to beat the immune regulatory systems and mount a compelling antitumor reaction. Such administrative components regularly keep up safe reactions inside an ideal physiological reach and shield the host from autoimmunity. Immunological resilience is accomplished through various processes that can be characterized as focal and fringe.

Focal resilience is intervened through clonal cancelation of high-fondness self-receptive clones during negative choice in the thymus (Peng et al., 2020). Notwithstanding, because self-reactivity is chosen for during positive determination in the thymus, different techniques are needed to limit autoreactivity. Fringe resistance intercedes through an assortment of systems, including administrative T cells (Treg), T-cell anergy, cell-outward tolerogenic signs, and fringe clonal erasure (Raïch-Regué et al., 2014). The immune system framework applies a particular substantial weight all through tumor movement, prompting insusceptible tumor altering. Subsequently, dangerous tumors regularly co-select immune suppressive and resistance systems to stay away from safe annihilation. The barricade system receptors additionally have a reiteration of opposite results which might be exacerbated antagonistically if it is continuously skirted for treating malignancy.

3 Nanotechnology and nanoparticles for vaccination

Nanotechnology stands at the pinnacle of modern vaccine innovation. Ideally, nanovaccines resemble the morphology of a particle whose size falls under a few hundred nanometers and can be effectively customized following the characteristics of size, hydrophobicity, surface charge, surface modification, and shape geometry. Table 2 helps us understand how different morphological shapes have different impacts for immunological usage.

TABLE 2   Morphology and usage of nanoparticles in the field of immunology.

Morphology of nanoparticles	Usage	References
Rods	Utilized for vaccine conveyance and specific tissues in body. One of the earliest methods of drug transport developed by nanotechnology	Malachowski and Hassel (2020)
Spheres	Hollow spheres are used to store vaccines, proteins, epitopes, antigens etc. Used in drug delivery as well as metallic spherical nanoparticles have their own property of antibacterial effects	Dykman and Khlebtsov (2017)
Hexagonal shapes	Hexagonal shapes nanoparticles can either be shaped so as to increase sensitivity and surface area. They possess higher surfaced enhanced Raman scattering, making them more effective for immunological purposes such as macrophage activation or vaccine delivery	Purwada et al. (2016)
Dendrimers	Dendrimer formation and shape ensures higher bioavailability and less toxicity for drug delivery	Kumar et al. (2015a)
Liposomes	Has better colloidal stability and more likely to be accepted by macrophages via phagocytosis as compared to normal nanoparticles	Nikkhoi et al. (2018)
Fullerene	Can be used for photodynamic therapy for treatment at specific tissues in conjunction with other metal nanoparticles	Petrovic et al. (2015)
Virus like particle arrangement	Promote immunogenic response at a safer level with its repetitive pattern and presence of PAMPs. Offers faster cellular uptake and DNA, RNA, and protein transfer to specific cells without the dangers of virus	Perotti and Perez (2019)

These particles can be tailored to stimulate an adequate antigenic response through modulated biodistribution and to-and-fro communication with the immune framework making it a competent medium of vaccination. The two main aspects that determine its effectiveness underlies in its composition and its back and forth interaction with the host immune system.

3.1 Immune system and its relationship with nanovaccine

From the perspective of immune interactions, the fundamental strategy of vaccination lies in the exploitation of the host immune system, tricking it into recognizing the vaccine as a naturally occurring foreign substance, thereby eliciting adequate immune response which then prepares the host immune system to counteract any subsequent attack by the same pathogen and its components of natural origin. The humoral immune defence is essentially responsible for the protection of the host from an antigen which can be self or nonself in nature and therefore a brief understanding is warranted of how the mechanism of immune response is ensued. The fundamental mechanism for detection of an antigen occurs through a cross-signaling in the midst of the various tissues of the innate and the adaptive immune system, the mutual communication between which drives the antigenic response. Macrophage, a cell which is fundamental to the innate immune system, contains pathogen recognition receptors which distinguish an exogenous pathogen, based on the associated molecular patterns present on its surface. Upon recognition, the lymphocytes of the innate immunity are deployed. By using the subsequent enzymatic activities, the pathogen is broken down into smaller aggregates and is thereby inactivated. The innate immune system possesses DCs known as Dendritic Cells, which simultaneously detects and disintegrates the invading pathogen through antigen processing. They further emigrate into the lymphatic regions where at the nodes they will provide the disintegrated aggregated proteins to the defenders of the adaptive immune system, which then subsequently activates the cells (Soares et al., 2017). The presentation of the external protein takes place through MHC-I as well as MHC-II pathways that are primarily associated with the activation of the latent thymus cells. Firstly, the MHC-I pathway involves the presentation of MHC-I and exogenous pathogen peptides to T cells while MHC-II pathway engages in MHC-II and endogenous peptide complex presentation (Neefjes et al., 2011). Upon activation, T cells regulate the pathogen clearance pathway and terminate the cells that had been damaged in the process of pathogen invasion. The diffused pathogens are also presented to B cells which on activation generate antibodies and further ameliorate the host immune system by diffusing pathogen lethality and by enhancing the phagocytic activity of the innate immune cells. These interactions also elicit memory B-cell response. Therefore, the innate and the adaptive immunity is altogether responsible for pathogen clearance.

So far, the artificial activation of the immune system is observed by using cytokines through lab as well as in vivo experiments along the use of components originating from the respective foreign pathogen. Nevertheless, due to size restrictions, the targeted delivery would often be inadequate, which would then render an insufficient immune response. This evoked the initiative of implementing nanoparticles as carriers for such small molecules. The advantages of having a higher volume, a better size compatibility and the ability of being released in a controlled manner have provided a room for nurturing its multifunctional nature as a carrier molecule, as well as a biochemical modulator. The physical and chemical modification of the surface of these particles allows for an efficient and targeted approach toward the delivery sites. In recent years, biomimetic approaches were prepared wherein the resonance and surface area of nanoparticles can be made to imitate the pathogen surface and be subsequently tested in vivo to determine the associated pathways and the degree to which the immune response is triggered in the model system. Therefore, by studying the cross-signaling of the cells of the immune system and the carrier nanoparticles is the key to determine the pathways associated with immune response induced by the pathogenic components. It has been found that a more robust B-cell response is evoked through a repetitious cluster of antigenic arrangement which resemble a natural pathogen surface. Accordingly, for activation of target immunity, providing a predefined density as well as a definite geometrical symmetry to the particle morphology is warranted, so that it can exhibit some pathogenic features. The importance of a defined symmetry has been demonstrated in a study wherein the stimulation of B-cell response to elicit antibody production by using a self-congregated protein nanoparticle arranged in an octahedral symmetry with Influenza HA antigen suggested a possible influence of viral geometry on the stimulation of immune response (Kanekiyo et al., 2013). Evidences also indicate that the optimal performance of monomeric CpG as a molecular adjuvant is further enhanced when the CpG ligand is assembled into nano-rings or by coupling CpG-C or CpG-B oligonucleotides with nanoparticles for dual-targeting of antigen and adjuvant (Soares et al., 2017). Therefore, the receptor-ligand interaction during the initial stages of communication between the immune cells and pathogens have relevance with respect to the spatial/geometrical aspect, which is therefore a fundamental factor to consider in designing nanoparticles as vaccine delivery agents.

In recent years, engineered nanoparticles have predominantly surfaced as artificially manufactured APCs, the design of which has been used to acquire a better understanding of the antigen presentation by DCs to the T cells (Neal et al., 2017). The strategy involves using the membrane proteins of DCs such as the MHC complex and other signaling molecules for bioconjugation or chemical crosslinking with the nanoparticle surface. The advantage of this construct over the use of free ligands is that it provides the nanoparticles with a biomimetic cloak which bears a similarity to the morphology of the native cells (Andreae et al., 2003). This is incredibly convenient because the T-cell signaling receptor domains are bundled to a minuscule scale of a few submicrons. Therefore, there is a high chance that the free ligands would escape the interaction with the targeted T-cell receptors. Besides this, these synthetically manufactured nanoparticles are tailored to avert an overblown immune response which is hazardous to the body, and to simultaneously emphasize on the antigen presentation for T-cell activation. Due to these advantages, recent studies on T-cell activation in model systems have been performed through these APCs. They have also been used for performing in vitro studies of T-cell interaction in relation to their physical requirements. One such example is a study on MHC presentation using ellipsoidal and spherical PLGA APCs particles with variable aspect ratios (known as ARs) which showed activated response caused by the high-AR APCs on a rodent model with melanoma, thereby highlighting the significance of spatial relation between the APCs particles and thymus cells and their role involved in activation of thymus cells. Studies have not only been restricted to spatial influence. The influence of surface MHC-peptide density of the nanoparticles on stimulation of T regulatory cells has been demonstrated in a separate study through nanoparticles engineered using iron oxide (IO) whose surface had been modified to mimic that of the immunosuppressive DCs using a complex of autoantigen peptides and MHC molecules. Therefore, alongside the geometrical aspect, density has a significant influence on the adaptive immune regulation.

3.2 Classification of nanovaccines

A variety of nanoformulations brings different opportunities of implementation. Each type of nanoparticle presents with a unique set of properties which can be exploited for ensuing the elicitation of the immune response, whether be used as vaccine delivery agents or as adjuvants to bolster immunogenicity. The classification of nanovaccines can be done in terms of its mode of action which relates directly to its physical nature, by extension, its material of formulation (Fig. 3).

3.2.1 Polymer-based nanovaccines

Polymer-based nanoparticles are deemed to be one of the most promising methods of vaccine delivery, primarily because of its fitting nature of biocompatibility, biodegradability, simplicity of fabrication, reduced toxicity, controlled vaccine release kinetics, and ease of modification of the particle surface properties (Demento et al., 2012). PLGA and PLA are considered as ubiquitously employed for the construction of polymeric carrier nanoparticles to

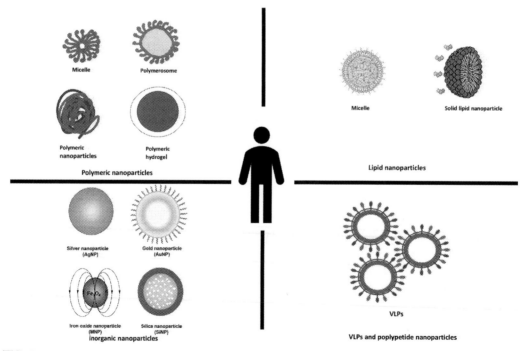

FIG. 3   Types of nanoparticles used for vaccination (created with biorender.com).

study the immune response using a large number of antigen types such as tetanus toxoid, *Mycobacterium tuberculosis* antigen, hepatitis B (HBV) antigen, ovalbumin, and *Bacillus anthracis* (Diwan et al., 2002). Alongside PLGA and PLA, various naturally occurring polymers have also been implemented as adjuvants, such as chitosan, inulin, pullulans, and alginate. Chitosan is a versatile polymer which has found its use in conjugation with alginate as HBV antigen carrier (Borges et al., 2008). In one particular study, it was found that chitosan nanoparticles used for delivery of recombinant DNA vaccine (nano-Esat-6/3e-FL) containing the above-mentioned T-cell epitopes and Esat-6/3e-FL genes of *Mycobacterium tuberculosis* elicited a higher degree of immune response in C57BL/6 mice (Feng et al., 2013). In another study, delivery of *Mycobacterium tuberculosis* lipid bound chitosan nanoparticles induced a higher secretion of Th1 and Th2 cytokines and also induced a higher degree of humoral immune response in mice models in comparison to the lipid molecules without chitosan (Pawar et al., 2013). Chitosan nanoparticles have also been used with PLGA polymers as vaccines to increase the immune response through adjuvant activity (Honda-Okubo et al., 2012). More polymers, such as inulin, have also been shown to produce higher antibody titers and robust cytokine response against influenza viruses and HBV along with enhanced site-targeting (Götze and Müller-Eberhard, 1971). Additionally, T-cell epitope P2 derived from tetanus toxoid has also been used with this formulation which functions as an immunostimulant. Therefore, the demonstration of multifunctionality of polymeric nanoparticles has hence been proved to be a potential tool for stretching the application of nanotechnology in the process of vaccination.

3.2.2 Liposomal nanovaccines

After polymeric nanoparticles, liposomes have perhaps been most extensively studied as vaccine conveyance vehicles, and also used for the drug delivery framework. The lipid bilayer of the liposomes is formed around the aqueous core through lipid hydration. In terms of biodegradability, immunogenicity and enhanced physico-chemical properties, liposomes prove to be more advantageous in comparison to the existing nanomaterials for vaccination, thus providing an ample scope for loading of different antigens with different surface charge. Variation in the properties of size, fluidity, bilayer composition, and surface modification such as PEGylation induces variable production of antibody titers in multiple model systems. The structural flexibility of liposome allows for the envelopment of the hydrophobic as well as hydrophilic molecular components, wherein hydrophobic molecules are bound by the phospholipid bilayer, and the hydrophilic molecules are engulfed into the core aqueous region (Moon et al., 2011). In multiple studies, it has been found that during the course of vaccine activity, the liposomal structure fuses with the cell membrane of a target cell. This has been demonstrated through various unilamellar and multilamellar liposomal vesicles composed of phosphatidylserine, cholesterol, phosphatidylcholine, and other biodegradable phospholipids (Moon et al., 2011). Delivery of DNA vaccines through liposomal nanovesicles have also been shown to have a potential implication for adjuvant therapy against fungal infection using DNAhps65 from *Mycobacterium leprae* (Ichihashi et al., 2013). Another existing implication of liposomal delivery is through DNA-cationic-liposomal complex formulated using *N*-(1[2,3dioleyloxy]propyl)*N,N,N*-trimethylammonium chloride which has now been paving a route for gene delivery for more than a decade. In case of other infections such as influenza, malaria, tuberculosis and chlamydia, application of nanoliposomes as both carriers and adjuvants have been shown to elicit high immune response.

More recently, many subsequent classes of liposomal nanovesicles have been distinguished, which offer a higher area for antigen/adjuvant/protein loading on the membrane surface. These are namely virosomes, hexosomes, and cubosomes. Virosomes are similar to liposomes with viral capsid proteins present (Blom et al., 2016) on its surface, which fuses more rapidly with the cell membrane of the targeted cell. In a particular study, phospholipid virosome (1,2-dioleoyl-sn-glycero-3-phosphocholine) known as DOPC fused with influenza A subtype H1N1 membrane proteins showcased a facilitated internalization in respiratory epithelial cell triple co-culture (Blom et al., 2016). In another study, subcutaneous employment of 40–60 nm virosomes as an adjuvant for chimera vaccine with peptides containing Tax, gp21, gp46, and gag epitopes of human T-cell lymphotropic virus type-1 enhanced the production of antibodies IgG2a and IGA and upscaled the release of cytokines IL-10 and IFN-γ in mice models (Rodrigues et al., 2018). The second subclass of liposomal nanovesicles are known as hexosomes which follow a nonlamellar framework under aqueous medium and produce strong antigen-specific antibody response on being used as a lipid adjuvant with other liposomal nanoparticles. The impact of hexosomes as adjuvants has been demonstrated in CB6/F1 mice through a comparative study between hexosomes containing immunopotentiator monomycoloyl glycerol-1 (MMG-1) and lipid phytantriol (Phy) respectively, wherein both hexosomes were administered with major outer membrane protein (MOMP) of *Chlamydia trachomatis*. Cubosomes, the third subtype of liposomal nanovesicles, are a continuous arrangement of cubic lattice linked with aqueous channels inside the lipid bilayers (Barriga et al., 2019). In one particular study, cubosomes based on Toll-like receptor (TLR) agonists monophosphoryl lipid A and imiquimod have been shown to elicit high proliferation of CD8+ and CD4+ T cells along with resulting in an elevated production of IFN-γ and antigen-specific antibody through subsequent release of antigen ovalbumin (OVA).

Because of its adjustable and modifiable properties of efficient antigen presentation, nanofabricated liposomes and different lipid-based nanostructures are evinced to be an effective method not only for better immune activation but also for delivery of DNA vaccines (Kumar et al., 2015b).

3.2.3 Inorganic particle-based nanovaccines

Nanoparticles based on various inorganic biocompatible materials, primarily iron, silica, carbon, and gold have been tested as vaccine delivery agents due to their easily modifiable morphology and surface chemistry, which allows for antigen functionalization. Due to their material nature, these nanoparticles prevent precipitation from early proteolytic degradation, thereby facilitating a steady antigen presentation process. For example, gold nanoparticles have been demonstrated as an antigen delivery vehicle for protection against infectious diseases such as influenza, tuberculosis, malaria, HIV/AIDS, and also against certain tumors (Molino et al., 2016). For studying cancer vaccine delivery against breast cancer, Tn-antigen mucin glycan conjugated gold nanoparticles were used to yield multivalent glycoconjugates which produced substantial antibody titers. Another study has shown that APCs were activated upon delivery of drug-modified gold nanorods (Au-NRs) as an adjuvant for HIV-1 Env encoding plasmid DNA vaccine which resulted in a significant promotion of humoral and cellular immunity along with proliferation of T cells (Xu et al., 2012). Other inorganic materials such as spherical carbon nanotubes and carbon nanoparticles have been demonstrated to induce increased immunogenicity when used as adjuvants and as peptide delivery agents against multiple viral infections. One study indicated that carbon nanoparticles loaded with bovine serum albumin (BSA) showed increased IgA mucosal production and also stimulated Th1 and Th2 reaction (Wang et al., 2011). In case of nanoparticles based on silica, a highly dense amount of silanol groups on its surface allows for the addition of distinct functional molecules for easier recognition of the vaccine by the targeted sites. The surface of silica (SiO_2 nanosphere) was conjugated covalently with antigen ovalbumin (OVA) to form OVA@SiO_2 nanovaccine (Pei et al., 2021). The testing of this nanovaccine was found to facilitate DC maturation and increase antigen uptake capacity. More examples of inorganic nanoparticles include iron nanoparticles (IO) which have been used to elicit immune response as vaccine carriers. As established in a study, a recombinant malaria antigen known as merozoite surface protein 1 (rMSP1) and IO nanoparticles were conjugated to form rMSP1-IO vaccine (Barriga et al., 2019). This formulation induced 80% antigen-specific antibody and also elevated the expression of cytokines such as TNF-α, IL-6, IL-12, and IFN-γ and chemokines such as CXCL1, CCL2, CXCL2, CCL4, CCL3, and CXCL10. In another study, magnetic IO nanoparticles were employed as cancer vaccines following phospholipid functionalization with tumor-associated carbohydrate antigen (TACAs) (Chen et al., 2016). This produced antibody response against tumor cells which expressed TACA glycopeptide. Therefore, inorganic nanoparticles present with exciting opportunities for studying immune activation.

3.2.4 VLPs and polypeptide based nanovaccines

VLPs or virus-like-particles are an arrangement of viral capsid proteins which contains a high density of epitopes that fall under the nanoscale range. The VLPs resemble a geometry which is similar to that of a symmetrical viral capsule with any nucleic acid being absent. To elicit an adequate immune response, its external structure can be enhanced through chemical and biological modification. These include the process of fusion of peptide molecules or coupling of nonpeptide antigens on the surface of the VLPs to render bioconjugates. In a particular study on a mice model, it was shown that the nasal vaccines which encapsulated antigens of respiratory syncytial virus (RSV) namely matrix (M) and matrix 2 (M2) inside VLPs derived from *Salmonella typhimurium* bacteriophage P22, induced a robust antigen-specific CD8+ and CD4+ T-cell response (Toth and Skwarczynski, 2014).

Other prokaryotic and eukaryotic forms of viral particles known as "cages" have also been demonstrated for stimulating an immune response. For example, cages from multiple sources such as bacterial or archaeal encapsulins, human ferritin heavy chain, and E2 from *Bacillus stearothermophilus*, have been used to demonstrate immunization upon antigen presentation.

In another study, a DC activating molecule CpG has been shown to manifest a much higher magnitude of DC activation upon in-vitro administration of CpG encapsulated inside E2 biomimetic nanoparticles in comparison to free CpG (Fig. 4). Dynamic light scattering analysis and transmission electron microscopy imaging presented uniform sHDL-Ag/CpG (10.5 nm 0.5 average diameter) available with nanodisc-like structure. Ag presentation was tested by flow-cytometry analysis of DCs. Co-culture of BMDCs with SIINFEKL-specific B3Z T-cell hybridoma was conducted and observed for another 24 h, followed by checking for T-cell activation (Molino et al., 2016; Kuai et al., 2017). Another type of viral carrier that has been used as

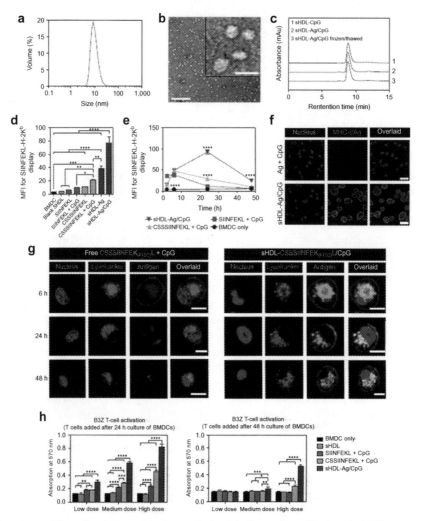

FIG. 4 Experimental data showing comparative analysis of CpG nanodiscs (Kuai et al., 2017).

a source for biomimetic inspiration is known as "vaults." These are made from eukaryotic major vault protein, constituting of an arrangement of ribonucleoproteins. In one study, vaults have been shown to elicit an increased antigen-specific CD4+ T-cell response toward *Chlamydia trachomatis* infection using polymorphic membrane protein G-1 (PmpG), a peptide derived from *Chlamydia muridarum* (Jiang et al., 2017). Therefore, the biomimetic approach using viral particles are a promising avenue of eliciting an antigen-specific immune response.

Besides viral particles, "nanofibrils" have also been used to demonstrate vaccine delivery. These are peptide sequences made of α-helices or β-sheets that arrange favorably in keeping the epitope site exposed from the surface. These nanofibrils have been studied for a plethora of antigens from strains such as Streptococcus aureus, Plasmodium falciparum, and

Mycobacterium tuberculosis. Among other notable means of vaccine delivery, "nanoclusters" have been designed using antigens and crosslinkers for an increased targeted response (Wang et al., 2014). "Micelles" are also another platform which features linked or conjugated antigenic molecules which can be tailored to suit the shape and size required to steer T-cell- and B-cell response.

4 Treatment of immunosuppressive diseases with nanoparticles

Nanotechnology has been utilized in the fields of anticipation, determination, and treatment of different ailments. Be that as it may, the connection of nanomaterials and the immune framework remains reasonably obscure. Past investigations have indicated that the nanomaterials can cause excitation or concealment of immune reactions through an official to blood proteins. Adsorption of these proteins bound to nanomaterials is perceived by different immunocyte cells.

Nanomaterials add to the movement of the adjuvant by expanding antigen introduction to the immune system just as the upgrade of the intrinsic resistant reactions. Deciding the level of biocompatibility of nanomaterials with the immune system is generally satisfied by their surface science. Today, the nanotechnology is generally used to improve focus on safe reactions to the anticipation and treatment of irresistible and nonirresistible illnesses. Local nano-immunotherapy through the decrease of harmful virulence improves the immunostimulatory particles. The uses of nanotechnology in medication and immunology are broad. Notwithstanding the clinical advances as of late, a few sicknesses, for example, AIDS, malignant growth, diabetes, and immune system infections have not been dealt with.

Especially when it comes to autoimmune diseases and others where the immune system is forced through an extensive wipe-out, nanoparticle therapy might be helpful in the following regard.

4.1 Potential of nanoparticle therapy

Since nanoparticles are the backbone of nanotechnology, their utilization in the clinical branch has opened new viewpoints in treatment. As needs are, the properties of nanoparticles should initially be assessed; and whenever endorsed, they will be then utilized for remedial purposes.

The utilization of nanomaterials in clinical intercessions has prompted direct contact of the nanomaterials with the human body. The nanomedicine can be refined by identifying, reestablishing, and recovering harmed tissues at the atomic levels. Considering the antimicrobial properties of various sorts of nanoparticles, for example, nanosilver, nanotitanium, and copper nanoparticles, one of their significant applications is to control an assortment of microorganisms.

Scientists, through the use of the outside surface of nanomaterials, have developed nanoscale associations among materials and natural systems to update their display and make new structures fundamentally. The usage of intelligent devices in prescription with negligible damage to including tissues is another utilization of the nanomaterials. Another utilization of nanomaterials in the clinical field is the production of feasible sections in sensor structures that can break down and thwart diseases. Adsorption of particles on the NPs known as nanoparticles in express microenvironments makes them be seen as new administrators by the characteristic immune system, achieving an inflammatory response. Nanotechnology is always considered for its capability to activate an inherent or adaptable safe feedback.

The adsorption of body particles outwardly of NPs causes their contorting, falling, and immunogenicity, achieving the adaptable, safe response. The acknowledgement of NPs intrudes with the subnuclear segments of DCs, impacts the peptides acquainted with T cells and thus changes the flexible, safe responses. In a research assessment by Gustafsson et al. (2014), TiO_2 NPs was injected intravenously to rodents regularly with a certain dosage for a few weeks, and it was observed that the following cytokines such as Interferon gamma, interleukin-4, and Interleukin-10 were activated. Such cytokine activated responses are very common for the varied forms of nanotechnology such as spheroids, colloidal powders, rods, etc. However, one major factor which helps to decide the efficacy of a nano-formulation in an immune environment includes the size of the nanoparticle. Without proper regulation of size, the specific response may be affected adversely.

4.2 Methodologies in discussion regarding the functionality of nanoparticles

Nanotechnology activated immune framework response has intrigued many researchers now. This is due to their primary efficacy in becoming highly viable adjuvants which strengthens the usually ineffective antigens in developing immunity by vaccination. The previous property has been appeared to rely upon molecule size and surface charge and can fundamentally add to the advancement of improved antibody features. As mentioned before, the dimensions of the nanoparticle molecule have been accounted for primary consideration required for deciding if antigens which are attached to nanoparticles. Often, depending upon the size and kind of antigen attached, the nanoparticles may help affect the interferon gamma or proinflammatory interleukin response. The main speculation on why nanotechnology inspired immunology is compelling for antibody improvement has been due to the fact that usually insoluble nanoparticles give a much-calibrated release of antigens, giving proper specific responses at target areas.

Nanotechnology empowers the creation of composite vehicles on a similar scale as individual cells and biomolecules, making a one-of-a-kind way to deal with imaging, detecting, drug conveyance, and portraying fundamental natural processes (Frey et al., 2018). Nanoparticles have in any event one measurement in the size of 0.1–100 nm. Liposomes, micelles, metallic nanoparticles, and polymeric nanoparticles are of the most ordinarily utilized nanoparticulate transporter frameworks for drug conveyance.

The following features are the basic requirements expected from nanoparticles to be feasible in the field of medicine:

(a) Embody and spare medications from degradation (Kroll et al., 2017).
(b) Improve drug conveyance.
(c) Control drug delivery (Kroll et al., 2017).
(d) Be created in enormous, reproducible, scale.

Nanoparticles can be focused by dynamic focusing on furthermore, detached focusing under in vivo conditions. Properties that intercede detached focusing on measure incorporates molecule structure, size, shape, and surface qualities. So, in most cases, whether it is the metallic nanoparticle or any organic nanoparticle or any nanocarrier like liposomes, nanotechnology is the perfect delivery system for bypassing the various challenging checkpoints.

The given table will help in the discussion of how the various methods of how nanotherapy is utilized against immunosuppressive diseases.

5 Cancer treatment with nanotechnology by immune modulation

5.1 Cancer immunology and immune evasion

As observed by Table 3, aberrant gene expression through random mutation and its unchecked replication results in the consistent growth and proliferation of cancer cells. These cancer cells express tumor associated antigens (TAA) on its surface which ideally, through cell-mediated immune response, is recognized by cytotoxic CD8+ T cells and natural killer

TABLE 3 Diseases and therapy.

S. no.	Disease	Process of therapy	References
1.	Rheumatoid arthritis	Macrophages carry nanocarriers to different inflamed areas	Yang et al. (2021)
2.	Diabetes	Various organic polymers such as *N*-isopropyl acrylamide, polyethyleneimine, polymethacrylic corrosive, poly (isobutyl cyanoacrylate), poly (-ε-caprolactone) are being utilized for oral insulin preparation as they help in better conveyance and protect the insulin from the proteolyzing effect within the body	Souto et al. (2019)
3.	Multiple sclerosis	Preparing the usage of nanosized carriers such as exosomes in order to get through the blood-brain barrier to deliver the target antibodies directly for the brain	Batrakova and Kim (2015)
4.	Lupus	Mycophenolic acid, usually used to treat lupus is packed in a biodegradable nanogel which helps in better conveyance and transport through the body	Sahu et al. (2017)
5.	Spondylitis	Various painkillers like diclofenac and other NSAIDS were stored in cationic liposomes or organic polymer liposomes for better transport and efficacy through the body	Xi et al. (2019)
6.	Coeliac disease	Gliadin, a necessary component is added via nanocapsules and provided to the patient	Adams (2021)
7.	Alzheimer	Nanoparticles act as carriers through the blood brain barrier. Certain nanoparticles made out of ceria can be utilized as oxidants which specifically target mitochondria	Chowdhury et al. (2017)
8.	Crohn's disease	Helps in delivery of nanomedicine and also to provide interference to defective genes via nanoparticles	Yang and Merlin (2019)

Continued

TABLE 3 Diseases and therapy—cont'd

S. no.	Disease	Process of therapy	References
9.	Ataxia-telangiectasia	Inhibition of the mutated gene with the help of various antisense oligonucleotide nanoparticles which makes it more receptive to gamma therapy	Chessa et al. (2021)
10.	Agammaglobulinemia	Antibodies conjugated to the specific nanoparticles are utilized for proper transport and subsequent treatment	Arruebo et al. (2009)

cells. These immune cells eliminate the cancer cells by direct apoptosis and simultaneous clearance of all cancer-related altercations with the release of IFN-γ (Wilhelm et al., 2016). These cell surface antigens are recognized by DCs along with the tumor-associated macrophages (known as TAMs) which act as antigen-presenting cells (APCs), activating TAA-specific CD4+T cells, which in turn results in the production of inflammatory cytokines TNF-α and IFN-γ. This communication between TAA and the immune system aids in constriction of the size of the tumor and subsequently upregulates antigen presentation, which in turn facilitates the activation of CD8+T cells resulting in checking of oncogenesis. However, with the use of immunosuppressive strategies, some oncogenic cells survive and continue to thrive inside the body bypassing immunosurveillance by creating a microenvironment which downregulates MHC expression, thereby restricting recognition by CD8+T cells and subsequently upregulates the expression of immune checkpoint proteins CD80/CD86 and PD-L1 (Sheikhpour et al., 2017), resulting in the inhibition of T-cell activation. In addition to this, the tumor microenvironment promotes the infiltration of fibroblasts and endothelial cells. These cells support the continued growth of tumor cells by aiding the homing of innate immune suppressor cells, along with infiltration of T regulatory cells (Treg) which inhibit CD8+ T cell function. Along with this, release of various antiinflammatory cytokines such as TGF-β as well as IL-10 which prevents the action of NK cells along with CD8+ T cells, which ultimately leads to malignancy.

In recent years, strategies of immunotherapy and immunomodulation have been found to be quintessential in enforcing the body's immune system against cancer, thereby increasing treatment effectiveness and decreasing the toxicity of existing chemotherapeutics. In this context, nanotechnology has proved to be pivotal in supplementing cancer immunomodulation.

5.2 Immunostimulants for cancer immunomodulation

5.2.1 Adjuvants

Alongside drug molecules, cancer immunomodulation is often studied through many other small co-stimulatory molecules. These are known as adjuvants, which are administered in conjugation with drug encapsulated nanoparticles to induce more effective humoral and antigen-specific immune response. As observed in Table 4, these are primarily used to target TLRs which are expressed on the surface of the innate immune cells like macrophages and DCs which recognize PAMPs expressed on the surface of pathogens and subsequently work through the lymphoid system. Synthetic adjuvants can also be targeted toward NOD-like

TABLE 4 Common adjuvants and their targeted TLRs and resulting immunostimulation.

Adjuvants	Targeted TLRs	Immunostimulation	References
Polyinosinic-polycytidylic acid-poly-L-lysine carboxymethylcellulose (poly-ICLC)	TLR3	• Activation of DCs • Activation of tumor-specific NK cells • Activation of antigen-specific CD8+ T-cell responses	Kyi et al. (2018)
Lipopolysaccharides (LPS)/monophosphoryl lipid A (MPLA)	TLR4	• Th1 dependent T-cell response • Increased production of IFN-β and IL-6 • Increased production of antibody titers	Duthie et al. (2011)
Flagellin	TLR5	• Tumor necrosis factor-α (TNFα) • Production of high antibody titers	Huleatt et al. (2007)
Imiquimod (R837)/imiquimod (R837), resiquimod (R848)	TLR7/TLR8	• Type-I interferon (IFN) production • IL-12 production	Shukla et al. (2010)
Unmethylated deoxycytidy l-deoxyguanosine oligodeoxynucleotides (CpG ODNs)	TLR9	• Activation of natural killer (NK) cells • Activation and proliferation of CD86 expressing B cells • IFN-α production • Strong Th1 response	Vollmer et al. (2004)

receptors of immune cells for activation of interferon genes. This stimulates the production of TNFα, IL6, IL1, and IL8 proinflammatory cytokines and also results in the activation of STING pathways to induce type-1 IFN and NF-κB-dependent production of cytokines (Duthie et al., 2011). In case of conventional cancer nano-medication, there can only be a limited amount of drug molecule that is encapsulated inside the vesicle, which on getting bioavailable may be subjected to nonspecific engulfment by phagocytic cells, further limiting its efficacy. Adjuvant co-delivery can make this process more proficient with an intrinsic focus on endogenous proteins, which can be aided through lymph nodes. For example, in-vivo delivery of amphiphilic adjuvants using lipid-based micelles drastically improves the proficiency of antigen presentation and subsequent T-cell antitumor response, because the nanoparticle itself disbands readily, which allows for quick binding to protein albumin.

5.2.2 Cytokines

Cytokines have also been well exploited for cancer immunomodulation because of its assistance in the MHC presentation process and for its favorable half-life in circulation. The cells targeted by cytokines express membrane receptors which sense the concentration of different cytokines and accordingly incites inflammatory and immune response. Therefore, this creates a room for the immune system to be manipulated through cytokine administration for cancer immunomodulation.

So far, many cytokines such as IFN-α, TNF-α, GM-CF, IL-2, IL-10, IL-12, IL-15, and IL-21 have been studied for potential cancer therapy as observed in Table 5. For antitumor therapy, these cytokines are mostly found to be effective when used in conjugation with one or more compounds. For example, when IFN-α is fused with apolipoprotein A-I, there is an improvement in the pharmacokinetics and anticancerous response of IFN-α. In another example, it has

TABLE 5 Common cytokines and their role in immune system modulation.

Cytokines	Impact on the immune system	References
IFN-α	• Activation of DCs • Activation of CD4+ and CD8+ T cells • Induction of JAK-STAT pathway • Activation of NK cells • Increased presentation of TAAs	Cauwels et al. (2018)
IL-2	• Induction of NK cells • Production of CD4+ and CD8+ T cells • Production of DCs and mast cells • Expansion of Treg cells • Activation of JAK-STAT pathway • Activation of (PI3K) AKT pathway • Activation of MAPK pathway	Choudhry et al. (2018)
IL-15	• Generation of cytotoxic lymphocytes (CTLs) • Generation of memory phenotype CD8 T cells • Maintenance and proliferation of NK cells • Increase in CD4 cells • Increase in levels of IL-6 and IFN-γ	Lodolce et al. (1998)
IL-21	• Differentiation and proliferation of B cells • Regulation of immunoglobulin production • Proliferation/effector function of CD4+ and CD8+ T cells • Induction of NK and NKT cells • Suppression FOXP3 expression • Regulation of expansion of Tregs • Stimulation of fibroblasts and epithelial cells • Production of inflammatory mediators	Santegoets et al. (2013)
IL-10	• Regulation of IL-1b, IL-6, VEGF, TNF-α, and MMP-9 • Promotion of IL-6 expression • Inhibition of NF-KB translocation • Inhibition of MHC expression and costimulatory molecules • Downregulation of Th1 cytokines expression • Induction of T-regulatory responses	Dennis et al. (2013) and Sheikhpour et al. (2017)
IL-12	• Activation of T and natural killer (NK) lymphocytes • Production of IFNγ • Signaling for Th1 differentiation via STAT-4 • Transformation and activation of M2 macrophages to antitumor M1 macrophages	Romagnani et al. (2001)

TABLE 5 Common cytokines and their role in immune system modulation—cont'd

Cytokines	Impact on the immune system	References
	• Expression of CXCR4 and CXCL9, 10, and 11 • Enhancement of MHC I tumor antigen presentation • CXCR3-mediated antiangiogenesis	
GM-CSF	• Differentiation and maturation of DCs • Priming of antitumor CTLs • Immunosuppression of tumor microenvironment • Stimulation of macrophage/monocyte activity • Enhanced CD4+ T cell and neutrophil activity	Simmons et al. (2007)
TNF-α	• Receptor mediated tumor cell apoptosis • Stimulation of T-effector cells, neutrophils and monocytes • Blocking of immune suppressor T-Reg cells • Modulation of endothelial cells • Disruption of tumor vasculature and neoangiogenesis • Promoting antitumor stage M1 • Downregulation of IL-13 expression • Inhibition of differentiation of tumor-induced monocyte into immunosuppressive phenotypes	Schaer et al. (2014)

been shown that PEGylation of IL-2 increases the half-life of IL-2 in circulation. Besides this, recombinant cytokines also prove to be more effective than regular variants (Charych et al., 2017). Administration of intravenous bolus from recombinant glycosylated IL-15 generated in *E. coli* to renal cell carcinoma or advanced melanoma patients showed increased levels of NK and CD8+ T cells in blood. These receptor-targeted administrations of cytokines along with chemotherapeutics are an effective means of combating cancer.

5.2.3 Nucleic acids

Delivery of nucleic acid therapeutics, such as plasmids, messenger RNA (mRNA), microRNA, antisense oligonucleotides (known as ASO), immunomodulatory RNA or DNA, for their ability to regulate the expression of targeted endogenous genes, can modulate immune response. For example, in vitro manufacturing of specific cancer-antigen encoding mRNA using DNA templates and T3 or T7 phage RNA polymerase have proved to be versatile in ex vivo approach of adoptive cell engineering for cancer vaccines (Sahin et al., 2014). A relatively new and exciting concept of genome editing has caught the attention of many researchers experimenting on nanoparticle-based drug delivery for immunomodulation. One of the examples include the siRNA-based nanoliposomes which are being used to target the innate immune system. Along with this, siRNA has also been conceptualized to be delivered to make transcriptional modification. For repairing specific genes associated with diseases, Cas9 mRNA has also been used for in vivo studies (Conde et al., 2015). One such example includes nanocarrier-based in vivo antitumor phenotyping of tumor-associated T cells through its nuclear insertion with leukemia-targeting CAR genes. This reprogramming strategy has proved to have an upper hand over traditional therapy using

CAR T cells which has in turn encouraged its ex vivo implementation in patient-specific T-cell therapy. Likewise, composite lipid nanoparticles have also been demonstrated to provide RNA protection against extracellular ribonuclease activity, thereby organizing easy recognition by antigen-presenting cells for in vivo expression of genetically modified antigenic peptide molecules. Nano-constructs containing viral and mutant-antigen coding RNA sequences induce strong memory and effector T-cell response through IFNα activation.

5.2.4 Monoclonal antibodies

As immunostimulants, monoclonal antibodies (mAbs) present with certain advantages such as resistance from degradation during circulation, higher target specificity and activation or suppression of cellular pathways which regulate immune response. For example, using anti-CD40 monoclonal antibodies, activation of humoral immune response (from the DC network) against cancer can be triggered by targeting the CD40 receptor present on the surface of macrophages, DCs and B cells. The examples of antibodies for receptor-mediated activation of signaling cascade include anti-4-1BB (anti-CD137) to target 4-1BB present on NKs, DCs, T cells and mast cells (Wilhelm et al., 2016). Anti-OX40 mAbs can be used to activate OX40 signaling pathway which result in increased production of cytokines and higher stimulation of CD8+ and CD4+ T cells.

As discussed earlier in the chapter, various types of nanoparticles have been formulated to be used as vaccine carrier agents and adjuvants to induce antigen-specific immune response. Similarly, for targeted delivery of various cancer therapeutics, primarily immunostimulatory agents like, cytokines, toll-like receptor agonists and nucleotides, numerous nanoformulations have been preclinically tested rendering favorable outcomes. Besides being a delivery platform, studying the nanoparticle-based approach from the perspective of material composition and surface property modulation, several immuno-nanomedicines have been shown to regulate many relevant signaling pathways which can be directly linked to immune response against cancer. Therefore, it is an interesting aspect to venture into by further exploring the various immunomodulatory properties of different formulations of nano-constructs which can regulate the immune system to produce an anticancerous response.

5.3 Nanovesicles for delivery of immunostimulants

A variety of nano-compositions have been tested for immunomodulation through targeted delivery of chemotherapeutic drugs and antigen-specific vaccines. In case of cancer treatment, the several advantages of nanoparticles in terms of size and tuneable physical and chemical properties can be exploited to specifically target the zone of oncogenesis. Cells of the innate immune system, namely, macrophages, monocytes and dendritic cells, can recognize and selectively uptake nanoparticles, which works for a distinct advantage. The traditional approach for administration of chemotherapeutic drugs usually presents with higher levels of toxicity and a multitude of off-targets. The material composition of the particle may vary in accordance with its targeted site of delivery, which is also a determinant factor in the modification of the surface chemistry in order to increase the specificity of drug delivery. In this regard, the particle surface can be embellished using peptides, recombinant proteins, and antibodies. As a matter of fact, various biological ligands which can be surface-docked for increasing the collection of the drug molecules encapsulated in the nanoparticles, distinctively at the site of the tumor, can be used to modify the nanoparticle surface. Therefore, it seems suitable for the exploration of nanofabrications for immunological approaches. Liposomal and synthetic nanoparticles have been investigated for immune response against cancer in most studies which have been well demonstrated through delivery of viral peptide antigens to evoke CD8+ T-cell response against cancer by upregulation of (IFN)-γ and (IL)-2 within these cells. Nanogels have also been shown to have a significant potential to partake in the process of immunomodulation. In the following subsections, these processes are discussed comprehensively.

5.3.1 Liposomal vesicles as immunomodulatory delivery platform

Liposomal nanoparticles, in particular, have been studied extensively for their capacity for immunomodulation, primarily due to its optimizable surface functionality but more so because of its flexibility of being able to be loaded with both hydrophilic and hydrophobic molecules. Such utility can be demonstrated through interlayer-crosslinked multilamellar vesicles (ICMVs). Innate immunomodulation using lipid-based nanoparticle has also been demonstrated using RNA-lipoplexes (RNA-LPXs) which are targeted toward IFN-α-dependent activity of macrophages and DCs (Phua, 2015). These lipid nanoparticles enclosed an RNA-encoded antigen and were tested for variable surface charge which induced strong memory and effector T-cell stimulation and resulted in steady decline of tumor. Liposomal carriers have also been used to deliver immunostimulant drugs for immune activation against tumor. For example, intravenous injection of liposomal construct, conjugated with prodrug indoximod (IND), followed by the loading of doxorubicin (DOX) in an orthotopic 4T1 tumor syngeneic mice model showed IFN-γ and perforin-dependent killing of tumor cells by CD8+ cytotoxic T cells, which was aided by reduced interference of Tregs and an overall increase in the ratio of CD8+/FOXP3+ T-cell (Nel et al., 2020). In another study using 4T1 breast cancer murine model, an increased expression of TNF-α, IL-1β and IFN-γ was recorded along with increase in the level of DCs, CD8+ T cells and F4/80+ macrophages upon intratumoral injection of IL-12 and monophosphoryl lipid A (MPL) which were encapsulated in 1,2-dioleoyl-3-trmethylammonium-propane (DOTAP) lipids (Meraz et al., 2012). Besides immunostimulants, delivery of nucleic acids such as CCR2-siRNA encapsulated in nano-liposomal vesicles showed efficient CCR2 mRNA degradation in monocytes which prevents inflammation and subsides tumor microenvironment in a mice model with lymphoma (Wilhelm et al., 2016).

5.3.2 Synthetic nanovesicles as immunomodulatory delivery platform

For synthetic nanoparticles, the virtue of its tuneable surface makes it fit for antitumor immunomodulation. The surface properties are favorable enough for chemical docking of serum proteins such as apolipoproteins and albumin, which facilitates its interaction with the complement and scavenger receptors of phagocytes, thereby extending its bioavailability. Synthetic polymers have been well explored as an adjuvant delivery platform. For example, delivery of Toll-like receptor agonists also known as TLR-7/8a attached with a polymeric scaffold resulted in enhanced pharmacokinetic profile of the polymer associated TLR-7/8a conjugate because of its prolonged activity in lymph nodes (Nuhn et al., 2018). This study also demonstrated how increasing polymer-TLR-7/8a density also resulted in increased levels of

CD8+ T cells and subsequently potentiated an antibody response through class switching of IgG to IgG2 with persistent formation and expression of local interferon as well as IL-12 (Lynn et al., 2015). In another study, imiquimod (R837) was used as an adjuvant wherein R837, Indocyanine green (also known as ICG) as well as a TLR7 agonist were encapsulated in PLGA. This formulation was also co-encapsulated with a near infrared dye which enhanced the permeation and retention (EPR) of PLGA-ICG-R837 nanoparticles selectively at the site of tumor, resulting in the photothermal clearance of the primary tumor while also producing tumor-specific antigens due to the presence of R837 (Mi et al., 2018). Along with adjuvant delivery, studies on the therapeutic benefit of combine checkpoint inhibitors have been pursued using polymeric nanovesicles. The PLGA-ICG-R837 nanoparticles discussed in the previous example were combined with the anti-CTLA4 checkpoint blockade for in vivo studies which showed inhibiting metastasis in multiple tumor models and reduction of secondary tumors along with a longstanding memory response. The administration of checkpoint inhibitors has proved to ameliorate some of the adverse effects caused by direct use of adjuvants. For example, methoxy-triethylene-glycol methacrylate and penta-fluoro-phenyl methacrylate-based nanoparticles were activated using TLR7 or TLR8 receptors and were demonstrated to locally activate DCs at the site of tumor. Additionally, through a combination treatment using checkpoint blockade anti-PDL1 in a B16 melanoma mice model, the tumor growth could be inhibited through activation of DCs under the influence of Flt3L growth factor. In another study, anti-PDL1 was combined with anti-OX40 and attached with PEGylated PLGA nanovesicles and were administered in two murine tumor models. This study showed an increase in T-cell activation and elevation in memory responses in comparison to the responses generated by administration of independent antibodies or with the use of nanoformulations with individual antibodies.

5.3.3 *Nanogels as immunomodulatory delivery platform*

Nanogels or nano-sized hydrogels (size 1–1000 nm) can be effectively used to encapsulate and deliver immunostimulants (Song et al., 2017). These are rather large and flexible polyionic networks made of polymers, hydrophilic or amphiphilic, which allow for an increased loading capacity and multivalent conjugation over a sizeable surface area, thus providing a scope for nurturing its immunomodulatory properties while also maintaining its structural stability and retaining high water content. For delivery of immunostimulants, nanogels are required to enact in a target-specific manner in order to avoid nonspecific accumulation in cellular compartments of different organs and bypassing the reticuloendothelial and glomerular filtration system. Therefore, the designing aspect of the respective nanogels are crucial (Ferreira et al., 2013). The common polymers for designing nanogels range from naturally occurring polymers just like alginate, dextran, chitosan, pullulan, hyaluronic acid, mannan, and heparin to synthetically derived polymeric substances like PGA, PLA, PMMA, PCL, and PLGA (Kabanov and Vinogradov, 2009).

When it comes to immunotherapy, different formulations of nanogels have been studied. For example, polymeric nanogels based on pentafluorophenyl reactive-esters derived from RAFT-polymerization allow for specific cellular targeting and can be further fine-tuned for personalized immunotherapy (Stickdorn and Nuhn, 2020). Since remodeling the TME is a traditional approach to develop antitumor immunity, a few nanogels have been explored to deliver such antitumor agents. Hydroxypropyl-β-cyclodextrin acrylate and chitosan derivative-based nanogels have been used to entrap and deliver paclitaxel to the TME which showed a significant enhancement in the drug release profile accompanied by an increased level of the presence of effector immunocytes and downregulation of other immune factors (Song et al., 2017), thereby resulting in an improved antitumor immunity.

Along with drugs, nanogel-based delivery of cytokines and adjuvants have also been used to generate antitumor activity. For example, in B16F10 melanoma mice model, the delivery of a combination of both IL-2 as well as TGF-β inhibiting factor into TME by utilizing the polymeric liposomal nanogels or nanolipogels (nLGs) remarkably stunted the tumor growth through tumor-activated infiltration of CD8+ thymus cell and expression of NK cell response (Park et al., 2012). Delivery of adjuvants using nanogels have also shown favorable results. For example, when β-glucan SPG nanogel is administered with a TLR-9 ligand CpG ODN, a much stronger Th1 response is generated in comparison to exclusive administration of β-glucan nanogel, which also delays the tumor growth when the mice is immunized with a formulation of antigen OVA and cross-linked CpG (Miyamoto et al., 2017). This could be due to improved cellular uptake as a result of increased size and therefore increased β-glucan recognition site inside the structure of the designed nanogel (Miyamoto et al., 2017). In another example, lymph node macrophages were activated against TAA encapsulated in cholesteryl pullulan (CHP) upon subcutaneous administration using a TLR agonist as an adjuvant (Muraoka et al., 2014). Even without the requirement of adjuvant co-delivery, nanogels can be used to target tumor sites. For example, a notable retardation in tumor growth was observed upon repeated delivery of recombinant IL12 encapsulated inside CHP nanogel (CHP/rmIL-12 complex), which resulted in subsequent increase in IL12 levels in the serum (Shimizu et al., 2008). In another patient-based study, it was observed that the protein known as NY-ESO-1 combined along with CHP nanogels elicited both CD8+ as well as CD4+ Thymus cell responses and production of antibody titers (Hasegawa et al., 2006).

Additionally, the material design for nanoparticles can significantly improve immunotherapy. Since nanoparticles can be made to mimic the local proteins through surface modification, engineered tumor-specific proteins can be coated on the surface of these particles with chemotherapeutic molecules entrapped in its core. Such an approach has been illustrated through lipoprotein mimicking nanoparticles which has shown a significant influence on increased targeting efficiency of the nanoparticles toward lymph nodes while also promoting the upregulation of T cells specific to the engineered protein molecules.

5.4 Nanoparticles as cancer immunomodulating agent

It has been well established that nanomaterials can be used as a vehicle for delivery of immunomodulatory agents. However, in recent years, some studies have highlighted that some nanomaterials can itself possess immunomodulatory abilities. Both liposomal and polymeric nanoparticles can be engineered and modified to elicit antigenic response upon interaction with APCs, followed by activation of tumor-specific T cells, thus promoting antitumor activity. For example, testing of neutrophil-targeted drug-loaded nanoparticles against brain tumor showed reduction in tumor size, implying that the nanoparticles were being carried to the tumor site when local cytokines were homing the local neutrophils. Besides using the innate immune cells for delivery of chemotherapeutics, these particles can itself be formulated to instigate similar immune reactions, such as activation and elevation of CD8+ and CD4+ T lymphocytes levels. In one particular study, it was demonstrated how particle size of an artificially fabricated nanoparticle mimicking APCs significantly increases the activation of T cells. Therefore, this goes to show that particle chemistry does generate immunomodulatory responses.

Some metallic nanoparticles also have been studied to evoke tumor immunomodulatory properties. For example, mesoporous silica-based nanoparticles (MSNs), most well-known for adjuvant-based combination therapies, have also been found to have co-stimulatory reactions which elevate Th1 and Th2 immune response (Ding et al., 2018). Large-pore upconversion MSNs or UCMSs (size <100 nm), in combination with model antigens OVA and Tumor This formulation was also used as a nanovaccine in colon cancer bearing murine model which showed a significant bit of inhibition in tumor growth, thereby indicating its immunotherapeutic potential of being used as an adjuvant (Ding et al., 2018). Another example of the use of metallic nanoparticles show reactive oxygen species-dependent elimination of cancer cells using a formulation of iron oxide against anemia which stunted tumorigenesis to a great extent. VLPs have also been demonstrated for anticancerous efficacy. In one particular study, it was shown that self-assembled cowpea mosaic viral nanoparticles, upon in situ administration yields higher volumes of (IFN)-γ and (IL)-2 inflammatory cytokines against B16F10 lung melanoma, while also bolstering the adaptive immune response (Lizotte et al., 2016). Alongside this, various immunostimulatory molecules, including growth factors, distinct cytokines as well as a blend of multiple stimulating molecules can be used to target the requisite site and thereby commence a higher immune cell-mediated activity against tumor.

An exciting aspect to nanoparticle-based immunomodulation is that the cancer cells can be used to embellish the nanoparticles loaded with adjuvants, especially the tumor cells which experience immunogenic cell death. This is particularly relevant, because chemotherapeutics causes immunogenic death of tumor cells and these tumor cells produce specific signals which cause its eradication. Hence, on amplification of these signals, more chemotherapeutic molecules will initiate antitumor activity at the site of the tumor. Therefore, through specific tumor cell-based embellishment on adjuvant loaded nanoparticles, the ease of the process can be increased efficiently as these nanoparticles can potentiate both external and internal signals for tumor death. Similarly, the death signal inducing molecules can be delivered to the site of the tumor through a target-specific approach. This method has provided a much better insight into the understanding of the elevated levels of tumor death signal production and an overall improved pharmacokinetic profile of these inducer molecules. Besides this, there are some intrinsic properties of nanoparticles which allow for the induction of tumor cell death upon administration. For example, nanoshell particles of gold are utilized to demonstrate for photothermal therapy against metastatic melanoma through dendritic cell activation by releasing immunostimulatory molecules.

To generate an even more robust immune response against tumor cells, a combinatorial strategy can be employed. As discussed in the previous paragraph, both tumor cell death-inducing chemotherapeutic molecules and photothermal therapy have shown to elicit high levels of antitumor response through innate immunomodulation and elevated T-cell response. A combinatorial approach involves the employment of both the approaches of death inducers and photothermal therapy. Such method stimulates the upregulation of the immune response generated by the immunomodulating agents and subsequently evokes antitumor activity coactively, which thereby causes eradication of malignant cells. In multiple model systems, this response has been recorded to be more significant than the response of individual therapies. As crucial as it is to modulate the immune system to bolster antitumor activity, it is also important to work against the systemic immunosuppression that is brought about with the onset of tumor (Sharpe and Mount, 2015). This is primarily due to the signals of immune suppression which are generated from the tumor, which in turn aids in its recruiting of local cells which help in its steady proliferation.

Therefore, with the increasing immunotherapeutic approaches, further studies are thus warranted for stretching the potential of cancer nanomedicines for immunomodulation. Using respective methods of administration or through combination therapy along with abolishment of immunosuppression is vital to immunotherapy. This avenue of medical biology strikes to be one of the most reassuring ventures that have the potential to be ubiquitously followed for cancer treatment shortly.

6 Conclusion

It is fair to say that while there have been many advances in the field of immunology when it comes to providing and redesigning new therapies for the betterment of humankind, nanotechnology helps to tap into a vast potential where it can help to augment and improve preexisting therapies or be synergistically developed with a new one (Toth and Skwarczynski, 2014). The utilization of nanoparticle-based conveyance frameworks is particularly appealing their acknowledgement by the immune system (Li et al., 2019).

Nanotechnology is utilized for delivery purposes, as mentioned multiple times in the book chapter. For example, various kinds of polymeric particles like liposomes can be utilized in the form of self-attacking adjuvants conveyance framework or in the form of a stage to convey antigen and adjuvant (Peng et al., 2020). PLA and PLGA polymers are among some of the best customarily contemplated nanotech carriers which have high compatibility with the human body. Utilizing many other methodologies like liposomal transport, VLPs, micelles, nanodiscs (Kuai et al., 2017) and rods, many different methods of immunomodulating pharmaceutical transport can be considered with higher efficacy. Nanotechnology permits the production of particles that utilize the inalienable capacity of immune defence to perceive little but significant detrimental behavior, for example, infections and poisons. In the mix with the defensive epitope plan, this allows the making of immunogenic nanoparticles that animate a reaction against the focused-on microbes (Jang, 2020). The very specific reaction of bodily defence to various forms of nanotechnology ensures that a large quantity of pharmacological agents at present is dependent on them. When it comes to the errant cells of the body or the external pathogens, there is no dearth of opportunities for usage of nanotechnology.

Thus, within the scope of safety and ethics, a wide range of research and planning can be conducted to ensure that nanotechnology-based interventions will be beneficial for augmenting and supporting the immune system.

Acknowledgment

Authors acknowledge DBT, Govt. of India for financial support to Ms. Shalini Dasgupta.

References

Adams, J., 2021. Can Gliadin Nanoparticles Cure Celiac Disease in Humans? Celiac.com. https://www.celiac.com/articles.html/can-gliadin-nanoparticles-cure-celiac-disease-in-humans-r5110/. (Accessed 14 January 2021).

Andreae, S., Buisson, S., Triebel, F., 2003. MHC class II signal transduction in human dendritic cells induced by a natural ligand, the LAG-3 protein (CD223). Blood 102 (6), 2130–2137. https://doi.org/10.1182/blood-2003-01-0273.

Arruebo, M., Valladares, M., González-Fernández, Á., 2009. Antibody-conjugated nanoparticles for biomedical applications. J. Nanomater. 2009. https://doi.org/10.1155/2009/439389.

Barriga, H.M.G., Holme, M.N., Stevens, M.M., 2019. Cubosomes: the next generation of smart lipid nanoparticles? Angew. Chem. Int. Ed. 58 (10), 2958–2978. https://doi.org/10.1002/anie.201804067.

Batrakova, E.V., Kim, M.S., 2015. Using exosomes, naturally-equipped nanocarriers, for drug delivery. J. Control. Release 219, 396–405. https://doi.org/10.1016/j.jconrel.2015.07.030.

Bertoletti, A., Tan, A.T., 2020. Challenges of CAR- and TCR-T cell-based therapy for chronic infections. J. Exp. Med. 217 (5). https://doi.org/10.1084/jem.20191663.

Bhatia, S., Afanasiev, O., Nghiem, P., 2011. Immunobiology of Merkel cell carcinoma: implications for immunotherapy of a polyomavirus-associated cancer. Curr. Oncol. Rep. 13 (6), 488–497. https://doi.org/10.1007/s11912-011-0197-5.

Blom, R.A.M., et al., 2016. A triple co-culture model of the human respiratory tract to study immune-modulatory effects of liposomes and virosomes. PLoS One 11 (9), e0163539. https://doi.org/10.1371/journal.pone.0163539.

Borges, O., et al., 2008. Immune response by nasal delivery of hepatitis B surface antigen and codelivery of a CpG ODN in alginate coated chitosan nanoparticles. Eur. J. Pharm. Biopharm. 69 (2), 405–416. https://doi.org/10.1016/j.ejpb.2008.01.019.

Boyiadzis, M.M., et al., 2018. Chimeric antigen receptor (CAR) T therapies for the treatment of hematologic malignancies: clinical perspective and significance. J. Immunother. Cancer 6 (1). https://doi.org/10.1186/s40425-018-0460-5.

Cauwels, A., et al., 2018. Delivering type I interferon to dendritic cells empowers tumor eradication and immune combination treatments. Cancer Res. 78 (2), 463–474. https://doi.org/10.1158/0008-5472.CAN-17-1980.

Charych, D., et al., 2017. Modeling the receptor pharmacology, pharmacokinetics, and pharmacodynamics of NKTR-214, a kinetically-controlled interleukin-2 (IL2) receptor agonist for cancer immunotherapy. PLoS One 12 (7). https://doi.org/10.1371/journal.pone.0179431.

Chen, Q., Xu, L., Liang, C., Wang, C., Peng, R., Liu, Z., 2016. Photothermal therapy with immune-adjuvant nanoparticles together with checkpoint blockade for effective cancer immunotherapy. Nat. Commun. 7 (1), 1–13. https://doi.org/10.1038/ncomms13193.

Chessa, L., Micheli, R., Molinaro, A., 2021. Focusing New Ataxia Telangiectasia Therapeutic Approaches. Insight Medical Publishing. https://raredisorders.imedpub.com/focusing-new-ataxia-telangiectasia-therapeutic-approaches.php?aid=8998. (Accessed 14 January 2021).

Chou, J., et al., 2008. Nasopharyngeal carcinoma—review of the molecular mechanisms of tumorigenesis. Head Neck 30 (7), 946–963. https://doi.org/10.1002/hed.20833.

Choudhry, H., et al., 2018. Prospects of IL-2 in cancer immunotherapy. Biomed. Res. Int. 2018. https://doi.org/10.1155/2018/9056173.

Chowdhury, A., Kunjiappan, S., Panneerselvam, T., Somasundaram, B., Bhattacharjee, C., 2017. Nanotechnology and nanocarrier-based approaches on treatment of degenerative diseases. Int. Nano Lett. 7 (2), 91–122. https://doi.org/10.1007/s40089-017-0208-0.

Chu, F., Cao, J., Neelalpu, S.S., 2017. Versatile CAR T-cells for cancer immunotherapy. Wspolczesna Onkol. 2 (1A), 73–80. https://doi.org/10.5114/wo.2018.73892.

Conde, J., et al., 2015. Dual targeted immunotherapy via in vivo delivery of biohybrid RNAi-peptide nanoparticles to tumor-associated macrophages and cancer cells. Adv. Funct. Mater. 25 (27), 4183–4194. https://doi.org/10.1002/adfm.201501283.

Davila, M.L., Brentjens, R.J., 2016. CD19-targeted CAR T cells as novel cancer immunotherapy for relapsed or refractory B-cell acute lymphoblastic leukemia. Clin. Adv. Hematol. Oncol. 14 (10), 802–808.

Demento, S.L., et al., 2012. Role of sustained antigen release from nanoparticle vaccines in shaping the T cell memory phenotype. Biomaterials 33 (19), 4957–4964. https://doi.org/10.1016/j.biomaterials.2012.03.041.

Dennis, K.L., Blatner, N.R., Gounari, F., Khazaie, K., 2013. Current status of interleukin-10 and regulatory T-cells in cancer. Curr. Opin. Oncol. 25 (6), 637–645. https://doi.org/10.1097/CCO.0000000000000006.

Ding, B., et al., 2018. Large-pore mesoporous-silica-coated upconversion nanoparticles as multifunctional immunoadjuvants with ultrahigh photosensitizer and antigen loading efficiency for improved cancer photodynamic immunotherapy. Adv. Mater. 30 (52). https://doi.org/10.1002/adma.201802479.

Diwan, M., Tafaghodi, M., Samuel, J., 2002. Enhancement of immune responses by co-delivery of a CpG oligodeoxynucleotide and tetanus toxoid in biodegradable nanospheres. J. Control. Release 85 (1–3), 247–262. https://doi.org/10.1016/S0168-3659(02)00275-4.

Duthie, M.S., Windish, H.P., Fox, C.B., Reed, S.G., 2011. Use of defined TLR ligands as adjuvants within human vaccines. Immunol. Rev. 239 (1), 178–196. https://doi.org/10.1111/j.1600-065X.2010.00978.x.

Dykman, L.A., Khlebtsov, N.G., 2017. Immunological properties of gold nanoparticles. Chem. Sci. 8 (3), 1719–1735. https://doi.org/10.1039/c6sc03631g.

Feng, G., et al., 2013. Enhanced immune response and protective effects of nano-chitosan-based DNA vaccine encoding T cell epitopes of Esat-6 and FL against *Mycobacterium tuberculosis* infection. PLoS One 8 (4), e61135. https://doi.org/10.1371/journal.pone.0061135.

Ferreira, S.A., Gama, F.M., Vilanova, M., 2013. Polymeric Nanogels as Vaccine Delivery Systems., https://doi.org/10.1016/j.nano.2012.06.001.

Frey, S., Castro, A., Arsiwala, A., Kane, R.S., 2018. Bionanotechnology for vaccine design. Curr. Opin. Biotechnol. 52, 80–88. https://doi.org/10.1016/j.copbio.2018.03.003.

Götze, O., Müller-Eberhard, H.J., 1971. The c3-activator system: an alternate pathway of complement activation. J. Exp. Med. 134 (3), 90–108.

Gustafsson, Å., Jonasson, S., Sandström, T., Lorentzen, J.C., Bucht, A., 2014. Genetic variation influences immune responses in sensitive rats following exposure to TiO_2 nanoparticles. Toxicology 326, 74–85. https://doi.org/10.1016/j.tox.2014.10.004.

Hasegawa, K., et al., 2006. In vitro stimulation of CD8 and CD4 T cells by dendritic cells loaded with a complex of cholesterol-bearing hydrophobized pullulan and NY-ESO-1 protein: identification of a new HLA-DR15-binding CD4 T-cell epitope. Clin. Cancer Res. 12 (6), 1921–1927. https://doi.org/10.1158/1078-0432.CCR-05-1900.

Honda-Okubo, Y., Saade, F., Petrovsky, N., 2012. Advax™, a polysaccharide adjuvant derived from delta inulin, provides improved influenza vaccine protection through broad-based enhancement of adaptive immune responses. Vaccine 30 (36), 5373–5381. https://doi.org/10.1016/j.vaccine.2012.06.021.

Huleatt, J.W., et al., 2007. Vaccination with recombinant fusion proteins incorporating Toll-like receptor ligands induces rapid cellular and humoral immunity. Vaccine 25 (4), 763–775. https://doi.org/10.1016/j.vaccine.2006.08.013.

Huye, L.E., Dotti, G., 2010. Designing T cells for cancer immunotherapy. Discov. Med. 9 (47), 297–303.

Ichihashi, T., Satoh, T., Sugimoto, C., Kajino, K., 2013. Emulsified phosphatidylserine, simple and effective peptide carrier for induction of potent epitope-specific T cell responses. PLoS One 8 (3), e60068. https://doi.org/10.1371/journal.pone.0060068.

Intlekofer, A.M., Thompson, C.B., 2013. At the bench: preclinical rationale for CTLA-4 and PD-1 blockade as cancer immunotherapy. J. Leukoc. Biol. 94 (1), 25–39. https://doi.org/10.1189/jlb.1212621.

Iwasaki, A., Foxman, E.F., Molony, R.D., 2017. Early local immune defences in the respiratory tract. Nat. Rev. Immunol. 17 (1), 7–20. https://doi.org/10.1038/nri.2016.117.

Jang, J., 2020. Nanomaterials and Immune System. [Online]. Available from: http://www.ksimm.or.krvolume. (Accessed 22 September 2020).

Jiang, J., et al., 2017. A protective vaccine against chlamydia genital infection using vault nanoparticles without an added adjuvant. Vaccine 5 (1). https://doi.org/10.3390/vaccines5010003.

Kabanov, A.V., Vinogradov, S.V., 2009. Nanogels as pharmaceutical carriers: finite networks of infinite capabilities. Angew. Chem. Int. Ed. 48 (30), 5418–5429. https://doi.org/10.1002/anie.200900441.

Kanekiyo, M., et al., 2013. Self-assembling influenza nanoparticle vaccines elicit broadly neutralizing H1N1 antibodies. Nature 499 (7456), 102–106. https://doi.org/10.1038/nature12202.

Kroll, A.V., et al., 2017. Nanoparticulate delivery of cancer cell membrane elicits multiantigenic antitumor immunity. Adv. Mater. 29 (47). https://doi.org/10.1002/adma.201703969.

Kuai, R., Ochyl, L.J., Bahjat, K.S., Schwendeman, A., Moon, J.J., 2017. Designer vaccine nanodiscs for personalized cancer immunotherapy. Nat. Mater. 16 (4), 489–498. https://doi.org/10.1038/NMAT4822.

Kumar, S., Anselmo, A.C., Banerjee, A., Zakrewsky, M., Mitragotri, S., 2015a. Shape and size-dependent immune response to antigen-carrying nanoparticles. J. Control. Release 220 (Pt A), 141–148. https://doi.org/10.1016/j.jconrel.2015.09.069.

Kumar, R., Ray, P.C., Datta, D., Bansal, G.P., Angov, E., Kumar, N., 2015b. Nanovaccines for malaria using Plasmodium falciparum antigen Pfs25 attached gold nanoparticles. Vaccine 33 (39), 5064–5071. https://doi.org/10.1016/j.vaccine.2015.08.025.

Kyi, C., et al., 2018. Therapeutic immune modulation against solid cancers with intratumoral poly-ICLC: a pilot trial. Clin. Cancer Res. 24 (20), 4937–4948. https://doi.org/10.1158/1078-0432.CCR-17-1866.

Li, Y., Ayala-Orozco, C., Rauta, P.R., Krishnan, S., 2019. The application of nanotechnology in enhancing immunotherapy for cancer treatment: current effects and perspective. Nanoscale 11 (37), 17157–17178. https://doi.org/10.1039/c9nr05371a.

Lim, W.A., June, C.H., 2017. The principles of engineering immune cells to treat cancer. Cell 168 (4), 724–740. https://doi.org/10.1016/j.cell.2017.01.016.

Lizotte, P.H., et al., 2016. In situ vaccination with cowpea mosaic virus nanoparticles suppresses metastatic cancer. Nat. Nanotechnol. 11 (3), 295–303. https://doi.org/10.1038/nnano.2015.292.

Lodolce, J.P., et al., 1998. IL-15 receptor maintains lymphoid homeostasis by supporting lymphocyte homing and proliferation. Immunity 9 (5), 669–676. https://doi.org/10.1016/S1074-7613(00)80664-0.

Lynn, G.M., et al., 2015. In vivo characterization of the physicochemical properties of polymer-linked TLR agonists that enhance vaccine immunogenicity. Nat. Biotechnol. 33 (11), 1201–1210. https://doi.org/10.1038/nbt.3371.

Malachowski, T., Hassel, A., 2020. Engineering nanoparticles to overcome immunological barriers for enhanced drug delivery. Eng. Regen. 1, 35–50. https://doi.org/10.1016/j.engreg.2020.06.001.

Maude, S.L., Teachey, D.T., Porter, D.L., Grupp, S.A., 2015. CD19-targeted chimeric antigen receptor T-cell therapy for acute lymphoblastic leukemia. Blood 125 (26), 4017–4023. https://doi.org/10.1182/blood-2014-12-580068.

Meraz, I.M., et al., 2012. Activation of the inflammasome and enhanced migration of microparticle-stimulated dendritic cells to the draining lymph node. Mol. Pharm. 9 (7), 2049–2062. https://doi.org/10.1021/mp3001292.

Mi, Y., et al., 2018. A dual immunotherapy nanoparticle improves T-cell activation and cancer immunotherapy. Adv. Mater. 30 (25). https://doi.org/10.1002/adma.201706098.

Miyamoto, N., Mochizuki, S., Fujii, S., Yoshida, K., Sakurai, K., 2017. Adjuvant activity enhanced by cross-linked CpG-oligonucleotides in β-glucan nanogel and its antitumor effect. Bioconjug. Chem. 28 (2), 565–573. https://doi.org/10.1021/acs.bioconjchem.6b00675.

Molino, N.M., Neek, M., Tucker, J.A., Nelson, E.L., Wang, S.W., 2016. Viral-mimicking protein nanoparticle vaccine for eliciting anti-tumor responses. Biomaterials 86, 83–91. https://doi.org/10.1016/j.biomaterials.2016.01.056.

Moon, J.J., et al., 2011. Interbilayer-crosslinked multilamellar vesicles as synthetic vaccines for potent humoral and cellular immune responses. Nat. Mater. 10 (3), 243–251. https://doi.org/10.1038/nmat2960.

Muraoka, D., et al., 2014. Nanogel-based immunologically stealth vaccine targets macrophages in the medulla of lymph node and induces potent antitumor immunity. ACS Nano 8 (9), 9209–9218. https://doi.org/10.1021/nn502975r.

Neal, L.R., et al., 2017. The basics of artificial antigen presenting cells in T cell-based cancer immunotherapies. J. Immunol. Res. Ther. 2 (1), 68–79.

Neefjes, J., Jongsma, M.L.M., Paul, P., Bakke, O., 2011. Towards a systems understanding of MHC class I and MHC class II antigen presentation. Nat. Rev. Immunol. 11 (12), 823–836. https://doi.org/10.1038/nri3084.

Nel, A.E., et al., 2020. Liposomal delivery of mitoxantrone and a cholesteryl indoximod prodrug provides effective chemo-immunotherapy in multiple solid tumors. ACS Nano 14 (10), 13343–13366. https://doi.org/10.1021/acsnano.0c05194.

Nikkhoi, S.K., Rahbarizadeh, F., Ranjbar, S., Khaleghi, S., Farasat, A., 2018. Liposomal nanoparticle armed with bivalent bispecific single-domain antibodies, novel weapon in HER2 positive cancerous cell lines targeting. Mol. Immunol. 96, 98–109. https://doi.org/10.1016/j.molimm.2018.01.010.

Nuhn, L., et al., 2018. Nanoparticle-conjugate TLR7/8 agonist localized immunotherapy provokes safe antitumoral responses. Adv. Mater. 30 (45), 1803397. https://doi.org/10.1002/adma.201803397.

Park, J., et al., 2012. Combination delivery of TGF-β inhibitor and IL-2 by nanoscale liposomal polymeric gels enhances tumour immunotherapy. Nat. Mater. 11 (10), 895–905. https://doi.org/10.1038/nmat3355.

Park, J.H., Geyer, M.B., Brentjens, R.J., 2016. CD19-targeted CAR T-cell therapeutics for hematologic malignancies: interpreting clinical outcomes to date. Blood 127 (26), 3312–3320. https://doi.org/10.1182/blood-2016-02-629063.

Pawar, D., Mangal, S., Goswami, R., Jaganathan, K.S., 2013. Development and characterization of surface modified PLGA nanoparticles for nasal vaccine delivery: effect of mucoadhesive coating on antigen uptake and immune adjuvant activity. Eur. J. Pharm. Biopharm. 85 (3 PART A), 550–559. https://doi.org/10.1016/j.ejpb.2013.06.017.

Pei, M., Xu, R., Zhang, C., Wang, X., Li, C., Hu, Y., 2021. Mannose-functionalized antigen nanoparticles for targeted dendritic cells, accelerated endosomal escape and enhanced MHC-I antigen presentation. Colloids Surf. B: Biointerfaces 197. https://doi.org/10.1016/j.colsurfb.2020.111378.

Peng, Q., et al., 2020. PD-L1 on dendritic cells attenuates T cell activation and regulates response to immune checkpoint blockade. Nat. Commun. 11 (1). https://doi.org/10.1038/s41467-020-18570-x.

Perica, K., Varela, J.C., Oelke, M., Schneck, J., 2015. Adoptive T cell immunotherapy for cancer. Rambam Maimonides Med. J. 6 (1), e0004. https://doi.org/10.5041/rmmj.10179.

Perotti, M., Perez, L., 2019. Virus-like particles and nanoparticles for vaccine development against HCMV. Viruses 12 (1). https://doi.org/10.3390/v12010035.

Petrovic, D., Seke, M., Srdjenovic, B., Djordjevic, A., 2015. Applications of anti/prooxidant fullerenes in nanomedicine along with fullerenes influence on the immune system. J. Nanomater., 2015. https://doi.org/10.1155/2015/565638.

Phua, K.K.L., 2015. Towards targeted delivery systems: ligand conjugation strategies for mRNA nanoparticle tumor vaccines. J Immunol Res 2015. https://doi.org/10.1155/2015/680620.

Powell, D.J., De Vries, C.R., Allen, T., Ahmadzadeh, M., Rosenberg, S.A., 2007. Inability to mediate prolonged reduction of regulatory T cells after transfer of autologous CD25-depleted PBMC and interleukin-2 after lymphodepleting chemotherapy. J. Immunother. 30 (4), 438–447. https://doi.org/10.1097/CJI.0b013e3180600ff9.

Purwada, A., et al., 2016. Self-assembly protein nanogels for safer cancer immunotherapy. Adv. Healthc. Mater. 5 (12), 1413–1419. https://doi.org/10.1002/adhm.201501062.

Raïch-Regué, D., Glancy, M., Thomson, A.W., 2014. Regulatory dendritic cell therapy: from rodents to clinical application. Immunol. Lett. 161 (2), 216–221. https://doi.org/10.1016/j.imlet.2013.11.016.

Redeker, A., Arens, R., 2016. Improving adoptive T cell therapy: the particular role of T cell costimulation, cytokines, and post-transfer vaccination. Front. Immunol. 7 (September), 1. https://doi.org/10.3389/fimmu.2016.00345.

Rodrigues, L., et al., 2018. Immune responses induced by nano-self-assembled lipid adjuvants based on a monomycoloyl glycerol analogue after vaccination with the *Chlamydia trachomatis* major outer membrane protein. J. Control. Release 285, 12–22. https://doi.org/10.1016/j.jconrel.2018.06.028.

Romagnani, P., et al., 2001. Cell cycle-dependent expression of CXC chemokine receptor 3 by endothelial cells mediates angiostatic activity. J. Clin. Invest. 107 (1), 53–63. https://doi.org/10.1172/JCI9775.

Romano, M., Fanelli, G., Albany, C.J., Giganti, G., Lombardi, G., 2019. Past, present, and future of regulatory T cell therapy in transplantation and autoimmunity. Front. Immunol. 10 (January), 43. https://doi.org/10.3389/fimmu.2019.00043.

Rosenberg, S.A., Restifo, N.P., Yang, J.C., Morgan, R.A., Dudley, M.E., 2008. Adoptive cell transfer: a clinical path to effective cancer immunotherapy. Nat. Rev. Cancer 8 (4), 299–308. https://doi.org/10.1038/nrc2355.

Rotte, A., 2019. Combination of CTLA-4 and PD-1 blockers for treatment of cancer. J. Exp. Clin. Cancer Res. 38 (1), 1–12. https://doi.org/10.1186/s13046-019-1259-z.

Ruffner, M.A., Sullivan, K.E., Henrickson, S.E., 2017. Recurrent and sustained viral infections in primary immunodeficiencies. Front. Immunol. 8 (June), 665. https://doi.org/10.3389/fimmu.2017.00665.

Sahin, U., Karikó, K., Türeci, Ö., 2014. MRNA-based therapeutics-developing a new class of drugs. Nat. Rev. Drug Discov. 13 (10), 759–780. https://doi.org/10.1038/nrd4278.

Sahu, P., Das, D., Kashaw, V., Iyer, A.K., Kashaw, S.K., 2017. Nanogels: a new dawn in antimicrobial chemotherapy. In: Antimicrobial Nanoarchitectonics: From Synthesis to Applications. Elsevier, pp. 101–137.

Sakai, T., et al., 2020. Artificial T cell adaptor molecule-transduced TCR-T cells demonstrated improved proliferation only when transduced in a higher intensity. Mol. Ther. Oncol. 18, 613–622. https://doi.org/10.1016/j.omto.2020.08.014.

Santegoets, S.J.A.M., Turksma, A.W., Powell, D.J., Hooijberg, E., Gruijl, T.D.D., 2013. IL-21 in cancer immunotherapy at the right place at the right time. Oncoimmunology 2 (6). https://doi.org/10.4161/onci.24522.

Schaer, D.A., Hirschhorn-Cymerman, D., Wolchok, J.D., 2014. Targeting tumor-necrosis factor receptor pathways for tumor immunotherapy. J. ImmunoTher. Cancer 2 (1), 7. https://doi.org/10.1186/2051-1426-2-7.

Sharpe, M., Mount, N., 2015. Genetically modified T cells in cancer therapy: opportunities and challenges. Dis. Model. Mech. 8 (4), 337–350. https://doi.org/10.1242/dmm.018036.

Sheikhpour, E., Noorbakhsh, P., Foroughi, E., Farahnak, S., Nasiri, R., Neamatzadeh, H., 2017. A survey on the role of interleukin-10 in breast cancer: a narrative. Rep. Biochem. Mol. Biol. 7 (1), 30–37.

Shimizu, T., et al., 2008. Nanogel DDS enables sustained release of IL-12 for tumor immunotherapy. Biochem. Biophys. Res. Commun. 367 (2), 330–335. https://doi.org/10.1016/j.bbrc.2007.12.112.

Shukla, N.M., Malladi, S.S., Mutz, C.A., Balakrishna, R., David, S.A., 2010. Structure-activity relationships in human toll-like receptor 7-active imidazoquinoline analogues. J. Med. Chem. 53 (11), 4450–4465. https://doi.org/10.1021/jm100358c.

Simmons, A.D., et al., 2007. GM-CSF-secreting cancer immunotherapies: preclinical analysis of the mechanism of action. Cancer Immunol. Immunother. 56 (10), 1653–1665. https://doi.org/10.1007/s00262-007-0315-2.

Soares, M.P., Teixeira, L., Moita, L.F., 2017. Disease tolerance and immunity in host protection against infection. Nat. Rev. Immunol. 17 (2), 83–96. https://doi.org/10.1038/nri.2016.136.

Song, Q., et al., 2017. Tumor microenvironment responsive nanogel for the combinatorial antitumor effect of chemotherapy and immunotherapy. Nano Lett. 17 (10), 6366–6375. https://doi.org/10.1021/acs.nanolett.7b03186.

Souto, E.B., et al., 2019. Nanoparticle delivery systems in the treatment of diabetes complications. Molecules 24 (23), 4209. https://doi.org/10.3390/molecules24234209.

Srivastava, S., Riddell, S.R., 2018. Chimeric antigen receptor T cell therapy: challenges to bench-to-bedside efficacy. J. Immunol. 200 (2), 459–468. https://doi.org/10.4049/jimmunol.1701155.

Stickdorn, J., Nuhn, L., 2020. Reactive-ester derived polymer nanogels for cancer immunotherapy. Eur. Polym. J. 124, 109481. https://doi.org/10.1016/j.eurpolymj.2020.109481.

Toth, I., Skwarczynski, M., 2014. The immune system likes nanotechnology. Nanomedicine 9 (17), 2607–2609. https://doi.org/10.2217/nnm.14.199.

Visser, J., Busch, V.J., de Kievit-van der Heijden, I.M., Ten Ham, A.M., 2012. Non-Hodgkin's lymphoma of the synovium discovered in total knee arthroplasty: a case report. BMC Res. Notes 5. https://doi.org/10.1186/1756-0500-5-449.

Vollmer, J., et al., 2004. Characterization of three CpG oligodeoxynucleotide classes with distinct immunostimulatory activities. Eur. J. Immunol. 34 (1), 251–262. https://doi.org/10.1002/eji.200324032.

Wang, T., Zou, M., Jiang, H., Ji, Z., Gao, P., Cheng, G., 2011. Synthesis of a novel kind of carbon nanoparticle with large mesopores and macropores and its application as an oral vaccine adjuvant. Eur. J. Pharm. Sci. 44 (5), 653–659. https://doi.org/10.1016/j.ejps.2011.10.012.

Wang, L., et al., 2014. Nanoclusters self-assembled from conformation-stabilized influenza M2e as broadly cross-protective influenza vaccines. Nanomed. Nanotechnol. Biol. Med. 10 (2), 473–482. https://doi.org/10.1016/j.nano.2013.08.005.

Wilhelm, S., et al., 2016. Analysis of nanoparticle delivery to tumours. Nat. Rev. Mater. 1 (5), 1–12. https://doi.org/10.1038/natrevmats.2016.14.

Xi, Y., et al., 2019. Advances in nanomedicine for the treatment of ankylosing spondylitis. Int. J. Nanomedicine 14, 8521–8542. https://doi.org/10.2147/IJN.S216199.

Xu, L., et al., 2012. Surface-engineered gold nanorods: promising DNA vaccine adjuvant for HIV-1 treatment. Nano Lett. 12 (4), 2003–2012. https://doi.org/10.1021/nl300027p.

Yang, C., Merlin, D., 2019. Nanoparticle-mediated drug delivery systems for the treatment of IBD: current perspectives. Int. J. Nanomedicine 14, 8875–8889. https://doi.org/10.2147/IJN.S210315.

Yang, Y., et al., 2021. Targeted silver nanoparticles for rheumatoid arthritis therapy via macrophage apoptosis and re-polarization. Biomaterials 264, 120390. https://doi.org/10.1016/j.biomaterials.2020.120390.

Yin, J., Retsch, M., Lee, J.-H., Thomas, E.L., Boyce, M.C., 2011. Mechanics of nanoindentation on a monolayer of colloidal hollow nanoparticles. Langmuir 27 (17), 10492–10500.

CHAPTER

15

Polycaprolactone-based shape memory polymeric nanocomposites for biomedical applications

Vaishnavi Hada[a], S.A.R. Hashmi[a,e], Medha Mili[a,e], Nikhil Gorhe[a,e], Sai Sateesh Sagiri[b], Kunal Pal[c], Rashmi Chawdhary[d], Manal Khan[d], Ajay Naik[a,e], N. Prashant[a,e], A.K. Srivastava[a,e], and Sarika Verma[a,e]

[a]Council of Scientific and Industrial Research—Advanced Materials and Processes Research Institute, Bhopal, Madhya Pradesh, India [b]Agro-Nanotechnology and Advanced Materials Research Center, Institute of Postharvest and Food Sciences, Agricultural Research Organization, The Volcani Center, Rishon LeZion, Israel [c]Department of Biotechnology and Medical Engineering, National Institute of Technology, Rourkela, Odisha, India [d]All India Institute of Medical Sciences (AIIMS), Bhopal, Madhya Pradesh, India [e]Academy of Council Scientific and Industrial Research (AcSIR)—Advanced Materials and Processes Research Institute (AMPRI), Hoshangabad Road, Bhopal, Madhya Pradesh, India

OUTLINE

HIGHLIGHTS

- Polycaprolactone-based nanocomposites are shape-memory polymers (SMPs) materials that can return to their original shape from the deformed state when activated by an external stimulus triggered by light, temperature, moisture, pH, radiations, chemicals, and magnetic field.

- The polycaprolactone-based nanocomposites are successfully used in various areas of the biomedical field like (a) drug delivery,

(b) cardiovascular diseases, (c) tissue engineering, (d) wound dressing, (e) scaffolds, and various other applications as they are being bio-friendly, biodegradable, biocompatible, and haemo-compatible and fulfills all the possible requirements for biomedical applications.

- The advancement of PCL as shape memory scaffolds for tissue engineering has gained much interest in the research field.

1 Introduction

Materials science is generally a new field of study that has arisen at the crossing point of physical science, chemical science, and designing. It includes the investigation of the properties of an actual substance that can be utilized in an application. The research tries to grasp the material's basic design, properties that act under different conditions, and adjust its properties while developing the desired material (ProvenProcess, n.d.). Practically, materials generally used in this universe are categorized into four types: metal, polymers, ceramic, and composite (ltc-proto, n.d.). Polymers are the most flexible of all materials, covering a broad spectrum of mechanical properties (Wnek, 2008) and thus have primary reasons for their widespread acceptance. From the past decades, polymer-based materials are generally being significant for various applications (Zhao et al., 2015). Biomaterials derived from polymers are divided into two categories: naturally occurring and synthetically created materials. In tissue engineering, biodegradable polymers have become very popular to avoid additional surgery to remove implants or scaffolds. Henceforth, much attention must be paid to the synthesis of biodegradable polymers (Khan et al., 2015). The use of polymers as biomaterials have enormously affected the progression of current medication. Specifically, polymeric biomaterials that are biodegradable give the critical benefit of having the option to be separated and taken out after they have served their capacity (Ulery et al., 2011). Distinctive manufactured polymeric materials, for example, polyethylene, polyurethane, poly dimethyl siloxane, perylene, polytetrafluoroethylene, polypropylene, polyamide, polyvinylidene fluoride, and poly (methyl methacrylate) were inspected (Teo et al., 2016). Despite the current wide range of polymers accessible in biomedicine, it is habitually hard to satisfy all material necessities simultaneously in an economically beneficial way (Maitz, 2015). Concerning this, "smart" polymers provide materials that react to various stimuli (e.g., temperature, pH, chemicals, solvents, electric and magnetic fields, light intensity, biological molecules), and research is still going on for the application in many different areas (Aguilar and San Román, 2019). These polymers are adjustive; for example, they may be disintegrating in aqueous solutions, adsorbed or grafted on solid aqueous interfaces, or cross-linked and forms the hydrogels.

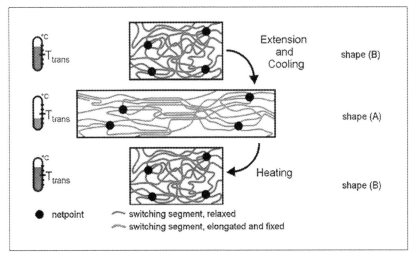

FIG. 1 Structure of SMP. *Retrieved by Lendlein, A., Langer, R., 2002. Biodegradable, elastic shape-memory polymers for potential biomedical applications. Science 296(5573), 1673–1676, under Creative Commons license 4.0.*

They may also be combined physically or chemically with other molecules, especially various bioactive molecules (Hoffman and Stayton, 2020). Owing to this, shape-memory polymers (SMPs) are excellent actively moving intelligent materials that can change themselves from a temporary shape to its permanent shape and hence are widely used in biomedical applications for a longer time (Mu et al., 2018).

The essential critical requirements are to achieve the proper design when it comes to biomedical applications. The material properties vary greatly depending on the design, particularly in the biomedical field (Gautrot and Zhu, 2009). Over the last several decades, there has been an increase in the demand for biodegradable polymers in healthcare applications (Lendlein and Langer, 2002). Polyesters such as poly (glycolic acid) (PGA), poly (lactic acid) (PLA), poly (trimethylene carbonate) (PTMC), and polycaprolactone (PCL) forms a large category of biodegradable materials (Wong et al., 2014). SMPs can be strengthened in mechanical strength and recovery stress by using high modulus inorganic or organic fillers. The degradation rate is determined by various factors, including chain length, crystallinity, branching extent, and the shape and medium in which the polymer is inserted (Leng et al., 2011). These can be accomplished by modifying the final polymer during production using various polymerization conditions, reactants, and monomers. This results in polymers with multiple properties, including rigidity, stretchability, impermeability, reduction in glass transition temperature (T_g), and melting temperature (T_m). Fig. 1 represents the structure of SMP in which the polymer expands itself when heated, and when cooled, it regains its original shape.

2 An insight of shape-memory polymers and shape memory effect

SMPs make a significant interest worldwide as these materials are smart and manifest intelligent behavior when triggered by external stimuli. They show high deforming nature and good recovery ability; therefore, they have multiple applications, as neoteric studies have described. The best feature of these materials is the shape recovery ability which makes them adaptable for numerous applications. These smart materials are used in many fields such as textile industries, biomedical applications, in the manufacturing of electro-conductive and magnetic composite materials, coatings for metal corrosion and many more. Recent studies conclude that SMP derived from epoxy resins has received significant attention since they have good mechanical and physical properties, low cost, and shape recovery features (Lendlein and Langer, 2002). Two phases are responsible for a polymer's shape memory ability, i.e., a hard segment and a soft segment. The hard segment is responsible for memorizing the material's permanent shape. Contrastingly, a need for a soft segment is obligatory, accountable for maintaining the temporary shape acquired by allylic groups derived from the new curing agents. SMPs from epoxy resins are primarily used in various protecting and fancy coatings because of their excellent adhesive, mechanical and chemical defence (Leng et al., 2011). They are predominantly utilized for layering cans and drums, coatings for car and cable paint, etc. In the electrical industry, they are utilized to preserve their effects in wet places or high humidity. Specific implementations here are high voltage conductors, switches, and transistor encapsulation. These can also be utilized to make coatings and matrix materials with reinforced fiber (Malikmammadov et al., 2017).

Nonetheless, shape memory is not an inherent feature but ascribed due to molecular structure, orientation, and polymer processing levels. However, shape memory's mechanism is based upon entropic elasticity (Behl et al., 2009). In shape-memory polymers, cross-linking prohibits the dislocation of polymeric chains. Chemical or physical crosslinks are reasonable for maintaining the original shape of the polymer. When chemical cross-linking is used to fix the original shape, a nonsoluble thermo-harder is acquired (Behl and Lendlein, 2007).

Contrarily, when physical cross-linking is applied, a thermoplastic material is obtained, which remains dissolvable in a suitable solvent, using intramolecular interactions (Leng et al., 2010). By using chemical or physical cross-links, a permanent shape is fixed in the first step of processing. Secondly, the original shape can be molded in a programmed shape. By using the shape fixing parts, the programmed shape are freezed by applying a particular trigger. When the force is removed, it will retain its temporary shape. However, the temporary shape cannot be obtained without programming. This stage is called programming. Although researchers typically refer to characterize the shape memory feature before the occurrence of material gets disrupted. Fig. 2 explains the concept behind shape-memory polymers, which shows an energy diagram of a thermo-responsive SMPs primary and secondary shape. A broader view of the thermodynamics behind the shape-memory mechanism is given by Lendlein and Kelch (2002).

The shape-memory polymers are differentiated into the thermal stimulus, electric stimulus, and solvent stimulus to retrieve the shape by a particular trigger. Some subclasses of glass transition temperature (T_g)-type and melting temperature (T_m)-kind, triple shape memory effect (Ahn and Kasi, 2011; Luo and Mather, 2010; Wang et al., 2013a), multistimuli (Wang et al., 2013b) multifunctional SMPs (Wang et al., 2014a), two-way SMPs (Behl et al., 2013a; Zhou et al., 2014), temperature memory polymers (Miaudet et al., 2007; Behl et al., 2013b; Wang et al., 2014b), etc. have also been studied for the SMPs.

The SMPs are beneficial because of their various properties like low density, more extensive strains, numerous scales of tailorable temperatures, easier processing, biocompatibility, biodegradation, and cost-effective characteristics. All these ease their implementations in

FIG. 2 Potential energy diagram of a thermoresponsive shape-memory polymer (A) shows the entropically favorable permanent shape, which can be programmed into the less favorable temporary state (C) through heating and deformation. The permanent shape (A) can be recovered through heating of the temporary shape (C). During switching between the temporary and permanent shape, the polymer goes through a transition state (B) where the polymer chains are mobile, allowing for the chains' reorganization. *Retrieved by Delaey, J., Dubruel, P., Van Vlierberghe, S., 2020. Shape-memory polymers for biomedical applications. Adv. Funct. Mater. 30(44), 1909047 with permission.*

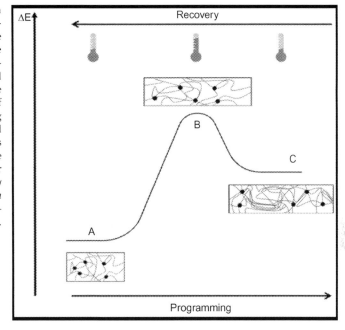

several areas such as textiles (Hu, 2007), aerospace technologies (Liu et al., 2014), biomedical applications (Baudis et al., 2014; Wache et al., 2003; Small et al., 2007; Zheng et al., 2017) and smart actuators (Kularatne et al., 2017), etc.

3 Significance of SMPs in biomedical applications

According to the biomedical perspective, SMP-based biomaterials have specific requirements for this application. SMPs fulfill all the basic needs such as nontoxicity, haemo-compatibility, noncarcinogenic, noninflammatory, and nonallergic properties. As per the second need, a biomaterial should be biocompatible, and it should trigger the desired function with its particular application (i.e., the material should be nonimmunogenic). Apart from this, a biomaterial should not disseminate its residues if it is not meant for it. Fig. 3 shows the requirements of biomaterials as per biomedical applications.

As compared to synthetic polymers, biopolymers show the beneficiary property as less inflammatory (Raghavendra et al., 2015). An explanation of these conditions can be found in polymers used for cardiovascular applications. When the blood comes in contact with the polymer, there will be no formation of a thrombus at the polymer's surface (Wang et al., 2014c), and hence, there will be no adsorption of proteins on the surface of the polymer (Siedlecki, 2017). For this analysis, haemo-compatibility studies on the biomaterial should be performed (Raghavendra et al., 2015; Hastings, 1985). Polymers with these conditions have been used for decades, such as PCL, poly(lactic acid) (PLA), poly(glycolic acid) (PGA), poly(dioxanone), along with various polyurethanes (McClure et al., 2011). The third need

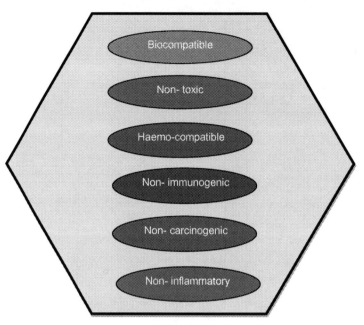

FIG. 3   Requirements of biomaterials as per biomedical applications.

for the biomaterial is not to give rise to an immunological response as a foreign body (Kularatne et al., 2017). Besides, a significant criterion for biomaterial that it should surely fulfill is, complementing the implant materials mechanical properties with those of the surrounding tissue (Kularatne et al., 2017; McClure et al., 2011; Safranski et al., 2013). The mechanical needs are specific to the application, which ultimately depends upon the site of implantation (Kularatne et al., 2017).

In some cases, biodegradation of the material plays a significant role. Depending on the application of the desired implant, the degradation period is mandatory. According to this concept, biodegradable sutures are pertinent (Wang et al., 2014c) as these can deteriorate after wound healing; hence, it prevents the removal of sutures. Regarding this concept, shape-memory behavior can develop biodegradable sutures that get self-tightened by them. This gives surgeons relief as it can be triggered by a stimulus that activates the shape memory effect and ultimately closes the wound. In some cases where degradation is not necessary, the biomaterial should stay in the body permanently.

And in some cases where biodegradation is mandatory, no traces of the biomaterial should remain in the body, and the degradation process should be nontoxic. Most often used biodegradable polymers include PGA, PLA, PCL, derivatives, and biopolymers such as cellulose or gelatine (Sternberg, 2009). Moreover, the polymer's degradation speed frequently clashes with the generation of new tissue depending upon the desired application. Hence, the polymer's degradation must be either slower or equal to the generation of the new tissue. Lastly, a biopolymer's final essential quality to fulfill biomedical needs is that it must have the sterilization capability. After all, polymers that cannot be sterilized cannot assist in vivo

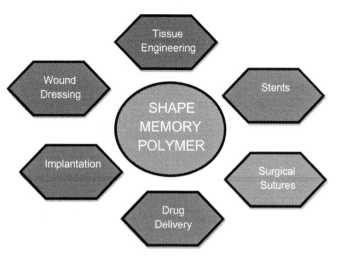

FIG. 4   Some applications of SMPs in the biomedical field.

applications or to investigate cellular response. Instead, the properties of biomaterials should not get modified by the sterilization process (Raghavendra et al., 2015).

Despite the enormous advances in smart materials preparation, it is still a significant obstacle to combine unique characteristics like self-healing and shape memory properties onto one material, which would have gained rapid interest because of their possible applications (Chen et al., 2020). Fig. 4 depicts some promising applications of SMPs in the biomedical field.

4 Synthesis and properties of PCL

4.1 Synthesis

PCL is an aliphatic polyester made out of repeated units of hexanoate. Synthesizing the PCL can be carried out by the following two systems: (1) use of 6-hydroxycaproic (6-hydroxyhexanoic), (2) the ring-opening polymerization (ROP) of PCL, as shown in Fig. 5 (Agarwal, 2010). The mechanism incorporates the advancement of aliphatic polyesters from hydroxyl carboxylic acids by the expulsion of water under vacuum (Braud et al., 1998) or then again catalyzing by naturally removed lipase protein eventually to increase the balance toward polymer arrangement (Woodruff and Hutmacher, 2010a; Kim and Kim, 2015). Some catalyst especially stannous octoate is used for polymerization and besides this; because alcohols have lower molecular weight, they can be utilized to control the polymer's molecular weight (Storey and Taylor, 1996). Several components influence the polymerization of PCL, which are anionic, cationic, and radical. Every strategy influences the subsequent subatomic weight, subatomic weight distribution, and synthetic design of the *co*-polymers (Okada, 2002). The average subatomic load of PCL tests ranges from 3000 to 80,000 g/mol, which may be evaluated by the subatomic weight (Hayash, 1994).

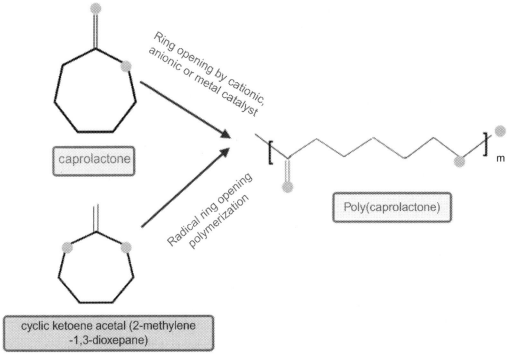

FIG. 5   Synthesis of PCL by radical ring-opening polymerization. Modified with permission from *Sternberg, K.,*
2009. Current requirements for polymeric biomaterials in otolaryngology. GMS Curr. Top. Otorhinolaryngol. Head Neck Surg.
8, under Creative Common Attribution 4.0.

4.2 Properties of PCL

PCL is broadly utilized in regenerative medication for different tissue engineering applications because of its attractive properties. Some properties have been mentioned below:

(a) Physicochemical properties

The average subatomic load of PCL tests may differ from 530 to 630,000, and it is evaluated
by the subatomic weight (Mondal et al., 2016). The melting point of this polymer ranges from
59°C to 64°C, and its T_g temperature is about 60°C, and subsequently, having semicrystalline
construction, with a crystalline rate up to 69%, brings about higher durability (Mishra et al.,
2008). PCL can be dissolved at different degrees in the majority of natural solvents when kept
at room temperature. For example, there are some soluble exceptions for the polymer, as it is
soluble in benzene, dichloromethane, 2-nitropropane, toluene, chloroform, cyclohexanone,
and carbon tetrachloride.

On the other hand, alcohols, oil ether, diethyl ether, and water cannot break up PCL (Sinha
et al., 2004a). The capacity of mixing with other polymers production to empower the adjustment of properties concurring the necessary application is one of the uncommon and most
beneficial aspects of PCL. This peculiar property of PCL is due to its well-known framework
forming nature.

(b) Surface properties

The surface properties of the created framework need to be viewed, depending on the given application. A framework is suggested to copy the extracellular network, wherein cells are cultivated within the fundamental boundaries of porosity and repulsiveness. A permeable microstructured framework with interconnected pores works well with cell multiplication and redesign. Likewise, the porosity of the platform assumes a pivotal part in regulating the simplicity of supplement dissemination, cellular movement, and mechanical firmness of the framework. As a rule, the acceptable surface unpleasantness range is somewhere in the range of 5 and 50 μm, while the pore size ought to be more prominent than 20 μm (Abedalwafa et al., 2013a). While designing scaffolds based on PCL nanofibers or cross-sections, water contact points must be considered a significant boundary. It illuminates the surface harshness and porosity just as the degree of achievement of adjustment measure (Cipitria et al., 2011). Part direction and thickness of layer are found to show the solid effect on superficial level features of the design. Subsequently, the technique for creation is picked depending on the typical attributes of the platform. Further, to improve these properties, mixing with various synthetic compounds is a viable methodology.

(c) Mechanical properties

Mechanical properties of the polymeric framework need to be considered during design, creation, and utility. The platform should detain mechanical solidarity to withstand the physical pressure at the implantation area. Further, it must have the option to help the tissue until it is fit for supporting itself (Cipitria et al., 2011; Bolaina-Lorenzo et al., 2016). Mechanical properties can be adjusted by creating and mixing with other polymers depending on cell growth and applications.

(d) Biocompatible properties

PCL polymer is well known in regenerative medicine because of its excellent property of defined boundaries. Nonetheless, the safe reaction created because of a foreign body is a significant issue to be handled. PCL has been applied in plenty of tissue engineering applications. The PCL has desired qualities of an ideal scaffold because of its hydrophobic nature. It is unfriendly to cell connection and multiplication; PCL has some constraints in vivo. These can be addressed by surface modification and mixing with other polymers (Ragaert et al., 2012).

(e) Degradation properties

The degradation conduct must be ideal for giving primary uprightness until the neotissue upholds itself. PCL corrupts most gradually among the polyesters because of five $-CH_2$ moieties as repeating units (Siddiqui et al., 2018). According to a portion of the exploration reports, it takes around 3–4 years for some PCL-based polymer frameworks to degrade totally (Hao et al., 2002). Having the lower pace of biodegradation, PCL is a favored possibility for delayed medication conveyance framework. Besides the synthetic construction of the polymer, surface change and platform structure likewise have pivotal influences during the time spent on debasement.

5 PCL-based shape memory polymeric nanocomposites

Biodegradable polymers have satisfied an expanding need for clinical use in the latest years. Among them, PCL is one of the polymers, which is generally utilized for its accessibility, cost-effectiveness, and appropriateness for alteration in tissue engineering (Pitt et al., 1981). PCL is a few recognized materials utilized in various applications, from manufacturing polymeric composites to biological applications, including tissue engineering, actuators, artificial skin, drug delivery system, etc. (Malikmammadov et al., 2018). Few of the PCL current applications in biomedical applications are described here.

(a) Scaffolds

Tissue engineering is a versatile field of science that makes tissue-like structures that uses living cells, biocompatible materials and chemicals, stimuli (electrical, physical) (e.g., cyclic mechanical stacking). It then fixes a biological tissue or replaces a weak organ (Abrisham et al., 2020). The scaffolds usually imitate the morphological structure of the target site for tissue engineering (Beatrice et al., 2021). For instance, scaffolds for bone tissue engineering (BTE) mainly comprise pore measurements between 200 and 600 µm and collagen and calcium phosphates. Like PCL, the organic compounds and manufactured materials restore the utilization of collagen owing to its cost-effective and controlling blend condition that put forward the expected and reproducible properties (rigidity, flexible modulus, crystallinity, and degradation rate) (Zhu et al., 2017).

Multifunctional scaffold structures are becoming increasingly essential in the field of tissue engineering. Fox et al. (2020) showed that the ND-PCL composites are effectively expelled to make a 3D scaffold exhibiting their potential as a composite material for tissue recovery. As the interest in TE increases, new materials are needed to improve scaffolds and their tissue interfaces. A platform made out of PCL and explosion nanodiamonds (ND-PCL), where ND incorporated at 10% (ND 10%) and 20% (ND 20%) in PCL composites were present. By adding ND into the PCL, the composite materials reported lower rigidity and lessening in crystallinity.

Their results showed that a debasement rate could be custom-made by utilizing composite ND-stacking to meet prerequisites. Further, the corruption profiles of the 10% ND and 20% ND composites are diverse, with the 20% ND composite ready to oppose disintegration for a more extended timeframe than the 10% ND composite. This is because NDs probably supplied different surface nucleation locales during film drying, expanding the surface harshness and the material's hydrophilicity. As ND is fused into the PCL material, the bio-interface gives an impression of a stronger osteoblast bond, with a reasonable improvement in cell connection with the expanded ND loads. Osteoblasts are a setup key in the bone recovery cycle, and as such, it is significant that the new composite material can uphold their connection and multiplication. Hydrophilicity is a substantial property for platform biomaterials. This adds to the expanded grip of osteoblast cells on ND-PCL composites over PCL alone. Ultimately, ND-PCL composite could be produced into 3D platforms using softening expulsion, making it ready to create cutting edge TE frameworks (Roseti et al., 2017). Fig. 6 depicts osteoblast adhesion on composite materials through 3D printing compatibility.

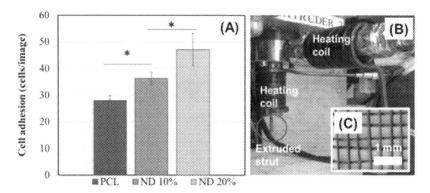

FIG. 6 (A) Osteoblast adhesion on composite materials depicting higher adhesion as compared to PCL alone (mean ± SEM, $n = 6$, student t-test ($P \leq .05$)). (B) The proof of concept of the 3D printing compatibility of PCL-ND 20% composite depicting the melt-extrusion setup and (C) also a scaffold printed using layer-by-layer deposition extruded struts (nozzle size: 20-gauge needle). *Retrieved by Fox, K., Ratwatte, R., Booth, M.A., Tran, H.M., Tran, P.A., 2020. High nanodiamond content-PCL composite for tissue engineering scaffolds. Nanomaterials. 10(5), 948, under Creative Commons 4.0.*

TE has been investigated as a supplemental method for treating craniomaxillofacial (CMF) bone deformities. Nail et al. carried work and described SMP structures orchestrated of biodegradable polycaprolactone-diacrylate (PCL-DA) by the photochemical fix projecting particulate-depleting procedure. To recognize bioactivity, they applied a polydopamine covering to the outside of the system. This bioactivity is expected to work with coordination and holding with enveloping bone tissue. Their work portrays self-fitting and shape memory rehearses, pore interconnectivity, and in vitro bioactivity.

PCL-based SMP systems were made with exceptionally interconnected pores subjected to SCPL methodology to achieve osteoconductivity. The portrayed tissue planning stage gains the specific course of action of properties essential for the powerful treatment of CMF (craniomaxillofacial) bone defects. The structure depends on osteointegration through its ability to "self-fit" inside a capricious CMF bone defect. Consequently, this stage tends to an alternative as opposed to autografting and standard bone substitutes for CMF bone distortion fix. Osteo-conductivity was expected along with the refined pore interconnectivity similarly to stage biodegradability. Finally, the system was bioactive due to the polydopamine covering, as shown by hydroxyapatite (HA) during in vitro tests (Nail et al., 2015). Fig. 7 shows a schematic diagram for the preparation of an SMP scaffold coated with polydopamine.

(b) Nanoparticles

Lately, various examinations exhibit some potential application in bioscience as the blending result of PCL/ PU/PLLA and HA (hydroxyapetite) nano-particles. Tian et al. aimed to make the network of PCL using HA nano-particles as substance net-points. This technique provides an advantage that each HA particle is considered a point convergence for PCL functionalized chains. At the same time, the materials became elastomeric over their melting

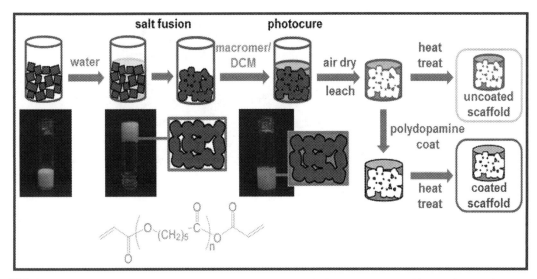

FIG. 7 Schematic for preparation of SMP scaffold coated with polydopamine. *Retrieved from Nail, L.N., Zhang, D., Reinhard, J.L., Grunlan, M.A., 2015. Fabrication of a bioactive, PCL-based "self-fitting" shape memory polymer scaffold. J. Vis. Exp. (104), with permission.*

temperature (61.7°C). In these situations, chilling the material to room temperature can deform the shape. Placing in the warm water again, the effect of shape-memory was set off; and could ceaselessly open till the holder recovers to its regular shape. The segment was progressed to orient the thermo-responsive SME of the cross-associated HA-PCL association to appreciate shape memory sway from the viewpoint on polymer associations. Initially, PCL had a semiglass state. HA nanoparticles probably work as cross-linkers, upheld the association framework, and limited the adaptability and risk of PCL chains. When the substance was heated up to its melting temperature, all the crystals start to melt. Simultaneously, PCL starts to arrange under an appropriate pressing factor. The substance gets changed from its hidden semigloss like state to a significantly adaptable state. In the wake of heating the model over its T_{trans} (switching temperature), the hardened PCL changed into a thick condensed state, and the substance shrinks and remits to its original state (Zhang et al., 2014). Considering the data, the HA-PCL composites may be the challenger for biomedical application in light material that stimulates bone cells (Tian et al., 2019a) (Fig. 8).

PCL composites have been utilized for different purposes such as drug delivery (Lee and Cho, 2016; Kumari et al., 2010), wound healing (Gamez et al., 2019), suture (Arora et al., 2019), tissue engineering (liver, breast, pancreas, cardiovascular, kidney, etc.) (Albrecht et al., 2016; Siqueira et al., 2017; Hewitt et al., 2019; Hsieh et al., 2018; Miao et al., 2019; Leong et al., 2015; Yang et al., 2016; Mohamadi et al., 2017; Centola et al., 2010; Chhaya et al., 2016; Serrano-Aroca et al., 2018; Kong and Mi, 2016; Grant et al., 2017; Burton et al., 2017; Ghorbani et al., 2017; Smink et al., 2017). Amidst these biomedical applications, PCL has mainly been implied for BTE due to its strength and low degradability, giving structural support for a longer duration (Le et al., 2019).

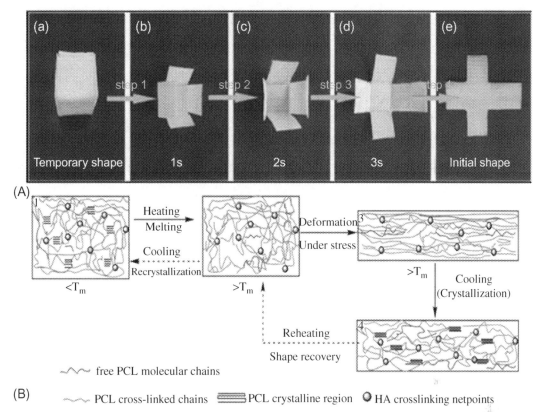

FIG. 8 (A) An example of the shape recovery process of a crosslinked HA-PCL 8000 material. (B) Schematic diagram delineating the shape memory mechanism of a covalently crosslinked HA-PCL network. *Retrieved by Tian, G., Zhu, G., Xu, S., Ren, T., 2019. A novel shape memory poly (ε-caprolactone)/hydroxyapatite nanoparticle networks for potential biomedical applications. J. Solid State Chem. 272, 78–86, under Creative Common Attrbution 3.0.*

(c) Cardiovascular diseases

Heart diseases influence all over the part of world populations and mainly affects the valves of the heart. Valve's substitution is a widely recognized therapy. Besides this, various factors influence the condition to choose between multiple natural or synthetic materials, depending upon the implantation. Ferreira et al. suggest using a polymeric coating to increase implant biocompatibility on titanium metal surfaces. PCL has several distinctive characteristics, especially biocompatibility, which accompanied greater efficiency and applicability for biomedical applications. Titanium plates were given alkaline and thermal reactions and covered with 1% PCL and 1% PCL + TiO$_2$. The union of TiO$_2$ was acted in two stages: Ultrasonic and microwave. The combination of titanium isopropoxide (98%), citrus extract, and isopropanol 20% (v/v) was mixed to develop metal complexes.

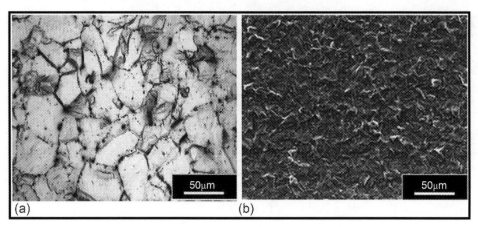

FIG. 9   Micrograph of Ti cp (commercially pure titanium) after alkaline and thermal treatments and reaction with Kroll solution. Micrographs obtained by (A) optical microscopy and (B) scanning electron microscopy. *Retrieved from Ferreira, C.C., Ricci, V.P., Sousa, L.L., Mariano, N.A., Campos, M.G., 2017. Improvement of titanium corrosion resistance by coating with poly-caprolactone and poly-caprolactone/titanium dioxide: potential application in heart valves. Mater. Res. 20, 126–133.*

Additionally, ethylene glycol was introduced to the combination exposed to microwave radiation to acquire a polymeric mass. At 300°C for 2h, the mass was calcined, and thus, the obtained material was mashed at 700°C for 2h to get powdered TiO_2. At that point, 0.4g of TiO_2 was scattered in 8.6M NaOH, sonicated for 2h and separated. Milli-Q water was utilized to rinse the drained solid, trailed by the expansion of 40mL of methanol followed by drying. Optical and SEM images of TiO_2 and PCL are shown in Fig. 9.

Covering of titanium plates with PCL and PCL with TiO_2 gave effective results, and no air bubble was formed. A transparent film was formed by PCL on the outside of the circle, while PCL+TiO_2 shaped a white film. The two films similarly covered the titanium plane and persisted attachment to the surface during all methods. They effectively covered titanium circles with PCL and PCL fused with titanium dioxide. In this manner, PCL and PCL incorporated with titanium dioxide are expected to be used as surface-covering titanium for heart valve applications (Ferreira et al., 2017).

Table 1 shows the potential applications of PCL in the biomedical sector. It also gives an overview of the chemical composition; the trigger is applied to actuate the developed material and some notable features that enhance the material's property.

6 Scope and future perspective

PCL is a bio-degradable polymer of the aliphatic polyester family (Sinha et al., 2004b), and its applications include wound healing materials, contraceptive devices, and drug delivery systems (Woodruff and Hutmacher, 2010b). PCL is a favorable biodegradable polymer

TABLE 1 Potential applications of PCL in biomedical area.

Sr. no.	Chemical composition	Trigger	Form	Application	Notable characteristics	References
1.	PLA/PCL-based poly(urethane) + iron oxide NP + PEG/gelatin	Thermal/moisture	Scaffold	Bone tissue engineering	(1) Osteogenic induction (2) Additive manufacturing	Wang et al. (2018)
2.	PCL based polyurethane	Ultrasound	2D shapes	Drug delivery	(1) Localized recovery (2) Temporal control over drug release	Han et al. (2013)
3.	PCL-poly(ethylene glycol)-urethanes + vancomycin	Thermal	2D films	Wound dressing	(1) Promote wound healing (2) Electroactive (3) Biocompatible (4) Free-radical scavenging (5) Antibacterial	Li et al. (2019)
4.	PCL-based poly(urethane) containing L-lysine	Thermal	TubesFilms	Vascular graft	(1) Biocompatible (2) Cellular alignment	Castillo-Cruz et al. (2019)
5.	Poly(ε-caprolactone-co-DL-lactide)	Thermal	Tubular stent	Esophageal stenosis	(1) Biodegradable (2) In vivo study in a dog	Yu et al. (2012)
6.	Poly(ε-caprolactone) acrylate + hydroxyapatite nanoparticles	Thermal	2D films	Bone defect repair	Presence of hydroxyapatite nanoparticles	Tian et al. (2019b)
7.	Hydroxypropyl cellulose-g-poly (ε-caprolactone)	Thermal	2D films	Drug delivery	Compares drug incorporation methods	Jahangiri et al. (2019)
8.	Poly(ε-caprolactone) + silver nanoparticle grafted cellulose nanocrystals	Photothermal	2D films	Self-tightening sutures	Antibacterial	Toncheva et al. (2018)

having lower degradability and has been widely used in vivo and in vitro. Compared with different biodegradable polymers, which possess very high toughness and flexibility, PCL offers superior mechanical properties. Compared to other biodegradable polymers, PCL contains excellent mechanical properties and very high strength and elasticity because of its molecular weight. Thus, PCL-based materials are appropriate for various applications such as bone, sutures, cartilage, tendons, etc., and different applications in which mechanical strength is required (Abedalwafa et al., 2013b). PCL has been extensively used for BTE (bone tissue engineering) among all biomedical applications because of its mechanical strength and lower degradation rate, giving structural support for a longer duration in the bone healing process (Le et al., 2019).

PCL has also been employed to fabric scaffold tissue engineering due to its superior viscoelastic and physicochemical properties (Woodruff and Hutmacher, 2010b; Palamà et al., 2017). So presumably, there is no other field where PCL has more extensively been used than nano-technology. This will emerge as a multidisciplinary field with a broad range of application areas (Singh et al., 2010). PCL has shown biodegradability, biocompatibility and FDA (Food and Drug Administration) approval for human use (Varan and Bilensoy, 2017).

PCL is a very multidisciplinary polymer and has ample uses in pharmaceutics and biomedicine. The property of biocompatibility; biodegradability, and mechanical strength of the polymer go along with the accessibility of its chemical modification and blending as well been the critical property of this polymer.

In the present review, PCL shows the positive, hopeful aspects of its use to manufacture polymeric nano-particles that can successfully be used for cancer treatment and possibly can enhance commercialization. Moreover, research is still going on for PCL to upgrade its availability in the near future (Espinoza et al., 2020).

7 Conclusion

PCL is a promising SMP that changes its shape from temporary to permanent when external stimuli are applied. PCL has become a mainstream polymer among bio-degradable polymers to create films, drug deliveries, and scaffolds for different applications of tissue engineering. This review gives an overview of the advancement of the ongoing PCL material properties, its composites, and its broad range of applications. It shows excellent features, like biodegradability, biocompatibility, 3D, and porous structures. Among all the biomedical applications, PCL has been extensively used for BTE (bone tissue engineering) due to its mechanical strength and lower degradation rate. This article presented a review of the PCL, its synthesis, properties, and applications in the healthcare sector, like bone tissue engineering (BTE), wound dressing, drug delivery, vascular graft, oesophagal stenosis, and bone defect repair, self-tightening sutures, etc.

Acknowledgment

The authors are grateful to CSIR, New Delhi, for supporting the grant of the present work, Project No. MLP 210 to author SAR Hashmi. Director CSIR-AMPRI Bhopal is also acknowledged for providing necessary institutional facilities and encouragement.

Conflicts of interest

The authors declare no conflict of interest related to this research work.

References

Abedalwafa, M., Wang, F., Wang, L., Li, C., 2013a. Biodegradable poly-epsilon-caprolactone (PCL) for tissue engineering applications: a review. Rev. Adv. Mater. Sci. 34 (2), 123–140.

Abedalwafa, M., Wang, F., Wang, L., Li, C., 2013b. Biodegradable poly-epsilon-caprolactone (PCL) for tissue engineering applications: a review. Rev. Adv. Mater. Sci. 34 (2), 123–140.

Abrisham, M., Noroozi, M., Panahi-Sarmad, M., Arjmand, M., Goodarzi, V., Shakeri, Y., Golbaten-Mofrad, H., Dehghan, P., Sahzabi, A.S., Sadri, M., Uzun, L., 2020. The role of polycaprolactone-triol (PCL-T) in biomedical applications: a state-of-the-art review. Eur. Polym. J. 131, 109701.

Agarwal, S., 2010. Chemistry, chances and limitations of the radical ring-opening polymerization of cyclic ketene acetals for the synthesis of degradable polyesters. Polym. Chem. 1 (7), 953–964.

Aguilar, M.R., San Román, J., 2019. Introduction to smart polymers and their applications. In: Smart Polymers and Their Applications. Woodhead Publishing, pp. 1–11.

Ahn, S.K., Kasi, R.M., 2011. Exploiting microphase-separated morphologies of side-chain liquid crystalline polymer networks for triple shape memory properties. Adv. Funct. Mater. 21, 4543–4549.

Albrecht, L.D., Sawyer, S.W., Soman, P., 2016. Developing 3D scaffolds in the field of tissue engineering to treat complex bone defects. 3D Print. Addit. Manuf. 3 (2), 106–112.

Arora, A., Aggarwal, G., Chander, J., Maman, P., Nagpal, M., 2019. Drug eluting sutures: a recent update. J. Appl. Pharm. Sci. 9 (7), 111–123.

Baudis, S., Behl, M., Lendlein, A., 2014. Smart polymers for biomedical applications. Macromol. Chem. Phys. 215, 2399–2402.

Beatrice, C.A., Shimomura, K.M., Backes, E.H., Harb, S.V., Costa, L.C., Passador, F.R., Pessan, L.A., 2021. Engineering printable composites of poly (ε-polycaprolactone)/β-tricalcium phosphate for biomedical applications. Polym. Compos. 42 (3), 1198–1213.

Behl, M., Lendlein, A., 2007. Shape-memory polymers. Mater. Today 10 (4), 20–28.

Behl, M., Zotzmann, J., Lendlein, A., 2009. Shape-memory polymers and shape-changing polymers. In: Shape-Memory Polymers. Springer, pp. 1–40.

Behl, M., Kratz, K., Zotzmann, J., Nöchel, U., Lendlein, A., 2013a. Reversible bidirectional shape memory polymers. Adv. Mater. 25, 4466–4469.

Behl, M., Kratz, K., Noechel, U., Sauter, T., Lendlein, A., 2013b. Temperature-memory polymer actuators. Proc. Natl. Acad. Sci. 110, 12555–12559.

Bolaina-Lorenzo, E., Martínez-Ramos, C., Monleón-Pradas, M., Herrera-Kao, W., Cauich-Rodríguez, J.V., Cervantes-Uc, J.M., 2016. Electrospun polycaprolactone/chitosan scaffolds for nerve tissue engineering: physicochemical characterization and Schwann cell biocompatibility. Biomed. Mater. 12 (1), 015008.

Braud, C., Devarieux, R., Atlan, A., Ducos, C., Vert, M., 1998. Capillary zone electrophoresis in normal or reverse polarity separation modes for the analysis of hydroxy acid oligomers in neutral phosphate buffer. J. Chromatogr. B Biomed. Sci. Appl. 706 (1), 73–82.

Burton, T.P., Corcoran, A., Callanan, A., 2017. The effect of electrospun polycaprolactone scaffold morphology on human kidney epithelial cells. Biomed. Mater. 13 (1), 015006.

Castillo-Cruz, O., Aviles, F., Vargas-Coronado, R., Cauich-Rodríguez, J.V., Chan-Chan, L.H., Sessini, V., Peponi, L., 2019. Mechanical properties of l-lysine based segmented polyurethane vascular grafts and their shape memory potential. Mater. Sci. Eng. C 102, 887–895.

Centola, M., Rainer, A., Spadaccio, C., De Porcellinis, S., Genovese, J.A., Trombetta, M., 2010. Combining electrospinning and fused deposition modeling for the fabrication of a hybrid vascular graft. Biofabrication 2 (1), 014102.

Chen, Y., Zhao, X., Luo, C., Shao, Y., Yang, M.B., Yin, B., 2020. A facile fabrication of shape memory polymer nanocomposites with fast light-response and self-healing performance. Compos. A: Appl. Sci. Manuf. 135, 105931.

Chhaya, M.P., Balmayor, E.R., Hutmacher, D.W., Schantz, J.T., 2016. Transformation of breast reconstruction via additive biomanufacturing. Sci. Rep. 6 (1), 1–2.

Cipitria, A., Skelton, A., Dargaville, T.R., Dalton, P.D., Hutmacher, D.W., 2011. Design, fabrication and characterization of PCL electrospun scaffolds—a review. J. Mater. Chem. 21 (26), 9419–9453.

Espinoza, S.M., Patil, H.I., San Martin Martinez, E., Casañas Pimentel, R., Ige, P.P., 2020. Poly-ε-caprolactone (PCL), a promising polymer for pharmaceutical and biomedical applications: focus on nanomedicine in cancer. Int. J. Polym. Mater. Polym. Biomater. 69 (2), 85–126.

Ferreira, C.C., Ricci, V.P., Sousa, L.L., Mariano, N.A., Campos, M.G., 2017. Improvement of titanium corrosion resistance by coating with poly-caprolactone and poly-caprolactone/titanium dioxide: potential application in heart valves. Mater. Res. 20, 126–133.

Fox, K., Ratwatte, R., Booth, M.A., Tran, H.M., Tran, P.A., 2020. High nanodiamond content-PCL composite for tissue engineering scaffolds. Nanomaterials. 10 (5), 948.

Gamez, E., Mendoza, G., Salido, S., Arruebo, M., Irusta, S., 2019. Antimicrobial electrospun polycaprolactone-based wound dressings: an in vitro study about the importance of the direct contact to elicit bactericidal activity. Adv. Wound Care 8 (9), 438–451.

Gautrot, J.E., Zhu, X.X., 2009. Shape memory polymers based on naturally-occurring bile acids. Macromolecules 42 (19), 7324–7331.

Ghorbani, F., Moradi, L., Shadmehr, M.B., Bonakdar, S., Droodinia, A., Safshekan, F., 2017. In-vivo characterization of a 3D hybrid scaffold based on PCL/decellularized aorta for tracheal tissue engineering. Mater. Sci. Eng. C 81, 74–83.

Grant, R., Hay, D.C., Callanan, A., 2017. A drug-induced hybrid electrospun poly-capro-lactone: cell-derived extracellular matrix scaffold for liver tissue engineering. Tissue Eng. A 23 (13–14), 650–662.

Han, J., Fei, G., Li, G., Xia, H., 2013. High intensity focused ultrasound triggered shape memory and drug release from biodegradable polyurethane. Macromol. Chem. Phys. 214 (11), 1195–1203.

Hao, J., Yuan, M., Deng, X., 2002. Biodegradable and biocompatible nanocomposites of poly (ε-caprolactone) with hydroxyapatite nanocrystals: thermal and mechanical properties. J. Appl. Polym. Sci. 86 (3), 676–683.

Hastings, G.W., 1985. Structural considerations and new polymers for biomedical applications. Polymer 26 (9), 1331–1335.

Hayash, T., 1994. Biodegradable polymers for biomedical uses. Prog. Polym. Sci. 19, 663–702.

Hewitt, E., Mros, S., McConnell, M., Cabral, J.D., Ali, A., 2019. Melt-electrowriting with novel milk protein/PCL biomaterials for skin regeneration. Biomed. Mater. 14 (5), 055013.

Hoffman, A.S., Stayton, P.S., 2020. Applications of "smart polymers" as biomaterials. In: Biomaterials Science. Academic Press, pp. 191–203.

Hsieh, Y.H., Shen, B.Y., Wang, Y.H., Lin, B., Lee, H.M., Hsieh, M.F., 2018. Healing of osteochondral defects implanted with biomimetic scaffolds of poly (ε-caprolactone)/hydroxyapatite and glycidyl-methacrylate-modified hyaluronic acid in a minipig. Int. J. Mol. Sci. 19 (4), 1125.

Hu, J., 2007. Shape Memory Polymers and Textiles. Elsevier.

Jahangiri, M., Kalajahi, A.E., Rezaei, M., Bagheri, M., 2019. Shape memory hydroxypropyl cellulose-g-poly (ε-caprolactone) networks with controlled drug release capabilities. J. Polym. Res. 26 (6), 136.

Khan, F., Tanaka, M., Ahmad, S., 2015. Fabrication of polymeric biomaterials: a strategy for tissue engineering and medical devices. J. Mater. Chem. B 3 (42), 8224–8249.

Kim, Y.B., Kim, G.H., 2015. PCL/alginate composite scaffolds for hard tissue engineering: fabrication, characterization, and cellular activities. ACS Comb. Sci. 17 (2), 87–99.

Kong, B., Mi, S., 2016. Electrospun scaffolds for corneal tissue engineering: a review. Materials 9 (8), 614.

Kularatne, R.S., Kim, H., Boothby, J.M., Ware, T.H., 2017. Liquid crystal elastomer actuators: synthesis, alignment, and applications. J. Polym. Sci. B Polym. Phys. 55, 395–411.

Kumari, A., Yadav, S.K., Yadav, S.C., 2010. Biodegradable polymeric nanoparticles based drug delivery systems. Colloids Surf. B: Biointerfaces 75 (1), 1–8.

Le, B.Q., Rai, B., Lim, Z.X., Tan, T.C., Lin, T., Lee, J.J., Murali, S., Teoh, S.H., Nurcombe, V., Cool, S.M., 2019. A polycaprolactone-β-tricalcium phosphate–heparan sulphate device for cranioplasty. J. Cranio-Maxillofac. Surg. 47 (2), 341–348.

Lee, H., Cho, D.W., 2016. One-step fabrication of an organ-on-a-chip with spatial heterogeneity using a 3D bioprinting technology. Lab Chip 16 (14), 2618–2625.

Lendlein, A., Kelch, S., 2002. Shape-memory polymers. Angew. Chem. Int. Ed. 41 (12), 2034–2057.

Lendlein, A., Langer, R., 2002. Biodegradable, elastic shape-memory polymers for potential biomedical applications. Science 296 (5573), 1673–1676.

Leng, J., Lan, X., Du, S., 2010. Shape-memory polymer composites. In: Shape-Memory Polymers and Multifunctional Composites. CRC Press, p. 203.

Leng, J.S., Lan, X., Liu, Y.J., Du, S.Y., 2011. Shape-memory polymers and their composites: stimulus methods and applications. Prog. Mater. Sci. 56 (7), 1077–1135.

Leong, N.L., Kabir, N., Arshi, A., Nazemi, A., Wu, B., Petrigliano, F.A., McAllister, D.R., 2015. Evaluation of polycaprolactone scaffold with basic fibroblast growth factor and fibroblasts in an athymic rat model for anterior cruciate ligament reconstruction. Tissue Eng. A 21 (11–12), 1859–1868.

Li, M., Chen, J., Shi, M., Zhang, H., Ma, P.X., Guo, B., 2019. Electroactive anti-oxidant polyurethane elastomers with shape memory property as non-adherent wound dressing to enhance wound healing. Chem. Eng. J. 375, 121999.

Liu, Y., Du, H., Liu, L., Leng, J., 2014. Shape memory polymers and their composites in aerospace applications: a review. Smart Mater. Struct. 23, 023001.

ltc-proto, n.d. https://www.ltc-proto.com/blog/4-types-of-materials-and-different-ways-of-their-manufacturing-processes/.

Luo, X., Mather, P.T., 2010. Triple-shape polymeric composites (TSPCs). Adv. Funct. Mater. 20, 2649–2656.

Maitz, M.F., 2015. Applications of synthetic polymers in clinical medicine. Biosurface Biotribol. 1 (3), 161–176.

Malikmammadov, E., Tanir, T., Kiziltay, A., Hasirci, V., Hasirci, N., 2017. PCL and PCL-based materials in biomedical applications. J. Biomater. Sci. Polym. Ed. 29 (7–9), 863–893.

Malikmammadov, E., Tanir, T.E., Kiziltay, A., Hasirci, V., Hasirci, N., 2018. PCL and PCL-based materials in biomedical applications. J. Biomater. Sci. Polym. Ed. 29 (7–9), 863–893.

McClure, M.J., Wolfe, P.S., Rodriguez, I.A., Bowlin, G.L., 2011. Bioengineered vascular grafts: improving vascular tissue engineering through scaffold design. J. Drug Deliv. Sci. Technol. 21 (3), 211–227.

Miao, S., Nowicki, M., Cui, H., Lee, S.J., Zhou, X., Mills, D.K., Zhang, L.G., 2019. 4D anisotropic skeletal muscle tissue constructs fabricated by staircase effect strategy. Biofabrication 11 (3), 035030.

Miaudet, P., Derre, A., Maugey, M., Zakri, C., Piccione, P.M., Inoubli, R., Poulin, P., 2007. Shape and temperature memory of nanocomposites with broadened glass transition. Science 318, 1294–1296.

Mishra, N., Goyal, A.K., Khatri, K., Vaidya, B., Paliwal, R., Rai, S., Mehta, A., Tiwari, S., Vyas, S., Vyas, S.P., 2008. Biodegradable polymer based particulate carrier (s) for the delivery of proteins and peptides. Anti-Inflammatory Anti-Allergy Agents Med. Chem. 7 (4), 240–251.

Mohamadi, F., Ebrahimi-Barough, S., Reza Nourani, M., Ali Derakhshan, M., Goodarzi, V., SadeghNazockdast, M., Farokhi, M., Tajerian, R., Faridi Majidi, R., Ai, J., 2017. Electrospun nerve guide scaffold of poly (ε-caprolactone)/collagen/nanobioglass: an in vitro study in peripheral nerve tissue engineering. J. Biomed. Mater. Res. A 105 (7), 1960–1972.

Mondal, D., Griffith, M., Venkatraman, S.S., 2016. Polycaprolactone-based biomaterials for tissue engineering and drug delivery: current scenario and challenges. Int. J. Polym. Mater. Polym. Biomater. 65 (5), 255–265.

Mu, T., Liu, L., Lan, X., Liu, Y., Leng, J., 2018. Shape memory polymer and its composite: function and application. Comprehensive Composite Materials II. Elsevier, pp. 454–486.

Nail, L.N., Zhang, D., Reinhard, J.L., Grunlan, M.A., 2015. Fabrication of a bioactive, PCL-based "self-fitting" shape memory polymer scaffold. J. Vis. Exp. (104), e52981.

Okada, M., 2002. Chemical syntheses of biodegradable polymers. Prog. Polym. Sci. 27, 87–133.

Palamà, I.E., Arcadio, V., D'Amone, S., Biasiucci, M., Gigli, G., Cortese, B., 2017. Therapeutic PCL scaffold for reparation of resected osteosarcoma defect. Sci. Rep. 7, 12672.

Pitt, C.G., Chasalow, F.I., Hibionada, Y.M., Klimas, D.M., Schindler, A., 1981. Aliphatic polyesters. I. The degradation of poly (ε-caprolactone) in vivo. J. Appl. Polym. Sci. 26 (11), 3779–3787.

ProvenProcess, n.d. https://provenprocess.com/medical-device-engineering/materials-science.

Ragaert, K., Van de Velde, S., Cardon, L., De Somer, F., 2012. Methods for improved flexural mechanical properties of 3D-plotted PCL-based scaffolds for heart valve tissue engineering. In: 5th International Conference on Polymers and Moulds Innovations (PMI 2012), pp. 49–60.

Raghavendra, G.M., Varaprasad, K., Jayaramudu, T., 2015. Nanotechnology Applications for Tissue Engineering. Elsevier Inc., Amsterdam.

Roseti, L., Parisi, V., Petretta, M., Cavallo, C., Desando, G., Bartolotti, I., Grigolo, B., 2017. Scaffolds for bone tissue engineering: state of the art and new perspectives. Mater. Sci. Eng. C 78, 1246–1262.

Safranski, D.L., Smith, K.E., Gall, K., 2013. Mechanical requirements of shape-memory polymers in biomedical devices. Polym. Rev. 53 (1), 76–91.

Serrano-Aroca, Á., Vera-Donoso, C.D., Moreno-Manzano, V., 2018. Bioengineering approaches for bladder regeneration. Int. J. Mol. Sci. 19 (6), 1796.

Siddiqui, N., Asawa, S., Birru, B., Baadhe, R., Rao, S., 2018. PCL-based composite scaffold matrices for tissue engineering applications. Mol. Biotechnol. 60 (7), 506–532.

Siedlecki, C.A., 2017. Hemocompatibility of Biomaterials for Clinical Applications: Blood-Biomaterials Interactions. Woodhead Publishing, Sawston, Cambridge.

Singh, M., Manikandan, S., Kumaraguru, A.K., 2010. Nanoparticles: a new technology with wide applications. Res. J. Nanosci. Nanotechnol. 1, 1–11.

Sinha, V.R., Bansal, K., Kaushik, R., Kumria, R., Trehan, A., 2004a. Poly-ε-caprolactone microspheres and nanospheres: an overview. Int. J. Pharm. 278 (1), 1–23.

Sinha, V.R., Bansal, K., Kaushik, R., Kumria, R., Trehan, A., 2004b. Poly-ε-caprolactone microspheres and nanospheres: an overview. Int. J. Pharm. 278, 1–23.

Siqueira, I.A., de Moura, N.K., de Barros Machado, J.P., Backes, E.H., Passador, F.R., de Sousa Trichês, E., 2017. Porous membranes of the polycaprolactone (PCL) containing calcium silicate fibers for guided bone regeneration. Mater. Lett. 206, 210–213.

Small, W., Buckley, P.R., Wilson, T.S., Benett, W.J., Hartman, J., Saloner, D., Maitland, D.J., 2007. Shape memory polymer stent with expandable foam: a new concept for endovascular embolization of fusiform aneurysms. IEEE Trans. Biomed. Eng. 54, 1157–1160.

Smink, A.M., Hertsig, D.T., Schwab, L., van Apeldoorn, A.A., de Koning, E., Faas, M.M., de Haan, B.J., de Vos, P., 2017. A retrievable, efficacious polymeric scaffold for subcutaneous transplantation of rat pancreatic islets. Ann. Surg. 266 (1), 149–157.

Sternberg, K., 2009. Current requirements for polymeric biomaterials in ear, nose and throat medicine. Laryngo-Rhino-Otologie 88, S1–S11.

Storey, R.F., Taylor, A., 1996. Effect of stannous octoate concentration on the ethylene glycolinitiated polymerization of epsiloncaprolactone. Abstr. Pap. Am. Chem. Soc. 211, 114.

Teo, A.J., Mishra, A., Park, I., Kim, Y.J., Park, W.T., Yoon, Y.J., 2016. Polymeric biomaterials for medical implants and devices. ACS Biomater. Sci. Eng. 2 (4), 454–472.

Tian, G., Zhu, G., Xu, S., Ren, T., 2019a. A novel shape memory poly (ε-caprolactone)/hydroxyapatite nanoparticle networks for potential biomedical applications. J. Solid State Chem. 272, 78–86.

Tian, G., Zhu, G., Xu, S., Ren, T., 2019b. A novel shape memory poly (ε-caprolactone)/hydroxyapatite nanoparticle networks for potential biomedical applications. J. Solid State Chem. 272, 78–86.

Toncheva, A., Khelifa, F., Paint, Y., Voue, M., Lambert, P., Dubois, P., Raquez, J.M., 2018. Fast IR-actuated shape-memory polymers using in situ silver nanoparticle-grafted cellulose nanocrystals. ACS Appl. Mater. Interfaces 10 (35), 29933–29942.

Ulery, B.D., Nair, L.S., Laurencin, C.T., 2011. Biomedical applications of biodegradable polymers. J. Polym. Sci. B Polym. Phys. 49 (12), 832–864.

Varan, C., Bilensoy, E., 2017. Cationic PEGylated polycaprolactone nanoparticles carrying post-operation docetaxel for glioma treatment. Beilstein J. Nanotechnol. 8, 1446–1456.

Wache, H.M., Tartakowska, D.J., Hentrich, A., Wagner, M.H., 2003. Development of a polymer stent with shape memory effect as a drug delivery system. J. Mater. Sci. Mater. Med. 14, 109–112.

Wang, L., Yang, X., Chen, H., Gong, T., Li, W., Yang, G., Zhou, S., 2013a. Design of triple-shape memory polyurethane with photo-cross-linking of cinnamon groups. ACS Appl. Mater. Interfaces 5 (21), 10520–10528.

Wang, L., Yang, X., Chen, H., Yang, G., Gong, T., Li, W., Zhou, S., 2013b. Multi-stimuli sensitive shape memory poly (vinyl alcohol)-graft-polyurethane. Polym. Chem. 4, 4461–4468.

Wang, L., Wang, W., Di, S., Yang, X., Chen, H., Gong, T., Zhou, S., 2014a. Silver-coordination polymer network combining antibacterial action and shape memory capabilities. RSC Adv. 4, 32276–32282.

Wang, L., Di, S., Wang, W., Chen, H., Yang, X., Gong, T., Zhou, S., 2014b. Tunable temperature memory effect of photo-cross-linked star PCL–PEG networks. Macromolecules 47, 1828–1836.

Wang, Y., Li, X., Pan, Y., Zheng, Z., Ding, X., Peng, Y., 2014c. High-strain shape memory polymers with movable cross-links constructed by interlocked slide-ring structure. RSC Adv. 4 (33), 17156–17160.

Wang, Y.J., Jeng, U.S., Hsu, S.H., 2018. Biodegradable water-based polyurethane shape memory elastomers for bone tissue engineering. ACS Biomater Sci. Eng. 4 (4), 1397–1406.

Wnek, G., 2008. Polymers. In: Encyclopedia of Biomaterials and Biomedical Engineering, second ed. vol. 4. CRC Press, pp. 2275–2281.

Wong, Y., Kong, J., Widjaja, L.K., Venkatraman, S.S., 2014. Biomedical applications of shape-memory polymers: how practically useful are they? Sci. China Chem. 57 (4), 476–489.

Woodruff, M.A., Hutmacher, D.W., 2010a. The return of a forgotten polymer—polycaprolactone in the 21st century. Prog. Polym. Sci. 35 (10), 1217–1256.

Woodruff, M.A., Hutmacher, D.W., 2010b. The return of a forgotten polymer—polycaprolactone in the 21st century. Prog. Polym. Sci. 35, 1217–1256.

Yang, G., Lin, H., Rothrauff, B.B., Yu, S., Tuan, R.S., 2016. Multilayered polycaprolactone/gelatinfiber-hydrogel composite for tendon tissue engineering. Acta Biomater. 35, 68–76.

Yu, X., Wang, L., Huang, M., Gong, T., Li, W., Cao, Y., Ji, D., Wang, P., Wang, J., Zhou, S., 2012. A shape memory stent of poly (ε-caprolactone-co-DL-lactide) copolymer for potential treatment of esophageal stenosis. J. Mater. Sci. Mater. Med. 23 (2), 581–589.

Zhang, J.L., Huang, W.M., Gao, G., Fu, J., Zhou, Y., Salvekar, A.V., Venkatraman, S.S., Wong, Y.S., Tay, K.H., Birch, W.-R., 2014. Shape memory/change effect in a double network nanocomposite tough hydrogel. Eur. Polym. J. 58, 41–51.

Zhao, Q., Qi, H.J., Xie, T., 2015. Recent progress in shape memory polymer: new behaviour, enabling materials, and mechanistic understanding. Prog. Polym. Sci. 49, 79–120.

Zheng, Y., Li, Y., Hu, X., Shen, J., Guo, S., 2017. Biocompatible shape memory blend for self-expandable stents with potential biomedical applications. ACS Appl. Mater. Interfaces 9, 13988–13998.

Zhou, J., Turner, S.A., Brosnan, S.M., Li, Q., Carrillo, J.M.Y., Nykypanchuk, D., Gang, O., Ashby, V.S., Dobrynin, A.V., Sheiko, S.S., 2014. Shapeshifting: reversible shape memory in semicrystalline elastomers. Macromolecules 47, 1768–1776.

Zhu, Y., Zhang, K., Zhao, R., Ye, X., Chen, X., Xiao, Z., Yang, X., Zhu, X., Zhang, K., Fan, Y., Zhang, X., 2017. Bone regeneration with micro/nano hybrid-structured biphasic calcium phosphate bioceramics at segmental bone defect and the induced immunoregulation of MSCs. Biomaterials 147, 133–144.

Nanoemulsions for antitumor activity

Soma Mukherjee[a], Darryl L. Holliday[a], Nabaraj Banjara[a], and Navam Hettiarachchy[b]

[a]Department of Biological and Physical Science, University of Holy Cross, New Orleans, LA, United States [b]Department of Food Science, University of Arkansas, Fayetteville, AR, United States

1 Introduction

Nanoemulsions are widely used to deliver water-insoluble or sparingly soluble drugs and constituted to protect nontoxic excipients (Ganta et al., 2014; McClements, 2012). The droplet of these emulsions are composed of heterogeneous colloidal dispersions of a nanometer size in another form of liquid (water or oil/oil in water) and exhibits greater stability and dispersibility (Gi et al., 1992). This encapsulation prevents the drug from functional degradation and also increases its functional life span inside the cellular plasma. To increase the solubility or to disperse the insoluble droplets in water, emulsifiers are added to stabilize the emulsion system. Emulsifiers are compounds of amphiphilic nature and possess the ability to decrease the interfacial tension within the two immiscible liquids (Maeda et al., 2000). Hydrophobic bicarbonate tails of the emulsifiers orient themselves in the oil phase and the polar head places themselves in the polar phase of the liquid (Tadros et al., 2004).

In the present decade, the research in cancer therapy is looking forward and focusing on using nanoemulsion formulation techniques as they have essential characteristics to reach the highest level of therapeutic effect of the drug. These particular important essential characteristics include greater surface area, outer ionic charge, expand half-life of translocation, specific cellular target sites, and the detectable imaging capacity of the system. The main barrier to reaching the target cancer cells are the vascularized tissues near the periphery of the cancer cells. Nanoemulsions can easily translocate in these tissues because their miniature size allows them to pass through the barriers. They can also be formulated for specific functionality and has the flexibility of encapsulating distinct types of cancer-inhibiting drugs with the specified target (Tiwari et al., 2006). Multifunctional nanoemulsion systems are the prime important delivery system in cancer therapy (Fig. 1, Table 1).

The extracellular matrix of tumor cell possess some surface markers and plays an important role in the subsequent progression of growth of the tumor. The surface markers of ECM are targeted by the drugs. Although the supply of oxygen and nutrients occurs through

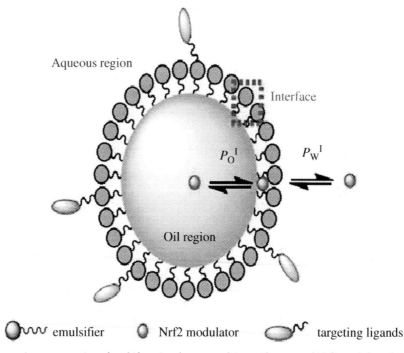

FIG. 1 Schematic representation of multifunctional nanoemulsion with potential Nrf2 modulator between two interphases. Pw′ and Po′ represent partition constant between water and oil interfacial regions respectively. *Adapted from Sezin-Bayindir et al. (2021) under Creative Commons Attribution License.*

TABLE 1   Nanoemulsion and different theranostic approach to detect cancer.

Nano-formulation	Chemotherapeutic, imaging agent	Cell line	Organismic model	Results	References
Hybrid liposomes	Imaging compound: ICG	MDA-MB-453 cancer cell	Orthotopic graft mouse model of breast cancer	Enhanced accumulation observed for HLs carrying ICG; growth inhibition of MDA-MB-453 cells and induced apoptosis	Ichihara et al. (2018)
Liposome	Chemotherapeutic Imaging compound: ruthenium polypyridine complex	MDA-MB-231 cells	Orthotopic mouse model with breast cancer	Enhanced accumulation in tumor area, reduction of growth	Shen et al. (2017)
Liposome	Chemotherapeutic agent PTX imaging agent: superparamagnetic iron oxide NPs	MDA-MB-231 cell line	Mouse model with breast cancer	Targeting achieved, MRI exhibited antitumor activity	Zheng et al. (2018)
NP-based sonophores	IR Dye QC1	4T1-breast cancer cell line	Athymic female mice Nude-Faxn1	Delivered dye offers excellent optoacoustic effect	Roberts et al. (2018)
PAMAM dendrimer	Imaging agent: ICG derivative, cypate, a clinically used NIR cyanine dye Chemotherapeutic: docetaxel	HepG2 malignant hepatic human cell	HepG2-induced mouse tumor model	Modified with iRGD, a tumor-targeting and penetrating peptide, enhanced antitumor efficiency with reduced risk of tumor recurrence	Ge et al. (2019)
PAMAM dendrimer	Chemotherapeutic: doxorubicin Imaging agent: Gd-doped ferrite NPs	Hela cell line	–	Maximum cumulative doxorubicin release; PAMAM-entrapped Gd-doped ferrite NP is potential noninvasive	Mekonnen et al. (2019)
Polydopamine-decorated mesoporous organosilica NPs	Perfluorohexane; ultrasound, ICG, NIRF imaging compounds	MCF-7 breast cancer cells	MCF-7 cell-induced breast tumor BALB/c nude mouse model	Long cellular uplate and retention; synergistic photothermal and photodynamic activity	Huang et al. (2019)
Mesoporous organosilica NPs	Imaging agent: perfluoropentane liquid Chemotherapeutic: doxorubicin	MDA-MB-231 breast cancer cells	MDA-MB-231 cell-induced tumor mouse model	pH sensitive coating helped in drug release and ultra-sonic imaging; combined effect in tumor suppression and photothermal treatment as well as chemotherapy	Xu et al. (2019)
Liposome, nano-droplet	Porphyrin-lipid stabilized paclitaxel MRI, in vivo fluorescence imaging	–	Mouse model	Higher accumulation of drug in tumor area and inhibited tumor growth (78%)	Chang et al. (2021)

simple diffusion, the larger surface area of the growing tumor inhibits the supply of oxygen causing hypoxia conditions to the microenvironment of the tumor. This triggers the angiogenic progress of new blood vessel formation (Arruebo et al., 2009; Keefe et al., 2010). So, by inhibiting the angiogenic process cellular growth can be inhibited. The commonly used anticancerous drugs also attribute marked toxicity, increase resistance and limit delivery of therapeutic compounds. Nanoemulsions can be a delivery vehicle encapsulating the compound inside the core of the micelles, decreasing toxicity and increasing target efficiency.

Another tumor biology plays important role in tumor development. The tumor cells are fed by the energy of glycolysis. Hypoxia in the cellular environment the final terminal metabolite of glycolysis is converted into lactate which is excluded by monocarboxylate transporter by adding H+ and this transforms the cellular ambiance acidic. Inadequate supply of oxygen also triggers the formation of carbonic anhydrase IX, resulting in the increase of bicarbonate pull from carbon dioxide. The bicarbonate ion is used up by the weakly alkaline cells and leading to a pH-responsive balance between the extracellular and intracellular sites of the tumor. The lipids that are stable at slightly alkaline pH can play a potential role by modifying chemical behavior in an acidic pH, still can release the drug efficiently (Ganta et al., 2014).

The uncontrolled and disordered proliferation of tumor development leads to an uncontrolled density of essential molecules in the intracellular microenvironment of the tumor. This triggers cellular hypoxia and the development of new vessels which leads to metastasis. When the lymphatic vessels exhibit instability this in turn also increases the retention time of the drugs since their elimination flow rate reduces. This affects the enhanced permeability and retention time of the therapeutics. Macromolecular water-insoluble drugs can be applied to take advantage of this EPR. Another approach to cancer treatment has emerged recently. This also involves the nanosized drug delivery systems (NDDs) that influence nuclear factor erythroid 2-related factor 2 (NRF2). NRF2 bases treatment modulators are encapsulated in nanoemulsion-based NDDs systems and produced encouraging results. The main advantage of using NDDs are passive targeting through enhanced permeability and retention (EPR) and specific active targeting via ligands (Sezgin-Bayindir et al., 2021) (Fig. 2). This type of cellular targeting is named passive targeting (Maeda et al., 2000). The size range also clears and increases the possibility of being engulfed by the mononuclear phagocytic system (MPS) (Keefe et al., 2010). This problem can be bypassed by coating nanoemulsions with hydrophilic compounds (Tan et al., 2011). The presence of negatively charged particles like phosphatidylserine promotes longer retention of negatively charged compounds by the malignant cells. The negativity of passive targeting is characterized by the inability to recognize and differentiate between normal healthy and malignant tissues (Gu et al., 2007).

The specific nature of antigen and antibody binding is commensurate with the conjugation of nanoemulsion and antibody. Many studies confirmed the delivery of drugs successfully by internalizing cancer cells. The stimuli-responsive nature of nanocarrier-Ab conjugation is more specific to malignant cells (Almeida et al., 2010). A library of ssDNA and ssRNA is used to design oligonucleotides of DNA and RNA to produce aptamer (Wooster et al., 2008; Ganta et al., 2008). Aptamers are smaller in size and have superior quality of no immunologic response, relatively easier to produce, and have a capacity for fast penetration. They can easily be combined with other molecules and can be transformed or folded into the secondary and tertiary structures of DNA/RNA (McClements, 2011; Wang and Huang, 1987). This process also allows the choice of specific aptamers specific to reduce tumor cells based on receptor

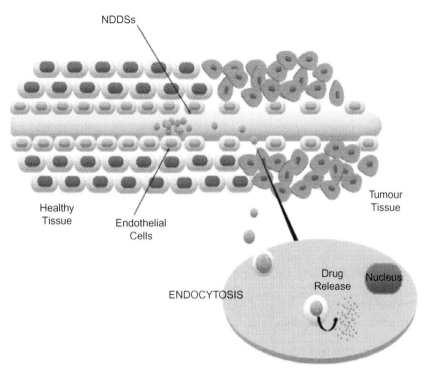

FIG. 2   The role of NDDs in drug targeting with increased permeability and retention (EPR) effect to carcinogenic cells. *Adapted from Sezin-Bayindir et al. (2021) under Creative Commons Attribution License.*

protein and specific recognizable biomarkers (McClements, 2011; Wang and Huang, 1987). Various malignant tissues exhibit high folate receptor expression of protein and higher affinity of folic acid and folate receptor makes folate an ideal candidate for targeting. In normal tissues, folate is located in the apical surface of the epithelial tissue (Ganta et al., 2008). It is confirmed that the number of folate receptor proteins increases with the progression of malignancy. Folic acid is inexpensive, nontoxic, nonimmunologic nanocarrier with an affinity for binding. Folic acid exhibits high circulating and storage ability. Another advantage of using nanoemulsion as a carrier in delivering chemotherapeutic drugs is its ability to combine with oligonucleotides. Although the oligonucleotides have several nonadvantage such as very short half-life, low penetrating ability over other carriers.

Active targeting with nanoemulsion however able to recognize particular molecules on the malignant tissue through the association of ligands. The advantage of active targeting lies in using the microenvironment of the tumor. It is a new and efficient strategy to deliver therapeutics to the particular cancer cell and also to a specific type of malignant type tissues (Parker et al., 2005). Different types of bonds/moieties can be created with specific ligand-receptor or antigen-antibody (Kumar and Divya, 2015). These ligand-receptor bonds can connect to specific overexpressed protein receptor molecules such as folate, transferrin, epidermal growth factor receptor, organ-specific membrane antigen (Elnakat and Ratnam, 2004; Low and Antony, 2004; Toub et al., 2006). The targeted therapeutic delivery induces specific toxic effect in malignant cells and thereby decrease side effects, also changing the molecular activity in order to increase stimulus effect to sensitivity (Bertrand et al., 2014). The mechanism of the drug efflux pump is controlled by the overexpressed protein of the ATP binding cassette (ABC) family. These proteins eliminate drugs from the cellular environment. The first gene revealed was the ABC1 gene encoded by P-glycoprotein. ABC1 gene was capable of transporting vinblastine, colchicine, etoposide, and paclitaxel (PVX) from the malignant cell (Zhang et al., 2006; Jain, 2005). Moreover, the MDR pathway is also responsible for inducing antiapoptotic Bcl 2 genes and nuclear factor-kappa B (NF-kB) (Al-Abd et al., 2009). Moreover, nanocarriers like passive and active targeting bonds and ligands can be used to reduce MDR by controlling the three important overexpressed genes that take part in the cellular mechanism of transporting drugs (McClements, 2012). Though the number of ABCs in humans is distributed for different organs of the human body and with several anatomical constraints (Lin et al., 2000).

Another auxiliary chemotherapeutic agent is small interfering RNA that is capable of regulating MDR by deregulating antiapoptotic genes. Only a few suitable carriers to co-deliver siRNA and drugs have been discovered. Nano-emulsions can be used as vectors as this can modulate P-gp and Bcl-2 both to control MDR (Fattal et al., 2004). For example, ceramide has been effectively (was capable of increasing apoptotic cell death) delivered by nanoemulsion with the specified target.

In this chapter, the major discoveries in the application of nanoemulsions in cancer therapy will be discussed. In order to utilize the physical and chemical characteristics of nanoemulsion and the facility that can be achieved in designing delivery systems have been focused in this field of research and development. The application of nanoemulsion has emerged a new horizon of drug delivery systems in cancer therapy.

2 Nanoemulsion and MDR

MDR contribute the highest number of death by cancer and was to apply cancer therapeutic drugs for the effective treatment of cancer patients. The overexpression of transporter protein responsible for the efflux of these transporters belonging to the ABC gene family has been reported to play an important role in inducing multidrug resistance to the malignant cells (Dean, 2009). Various intracellular ligands like proteins, lipids, and different metabolites, antibiotics also play an important role and remain active by gaining energy from ATP hydrolysis.

The most widely expressed transporter was MRP1 is responsible for a wide range of drugs such as vincristine, etoposide, colchicine, camptothecin, and methotrexate. The encoding gene of MRP1 protein was ABCC1. MRP2 was identified from the cell membrane of polarized cells of the kidney, liver, and intestinal epithelium. Gradually MRP3, MRP4, MRP 8 were discovered and shown to be responsible for various drugs. These data create the search for novel compounds to inactivate ABC pumps to effectively deliver cancer therapeutic drugs. Nanoemulsions with ABC pumps inhibitor have been researched to address MDR resistance and cancer treatment (Mohammad et al., 2018). The first docetaxel was delivered for ovarian cancer cells by surpassing docetaxel resistance binding with folate. The target was made with folate as folate is weakly expressed in normal tissues and this facilitates targeting cancer cells with overexpressed folate sensitive receptor proteins. In addition, the author was also able to explore P-gp expression and with folate receptors could be mediated by endocytosis that can exhibit an increased level of cytotoxicity by suppressing ABC transporters (Chanamai and McClements, 2000; Ganta et al., 2016). The taxanes drugs (paclitaxel and docetaxel) have been used as a fundamental drug in the treatment of breast cancer although inhibiting MRD is still a vital issue. In order to run over this issue, baicalein is used to decrease P-gp and also to trigger oxidative stress by Meng and his team. Inducing oxidative stress in the cellular environment is claimed as another bolstered approach to sensitize paclitaxel drugs that have the potentiality to induce reactive oxygen species and decrease the formation of glutathione. Using this strategy, a novel formulation of nanoemulsion with paclitaxel and baicalein was delivered to treat breast cancer cells. This formulation was effective in inducing ROS, inhibiting cellular GSH, and triggering caspase-3 activity in breast cancer cells (MCF-7/tax. The most beneficial result was this co-encapsulation exhibited highly effective compared to other paclitaxel drug formulations. The result of this co-encapsulation delivery suggests that co-delivery of paclitaxel and baicalein might be used to suppress MDR in cancer therapy (Meng et al., 2016). In another approach, it has the potentiality to change the expression level of Bax and Bcl 2 and also can inhibit transport protein P-ag. Derivative of vitamin A has the potentiality of working as an antioxidant and the antioxidating mechanism involves the reduction of peroxyl and fatty acid radicals (Zheng et al., 2016). Vitamin E was also used in a study with TPGS was a potent surfactant that can inhibit p-gp and can effectively alter MDR in cancer therapy (Tang et al., 2013). Vitamin E can interfere or break Bcl-xL-Bax interaction, stimulate Bax, can intervene mitochondrial apoptotic cell cycle. Therefore, vitamin E was a potential candidate to formulate nanoemulsion with paclitaxel to study MDR resistance in ovarian cell lines (Zheng et al., 2016).

3 Application and different types of cancer therapy

In the last decades, nanoemulsion has shown to be an effective approach in cancer medication delivery and therefore major research has been focused on finding remedies and treatment of different types of cancer. These recent efforts and advances in this emulsion technology and antitumor activity have been discussed in the following sections.

3.1 Colon cancer therapy with nanoemulsion

A large number of cancer death occur in the world due to colon cancer (Hu and Zhang, 2012). Four subcategories within this type of cancer include adenomatous polyposis (familial), nonpolyposis (hereditary), colon cancer (sporadic), and colitis-associated cancer (Yu et al., 2015). The traditional best treatment choices were followed by surgery combined with immunotherapy, radiotherapy, and/or chemotherapy and also less significant herbal therapy. After 5 years of the postsurgery rate of survival decreases as metastasis and recurrence of the malignant cells still continue, which signifies that the primary cause of death is not the tumor itself but the invasion and migration of the mechanism of malignancy

(Huang et al., 2015). The mechanism of invasion and migration of malignancy involves epithelial-mesenchymal transition (EMT). In EMT, epithelial cells change into mesenchymal cells by changing the structure and function of the cell that triggers adhesion and migration (Cunningham et al., 2010).

Lycopene (LP), naturally found in tomatoes, exhibit several functional properties—it has the ability to protect against incurable diseases, antiproliferation against certain blood and colon cancer cells and can further enhance cell cycle arrest on certain tumor cells (Winawer et al., 1996)—and its antiproliferative activity could be of use in cancer treatment if the low stability and availability are not of major concern (Thiery and Sleeman, 2006). A nanoemulsion with LP has been developed to deliver this bioactive competence along with gold nanoparticles (AN). AN is not only a wonderful drug carrier but also can be attached with ligands which is the receptor of the cell and can target specific cell (Mein et al., 2008). This specific cell targeting mechanism seems beneficial but the high dose administration promotes fibroblast cell migration in humans (Chen et al., 2014). This toxic effect can be eliminated by introducing liposomes, some polymeric components, some lipid-based particles, for example, LP-derived substances (Lu et al., 2010). A specific emulsion-based formulation has been created with the oil phase with LP, the water phase is with AN solution, and Tween 80 is used as an emulsifier. The study was conducted in colon cancer cell line (human) HT-29. The efficacy of the formulation has been evaluated AN and LP alone, and the combined effect with AN and LP on the cell lone exhibited the efficacy of the emulsion carrying LP and AN (Lu et al., 2010).

The smaller droplet size of the emulsion showed a stronger effect on the human colon cancer cells (HT-29). The presence of a higher number of advance or proapoptotic and necrotic cells is determined by the increasing volume of the delivered LP in the emulsion droplet. Both the combined application of AN and LP; and the nanoemulsion enhance the levels of advance apoptotic, late apoptotic, and necrotic cell development. In spite of that, the application of nanoemulsion in cancer treatment instigates apoptotic and necrotic cell development. No significant effect by the emulsifier has been observed to the cell except this confers the stability of the nanoemulsion. The application of nanoemulsion with a lower concentration of AN and LP doses induce less differentially expressed proteins such as procaspases 3 and 8, and tumor marker protein Bcl-2 and can also eliminate the expression of Bax and PARP-1 along with initiation of apoptotic on cells. In another similar experiment, fivefold lower migration capability of HT-29 cells when compared to control, the markers showed significantly reversed through upregulation of E-cadherin and just the opposite regulation of Akt, nuclear kappa factor B, pro-matrix metalloproteinase (MMP)-2 and MMP-9 (Huang et al., 2015).

3.2 Ovarian cancer therapy and nanoemulsion

Platinum (Pt) is used widely in the vast majority of cancer therapy as a chemotherapeutic agent combined with other compounds. The two important drug molecules carboplatin and cisplatin while containing Pt in their formulation-enhance survival rate better than any other cancer therapeutics (Leu et al., 2012). Pt compounds can form both intrastrand and interstrand cross-links and thus shatter the structure of cellular DNA (Zhao et al., 2013).

The major problem that arises with the application of Pt is that it all demolishes healthy cells along with cancer cells. Furthermore, cancer cells can transform into highly Pt resistant by developing a resistant mechanism like advocating membrane pump functionality, or recruiting enzymes, or DNA renovating mechanism. Therefore, the foremost answer required for cancer treatment is whether the ovarian tumor is Pt-sensitive or resistant (Neijt et al., 1991).

The evolution of nanomedicine granted the design of formulation to eliminate toxicity and resistance of Pt-sensitivity or resistance. It was difficult to address this issue as Pt is a highly lipophilic metal (Jamieson and Lippard, 1999). However, nanoemulsion is able to deliver Pt-combined drugs that are capable of carrying large amounts of hydrophobic compounds and also allow the adherence of specific ligands to their surface that grant their targeted delivery. A Pt-based drug myrisplatin and one proapoptotic substance C6-ceramide have been delivered by encapsulating in nanoemulsion. This emulsion was designed with a surface ligand peptide (EGER) and one imaging agent gadolinium and applied on ovarian cancer dell lines SKOV3, A2780, and A2780CP (Neijt et al., 1991).

The cytotoxicity study of cisplatin revealed that the cells (SKOV3) those expressing epidermal growth factor receptor (EGFR) also were resistant to the drug at a certain concentration (IC50, of 18uM). Also targeted and nontargeted delivery of encapsulated myrisplatin exhibited an increased level of cytotoxicity and cytotoxicity of targeted nanoemulsion was twice more toxic than nontargeted emulsion formulation. But when ceramide is encapsulated along with the combination, cytotoxic level changes which confirm its positive synergistic functionality. The combination of ceramide and myrisplatin encapsulation showed 0.5-fold positive effect than cisplatin alone. Nanoemulsion with vitamin E with paclitaxel was able to modulate the expression of bac and BCL-2 protein those are connected to drug resistance in cancer cells and also suppress P-gp transport (Zheng et al., 2016). The same team of researchers also assessed the same drug-resistance into ovarian cancer cell line A2780. The drug Taxol decreased proliferation effect and mitochondrial potential. The authors declared their claim that a combination of anticancer therapeutics with vitamin E and its derivative has the potential to function in a multifunctional direction and it could be served as an alternative solution to multidrug resistance to cancer treatment (Cronin et al., 2002).

3.3 Prostate cancer therapy and nanoemulsion

Prostate cancer therapy faces the recurring effect of malignancy and patients suffer in an untreatable state. The initiation of malignancy starts from cancer stem cells (CSCs) and tumor-initiating cells (TICs) and these calls are the origins of cancer progression, metastasis, and in developing resistance to drugs (Zheng et al., 2016). The two cancer stem cell markers CD133 and CD4 (expressed proteins) are connected with drug resistance and also play a vital role in proliferating cancer cells after treatment (Jemal et al., 2011). Drug resistance in cancer cells is strongly associated with upregulation of transporters, arousal of antiapoptotic pathways, inhibition of apoptotic mechanism, and responsiveness in DNA damage and recovery processes (Reya et al., 2001).

The obstacle with prostate cancer treatment is that it aims at populations of rapidly growing malignant cells but not cancer stem cells. Also, prostate cancer drug discovery relies on the study with cancer cell lines with a higher number of preclinical trials. In many cases, the cell line study gathering genomic and epigenomic data has no or low similarity with the original tumor. Most of the time now the cell lines (PPT2) are collected from the patient with prostate cancer, with a very small passage number and the cells are immature and the stem-like properties are preserved. These PPT2 cells have genomic properties associated with antiapoptotic signaling and drug resistance and represent an ideal model for CSC-targeted drug delivery studies (Ahmad et al., 2017).

Abraxane is the most frequently used drug for cancer therapy. It is a paclitaxel prodrug, emerged to enhance its solubility with human serum albumin protein-bound nano-formulation. This drug encountered some problems with MDR cancer cells. The new generation cancer drug SBT-1214 is efficient in overcoming the drug-resistance problem. This drug can be efficiently delivered to induce toxicity to cancer cells by combining with DHA and PUFA. Combined delivery with DHA and paclitaxel showed a mild decrease in P-gp and ABC protein transporters. Nanoemulsion with DHA-SBT-1214, phospholipid, and fish oil combined formulation was studied and improved encapsulation has been reported. The theory developed after these studies concluded that nanoemulsion will be effective first on the CSC-initiated cell line, taking over the EPR effect and finally inhibiting apoptosis (Reya et al., 2001).

The collection of patient-derived CSC with PPT2 cells can be applied in the development of therapeutics that can target cells for tumor development. The combined formulation of DHA with SBT-1214 can stay longer duration in blood circulation. The hydrophobic drug can be conjugated and delivered through nanoemulsion effectively. This also allows surface modification (Fig. 3) with PEG that also increases the circulation time, which enhances accumulation. The success of nanoemulsion in drug delivery depends on its payload compared to drug solution (Reya et al., 2001).

3.4 Liver cancer and nanoemulsion

Nanotechnology is an important means for the delivery and targeting of therapeutics in hepatocellular carcinoma (Depalo et al., 2017; Lamprecht, 2015; Hu et al., 2013). Nanoparticles, micelles, and liposomes have been used for both passively and designed receptor-mediated specific targeting. Nanosized droplets carrying drugs may reach the tumor cells via spongy vascular tissues and exhibit improved site-specific target delivery. Angiogenic related molecules VEGFR (vascular endothelial growth factor), RAF (rapidly accelerated fibrosarcoma), and EGFR (epidermal growth factor receptor) are effective targets in the progress of therapeutics. It has been observed that specific surface ligands and receptors are expressed differentially in hepatocellular carcinoma cells and they are very selected as important targets in the development of selective drugs (Upadhyay et al., 2020) (Table 2).

3.5 Leukemia and nanoemuslion

Biocompatible nanoemulsion the system also has been used to encapsulate to deliver drugs and to enhance the therapeutic effect by eliminating adverse toxic side effects. Lipid-based nanoemulsion has been used by several researchers as a suitable encapsulating and delivery method for cancer therapy. Lipid nanoemulsion is able to bind LDL receptors and the emulsion was designed with the target of concentrating anticancer drugs in tissues with LDL

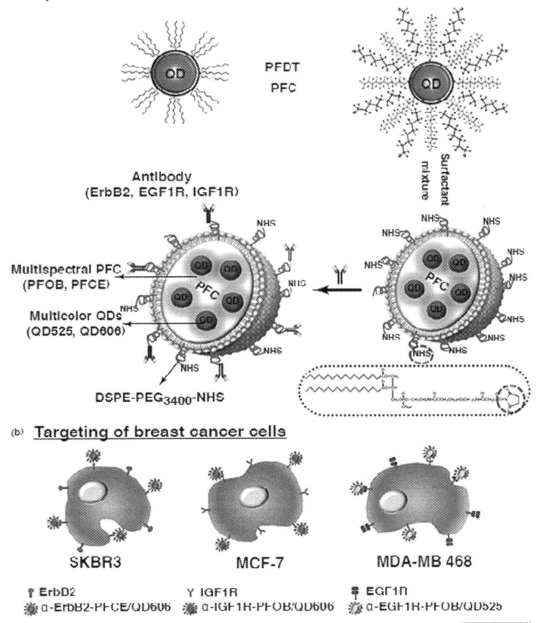

FIG. 3 Schematic model of three types of nanoemulsion: (A) perfluorocarbon (PFC)/quantum dot (AD) and (B) recognition of growth factors by receptors of breast cancer cell (Bae and Chung, 2014).

TABLE 2 Compounds delivered through nanoemulsion for liver cancer treatment.

	Cell lines	Specific functionality	References
Natural			
Curcumin	N/A	Increased bioavailability and found in hepatic region	Ahmed et al. (2012)
Silymarin	HepG2	Bioavailability was high and ROS Condensation was increased in hepatic cells	Ahmad et al. (2018)
Eugenol	HB8065	Showed higher apoptosis (69%) in HB8065 hepatic cell lines	Majeed et al. (2014)
Resveratrol	HepG2 and Rat liver cells	Stopped S and G2/M cell cycle, gown regulated hepatic growth factors	Herneisey et al. (2018)
Crocin	HepG2	Antiproliferative, 59% decrease in telomerase activity	Noureini and Wink (2012)
Gingko biloba	HepG2	Low genotoxicity and cytotoxicity	Tao et al. (2018)
Nigella sativa	HepG2	Upregulated apoptotic marker Genes (p53, bax, caspase-3, and caspase-9) and antiapoptotic Gene (bcl-2) was downregulated	Usmani et al. (2019)
Synthetic			
Sorofenib	MTT tests	Higher efficacy and not affecting normal cells	Cao et al. (2015)
Cisplatin	N/A	Higher amount of drug delivery in cancer cells	Abu-Fayyad and Nazzal (2017)
Gemcitabine	Mice	Gemcitabine chlorambucil Conjugated nanoparticle exhibited Enhanced anticancerous activity	Fan et al. (2015)
Doxorubicin	Swiss albino mice	Showed enhanced efficacy with partial toxicity in heart tissues	Alkhatib et al. (2016)
Vitamin D	Mice	Showed hepatoprotective effect and also decrease TNF alpha level (20%)	El-Sherbiny et al. (2018)
Combined			
Doxoruibcin and gemcitabine	Mice	Combined nanoemulsion administered mice survived longer (32 days) than the control group	Upadhyay et al. (2020)
Curcumin and sorafenib	HepG2	Enhanced cell cytotoxicity and cell apoptosis	Cao et al. (2015)
5-Fluorouracil and oxaliplatin	Caco-2	Enhanced permeability and efficacy	Pangeni et al. (2016)

receptor in tumor tissues. The methotrexate has been encapsulated and evaluated the efficacy in vitro and it has been observed that cellular uptake of nanoemulsion with the drug was higher than the nonencapsulated drug and was able to induce toxicity to cancer cells (Moura et al., 2011).

Chalcones have been encapsulated in nanoemulsion formulation and evaluated both in vitro and in vivo. The authors reported that the formulated nanoemulsion and free chalcones induce apoptosis to the malignant cells in vitro and exhibited an antileukemic effect. However, nonencapsulated chalcones showed more toxicity than nanoemulsion-loaded chalcones and similar observation has been reported in vivo. Nonencapsulated chalcones also reduce weight gain and hepatic injuries confirmed by oxidative stress and inflammatory response (Jones et al., 2008).

3.6 Breast cancer and nanoemulsion

Breast cancer incidence accounts for 11.7% of new cases and 6.7% of breast cancer death in women worldwide (Sung et al., 2021). The available chemotherapeutic agents have limitations as these accumulate in lesser quantity at the tumors rather their increased concentration in other organs leads to toxicities. Several measures have emerged in order to enhance the efficiency of the treatment plan in breast cancer patients. The use of natural compounds to reduce inflammation due to chemotherapy has been developed and nanoemulsion delivery system with these natural compounds could be a suitable strategy for breast cancer management. A nanoemulsion system with essential oil *Nigella sativa* has been developed and it exhibited anticancer functionality in vitro in MCF-7 breast cancer cell lines. This type of nanoemulsion could be beneficial for the entrapment of active natural compounds in order to control breast cancer treatment-related side effects (Moura et al., 2011).

Ceremide encapsulated nanoemulsion had been administered locally on breast cancer tumors. The research aimed at using the encapsulated nanoemulsion to see and apply in localized areas of the tumor and pretumor lesions to observe the systematic adverse outcome. The developed nanoemulsion was loaded with bioadhesive ceramide and the surface of the droplet particles was modified with hydrophobic chitosan. The concentration of C6 necessary to reduce the MCF-7 cell viability to 50% (EC50) and 4.5% fold reduction was observed compared to the nonencapsulated solution (Fig. 3). Further reduction (2.6-fold) was achieved when tributyrin was co-delivered in the oil phase of the nanoemulsion. Intraductal localized administration prolonged drug accumulation for more than 120h in the breast tissues compared to nonencapsulated drug solution. Camptothecin was also encapsulated in nanoemulsion and evaluated the efficacy of the encapsulated formulation in vivo and in vitro compared to the free drug solution (Periasamy et al., 2016).

3.7 Melanoma and nanoemulsion

Melanoma is one of the major types of skin cancer that affects about 80% of skin cancer-related death. The important problem of skin cancer treatment is the low effective rate of the currently available treatment regimen. The currently existing chemotherapeutics receive a relatively slow or incomplete response due to the inherent resistance to this type of cancer cells. Standard chemotherapeutics for late-stage and after the development of metastasis usually exhibit unsatisfied results, which leads to adverse side effects and an increased mortality rate. As a result, a novel combination of skin cancer drugs and formulation for effective delivery systems such as nanoemulsion was being explored.

The association of simvastatin was evaluated in the mice model, revealing that the association of these two drugs in nanoemulsion increased antitumoral efficiency compared to free paclitaxel. The reason of this increased activity is assumed to be the fact that statins increase some LDL receptor protein expression and these protein receptors are in charge of internalization of the lipid-based nanoemulsion (Migotto et al., 2018). Other researchers encapsulated 7-ketocholesterol which is a cholesterol derivative inside the core of the nanoemulsions and evaluated them in a melanoma cell line. More than 50% size reduction, expansion of necrotic lesions, reduction of intratumoral vascular tissue has been reported. The report of 10% death of melanoma cells has also been reported by the administration of a single dose of cholesterol encapsulated nanoemulsion (Natesan et al., 2017).

Monge-Fuentes and his team also followed a different approach where acai oil was encapsulated in nanoemulsion and the nanoemulsion were used as a photosensitizer in both in vitro and in vivo the study, normal (NIH/3te) and melanoma (B16F10) cell lines were treated with photosensitized nanoemulsion and 85% melanoma cell death was reported, while the normal cells were unaffected and could maintain the viability after treatment. Administration of acai oil nanoemulsion on malignant tumor-bearing mice exhibited a significant reduction (82%) of tumor size (Kretzer et al., 2016).

Paclitaxel is used as a multicancer therapeutic agent to treat all sorts of cancer. It can effectively take part in the breakdown mechanism of microtubules during the cell cycle and can inhibit cellular apoptosis, arrest mitotic division and inhibit total cellular functionality. PTX is a hydrophobic drug and exhibits low solubility in water. PTX is co-delivered with Cremophore-EL and ethanol, though Cremophore exhibits toxicity. To solve this problem, hyaluronic acid has been investigated as a co-delivery compound. Hyaluronic acid is negatively charged and has an affinity for highly expressed tumor protein markers (CD44) (Journo-Gershfeld et al., 2012). Two hyaluronic acid and PTX complexed nanoemulsion (HPNs) formulation has been developed and evaluated against molecular marker (CD44) in a nonsmall lung carcinoma cell line (NCI-H460).

HPN complexed nanoemulsion demonstrated superior physiochemical emulsion properties. The half-life of the particles and the zeta potential supported the stability of the emulsion formulation, low polydispersity index support the homogeneity of the emulsion droplet and the desired globular distribution.

Efficacy evaluation of both formulation HPN and HA alone showed the reduced size of the tumor confirming enhanced efficacy. No significant change has been observed in body weight loss in a treated individual with PTX nanoemulsion and combined formulation nanoemulsion (HPN), but both confirmed less toxic effect for healthy tissues (Liebmann et al., 1993). Curcuminoid extract has been encapsulated in nanoemulsion and anticancer activity has been evaluated in lung cancer cell lines to study the mechanism of inhibition action. The cell cycle was arrested in the lung cancer cell line at the G2/M phase in both curcuminoid extract and in encapsulated nanoemulsion, though the author declared pathway differentiation in both treatments. Among two cell lines, H460 demonstrated higher susceptibility to cell division compared to A549 for both treatments (Matsubara et al., 2000).

4 Theragonostic application of nanoemulsion

Detection of cancer in an early stage was problematic and challenging. The multimodal application of nanotechnology has opened up new dimensions toward imaging the microenvironment of the cellular cancer lesions, where persistent dispersion of hydrophobic dyes can be easily administered in the oil phase of a nanoemulsion. Higher accumulation of nanoparticles can be achieved by passive targeting (Choudhury et al., 2019). The incorporation of hydrophobic contrast dyes such as fluorescent or dark quencher in nanoemulsions clears detection specificity by improving the sensitivity of the dyes. For example, a gold nano-spere coated perflurohexane core produces a photoacoustic signal when observed through very low laser fluence (Wei et al., 2014) within medical safety limits. The surface dynamics of nanoemulsion generated nonlinear increased contrast in signaling by forming microbubbles (Wei et al., 2014). The same technology has been used to construct nanoemulsion with 19F probes (Lim et al., 2010) perfluorocarbon based nanoemulsion for magnetic resonance imaging. PFC/perfluoropolyether is hydrophobic compounds that were ligated with p-anisyl fluorinated b-diketones and exposed to a higher amount of fluorine content that improved image quality.

In another application, nanoemulsion was able to improve the quality of the image by quantum dots. Quantum dots of indium-gallium phosphide (InGaP)/zinc oxide and 19F in perluorodecalin were combined in a nanoemulsion and formed a dual-mode imaging nano-probe. This fabricated nanoemulsion probe was able to identify phagocytotic and nonphagocytotic immune cells. Application of this technology can diagnose the early stage of cancer development as it allows to manipulate magnetic field for fluorescence detection visualization (Hingorani et al., 2020). Target-specific cellular imaging was made possible with nanoemulsion technology while PFC was introduced to display trans-activator of transcription component on the surface of engineered T (chimeric antigen receptor—CAR) cells. This also facilitates the surface-specific ligand binding and allows nanoprobe in improving image diagnosis. In a similar way, monoclonal antibodies were ligated onto the emulsion droplet surface of cadmium-selenide (cdSe)/zinc sulfide (ZnS) quantum dot/PFC nanoemulsions to target breast cancer cells (Bae and Chung, 2014).

5 Future prospects

The major advantage of nanoemulsion lies in the co-delivery of hydrophobic and hydrophilic molecules in order to facilitate a variety of targets by surface modification.

The primary challenge in future development is to pursue finding effective co-delivery of efficient therapeutics to improve functionality and find differences from other formulations. In this effort, the major points to keep in mind are the interaction of the drug with other ingredients of the whole system, the impact of processing technology, and the stability of the target drug. Along with the development, the reaction of the nanoemulsion with target cells and exploring different ways of drug release and cellular uptake and circulation.

Different routes of emulsion administration for nanoemulsions loaded with chemotherapeutics can be investigated. The core focus should be on new insights on nano-formulation, developing new opportunities for anticancer agent delivery. The development of novel imaging agents in nanoemulsion systems to improve the diagnostic technology even at a very early stage of cancer development. This will open a wide range of scope to early detection molecular probe technology that can be commensurate with imaging, as it can provide real-time monitoring of cancer patients with minimum pain and invasive diagnostic procedure. Traditional imaging technology also involves radioactive isotope or a fluorophore. This needs more improvement to reduce the adverse effects and spread of malignancy.

The current development can also focus on the delivery of antiinflammatory natural compounds for cancer patients that can reduce inflammation due to chemotherapy and soothe the recovery after chemotherapy.

6 Conclusion

Nanoemulsions constitute a novel and propitious strategy in the therapeutic application of anticancer drugs. The encapsulation of lipid-soluble drugs inside the hydrophobic center allows a solution to the delivery problems in anticancer therapeutics. The addition of an emulsifier and GRAS compounds allows the construction of a stable and guarded route of drug delivery. The small size of the particle allows longer the retention and circulatory cycle in the metabolic system.

The superiority of nanoemulsion formulation compared to other drug delivery strategies is that surface modification and design facilitated to target malignant cells and that way effect of MDR can be avoided. This significant contribution has a remarkable development in cancer treatment, as its vital issue was the fact that other drugs cause cellular normal cell toxicity and cancer cells develop resistance to the drugs. The application of passive targeting has also a superior effect, specifically in tumor tissues. However, active research in targeting might formulate a more positive physiological effect on the formulation. Also, co-encapsulation, ligand binding, surface bonding of other excipient compounds can stop the MDR mechanism.

All these positive outcomes are less worthy if the production process and the metabolic effect are not specifically tested. The approved clinical trials are necessary for every phase of the therapy, all parameters must be scrutinized and scientific clarification has to be investigated to create therapeutic use of such delivery systems. This development in cancer therapy might result in various factors of physiological response that can cause unwanted deaths and no risk viable therapy has been established till today.

References

Abu-Fayyad, A., Nazzal, S., 2017. Gemcitabine-vitamin E conjugates: synthesis, characterization, entrapment into nanoemulsions, and in-vitro deamination and antitumor activity. Int. J. Pharm. 528 (1–2), 463–470.

Ahmad, G., El Sadda, R., Botchkina, G., Ojima, I., Egan, J., Amiji, M., 2017. Nanoemulsion formulation of a novel taxoid DHA-SBT-1214 inhibits prostate cancer stem cell-induced tumor growth. Cancer Lett. 406, 71–80.

Ahmad, U., Akhtar, J., Singh, S.P., Badruddeen, Ahmad, F.J., Siddiqui, S., Wahajuddin, 2018. Silymarin nanoemulsion against human hepatocellular carcinoma: development and optimization. Artif. Cells Nanomed. Biotechnol. 46 (2), 231–241.

Ahmed, K., Li, Y., McClements, D.J., Xiao, H., 2012. Nanoemulsion- and emulsion-based delivery systems for curcumin: encapsulation and release properties. Food Chem. 132 (2), 799–807.

Al-Abd, A.M., Lee, S.H., Kim, S.H., Cha, J.H., Park, T.G., Lee, S.J., Kuh, H.J., 2009. Penetration and efficacy of VEGF siRNA using polyelectrolyte complex micelles in a human solid tumor model in-vitro. J. Control. Release 137 (2), 130–135.

Alkhatib, M.H., Alkreathy, H.M., Balamash, K.S., Abdu, F., 2016. Antitumor activity of doxorubicine-loaded nanoemulsion against Ehrlich ascites carcinoma-bearing mice. Trop. J. Pharm. Res. 15 (5), 937–943.

Almeida, C.P., Vital, C.G., Contente, T.C., Maria, D.A., Maranhão, R.C., 2010. Modification of composition of a nanoemulsion with different cholesteryl ester molecular species: effects on stability, peroxidation, and cell uptake. Int. J. Nanomedicine 5, 679.

Arruebo, M., Valladares, M., González-Fernández, Á., 2009. Antibody-conjugated nanoparticles for biomedical applications. J. Nanomater. 2009, 439389.

Bae, P.K., Chung, B.H., 2014. Multiplexed detection of various breast cancer cells by perfluorocarbon/quantum dot nanoemulsions conjugated with antibodies. Nano Converg. 1 (1), 1–8.

Bertrand, N., Wu, J., Xu, X., Kamaly, N., Farokhzad, O.C., 2014. Cancer nanotechnology: the impact of passive and active targeting in the era of modern cancer biology. Adv. Drug Deliv. Rev. 66, 2–25.

Cao, H., Wang, Y., He, X., Zhang, Z., Yin, Q., Chen, Y., et al., 2015. Codelivery of sorafenib and curcumin by directed self-assembled nanoparticles enhances therapeutic effect on hepatocellular carcinoma. Mol. Pharm. 12 (3), 922–931.

Chanamai, R., McClements, D.J., 2000. Impact of weighting agents and sucrose on gravitational separation of beverage emulsions. J. Agric. Food Chem. 48 (11), 5561–5565.

Chang, E., Bu, J., Ding, L., Lou, J.W., Valic, M.S., Cheng, M.H., et al., 2021. Porphyrin-lipid stabilized paclitaxel nanoemulsion for combined photodynamic therapy and chemotherapy. J. Nanobiotechnol. 19 (1), 1–15.

Chen, Y.J., Inbaraj, B.S., Pu, Y.S., Chen, B.H., 2014. Development of lycopene micelle and lycopene chylomicron and a comparison of bioavailability. Nanotechnology 25 (15), 155102.

Choudhury, H., Pandey, M., Yin, T.H., Kaur, T., Jia, G.W., Tan, S.L., et al., 2019. Rising horizon in circumventing multidrug resistance in chemotherapy with nanotechnology. Mater. Sci. Eng. C 101, 596–613.

Cronin, M.T.D., Dearden, J.C., Duffy, J.C., Edwards, R., Manga, N., Worth, A.P., Worgan, A.D.P., 2002. The importance of hydrophobicity and electrophilicity descriptors in mechanistically-based QSARs for toxicological endpoints. SAR QSAR Environ. Res. 13 (1), 167–176.

Cunningham, D., Atkin, W., Lenz, H.J., Lynch, H.T., Minsky, B., Nordlinger, B., Starling, N., 2010. Colorectal cancer. Lancet 20, 1030–1047.

Dean, M., 2009. ABC transporters, drug resistance, and cancer stem cells. J. Mammary Gland Biol. Neoplasia 14 (1), 3–9.

Depalo, N., Iacobazzi, R.M., Valente, G., Arduino, I., Villa, S., Canepa, F., et al., 2017. Sorafenib delivery nanoplatform based on superparamagnetic iron oxide nanoparticles magnetically targets hepatocellular carcinoma. Nano Res. 10 (7), 2431–2448.

Elnakat, H., Ratnam, M., 2004. Distribution, functionality and gene regulation of folate receptor isoforms: implications in targeted therapy. Adv. Drug Deliv. Rev. 56 (8), 1067–1084.

El-Sherbiny, M., Eldosoky, M., El-Shafey, M., Othman, G., Elkattawy, H.A., Bedir, T., Elsherbiny, N.M., 2018. Vitamin D nanoemulsion enhances hepatoprotective effect of conventional vitamin D in rats fed with a high-fat diet. Chem. Biol. Interact. 288, 65–75.

Fan, M., Liang, X., Li, Z., Wang, H., Yang, D., Shi, B., 2015. Chlorambucil gemcitabine conjugate nanomedicine for cancer therapy. Eur. J. Pharm. Sci. 79, 20–26.

Fattal, E., Couvreur, P., Dubernet, C., 2004. "Smart" delivery of antisense oligonucleotides by anionic pH-sensitive liposomes. Adv. Drug Deliv. Rev. 56 (7), 931–946.

Ganta, S., Devalapally, H., Shahiwala, A., Amiji, M., 2008. A review of stimuli-responsive nanocarriers for drug and gene delivery. J. Control. Release 126 (3), 187–204.

Ganta, S., Talekar, M., Singh, A., Coleman, T.P., Amiji, M.M., 2014. Nanoemulsions in translational research—opportunities and challenges in targeted cancer therapy. AAPS PharmSciTech 15 (3), 694–708.

Ganta, S., Singh, A., Rawal, Y., Cacaccio, J., Patel, N.R., Kulkarni, P., et al., 2016. Formulation development of a novel targeted theranostic nanoemulsion of docetaxel to overcome multidrug resistance in ovarian cancer. Drug Deliv. 23 (3), 958–970.

Ge, R., Cao, J., Chi, J., Han, S., Liang, Y., Xu, L., et al., 2019. NIR-guided dendritic nanoplatform for improving antitumor efficacy by combining chemo-phototherapy. Int. J. Nanomedicine 14, 4931.

Gi, H.J., Chen, S.N., Hwang, J.S., Tien, C., Kuo, M.T., 1992. Studies of formation and interface of oil-water microemulsion. Chin. J. Phys. 30 (5), 665–678.

Gu, F.X., Karnik, R., Wang, A.Z., Alexis, F., Levy-Nissenbaum, E., Hong, S., et al., 2007. Targeted nanoparticles for cancer therapy. Nano Today 2 (3), 14–21.

Herneisey, M., Mejia, G., Pradhan, G., Dussor, G., Price, T., Janjic, J., 2018. Resveratrol nanoemulsions target inflammatory macrophages to prevent. J. Pain 19 (3), S75–S76.

Hingorani, D.V., Chapelin, F., Stares, E., Adams, S.R., Okada, H., Ahrens, E.T., 2020. Cell penetrating peptide functionalized perfluorocarbon nanoemulsions for targeted cell labeling and enhanced fluorine-19 MRI detection. Magn. Reson. Med. 83 (3), 974–987.

Hu, C.M.J., Zhang, L., 2012. Nanoparticle-based combination therapy toward overcoming drug resistance in cancer. Biochem. Pharmacol. 83 (8), 1104–1111.

Hu, X., Hu, J., Tian, J., Ge, Z., Zhang, G., Luo, K., Liu, S., 2013. Polyprodrug amphiphiles: hierarchical assemblies for shape-regulated cellular internalization, trafficking, and drug delivery. J. Am. Chem. Soc. 135 (46), 17617–17629.

Huang, R.F.S., Wei, Y.J., Inbaraj, B.S., Chen, B.H., 2015. Inhibition of colon cancer cell growth by nanoemulsion carrying gold nanoparticles and lycopene. Int. J. Nanomedicine 10, 2823.

Huang, C., Zhang, Z., Guo, Q., Zhang, L., Fan, F., Qin, Y., et al., 2019. A dual-model imaging theragnostic system based on mesoporous silica nanoparticles for enhanced cancer phototherapy. Adv. Healthc. Mater. 8 (19), 1900840.

Ichihara, H., Okumura, M., Tsujimura, K., Matsumoto, Y., 2018. Theranostics with hybrid liposomes in an orthotopic graft model mice of breast cancer. Anticancer Res. 38 (10), 5645–5654.

Jain, R.K., 2005. Normalization of tumor vasculature: an emerging concept in antiangiogenic therapy. Science 307 (5706), 58–62.

Jamieson, E.R., Lippard, S.J., 1999. Structure, recognition, and processing of cisplatin–DNA adducts. Chem. Rev. 99 (9), 2467–2498.

Jemal, A., Bray, F., Center, M.M., Ferlay, J., Ward, E., Forman, D., 2011. Global cancer statistics. CA Cancer J. Clin. 61 (2), 69–90.

Jones, R.J., Hawkins, R.E., Eatock, M.M., Ferry, D.R., Eskens, F.A., Wilke, H., Evans, T.J., 2008. A phase II open-label study of DHA-paclitaxel (Taxoprexin) by 2-h intravenous infusion in previously untreated patients with locally advanced or metastatic gastric or oesophageal adenocarcinoma. Cancer Chemother. Pharmacol. 61 (3), 435–441.

Journo-Gershfeld, G., Kapp, D., Shamay, Y., Kopeček, J., David, A., 2012. Hyaluronan oligomers-HPMA copolymer conjugates for targeting paclitaxel to CD44-overexpressing ovarian carcinoma. Pharm. Res. 29 (4), 1121–1133.

Keefe, A.D., Pai, S., Ellington, A., 2010. Aptamers as therapeutics. Nat. Rev. Drug Discov. 9 (7), 537–550.

Kretzer, I.F., Maria, D.A., Guido, M.C., Contente, T.C., Maranhão, R.C., 2016. Simvastatin increases the antineoplastic actions of paclitaxel carried in lipid nanoemulsions in melanoma-bearing mice. Int. J. Nanomedicine 11, 885.

Kumar, G.P., Divya, A., 2015. Nanoemulsion based targeting in cancer therapeutics. Med. Chem. 5 (5), 272–284.

Lamprecht, A., 2015. Nanomedicines in gastroenterology and hepatology. Nat. Rev. Gastroenterol. Hepatol. 12 (4), 195–204.

Leu, J.G., Chen, S.A., Chen, H.M., Wu, W.M., Hung, C.F., Yao, Y.D., et al., 2012. The effects of gold nanoparticles in wound healing with antioxidant epigallocatechin gallate and α-lipoic acid. Nanomedicine 8 (5), 767–775.

Liebmann, J., Cook, J., Mitchell, J., 1993. Cremophor EL, solvent for paclitaxel, and toxicity. Lancet 342 (8884), 1428.

Lim, Y.T., Cho, M.Y., Kang, J.H., Noh, Y.W., Cho, J.H., Hong, K.S., et al., 2010. Perfluorodecalin/[InGaP/ZnS quantum dots] nanoemulsions as 19F MR/optical imaging nanoprobes for the labeling of phagocytic and nonphagocytic immune cells. Biomaterials 31 (18), 4964–4971.

Lin, A., Slack, N., Ahmad, A., Koltover, I., George, C., Samuel, C., Safinta, C., 2000. Structure and structure—function studies of lipid/plasmid DNA complexes. J. Drug Target. 8 (1), 13–27.

Low, P.S., Antony, A.C., 2004. Folate receptor-targeted drugs for cancer and inflammatory diseases. Adv. Drug Deliv. Rev. 56 (8), 1055–1058.

Lu, W., Zhang, G., Zhang, R., Flores, L.G., Huang, Q., Gelovani, J.G., Li, C., 2010. Tumor site–specific silencing of NF-κ B p65 by targeted hollow gold nanosphere–mediated photothermal transfection. Cancer Res. 70 (8), 3177–3188.

Maeda, H., Wu, J., Sawa, T., Matsumura, Y., Hori, K., 2000. Tumor vascular permeability and the EPR effect in macromolecular therapeutics: a review. J. Control. Release 65 (1–2), 271–284.

Majeed, H., Antoniou, J., Fang, Z., 2014. Apoptotic effects of eugenol-loaded nanoemulsions in human colon and liver cancer cell lines. Asian Pac. J. Cancer Prev. 15 (21), 9159–9164.

Matsubara, Y., Katoh, S., Taniguchi, H., Oka, M., Kadota, J., Kohno, S., 2000. Expression of CD44 variants in lung cancer and its relationship to hyaluronan binding. J. Int. Med. Res. 28 (2), 78–90.

McClements, D.J., 2011. Edible nanoemulsions: fabrication, properties, and functional performance. Soft Matter 7 (6), 2297–2316.

McClements, D.J., 2012. Nanoemulsions versus microemulsions: terminology, differences, and similarities. Soft Matter 8 (6), 1719–1729.

Mein, J.R., Lian, F., Wang, X.D., 2008. Biological activity of lycopene metabolites: implications for cancer prevention. Nutr. Rev. 66 (12), 667–683.

Mekonnen, T.W., Birhan, Y.S., Andrgie, A.T., Hanurry, E.Y., Darge, H.F., Chou, H.Y., et al., 2019. Encapsulation of gadolinium ferrite nanoparticle in generation 4.5 poly (amidoamine) dendrimer for cancer theranostics applications using low frequency alternating magnetic field. Colloids Surf. B: Biointerfaces 184, 110531.

Meng, L., Xia, X., Yang, Y., Ye, J., Dong, W., Ma, P., et al., 2016. Co-encapsulation of paclitaxel and baicalein in nanoemulsions to overcome multidrug resistance via oxidative stress augmentation and P-glycoprotein inhibition. Int. J. Pharm. 513 (1–2), 8–16.

Migotto, A., Carvalho, V.F., Salata, G.C., da Silva, F.W., Yan, C.Y.I., Ishida, K., et al., 2018. Multifunctional nanoemulsions for intraductal delivery as a new platform for local treatment of breast cancer. Drug Deliv. 25 (1), 654–667.

Mohammad, I.S., He, W., Yin, L., 2018. Understanding of human ATP binding cassette superfamily and novel multidrug resistance modulators to overcome MDR. Biomed. Pharmacother. 100, 335–348.

Moura, J.A., Valduga, C.J., Tavares, E.R., Kretzer, I.F., Maria, D.A., Maranhão, R.C., 2011. Novel formulation of a methotrexate derivative with a lipid nanoemulsion. Int. J. Nanomedicine 6, 2285.

Natesan, S., Sugumaran, A., Ponnusamy, C., Thiagarajan, V., Palanichamy, R., Kandasamy, R., 2017. Chitosan stabilized camptothecin nanoemulsions: development, evaluation and biodistribution in preclinical breast cancer animal mode. Int. J. Biol. Macromol. 104, 1846–1852.

Neijt, J.P., ten Bokkel Huinink, W.W., Van der Burg, M.E.L., Van Oosterom, A.T., Willemse, P.H.B., Vermorken, J.B., et al., 1991. Long-term survival in ovarian cancer: mature data from The Netherlands Joint Study Group for Ovarian Cancer. Eur. J. Cancer Clin. Oncol. 27 (11), 1367–1372.

Noureini, S.K., Wink, M., 2012. Antiproliferative effects of crocin in HepG2 cells by telomerase inhibition and hTERT down-regulation. Asian Pac. J. Cancer Prev. 13 (5), 2305–2309.

Pangeni, R., Choi, S.W., Jeon, O.C., Byun, Y., Park, J.W., 2016. Multiple nanoemulsion system for an oral combinational delivery of oxaliplatin and 5-fluorouracil: preparation and in vivo evaluation. Int. J. Nanomedicine 11, 6379.

Parker, N., Turk, M.J., Westrick, E., Lewis, J.D., Low, P.S., Leamon, C.P., 2005. Folate receptor expression in carcinomas and normal tissues determined by a quantitative radioligand binding assay. Anal. Biochem. 338 (2), 284–293.

Periasamy, V.S., Athinarayanan, J., Alshatwi, A.A., 2016. Anticancer activity of an ultrasonic nanoemulsion formulation of Nigella sativa L. essential oil on human breast cancer cells. Ultrason. Sonochem. 31, 449–455.

Reya, T., Morrison, S.J., Clarke, M.F., Weissman, I.L., 2001. Stem cells, cancer, and cancer stem cells. Nature 414 (6859), 105–111.

Roberts, S., Andreou, C., Choi, C., Donabedian, P., Jayaraman, M., Pratt, E.C., et al., 2018. Sonophore-enhanced nanoemulsions for optoacoustic imaging of cancer. Chem. Sci. 9 (25), 5646–5657.

Sezgin-Bayindir, Z., Losada-Barreiro, S., Bravo-Díaz, C., Sova, M., Kristl, J., Saso, L., 2021. Nanotechnology-based drug delivery to improve the therapeutic benefits of NRF2 modulators in cancer therapy. Antioxidants 10 (5), 685.

Shen, J., Kim, H.C., Wolfram, J., Mu, C., Zhang, W., Liu, H., et al., 2017. A liposome encapsulated ruthenium polypyridine complex as a theranostic platform for triple-negative breast cancer. Nano Lett. 17 (5), 2913–2920.

Sung, H., Ferlay, J., Siegel, R.L., Laversanne, M., Soerjomataram, I., Jemal, A., Bray, F., 2021. Global cancer statistics 2020: GLOBOCAN estimates of incidence and mortality worldwide for 36 cancers in 185 countries. CA Cancer J. Clin. 71 (3), 209–224.

Tadros, T., Izquierdo, P., Esquena, J., Solans, C., 2004. Formation and stability of nano-emulsions. Adv. Colloid Interf. Sci. 108, 303–318.

Tan, W., Wang, H., Chen, Y., Zhang, X., Zhu, H., Yang, C., et al., 2011. Molecular aptamers for drug delivery. Trends Biotechnol. 29 (12), 634–640.

Tang, J., Fu, Q., Wang, Y., Racette, K., Wang, D., Liu, F., 2013. Vitamin E reverses multidrug resistance in vitro and in vivo. Cancer Lett. 336 (1), 149–157.

Tao, R., Wang, C., Zhang, C., Li, W., Zhou, H., Chen, H., Ye, J., 2018. Characterization, cytotoxicity, and genotoxicity of TiO_2 and folate-coupled chitosan nanoparticles loading polyprenol-based nanoemulsion. Biol. Trace Elem. Res. 184 (1), 60–74.

Thiery, J.P., Sleeman, J.P., 2006. Complex networks orchestrate epithelial–mesenchymal transitions. Nat. Rev. Mol. Cell Biol. 7 (2), 131–142.

Tiwari, S., Tan, Y.M., Amiji, M., 2006. Preparation and in vitro characterization of multifunctional nanoemulsions for simultaneous MR imaging and targeted drug delivery. J. Biomed. Nanotechnol. 2 (3–4), 217–224.

Toub, N., Malvy, C., Fattal, E., Couvreur, P., 2006. Innovative nanotechnologies for the delivery of oligonucleotides and siRNA. Biomed. Pharmacother. 60 (9), 607–620.

Upadhyay, T., Ansari, V.A., Ahmad, U., Sultana, N., Akhtar, J., 2020. Exploring nanoemulsion for liver cancer therapy. Curr. Cancer Ther. Rev. 16 (4), 260–268.

Usmani, A., Mishra, A., Arshad, M., Jafri, A., 2019. Development and evaluation of doxorubicin self nanoemulsifying drug delivery system with Nigella Sativa oil against human hepatocellular carcinoma. Artif. Cells Nanomed. Biotechnol. 47 (1), 933–944.

Wang, C.Y., Huang, L., 1987. pH-sensitive immunoliposomes mediate target-cell-specific delivery and controlled expression of a foreign gene in mouse. Proc. Natl. Acad. Sci. 84 (22), 7851–7855.

Wei, C.W., Lombardo, M., Larson-Smith, K., Pelivanov, I., Perez, C., Xia, J., et al., 2014. Nonlinear contrast enhancement in photoacoustic molecular imaging with gold nanosphere encapsulated nanoemulsions. Appl. Phys. Lett. 104 (3), 033701.

Winawer, S.J., Zauber, A.G., Gerdes, H., O'Brien, M.J., Gottlieb, L.S., Sternberg, S.S., et al., 1996. Risk of colorectal cancer in the families of patients with adenomatous polyps. N. Engl. J. Med. 334 (2), 82–87.

Wooster, T.J., Golding, M., Sanguansri, P., 2008. Impact of oil type on nanoemulsion formation and Ostwald ripening stability. Langmuir 24 (22), 12758–12765.

Xu, C., Gao, F., Wu, J., Niu, S., Li, F., Jin, L., et al., 2019. Biodegradable nanotheranostics with hyperthermia-induced bubble ability for ultrasound imaging-guided chemo-photothermal therapy. Int. J. Nanomedicine 14, 7141.

Yu, P., Yu, H., Guo, C., Cui, Z., Chen, X., Yin, Q., et al., 2015. Reversal of doxorubicin resistance in breast cancer by mitochondria-targeted pH-responsive micelles. Acta Biomater. 14, 115–124.

Zhang, C., Tang, N., Liu, X., Liang, W., Xu, W., Torchilin, V.P., 2006. siRNA-containing liposomes modified with polyarginine effectively silence the targeted gene. J. Control. Release 112 (2), 229–239.

Zhao, C., Feng, Q., Dou, Z., Yuan, W., Sui, C., Zhang, X., Xia, G., Sun, H., Ma, J., 2013. Local targeted therapy of liver metastasis from colon cancer by galactosylated liposome encapsulated with doxorubicin. PLoS One 8 (9), e73860.

Zheng, N., Gao, Y., Ji, H., Wu, L., Qi, X., Liu, X., Tang, J., 2016. Vitamin E derivative-based multifunctional nanoemulsions for overcoming multidrug resistance in cancer. J. Drug Target. 24 (7), 663–669.

Zheng, X.C., Ren, W., Zhang, S., Zhong, T., Duan, X.C., Yin, Y.F., et al., 2018. The theranostic efficiency of tumor-specific, pH-responsive, peptide-modified, liposome-containing paclitaxel and superparamagnetic iron oxide nanoparticles. Int. J. Nanomedicine 13, 1495.

17

Nanomaterials for aging and cosmeceutical applications

*Mh Busra Fauzi[a], Ali Smandri[a], Ibrahim N. Amirrah[a],
Nurkhuzaiah Kamaruzaman[a], Atiqah Salleh[a], Zawani Mazlan[a],
Nusaibah Sallehuddin[a], Izzat Zulkiflee[a], Law Xia Jian[a],
and Fatimah Mohd Nor[b]*

[a]Centre for Tissue Engineering and Regenerative Medicine, Faculty of Medicine, National University of Malaysia, Kuala Lumpur, Malaysia [b]KPJ Ampang Puteri Specialist Hospital, Ampang, Selangor, Malaysia

O U T L I N E

1 Introduction

1.1 Overview

Skin is a multilayered organ that constitutes 15% of the body's mass, barriers the body from the environment, and assists in multiple functions such as temperature regulations, sensations, and protection against external threats (Dąbrowska et al., 2018). Skin characteristics and conditions vary among individuals and depend significantly on many factors, such as site, ethnicity, gender, age, or lifestyle and body mass index (Dąbrowska et al., 2018; Wong et al., 2016). Through life, the skin experiences many changes in shape, performance, and formulation, where the skin becomes thinner, rigid, and less flexible, thereby changing the defensive function against mechanical injuries and other triggers (Pawlaczyk et al., 2013). Changes in aged skin architecture may be represented by a decrease in the number of dermal channels contributing to the presence of more plateau areas, allowing pulling and intensifying of the visible skin lines. Some wrinkle formation causes are correlated with dermis-epidermis thinning, stratum spongiosum thinning, stratum corneum thickening, collagen IV and VII deficiency at the dermal-epidermal junction under the wrinkle base (Wong et al., 2016). Furthermore, many other intrinsic factors (i.e., free radicals formation, hormone decline, DNA damage, and telomere shortening) and extrinsic factors (i.e., smoking, ultraviolet radiations, and lifestyle) can accelerate skin aging (Shanbhag et al., 2019).

After many innovations in the cosmetics field, nanocosmeceutical approaches that incorporate novel cosmetic formulations utilizing nanotechnology have been adopted to improve the efficiency of bioactive components to have better product effectivity by adding small nanoparticles of cosmetic ingredients to the damaged skin to repair (Aziz et al., 2019; Bilal and Iqbal, 2020). The use of nanomaterials in cosmeceuticals is rapidly increasing with better efficiency, active ingredient delivery, enhanced entrapment, and occlusive property than standard methods. Nanocosmeceuticals are applicable in repairing various body parts besides the skin such as nails, hair, and lips, and many cosmetic manufacturers invest heavily in new and improved products with this technology. This chapter discusses many aspects of nanocosmeceuticals such as types of nanoparticles in use, nanocosmeceuticals applications, antiaging mechanism, nanocosmeceuticals toxicity and environmental concerns, current regulations, as well as the prospects and recommendations of nanomaterials in cosmetics.

1.2 Advantages and disadvantages of nanomaterials in cosmetics

Using nanotechnology to carry cosmetic products have many benefits including controlled drug release mechanisms, improving efficacy, having occlusive properties which protect the skin surface and prevent hydration loss, ability to manipulate physical stability, high entrapment capability, and specific site targetability compared to previous conventional cosmetic products (Aziz et al., 2019). Also, the small size enhances skin penetration and allows a larger surface area in more quantity enabling these nanocarriers to be active transport of active drugs in nanocosmeceuticals for both hydrophilic and lipophilic delivery (Kaul et al., 2018). The new technology can replace conventional methods with enhanced delivery systems like liposomes, niosomes, solid lipid nanoparticles (SLNs), cubosomes, nanostructured lipid carriers (NLCs), and nanoparticles such as fullerenes, nanosilver, nanogold, and nanopigments (Srinivas, 2016). Thus, nanomaterials can be used extensively in producing

TABLE 1   List of advantages and disadvantages of nanocosmeceuticals.

Using nanomaterials in cosmeceuticals	
Advantages	**Disadvantages**
Enhanced drug delivery transport	Potentially toxic
Controlled drug release	Reactive small particles
Occlusive characteristics	Potentially harmful to the environment
Specific site targeting	Clinical trials not necessary
High drug loading and entrapment	Clinical safety regulations not standardized
Stable physicality	High risk of exposure

antiaging and antiwrinkle creams, moisturizers, skin whitening products, and hair repair serums, conditioning, and washing.

Despite the many advantages, there are drawbacks regarding toxicology and health risks for consumers, production workers, as well as the environment associated with using nanomaterials for cosmeceutical applications. This is especially concerning as the rate of nanocosmeceuticals rapidly increases pushed forward by their global commercialization potential. Although their small size scale allows efficient drug delivery, it also enables a wide range of exposure risks to consumers and production workers such as inhalation, ingestion, and skin penetration (Dhawan et al., 2020). Besides, these nanomaterials can contaminate water, air, and oil during manufacturing, use, or disposal, which may cause environmental harm, the extent of which remains largely unknown (Santos et al., 2019). Some nanoparticles can easily reach the brain via nasal nerves, while others can potentially enter the bloodstream and be transported to other body organs which may be hazardous (Monsé et al., 2018). The production of nanoparticles may cause expansive production of reactive oxygen species, inflammation, oxidation stress, DNA damage, protein breakdown, and membrane ruptures (Singh et al., 2020). For example, ultrafine nanomaterials like carbon nanotubes, titanium dioxide (TiO_2), fullerenes which are carbon-based, and nanoparticles from copper and silver may be toxic to human cells and tissues in certain doses (Bahadar et al., 2016; Subramaniam et al., 2019). It has been reported that TiO_2 damages DNA, RNA, and fat cells in sunscreens (Tyagi et al., 2016). Also, clinical trials are unnecessary for the approval of nanocosmeceutical products, and there is a lack of stringent laws by regulatory bodies for the regulation, license, and use of nanomaterials in nanocosmeceuticals (Kumud and Sanju, 2018). The advantages and disadvantages of nanomaterials in cosmeceuticals are shown in Table 1.

2 Classifications of nanocosmeceuticals

2.1 Types of nanoparticles used in nanocosmeceuticals

Nanotechnology has been widely used in cosmetic applications and involved in the cosmeceutical industries for over the decades, as there have been increasing skincare product manufacturers that use various methods such as pressurized emulsification to decrease the size of their active ingredients into nanosized to induce skin permeability and enhance their ability to absorbs into the skin (Yapar and Inal, 2012). Nanotechnologies in the cosmetic industry are high in demand as these nanomaterials are applied in various products such as shampoos, conditioners, face creams, moisturizers, toothpaste, soaps, deodorants as well as sunscreens. Nanomaterials in cosmetics could be their active ingredients or act as delivery agents which encapsulate or carry the active ingredients into the desired target (Melo et al., 2015). Fig. 1 shows the different types of nanomaterials that have been used in nanocosmetics. These nanomaterials could be classified as inorganic and organic materials following their physicochemical properties.

Inorganic nanomaterials, generally called hard particles, are composed of any inorganic substances which include metal, metal oxides, and ceramics. Inorganic nanoparticles usually consist of low to none toxicity, hydrophobic, and very stable in comparison with their bulk counterparts (Fytianos et al., 2020). TiO$_2$ and ZnO are widely used inorganic nanoparticles for

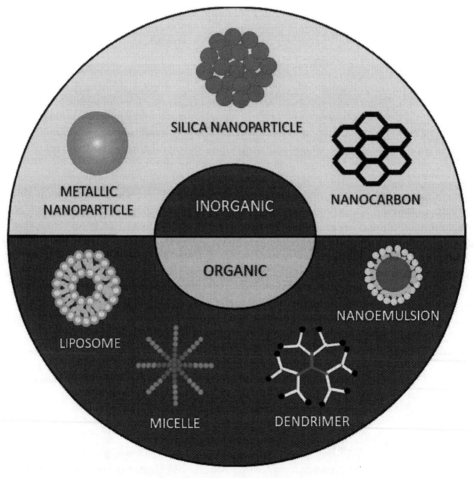

FIG. 1   Different types of nanoparticles used in the cosmeceutical industry.

sunscreen applications. Due to their nanoscale size, the nanoparticles are able to produce a higher sun protection factor (SPF) more effectively and provide a better result for the consumer because of their transparency properties. These nanoparticles also show better Ultraviolet A (UVA) and Ultraviolet B (UVB) absorption ability which is crucial in sunscreen production (Khan et al., 2018). Other examples of inorganic nanomaterials are the uses of silica nanoparticles which are used in a wide range of products such as makeup, nail care, oral care, and hair care products. Silica nanoparticles are favorable in the cosmetic industry due to their unique physicochemical properties and their low production cost. It has been shown that these silica nanoparticles were able to improve the texture and shelf-life of cosmetic products.

Metallic nanoparticles which could be classified under inorganic nanomaterials have also been shown to be beneficial in the cosmeceutical industry as their proven to have antibacterial and antifungal properties. Colloidal metallic nanoparticles such as gold and silver nanoparticles are intensely used in personal hygiene products such as oral and body care as these products can remain stable for a long time. There also have been increasing demands of antimicrobial properties in hygiene products as consumers are starting to be more focused on ensuring safety during a pandemic such as the COVID-19 pandemic. Silver nanoparticles are proven to have mechanisms of action when dealing with a wide range of e bacteria and viruses, which made them favorable in hygiene care products (Salleh et al., 2020). At the same time, gold nanoparticles are often used in skincare products. Cosmeceutical company introduces the use of gold nanoparticles in skincare products as they claimed to affect skin elasticity, and various studies could support the claims (Bilal and Iqbal, 2020). However, the safety of these metallic nanoparticles is still debatable as these nanoparticles are harmful in high doses. Therefore, organic nanomaterials are manufacturers to counteract the disadvantages of inorganic nanomaterials.

Organic nanomaterials are generally called soft particles, are composed of organic materials such as lipids, proteins, or polymers that can be altered structurally through stress and surface contact. There are a few examples of organics nanomaterials which are liposomes, nanoemulsions, micelles, and dendrimers. Liposomes are known as tiny bubbles which consist of a phospholipid bilayer and a hydrophilic core (vesicular bilayer system) (Mishra et al., 2018). These nanomaterials are often used as a carrier in cosmetics products to improve the active ingredient deposition within the skin at the targeted site and thus reduce the systemic absorption and minimize the side effect, therefore providing a long-lasting effect (Bochicchio et al., 2020). Nanoemulsions are emulsion that has nano-droplet ranging from 20 to 300 nm. The nanoemulsions are often in oil-in-water or water-in-oil colloidal dispersions (Aziz et al., 2019). These nanosized droplets can notably influence the permeability of molecules, especially active ingredients that exhibited apolar characteristics.

2.2 Application of nanocosmeceuticals

2.2.1 Nano antiaging

Skin aging is identified by characteristics such as wrinkling, loss of elasticity, laxity, and a rough-textured look. Various antiaging products and procedures are commercially available currently, such as antiaging cream, healthy diet programs, antiaging supplements, hormone replacement therapy, and antiprogeria strategies (Zhang and Duan, 2018). Presently, nanosized synthetic and natural antiaging creams are already available in the market with improved skin protection to moisturize, lift, and whiten skin such as Hydra Zen Cream, Revitalift, and Skin Caviar Ampoules (Ganesan and Choi, 2016). Other than nanosizing, few delivery systems have been used to fabricate nano antiaging creams and lotions to enhance the activity of bioactive compounds to deeper layers of skin. Che Sulaiman et al. synthesized nanoemulsion containing leaf extracts of *Clinacanthus nuthans* Lindau optimized by an artificial neural network aimed to prevent aging. The fabricated product demonstrated nonirritant properties and biocompatible with human skin cells. In vivo ultrasound showed collagen content increased significantly with the application of *Clinacanthus nuthans* Lindau, further consolidating its antiaging property (Che Sulaiman et al., 2017).

2.2.2 Nano moisturizers

Adequate moisturizing is a basic process to improve skin health. Moisturizer forms a thin layer on the surface of the skin maintaining skin moisture. Presently, nanoemulsions, nanoliposomes, solid lipid nanocarriers are successfully used in moisturizer formulations with active phytoingredients to improve skin silkiness. Noor et al. synthesized solid lipid particles carrying virgin coconut oil by ultrasonication of molten stearic acid and virgin coconut oil in an aqueous solution aimed for better dermal distribution. The fabricated product demonstrated better skin penetration by penetration study through rat skin and hence, proven to be one of the promising cosmeceutical carrying scaffolds for improved dermal delivery (Noor et al., 2016).

2.2.3 Nano facial creams

Skin creams with natural bioactive ingredients are the trend nowadays in the cosmetic industry. These natural ingredients such as *Nigella sativa*, *Hypericum perforatum*, and *Aloe vera* possess excellent properties in skincare such as antioxidant, antibacterial and antifungal properties. However, natural extracts may display unpleasant sensory effects due to their formation. To overcome this, Kalouta et al. synthesized nano encapsulated pomegranate and tea tree oil extract via an electrohydrodynamic process. Dynamic mechanical analysis of the creams showed that the creams are suitable for application in the cosmetic industry and their rheological properties were better compared to commercially available products. Pomegranate in the cream softens its consistency while tea tree extract showed a hardening effect (Kalouta et al., 2020).

2.2.4 Nano sunscreen

Continuous skin exposure to UV rays leads to disastrous effects such as premature skin aging and skin cancer. Sunscreens are essential to shield the skin from damaging sun rays. Gollavilli et al. synthesized novel naringin-loaded nanoethosomal sunscreen aimed to achieve sunscreen with enhanced skin penetration, skin retention, and excellent antioxidant property. Organoleptic evaluation, ABTS assay, as well as in vitro and in vivo skin permeation studies showed that the optimized sunscreen creams enhanced naringin retention within the skin with minimum permeation and provide an SPF value of 21.21 ± 0.62 (Gollavilli et al., 2020).

2.2.5 Nanofibrous cosmetic face mask

Nanofibrous face masks, a promising novel cosmetic sector, are formed of fiber mats or membranes capable of delivering bioactive ingredients to the skin. Nanofibrous cosmetic face masks were developed using electrospinning with active compounds which could provide cosmeceutical benefits for the user. Recently, a nanofibrous face mask was created with antiaging, whitening, antioxidant, and antiwrinkling effects. Gold nanoparticles, polyethylene oxide, gelatin, collagen, and orange peel extract were synthesized to form a paper-like thin hydrogel face mask. Releasing study, skin permeation study, radical scavenging activity, a toxicological study confirmed the cosmetic benefits of the product (Manatunga et al., 2020).

2.2.6 Nano perfumes

Ethanol is an essential ingredient for traditional perfumes since it is the solvent of a hydrophobic aromatic compound. However, sensitive-skinned individuals may have skin irritation as ethanol is known to be an irritative solvent. Miastkowska et al. prepared a stable oil-in-water nanoemulsion system as carriers for fragrances, without ethanol. Stable and transparent nano perfumes were achieved with fragrance composition concentration within the 6%–15% range. The achieved result confirmed the protective role of nanoemulsions. The dermatological tests confirmed the safety of developed preparations (Miastkowska et al., 2018).

2.2.7 Nanonail care

Nail polishes are currently available in countless colors and effects such as glittering, matte, and shiny materials. Numerous studies demonstrated the usage of nanoparticles in nail polishes can increase resistance with antibacterial and antifungal properties. Lau et al. used laser ablation to integrate metal nanoparticles directly into the nail polish with no chemical additives, producing a highly pure colloid in the polish. This method also can easily apply bioactive nanoparticles with antibacterial and antifungal effects to any solid surface (Lau et al., 2017).

2.2.8 Nano hair care

Haircare is one of the major cosmetic concerns of the market. Individuals can start to lose hair at a very young age, causing a decrease in self-esteem. Drug delivery into hair follicles is gaining more exposure to hair loss prevention. Główka et al. successfully synthesized encapsulated antiapoptotic roxithromycin-loaded nanoparticles for hair's follicular targeting. Ex vivo penetration study on human scalp skin proved that preferential targeting to the pilosebaceous unit by using polymeric nanoparticles is achievable. Follicular penetration behavior in water suspensions and organogel was significantly improved than that in an oily solution. The particles in the water dispersion able to reach the hair bulb with the dermal papilla suggested that the formulation may be influential in preventing hair loss (Główka et al., 2014).

3 Nanocosmeceuticals mechanisms of action

3.1 Penetration in skincare products

Nanotechnology shows advancement in the area of research and development by increasing the effectiveness of the product through the delivery of creative solutions. To address such limitations associated with conventional goods, the use of nanotechnology is escalating in the field of cosmetics. Cosmeceuticals are known to be the fastest expanding sector of the personal care market and have grown significantly over the years. Cosmeceuticals have high entrapment quality and strong sensory properties and are more durable than conventional or traditional cosmetics. Most nanoparticles are suitable for both lipophilic and hydrophilic drug delivery. Nanocosmeceuticals used in the treatment of skin, hair, nails, and lips for conditions such as wrinkles, photoaging, hyperpigmentation, dandruff, and hair damage have been commonly used.

The skin consists predominantly of three layers: epidermis, dermis, and subcutaneous tissue. The epidermis comprises five distinct cell strata. The stratum corneum (SC), which is the outermost layer of the epidermis, usually consists of 10–15 histologically distinguishable corneocyte layers, each of which is ~1 μm thick (Scheuplein, 1967; Katz and Poulsen, 1971; Anderson and Cassidy, 1973; Holbrook and Odland, 1974). Corneocytes are typically flattened, tightly packed, and metabolically inert cells creating a waterproof barrier. Cell membranes are so closely bound together that there is barely any intercellular space from which polar nonelectrolyte molecules and ions can disperse (Fox et al., 2011).

There are three potential mechanisms of epidermal penetration of active compounds (Ghaffarian and Muro, 2013). There is appendageal (intercellular) penetration through the hair follicle, sebaceous, sweat glands, and transcellular (intracellular) permeation through the corneocytes and intercellular lipid matrix. The latter route makes a direct line across the SC to the lower layers of the epidermis, eventually to the dermis below. It is widely agreed that the appendices (hair follicles and glands) contain just a fractional region for permeation. It concludes that this transcellular pathway is the primary transepidermal pathway (Kim et al., 2020). A molecule that moves through the transcellular path must be partitioned and diffused through corneocytes. Still, to travel between the corneocytes, the molecule must also be partitioned and diffused through approximately 4–20 lipid lamellae in each of these cells. Therefore, the transcellular path involves the negotiation of several hydrophilic and hydrophobic domains (Fig. 2). Thus, the intercellular lipid matrix plays a crucial role in the SC barrier mechanism. As a result, most of the research aimed at improving successful skin permeation focused on the contributions of lipids to barrier function, the modulation of lipid solubility, and the modification of the ordered SC structure.

3.2 Antiaging activity

Currently, cosmetic manufacturers are concentrating their attention on producing groundbreaking natural cosmetics to counter symptoms of skin aging, thereby satisfying the safe appearance and well-being needs of customers. Skin aging is a complex biological process that can be caused by a mix of both endogenous/intrinsic and exogenous/extrinsic factors. For

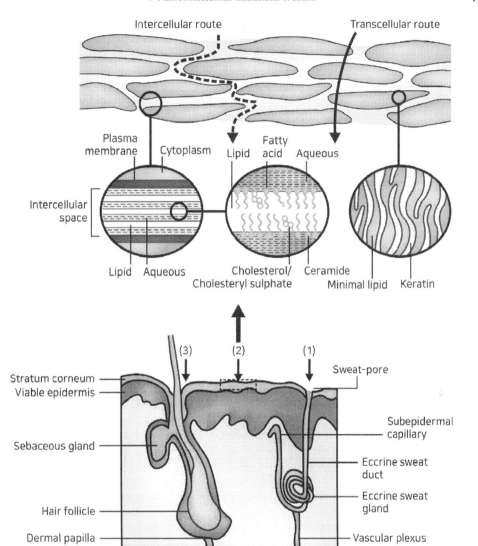

FIG. 2 Permeation pathways of Stratum Corneum model (Kim et al., 2020).

example, oxidative stress has been associated with loss of skin elasticity. The presence of excessive reactive oxygen species (ROS) is the primary cause of oxidative stress (Silva et al., 2017) and leads to the two immediate signs of skin aging, which are wrinkles and sagging. Antiaging products currently in the market have been developed with claims of antiwrinkle and firming, moisturizing, and whitening effects (Wu et al., 2016).

Different approaches are available for the prevention and delay of skin aging. Topical antioxidant supplementation is one of the most effective methods for preventing or treating skin aging (Mohiuddin, 2019). Most of the antioxidants commonly used in skincare formulations, however, have some drawbacks such as excessive skin irritation, chemical instability, and low skin penetration, which actively limit their effectiveness following topical application (Sharadha et al., 2020). Therefore, loading cosmetic antioxidants into suitable nanocarriers delivery systems to improve the efficacy of antiaging products by enhancing the solubility, permeability, and stability of antioxidants can be a strategy to resolve these problems and mechanisms for antiaging activity (Fig. 3). Examples of nanocarriers that are widely explored and studied are liposomes, micelle, nanostructured lipid carriers, and nanoemulsions (Vinardell and Mitjans, 2015).

Several types of nanomaterials possess intrinsic antioxidant properties due to the surface properties of the material itself. Antioxidants interact with free radicals, terminating the adverse chain reactions and converting them to harmless products (Fig. 4). Examples of antioxidant functionalized nanoparticles derived from various biological extracts are silver nanoparticles (AgNPs), gold nanoparticles (AuNPs), copper oxide nanoparticles (Cu_2ONPs), iron nanoparticles (INPs), zinc oxide (ZnONPs), selenium (SeNPs) and nickel oxide nanoparticles (NiONPs) (Valgimigli et al., 2018). These antioxidant nanoparticles have been reported to exhibit potent

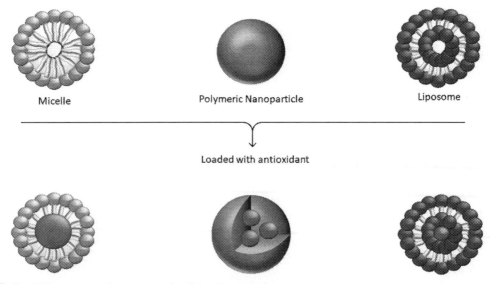

FIG. 3   Different types of nanocarriers loaded with antioxidant.

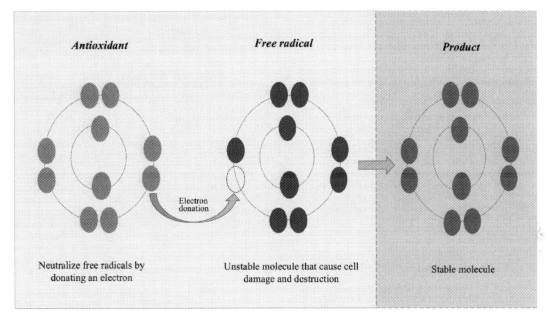

FIG. 4 Antioxidant mechanism of action.

and good antioxidant properties. They are known as preventive and chain-breaking antioxidants. Preventive antioxidants minimize the formation of initiating radicals, whereas chain-breaking antioxidants slow down the propagative phase by taking out the free radicals from the reaction (Valgimigli et al., 2018).

To sum up, by acting as carriers and antioxidants, nanomaterials have been widely used in cosmetic formulations as antiaging. By displaying superior bioavailability, higher durability, and the probability of meeting specific targets, these materials can theoretically overcome several weaknesses of small-molecule antioxidants.

4 Toxicity of nanoparticles for cosmeceuticals

4.1 Nanocosmeceuticals products toxicity

The revolution of nanotechnology throughout the years has led to an increase in demand, production, and application in many areas, particularly in the cosmeceuticals industry. Despite the many substantial potentials and unique benefits that they offer, several alarming health and safety issues have become a significant concern regarding the possible dangers of these nanomaterials toward the consumers, workforce, and environment. There are risks and uncertainties of these nanoparticles which are in need of a greater scientific understanding.

The toxicity of these nanocosmeceuticals products immensely depends on nanoparticles' chemical composition, particle size, surface properties (i.e., surface coating), and aggregation ability which can be engineered during the process of manufacturing. This simple explanation can picture the correlation between the toxicity of these particles with its size: the smaller the nanometers dimensions, the greater the ratio of surface area to volume. Hence, it will cause greater biological and chemical reactivity. Nanoparticles with low solubility have been proved to be cancerous and may manifest more prominent toxicity (Kaul et al., 2018). In vitro studies showed that free radicals which were produced by titanium dioxide nanoparticles (TiO_2) caused cell toxicity when exposed to UV rays. Other studies demonstrated that cobalt-chromium nanoparticles could destroy human fibroblast across a cellular barrier (Lohani et al., 2018). These nanoparticles could cause a bizarre health hazard, which includes organ toxicity, lung and skin damage, and even harm unborn fetus (Lohani et al., 2018).

4.2 Postapplication adverse events

4.2.1 Postapplication exposure through skin penetration

The human skin consists of three layers: the epidermis, dermis, and hypodermis, as the largest organ of the body it acts as a significant barrier toward contaminants and harmful foreign particles. However, the presence of sweat glands and hair follicles facilitates and enables the penetration of these nanomaterials into the skin barrier. AgNPs is one of the most utilized nanoparticles in the cosmetics industries as it possesses antimicrobial properties (Burduşel et al., 2018). Have been reported that AgNPs toxicity was mediated by inducing oxidative stress which is associated with low viability, mitochondrial activity inhibition, and the initiation of cell death and apoptosis (Sajid et al., 2015). A study revealed that sunscreens, which consist of zinc oxide nanoparticles, can penetrate through the barrier when applied to healthy skin and can be observed in the urine and blood (Lohani et al., 2018).

4.2.2 Respiratory system postapplication exposure through inhalation

Inhalation is the primary route for airborne nanoparticles exposure. The production workforce and product consumers may inhale the nanomaterials while producing or applying the cosmeceuticals product, which is likely to be in the form of aerosol or mist. The nanoparticles accumulation in the respiratory system depends on the interaction between the respiratory epithelium membrane. Upon inhalation, nanoparticles can penetrate the lungs hence interact with the epithelium which may cause inflammation, and more profound penetration into the interstitium may cause chronic effects and will gain access to the lymph nodes (Sajid et al., 2015). These nanoparticles travel to the brain through the nasal nerves (transsynaptic transport) which then enter the nervous system, in conjunction with their nanometers dimensions these particles are also able to access the bloodstream hence gaining access to the various organ via the blood flow (Lohani et al., 2018). Nanoparticles with sizes of less than 10 nm were reported to prolonged eschar formation, erythema, and edema (Kaul et al., 2018).

4.2.3 Digestive system postapplication exposure through ingestion

Ingestion of nanoparticles into the digestive system is usually transferred from hand to mouth either intentionally or unintentionally. These nanomaterials can also be ingested through the application of cosmeceuticals product which is commonly used on the lips or mouth such as lipstick, lip mask, lip balms, and other cosmetic products that incorporate nanoparticles. Numbers of commercial cosmeceuticals products contain copper nanoparticles, in vivo this material exhibits toxicological effects and injuries to the internal organs of the mice when exposed to the copper particles (Kaul et al., 2018). When ingested, TiO_2 nanoparticles will damage the cell membranes of the digestive gland via the mechanism of oxidative stress and are also able to induce nasal pathology. When the cells are exposed, they eventually become injured, hence damaging the mucous membranes. Thus, the ability to smell and the humidification of the nasal are reduced (Sajid et al., 2015). Silver nanoparticles were widely used in cosmeceuticals as antimicrobial agents. However, the same silver concentrations, which were found lethal for bacteria, were reported to be lethal for both keratinocytes and fibroblasts (Kaul et al., 2018).

4.3 Environmental concerns

Environmental factors may also alter the toxicity of nanoparticles. The weather conditions, which include geographical latitude, humidity, wind flow rate, nature of light, and temperature, may drive certain properties of nanoparticles to increase their toxicity. Nanoparticles may be dispersed at a greater rate in higher temperatures, compared to lower or room temperature. Nanoparticles also are known to behave differently under UV light compared to visible lights. Nanoparticles with small dimensions penetrate easily through animal tissue and plant in a high wind speed environment (Sajid et al., 2015).

The aquatic habitat is likely to be polluted with nanoparticles as the use of nanoparticles in cosmetics products such as sunscreen, moisturizer, lotion, and even deodorants are immensely used and applied. The most common nanoparticle contaminations are ZnO and TiO_2 as they are commonly found in sunscreen. A study was done on zebrafish using these nanoparticles and found that the long-term exposure of these particles led to significant damage of different organs as the gills, brain, and liver growth rates were affected. ZnO nanoparticles were also found to be harmful to algae either in bulk or the form of small particles while the nanoparticles of CuO and TiO_2 were found to be much more damaging for algae species (Sajid et al., 2015). In another study, AgNPs were found to be extremely lethal to aquatic life, an analysis of zebrafish liver tissue was done after AgNps exposure, and the result indicates various cellular changes which include disruption of hepatic cells and apoptotic changes on the tissue (Sajid et al., 2015).

5 Safety assessment of nanomaterials in cosmetic industry

Safety assessment of nanomaterials is very important as they are very small in size and can easily penetrate through the skin, inhaled and swollen, thus, may be toxic to the users. European Union and United States have come out with their own regulation for the usage of nanomaterials in the cosmetic industry.

5.1 European Union

In the EU, the Cosmetics Regulation (EC) No 1223/2009 has prohibited the use of animals in cosmetic testing. In addition, the regulation also requires that the list of nanomaterials used in cosmetic products be made available to the consumers. More recently, the Guidance on the Safety Assessment of Nanomaterials in Cosmetics (SCCS/1611/19) updated the safety assessment needed for nanomaterials used in cosmetics. In July 2017, the EC published a list of nanomaterials used in cosmetic products in the EU. The list was last updated in December 2019 (European Commission, 2021). Starting from 2020, the European Union Observatory for Nanomaterials (EUON) requires that nanomaterials used in cosmetics should have a REACH (Registration, Evaluation, Authorization, and Restriction of Chemicals) registration compliant to assessment the risk to human health and the environment. The nanomaterials must be registered to be used legally in the EU (Fytianos et al., 2020).

5.2 United States

In the United States, the use of nanomaterials in cosmetics is regulated by the Guidance for Industry: Safety of Nanomaterials in Cosmetic Products. According to the guideline, the safety assessment of nanomaterials should include physicochemical properties, size, morphology, density, solubility, porosity, stability, agglomeration, and impurities. In addition, the potential route of exposure, e.g., dermal, ingestion and inhalation, should be identified and the potential toxicity should be evaluated in vitro and in vivo. Furthermore, data on skin penetration, potential inhalation, genotoxicity, and skin and eye irritation should be made available (FDA, 2011).

6 Future perspective and recommendations

Using nanomaterials in cosmetics has escalated in interest to increase the efficacy of products using innovative technology. Cosmeceuticals are personal care products that combine biologically active components bringing remedial benefits on the surfaces applied. Nanocosmeceutical involves developing or manipulating nanomaterials, which are small particles in the size range of 1 to 100 nm, for cosmeceutical solutions (Nahar and Sarker, 2017). Nanomaterials in cosmeceutical products are used worldwide to treat aging, pigmentation, and damage to skin, body, and hair. Cosmeceuticals are the fastest growing division of the personal care industry with a drastic rise of interest over the years. Products with nanotechnology have risen almost 516% within a mere 7 years between 2006 and 2013 (Ajazzuddin et al., 2015). The sale of nanomaterial-containing cosmeceutical products is projected to reach above US$55.3 billion by 2022 (Shokri, 2017). The rapid increase of nanocosmeceutical production gives growing concerns regarding potential toxic effects on consumers, manufacturers, and the environment. Issues also apply to the lack of stringent scrutiny and clinical trials for production and marketing approval in regulatory bodies.

Because cosmeceuticals are in between pharmaceuticals and personal care products, the Federal Food, Drug and Cosmetics Act (FDA) does not recognize the term "cosmeceuticals," and products enjoy the functional benefits without stringent requirements for approval as they do for drugs (US Food and Drug Administration, 2020). Although cosmeceuticals can alter skin physiological processes, clinical trials are avoided by manufacturers with specific claims to prevent being subjected to lengthy and expensive approval processes by the FDA. The issue of nanomaterial safety in consumer goods has been raised by many organizations, including the World Health Organization (WHO), political institutions, and nongovernmental agencies (Fytianos et al., 2020). Most countries have independent regulations and definitions of controlled and approved ingredients in cosmetics, some more strict than others regarding the use of nanomaterials, but they are not globally standardized (Melo et al., 2015). In some cases, an additional category is created to include items that are in the boundary between medical drugs and personal care products (Raj and Chandrul, 2016).

The US Environmental Protection Agency (EPA) has focused research on several types of nanomaterials that had reports of toxicity, including TiO_2, nanotubes, silver nanoparticles, fullerenes, zero-valent, and fullerenes (US Environmental Protection Agency Federal Facilities Restoration and Reuse Office, 2017). The European Commission (EC) Scientific Committee of Consumer Safety (SCCS) has published updated guidance on the safety assessment of nanomaterials in cosmetics in 2019 (SCCS, 2019). Also, the FDA produced an independent guideline for applying nanotechnology in industrial products (US Department of Health and Human Services Food and Drug Administration Guidance for Industry, 2017) and has collaborated with various research centers to improve regulatory standards for nanotechnology products which encompass nanomedical drugs, biologics, devices, food, and cosmetics. Currently, there are about 54 FDA-approved nanoparticles which may or may not be included in the nanocosmeceutical categories (Anselmo and Mitragotri, 2016; Choi and Han, 2018). The European Observatory for Nanomaterials (EUON) has started regulating toxicology data which may not involve animal testing, as well as standardized compliance from all companies that develop, use, or import nanoforms to have REACH registration application from 2020 (Fytianos et al., 2020). REACH stands for Registration, Evaluation, Authorization, and Restriction of Chemicals as a regulation method for the European Union (EU) (Catalogue of cosmetic ingredients—European Observatory for Nanomaterials. Understanding REACH—ECHA, 2021) whereby any nanomaterials that are categorized by REACH but not registered are deemed illegal.

There is no doubt that nanotechnology will grow rapidly in cosmetics and personal care products in the near future. Current researchers focus on solubility, biocompatibility, and bio-effectiveness carefully for product enhancement. The rising demand for nanocosmeceuticals necessitates concentrating on the health and safety of consumers and the environment. The accountability falls on both the manufacturer and the regulatory bodies to answer raising concerns. As such, increased transparency is called. More stringent laws are recommended regarding using nanomaterials specifically for cosmeceuticals, including a standardized global definition of nanomaterials. Mandatory reports regarding the nanoparticles used in cosmetics should include physicochemical characterization, toxicology report, assess the presence of impurities, and the pharmacodynamic such as distribution, absorption, metabolism, and excretion. The toxicity and pharmacology and toxicology profile should be made available along with the risk estimation of products. A standardized testing method that is continually improved is suggested for nanomaterial-loaded cosmeceuticals with reported biological and change in characteristic properties. Information and data such as proof of product safety as well as instructions of use should be prepared by manufacturers and made available to the consumers.

The significant impact of nanotechnology is undeniable. It is useful in almost every field of industry and commercialization in biology, engineering, chemistry, physics including cosmetics, devices, and medical preparations. The benefits of this revolutionary technological development in cosmetics outweigh the negative aspects, mainly due to its diversified market, with products emerging from small and major manufacturers worldwide. This pushes forward the rise of using nanomaterials in cosmeceuticals which offer excellent technical and economic aspirations. However, questions and caution are commended for the future involving manufacturers and regulatory bodies to ensure that there will not be prolonged damaging effects to consumers as well as the environment.

References

Ajazzuddin, M., Jeswani, G., Jha, A., 2015. Nanocosmetics: past, present and future trends. Recent Pat. Nanomed. 5, 3–11. https://doi.org/10.2174/1877912305666150417232826.

Anderson, R.L., Cassidy, J.M., 1973. Variations in physical dimensions and chemical composition of human stratum corneum. J. Invest. Dermatol. 61, 30–32. https://doi.org/10.1111/1523-1747.ep12674117.

Anselmo, A.C., Mitragotri, S., 2016. Nanoparticles in the clinic. Bioeng. Transl. Med. 1, 10–29. https://doi.org/10.1002/btm2.10003.

Aziz, Z.A.A., Mohd-Nasir, H., Ahmad, A., Siti, S.H., Peng, W.L., Chuo, S.C., Khatoon, A., Umar, K., Yaqoob, A.A., Mohamad Ibrahim, M.N., 2019. Role of nanotechnology for design and development of cosmeceutical: application in makeup and skin care. Front. Chem. 7, 1–15. https://doi.org/10.3389/fchem.2019.00739.

Bahadar, H., Maqbool, F., Niaz, K., Abdollahi, M., 2016. Toxicity of nanoparticles and an overview of current experimental models. Iran. Biomed. J. 20, 1–11. https://doi.org/10.7508/ibj.2016.01.001.

Bilal, M., Iqbal, H.M.N., 2020. New insights on unique features and role of nanostructured materials in cosmetics. Cosmetics 7, 24. https://doi.org/10.3390/cosmetics7020024.

Bochicchio, S., Dalmoro, A., De Simone, V., Bertoncin, P., Lamberti, G., Barba, A.A., 2020. Simil-microfluidic nanotechnology in manufacturing of liposomes as hydrophobic antioxidants skin release systemsc. Cosmetics 7. https://doi.org/10.3390/COSMETICS7020022.

Burduşel, A.C., Gherasim, O., Grumezescu, A.M., Mogoantă, L., Ficai, A., Andronescu, E., 2018. Biomedical applications of silver nanoparticles: an up-to-date overview. Nanomaterials 8, 1–25. https://doi.org/10.3390/nano8090681.

Anon., 2021. Catalogue of cosmetic ingredients—European Observatory for Nanomaterials. Understanding REACH—ECHA.

Che Sulaiman, I.S., Basri, M., Fard Masoumi, H.R., Ashari, S.E., Basri, H., Ismail, M., 2017. Predicting the optimum compositions of a transdermal nanoemulsion system containing an extract of Clinacanthus nutans leaves (L.) for skin antiaging by artificial neural network model. J. Chemom. 31, 1–13. https://doi.org/10.1002/cem.2894.

Choi, Y.H., Han, H.K., 2018. Nanomedicines: current status and future perspectives in aspect of drug delivery and pharmacokinetics. J. Pharm. Investig. 48, 43–60. https://doi.org/10.1007/s40005-017-0370-4.

Dąbrowska, A.K., Spano, F., Derler, S., Adlhart, C., Spencer, N.D., Rossi, R.M., 2018. The relationship between skin function, barrier properties, and body-dependent factors. Skin Res. Technol. 24, 165–174. https://doi.org/10.1111/srt.12424.

Dhawan, S., Sharma, P., Nanda, S., 2020. Cosmetic nanoformulations and their intended use. Nanocosmetics, 141–169. https://doi.org/10.1016/b978-0-12-822286-7.00017-6.

European Commission, 2021. Catalogue of Nanomaterials Used in Cosmetic Products Placed on the EU Market. [Internet]. Available from: https://euon.echa.europa.eu/catalogue-of-cosmetic-ingredients.

FDA, 2011. Final Guidance for Industry—Safety of Nanomaterials in Cosmetic Products. [Internet]. Available from: https://www.fda.gov/media/83957/download.

Fox, L.T., Gerber, M., Du Plessis, J., Hamman, J.H., 2011. Transdermal drug delivery enhancement by compounds of natural origin. Molecules 16, 10507–10540. https://doi.org/10.3390/molecules161210507.

Fytianos, G., Rahdar, A., Kyzas, G.Z., 2020. Nanomaterials in cosmetics: recent updates. Nanomaterials 10, 1–16. https://doi.org/10.3390/nano10050979.

Ganesan, P., Choi, D.K., 2016. Current application of phytocompound-based nanocosmeceuticals for beauty and skin therapy. Int. J. Nanomedicine 11, 1987–2007. https://doi.org/10.2147/IJN.S104701.

Ghaffarian, R., Muro, S., 2013. Models and methods to evaluate transport of drug delivery systems across cellular barriers. J. Vis. Exp., 1–13. https://doi.org/10.3791/50638.

Główka, E., Wosicka-Frąckowiak, H., Hyla, K., Stefanowska, J., Jastrzębska, K., Klapiszewski, Ł., Jesionowski, T., Cal, K., 2014. Polymeric nanoparticles-embedded organogel for roxithromycin delivery to hair follicles. Eur. J. Pharm. Biopharm. 88, 75–84. https://doi.org/10.1016/j.ejpb.2014.06.019.

Gollavilli, H., Hegde, A.R., Managuli, R.S., Bhaskar, K.V., Dengale, S.J., Reddy, M.S., Kalthur, G., Mutalik, S., 2020. Naringin nano-ethosomal novel sunscreen creams: development and performance evaluation. Colloids Surf. B: Biointerfaces 193, 111122. https://doi.org/10.1016/j.colsurfb.2020.111122.

Holbrook, K.A., Odland, G.F., 1974. Regional differences in the thickness (cell layers) of the human stratum corneum: an ultrastructural analysis. J. Invest. Dermatol. 62, 415–422. https://doi.org/10.1111/1523-1747.ep12701670.

Kalouta, K., Eleni, P., Boukouvalas, C., Vassilatou, K., Krokida, M., 2020. Dynamic mechanical analysis of novel cosmeceutical facial creams containing nano-encapsulated natural plant and fruit extracts. J. Cosmet. Dermatol. 19, 1146–1154. https://doi.org/10.1111/jocd.13133.

Katz, M., Poulsen, B.J., 1971. Absorption of drugs through the skin. In: Brodie, B.B., Gillette, J.R., Ackerman, H.S. (Eds.), Concepts in Biochemical Pharmacology. Handbuch der Experimentellen Pharmakologie/Handbook of Experimental Pharmacology. 28/1. Springer, Berlin, Heidelberg, pp. 103–174, https://doi.org/10.1007/978-3-642-65052-9_7.

Kaul, S., Gulati, N., Verma, D., Mukherjee, S., Nagaich, U., 2018. Role of nanotechnology in cosmeceuticals: a review of recent advances. J. Pharm. 2018, 1–19. https://doi.org/10.1155/2018/3420204.

Khan, M.Z., Baheti, V., Ashraf, M., Hussain, T., Ali, A., Javid, A., Rehman, A., 2018. Development of UV protective, superhydrophobic and antibacterial textiles using ZnO and TiO_2 nanoparticles. Fibers Polym. 19, 1647–1654. https://doi.org/10.1007/s12221-018-7935-3.

Kim, B., Cho, H.-E., Moon, S.H., Ahn, H.-J., Bae, S., Cho, H.-D., An, S., 2020. Transdermal delivery systems in cosmetics. Biomed. Dermatol. 4, 1–12. https://doi.org/10.1186/s41702-020-0058-7.

Kumud, M., Sanju, N., 2018. Nanotechnology driven cosmetic products: commercial and regulatory milestones. Appl. Clin. Res. Clin. Trials Regul. Aff. 5, 112–121. https://doi.org/10.2174/2213476x05666180530093111.

Lau, M., Waag, F., Barcikowski, S., 2017. Direct integration of laser-generated nanoparticles into transparent nail polish: the plasmonic "goldfinger". Ind. Eng. Chem. Res. 56, 3291–3296. https://doi.org/10.1021/acs.iecr.7b00039.

Lohani, A., Verma, A., Joshi, H., Yadav, N., Karki, N., 2018. Nanotechnology-based cosmeceuticals. ISRN Dermatol. 2014, 843687.

Manatunga, D.C., Godakanda, V.U., Herath, H.M.L.P.B., de Silva, R.M., Yeh, C.-Y., Chen, J.-Y., Akshitha de Silva, A.A., Rajapaksha, S., Nilmini, R., Nalin de Silva, K.M., 2020. Nanofibrous cosmetic face mask for transdermal delivery of nano gold: synthesis, characterization, release and zebra fish employed toxicity studies. R. Soc. Open Sci. 7, 201266. https://doi.org/10.1098/rsos.201266.

Melo, A., Amadeu, M.S., Lancellotti, M., De Hollanda, L.M., Machado, D., 2015. The role of nanomaterials in cosmetics: national and international legislative aspects. Quim Nova 38, 599–603. https://doi.org/10.5935/0100-4042.20150042.

Miastkowska, M., Lasoń, E., Sikora, E., Wolińska-Kennard, K., 2018. Preparation and characterization of water-based nano-perfumes. Nanomaterials 8. https://doi.org/10.3390/nano8120981.

Mishra, H., Chauhan, V., Kumar, K., Teotia, D., 2018. A comprehensive review on liposomes: a novel drug delivery system. J. Drug Deliv. Ther. 8, 400–404. https://doi.org/10.22270/jddt.v8i6.2071.

Mohiuddin, A.K., 2019. Skin care creams: formulation and use. Am. J. Dermatol. Res. Rev. 2, 1–45.

Monsé, C., Hagemeyer, O., Raulf, M., Jettkant, B., van Kampen, V., Kendzia, B., Gering, V., Kappert, G., Weiss, T., Ulrich, N., et al., 2018. Concentration-dependent systemic response after inhalation of nano-sized zinc oxide particles in human volunteers. Part. Fibre Toxicol. 15, 1–11. https://doi.org/10.1186/s12989-018-0246-4.

Nahar, L., Sarker, S.D., 2017. Importance of nanotechnology in drug delivery. Open Access J. Pharm. Res. 1. https://doi.org/10.23880/oajpr-16000130.

Noor, N.M., Khan, A.A., Hasham, R., Talib, A., Sarmidi, M.R., Aziz, R., Abd, A., 2016. Empty nano and microstructured lipid carriers of virgin coconut oil for skin moisturisation. IET Nanobiotechnol. 10, 195–199. https://doi.org/10.1049/iet-nbt.2015.0041.

Pawlaczyk, M., Lelonkiewicz, M., Wieczorowski, M., 2013. Age-dependent biomechanical properties of the skin. Postep. Dermatol. Alergol. 30, 302–306. https://doi.org/10.5114/pdia.2013.38359.

Raj, R.K., Chandrul, K.K., 2016. Regulatory requirements for cosmetics in relation with regulatory authorities in India against US, Europe, Australia and Asean countries. Int. J. Pharma Res. Health Sci. 4, 1332–1341. https://doi.org/10.21276/ijprhs.2016.05.01.

Sajid, M., Ilyas, M., Basheer, C., Tariq, M., Daud, M., Baig, N., Shehzad, F., 2015. Impact of nanoparticles on human and environment: review of toxicity factors, exposures, control strategies, and future prospects. Environ. Sci. Pollut. Res. 22, 4122–4143. https://doi.org/10.1007/s11356-014-3994-1.

Salleh, A., Naomi, R., Utami, N.D., Mohammad, A.W., Mahmoudi, E., Mustafa, N., Fauzi, M.B., 2020. The potential of silver nanoparticles for antiviral and antibacterial applications: a mechanism of action. Nanomaterials 10, 1–20. https://doi.org/10.3390/nano10081566.

Santos, A.C., Morais, F., Simões, A., Pereira, I., Sequeira, J.A.D., Pereira-Silva, M., Veiga, F., Ribeiro, A., 2019. Nanotechnology for the development of new cosmetic formulations. Expert Opin. Drug Deliv. 16, 313–330. https://doi.org/10.1080/17425247.2019.1585426.

SCCS, 2019. Guidance on the Safety Assessment of Nanomaterials in Cosmetics. SCCS, Brussels, Belgium.

Scheuplein, R.J., 1967. Mechanism of percutaneous absorption. II. Transient diffusion and the relative importance of various routes of skin penetration. J. Invest. Dermatol. 48, 79–88. https://doi.org/10.1038/jid.1967.11.

Shanbhag, S., Nayak, A., Narayan, R., Nayak, U.Y., 2019. Anti-aging and sunscreens: paradigm shift in cosmetics. Adv. Pharm. Bull. 9, 348.

Sharadha, M., Gowda, D.V., Vishal Gupta, N., Akhila, A.R., 2020. An overview on topical drug delivery system—updated review. Int. J. Res. 11, 368–385.

Shokri, J., 2017. Nanocosmetics: benefits and risks. BioImpacts 7, 207–208. https://doi.org/10.15171/bi.2017.24.

Silva, S.A.M., Michniak-Kohn, B., Leonardi, G.R., 2017. An overview about oxidation in clinical practice of skin aging. An. Bras. Dermatol. 92, 367–374.

Singh, R., Cheng, S., Singh, S., 2020. Oxidative stress-mediated genotoxic effect of zinc oxide nanoparticles on *Deinococcus radiodurans*. 3 Biotech 10, 1–13. https://doi.org/10.1007/s13205-020-2054-4.

Srinivas, K., 2016. The current role of nanomaterials in cosmetics. J. Chem. Pharm. Res. 8, 906–914.

Subramaniam, V.D., Prasad, S.V., Banerjee, A., Gopinath, M., Murugesan, R., Marotta, F., Sun, X.F., Pathak, S., 2019. Health hazards of nanoparticles: understanding the toxicity mechanism of nanosized ZnO in cosmetic products. Drug Chem. Toxicol. 42, 84–93. https://doi.org/10.1080/01480545.2018.1491987.

Tyagi, N., Srivastava, S.K., Arora, S., Omar, Y., Ijaz, Z.M., Al-Ghadhban, A., Deshmukh, S.K., Carter, J.E., Singh, A.P., Singh, S., 2016. Comparative analysis of the relative potential of silver, zinc-oxide and titanium-dioxide nanoparticles against UVB-induced DNA damage for the prevention of skin carcinogenesis. Cancer Lett. 383, 53–61. https://doi.org/10.1016/j.canlet.2016.09.026.

US Department of Health and Human Services Food and Drug Administration Guidance for Industry, 2017. Drug Products, Including Biological Products, That Contain Nanomaterials—Guidance for Industry. Food Drug Administration, https://doi.org/10.1002/jgm.

US Environmental Protection Agency Federal Facilities Restoration and Reuse Office, 2017. EPA Technical Fact Sheet—Nanomaterials., p. 9.

US Food & Drug Administration, 2020. Is It a Cosmetic, a Drug, or Both? (Or Is It Soap?). FDA.

Valgimigli, L., Baschieri, A., Amorati, R., 2018. Antioxidant activity of nanomaterials. J. Mater. Chem. B 6, 2036–2051. https://doi.org/10.1039/c8tb00107c.

Vinardell, M.P., Mitjans, M., 2015. Nanocarriers for delivery of antioxidants on the skin. Cosmetics 2, 342–354. https://doi.org/10.3390/cosmetics2040342.

Wong, R., Geyer, S., Weninger, W., Guimberteau, J.C., Wong, J.K., 2016. The dynamic anatomy and patterning of skin. Exp. Dermatol. 25, 92–98. https://doi.org/10.1111/exd.12832.

Wu, Y., Choi, M.H., Li, J., Yang, H., Shin, H.J., 2016. Mushroom cosmetics: the present and future. Cosmetics 3, 1–13. https://doi.org/10.3390/cosmetics3030022.

Yapar, E.A., Inal, Ö., 2012. Nanomaterials and cosmetics. J. Pharm. Istanbul Univ. 42, 43–70. https://doi.org/10.2991/978-94-6239-012-6_18.

Zhang, S., Duan, E., 2018. Fighting against skin aging: the way from bench to bedside. Cell Transplant. 27, 729–738. https://doi.org/10.1177/0963689717725755.

Nano-formulations in drug delivery

Melissa Garcia-Carrasco[a], Itzel F. Parra-Aguilar[a],
Erick P. Gutiérrez-Grijalva[b], Angel Licea-Claverie[c],
and J. Basilio Heredia[a]

[a]Research Center for Food and Development (CIAD), Nutraceuticals and Functional Foods
Laboratory, Culiacán, Sinaloa, Mexico [b]CATEDRAS CONACYT-Research Center for Food and
Development (CIAD), Nutraceuticals and Functional Foods Laboratory, Culiacán, Sinaloa,
Mexico [c]TNM, Tijuana Technological Institute, Graduate Studies and Chemistry Research
Center, Tijuana, Baja California, México

OUTLINE

1 Nanotechnology in nano-formulations in drug delivery

1.1 Smart delivery systems

The delivery systems used in the different treatments until a few years ago brought with them a series of deficiencies such as a low drug solubility, the appearance of different adverse effects with the treatment, the accumulation of the drug in areas where it is not needed, as well as being excreted without performing its function (Alvarez-Lorenzo and Concheiro, 2014). For many years, different delivery systems have been investigated with nanotechnology (nanomedicine), which has opened a new door to the controlled release of drugs. These systems have been improved with the help of new technologies based on the needs of the population. A clear example is the decrease of the adverse effects in a treated patient with a chronic degenerative disease such as cancer. That is why the synthesis of new release systems helps to a large extent to have better bioavailability, reduce adverse effects, obtain a prolonged effect of the drug, and greater specificity in the target organ where this drug is released (Khan et al., 2020). Within the different release systems, some can change their conformation when there are certain stimuli or variations in the environment, such as changes in pH, temperature, light, and radiation, which have made these systems in some way called "smart" (Liu et al., 2016b). Only at the site where the stimulus is received will the drug be released. The main characteristic of these systems is that they have a size <300 nm to pass through the different tissues (Patra et al., 2018). However, macro-scale systems can also exist, which can be synthesized by different organic and inorganic components.

1.2 Organic and inorganic nanoparticles-polymers

These nanoparticles (NPs) can be classified depending on the different materials they are composed of (Fig. 1). The organic NPs can be synthesized based on polymers, while others are based on carbon, such as nanotubes and carbon dots. On the other hand, inorganic NPs are synthesized based on metals such as gold and silver and mesoporous silica (Lombardo et al., 2019).

1.2.1 Polymers

Polymeric systems such as liposomes, micelles, or dendrimers, have been used for drug delivery at the nanoscale and have been approved by the Food and Drug Administration (FDA) (Khan et al., 2020). These systems have been characterized by having high compatibility and having a large drug storage capacity. Due to that versatility, different morphologies and chemical compositions of the polymers can be obtained under controlled conditions. Most of these systems are made up of hydrophilic/hydrophobic (amphiphilic) systems (Lombardo et al., 2019).

1.2.2 Nanoparticles

Metallic nanoparticles (NPs) have been investigated in clinical and preclinical studies to diagnose and detect different diseases. Gold (AuNPs) and silver (AgNPs) NPs have gained great interest because they have certain electronic and chemical properties, which can be given different types of applications. AuNPs and AgNPs have been specially studied since

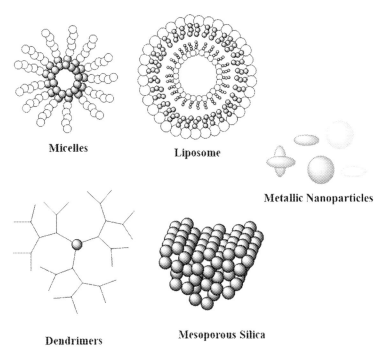

Micelles   **Liposome**

Metallic Nanoparticles

Dendrimers   **Mesoporous Silica**

FIG. 1   Different morphologies of the most commonly used organic and inorganic nanocarriers.

they present remarkable photoactivity irradiated with ultraviolet-visible light (UV-Vis). This activity varies depending on the nanoparticles' size and morphology, determining the chemical, physical, optical, electronic, and catalytic properties that the nanoparticles will have. The shape of nanoparticles normally tends to be spherical but can be obtained as rods, and tubes, contributing to their modified properties (Liang et al., 2014).

Gold nanoparticles can confine resonant photons in their periphery, which in some way induce a coherent oscillation of the band of electrons present on their surface, showing a surface plasmon, which confers two important properties. The first is that the confinement of the photon in the dimensions of the nanoparticles leads to an increase in the electromagnetic field, which helps the radiation properties such as absorption and radiation to be improved, so there is an improvement in the image resolution of these nanoparticles (Yang et al., 2019b). The second is related to the light that is strongly absorbed by these nanoparticles, followed by a rapid phase shift of the electronic movement, which causes a large energy transfer, giving its photothermic properties. These two properties give a great application to these nanoparticles since they can be used as a diagnosis and theranostic therapy against cancer (Elzoghby et al., 2016).

1.2.3 Silice mesoporosa

Mesoporous silica is a material in which its pore diameters vary between 2 and 50 nm. In recent decades this material has been studied for medical applications. The pore's uniformity and the large surface area also present stability to temperature, pH changes, and mechanical stress (Lombardo et al., 2019). This is one of the best-known release systems; these can be differentiated depending on the different properties that concern the texture (surface area, pore size, and volume), mesostructures (messy, hexagonal, and cubic), and its morphology (spherical, nanofibers, and nanotubes) (Chen et al., 2016b).

2 Morphologies and their properties in drug delivery

For many years, different types of morphologies have been investigated for the release of drugs in a controlled way, these nano-formulations in some cases can be stimulated with external factors that help them to release the content within them, such as infrared irradiation, alternating the magnetic field and in some cases even vibrations.

2.1 Polymers

2.1.1 Micelles

Polymeric micelles are nanoparticles that have an amphiphilic property, which is attributed to the type of polymers that is developed, having a hydrophobic polymer at one end and a hydrophilic one at the other, which causes their self-assembly when are in the right environment (Zhou et al., 2018). Those features are the main reasons which these micelles have gained high relevance in recent years. Besides their simple morphology and easy preparation, they increase the drug's solubility while lowering their toxicity, reducing the number of adverse effects; size can vary from 15 to 300 nm, where 300 nm is the maximum value to synthesize the micelles since it is expected they can pass through the cracks of the tissues and thus reach their objective (Wang et al., 2015a; Andrade et al., 2021; Su et al., 2021).

The combination of different polymers for the micelles synthesis also causes variations in the properties of these nanoparticles since when using a cationic polymer would cause the system to modify its morphology due to changes in the pH of the medium in which it is found as the micelles blocks are of the electrostatic type (Bazban-Shotorbani et al., 2017). Therefore, this is one reason this type of system would have to be administered intravenously (IV) since variations in the digestive system's pH could cause the system to lose stability. Some of the polymers that are used are more frequently in the synthesis of these systems is poly (ethylene glycol) (PEG) (Biswas et al., 2016), and this gives greater solubility and stability to the system (Ahmed et al., 2021). Furthermore, it has been found that it is biocompatible. The FDA already approves some systems of this type. Table 1 shows some of them.

2.1.2 Liposomes

Liposomes are synthetic structures containing a phospholipid bilayer, which can vary depending on the application to be given to the produced liposome, mostly synthesized from cholesterol. Due to the similarity that these have with extracellular vesicles, the interest in this type of nano-formulations for different drug loading has increased (Johnsen et al., 2018). The chemical properties that its components give it, such as its hydrophobicity, its size, and responding to physical and chemical stimuli such as changes in pH or light irradiation. In a wide variety of work done with these systems, it has been found that the ideal size for

TABLE 1 Micelle nano-formulations clinical approved and clinical trials (Zhou et al., 2018).

Name	Phase	Drug
Genexol-PM	II–IV	Paclitaxel
NK105	III	Paclitaxel
NC-6004	I–III	Cisplatin
NK012	I and II	7-Ethyl-10-hydroxy-camptothecin
SPI1049C	III	Doxorubicin
BIND-014	I and II	Docetaxel
NC-4016	I	Oxaliplatin
NanoxelM	II and III	Docetaxel
NK911	II	Doxorubicin

the release of different anticancer, antiinflammatory, antifungal drugs and some genes are around 100 nm or less (Liu et al., 2016b). It should be noted that this type of system was the first to be approved for commercial use and that, like most micelles, it contained PEG, thus helping it remain in the bloodstream for a longer time (Liu et al., 2016b; Zylberberg and Matosevic, 2016). Some of the response stimuli investigated with the different systems are thermosensitivity, magnetic variations, and sensitivity to pH changes and enzymatic reactions (Lamichhane et al., 2018) (Table 2).

TABLE 2 Liposomal nano-formulations clinical approved (Alavi et al., 2017; Johnsen et al., 2018; Lamichhane et al., 2018).

Name	Phase	Drug
Doxil	IV	Doxorubicin
Amphotec	IV	Amphotericin B
DaunoXome	IV	Daunorubicin
AmBisome	IV	Amphotericin B
Depocyt	IV	Cytarabine
Myocet	IV	Doxorubicin
Visudyne	IV	Verteporfin
Lipusu	IV	Paclitaxel
MARQIBO KIT	IV	Vincristine
ELA-MAX	IV	Lidocaine
Epaxal	IV	Hepatitis-A Vaccine

2.1.3 Dendrimers

Dendrimers are nanometric-scale structures, which start from a center and grow in a well-defined branched and symmetrical way. These can be conjugated to different chemical species and detect different analytes; they can also act as contrast agents, among other applications. This variety of applications is due to its structure, since depending on the type of agents with which it is synthesized, it can present hydrophilic or hydrophobic cavities. Therefore, dendrimers can encapsulate different types of drugs or molecules and expand their application window in different areas. In addition to having a very low dispersity, the variation in these formulations' sizes is very small (Parajapati et al., 2016; Wang et al., 2016).

It has been found that dendrimers have a high solubility in water and biocompatibility, as compared to other polymeric systems, emphasizing the use of biodegradable compounds for their synthesis and avoiding the accumulation in some organs (Huang and Wu, 2018). The terminals that this type of nano-formulation presents can be easily functionalized with different molecules, increasing their possible applications. So far, only some dendrimer systems are in clinical studies, such as Starpharma Holding Limited (Melbourne, Australia) and Starpharma/AstraZeneca. Table 3 shows the systems that are being studied with dendrimers.

2.2 Metallic nanoparticles

Metallic nanoparticles (MNPs) are synthesized only by metallic precursors, and these will present different properties depending on the size and shape that they present; their synthesis is given by noble metals that provide them great chemical stability, good conductivity, catalytic properties, and the possibility of conjugating with biomolecules (Baranwal et al., 2016). These commonly used metals are copper, gold, and silver, where the latter shows antimicrobial properties (Amini, 2019). In recent years, NPs have gained the great interest for their application in chemotherapy as they can be covered with a polymeric coating (Qiu et al., 2018). Therefore, they can perform a dual activity, thus being able to store a greater amount of drug to be released; one of the mechanisms by which this release occurs is by infrared radiation, which makes these NPs radiate heat, breaking the shell in which are wrapped releasing the drug at a specific site (McNamara and Tofail, 2016).

NPs physical characteristics, such as its magnetic field, can attract different particles or drugs and help as contrast agents due to the same molecules' density (McNamara and Tofail, 2016). Based on all of these properties, NPs are very useful in different biological, biomedical,

TABLE 3 Dendrimers nano-formulation clinical approved and trials (Mignani et al., 2018).

Name	Phase	Drug
VivaGel	IV	Antiviral and block bacteria[a]
DEPdocetaxel	III	Docetaxel
DEPcabazitaxel	II	Jevtana[a]
DEPirinotecan	II	SN-38[a]

[a] *Starpharma (2020).*

and pharmaceutical areas (Baetke et al., 2015; Han et al., 2019). In the controlled release of drugs, it has been found that these nanoparticles can optimize the dose of the necessary drug and thus obtain an increase in therapeutic efficiency (Khan et al., 2019).

2.3 Mesoporous silica

Mesoporous silica has gained great interest since it can store large volumes in its pores, has a large surface area, low costs, and low toxicity (Manzano and Vallet-Regí, 2019). One of the main attractions of these types of NPs is their porosity and the type of synthesis conditions this is being carried out. They are very easy to functionalize, showing a size that ranges from 10 to 300 nm with varied applications in the biological field; these areas are cell imaging, diagnostics, bio-analysis, and as drug/gene/protein distributor (Lv et al., 2016). The pores' size and shape vary depending on the type of surfactant used for their synthesis. Within the pores can contain an endless number of biomolecules that can vary from metallic particles to drugs and can be released in a controlled manner (Teruel et al., 2018). This characteristic may be due to different reasons; one of them would be the pore size that our molecule contains, the number of pores, or if it is modified in the periphery with other particles that cause the drug to be released under different stimuli such as pH, enzymes, and temperature (Wang et al., 2015b).

3 Preparation of nano-formulations

3.1 Micelles

Polymeric micelles typically exhibit core-shell architectures, and the inner hydrophobic core and outer shells are composed of hydrophobic blocks and hydrophilic moieties of the amphiphiles, respectively (Kwon and Kataoka, 2012). The hydrophobic cores usually serve

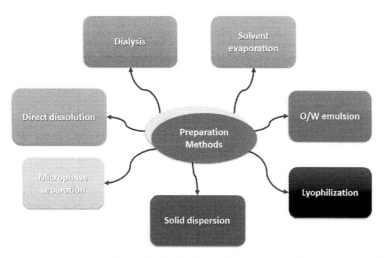

FIG. 2   Most common preparation methods for the loading of drugs or natural active compounds.

as reservoirs for bioactive agents. Fig. 2 shows the most common methods for the preparation of micelles that act as reservoirs. In contrast, the hydrophilic shell can provide the necessary steric protection to stabilize the cores and maintain micelles' dispersion and prevent rapid phagocytosis by macrophages in vivo. A very small micelles particle size with an average hydrodynamic diameter range of 5–100 nm is easily achieved by simple dialysis. This small particle size helps nanocarriers avoid surveillance by the in vivo immune system and allows them to effectively target tumor tissues because of their enhanced permeability and retention (EPR) (Dufort et al., 2012).

Spherical micelles are small, monodisperse structures with a radius of approximately the extended length of the surfactant tail, while wormlike micelles can constitute long and flexible aggregate structures, which can entangle into a dynamic reversible network, resulting in remarkable viscoelastic properties (de Vries and Ravoo, 2019).

The structure of micelles has two parts; a hydrophobic core that acts as a depot for the poorly water-soluble drugs and a hydrophilic shell that plays an important role in the pharmacokinetic part of the micelles by decreasing opsonization and thus increasing their circulation time in vivo (Farokhzad and Langer, 2009; Alexis et al., 2010).

These are the methods for the Preparation of Polymeric Micelles (Kesharwani et al., 2019):

A. Solvent evaporation method/solution casting method. It is the widely employed method for micelles preparation. In this method, the drug and the polymer are dissolved in an organic solvent, followed by the formation of the drug impregnated on thin polymer film after the organic solvent's evaporation (usually 12–24 h). After that, the thin film is reconstituted using an aqueous phase to obtain drug-loaded polymeric micelles. The nonentrapped drug is eliminated using a dialysis technique (Rapoport, 2007; Mourya et al., 2011; Kulthe et al., 2012). This method has been successfully utilized for entrapment of paclitaxel and amphotericin B in poly(ethylene oxide-*b*-D, L-lactide) (PEOPDLLA) micelles and poly(ethylene oxide)-*block*-poly(*N*-hexyl stearate-L-aspartamide) (PEOPHSA) micelles, respectively (Rapoport, 2007). However, this method is unsuitable when long-chain and more hydrophobic core-forming blocks are used (Rajeshwar et al., 2011).

B. Direct dissolution method. It is the simplest technique used to prepare polymeric micelles when polymers possess short-chain and low molecular weight insoluble blocks. Briefly, amphiphilic polymers and drugs are directly dissolved in water at or above critical micellar concentration (CMC) for micelle formation with or without stirring, thermal and ultrasound treatments (Kulthe et al., 2012; Atanase et al., 2017). This method's main limitation is low drug loading (Mourya et al., 2011; Rajeshwar et al., 2011).

C. Dialysis method. This method is usually applied when core-forming blocks are highly hydrophobic and possess long sequences (Mourya et al., 2011). It involves dissolving both the drug and the polymer in a water-miscible organic phase such as dimethylformamide (DMF). The dialysis bag filled with the above solution is immersed in water for several hours, inducing micelle assembly through a solvent exchange (Kedar et al., 2010; Atanase et al., 2017). Doxorubicin was successfully encapsulated in poly his-PEG/poly (L-lactic acid) (PLLA-PEG) and NIPAAm with amphiphilic blocks (was dissolved in THF) mixed micelles using this technique (Cagel et al., 2017; Picos-Corrales et al., 2019). Although extensively used, dialysis is a time-consuming methodology and may alter morphology due to the gradual change in solvent properties (Kulthe et al., 2012; Zhu et al., 2017).

D. Oil/water emulsion solvent evaporation method. The method involves dissolving the copolymer in organic or the aqueous solvent and the drug in a water-immiscible volatile organic phase. The solvents employed mainly include acetone, tetrahydrofuran, ethyl acetate, chloroform, and methylene chloride. Drug-loaded polymeric micelles are obtained under vigorous and constant stirring by adding the organic phase to the aqueous phase (Jones and Leroux, 1999; Gaucher et al., 2005). Paclitaxel was loaded in copolymer micelles of methoxy poly(ethylene glycol)-*b* poly(L-lactide) (PEG-PLA) using the O/W emulsion method (Palao-Suay et al., 2016). However, this method involves the usage of toxic chlorinated solvents (Jones and Leroux, 1999).

E. Lyophilization/lyophilization method. It is the most simple and cost-effective one-step technique. Initially, the drug and the polymer are dissolved in a mixture of the organic phase (tert-butanol) and water, followed by lyophilization. The lyophilized cake is reconstituted in an aqueous medium, and the unencapsulated drug is removed by dialysis (Rapoport, 2007; Kedar et al., 2010). This method has been described for poly (N-vinyl-pyrrolidone)-*block*-poly (D, L-lactide) (PVPPDLLA) micelles loaded with paclitaxel (Rapoport, 2007).

F. Solid dispersion method. This method involves forming a solid drug-polymer matrix after evaporation of organic solvent containing both the drug and the polymer in a dissolved state. The preheated polymer-drug matrix is exposed to an aqueous phase to facilitate the formation of drug encapsulated polymeric micelles (Zhang et al., 1996; Taillefer et al., 2000; Kedar et al., 2010).

G. Microphase separation method. Briefly, in this method, the drug and polymer are normally dissolved in tetrahydrofuran, followed by the dropwise addition of the aqueous phase with constant stirring. Subsequently, the organic phase is evaporated under pressure to obtain an aqueous polymeric micelles solution (Zhang et al., 2009; Kedar et al., 2010).

3.2 Liposomes

Liposomes are spherical structures composed of blocks of surfactants (i.e., phospholipids), which can spontaneously form bilayers in aqueous solutions consisting of two layers of molecules that have their nonpolar groups facing toward each other (McClements, 2014). Therefore, the liposome constitutes an outer hydrophobic layer and an inner hydrophilic layer with sizes varying from 50 to 100 nm depending on the cholesterol and phospholipid composition (Yhee et al., 2016). Liposome formation enables the amphiphilic character of lipids which, with the force of hydrophobic interaction, assemble into bilayers. Three groups of liposomes exist by their morphology, using conventional classification: multilamellar vesicle (MLV), small unilamellar vesicle (SUV), large unilamellar vesicle (LUV), and multivesicular vesicle (MVV) (Elizondo et al., 2011). The liposome can be made from natural substances and are therefore nontoxic, biocompatible, biodegradable, and nonimmunogenic. Phosphatidylcholine(PC), phosphatidylglycerol (PG), di-palmitoyl phosphatidylcholine (DPPC), di-stearoyl phosphatidylcholine (DSPC), cholesterol (CH), stearylamine, dicetylphosphate, or a mixture of these are usually the composites of the bilayer (Đorđević et al., 2015).

The loading of ingredients in the liposomes can be classified as passive or active for a wide variety of approaches available. Passive loading is the incorporation of the compounds by mixing them with the liposome before or after the vesicle formation. On the other hand, the active loading consists of a specific physicochemical mechanism to incorporate the ingredient in the liposomes, including the electrostatic attraction between the active and the liposome-when they have different charged structures (Esposto et al., 2020).

Besides, various materials have been used to modify liposomes to improve their functionality, such as polysaccharides (e.g., chitosan, pectin) (Zhou et al., 2014; Liu et al., 2016a), protein (e.g., whey protein, silk fibroin) (Dong et al., 2015; Frenzel and Steffen-Heins, 2015), polymers (e.g., PEG) (Oberoi et al., 2016), and others (e.g., silica, gold) (Mohanraj et al., 2010; Rengan et al., 2014).

For their amphiphilic nature, the liposomes can introduce both lipid-soluble and water-soluble drugs (Torchilin, 2005). This encapsulation protects the drug from rapid degradation and reduces drug toxicity by making it unavailable to the systemic circulation. Liposomes can also improve the therapeutic index of a new or established drug by changing its pharmacokinetics, such as absorption and metabolism, increasing its biological half-life, or reducing its elimination (Sonju et al., 2020).

3.3 Dendrimers

Dendrimers are defined as hyperbranched polymeric macromolecules, having well-defined radical branching (Kesharwani et al., 2014). Their basic dendrimer structure comprises three main components: an initiating core molecule, interior repetitive branching units attached to the central core, and terminal modifiable functional groups (Singh et al., 2016a). Dendrimers of small size, spherical shape, and lipophilicity are better than linear polymers, as they have better penetration to the cell membrane. The loading on dendrimers depends on the type and number of active sites, loading capacity, external functional groups, and lipophilicity (Khan et al., 2015). The synthesis of dendrimers could be done by the divergent approach, which is when the synthesis starts from the interior core; dendrimers are synthesized layer by layer to the peripheral surface. The convergent approach is done when dendrimer segments are synthesized separately and then attached to the dendrimer core molecule (Pana et al., 2020).

3.4 Metallic nanoparticles

The key point of the synthesis of metallic nanoparticles is to reduce the used metal in the first instance, whether it is Au, Ag, or Cu. For the reduction of these metals, different methods are used, among which the use of some polymers such as chitosan, phytochemical agents with antioxidant properties, chemical reduction, irradiation, electrochemistry, assistance through microwaves, among others (Yeh et al., 2012; Uehara et al., 2015; Rao et al., 2016).

The synthesis of these nanoparticles can be classified mainly into three types:

A. Synthesis by chemical methods. In this method, a reduction of tetrachloroauric acid ($HAuCl_4$) is carried out for the synthesis of AuNPs, and silver nitrate ($AgNO_3$) for the case of AgNPS. This chemical reduction is carried out with sodium borohydride ($NaBH_4$), where particles sizes of 10–100 nm can be obtained (Yeh et al., 2012). Different polymers that act as reducers have also been used, such as chitosan and ethylene glycol (Sun and Xia, 2002; Wei and Qian, 2008; Rubina et al., 2016). Also, the use of different phytochemical

TABLE 4 Some organisms used for the synthesis of Au and AgNPs (Hassan, 2015; Velusamy et al., 2016).

Organisms	NPs
Pseudomonas aeruginosa	Au
Pseudomonas stutzeri	Ag and Au
Lactobacillus sp.	Ag and Au
Klebsiella pneumonia	Ag
Aspergillus fumigates	Ag
Pichia jadinii	Au

agents have provided the same function (alkaloids, phenolic acids, terpenoids, tannins, etc.) (Baranwal et al., 2016; Jayaprakash et al., 2017; Amini, 2019; Ahmad et al., 2020).

B. Synthesis by biological methods. This type of synthesis falls within those classified as green syntheses since no toxic substance is used, which gives a plus to this type of nanoparticles for use in the biological and pharmaceutical area, ensuring that these nanoparticles are as innocuous as possible. For the synthesis, different organisms were used: bacteria, viruses, fungi, yeasts, among other organisms. The reduction of HAuCl4 occurred, and the particle sizes obtained vary from 2 to 100 nm (Alaqad and Saleh, 2016; Velusamy et al., 2016) (Table 4).

C. Synthesis by physical and mechanical methods. Some of the physical methods used to obtain AuNPs is evaporation-condensation, which can obtain NPs with high purity without using chemicals or toxic substances. It should be noted that obtaining these NPs is very expensive since very complex equipment is needed to obtain them. Another widely used physical method is laser-assisted synthesis (e.g., gamma); this type of synthesis starts from irradiating a small metal block (Au, Ag). Thus obtaining the NPs, like the previous method, the size of these is very homogeneous (Alaqad and Saleh, 2016; Lee and Jun, 2019).

3.5 Sílice mesoporosa

Mesoporous silica is classified as one of the best-studied inorganic release systems with an infinite number of applications. This type of system appeared for the first time in the '70s. After this, a group of researchers managed to synthesize the first mesoporous nanostructure based on aluminosilicate gels using the liquid crystal template mechanism; these NPs received the name MCM-41 (Narayan et al., 2018). The pores' size varied from 2 to 50 nm, and this variation was attributed to the use of different surfactants for its synthesis. There are many methodologies for obtaining these NPs, such as sol-gel, microwave synthesis, hydrothermal synthesis, and template synthesis, among others (Farjadian et al., 2019). The synthesis of these NPs can be classified into three different types (Li et al., 2019):

A. Based on the Stöber method. This method is based on the production of silica NPs. The precursors of the alkoxide silicone are hydrolyzed and condensed under acid or basic catalysis; for this, the surfactants have the function of giving the NPs structure (Ghaferi et al., 2021). Pores are formed after calcining the silica sample.

B. Solution based on the synthesis of mesoporous silica. In this method, cetyltrimethylammonium bromide (CTAB) is used as a template when it presents a concentration similar to what is called critical micellar concentration (CMC) (which consists of the minimum concentration of product necessary for the micelles to form). There is self-assembly in spherical micelles; in the solution, a mesophase is produced, and through precursors for the formation of silica, the sample is calcined (Sun et al., 2016; Yang et al., 2019a).

C. EISA. In this method, a homogeneous solution of silica is used with ethanol as the surfactant. As the ethanol evaporates, the surfactant's concentration increases, and the micelles are formed within the silica. The size of the pores obtained by this method range from 25 to 250 nm. Some other synthesis routes that can be used are micelles formation due to changes in pH, time and speed of stirring, and temperature (Baeza et al., 2016; Narayan et al., 2018; Ghaferi et al., 2021).

4 Different applications of nano-formulations

The applications given to the different nano-formulations are very varied; they can be used as sensors, catalysts, and waste absorbents. However, the demand is presented by some morbid diseases such as cancer, diabetes, and Alzheimer's. These have been mainly used to improve poorly soluble drugs' solubility, so their effect could significantly increase if they are marketed. There are some nano-formulations on the market for some years, and these are focused on the loading and release of antineoplastic drugs (liposomes and micelles). The administration of these nano-formulations can be oral, topical, inhaled, and intravenous (IV). Their administration type will be decided largely depending on the type of nano-formulations and the components used for their synthesis. Table 5 shows a small summary of the application of NPs in different areas.

5 Biocompatibility and mechanism of some system drug delivery

The biocompatibility of these nano-formulations is one of the most important factors that must be considered when developing a nanocarrier (Fig. 3 shows a resume of these characteristics). It depends on whether it can be used for biological applications. Organic NPs have a lower probability of presenting toxicity, but this can be presented depending on the type of polymer with which it is synthesized (Cho et al., 2015; Biswas et al., 2016; Choi and Han, 2018). For dendrimers, this toxicity will be reflected by the number of generations (branches) that the molecule contains (Parajapati et al., 2016; Singh et al., 2016; Davaran et al., 2018; Huang and Wu, 2018).

Depending on the size of these NPs, they can enter through the passive or active route. To activate the route, they have to be functionalized on their surface with different biomolecules (e.g., peptides) to be recognized by membrane receptors and enter through endocytosis/phagocytosis (Paliwal et al., 2015; Pattni et al., 2015; Zhu and Liao, 2015; Daraee et al.,

TABLE 5 Some applications of nano-formulations.

NPs	Drug/ substance	Application	Author
Liposomes	Doxorubicin Morphine sulfate Inactivated Vaccine Vincristine 5-Fluorouracil Nucleotides	Cancer therapy Pain Vaccine shot Molecular imaging	Paliwal et al. (2015), Pattni et al. (2015), Sercombe et al. (2015), Daraee et al. (2016), Liu et al. (2016b), Lamichhane et al. (2018), Zangabad et al. (2018), Camilo et al. (2020)
Micelles	Doxorubicin Docetaxel Paclitaxel MNPs Camptothecin	Cancer therapy Bioimaging	Cho et al. (2015), Movassaghian et al. (2015), Wang et al. (2015a), Biswas et al. (2016), Gothwal et al. (2016), Liu et al. (2016b), Sheikhpour et al. (2017), Jo et al. (2020)
Dendrimers	Camptothecin Doxorubicin Methotrexate Paclitaxel Cisplatin	Cancer therapy Bioimaging Photothermal therapy Vaccine shot	Wang et al. (2016), Santos et al. (2017), Sharma et al. (2017), Davaran et al. (2018), Huang and Wu (2018)
MNPs	Methotrexate Phthalocyanine Doxorubicin Cefotaxime	Cancer therapy Colorimetric sensor Bioassays Bioimaging Hyperthermia Therapy Theranostics Antimicrobial	Baetke et al. (2015), Meena Kumari et al. (2015), Alaqad and Saleh (2016), Chen et al. (2016a), McNamara and Tofail (2016), Abdal Dayem et al. (2018), Freitas de Freitas et al. (2018), Darabdhara et al. (2019), Han et al. (2019), Lee and Jun (2019), Shaikh et al. (2019)
Silica mesoporous	Cytochrome c Camptothecin Doxorubicin Methotrexate 5-Fluorouracil Alendronate	Cancer therapy Osteogenic Antiinflammatory Bioimaging	Bharti et al. (2015), Wang et al. (2015b), Baeza et al. (2016), Watermann and Brieger (2017), Teruel et al. (2018), Farjadian et al. (2019), Li et al. (2019), Manzano and Vallet-Regí (2019), Yang et al. (2019a)

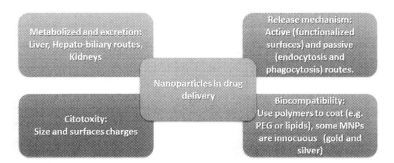

FIG. 3 Factors and their characteristics that must be considered when developing a nanocarrier.

2016; Sharma et al., 2017; Huang and Wu, 2018). Due to these nanoparticles are not quickly opsonized when entering our body, polymers such as polyethylene glycol (PEG) are used to coat the molecule or to form part of the outer part of the NP, which helps it to pass longer through the bloodstream (Biffi et al., 2015). In the case of liposomes, because they are synthesized with lipids similar to those contained in our cell membranes, they present very low cytotoxicity (Paliwal et al., 2015; Zangabad et al., 2018). Different studies have been carried out for micelles in which it has been observed that the ideal size for the penetration of this type of particle varies between 100 and 160nm. The excretion of these NPs occurs mainly by the liver, where the different components of lysosomes, micelles, and dendrimers are metabolized (Wang et al., 2015a; Huang and Wu, 2018).

Inorganic NPs, in particular, have been subjected to biocompatibility tests due to their nature. In the case of metallic NPs, it has been found that despite being innocuous and not presenting any harm to our body (Alaqad and Saleh, 2016). The morphology they present, their size, surface charge, coating, and their functionalization can make them present a certain degree of cytotoxicity. In recent studies, it has been found that particle sizes smaller than 37nm cause toxicity while 50–100nm have zero toxicity (Wozniak et al., 2017; Freitas de Freitas et al., 2018). This could be since the smaller SPNs can pass through the membrane without a problem, causing their accumulation within the different systems.

In contrast, when the size is slightly larger, there is a lower probability of entering the system passively. Molecules around 100nm enter by endocytosis with clathrin while using caveolae for 60–80nm (Wozniak et al., 2017). The excretion of these MNPs (e.g., AuNPs) occurs through the kidney and hepato-biliary routes, depending on their form and coating material, while NPs are excreted mostly through feces (Lin et al., 2015). Mesoporous silica has caused great controversy regarding this issue since these nanoparticles are synthesized with inorganic compounds. However, the Administration of Food and Drug (FDA) approved the use of hybrid silica nanoformulations as image biomarkers, using a fluorescent marker (124I) and functionalized with a peptide (cRGDY) which is selective for tumors.

In pharmacokinetic studies, it was found that these nanoparticles are rapidly excreted via the kidneys (Bharti et al., 2015; Baeza et al., 2016; Narayan et al., 2018; Farjadian et al., 2019). The process by which these can enter the cell is by endocytosis; once inside, the substances within the endosome cause the release of the content within the mesoporous silica. The biocapacity of this nano-formulation is given by different factors such as pore size, surface properties, and the pores' morphology and structure (Farjadian et al., 2019; Yang et al., 2019a). The first mesoporous silica-based on NPs, called "Cornell dots," was approved in its first phase of the study, proving that these NPs are biocompatible and do not produce cytotoxicity (Bharti et al., 2015; Ghaferi et al., 2021).

6 Perspectives

NPs have a great potential for application regardless of whether they are organic or inorganic. The combination of two different systems could have a dual effect of activity, as in hybrid nanoparticles (metallic/polymers). The combination of the different nano-formulations gives rise to theragnostic. These systems must continue to be investigated since it has been observed that there are high biocompatibility and an increase in the therapeutic effect of the administered drugs compared to the drugs released freely.

References

Abdal Dayem, A., Lee, S.B., Cho, S.G., 2018. The impact of metallic nanoparticles on stem cell proliferation and differentiation. Nanomaterials 8 (10), 761.

Ahmad, T., Bustam, M.A., Zulfiqar, M., Moniruzzaman, M., Idris, A., Iqbal, J., Asghar, H.M.A., Ullah, S., 2020. Controllable phytosynthesis of gold nanoparticles and investigation of their size and morphology-dependent photocatalytic activity under visible light. J. Photochem. Photobiol. A Chem. 392, 112429.

Ahmed, A., Sarwar, S., Hu, Y., Munir, M.U., Nisar, M.F., Ikram, F., Asif, A., Rahman, S.U., Chaudhry, A.A., Rehman, I.U., 2021. Surface-modified polymeric nanoparticles for drug delivery to cancer cells. Expert Opin. Drug Deliv. 18 (1), 1–24.

Alaqad, K., Saleh, T.A., 2016. Gold and silver nanoparticles: synthesis methods, characterization routes and applications towards drugs. J. Environ. Anal. Toxicol. 6 (4), 1000384.

Alavi, M., Karimi, N., Safaei, M., 2017. Application of various types of liposomes in drug delivery systems. Adv. Pharm. Bull. 7 (1), 3–9.

Alexis, F., Pridgen, E.M., Langer, R., Farokhzad, O.C., 2010. Nanoparticle technologies for cancer therapy. Drug Deliv. (197), 55–86.

Alvarez-Lorenzo, C., Concheiro, A., 2014. Smart drug delivery systems: from fundamentals to the clinic. Chem. Commun. 50 (58), 7743–7765.

Amini, S.M., 2019. Preparation of antimicrobial metallic nanoparticles with bioactive compounds. Mater. Sci. Eng. C 103, 109809.

Andrade, F., Rafael, D., Vilar-Hernandez, M., Montero, S., Martinez-Trucharte, F., Seras-Franzoso, J., Diaz-Riascos, Z.V., Boullosa, A., Garcia-Aranda, N., Camara-Sanchez, P., Arango, D., Nestor, M., Abasolo, I., Sarmento, B., Schwartz Jr., S., 2021. Polymeric micelles targeted against CD44v6 receptor increase niclosamide efficacy against colorectal cancer stem cells and reduce circulating tumor cells in vivo. J. Control. Release 331, 198–212.

Atanase, L., Desbrieres, J., Riess, G., 2017. Micellization of synthetic and polysaccharides-based graft copolymers in aqueous media. Prog. Polym. Sci. 73, 32–60.

Baetke, S.C., Lammers, T., Kiessling, F., 2015. Applications of nanoparticles for diagnosis and therapy of cancer. Br. J. Radiol. 88 (1054), 20150207.

Baeza, A., Ruiz-Molina, D., Vallet-Regí, M., 2016. Recent advances in porous nanoparticles for drug delivery in antitumoral applications: inorganic nanoparticles and nanoscale metal-organic frameworks. Expert Opin. Drug Deliv. 14 (6), 783–796.

Baranwal, A., Mahato, K., Srivastava, A., Maurya, P.K., Chandra, P., 2016. Phytofabricated metallic nanoparticles and their clinical applications. RSC Adv. 6 (107), 105996–106010.

Bazban-Shotorbani, S., Hasani-Sadrabadi, M.M., Karkhaneh, A., Serpooshan, V., Jacob, K.I., Moshaverinia, A., Mahmoudi, M., 2017. Revisiting structure-property relationship of pH-responsive polymers for drug delivery applications. J. Control. Release 253, 46–63.

Bharti, C., Nagaich, U., Pal, A.K., Gulati, N., 2015. Mesoporous silica nanoparticles in target drug delivery system: a review. Int. J. Pharm. Investig. 5 (3), 124–133.

Biffi, S., Voltan, R., Rampazzo, E., Prodi, L., Zauli, G., Secchiero, P., 2015. Applications of nanoparticles in cancer medicine and beyond: optical and multimodal in vivo imaging, tissue targeting and drug delivery. Expert Opin. Drug Deliv. 12 (12), 1837–1849.

Biswas, S., Kumari, P., Lakhani, P.M., Ghosh, B., 2016. Recent advances in polymeric micelles for anti-cancer drug delivery. Eur. J. Pharm. Sci. 83, 184–202.

Cagel, M., Tesan, F.C., Bernabeu, E., Salgueiro, M.J., Zubillaga, M.B., Moretton, M.A., Chiappetta, D.A., 2017. Polymeric mixed micelles as nanomedicines: achievements and perspectives. Eur. J. Pharm. Biopharm. 113, 211–228.

Camilo, C.J.J., Leite, D.O.D., Silva, A.R.A., Menezes, I.R.A., Coutinho, H.D.M., Costa, J.G.M., 2020. Lipid vesicles: applications, principal components and methods used in their formulations: a review. Acta Biol. Colomb. 25 (2), 339–352.

Chen, C.W., Chan, Y.C., Hsiao, M., Liu, R.S., 2016a. Plasmon-enhanced photodynamic cancer therapy by upconversion nanoparticles conjugated with Au nanorods. ACS Appl. Mater. Interfaces 8 (47), 32108–32119.

Chen, S., Hao, X., Liang, X., Zhang, Q., Zhang, C., Zhou, G., Shen, S., Jia, G., Zhang, J., 2016b. Inorganic nanomaterials as carriers for drug delivery. J. Biomed. Nanotechnol. 12 (1), 1–27.

Cho, H., Lai, T.C., Tomoda, K., Kwon, G.S., 2015. Polymeric micelles for multi-drug delivery in cancer. AAPS PharmSciTech 16 (1), 10–20.

Choi, Y.H., Han, H.K., 2018. Nanomedicines: current status and future perspectives in aspect of drug delivery and pharmacokinetics. J. Pharm. Investig. 48 (1), 43–60.

Darabdhara, G., Das, M.R., Singh, S.P., Rengan, A.K., Szunerits, S., Boukherroub, R., 2019. Ag and Au nanoparticles/reduced graphene oxide composite materials: synthesis and application in diagnostics and therapeutics. Adv. Colloid Interf. Sci. 271, 101991.

Daraee, H., Etemadi, A., Kouhi, M., Alimirzalu, S., Akbarzadeh, A., 2016. Application of liposomes in medicine and drug delivery. Artif. Cells Nanomed. Biotechnol. 44 (1), 381–391.

Davaran, S., Nasibova, A., Saghfi, S., Kavetskyy, T., Herizchi, R., Abasi, E., Kafshdooz, T., Annabi, N., Mostafavi, E., Khalilov, R., Akbarzadeh, A., 2018. Role of dendrimers in advanced drug delivery and biomedical applications: a review. Exp. Oncol. 40 (3), 178–183.

de Vries, W.C., Ravoo, B.J., 2019. Vesicles and micelles. In: Supramolecular Chemistry in Water. Wiley, pp. 375–411.

Dong, Y., Dong, P., Huang, D., Mei, L., Xia, Y., Wang, Z., Pan, X., Li, G., Wu, C., 2015. Fabrication and characterization of silk fibroin-coated liposomes for ocular drug delivery. Eur. J. Pharm. Biopharm. 91, 82–90.

Đorđević, V., Balanč, B., Belščak-Cvitanović, A., Lević, S., Trifković, K., Kalušević, A., Kostić, I., Komes, D., Bugarski, B., Nedović, V., 2015. Trends in encapsulation technologies for delivery of food bioactive compounds. Food Eng. Rev. 7 (4), 452–490.

Dufort, S., Sancey, L., Coll, J.-L., 2012. Physico-chemical parameters that govern nanoparticles fate also dictate rules for their molecular evolution. Adv. Drug Deliv. Rev. 64 (2), 179–189.

Elizondo, E., Moreno, E., Cabrera, I., Córdoba, A., Sala, S., Veciana, J., Ventosa, N., 2011. Liposomes and other vesicular systems: structural characteristics, methods of preparation, and use in nanomedicine. Prog. Mol. Biol. Transl. Sci. 104, 1–52.

Elzoghby, A.O., Hemasa, A.L., Freag, M.S., 2016. Hybrid protein-inorganic nanoparticles: from tumor-targeted drug delivery to cancer imaging. J. Control. Release 243, 303–322.

Esposto, B.S., Jauregi, P., Tapia-Blácido, D.R., Martelli-Tosi, M., 2020. Liposomes vs. chitosomes: encapsulating food bioactives. Trends Food Sci. Technol. 108, 40–48.

Farjadian, F., Roointan, A., Mohammadi-Samani, S., Hosseini, M., 2019. Mesoporous silica nanoparticles: synthesis, pharmaceutical applications, biodistribution, and biosafety assessment. Chem. Eng. J. 359, 684–705.

Farokhzad, O.C., Langer, R., 2009. Impact of nanotechnology on drug delivery. ACS Nano 3 (1), 16–20.

Freitas de Freitas, L., Varca, G.H.C., Dos Santos Batista, J.G., Benevolo Lugao, A., 2018. An overview of the synthesis of gold nanoparticles using radiation technologies. Nanomaterials 8 (11), 939.

Frenzel, M., Steffen-Heins, A., 2015. Whey protein coating increases bilayer rigidity and stability of liposomes in food-like matrices. Food Chem. 173, 1090–1099.

Gaucher, G., Dufresne, M.-H., Sant, V.P., Kang, N., Maysinger, D., Leroux, J.-C., 2005. Block copolymer micelles: preparation, characterization and application in drug delivery. J. Control. Release 109 (1–3), 169–188.

Ghaferi, M., Koohi Moftakhari Esfahani, M., Raza, A., Al Harthi, S., Ebrahimi Shahmabadi, H., Alavi, S.E., 2021. Mesoporous silica nanoparticles: synthesis methods and their therapeutic use-recent advances. J. Drug Target. 29 (2), 131–154.

Gothwal, A., Khan, I., Gupta, U., 2016. Polymeric micelles: recent advancements in the delivery of anticancer drugs. Pharm. Res. 33 (1), 18–39.

Han, X., Xu, K., Taratula, O., Farsad, K., 2019. Applications of nanoparticles in biomedical imaging. Nanoscale 11 (3), 799–819.

Hassan, S., 2015. A review on nanoparticles: their synthesis and types. Res. J. Recent Sci. 4, 1–3.

Huang, D., Wu, D., 2018. Biodegradable dendrimers for drug delivery. Mater. Sci. Eng. C 90, 713–727.

Jayaprakash, N., Vijaya, J.J., Kaviyarasu, K., Kombaiah, K., Kennedy, L.J., Ramalingam, R.J., Munusamy, M.A., Al-Lohedan, H.A., 2017. Green synthesis of Ag nanoparticles using Tamarind fruit extract for the antibacterial studies. J. Photochem. Photobiol. B Biol. 169, 178–185.

Jo, M.J., Lee, Y.J., Park, C.W., Chung, Y.B., Kim, J.S., Lee, M.K., Shin, D.H., 2020. Evaluation of the physicochemical properties, pharmacokinetics, and in vitro anticancer effects of docetaxel and osthol encapsulated in methoxy poly(ethylene glycol)-b-poly(caprolactone) polymeric micelles. Int. J. Mol. Sci. 22 (1), 231.

Johnsen, K.B., Gudbergsson, J.M., Duroux, M., Moos, T., Andresen, T.L., Simonsen, J.B., 2018. On the use of liposome controls in studies investigating the clinical potential of extracellular vesicle-based drug delivery systems – a commentary. J. Control. Release 269, 10–14.

Jones, M.-C., Leroux, J.-C., 1999. Polymeric micelles—a new generation of colloidal drug carriers. Eur. J. Pharm. Biopharm. 48 (2), 101–111.

Kedar, U., Phutane, P., Shidhaye, S., Kadam, V., 2010. Advances in polymeric micelles for drug delivery and tumor targeting. Nanomedicine 6 (6), 714–729.

Kesharwani, P., Jain, K., Jain, N.K., 2014. Dendrimer as nanocarrier for drug delivery. Prog. Polym. Sci. 39 (2), 268–307.

Kesharwani, S.S., Kaur, S., Tummala, H., Sangamwar, A.T., 2019. Multifunctional approaches utilizing polymeric micelles to circumvent multidrug resistant tumors. Colloids Surf. B: Biointerfaces 173, 581–590.

Khan, O.F., Zaia, E.W., Jhunjhunwala, S., Xue, W., Cai, W., Yun, D.S., Barnes, C.M., Dahlman, J.E., Dong, Y., Pelet, J.M., 2015. Dendrimer-inspired nanomaterials for the in vivo delivery of siRNA to lung vasculature. Nano Lett. 15 (5), 3008–3016.

Khan, I., Saeed, K., Khan, I., 2019. Nanoparticles: properties, applications and toxicities. Arab. J. Chem. 12 (7), 908–931.

Khan, A.U., Khan, M., Cho, M.H., Khan, M.M., 2020. Selected nanotechnologies and nanostructures for drug delivery, nanomedicine and cure. Bioprocess Biosyst. Eng. 43 (8), 1339–1357.

Kulthe, S.S., Choudhari, Y.M., Inamdar, N.N., Mourya, V., 2012. Polymeric micelles: authoritative aspects for drug delivery. Des. Monomers Polym. 15 (5), 465–521.

Kwon, G.S., Kataoka, K., 2012. Block copolymer micelles as long-circulating drug vehicles. Adv. Drug Deliv. Rev. 64, 237–245.

Lamichhane, N., Udayakumar, T.S., D'Souza, W.D., Simone 2nd, C.B., Raghavan, S.R., Polf, J., Mahmood, J., 2018. Liposomes: clinical applications and potential for image-guided drug delivery. Molecules 23 (2), 288.

Lee, S.H., Jun, B.H., 2019. Silver nanoparticles: synthesis and application for nanomedicine. Int. J. Mol. Sci. 20 (4), 865.

Li, Z., Zhang, Y., Feng, N., 2019. Mesoporous silica nanoparticles: synthesis, classification, drug loading, pharmacokinetics, biocompatibility, and application in drug delivery. Expert Opin. Drug Deliv. 16 (3), 219–237.

Liang, R., Wei, M., Evans, D.G., Duan, X., 2014. Inorganic nanomaterials for bioimaging, targeted drug delivery and therapeutics. Chem. Commun. 50 (91), 14071–14081.

Lin, Z., Monteiro-Riviere, N.A., Riviere, J.E., 2015. Pharmacokinetics of metallic nanoparticles. WIREs Nanomed. Nanobiotechnol. 7 (2), 189–217.

Liu, W., Liu, W., Ye, A., Peng, S., Wei, F., Liu, C., Han, J., 2016a. Environmental stress stability of microencapsules based on liposomes decorated with chitosan and sodium alginate. Food Chem. 196, 396–404.

Liu, D., Yang, F., Xiong, F., Gu, N., 2016b. The smart drug delivery system and its clinical potential. Theranostics 6 (9), 1306–1323.

Lombardo, D., Kiselev, M.A., Caccamo, M.T., 2019. Smart nanoparticles for drug delivery application: development of versatile nanocarrier platforms in biotechnology and nanomedicine. J. Nanomater. 2019, 1–26.

Lv, X., Zhang, L., Xing, F., Lin, H., 2016. Controlled synthesis of monodispersed mesoporous silica nanoparticles: particle size tuning and formation mechanism investigation. Microporous Mesoporous Mater. 225, 238–244.

Manzano, M., Vallet-Regí, M., 2019. Mesoporous silica nanoparticles for drug delivery. Adv. Funct. Mater. 30 (2), 47.

McClements, D.J., 2014. Nanoparticle- and Microparticle-Based Delivery Systems: Encapsulation, Protection and Release of Active Compounds. CRC Press.

McNamara, K., Tofail, S.A.M., 2016. Nanoparticles in biomedical applications. Adv. Phys. 2 (1), 54–88.

Meena Kumari, M., Jacob, J., Philip, D., 2015. Green synthesis and applications of Au-Ag bimetallic nanoparticles. Spectrochim. Acta A Mol. Biomol. Spectrosc. 137, 185–192.

Mignani, S., Rodrigues, J., Tomas, H., Roy, R., Shi, X., Majoral, J.P., 2018. Bench-to-bedside translation of dendrimers: reality or utopia? A concise analysis. Adv. Drug Deliv. Rev. 136–137, 73–81.

Mohanraj, V.J., Barnes, T.J., Prestidge, C.A., 2010. Silica nanoparticle coated liposomes: a new type of hybrid nanocapsule for proteins. Int. J. Pharm. 392 (1–2), 285–293.

Mourya, V., Inamdar, N., Nawale, R., Kulthe, S., 2011. Polymeric micelles: general considerations and their applications. Indian J. Pharm. Educ. Res. 45 (2), 128–138.

Movassaghian, S., Merkel, O.M., Torchilin, V.P., 2015. Applications of polymer micelles for imaging and drug delivery. WIREs Nanomed. Nanobiotechnol. 7 (5), 691–707.

Narayan, R., Nayak, U.Y., Raichur, A.M., Garg, S., 2018. Mesoporous silica nanoparticles: a comprehensive review on synthesis and recent advances. Pharmaceutics 10 (3), 118.

Oberoi, H.S., Yorgensen, Y.M., Morasse, A., Evans, J.T., Burkhart, D.J., 2016. PEG modified liposomes containing CRX-601 adjuvant in combination with methyl glycol chitosan enhance the murine sublingual immune response to influenza vaccination. J. Control. Release 223, 64–74.

Palao-Suay, R., Gómez-Mascaraque, L., Aguilar, M.R., Vázquez-Lasa, B., San Román, J., 2016. Self-assembling polymer systems for advanced treatment of cancer and inflammation. Prog. Polym. Sci. 53, 207–248.

Paliwal, S.R., Paliwal, R., Vyas, S.P., 2015. A review of mechanistic insight and application of pH-sensitive liposomes in drug delivery. Drug Deliv. 22 (3), 231–242.

Pana, J., Attiaa, S.A., Filipczaka, N., Torchilina, V.P., 2020. 10 Dendrimers for drug delivery purposes. In: Nanoengineered Biomaterials for Advanced Drug Delivery. Elsevier, p. 201.

Parajapati, S.K., Maurya, S.D., Das, M.K., Tilak, V.K., Verma, K.K., Dhakar, R.C., 2016. Potential application of dendrimers in drug delivery: a concise review and update. J. Drug Deliv. Ther. 6 (2), 71–88.

Patra, J.K., Das, G., Fraceto, L.F., Campos, E.V.R., Rodriguez-Torres, M.D.P., Acosta-Torres, L.S., Diaz-Torres, L.A., Grillo, R., Swamy, M.K., Sharma, S., Habtemariam, S., Shin, H.S., 2018. Nano-based drug delivery systems: recent developments and future prospects. J. Nanobiotechnol. 16 (1), 71.

Pattni, B.S., Chupin, V.V., Torchilin, V.P., 2015. New developments in liposomal drug delivery. Chem. Rev. 115 (19), 10938–10966.

Picos-Corrales, L.A., Garcia-Carrasco, M., Licea-Claveríe, Á., Chávez-Santoscoy, R.A., Serna-Saldívar, S.O., 2019. NIPAAm-containing amphiphilic block copolymers with tailored LCST: aggregation behavior, cytotoxicity and evaluation as carriers of indomethacin, tetracycline and doxorubicin. J. Macromol. Sci. A 56, 759–772.

Qiu, M., Wang, D., Liang, W., Liu, L., Zhang, Y., Chen, X., Sang, D.K., Xing, C., Li, Z., Dong, B., Xing, F., Fan, D., Bao, S., Zhang, H., Cao, Y., 2018. Novel concept of the smart NIR-light-controlled drug release of black phosphorus nanostructure for cancer therapy. PNAS 115 (3), 501–506.

Rajeshwar, B.R., Gatla, A., Rajesh, G., Arjun, N., Swapna, M., 2011. Polymeric micelles: a nanoscience technology. Am. J. Pharm. Res. 1 (4), 351–363.

Rao, P.V., Nallappan, D., Madhavi, K., Rahman, S., Jun Wei, L., Gan, S.H., 2016. Phytochemicals and biogenic metallic nanoparticles as anticancer agents. Oxidative Med. Cell. Longev. 2016, 3685671.

Rapoport, N., 2007. Physical stimuli-responsive polymeric micelles for anti-cancer drug delivery. Prog. Polym. Sci. 32 (8–9), 962–990.

Rengan, A.K., Jagtap, M., De, A., Banerjee, R., Srivastava, R., 2014. Multifunctional gold-coated thermo-sensitive liposomes for multimodal imaging and photo-thermal therapy of breast cancer cells. Nanoscale 6 (2), 916–923.

Rubina, M.S., Kamitov, E.E., Zubavichus, Y.V., Peters, G.S., Naumkin, A.V., Suzer, S., Vasil'kov, A.Y., 2016. Collagen-chitosan scaffold modified with Au and Ag nanoparticles: synthesis and structure. Appl. Surf. Sci. 366, 365–371.

Santos, S.S., Gonzaga, R.V., Silva, J.V., Savino, D.F., Prieto, D., Shikay, J.M., Paulo, L.H.A., Ferreira, E.I., Giarolla, J., 2017. Peptide dendrimers: drug/gene delivery and others approaches. Can. J. Chem. 95, 907–916.

Sercombe, L., Veerati, T., Moheimani, F., Wu, S.Y., Sood, A.K., Hua, S., 2015. Advances and challenges of liposome assisted drug delivery. Front. Pharmacol. 6, 286.

Shaikh, S., Nazam, N., Rizvi, S.M.D., Ahmad, K., Baig, M.H., Lee, E.J., Choi, I., 2019. Mechanistic insights into the antimicrobial actions of metallic nanoparticles and their implications for multidrug resistance. Int. J. Mol. Sci. 20 (10), 2468.

Sharma, A.K., Gothwal, A., Kesharwani, P., Alsaab, H., Iyer, A.K., Gupta, U., 2017. Dendrimer nanoarchitectures for cancer diagnosis and anticancer drug delivery. Drug Discov. Today 22 (2), 314–326.

Sheikhpour, M., Barani, L., Kasaeian, A., 2017. Biomimetics in drug delivery systems: a critical review. J. Control. Release 253, 97–109.

Singh, J., Jain, K., Mehra, N.K., Jain, N., 2016a. Dendrimers in anticancer drug delivery: mechanism of interaction of drug and dendrimers. Artif. Cells Nanomed. Biotechnol. 44 (7), 1626–1634.

Sonju, J.J., Dahal, A., Singh, S.S., Jois, S.D., 2020. Peptide-functionalized liposomes as therapeutic and diagnostic tools for cancer treatment. J. Control. Release 329, 624–644.

Starpharma, 2020. DEP®. Retrieved 30, 2021, from: https://www.starpharma.com/drug_delivery/dep-posters.

Su, T., Cheng, F., Pu, Y., Cao, J., Lin, S., Zhu, G., He, B., 2021. Polymeric micelles amplify tumor oxidative stresses through combining PDT and glutathione depletion for synergistic cancer chemotherapy. Chem. Eng. J. 411 (128561), 1–12.

Sun, Y., Xia, Y., 2002. Shape-controlled synthesis of gold and silver nanoparticles reports. Science 298, 2176–2179.

Sun, B., Zhou, G., Zhang, H., 2016. Synthesis, functionalization, and applications of morphology-controllable silica-based nanostructures: a review. Prog. Solid State Chem. 44 (1), 1–19.

Taillefer, J., Jones, M.C., Brasseur, N., Van Lier, J., Leroux, J.C., 2000. Preparation and characterization of pH-responsive polymeric micelles for the delivery of photosensitizing anticancer drugs. J. Pharm. Sci. 89 (1), 52–62.

Teruel, A.H., Perez-Esteve, E., Gonzalez-Alvarez, I., Gonzalez-Alvarez, M., Costero, A.M., Ferri, D., Parra, M., Gavina, P., Merino, V., Martinez-Manez, R., Sancenon, F., 2018. Smart gated magnetic silica mesoporous particles for targeted colon drug delivery: new approaches for inflammatory bowel diseases treatment. J. Control. Release 281, 58–69.

Torchilin, V.P., 2005. Recent advances with liposomes as pharmaceutical carriers. Nat. Rev. Drug Discov. 4 (2), 145–160.

Uehara, A., Booth, S.G., Chang, S.Y., Schroeder, S.L., Imai, T., Hashimoto, T., Mosselmans, J.F., Dryfe, R.A., 2015. Electrochemical insight into the Brust-Schiffrin synthesis of Au nanoparticles. J. Am. Chem. Soc. 137 (48), 15135–15144.

Velusamy, P., Kumar, G.V., Jeyanthi, V., Das, J., Pachaiappan, R., 2016. Bio-inspired green nanoparticles: synthesis, mechanism, and antibacterial application. Toxicol. Res. 32 (2), 95–102.

Wang, J., Mao, W., Lock, L.L., Tang, J., Soi, M., Sun, W., Cui, H., Xu, D., Shen, Y., 2015a. The role of micelle size in tumor accumulation, penetration and treatment. ACS Nano 9, 7195–7206.

Wang, Y., Zhao, Q., Han, N., Bai, L., Li, J., Liu, J., Che, E., Hu, L., Zhang, Q., Jiang, T., Wang, S., 2015b. Mesoporous silica nanoparticles in drug delivery and biomedical applications. Nanomedicine 11 (2), 313–327.

Wang, H., Huang, Q., Chang, H., Xiao, J., Cheng, Y., 2016. Stimuli-responsive dendrimers in drug delivery. Biomater. Sci. 4 (3), 375–390.

Watermann, A., Brieger, J., 2017. Mesoporous silica nanoparticles as drug delivery vehicles in cancer. Nanomaterials 7 (7), 189.

Wei, D., Qian, W., 2008. Facile synthesis of Ag and Au nanoparticles utilizing chitosan as a mediator agent. Colloids Surf. B: Biointerfaces 62 (1), 136–142.

Wozniak, A., Malankowska, A., Nowaczyk, G., Grzeskowiak, B.F., Tusnio, K., Slomski, R., Zaleska-Medynska, A., Jurga, S., 2017. Size and shape-dependent cytotoxicity profile of gold nanoparticles for biomedical applications. J. Mater. Sci. Mater. Med. 28 (6), 92.

Yang, B., Chen, Y., Shi, J., 2019a. Mesoporous silica/organosilica nanoparticles: synthesis, biological effect and biomedical application. Mater. Sci. Eng. R 137, 66–105.

Yang, W., Veroniaina, H., Qi, X., Chen, P., Li, F., Ke, P.C., 2019b. Soft and condensed nanoparticles and nanoformulations for cancer drug delivery and repurpose. Adv. Ther. 3 (1), 1900102.

Yeh, Y.C., Creran, B., Rotello, V.M., 2012. Gold nanoparticles: preparation, properties, and applications in bionanotechnology. Nanoscale 4 (6), 1871–1880.

Yhee, J.Y., Im, J., Nho, R.S., 2016. Advanced therapeutic strategies for chronic lung disease using nanoparticle-based drug delivery. J. Clin. Med. 5 (9), 82.

Zangabad, P.S., Mirkiani, S., Shahsavari, S., Masoudi, B., Masroor, M., Hamed, H., Jafari, Z., Taghipour, Y.D., Hashemi, H., Karimi, M., Hamblin, M.R., 2018. Stimulus-responsive liposomes as smart nanoplatforms for drug delivery applications. Nanotechnol. Rev. 7 (1), 95–122.

Zhang, X., Jackson, J.K., Burt, H.M., 1996. Development of amphiphilic diblock copolymers as micellar carriers of taxol. Int. J. Pharm. 132 (1–2), 195–206.

Zhang, J., Wu, M., Yang, J., Wu, Q., Jin, Z., 2009. Anionic poly (lactic acid)-polyurethane micelles as potential biodegradable drug delivery carriers. Colloids Surf. A Physicochem. Eng. Asp. 337 (1–3), 200–204.

Zhou, W., Liu, W., Zou, L., Liu, W., Liu, C., Liang, R., Chen, J., 2014. Storage stability and skin permeation of vitamin C liposomes improved by pectin coating. Colloids Surf. B: Biointerfaces 117, 330–337.

Zhou, Q., Zhang, L., Yang, T., Wu, H., 2018. Stimuli-responsive polymeric micelles for drug delivery and cancer therapy. Int. J. Nanomedicine 13, 2921–2942.

Zhu, Y., Liao, L., 2015. Applications of nanoparticles for anticancer drug delivery: a review. J. Nanosci. Nanotechnol. 15 (7), 4753–4773.

Zhu, Y., Yang, B., Chen, S., Du, J., 2017. Polymer vesicles: mechanism, preparation, application, and responsive behavior. Prog. Polym. Sci. 64, 1–22.

Zylberberg, C., Matosevic, S., 2016. Pharmaceutical liposomal drug delivery: a review of new delivery systems and a look at the regulatory landscape. Drug Deliv. 23 (9), 3319–3329.

CHAPTER

19

Nano-materials as biosensor for heavy metal detection

Samprit Bose, Sourav Maity, and Angana Sarkar

Department of Biotechnology and Medical Engineering, National Institute of Technology, Rourkela, Odisha, India

1 Introduction

One of the most important biosensor components is the sensing element which is also termed as the receptor unit. The main objective of a receptor is to properly select an analyte that has to be determined. The next component is the transducer system which converts the biological response of an analyte into an electrical form. The transducer is responsible for energy conversion. The detector is the third component that amplifies the received electrical signal so that the response can be analyzed properly (Malik et al., 2013). The surface-to-volume ratio of a nanomaterial is very high. The range of a nanomaterial size is 1 to 100 nm. Due to their smaller size, nanomaterials possess unique characteristics that are highly distinct from the same materials in a bulk state, thus making the nanomaterials suitable for the fabrication of biosensors. The nanostructures which are being created using the concept of nanotechnology for biosensing purposes have brought a revolution in the field of molecular biology and have allowed the opportunity to manipulate at the molecular level and monitor the biological activities occurring at the anatomical level with greater accuracy. For the improvement of biological signaling and transduction mechanism, the electronic and magnetic properties of several nanomaterials have been studied. Some of the nanomaterials that are used to fabricate biosensors include nanoparticles, nanotubes, nanorods, nanowires, and thin films of nanocrystalline materials among which nanoparticles are the most analyzed nanomaterial till date (Lee et al., 2012). The nanomaterials used to fabricate biosensors can perform a wide range of functions. Nanomaterial-based biosensors are used as an amperometric device for the enzymatic determination of glucose. Quantum dots are being used as fluorescent agents for the determination of biomolecules. Nanomaterials are used to determine the presence of heavy metals (Maity et al., 2020a). Also, the detection of nucleic acid sequences is being done by metal-based nanoparticles. Several factors have to be taken care of while using a certain type of nanomaterial for biosensing purposes. Different types of properties such as electrical and electromechanical properties are being engineered into different materials along with nanoelectromechanical systems (NEMS). This has enabled the materials to possess several complex electrical, optic, thermal, fluidic, and magnetic properties that are unique in nature. Several novel properties are being incorporated into different materials as they operate at the nanoscale level by NEMS technology. Improvements in the field of bio-adhesion properties and response to a large range of stimuli have been achieved by coupling the NEMS devices with biological molecules and systems. Several surface forces like frictional, adhesive, and cohesive forces can be regulated more accurately using NEMS technology. The implementation of NEMS has improved the biochemical interactions which occur in biosensing technology. Nanofabrication is the process of design and manufacture of devices having nanoscale dimensions. The technique of nanofabrication used for biosensing involves four main processes, photolithography, surface etching strategies, thin-film growth, and several chemical bonding parameters. Nanomaterial-based biosensors have made the process of biosensing much easier, smarter, quicker, and cost-effective. Such types of biosensors made up of nanomaterials and nanostructures like nanoparticles, nanotubes and quantum dots are much more user-friendly and have multiple functionalities. Nanomaterial-based biosensors are recently used to merge the chemical and biological sensors which helps to make the overall process much more efficient and better in terms of performance. Heavy metals like mercury, arsenic, chromium, lead, and cadmium are highly toxic in nature. Increasing industrial activities are responsible for heavy metals' discharge into the environment (Maity et al., 2021a). The nonbiodegradable nature of these heavy metals contaminates the environment. They also enter the body through the foods we consume and possess several health-related problems to a human being (Maity et al., 2020b). Thus, pollution through heavy metals has emerged as a serious threat globally, and hence it has become of the utmost importance to monitor and detect heavy metals present in the environment, drinking water, biological fluids, and food for their remediation (Maity et al., 2021b). There are several conventional techniques for heavy metals detection which include ultraviolet-visible (UV) spectroscopy, atomic absorption spectroscopy (AAS), and atomic emission spectroscopy. These methods are tedious, costly, highly sensitive, and require trained personnel for implementation. They are also very difficult to carry from one place to another for on-site detection. Biosensors have become a relevant substitute for these conventional methods. Advancement in the study area of nanotechnology has given a great opportunity for refining the quality and execution of these biosensors for the on-site determination of multiple heavy metals. Though there are several technical complexities involved in the fabrication of a particular biosensor, the incorporation of nanomaterials in biosensors has proven to be a big success due to its result-oriented experimental support. Nanomaterial-based biosensors have drastically improved the quality of biosensors in respect to sensitivity, reproducibility, and detection range also made miniaturization of devices possible with the help of lab-on-chip (LOC) technology (Alhadrami, 2018). In this chapter, we discussed various types of nanomaterial-based biosensors responsible for heavy metals detection.

2 Biosensor

A biosensor is a scientific device where a response that is biological in nature is converted to an electrical signal. A biosensor integrates a biological element with the help of a transducer that is physicochemical in nature. The signal generated is in proportion to a single analyte which is then finally transmitted to a detector (Malik et al., 2013).

Biosensors provide better sensitivity and stability in respect to other standard methods. The different applications of a biosensor are discussed below:

- Food processing industries: Biosensors are used to detect pathogens in food. In food processing industries, biosensors are used for constant monitoring and maintaining the quality of food products in an economical manner thereby replacing the expensive and time-consuming traditional techniques.
- Medical field: Applications of biosensors to diagnose harmful diseases. Glucose biosensors are used for diabetes mellitus detection. Various other applications of biosensors in the field of medicine include cardiac markers quantification in undiluted serum, biological chips for efficient and precise determination of numerous neurochemical and cancer markers detection.
- Drug discovery: Biosensors are also used for fluorescence purposes for the identification of drugs and cancer cells.
- Biodefence: Biosensors are used for sensitive and selective detection of viruses during biological attacks.
- Plant biology: Biosensors have replaced the traditional methods and have developed advanced technologies for the detection of enzyme substrates, receptors, and transporters in plants.

2.1 Types of biosensor

The different types of biosensors (Fig. 1) are discussed below:

- The biological element is the first component of a biosensor. It is used to hold together the molecules that are being targeted and should be immobilized, highly particular, and steady when they are under storage conditions. The biological element may include antibody, DNA, enzyme, biomimetic, and phage (Ali et al., 2017).
- The transducing element is divided into four categories (Alhadrami, 2018):

  1. Optical Biosensors: They are the most common analytical device consisting of a biorecognition element and an optical transducer. The optical biosensor produces a

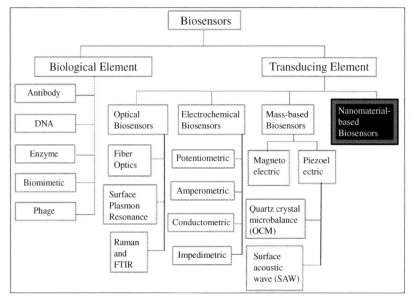

FIG. 1   Types of biosensors.

signal with respect to the measured analyte concentration. Optical biosensors can be
further subcategorized into:
a. Fiber optics: They are mainly derived from the principle of total internal reflection.
    Fiber-optic biosensors make use of the optical field for the measurement of cells,
    proteins, and DNA.
b. Surface Plasmon Resonance (SPR): When a plane-polarized light, hits a thin film
    made of metal, under the conditions of total internal reflection, SPR takes place.
c. Raman and FTIR: Cellular biochemistry study is successfully achieved through the
    technique of Raman spectroscopy. The ability of the Raman spectroscopy cell-based
    system to respond freely to many biologically active compounds instead of a single
    toxic agent is a potential advantage. Additional information regarding the time-
    dependent changes of cellular biochemistry is also provided by Raman
    spectroscopy biosensors.
2. Electrochemical biosensors: A self-contained integrated device, an electrochemical
    biosensor provides particular measurable or semi measurable scientific information
    employing a biochemical receptor, kept in direct vicinity with an electrochemical
    transducer. They can be further subcategorized into (Girigoswami and Akhtar, 2019):
    a. Potentiometric biosensor: It is an analytical device that makes use of ion-selective
        electrodes, such as pH meters to convert biological reactions into electrical signal.
    b. Amperometric biosensor: It is an integrated, analytical device in which current is
        measured based on oxidation or reduction of the biological element which are
        electroactive in nature.
    c. Conductometric biosensor: It measures the electrical conductivity of a sample
        solution present between two adjacent electrodes.

    d.   Impedimetric biosensor: The biological receptors, immobilized on the electrode surface result in the formation of an impedimetric biosensor.

  3. Mass-based biosensors: The mass-based biosensors are subcategorized into:

    a.   Magnetoelectric biosensor: It produces a magnetoelectric (ME) effect which generates an electrical output when a magnetic field is applied.

    b.   Piezoelectric biosensor: It is an analytical device that works on the principle of affinity interaction recording. Some piezoelectric devices use crystals, like quartz, which oscillate under the influence of an electric field.

  4. Nanomaterial-based biosensor: It is a biosensor operating at a nano-scale level used for the detection of an analyte.

2.2 Nanomaterial-based biosensors

- A nano-biosensor is a biosensor that operates at a nano-scale level for the detection of an analyte with the help of a bioreceptor, transducer, and a physicochemical detector (Malik et al., 2013).
- Nanomaterials have dimensions within 1 and 100 nm and due to their small size, most of the atoms are present in the vicinity of the surface.
- Nanomaterials have several physical and chemical characteristics that do not match with the characteristics of the same materials when they are present at a large scale thus making them perfectly suitable for the manufacturing of biosensors and monitor every occurrence at the physiological level with far great accuracy.
- The surface-to-volume ratios of nanomaterials are very high, thus allowing the surface to be used more properly and efficiently.
- One of the important factors for considering nanomaterials for sensing function is due to their excellent optical properties. The unique optical characteristics of nanomaterials provide them photonic properties and make them highly potential to be used as fluorophores.
- Several nanomaterials like nanoparticles, quantum dots, carbon nanotubes, nanorods, and nanowires are used as biosensors due to their several unique properties like large aspect ratios, potential to be functionalized, fine catalytic properties, better electrical conductivity, outstanding fluorescence, good charge conduction, and good electrical and sensing properties.

The different types of nano-based biosensors are outlined in Fig. 2.

2.3 Basic principles of nanomaterial-based biosensors

Heavy metals like chromium, mercury, cadmium, arsenic, and lead are nonbiodegradable and highly toxic in nature. These heavy metals are being detected using nanomaterial-based biosensors. Nanomaterial-based biosensors have made the overall process of detection of heavy metals much easier, smarter, quicker, and cost-effective thereby replacing the costly, tedious conventional methods of heavy metal detection. The nanomaterials that are being used for the detection are mainly nanoparticles, quantum dots (QDs), and carbon nanotubes (CNTs) (Fig. 3).

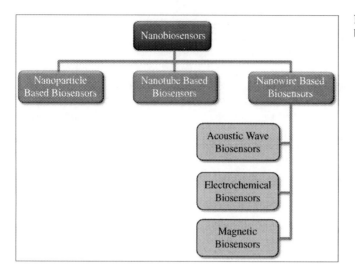

FIG. 2   Types of nanomaterial-based biosensors.

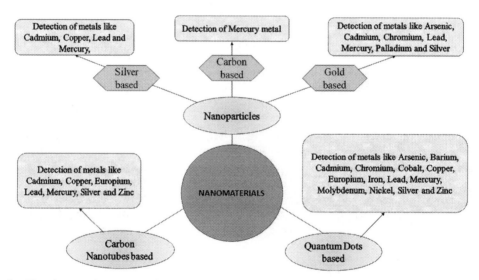

FIG. 3   Classification of nanomaterials.

Unmodified silver nanoparticles are applied to determine the presence of Hg^{2+} ions and have a detection range of 25–500 nM (Wang et al., 2010). Nanoparticles made of silver are used to manufacture a colorimetric sensor and electrochemical biosensor for the determination of Hg^{2+} (Duan et al., 2014) and Pb^{2+} (Tang et al., 2016) respectively. Biosynthesized silver nanoparticles detect Hg^{2+} and Cu^{2+} (Maiti et al., 2016). Different biosensors like colorimetric and resonance scattering-based biosensors, DNA biosensors are constructed using gold nanoparticles to detect As (III) (Wu et al., 2012a,b) and Hg^{2+} (Mashhadizadeh and Talemi, 2014) respectively. The pairing of fluorescence donors with gold nanoparticles improves Pb^{2+} detection (Kim et al., 2011). Gold nanoparticle-modified optical fiber is used to develop localized surface plasmon resonance (LSPR) optical fiber to detect Cd^{2+} ions (Lin and Chung, 2009). Development of potentiometric biosensor using self-assembled gold nanoparticles detects the presence of mercury ions (Yang et al., 2006). Green synthesis of fluorescent carbon nanoparticles is done to detect Hg^{2+} ions (Lu et al., 2012). Carbon nanoparticles are used for the development of a ratiometric fluorescent biosensor for Hg^{2+} ions detection in an aqueous medium (Lan et al., 2014).

One of the key nanomaterials used for heavy metals detection is quantum dots. Functionalized quantum dots of Cadmium Sulfide (CdS) are applied to develop a luminescence probe, helpful for heavy metals determination like Ag^+, Hg^{2+}, Cu^{2+}, Co^{2+}, Ni^{2+} in aqueous solution (Chen et al., 2008). Detection of Cu^{2+} ions using fluorescent probes made of DDTC-functionalized CdS/CdSe quantum dots (Wang et al., 2011b), fluorescent sensor made of CdS quantum dots (Lai et al., 2006), glyphosphate-functionalized CdS quantum dots (Liu et al., 2012), and cysteamine capped CdS quantum dots (Boonmee et al., 2016). To detect heavy metals like Cu^{2+} ions (Lai et al., 2006) and Pb^{2+} ions (Li et al., 2013b) cadmium selenide (CdSe) quantum dots are being used. Detection of Cd^{2+} ions is done by developing Thiogycerol-functionalized CdSe quantum dots (Brahim et al., 2015). Application of cadmium telluride (CdTe) quantum dots to determine heavy metals like Pb^{2+} ions (Wu et al., 2008), Fe^{2+} ions (Wu et al., 2009), Cu^{2+} ions (Wang et al., 2009; Guo et al., 2012; Rao et al., 2016), Hg^{2+} ions (Li et al., 2011b; Tao et al., 2014) and Ag^+ ions (Feng et al., 2016). Determination of Hg^{2+} ions are done by using functionalized zinc sulfide (ZnS) quantum dots (Li et al., 2008), N-acetyl-L-cysteine which is capped to ZnS quantum dots (Duan et al., 2011), and cetyltrimethyammonium bromide(CTAB)-capped ZnS quantum dots that are manganese doped (Xie et al., 2012), protein-functionalized ZnS quantum dots that are manganese doped, are used to detect Cr^{3+} ions (Zhao et al., 2013a). Detection of Pb^{2+} ions in water is done by developing a room temperature phosphorescence (RTP) sensor using glutathione-capped ZnS quantum dots doped with manganese (Chen et al., 2016). Zinc selenide (ZnSe) quantum dots are applied for the detection of heavy metals like Hg^{2+} (Ke et al., 2012) and Cu^{2+} (Wu et al., 2014). Fluorescent sensors are also being developed using zinc telluride (ZnTe) quantum dots to detect Fe^{3+} ions (Xing et al., 2018). The graphene quantum dots are used to detect heavy metals like Hg^{2+} (Li et al., 2015), Pb^{2+} (Qian et al., 2015; Niu et al., 2018), and Fe^{3+} (Xu et al., 2016).

Nano adsorbents like titanium dioxide, alumina, and oxides of iron which are based on metals are being used for heavy metal detection such as arsenic, copper, chromium, lead, nickel, mercury, and cadmium (Bose et al., 2021; Maity et al., 2020a).

Another type of nanomaterial used for heavy metal detection is carbon nanotubes (CNTs). CNT threads are used for the determination and analysis of various trace metals or organic pollutants (Maity et al., 2020a) like Cu^{2+}, Pb^{2+}, Cd^{2+}, Zn^{2+} ions (Zhao et al., 2014), three dimensional graphene-carbon nanotube hybrid electrodes are used to determine Pb^{2+}, Cd^{2+} ions (Huang et al., 2014), single-walled and multiwalled CNTs are used to detect Pb^{2+} (Lian et al., 2014) and Eu^{3+} (Wang et al., 2014a) ions respectively. The development of a sensor was done using a composite electrode made of carbon nanotube for the determination of heavy metals Pb^{2+}, Cd^{2+} (Xuan and Park, 2018).

Table 1 represents a brief analysis of the different kinds of nanomaterials used for heavy metal detection, the basic principles behind each detection, and their respective range of detection.

TABLE 1 Different nanomaterials used heavy metal detection.

Sl. no.	Nanomaterials	Basic principles	Metal detected	Range of detection	References
A.	*Nanoparticles*				
1.	Unmodified silver nanoparticles	Change in the shape of mercury-specific oligonucleotides (MSO) from random curly design to hairpin structure due to the incorporation of Hg^{+2} and salt-induced unmodified silver nanoparticles aggregation	Hg^{+2}	25–500 nM	Wang et al. (2010)
2.	Cysteine-functionalized silver nanoparticles	Surface-enhanced Raman scattering (SERS) with the help of cysteine-functionalized silver nanoparticles attached with Raman-labeling molecules	Hg^{+2}, Cu^{+2}	1 pM for Hg^{+2} 10 pM for Cu^{+2}	Zhang et al. (2018)
3.	Silver nanoparticles	Development of a colorimetric sensor by adding 6-thioguanine to silver nanoparticle solution to detect Hg^{2+} ions	Hg^{2+}	0 nM–333 nM	Duan et al. (2014)
4.	Biosynthesized silver nanoparticles	Functionalization of silver nanoparticles with 3-mercapto-1, 2-propanediol (MPD) to determine the presence of Hg^{+2} in water by colorimetric method Detection of Cu^{+2} is done by observing the change in absorbance due to the complex formation of metal ions	Hg^{+2}, Cu^{+2}	–	Maiti et al. (2016)
5.	Silver nanoparticles	Fabrication of a novel electrochemical biosensor using graphene oxide as a catalytic probe and deposition substrate to detect Pb^{2+} ions	Pb^{2+}	0.1 nM to 10 μM	Tang et al. (2016)

#	Material	Description	Ion	Detection limit	Reference
6.	Silver nanoparticles	Development of a colorimetric sensor to detect mercury ions using nanoparticles of silver in an aqueous medium	Hg^{2+}	Quantification limit of 8.5×10^{-7} M	Firdaus et al. (2017)
7.	Thiamine-functionalized silver nanoparticles	Development of colorimetric sensor to determine mercury ions using an aqueous medium using thiamine-functionalized nanoparticles made of silver	Hg^{2+}	1×10^{-8} M to 5×10^{-6} M	Khan et al. (2018)
8.	Graphene oxide/silver nanoparticle composite	To detect heavy metal ions by reduced nanoparticle composite of graphene oxide/silver through facile synthesis	Cu^{2+}, Cd^{2+}, Hg^{2+}	Limit of detection: 10^{-15} M for Cu^{2+}, 10^{-21} M for Cd^{2+} and 10^{-29} M for Hg^{2+}	Cheng et al. (2019)
9.	Gold nanoparticles	Colorimetric biosensor development based upon DNAzyme-directed assembly of gold nanoparticles to detect lead	Pb^{2+}	100nM–200µM	Liu and Lu (2003)
10.	Self-assembled gold nanoparticles	Development of a renewable potentiometric biosensor that inhibits urease and detects the presence of mercury ions	Hg^{2+}	0.09–1.99 µmol/L	Yang et al. (2006)
11.	DNA-functionalized gold nanoparticles	To detect Hg^{2+} ions applying DNA-functionalized nanoparticles made of gold by colorimetric method	Hg^{2+}	100nM (20ppb) approx.	Lee et al. (2007)
12.	Optical fiber modified with nanoparticles made of gold ($NM_{Au}OF$)	Development of a localized surface plasmon resonance (LSPR) optical fiber to detect Cd(II) concentration	Cd(II)	1–8ppb	Lin and Chung (2009)
13.	Label-free gold nanoparticles	Development of a colorimetric biosensor by applying label-free nanoparticles made of gold and alkanethiols to detect Hg^{2+}, Ag^+, Pb^{2+} ions	Hg^{2+}, Ag^+, Pb^{2+}	—	Hung et al. (2010)

Continued

TABLE 1 Different nanomaterials used heavy metal detection—cont'd

Sl. no.	Nanomaterials	Basic principles	Metal detected	Range of detection	References
14.	Glutathione functionalized gold nanoparticles	A cost-efficient and effective method of colorimetric determination of Pb^{2+} in lake water using nanoparticles made of gold and functionalized using glutathione	Pb^{2+}	0.1–10 μM	Chai et al. (2010)
15.	Gold nanoparticles	Pairing of fluorescence donors with gold nanoparticles to improve Pb^{2+} detection	Pb^{2+}	Detection limit of 5nM	Kim et al. (2011)
16.	Gold nanoparticles	Development of a colorimetric and resonance scattering-based biosensor to detect As(III) by aggregating gold nanoparticles and forming a special bond between arsenic-binding aptamer, target, and cationic surfactant in aqueous solution	As(III)	1–1500 ppb	Wu et al. (2012a)
17.	Gold nanoparticles	Development of a colorimetric biosensor to detect arsenic (III)	Arsenic (III)	Quantification limit of 5.3 ppb	Wu et al. (2012b)
18.	Gold nanoparticles	Development of a DNA biosensor for quantifying the concentration of mercury ions using nanoparticles made of gold on a modified glass surface	Hg^{2+}	10nM–10 μM	Mashhadizadeh and Talemi (2014)
19.	Gold nanoparticles	Development of colorimetric sensor using nanoparticles made of gold on a paper-based scientific device to find out the concentration of Hg^{2+} in water	Hg^{2+}	Limit of detection of 50nM	Chen et al. (2014)

20.	DNA-functionalized gold nanoparticles	Development of a fluorescence spectrometric sensor using thiol-DNA-functionalized nanoparticles made of gold to detect Hg^{2+}	Hg^{2+}	20–90 nM	Wang et al. (2015)
21.	Thymine modified gold nanoparticles	Development of a reusable electrochemical biosensor based on nanoparticles made of gold and modified using thymine for the determination of Hg^{2+} ions	Hg^{2+}	10 ng/L to 1.0 µg/L	Wang et al. (2016)
22.	Gold nanoparticles	Development of a turn-on fluorescent sensor to detect Pb^{2+} ions using gold nanoparticles and graphene quantum dots	Pb^{2+}	50 nM–4 µM	Niu et al. (2017)
23.	Mecaptosuccinic acid-modified nanoparticles made of gold	Colorimetric sensor development using mecaptosuccinic acid-modified gold nanoparticles to detect Cr^{3+} ions	Cr^{3+}	0.6–1.4 µM	Yu et al. (2017)
24.	Gold nanoparticles	Colorimetric nanosensor development to detect Pd^{2+} ions using synthesized gold nanoparticles	Pd^{2+}	Quantification limit of 4.3 µM	Anwar et al. (2018)
25.	Carbon nanoparticles	Development of a fluorescence sensor using carbon nanoparticles to detect Hg^{2+} ions	Hg^{2+}	Quantification limit of 10 nM	Li et al. (2011a)
26.	Fluorescent carbon nanoparticles	Detection of Hg^{2+} ions by fluorescent carbon nanoparticles through green synthesis	Hg^{2+}	Detection limit 0.23 nM	Lu et al. (2012)
27.	Carbon nanoparticles	Development of a ratiometric fluorescent biosensor using carbon nanoparticles to detect Hg^{2+} ions	Hg^{2+}	0–6 µM	Lan et al. (2014)

Continued

TABLE 1 Different nanomaterials used heavy metal detection—cont'd

Sl. no.	Nanomaterials	Basic principles	Metal detected	Range of detection	References
B.	*Quantum dots*				
1.	Functionalized cadmium sulfide quantum dots	Fluorescence probe development using functionalized cadmium sulfide quantum dots to detect silver ions	Ag^+	2.0×10^{-8}–1.0×10^{-6} mol/L	Chen and Zhu (2005)
2.	Functionalized cadmium sulfide quantum dots	Luminescence probe development using functionalized cadmium sulfide quantum dots to detect heavy metals in sample aqueous medium	Ag^+, Hg^{2+}, Cu^{2+}, Co^{2+}, Ni^{2+}	1.25×10^{-7}–5×10^{-6} mol/L for Ag^+, 1.5×10^{-8}–7.5×10^{-7} mol/L for Hg^{2+}, 3.0×10^{-7}–1.0×10^{-5} mol/L for Ni^{2+}, 4.59×10^{-8}–2.29×10^{-6} mol/L for Cu^{2+} ions	Chen et al. (2008)
3.	Diethyldithiocarbamate (DDTC) functionalized cadmium sulfide quantum dots	Development of a fluorescent probe using DDTC-functionalized cadmium sulfide quantum dots to detect Cu^{2+} ions	Cu^{2+}	0–100 μg/L	Wang et al. (2011b)
4.	Cadmium sulfide quantum dots	Development of a fluorescent sensor using cadmium sulfide quantum dots to detect Cu^{2+} ions	Cu^{2+}	—	Lai et al. (2006)
5.	Glyphosphate-functionalized cadmium sulfide quantum dots	Fluorescence probe development using glyphosphate functionalized cadmium sulfide quantum dots to detect Cu^{2+} ions	Cu^{2+}	2.4×10^{-2}–28 μg/mL	Liu et al. (2012)
6.	Dithizone functionalized cadmium sulfide quantum dots	Development of a fluorescence probe using dithizone-functionalized cadmium sulfide quantum dots to detect Pb^{2+} ions	Pb^{2+}	0.01 nmol/L to 20 μmol/L	Zhao et al. (2013b)

No.	Quantum dots	Description	Ion	Range/Limit	Reference
7.	Thiourea functionalized cadmium sulfide quantum dots	Development of a fluorescent probe using thiourea functionalized cadmium sulfide quantum dots to detect Hg^{2+} ions	Hg^{2+}	1–300 μg/L	Xi et al. (2016)
8.	Cysteamine capped cadmium sulfide quantum dots	Fluorescence probe development using cysteamine capped cadmium sulfide quantum dots to detect Cu^{2+} ions	Cu^{2+}	2 to 10 μM	Boonmee et al. (2016)
9.	Mercaptopropionic (MPA)-capped cadmium sulfide quantum dots	Synthesis and fluorescence of MPA-capped cadmium sulfide quantum dots to detect Mo^{2+} ions present in an aqueous solution	Mo^{2+}	0.04–2 μM	Mohamed et al. (2018)
10.	Citric acid capped cadmium sulfide quantum dots	Development of fluorescent sensor using cadmium sulfide quantum dots crowned with citric acid to detect Cu^{2+} ions present in an aqueous solution	Cu^{2+}	1.0×10^{-8} M to 5.0×10^{-5} M	Wang et al. (2018)
11.	Functionalized CdSe quantum dots	Development of a luminescent nanosensor to detect Hg^{2+} ions applying functionalized CdSe quantum dots	Hg^{2+}	Quantification limit of 1.8×10^{-7} M	Li et al. (2008)
12.	16-Mercaptohexadecanoic acid (16-MHA) capped CdSe quantum dots	Detection of Cu^{2+} ions using photoluminescence of CdSe quantum dots	Cu^{2+}	Quantification limit of 5nM and a dynamic range up to 100 μM	Chan et al. (2010)
13.	CdSe quantum dots	Spectrofluorometric determination of Cu^{2+} ions using CdSe quantum dots as fluorescent sensors	Cu^{2+}	Detection limit of 8.5 μg/L	Lai et al. (2006)
14.	Diethyldithiocarbamate (DDTC) functionalized CdSe quantum dots	Development of a fluorescent probe using CdSe quantum dots and functionalized with DDTC to detect Cu^{2+} ions	Cu^{2+}	0–100 μg/L	Wang et al. (2011b)

Continued

TABLE 1 Different nanomaterials used heavy metal detection—cont'd

Sl. no.	Nanomaterials	Basic principles	Metal detected	Range of detection	References
15.	Functionalized mercaptoethanol (ME) capped CdSe quantum dots	Detection of Ba^{2+} ions by developing luminescence probe based on functionalized Mercaptoethanol (ME) capped CdSe quantum dots	Ba^{2+}	1×10^{-7} to 1.2×10^{-6} mol/L	Mahmoud (2012)
16.	CdSe quantum dots	Biosensor development is based on the transfer of energy to graphene oxide from CdSe quantum dots to detect Pb^{2+} ions	Pb^{2+}	Quantification limit of 90 pM	Li et al. (2013b)
17.	Aqueous CdSe quantum dots	To detect Cu^{2+} ions by synthesis of CdSe quantum dots in an aqueous form directly	Cu^{2+}	10 nM to 7.5 μM	Bu et al. (2013)
18.	Fluorescence enhanced CdSe quantum dots	Development of fluorescence enhanced CdSe quantum dot to detect Cu^{2+} ions in sample aqueous solution	Cu^{2+}	0.13–0.16 ppb (2.0–2.5 nM)	Zhang et al. (2014)
19.	CdSe quantum dots functionalized with thiogycerol	Detection of Cd^{2+} ions by developing thiogycerol-functionalized CdSe quantum dots	Cd^{2+}	1 to 22 μM	Brahim et al. (2015)
20.	CdSe quantum dots functionalized with thiourea	Development of a fluorescent probe using thiourea functionalized CdSe dots for the determination of Hg^{2+} ions	Hg^{2+}	1–300 μg/L	Xi et al. (2016)
21.	P-Nitrophenyldiazenyphenyloxadiazole (NDPO) capped CdSe quantum dots	Development of a nanosensor using NDPO capped CdSe quantum dots to detect Cd^{2+} ions	Cd^{2+}	0.05–1.0 mmol/L	Eftekhari-Sis et al. (2018)

22.	CdTe quantum dots	Method development to detect Pb^{2+} ions based on the fluorescence quenching of CdTe quantum dots	Pb^{2+}	2.0×10^{-6} to 1.0×10^{-4} mol/L	Wu et al. (2008)
23.	CdTe quantum dots capped with glutathione	Development of fluorescent probes to detect Cr(IV) applying CdTe quantum dots capped with glutathione	Cr(VI)	0.01 to 1.00 μg/mL	Zhang et al. (2009)
24.	CdTe quantum dots	Detection of Fe^{2+} by photoluminescence using CdTe quantum dots-fenton hybrid system	Fe^{2+}	Detection limit of 5 nM	Wu et al. (2009)
25.	CdTe quantum dots	To detect Cu^{2+} applying CdTe quantum dots by fluorescence quenching method	Cu^{2+}	20–300 μg/L	Wang et al. (2009)
26.	Glutathione-capped CdTe quantum dots	Development of fluorescent probes to detect As(III) using glutathione-capped CdTe quantum dots	As(III)	5.0×10^{-6} to 25×10^{-5} mol L^{-1}	Wang et al. (2011a)
27.	CdTe quantum dots	Development of fluorescent probes using sol-gel extracted silica spheres surfaced with calix [6]arene to detect Hg^{2+} ions	Hg^{2+}	2.0–14.0 nmol/L	Li et al. (2011b)
28.	Multicolor CdTe quantum dots	To detect Pb^{2+} ions applying multicolor CdTe quantum with thioglycolic acid as a stabilizer	Pb^{2+}	Detection limit of 4.7 nmol/L.	Zhong et al. (2012)
29.	CdTe quantum dots	To detect Cu^{2+} ions applying CdTe quantum dots combined with enzyme inhibition	Cu^{2+}	Quantification limit of 0.176 ng/mL (2075 nM)	Guo et al. (2012)
30.	Cysteamine-quoted CdTe quantum dots	To detect Hg^{2+} ions by the synthesis of cysteamine-quoted CdTe quantum dots	Hg^{2+}	0.08–3.33 μM	Pei et al. (2012)

Continued

TABLE 1 Different nanomaterials used heavy metal detection—cont'd

Sl. no.	Nanomaterials	Basic principles	Metal detected	Range of detection	References
31.	Mercaptopropionic acid stabled CdTe quantum dots	Development of fluorescent probe to detect Ag^+ using mercaptopropionic acid stabled CdTe quantum dots	Ag^+	4×10^{-7} to 32×10^{-7} mol/L	Gan et al. (2012)
32.	CdTe quantum dots	Transfer of fluorescence resonance energy between fluorescent brightener and CdTe quantum dots to detect Hg^{2+} ions in water samples	Hg^{2+}	8.0×10^{-9} to 8.0×10^{-7} g/L	Tao et al. (2014)
33.	Mercaptosuccinic acid-modified CdTe quantum dots	Development of fluorescent sensor to detect Ag^+ using CdTe quantum dots crowned with mercaptosuccinic acid in aqueous solution	Ag^+	0.4 to 8 μmol/L	Jiao et al. (2014)
34.	Homocysteine capped CdTe quantum dots	Fluorometric detection of Ag^+ using CdTe quantum dots crowned with homocysteine	Ag^+	Quantification limit of 8.3 nM	Cai et al. (2014)
35.	CdTe quantum dots	Development of ratiometric fluorescent probe to detect Cu^{2+} ion applying CdTe quantum dots	Cu^{2+}	0.1 to 1 μM	Rao et al. (2016)
36.	CdTe quantum dots	Development of a fluorescence resonance energy transfer (FRET) system between fluorescein isothiocyanate and CdTe quantum dots	Ag^+	0.01–8.96 nmol/L	Feng et al. (2016)
37.	Thiol-capped CdTe quantum dots	Development of a fluorescence quenching technique to detect the Cu^{2+} ions in water samples	Cu^{2+}	0.10–4.0 μg/mL	Nurerk et al. (2016)
38.	CdTe quantum dots	Visual detection of Cu^{2+} ions by developing a ratiometric fluorescent paper sensor applying CdTe quantum dots	Cu^{2+}	Quantification limit of 0.36 nM	Wang et al. (2016)

No.	Material	Study	Ion	Range/Limit	Reference
39.	Bovine serum albumin (BSA)-capped CdTe quantum dots	Fluorescent probe development to detect Hg^{2+} using bovine serum albumin(BSA)-capped CdTe quantum dots	Hg^{2+}	0.001 to 1 μmol/L	Zhu et al. (2017)
40.	CdTe quantum dots coated with ligand	Development of a nanosensor using CdTe quantum dots coated with ligand to determine Cu^{2+} ions in environmental water samples	Cu^{2+}	$5.16 \pm 0.07 \times 10^{-8}$ to $1.50 \pm 0.03 \times 10^{-5}$ mol/L	Elmizadeh et al. (2017)
41.	CdTe quantum dots	Development of ratiometric fluorescent sensor using CdTe quantum dots to detect Fe^{3+} ions	Fe^{3+}	0–3.5 μM	Zhou et al. (2018a)
42.	Ligand-coated CdTe quantum dots	Development of a nanosensor to detect Cr^{3+} ions using ligand-coated CdTe quantum dots	Cr^{3+}	$6.78 \pm 0.05 \times 10^{-8}$ to $3.7 \pm 0.02 \times 10^{-6}$ mol/L	Elmizadeh et al. (2018)
43.	Functionalized ZnS quantum dots	Development of a luminescent nanosensor to detect Hg^{2+} ions applying functionalized ZnS quantum dots	Hg^{2+}	Quantification limit of 1.8×10^{-7} M	Li et al. (2008)
44.	ZnS quantum dots surfaced with N-acetyl-L-cysteine	An eco-friendly sensor development to detect Hg^{2+} ions using ZnS quantum dots surfaced with N-acetyl-L-cysteine	Hg^{2+}	0–2.4 μmol/L	Duan et al. (2011)
45.	ZnS quantum dots	Detection of Hg^{2+} ions by developing a highly sensitive probe using ZnS quantum dots	Hg^{2+}	0–40nM	Ke et al. (2012)
46.	Cetyltrimethyammonium bromide (CTAB) crowned with manganese doped ZnS quantum dots	Development of a room temperature phosphorescence (RTP) mercury ions sensor to detect mercury ions using CTAB/Mn-ZnS quantum dots	Hg^{2+}	Quantification limit of 1.5nM	Xie et al. (2012)

Continued

TABLE 1 Different nanomaterials used heavy metal detection—cont'd

Sl. no.	Nanomaterials	Basic principles	Metal detected	Range of detection	References
47.	Protein-functionalized manganese doped ZnS quantum dots	Development of a phosphorescent sensor using protein-functionalized manganese doped ZnS quantum dots for the detection of Cr^{3+} ions	Cr^{3+}	10–300 nM with a limit of detection of 3 nM	Zhao et al. (2013a)
48.	ZnS quantum dots capped with thioglycolic acid	Development of a fluorescent probe to detect Co^{2+} ions using ZnS quantum dots capped with thioglycolic acid	Co^{2+}	0.3012–90.36 μmol/L	Zi et al. (2014)
49.	ZnS quantum dots coated with silica	Development of a fluorescent probe using ZnS quantum dots coated with silica to detect Pb^{2+} ions	Pb^{2+}	10^{-9} to 2.6×10^{-4} M	Qu et al. (2014)
50.	3-Mercaptopropionic acid capped core-shell ZnS quantum dots	Development of a fluorescent probe using 3-mercaptopropionic acid capped core-shell ZnS quantum dots for the determination of Cu^{2+} ions	Cu^{2+}	2.5×10^{-9} to 17.5×10^{-7} M	Bian et al. (2015)
51.	Cystamine-stabilized ZnS quantum dots	Detection of As(III) by developing a fluorimetricapta sensor using cystamine-stabilized ZnS quantum dots	As(III)	1.0×10^{-11} to 1.0×10^{-6} mol/L	Ensafi et al. (2016)
52.	Chitosan-capped ZnS quantum dots	Sensor development for heavy metal ions using chitosan-capped ZnS quantum dots to detect Zn^{2+}, Cu^{2+} ions	Zn^{2+}, Cu^{2+}	4–400 ppm	Borgohain et al. (2016)
53.	Glutathione-capped manganese doped ZnS quantum dots	To detect Pb^{2+} ions in water by developing a room temperature phosphorescence (RTP) sensor using glutathione-capped manganese doped ZnS quantum dots	Pb^{2+}	1.0 to 100 μg/L	Chen et al. (2016)

No.	Quantum dots	Description	Ion	Detection	Reference
54.	ZnS quantum dots	Development of a PET sensor using ZnS quantum dots to detect Hg^{2+} ion aqueous medium	Hg^{2+}	Detection limit of 1 pM	Saikia et al. (2016)
55.	Mercaptopropionic acid capped Mn-doped ZnS quantum dots	Phosphorescence detection of Pb^{2+} ions in aqueous samples using mercaptopropionic acid capped Mn-doped ZnS quantum dots	Pb^{2+}	$3.69{-}10^{-8}$ mol/L	Gan et al. (2017)
56.	ZnS quantum dots	Development of a photoelectrochemical sensor using ZnS quantum dots to detect Hg^{2+} ions	Hg^{2+}	1×10^{-11} to 1×10^{-6} M	Wang et al. (2018)
57.	ZnSe quantum dots	Detection of Hg^{2+} ions by developing a highly sensitive probe using ZnSe quantum dots	Hg^{2+}	0–40 nM	Ke et al. (2012)
58.	ZnSe quantum dots	Development of a fluorescence sensor using ZnSe quantum dots to detect Cu^{2+} ions	Cu^{2+}	Quantification limit of 4.7×10^{-7} mol/L with an optimal pH of 7.0	Wu et al. (2014)
59.	ZnSe quantum dots capped with L-glutathione	Determination of Cu^{2+} ions using ZnSe quantum dots capped with L-glutathione	Cu^{2+}	Quantification limit of 2×10^{-10} mol/L	Ding et al. (2014)
60.	Manganese doped ZnSe quantum dots	To detect Hg^{2+} ions by fluorescence enhancement of manganese doped ZnSe quantum dots	Hg^{2+}	Limit of detection of 7 nM	Zhou et al. (2018b)
61.	ZnTe quantum dots	Development of a fluorescence sensor using ZnTe quantum dots to detect Fe^{3+} ions	Fe^{3+}	2.0×10^{-6} to 1.0×10^{-4} mol/L	Xing et al. (2018)
62.	Fluorescent graphene quantum dots	Development of nanoprobe using fluorescent graphene quantum dots to detect Hg^{2+} ions	Hg^{2+}	8×10^{-7} to 9×10^{-6} M	Wang et al. (2014b)

Continued

TABLE 1 Different nanomaterials used heavy metal detection—cont'd

Sl. no.	Nanomaterials	Basic principles	Metal detected	Range of detection	References
63.	Graphene quantum dots doped with sulfur	Fluorescent probe development to detect Fe^{3+} ions using sulfur-doped graphene quantum dots	Fe^{3+}	0–0.7 μM	Li et al. (2014)
64.	Graphene quantum dots	To detect Hg^{2+} ions by the process of fluorescence switching of graphene quantum dots	Hg^{2+}	Quantification limit of 0.439 nmol/L	Li et al. (2015)
65.	Graphene quantum dots	Development of a fluorescent nanosensor based on graphene quantum dots to detect Pb^{2+} ions	Pb^{2+}	9.9–435 nM	Qian et al. (2015)
66.	Graphene quantum dots doped with nitrogen	Development of a fluorescent probe to detect Ag^+ ions using graphene quantum dots doped with nitrogen	Ag^+	0.2–40 μM	Tabaraki and Nateghi (2016)
67.	Graphene quantum dots	Synthesis of graphene quantum dots with nanoparticles made of gold to detect heavy metal ions	Hg^{2+}, Cu^{2+}	Quantification limit of 0.02 nM for Hg^{2+} and 0.05 nM for Cu^{2+}	Ting et al. (2015)
68.	Graphene quantum dots	Development of a fluorescent sensing platform for the determination of Fe^{3+} ions using graphene quantum dots	Fe^{3+}	0.005 to 500 μM	Xu et al. (2016)
69.	Graphene quantum dots doped with sulfur	Development of a fluorescent sensing probe for the determination of Ag^+ ions applying graphene quantum dots doped with sulfur	Ag^+	0.1–130 μM	Bian et al. (2017)
70.	Graphene quantum dots	Development of a ratiometric fluorescent nanosensor for the determination of Ag^+ ions applying graphene quantum dots	Ag^+	0–115.2 μM	Zhao et al. (2017)

71.	Graphene quantum dots	Development of fluorescent sensor using quantum dots made of graphene and nanoparticles using gold to detect Pb^{2+} ions	Pb^{2+}	50nM to 4 μM	Niu et al. (2018)
72.	Graphene quantum dots doped with nitrogen	Development of an "on-off-on" fluorescent sensor to detect Hg^{2+} ions graphene quantum dots doped with nitrogen	Hg^{2+}	0–4.31 μM	Du et al. (2019)
C.	*Carbon nanotubes*				
1.	Carbon nanotube nanoelectrode array	Development of a carbon nanotube nanoelectrode array to detect trace metal ions	Cd^{2+}, Pb^{2+}	Quantification limit of 0.04 μg/L	Liu et al. (2005)
2.	Bismuth-modified carbon nanotube electrode	Development of a bismuth-modified carbon nanotube electrode to detect trace metal ions	Pb^{2+}, Cd^{2+}, Zn^{2+}	Quantification limit of 1.3 μg/L for Pb^{2+}, 0.7 μg/L for Cd^{2+} and 12 μg/L for Zn^{2+}	Hwang et al. (2008)
3.	Carbon nanotube-modified electrodes	Development of a voltammetry process using carbon nanotube-modified electrodes for the determination of trace heavy metals	Cu^{2+}, Pb^{2+}	Quantification limit of 15 ppb for Cu^{2+} and 1 ppb for Pb^{2+}	Morton et al. (2009)
4.	Carbon nanotube grafted with quinoline group	Development of a fluorescent sensor using carbon nanotube grafted with quinoline group to detect Cu^{2+} ions	Cu^{2+}	Quantification limit of 10^{-7} M	Dong et al. (2009)
5.	Single-walled carbon nanotube	Development of a reusable single-walled carbon nanotube fluorescent sensor to detect Ag^{+} ions in samples of water	Ag^{+}	Quantification limit of 1 nM	Zhao et al. (2010)

Continued

TABLE 1 Different nanomaterials used heavy metal detection—cont'd

Sl. no.	Nanomaterials	Basic principles	Metal detected	Range of detection	References
6.	Carbon nanotube tower electrode	Detection of heavy metals using carbon nanotube tower electrode by anodic stripping voltammetry method	Pb^{2+}, Cd^{2+}, Cu^{2+}, Zn^{2+}	Quantification limit of 12 nM for Pb^{2+}, 25 nM for Cd^{2+}, 44 nM for Cu^{2+}, 67 nM for Zn^{2+}	Guo et al. (2011)
7.	Pristine single-walled carbon nanotube electrodes	Development of pristine single-walled carbon nanotube electrodes to detect Cd^{2+} and Pb^{2+}	Cd^{2+}, Pb^{2+}	Quantification limit of 0.7 ppb for Cd^{2+} and 0.8 ppb for Pb^{2+}	Bui et al. (2012a)
8.	Carbon nanotube with single wall	Development of single-walled carbon nanotube using nanoparticles made of gold for the detection of lead and copper	Cu^{2+}, Pb^{2+}	3 Quantification limit of 0.613 ppb for Cu^{2+} and 0.546 ppb for Pb^{2+}	Bui et al. (2012b)
9.	Antimony nanoparticle-multiwalled carbon nanotube	Detection of heavy metals using antimony nanoparticle-multiwalled carbon nanotube	Cd^{2+}, Pb^{2+}	Quantification limit of 0.77 µg/L for Cd^{2+} and 0.65 µg/L for Pb^{2+}	Ashrafi et al. (2014)
10.	Carbon nanotube thread	Development of an anodic stripping voltammetry process using carbon nanotube threads to detect heavy metals	Cu^{2+}, Pb^{2+}, Cd^{2+}, Zn^{2+}	Quantification limit of 0.27 nM for Cu^{2+}, 1.5 nM for Pb^{2+}, 1.9 nM for Cd^{2+}, 1.4 nM for Zn^{2+}	Zhao et al. (2014)
11.	Three-dimensional graphene-carbon nanotube hybrid electrode	To detect heavy metal ions by developing a three-dimensional graphene-carbon nanotube hybrid electrode	Pb^{2+}, Cd^{2+}	Quantification limit of 0.2 µg/L for Pb^{2+} and 0.1 µg/L for Cd^{2+}	Huang et al. (2014)
12.	Single-walled carbon nanotube	Detection of Pb^{2+} ions by developing a sensor using DNA wrapped metallic single-walled carbon nanotube	Pb^{2+}	1.0×10^{-9} to 1.0×10^{-8} mol/L	Lian et al. (2014)

No.	Material	Description	Ion	Range	Note	Reference
13.	Multiwalled carbon nanotube	Development of an electrochemical sensor using carbon nanotube-loaded film of Nafion for the determination of Eu^{3+} ions	Eu^{3+}	1–100 nM		Wang et al. (2014a)
14.	Carbon nanotube	Development of carbon nanotube-based sensor to detect mercury ions in an aqueous medium	Hg^{2+}	5–90 nM		Pokhrel et al. (2017)
15.	Carbon nanotube thread	Development of an electrochemical cell using a thread of carbon nanotube to detect heavy metals	Hg^{2+}, Cu^{2+}, Pb^{2+}		Quantification limit of 1.05 nM for Hg^{2+} .0.53 nM for Cu^{2+}, 0.57 nM for Pb^{2+}	Zhao et al. (2017)
16.	Carbon nanotube nanocomposite electrode	Sensor development using carbon nanotube composite electrode for heavy metal determination	Pb^{2+}, Cd^{2+}		Quantification limit of 0.2 ppb for Pb^{2+} and 0.6 ppb for Cd^{2+}	Xuan and Park (2018)

3 Advancement on nanomaterial-based biosensor

Biosensors of nanomaterial-based are being used for the determination of heavy metals. The nanomaterials that are being used can be broadly classified into nanoparticles, quantum dots, and carbon nanotubes.

- Nanoparticles play a pivotal role to detect heavy metals. Heavy metals like cadmium, arsenic (III), mercury, copper, silver, mercury, and lead have been detected using silver, gold, and carbon nanoparticles.

    i. Unmodified silver nanoparticles were used to determine Hg^{2+} ions, having a range of detection of 25–500 nM (Wang et al., 2010). Surface-enhanced Raman scattering (SERS) with the help of cysteine-functionalized nanoparticles made of silver to detect Hg^{2+} ions (Li et al., 2013a) brought down the quantification limit for Hg^{2+} ions to 1 pM. A colorimetric sensor was developed by adding 6-thioguanine to silver nanoparticle solution to determine Hg^{2+} ions and the quantification limit was found to be 4 nM (Duan et al., 2014). Another colorimetric sensor was designed to detect Hg^{2+} ions applying silver nanoparticles in an aqueous medium having a quantification limit of 8.5×10^{-7} M (Firdaus et al., 2017). Facile synthesis was applied to detect Hg^{2+} using reduced Graphene oxide/Silver nanoparticle composite having a quantification limit of 10^{-29} M for Hg^{2+} (Cheng et al., 2019).

    ii. Self-assembled Gold Nanoparticles were used to develop a renewable potentiometric biosensor that inhibits urease and detects the presence of Hg^{2+} ions having a quantification limit of 0.05 µmol/L (Yang et al., 2006). Determination of Hg^{2+} ions was done using Gold Nanoparticles functionalized with DNA by the colorimetric method enhanced the quantification limit to 100 nM (Lee et al., 2007). Gold Nanoparticles functionalized with DNA were applied to detect Hg^{2+} ion in an aqueous medium further reducing the quantification limit of 8 nM (Wang et al., 2015).

    iii. Glutathione Functionalized Gold Nanoparticles were used to detect Pb^{2+} ions having the lowest detectable limit of 100 nM (Chai et al., 2010). The pairing of fluorescence donors with gold nanoparticles improved Pb^{2+} detection and lowered the detection limit to 5 nM (Kim et al., 2011). A turn-on fluorescent sensor was designed to determine Pb^{2+} ions using gold nanoparticles and graphene quantum dots which had a quantification limit of 16.7 nM (Niu et al., 2018).

    iv. Development of a colorimetric biosensor and resonance scattering-based biosensor to determine As(III) by aggregating gold nanoparticles with a quantification limit of 40 ppb (Wu et al., 2012a). The development of a colorimetric biosensor to detect Arsenic (III) in an aqueous solution reduced the quantification limit to 5.3 ppb (Wu et al., 2012b). A colorimetric nano-based sensor was designed for the determination of Pd^{2+} ions by making the use of synthesized gold nanoparticles having a quantification limit of 4.3 µM (Anwar et al., 2018).

    v. Development of a fluorescence sensor using carbon nanoparticles to detect Hg^{2+} ions in an aqueous medium with a quantification limit of 10 nM (Li et al., 2011a). Determination of Hg^{2+} ions by green synthesis of fluorescent nanoparticles of carbon molecules with a quantification limit of 0.23 nM (Lu et al., 2012). A ratiometric fluorescent biosensor using nanoparticles made of carbon was developed to detect Hg^{2+} ions in an aqueous medium which had a quantification limit of 42 nM (Lan et al., 2014).

Some nanoparticles-derived sensors for the detection of heavy metals are outlined in Fig. 4.

- Quantum dots are applied for heavy metal's detection like Barium, Arsenic, Europium, Cadmium, Chromium, Lead, Copper, Cobalt, Iron, Nickel, Mercury, Molybdenum, Zinc, and Silver.

  i. Development of a luminescence probe using functionalized Cadmium Sulfide (CdS) quantum dots for the determination of heavy metals like Cu^{2+} in an aqueous solution which as a range of detection of 4.59×10^{-8} to 2.295×10^{-6} mol/L (Chen et al., 2008). A fluorescent probe using DDTC-functionalized Cadmium Sulfide Quantum dots to determine Cu^{2+} ions has a quantification limit of 0.29 μg/L (Wang et al., 2011a,b). A fluorescence probe using glyphosphate-functionalized Cadmium Sulfide quantum dots developed for the determination of Cu^{2+} ions has a quantification limit of 1.3 μg/mL (Liu et al., 2012). Cysteamine-capped Cadmium Sulfide quantum dots were applied to develop a fluorescence probe for the determination of Cu^{2+} ions with a quantification limit of 1.5 μM (Boonmee et al., 2016). Citric acid-capped Cadmium Sulfide Quantum Dots were applied to develop a fluorescent sensor to detect Cu^{2+} ions in an aqueous medium and has a quantification limit of 9.2 nM (Wang et al., 2018).

  ii. A luminescent nano-sensor to detect Hg^{2+} ions applying functionalized Cadmium Selenide (CdSe) quantum dots was developed having a quantification limit of 1.8×10^{-7} M (Li et al., 2008). A fluorescent probe using Thiourea functionalized CdSe quantum dots was developed to detect Hg^{2+} ions and had a quantification limit of 0.56 μg/L (Xi et al., 2016).

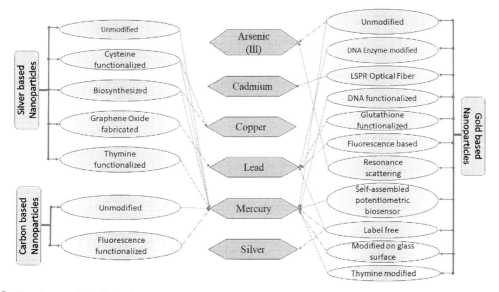

FIG. 4   Nanoparticles derived sensors.

iii. Photoluminescence of CdSe Quantum Dots was done to determine Cu^{2+} ions and the quantification limit was observed to be 5 nM (Chan et al., 2010). Fluorescent sensors were developed using CdSe quantum for spectro-fluorometrically detecting Cu^{2+} ions and have a quantification limit of 8.5 μg/L (Lai et al., 2006). Aqueous CdSe Quantum Dots have been synthesized directly to determine Cu^{2+} ions and the quantification limit was lowered to 5 nM (Bu et al., 2013). Fluorescence enhanced CdSe Quantum dot to detect Cu^{2+} ions in fluidic solution was designed which had a range of detection of 2.0–2.5 nM (Zhang et al., 2014).

iv. Cd^{2+} ions were detected by developing thiogycerol-functionalized CdSe Quantum Dots and the quantification limit was observed to be 0.32 μM (Brahim et al., 2015). Development of a nano-sensor using CdSe quantum dots crowned with NDPO to detect Cd^{2+} ions having a quantification range of 0.05–1.0 mmol/L (Eftekhari-Sis et al., 2018).

v. Development of a method to determine Pb^{2+} ions derived from the quenching property of fluorescence from CdTe Quantum Dots and the quantification limit was found to be 2.7×10^{-7} mol/L (Wu et al., 2008). Pb^{2+} ions were detected using Multicolor CdTe Quantum with thioglycolic acid as a stabilizer. The quantification limit was found to be 4.7 nmol/L (Zhong et al., 2012).

vi. Cu^{2+} ions were detected using CdTe Quantum Dots by fluorescence quenching method and the detection limit was 9.3 μg/L (Wang et al., 2009). CdTe Quantum Dots coupled with enzyme inhibition were applied to detect Cu^{2+} ions having a quantification limit of 0.176 ng/mL (Guo et al., 2012). A ratiometric illuminating probe was developed to detect Cu^{2+} applying CdTe Quantum Dots. The detection limit was found to be 0.096 μM (Rao et al., 2016). A nanosensor using Ligand-capped CdTe Quantum Dots was designed to determine Cu^{2+} ions in environmental aqueous samples and the quantification limit was found to be $1.55 \pm 0.05 \times 10^{-8}$ mol/L (Elmizadeh et al., 2017).

vii. An illuminating probe to detect Ag^{+} using Mercaptopropionic acid stabled CdTe Quantum Dots was developed and the quantification limit was 4.106×10^{-8} mol/L (Gan et al., 2012). Fluorometric detection of Ag^{+} applying CdTe Quantum Dots covered with Homocysteine was done and the quantification limit was observed to be 8.3 nM (Cai et al., 2014). A fluorescence resonance energy transfer (FRET) system was developed between fluorescein isothiocyanate and CdTe Quantum Dots to detect Ag^{+} ions having a range of detection between 0.01 and 8.96 nmol/L (Feng et al., 2016).

viii. A luminescent nanosensor was designed to detect Hg^{2+} ions by applying functionalized ZnS quantum dots with a quantification limit of 1.8×10^{-7} M (Li et al., 2008). An eco-friendly sensor was designed to detect Hg^{2+} ions applying ZnS quantum dots crowned with *N*-acetyl-L-cysteine and has a quantification limit of 5 nmol/L (Duan et al., 2011). Hg^{2+} ions were detected by developing a highly sensitive probe using ZnS quantum dots and the quantification limit was observed to be 2.5 nM (Ke et al., 2012). Development of a PET sensor using ZnS quantum dots to detect Hg^{2+} ion with a quantification limit of 1 pM in the aqueous medium (Saikia et al., 2016). Hg^{2+} ions were detected by developing a photoelectrochemical sensor using ZnS quantum dots with a quantification limit of 2 pM (Wang et al., 2018).

ix. Detection of Hg^{2+} ions by developing a highly sensitive probe using ZnSe quantum dots with a quantification limit of 2.5 nM (Ke et al., 2012). Hg^{2+} ions were detected by fluorescence enhancement of ZnSe Quantum dots doped with Manganese and the quantification limit was observed to be 7 nM (Zhou et al., 2018b).

x. A fluorescent nanosensor was developed based upon graphene quantum dots to detect Pb^{2+} ions with a quantification limit of 0.6 nM (Qian et al., 2015). Pb^{2+} ions were detected by developing a fluorescent sensor using quantum dots made of graphene and gold nanoparticles and the quantification limit was 16.7 nM (Niu et al., 2018).

xi. Development of nanoprobe using Fluorescent graphene Quantum dots to determine Hg^{2+} ions with a quantification limit of 1.0×10^{-7} M (Wang et al., 2014b). Hg^{2+} ions were detected by the process of fluorescence switching of graphene quantum dots having a quantification limit of 0.439 nmol/L (Li et al., 2015). Development of an "on-off-on" fluorescent sensor to detect Hg^{2+} ions applying Nitrogen-doped Graphene Quantum Dots. The quantification limit was found to be 23 nM (Du et al., 2019).

- Carbon nanotubes are applied as a biomaterial for heavy metals detection like Cadmium, Copper, Europium, Lead, Mercury, Silver, and Zinc.

   i. Development of a Carbon Nanotube nanoelectrode array to detect trace metal ions like Cd^{2+} and Pb^{2+} having quantification limit of 0.04 µg/L (Liu et al., 2005). Development of a Bismuth-modified carbon nanotube electrode for trace metal ions determination like Pb^{2+}, Cd^{2+}, Zn^{2+} having quantification limit of 1.3 µg/L for Pb^{2+}, 0.7 µg/L for Cd^{2+}, 12 µg/L for Zn^{2+} (Hwang et al., 2008). Heavy metals detection like Pb^{2+} and Cd^{2+} ions using carbon nanotube tower electrode by anodic stripping voltammetry method with a quantification limit of 12 nM for Pb^{2+} and 25 nM for Cd^{2+} (Guo et al., 2011). Heavy metals detection like Pb^{2+} and Cd^{2+} by developing a three-dimensional graphene-carbon nanotube hybrid electrode with a quantification limit of 0.2 µg/L for Pb^{2+} and 0.1 µg/L for Cd^{2+} (Huang et al., 2014). Sensor development using composites of carbon nanotube electrode for heavy metals detection like Pb^{2+} and Cd^{2+} with a quantification limit of 0.2 ppb for Pb^{2+} and 0.6 ppb for Cd^{2+} (Xuan and Park, 2018).

   ii. Development of a voltammetry process using carbon nanotube-modified electrodes to detect trace heavy metals Cu^{2+} and Pb^{2+} with a quantification limit of 15 ppb for Cu^{2+} and 1 ppb for Pb^{2+} (Morton et al., 2009). Designing a fluorescent sensor using Quinoline group grafted Carbon nanotube to detect Cu^{2+} ions with a quantification limit of 10^{-7} M (Dong et al., 2009). Development of gold nanoparticle patterned single-walled carbon nanotube to detect Cu^{2+} and Pb^{2+} with a quantification limit of 0.613 ppb for Cu^{2+} and 0.546 ppb for Pb^{2+} (Bui et al., 2012b). Development of an anodic stripping voltammetry process exploiting carbon nanotube threads to detect heavy metals like Cu^{2+} and Pb^{2+} with a quantification limit of 0.27 nM for Cu^{2+}, 1.5 nM for Pb^{2+} (Zhao et al., 2014). Development of an electrochemical cell using carbon nanotube thread to detect heavy metals like Cu^{2+} and Pb^{2+} with a quantification limit of 0.53 nM for Cu^{2+}, 0.57 nM for Pb^{2+} (Zhao et al., 2017).

    iii. Development of carbon nanotube-based sensor to detect Hg^{2+} ions in water with a quantification limit of 3 nM (Pokhrel et al., 2017). The development of an electrochemical cell using carbon nanotube thread to detect heavy metals like Hg^{2+}, Cu^{2+}, Pb^{2+} improved the quantification limit of Hg^{2+} to 1.05 nM (Zhao et al., 2017).

4 Pros and cons

Nanotechnology has proven to be a very significant blessing in the field of developing a biosensor. There are several advantages of designing a biosensor using nanoscale materials which are discussed below:

- Nanomaterials have a dimension between 1 and 100 nm and owing to the tiny dimension and big surface area of these nanomaterials they possess some unique properties which are absent in the same materials in bulk state, hence making them suitable for designing biosensors.
- A lower detection limit can be achieved in a nano-biosensor by making use of nanomaterials like nanoparticles, nanotubes, nanowires, and quantum dots.
- Carbon nanotubes have higher aspect ratios and can be functionalized easily. Nanoparticles have excellent catalytic properties. Nanowires are highly versatile and have good charge conduction and quantum dots have size tuneable band energy.
- Nanoparticles have excellent and unique optical properties and hence nano-biosensors can be used as fluorophores.
- By using nanomaterial-based biosensors, high sensitivity and shorter response time can be achieved.

Some of the disadvantages related to nano-biosensors are discussed below:

- Nano-biosensors are highly sensitive and the chances of error are very high.
- Nanomaterial-based biosensors are still in the developing stage.
- In a few cases, nano-biosensors are very difficult to fabricate and the cost of production is also high.

5 Future prospects

The main objective of a nano-biosensor is to determine any kind of signal which is biochemical in nature at the grassroots levels and can be integrated into technologies for molecular diagnostics. Some of the applications of nano-biosensors incorporate the determination of microbes and heavy metals in different samples, controlling the regulation of body fluids, and finding the location of cancer cells. Nano-biosensors with high sensitivity properties and reduced size have immense potential to be used in near future for bio diagnosis and other medical purposes. Due to their small size, they can be easily implanted into several medical devices. The prospect of nano-biosensors application in the field of the medical diagnostic will be unmatched. The use of nano-biosensors, in the future, will outrun the traditional medical practices and ensure the advanced diagnosis of various infections, determination of pathogens, and more advanced patient monitoring instruments.

References

Alhadrami, H.A., 2018. Biosensors: classifications, medical applications, and future prospective. Biotechnol. Appl. Biochem. 65 (3), 497–508.

Ali, J., Najeeb, J., Ali, M.A., Aslam, M.F., Raza, A.J.J.B.B., 2017. Biosensors: their fundamentals, designs, types and most recent impactful applications: a review. J. Biosens. Bioelectron. 8 (1), 1–9.

Anwar, A., Minhaz, A., Khan, N.A., Kalantari, K., Afifi, A.B.M., Shah, M.R., 2018. Synthesis of gold nanoparticles stabilized by a pyrazinium thioacetate ligand: a new colorimetric nanosensor for detection of heavy metal Pd (II). Sensors Actuators B Chem. 257, 875–881.

Ashrafi, A.M., Cerovac, S., Mudrić, S., Guzsvány, V., Husáková, L., Urbanová, I., Vytřas, K., 2014. Antimony nanoparticle-multiwalled carbon nanotubes composite immobilized at carbon paste electrode for determination of trace heavy metals. Sensors Actuators B Chem. 191, 320–325.

Bian, S., Shen, C., Qian, Y., Liu, J., Xi, F., Dong, X., 2017. Facile synthesis of sulfur-doped graphene quantum dots as fluorescent sensing probes for Ag^+ ions detection. Sens. Actuators B Chem. 242, 231–237.

Bian, W., Wang, F., Zhang, H., Zhang, L., Wang, L., Shuang, S., 2015. Fluorescent probe for detection of $Cu2+$ using core-shell CdTe/ZnS quantum dots. Luminescence 30 (7), 1064–1070.

Boonmee, C., Noipa, T., Tuntulani, T., Ngeontae, W., 2016. Cysteamine capped CdS quantum dots as a fluorescence sensor for the determination of copper ion exploiting fluorescence enhancement and long-wave spectral shifts. Spectrochim. Acta A Mol. Biomol. Spectrosc. 169, 161–168.

Borgohain, R., Boruah, P.K., Baruah, S., 2016. Heavy-metal ion sensor using chitosan capped ZnS quantum dots. Sensors Actuators B Chem. 226, 534–539.

Bose, S., Maity, S., Sarkar, A., 2021. Review of microbial biosensor for the detection of mercury in water. Environ. Qual. Manag., 1–12. https://doi.org/10.1002/tqem.21742.

Brahim, N.B., Mohamed, N.B.H., Echabaane, M., Haouari, M., Chaâbane, R.B., Negrerie, M., Ouada, H.B., 2015. Thioglycerol-functionalized CdSe quantum dots detecting cadmium ions. Sensors Actuators B Chem. 220, 1346–1353.

Bu, X., Zhou, Y., He, M., Chen, Z., Zhang, T., 2013. Bioinspired, direct synthesis of aqueous CdSe quantum dots for high-sensitive copper (II) ion detection. Dalton Trans. 42 (43), 15411–15420.

Bui, M.P.N., Li, C.A., Han, K.N., Pham, X.H., Seong, G.H., 2012a. Electrochemical determination of cadmium and lead on pristine single-walled carbon nanotube electrodes. Anal. Sci. 28 (7), 699–704.

Bui, M.P.N., Li, C.A., Han, K.N., Pham, X.H., Seong, G.H., 2012b. Simultaneous detection of ultratrace lead and copper with gold nanoparticles patterned on carbon nanotube thin film. Analyst 137 (8), 1888–1894.

Cai, C., Cheng, H., Wang, Y., Bao, H., 2014. Mercaptosuccinic acid modified CdTe quantum dots as a selective fluorescence sensor for $Ag+$ determination in aqueous solutions. RSC Adv. 4 (103), 59157–59163.

Chai, F., Wang, C., Wang, T., Li, L., Su, Z., 2010. Colorimetric detection of $Pb2+$ using glutathione functionalized gold nanoparticles. ACS Appl. Mater. Interfaces 2 (5), 1466–1470.

Chan, Y.H., Chen, J., Liu, Q., Wark, S.E., Son, D.H., Batteas, J.D., 2010. Ultrasensitive copper (II) detection using plasmon-enhanced and photo-brightened luminescence of CdSe quantum dots. Anal. Chem. 82 (9), 3671–3678.

Chen, J.L., Zhu, C.Q., 2005. Functionalized cadmium sulfide quantum dots as fluorescence probe for silver ion determination. Anal. Chim. Acta 546 (2), 147–153.

Chen, J., Zheng, A., Gao, Y., He, C., Wu, G., Chen, Y., Kai, X., Zhu, C., 2008. Functionalized CdS quantum dots-based luminescence probe for detection of heavy and transition metal ions in aqueous solution. Spectrochim. Acta A Mol. Biomol. Spectrosc. 69 (3), 1044–1052.

Chen, G.H., Chen, W.Y., Yen, Y.C., Wang, C.W., Chang, H.T., Chen, C.F., 2014. Detection of mercury (II) ions using colorimetric gold nanoparticles on paper-based analytical devices. Anal. Chem. 86 (14), 6843–6849.

Chen, J., Zhu, Y., Zhang, Y., 2016. Glutathione-capped Mn-doped ZnS quantum dots as a room-temperature phosphorescence sensor for the detection of $Pb2+$ ions. Spectrochim. Acta A Mol. Biomol. Spectrosc. 164, 98–102.

Cheng, Y., Li, H., Fang, C., Ai, L., Chen, J., Su, J., Zhang, Q., Fu, Q., 2019. Facile synthesis of reduced graphene oxide/silver nanoparticles composites and their application for detecting heavy metal ions. J. Alloys Compd. 787, 683–693.

Ding, Y., Shen, S.Z., Sun, H., Sun, K., Liu, F., 2014. Synthesis of L-glutathione-capped-ZnSe quantum dots for the sensitive and selective determination of copper ion in aqueous solutions. Sens. Actuators B Chem. 203, 35–43.

Dong, Z., Jin, J., Zhao, W., Geng, H., Zhao, P., Li, R., Ma, J., 2009. Quinoline group grafted carbon nanotube fluorescent sensor for detection of $Cu2+$ ion. Appl. Surf. Sci. 255 (23), 9526–9530.

Du, F., Sun, L., Zen, Q., Tan, W., Cheng, Z., Ruan, G., Li, J., 2019. A highly sensitive and selective "on-off-on" fluorescent sensor based on nitrogen doped graphene quantum dots for the detection of Hg2+ and paraquat. Sensors Actuators B Chem. 288, 96–103.

Duan, J., Jiang, X., Ni, S., Yang, M., Zhan, J., 2011. Facile synthesis of N-acetyl-l-cysteine capped ZnS quantum dots as an eco-friendly fluorescence sensor for Hg2+. Talanta 85 (4), 1738–1743.

Duan, J., Yin, H., Wei, R., Wang, W., 2014. Facile colorimetric detection of Hg2+ based on anti-aggregation of silver nanoparticles. Biosens. Bioelectron. 57, 139–142.

Eftekhari-Sis, B., Malekan, F., Younesi Araghi, H., 2018. CdSe quantum dots capped with p-nitrophenyldiazenylphenyloxadiazole: a nanosensor for Cd2+ ions in aqueous media. Can. J. Chem. 96 (4), 371–376.

Elmizadeh, H., Soleimani, M., Faridbod, F., Bardajee, G.R., 2017. Ligand-capped CdTe quantum dots as a fluorescent nanosensor for detection of copper ions in environmental water sample. J. Fluoresc. 27 (6), 2323–2333.

Elmizadeh, H., Soleimani, M., Faridbod, F., Bardajee, G.R., 2018. A sensitive nano-sensor based on synthetic ligand-coated CdTe quantum dots for rapid detection of Cr (III) ions in water and wastewater samples. Colloid Polym. Sci. 296 (9), 1581–1590.

Ensafi, A.A., Kazemifard, N., Rezaei, B., 2016. A simple and sensitive fluorimetric aptasensor for the ultrasensitive detection of arsenic (III) based on cysteamine stabilized CdTe/ZnS quantum dots aggregation. Biosens. Bioelectron. 77, 499–504.

Feng, Y., Liu, L., Hu, S., Zou, P., Zhang, J., Huang, C., Wang, Y., Wang, S., Zhang, X., 2016. Efficient fluorescence energy transfer system between fluorescein isothiocyanate and CdTe quantum dots for the detection of silver ions. Luminescence 31 (2), 356–363.

Firdaus, M.L., Fitriani, I., Wyantuti, S., Hartati, Y.W., Khaydarov, R., McAlister, J.A., Obata, H., Gamo, T., 2017. Colorimetric detection of mercury (II) ion in aqueous solution using silver nanoparticles. Anal. Sci. 33 (7), 831–837.

Gan, T.T., Zhang, Y.J., Zhao, N.J., Xiao, X., Yin, G.F., Yu, S.H., Wang, H.B., Duan, J.B., Shi, C.Y., Liu, W.Q., 2012. Hydrothermal synthetic mercaptopropionic acid stabled CdTe quantum dots as fluorescent probes for detection of Ag+. Spectrochim. Acta A Mol. Biomol. Spectrosc. 99, 62–68.

Gan, T., Zhao, N., Yin, G., Tu, M., Liu, J., Liu, W., 2017. Mercaptopropionic acid-capped Mn-doped ZnS quantum dots as a probe for selective room-temperature phosphorescence detection of Pb 2+ in water. New J. Chem. 41 (22), 13425–13434.

Girigoswami, K., Akhtar, N., 2019. Nanobiosensors and fluorescence based biosensors: an overview. Int. J. Nano Dimens. 10 (1), 1–17.

Guo, C., Wang, J., Cheng, J., Dai, Z., 2012. Determination of trace copper ions with ultrahigh sensitivity and selectivity utilizing CdTe quantum dots coupled with enzyme inhibition. Biosens. Bioelectron. 36 (1), 69–74.

Guo, X., Yun, Y., Shanov, V.N., Halsall, H.B., Heineman, W.R., 2011. Determination of trace metals by anodic stripping voltammetry using a carbon nanotube tower electrode. Electroanalysis 23 (5), 1252–1259.

Huang, H., Chen, T., Liu, X., Ma, H., 2014. Ultrasensitive and simultaneous detection of heavy metal ions based on three-dimensional graphene-carbon nanotubes hybrid electrode materials. Anal. Chim. Acta 852, 45–54.

Hung, Y.L., Hsiung, T.M., Chen, Y.Y., Huang, Y.F., Huang, C.C., 2010. Colorimetric detection of heavy metal ions using label-free gold nanoparticles and alkanethiols. J. Phys. Chem. C 114 (39), 16329–16334.

Hwang, G.H., Han, W.K., Park, J.S., Kang, S.G., 2008. Determination of trace metals by anodic stripping voltammetry using a bismuth-modified carbon nanotube electrode. Talanta 76 (2), 301–308.

Jiao, H., Zhang, L., Liang, Z., Peng, G., Lin, H., 2014. Size-controlled sensitivity and selectivity for the fluorometric detection of Ag+ by homocysteine capped CdTe quantum dots. Microchim. Acta 181 (11–12), 1393–1399.

Ke, J., Li, X., Shi, Y., Zhao, Q., Jiang, X., 2012. A facile and highly sensitive probe for Hg (II) based on metal-induced aggregation of ZnSe/ZnS quantum dots. Nanoscale 4 (16), 4996–5001.

Khan, U., Niaz, A., Shah, A., Zaman, M.I., Zia, M.A., Iftikhar, F.J., Nisar, J., Ahmed, M.N., Akhter, M.S., Shah, A.H., 2018. Thiamine-functionalized silver nanoparticles for the highly selective and sensitive colorimetric detection of Hg 2+ ions. New J. Chem. 42 (1), 528–534.

Kim, J.H., Han, S.H., Chung, B.H., 2011. Improving Pb2+ detection using DNAzyme-based fluorescence sensors by pairing fluorescence donors with gold nanoparticles. Biosens. Bioelectron. 26 (5), 2125–2129.

Lai, Y., Yu, Y., Zhong, P., Wu, J., Long, Z., Liang, C., 2006. Development of novel quantum dots as fluorescent sensors for application in highly sensitive spectrofluorimetric determination of Cu2+. Anal. Lett. 39 (6), 1201–1209.

Lan, M., Zhang, J., Chui, Y.S., Wang, P., Chen, X., Lee, C.S., Kwong, H.L., Zhang, W., 2014. Carbon nanoparticle-based ratiometric fluorescent sensor for detecting mercury ions in aqueous media and living cells. ACS Appl. Mater. Interfaces 6 (23), 21270–21278.

Lee, J.S., Han, M.S., Mirkin, C.A., 2007. Colorimetric detection of mercuric ion (Hg2+) in aqueous media using DNA-functionalized gold nanoparticles. Angew. Chem. Int. Ed. 46 (22), 4093–4096.

Lee, S.H., Sung, J.H., Park, T.H., 2012. Nanomaterial-based biosensor as an emerging tool for biomedical applications. Ann. Biomed. Eng. 40 (6), 1384–1397.

Li, H., Zhang, Y., Wang, X., Gao, Z., 2008. A luminescent nanosensor for Hg (II) based on functionalized CdSe/ZnS quantum dots. Microchim. Acta 160 (1–2), 119–123.

Li, H., Zhai, J., Tian, J., Luo, Y., Sun, X., 2011a. Carbon nanoparticle for highly sensitive and selective fluorescent detection of mercury (II) ion in aqueous solution. Biosens. Bioelectron. 26 (12), 4656–4660.

Li, T., Zhou, Y., Sun, J., Tang, D., Guo, S., Ding, X., 2011b. Ultrasensitive detection of mercury (II) ion using CdTe quantum dots in sol-gel-derived silica spheres coated with calix [6] arene as fluorescent probes. Microchim. Acta 175 (1–2), 113.

Li, F., Wang, J., Lai, Y., Wu, C., Sun, S., He, Y., Ma, H., 2013a. Ultrasensitive and selective detection of copper (II) and mercury (II) ions by dye-coded silver nanoparticle-based SERS probes. Biosens. Bioelectron. 39 (1), 82–87.

Li, M., Zhou, X., Guo, S., Wu, N., 2013b. Detection of lead (II) with a "turn-on" fluorescent biosensor based on energy transfer from CdSe/ZnS quantum dots to graphene oxide. Biosens. Bioelectron. 43, 69–74.

Li, S., Li, Y., Cao, J., Zhu, J., Fan, L., Li, X., 2014. Sulfur-doped graphene quantum dots as a novel fluorescent probe for highly selective and sensitive detection of Fe3+. Anal. Chem. 86 (20), 10201–10207.

Li, Z., Wang, Y., Ni, Y., Kokot, S., 2015. A rapid and label-free dual detection of Hg (II) and cysteine with the use of fluorescence switching of graphene quantum dots. Sensors Actuators B Chem. 207, 490–497.

Lian, Y., Yuan, M., Zhao, H., 2014. DNA wrapped metallic single-walled carbon nanotube sensor for Pb (II) detection. Fullernes, Nanotubes, Carbon Nanostruct. 22 (5), 510–518.

Lin, T.J., Chung, M.F., 2009. Detection of cadmium by a fiber-optic biosensor based on localized surface plasmon resonance. Biosens. Bioelectron. 24 (5), 1213–1218.

Liu, G., Lin, Y., Tu, Y., Ren, Z., 2005. Ultrasensitive voltammetric detection of trace heavy metal ions using carbon nanotube nanoelectrode array. Analyst 130 (7), 1098–1101.

Liu, Z., Liu, S., Yin, P., He, Y., 2012. Fluorescence enhancement of CdTe/CdS quantum dots by coupling of glyphosate and its application for sensitive detection of copper ion. Anal. Chim. Acta 745, 78–84.

Liu, J., Lu, Y., 2003. A colorimetric lead biosensor using DNAzyme-directed assembly of gold nanoparticles. J. Am. Chem. Soc. 125 (22), 6642–6643.

Lu, W., Qin, X., Liu, S., Chang, G., Zhang, Y., Luo, Y., Asiri, A.M., Al-Youbi, A.O., Sun, X., 2012. Economical, green synthesis of fluorescent carbon nanoparticles and their use as probes for sensitive and selective detection of mercury (II) ions. Anal. Chem. 84 (12), 5351–5357.

Mahmoud, W.E., 2012. Functionalized ME-capped CdSe quantum dots based luminescence probe for detection of Ba2+ ions. Sensors Actuators B Chem. 164 (1), 76–81.

Maiti, S., Barman, G., Laha, J.K., 2016. Detection of heavy metals (Cu+2, Hg+2) by biosynthesized silver nanoparticles. Appl. Nanosci. 6 (4), 529–538.

Maity, S., Sinha, D., Sarkar, A., 2020a. Wastewater and industrial effluent treatment by using nanotechnology. In: Bhushan, I., Singh, V., Tripathi, D. (Eds.), Nanomaterials and Environmental Biotechnology. Nanotechnology in the Life Sciences, Springer, Cham, https://doi.org/10.1007/978-3-030-34544-0_16.

Maity, S., Biswas, R., Sarkar, A., 2020b. Comparative valuation of groundwater quality parameters in Bhojpur, Bihar for arsenic risk assessment. Chemosphere 259, 127398. https://doi.org/10.1016/j.chemosphere.2020.127398.

Maity, S., Biswas, R., Verma, S.K., Sarkar, A., 2021a. Natural polysaccharides as potential biosorbents for heavy metal removal. In: Food, Medical, and Environmental Applications of Polysaccharides. Elsevier, pp. 627–665, https://doi.org/10.1016/B978-0-12-819239-9.00012-9.

Maity, S., Nanda, S., Sarkar, A., 2021b. Colocasia esculenta stem as novel biosorbent for potentially toxic metals removal from aqueous system. Environ. Sci. Pollut. Res. https://doi.org/10.1007/s11356-021-13026-1.

Malik, P., Katyal, V., Malik, V., Asatkar, A., Inwati, G., Mukherjee, T.K., 2013. Nanobiosensors: concepts and variations. Int. Sch. Res. Notices 2013, 01–09.

Mashhadizadeh, M.H., Talemi, R.P., 2014. A novel optical DNA biosensor for detection of trace amounts of mercuric ions using gold nanoparticles introduced onto modified glass surface. Spectrochim. Acta A Mol. Biomol. Spectrosc. 132, 403–409.

Mohamed, N.B.H., Brahim, N.B., Mrad, R., Haouari, M., Chaâbane, R.B., Negrerie, M., 2018. Use of MPA-capped CdS quantum dots for sensitive detection and quantification of Co2+ ions in aqueous solution. Anal. Chim. Acta 1028, 50–58.

Morton, J., Havens, N., Mugweru, A., Wanekaya, A.K., 2009. Detection of trace heavy metal ions using carbon nanotube-modified electrodes. Electroanalysis 21 (14), 1597–1603.

Niu, X., Zhong, Y., Chen, R., Wang, F., Liu, Y., Luo, D., 2018. A "turn-on" fluorescence sensor for Pb2+ detection based on graphene quantum dots and gold nanoparticles. Sensors Actuators B Chem. 255, 1577–1581.

Nurerk, P., Kanatharana, P., Bunkoed, O., 2016. A selective determination of copper ions in water samples based on the fluorescence quenching of thiol-capped CdTe quantum dots. Luminescence 31 (2), 515–522.

Pei, J., Zhu, H., Wang, X., Zhang, H., Yang, X., 2012. Synthesis of cysteamine-coated CdTe quantum dots and its application in mercury (II) detection. Anal. Chim. Acta 757, 63–68.

Pokhrel, L.R., Ettore, N., Jacobs, Z.L., Zarr, A., Weir, M.H., Scheuerman, P.R., Kanel, S.R., Dubey, B., 2017. Novel carbon nanotube (CNT)-based ultrasensitive sensors for trace mercury (II) detection in water: a review. Sci. Total Environ. 574, 1379–1388.

Qian, Z.S., Shan, X.Y., Chai, L.J., Chen, J.R., Feng, H., 2015. A fluorescent nanosensor based on graphene quantum dots–aptamer probe and graphene oxide platform for detection of lead (II) ion. Biosens. Bioelectron. 68, 225–231.

Qu, H., Cao, L., Su, G., Liu, W., Gao, R., Xia, C., Qin, J., 2014. Silica-coated ZnS quantum dots as fluorescent probes for the sensitive detection of Pb2+ ions. J. Nanopart. Res. 16 (12), 2762.

Rao, H., Liu, W., Lu, Z., Wang, Y., Ge, H., Zou, P., Wang, X., He, H., Zeng, X., Wang, Y., 2016. Silica-coated carbon dots conjugated to CdTe quantum dots: a ratiometric fluorescent probe for copper (II). Microchim. Acta 183 (2), 581–588.

Saikia, D., Dutta, P., Sarma, N.S., Adhikary, N.C., 2016. CdTe/ZnS core/shell quantum dot-based ultrasensitive PET sensor for selective detection of Hg (II) in aqueous media. Sensors Actuators B Chem. 230, 149–156.

Tabaraki, R., Nateghi, A., 2016. Nitrogen-doped graphene quantum dots: "turn-off" fluorescent probe for detection of Ag+ ions. J. Fluoresc. 26 (1), 297–305.

Tang, S., Tong, P., You, X., Lu, W., Chen, J., Li, G., Zhang, L., 2016. Label free electrochemical sensor for Pb2+ based on graphene oxide mediated deposition of silver nanoparticles. Electrochim. Acta 187, 286–292.

Tao, H., Liao, X., Xu, M., Li, S., Zhong, F., Yi, Z., 2014. Determination of trace Hg2+ ions based on the fluorescence resonance energy transfer between fluorescent brightener and CdTe quantum dots. J. Lumin. 146, 376–381.

Ting, S.L., Ee, S.J., Ananthanarayanan, A., Leong, K.C., Chen, P., 2015. Graphene quantum dots functionalized gold nanoparticles for sensitive electrochemical detection of heavy metal ions. Electrochim. Acta 172, 7–11.

Wang, N., Lin, M., Dai, H., Ma, H., 2016. Functionalized gold nanoparticles/reduced graphene oxide nanocomposites for ultrasensitive electrochemical sensing of mercury ions based on thymine–mercury–thymine structure. Biosens. Bioelectron. 79, 320–326.

Wang, Y., Lu, J., Tong, Z., Huang, H., 2009. A fluorescence quenching method for determination of copper ions with CdTe quantum dots. J. Chil. Chem. Soc. 54 (3), 274–277.

Wang, Y., Yang, F., Yang, X., 2010. Colorimetric detection of mercury (II) ion using unmodified silver nanoparticles and mercury-specific oligonucleotides. ACS Appl. Mater. Interfaces 2 (2), 339–342.

Wang, G., Lu, Y., Yan, C., Lu, Y., 2015. DNA-functionalization gold nanoparticles based fluorescence sensor for sensitive detection of Hg2+ in aqueous solution. Sens. Actuators B Chem. 211, 1–6.

Wang, X., Lv, Y., Hou, X., 2011a. A potential visual fluorescence probe for ultratrace arsenic (III) detection by using glutathione-capped CdTe quantum dots. Talanta 84 (2), 382–386.

Wang, J., Zhou, X., Ma, H., Tao, G., 2011b. Diethyldithiocarbamate functionalized CdSe/CdS quantum dots as a fluorescent probe for copper ion detection. Spectrochim. Acta A Mol. Biomol. Spectrosc. 81 (1), 178–183.

Wang, T., Zhao, D., Guo, X., Correa, J., Riehl, B.L., Heineman, W.R., 2014a. Carbon nanotube-loaded nafion film electrochemical sensor for metal ions: europium. Anal. Chem. 86 (9), 4354–4361.

Wang, B., Zhuo, S., Chen, L., Zhang, Y., 2014b. Fluorescent graphene quantum dot nanoprobes for the sensitive and selective detection of mercury ions. Spectrochim. Acta A Mol. Biomol. Spectrosc. 131, 384–387.

Wang, Y., Zhang, C., Chen, X., Yang, B., Yang, L., Jiang, C., Zhang, Z., 2016. Ratiometric fluorescent paper sensor utilizing hybrid carbon dots–quantum dots for the visual determination of copper ions. Nanoscale 8 (11), 5977–5984.

Wang, Y., Wang, P., Wu, Y., Di, J., 2018. A cathodic "signal-on" photoelectrochemical sensor for Hg2+ detection based on ion-exchange with ZnS quantum dots. Sensors Actuators B Chem. 254, 910–915.

Wu, H., Liang, J., Han, H., 2008. A novel method for the determination of Pb2+ based on the quenching of the fluorescence of CdTe quantum dots. Microchim. Acta 161 (1–2), 81–86.

Wu, P., Li, Y., Yan, X.P., 2009. CdTe quantum dots (QDs) based kinetic discrimination of Fe2+ and Fe3+, and CdTe QDs-fenton hybrid system for sensitive photoluminescent detection of Fe2+. Anal. Chem. 81 (15), 6252–6257.

Wu, Y., Liu, L., Zhan, S., Wang, F., Zhou, P., 2012a. Ultrasensitive aptamer biosensor for arsenic (III) detection in aqueous solution based on surfactant-induced aggregation of gold nanoparticles. Analyst 137 (18), 4171–4178.

Wu, Y., Zhan, S., Wang, F., He, L., Zhi, W., Zhou, P., 2012b. Cationic polymers and aptamers mediated aggregation of gold nanoparticles for the colorimetric detection of arsenic (III) in aqueous solution. Chem. Commun. 48 (37), 4459–4461.

Wu, D., Chen, Z., Huang, G., Liu, X., 2014. ZnSe quantum dots based fluorescence sensors for Cu2+ ions. Sensors Actuators A Phys. 205, 72–78.

Xi, L.L., Ma, H.B., Tao, G.H., 2016. Thiourea functionalized CdSe/CdS quantum dots as a fluorescent sensor for mercury ion detection. Chin. Chem. Lett. 27 (9), 1531–1536.

Xie, W.Y., Huang, W.T., Luo, H.Q., Li, N.B., 2012. CTAB-capped Mn-doped ZnS quantum dots and label-free aptamer for room-temperature phosphorescence detection of mercury ions. Analyst 137 (20), 4651–4653.

Xing, X., Wang, D., Chen, Z., Zheng, B., Li, B., Wu, D., 2018. ZnTe quantum dots as fluorescence sensors for the detection of iron ions. J. Mater. Sci. Mater. Electron. 29 (16), 14192–14199.

Xu, L., Mao, W., Huang, J., Li, S., Huang, K., Li, M., Xia, J., Chen, Q., 2016. Economical, green route to highly fluorescence intensity carbon materials based on ligninsulfonate/graphene quantum dots composites: application as excellent fluorescent sensing platform for detection of Fe3+ ions. Sensors Actuators B Chem. 230, 54–60.

Xuan, X., Park, J.Y., 2018. A miniaturized and flexible cadmium and lead ion detection sensor based on micro-patterned reduced graphene oxide/carbon nanotube/bismuth composite electrodes. Sensors Actuators B Chem. 255, 1220–1227.

Yang, Y., Wang, Z., Yang, M., Guo, M., Wu, Z., Shen, G., Yu, R., 2006. Inhibitive determination of mercury ion using a renewable urea biosensor based on self-assembled gold nanoparticles. Sensors Actuators B Chem. 114 (1), 1–8.

Yu, Y., Hong, Y., Wang, Y., Sun, X., Liu, B., 2017. Mecaptosuccinic acid modified gold nanoparticles as colorimetric sensor for fast detection and simultaneous identification of Cr3+. Sensors Actuators B Chem. 239, 865–873.

Zhang, W., Jiang, L., Piper, J.A., Wang, Y., 2018. SERS nanotags and their applications in biosensing and bioimaging. J. Anal. Test. 2 (1), 26–44.

Zhang, L., Xu, C., Li, B., 2009. Simple and sensitive detection method for chromium (VI) in water using glutathione—capped CdTe quantum dots as fluorescent probes. Microchim. Acta 166 (1–2), 61–68.

Zhang, K., Guo, J., Nie, J., Du, B., Xu, D., 2014. Ultrasensitive and selective detection of Cu2+ in aqueous solution with fluorescence enhanced CdSe quantum dots. Sensors Actuators B Chem. 190, 279–287.

Zhao, C., Qu, K., Song, Y., Xu, C., Ren, J., Qu, X., 2010. A reusable DNA single-walled carbon-nanotube-based fluorescent sensor for highly sensitive and selective detection of Ag+ and cysteine in aqueous solutions. Chem. Eur. J. 16 (27), 8147–8154.

Zhao, T., Hou, X., Xie, Y.N., Wu, L., Wu, P., 2013a. Phosphorescent sensing of Cr3+ with protein-functionalized Mn-doped ZnS quantum dots. Analyst 138 (21), 6589–6594.

Zhao, Q., Rong, X., Ma, H., Tao, G., 2013b. Dithizone functionalized CdSe/CdS quantum dots as turn-on fluorescent probe for ultrasensitive detection of lead ion. J. Hazard. Mater. 250, 45–52.

Zhao, D., Guo, X., Wang, T., Alvarez, N., Shanov, V.N., Heineman, W.R., 2014. Simultaneous detection of heavy metals by anodic stripping voltammetry using carbon nanotube thread. Electroanalysis 26 (3), 488–496.

Zhao, D., Siebold, D., Alvarez, N.T., Shanov, V.N., Heineman, W.R., 2017. Carbon nanotube thread electrochemical cell: detection of heavy metals. Anal. Chem. 89 (18), 9654–9663.

Zhong, W., Zhang, C., Gao, Q., Li, H., 2012. Highly sensitive detection of lead (II) ion using multicolor CdTe quantum dots. Microchim. Acta 176 (1–2), 101–107.

Zhou, M., Guo, J., Yang, C., 2018a. Ratiometric fluorescence sensor for Fe3+ ions detection based on quantum dot-doped hydrogel optical fiber. Sensors Actuators B Chem. 264, 52–58.

Zhou, Z.Q., Yan, R., Zhao, J., Yang, L.Y., Chen, J.L., Hu, Y.J., Jiang, F.L., Liu, Y., 2018b. Highly selective and sensitive detection of Hg2+ based on fluorescence enhancement of Mn-doped ZnSe QDs by Hg2+-Mn2+ replacement. Sensors Actuators B Chem. 254, 8–15.

Zhu, J., Zhao, Z.J., Li, J.J., Zhao, J.W., 2017. CdTe quantum dot-based fluorescent probes for selective detection of Hg (II): the effect of particle size. Spectrochim. Acta A Mol. Biomol. Spectrosc. 177, 140–146.

Zi, L., Huang, Y., Yan, Z., Liao, S., 2014. Thioglycolic acid-capped CuInS2/ZnS quantum dots as fluorescent probe for cobalt ion detection. J. Lumin. 148, 359–363.

CHAPTER

20

Smart nano-biosensors in sustainable agriculture and environmental applications

Rani Puthukulangara Ramachandran[a], Chelladurai Vellaichamy[b], and Chyngyz Erkinbaev[a]

[a]Department of Biosystems Engineering, University of Manitoba, Winnipeg, MB, Canada
[b]Department of Agricultural Engineering, Bannariamman Institute of Technology, Sathyamangalam, Tamil Nadu, India

OUTLINE

1 Introduction

The shift in global focus toward sustainability had led to a constant search for a resilient management system of all natural resources that we are privileged with including land, food systems, and the environment. As agriculture plays a vital role in reforming land, ecosystem, and environmental resources, there have been controversial conversations on emphasizing sustainability in agricultural systems considering it as a major contributor to greenhouse gases. This led to the use of various technologies such as artificial intelligence and sensor technology to increase the quantity, quality, and economy of agricultural production with a focus on preserving natural resources and ecosystem functionality. With the advancement in nanotechnology, new solutions were developed and in physical, chemical, and biological sensing techniques with extended applications in agriculture and the environment. The term nanotechnology refers to a technology that uses molecules, atoms, or submicron dimensions to apply in a larger complex system. This technology has its applications in biological systems and all applied sciences. The intervention of nanotechnology and engineered nanomaterials to advance sensor development had played an indispensable role in minimizing the challenges in developing sensors with suitable size, compatibility, and application-specific features. Therefore, any sensors capable of detecting/monitoring a chemical species, molecules, and nanoparticles, or quantifying physical parameters such as pressure, temperature, and relative humidity, on the nanoscale are called a nano-sensor. The nano-sensors that target certain interesting molecules with immobilized receptors are called nano-biosensors. Usually, these sensors consist of a biological sensing unit, which is integrated with a transducer and activator and they are developed in nanoscales, targeting analyte that is in atomic scales. This chapter gives an overview of nano-biosensors, fabrication of different types of nano-biosensors, its applications in environmental science and agriculture.

There are some advantages of using a smart nano-biosensor over conventional biosensors. The sensitivity and selectivity of these sensors is high, capable of detecting a single organism like viruses or trace amounts of harmful chemicals by providing accurate analytical signals. Moreover, efficiency is also relatively high, as they are working at an atomic scale. Smart nano-biosensors also have an increased surface-to-volume ratio ensuring higher sensitivity and precision in sensing. However, despite all these advantages, nano-biosensors can sometimes be prone to faulty measures with high sensitivity. A typical nano-biosensor consists of three components, a probe or biological sensory element, a transducer, and a detector. The probe is a sensor that is capable of sensing an analyte by generating a signal and is capable of transmitting those signals to the transmitter. These can be either enzymes, receptors, nucleic acid, lectins, tissues, or microorganisms.

2 Principle of nano-biosensors

A nano-biosensor primarily targets biological agents such as antibodies, proteins, nucleic acids, and metabolites, etc. The biosensor element's working principle is based on the selective binding of bio-analytes onto receptors, which generate/modulate the physiochemical signal associated with the binding (Malik et al., 2013; Prasad, 2014; White et al., 2009). These physiochemical signals are then captured by a transducer and converted into an electrical signal. These electrical signals with varying electrical properties such as potential, current, conductance, impedance, intensity, and resonance frequency, etc. (Huang et al., 2021) are monitored/recorded. In the cases where multiple analytes are measured simultaneously, a multiarray biosensor can be used as a separate system in a single assembled unit called a bio-chip (Huang et al., 2021; Min et al., 2009). The measurable electrical signal output from the transducer is then sent to a detector. The detector captures this electrical output and amplifies them and transfers them into a microprocessor which then analyzes the signal. In addition to these components, nano-biosensors also may have an immobilizer to immobilize the bioreceptor so that it reacts/binds with bio-analyte more efficiently to facilitate the electrochemical sensing mechanism (Huang et al., 2021; Malik et al., 2013). The two main immobilization methods which are most popular are: Physical (reversible) is done by attaching various types of enzymes to the surface of the sensing element without chemical connection. Whereas the chemical (irreversible) immobilization techniques require strong chemical bonds (i.e., covalent) (Naresh and Lee, 2021). The high sensitivity and reliability of these smart nano-biosensors justify its successful integration with the artificial intelligence, machine learning, big data analytics, etc. for real-time monitoring, and prediction modeling.

There are four types of nano-sensors based on the material used, (1) metal-based like gold, palladium, silver, copper, and metal oxides (2) carbon-based like CNTs (3) nano-sized polymers (4) composites combining nanoparticles. Nano-biosensors associated with very small size and high surface and volume were able to provide new technology and a new tool for real bioanalytical applications which was not available in the past (Abdel-Karim et al., 2020).

3 Types of nano-bio sensors

Typically, nano-biosensors are classified based on the interaction between the probe and the analyte element, identification method uses to identify the interactions, transduction system, and nature of the recognized compound. The common classification of nano-biosensors in environmental and agricultural applications is given in Fig. 1.

In general, the biosensors are classified into enzyme-based antibody-based and aptamer-based biosensors based on the biological element/receptors (Chatterjee et al., 2020; Han et al., 2010; Liu and Liu, 2021; Negi and Choephel, 2020; Piro et al., 2016; Seo and Gu, 2017; Yoo et al., 2020). Enzyme-based biosensors are operate based on the redox reactions of various enzymes and have proven applications in the food quality and safety, environmental, agricultural, and pharmaceutical industries (Dhar et al., 2019; Ghorbanpour et al., 2020; Rocchitta et al., 2016; Verma and Bhardwaj, 2015; Zhu et al., 2019). Antibody-based biosensors utilize the capability of antibodies as recognition segments for an individual amino acid. These sensors have been utilized in clinical and environmental applications (Wujcik et al., 2014) aptamer-based biosensors contain the DNA/RNA oligonucleotides that react with their target ligands. These aptamers can recognize small molecules of interest and bind to them and change their structures (Chiu and Huang, 2009). Aptamer-based biosensors with applications in environmental monitoring especially in water contaminants such as the ones for detecting bacterial cells, bacterial toxins, viruses, mycotoxins, or heavy metals. It also targets aquatic toxins, pesticides, industrial by-products, antibiotics, and pharmaceuticals (McConnell et al., 2020).

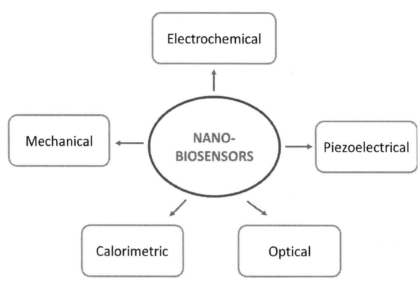

FIG. 1 Type of common nano-biosensors in environmental and agricultural applications.

Most commonly the nano-biosensors are classified based on the transducer. This classification includes mechanical, electrochemical, optical, piezoelectric, calorimetric, etc. (Naresh and Lee, 2021; Negi and Choephel, 2020; Purohit et al., 2020).

3.1 Mechanical nano-biosensors

As the name implies mechanical nano-biosensors measure the nanoscale forces created through biomolecular interactions or variations in mechanical (i.e., surface stresses, expansion, deformation, motion) and physical (weights, forces) (Purohit et al., 2020). The magnitude of the stress or the force is depending on the type of bond created between the analyte and the probe, like hydrogen bonding, van der Waals, and electrostatic bonding (Naresh and Lee, 2021). The mechanical biosensors can be further classified into the following classes: (1) quartz microbalance (mass sensitive sensors), acoustic wave (longitudinal and shear wave sensitive), and nano-mechanical systems (mechanical motion-sensitive) (Chalklen et al., 2020). Usually, mechanical biosensors comprise a microscale cantilever beam to identify/detect the analyte, a mechanical transducer, and the processor. The changes in mechanical properties of the cantilever are usually proportional to mass change and create a force that is unique and measurable to the specific molecule. These forces can be quantified by transformation and/or vibration/resonance (Chalklen et al., 2020; Naresh and Lee, 2021). Depending on the cause for the bending these mechanical nano-biosensors can be further divided into three categories including bending in response to the mechanical weight, bending in response to the increased weight or load, and bending in response to a temperature change. The main advantage of this type of sensor is the high sensitivity toward the mass (Arlett et al., 2011). A relatively light weighed sensor can be highly sensitive to the additional weight added by the binding molecules compared to a normally weighed sensor.

Mechanical nano-biosensors have many advantages over other types of biosensors including their capability of detecting the smallest amount of analyte, ability to quantify forces at the cellular level, fast sensing time, and ability to detect an analyte in liquids and gaseous phases (Arlett et al., 2011; Naresh and Lee, 2021).

3.2 Optical nano-biosensors

Unlike mechanical nano-biosensors, optical biosensors are based on an oscillating light beam in a closed space with a high-affinity element coupled with an optical system. The detection principle of an optical biosensor is based on the light beam traveling with the optical arrangements through a closed path where the analyte combines with a resonator to create a resonant frequency. The generated electrical signals with magnitudes or frequencies are proportional to the chemical analyte and its concentration resulting in nondestructive and real-time detection. The linear resonator bounces the light between two mirrors and the circular resonator bounces light into two directions. Moreover, optical nano-biosensors have been widely used in pathogen detection and clinical detections such as cancer cells (Xu et al., 2020). Optical biosensors use various types of biological substances (i.e., antibodies, aptamers, enzymes, nucleotides, cells, tissue, etc.) as a biosensing element that can create a respective form of optocoupled systems.

The optical nano-biosensors are based on the path of light in a closed path. The traveling light hits the resonator where the analytes are present. The two types of commonly available resonators are linear and circular. Surface plasmon resonance (SPR) is a phenomenon, which has been used in these optical nano-biosensors. The surface plasmon resonance (SPR) is a technique based on an electro-optical interaction where photon energy of the lights is converted into an electric signal, detecting the changes. The SPR chip contains a functional adhesive layer that immobilizes interacting molecules via chemical coupling (Damborský et al., 2016). The energy carried by the photons in the light beam has transferred into the plasma (a group of electrons) in the metal. The main advantages associated with these sensors are the minimal damages to the target cells and can penetrate the deeper parts of the cells. A fiber-optic nano-biosensor designed with optical fibers, onto which biochemical target molecules such as antibodies, peptides, and DNA, etc. inside the living cells can be immobilized (Long et al., 2013).

3.3 Electrochemical nano-biosensors

Electrochemical sensors are based on the reaction of the biomolecules/analyte to generate ions and electrons forming an electrical signal with varying electric current and potential proportional to the analyte concentration. An electrochemical sensor consists of a sensing electrode that serves as the transduction element in the biochemical reaction, and a reference electrode separated by an electrolyte to enable electron flow (Hammond et al., 2016). Electrochemical biosensors are relatively simple, inexpensive, and easy to miniaturize into a compact portable system and hence can be easily integrated with simple electronics to rapidly and accurately measure target analytes. Therefore, they are very attractive for medical applications where quick diagnostic and screening methods as well as in environmental monitoring (Hammond et al., 2016; Pumera et al., 2007). There are different types of electrochemical sensors and can be classified as amperometric, photo-electrochemical, impedance, and electrochemical luminescence. These electrochemical sensors have been used in the detection of biological and chemical substances and provided a wide range of applications in environmental and biomedical domains (Bakirhan et al., 2020; Hammond et al., 2016; Zhu et al., 2015).

Amperometric biosensors are based on the detection of change in electrical current and electron motion resulting from enzyme-catalyzed redox reactions (Huang et al., 2021). These sensors are widely used in low-cost gas monitoring sensors for CO_2 and CH_4 and volatile organic compounds detections (Baron and Saffell, 2017). Electrochemical impedance spectroscopy (EIS) sensors are used to monitor biomolecular events occurring at the electrode surface based on the impedance properties of the electrochemical system using a large range of frequencies (Zamfir et al., 2020). Electrogenerated chemiluminescence which combines the advantages of the electrochemical as well as the photo-luminescence techniques is the basic principle of electrochemical luminescence sensors (Shen et al., 2019). Photo-electrochemical sensors contain semiconductor materials capable of converting chemical energy to electrical signals under irradiation conditions. These materials interact with the electrode by oxidizing the molecules remaining on the electrode surface (Syrek et al., 2019).

3.4 Piezoelectric nano-biosensor

A piezoelectric material or piezoelectric crystal such as lead zirconate titanate, Barium Titanate, etc. is the core sensor part in a piezoelectric nano-biosensor (Selvarajan et al., 2017). The piezoelectric sensor is based on sensing changes in mechanical stress due to a mass-bound biomolecular interaction on the piezoelectric crystal surface and the voltage produced when the material is mechanically stressed (Pohanka, 2018). These sensors have been studied for their application in various environmental and healthcare, such as monitoring foodborne pathogens and toxins (Malekzad et al., 2017).

3.5 Calorimetric nano-biosensor

Colorimetric biosensors determine an analyte of choice by detecting small changes in color using spectrophotometer detectors for quantitative measurement utilizing the optical properties of the nanoparticles (Zhao et al., 2020). Gold nanoparticles are commonly used in colorimetric nano-sensors attributed to their simplicity, controllable size, optical properties, and high extinction coefficients (Chang et al., 2019; Mondal et al., 2018; Tavakkoli Yaraki and Tan, 2020).

4 Nanostructures used in sensors

Nano-biosensors use various nanostructures such as nanoparticles, nanowires, and nanocomposites. The combination of this nanostructure and the biomolecule with an increased surface-to-volume ratio of nanomaterials had improved the sensitivity of the present-day biosensors, etc. (Huang et al., 2021). A nanowire-nano-biosensor is a cylindrical wire like a sensor with diameters in nanoscale and various lengths ranging from micrometers to centimeters. The nanowire nano-sensor is coated with a sensing biomolecule such as a DNA strand, polypeptides, and proteins. This biomolecule acts as a detector in the sensor while a carbon nanotube is serving as the transmitter. These sensors are useful in creating specific sensors for certain species. The simplest form of nanotube nanostructure is composed of conductive polymers and metal oxides (e.g. carbon nanotube which is used commonly in biosensors and has sensors) (Ziegler et al., 2021).

5 Nano-biosensors for environmental and agricultural application

Application of nano-biosensors in agriculture allows farmers to closely monitor environmental conditions associated with crop production and indirect environmental monitoring are interrelated. Nano-biosensors can be used in the detection of a wide variety of effective environmental factors like fertilizers, moisture, pH, pathogens, herbicides, and pesticides (Fig. 2).

Compared to the traditional detection methods like chromatography, these sensors are rapid in detection, accurate, highly sensitive, and selective. Nano-biosensors combined with other technologies like GPS can be useful to in-site monitoring plant growth conditions like soil moisture depletion or fertilizer level depletion or plant health and diseases. These data can be useful in deciding the optimal time to harvest the crops or deciding the amount of water and other fertilizer need for optimal plant growth avoiding the excess usage of these. Moreover, this helps in environmental protection avoiding the fertilizer leach and contaminating the water bodies and excess water usage. These factors are further discussed in the below sections.

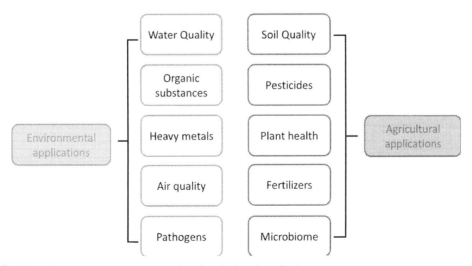

FIG. 2 Nano-biosensors for environmental and agricultural applications.

As mentioned above, apart from the agricultural application of these sensors they can serve for pure environmental monitoring of various gaseous or trace elements in the environment such as pollutant levels, soil condition, pH, etc. These nano-biosensors are highly portable and efficient in detecting the target molecules in complex systems. Further, one of the main advantages of these sensors is the ability to monitor the desired molecules in real-time, on-site, from a minimum of sample amount. Furthermore, the application of these sensors as environmental quality monitoring tools is commonly available in the field as pesticides, herbicides, pH, insecticides, and fertilizer level detecting nano-biosensors (Mandal et al., 2020; Salouti and Khadivi Derakhshan, 2020). The major environment applications for the nano-biosensors are discussed below.

5.1 Pesticides and biological contaminants in soil and water

Nano-biosensors demonstrated accurate and enhanced detection with high sensitivity and specificity of various biochemical elements and shows strong potential for environmental application, including areas of monitoring soil and water. The rapid, precise, and accurate detection of pathogens and pollutants is extremely important to maintain healthy soil and the environment. The metal and metal oxide-based nano-sensors are used extensively in nutrient detection in soil because of their ability in rapid electron transfer or electron conductivity and electroactivity. For example, the presence of urea in soil by measuring urease activity is possible with the help of metal oxide nanoparticles-chitosan-based nano-sensor (Kaushik et al., 2009). Nano-biosensors also plays role in the diagnosis of soil quality caused by undesirable soil microbes (i.e., fungi and bacteria). The ability of nano-biosensor to quantify these microbial contaminations is based on accurate estimation of the oxygen consumption rate of useful/undesirable microbes in the soil (Mandal et al., 2020). Further, a bi-enzyme-based electrochemical biosensing carbon electrode has been used to detect methyl salicylate a chemical released by the plants especially during fungal infection (Fang et al., 2016).

Recently, many types of research have been conducted to identify various pesticides residues in soil using nano-biosensors. The application of various nano-biosensor for detection of pesticide in biological samples are listed in Table 1. The organophosphorus, carbamates, neonicotinoids, dichlorvos and paraoxon, and triazines pesticides were detected using nano-biosensor (Willner and Vikesland, 2018). The acetylcholinesterase enzyme and pyranine (a pH-sensitive fluorescent indicator) were stabilized in the nano-size liposomes and used as a fluorescent nano-biosensor. The enzyme activity of hydrolysis of acetylcholine was reduced with incoming quantities of pesticides, reducing the fluorescent effect of the pH indicator. This system has been used in detecting the toxicity level of drinking water (di Tuoro et al., 2011; Martinazzo et al., 2018; Xia et al., 2015). Researchers also claim that direct detection of pesticides using aptamers biosensors could serve as an excellent option over traditional methods for the analysis of pesticides (Saini et al., 2017). Attempts were already done by researchers for detection and quantification of the pesticides such as carbofuran, carbaryl, methyl paraoxon, and dichlorvos using amperometric acetylcholinesterase biosensors. The combination of commercial and genetically modified acetylcholinesterase enzymes in amperometric biosensors enabled improved sensitivity of the sensors toward these pesticides (Valdés-Ramírez et al., 2008). Gold nanoparticle-based biosensors were reported to have

TABLE 1 Summary of application of nano-biosensor in environmental and agricultural application.

Type of contaminants	Sensing method	Limit of detection	Reference
Urea and urease activity in soil	Gold nanoparticles-based colorimetric-pH sensing	5 μM and 1.8 U/L	Deng et al. (2016)
Nitrate in soil	Electrochemical sensing using graphene oxide (GO) nanosheets and poly(3,4-ethylenedioxythiophene) nanofibers	0.68 mg/L	Ali et al. (2017)
Dichlorvos, methyl paraoxon, carbofuran, and carbaryl	Amperometric acetylcholinesterase biosensors	1×10^{-7}, 2.6×10^{-9}, 4.5×10^{-9}, 1.6×10^{-7} mol/L	Valdés-Ramírez et al. (2008)
Neonicotinoids (neuro-active insecticide)	Colorimetric biosensor with gold nanoparticles	0.1 ppm	Weerathunge et al. (2014)
Triazines (herbisides)	Electrochemical immunosensor with gold nanoparticles	0.016 ng/mL	Liu et al. (2014)
Glyphosate and glufosinate	Electrochemical biosensor with graphite electrode	0.35 ng/mL 0.19 ng/mL	Prasad et al. (2014)
Organophosphorus pesticides-paraoxon	Acetylcholinesterase biosensor with gold and graphene oxide nanocomposite	0.5 nM	Yang et al. (2014)
Methyl parathion-organophosphorus insecticides	Amperometric biosensor with nanotubes-chitosan	7.5×10^{-13} M	Dong et al. (2013)
Phorate, profenofos, isocarbophos, omethoate	Aptamer-type DNA-based biosensor	19.2, 13.4, 17.2, and 23.4 nM	Zhang et al. (2014)
Trichoderma harzianum	Electrochemical sensor with gold electrode and chitosan nanocomposite	1.0×10^{-13} M mol/L	Siddiquee et al. (2014)
E. coli bacteria	Chemi-resistive biosensor with graphene, AuNPs, and streptavidin-antibody system	12 cfu/mL	Zhao et al. (2020)
Virus (plant and soil)	Colorimetric/chemiluminescent DNA-based biosensor	3.3 μM	Ortiz-Tena et al. (2018)
Soil and plant fungi including Fusarium spp.	qPCR-based microarray biosensor	0.6–43.5 pg of DNA	Nikitin et al. (2018)
Trichoderma harzianum Fungicide	Electrochemical biosensor with gold electrode and ZnO nanoparticles	10×10^{-19} M	Siddiquee et al. (2014)
Lead	Fluorescent aptamer DNA silver nanoclusters sensor	5–50 nM	Zhang and Wei (2018)
Mercury and lead	Colorimetric sensor with graphene oxide–gold	0–50 μM	Chen et al. (2015)
Arsenic	Electrochemical aptamer biosensor	0.003×10^{-3} μg/L	Yadav et al. (2020)
Mercury	Electrochemical biosensor	0.005 nM	Liu et al. (2018)

Continued

TABLE 1 Summary of application of nano-biosensor in environmental and agricultural application—cont'd

Type of contaminants	Sensing method	Limit of detection	Reference
Cadmium	Label-free electrochemical aptasensor	0.05 ng/mL	Li et al. (2019)
Metals (As^{5+}, Cr^{6+}, Cu^{2+}, Na^+, Mg^{2+}, Al^{3+}, K^+, Sr^{2+}, Mn^{2+}, Fe^{3+}, Fe^{2+}, Zn^{2+})	Electrochemical sensors using aniline, N-phenylglycine, and graphene oxide nanocomposites	2.19×10^{-7} M	Dutta and Panda (2019)

accurate detection of heavy metals such as mercury, lead, etc., and the dithiocarbamate pesticide to a concentration. Zinc bis (dimethyl dithiocarbamate), a fungicide is reported to be linked to the development of Parkinson's disease in human beings. The traditional method of laboratory analysis of these chemicals is time-consuming and expensive. A goldnanoparticle-based lab-on-a-chip device technology enabled this detection portable and faster using colorimetric and fluorescent detection methods (Lafleur et al., 2012).

5.2 Heavy metals in soil and water detection

Heavy metals and respective ions can cause detrimental damage to the environment and living beings, especially to human health as they are nondegradable and can accumulate in the organism in contact. The demand for a low-cost rapid detection method to analyze the heavy metal content in drinking water and other biological samples are high. The introduction of nano-biosensors to detect the heavy metal concentration of water has been an interesting topic among researchers recently (Liu et al., 2020; Maghsoudi et al., 2021). Different types of nanomaterials such as metal and metal oxides, carbon or silicon-based nanomaterials provided a basis for the development of these nano-biosensors. Detection of heavy metals in any biological sample is achieved by the determination and quantification of the Optochemical properties of the nanomaterial in the sensor. Optical chemical sensors based on Raman scattering and plasmon resonance, including surface-enhanced Raman scattering (SERS) and surface plasmon resonance (SPR) sensors along with fluorescent and colorimetric detection method has exhibited promising potential for the detection of metallic ions, and anions with many advantages including fast detection, high sensitivity, etc. (Chang et al., 2019; Liu et al., 2020; Ullah et al., 2018).

Fluorescence-based nano-biosensors are also highly important in determining heavy metal concentration in biological samples. The basic mechanism of these types of sensors is based on the change in fluorescence properties when binds with heavy metal particles. The main advantages associated with these types of sensors are sensitivity and selectivity due to the characteristic optical properties of fluorophores (i.e., ultraviolet, visible, and infrared absorption spectra, short range of emission and high quantum yield, etc.) (Li and Zhu, 2013; Liu et al., 2017; Zhang and Chen, 2014; Zhang et al., 2013). SERS has a greater potential of optical probing as analyte molecules are absorbed the Raman scattering area is increased and provided qualitative and quantitative information. The narrow peak widths in the enhanced Raman spectra enable the better application of SERS in multiplex analysis with the ability to detect a single molecule species from highly similar one than fluorescent techniques (Frost et al., 2015).

Studies revealed that electrochemical sensors with different sensing techniques such as voltammetry, potentiometry, impedimetric, amperometry, and conductometry are amendable for heavy metal detection and quantification in biological samples (Li et al., 2013). Despite the advancement in the nano-sensing techniques for heavy metal detection, there is still room for improvements in the design and development of novel nanomaterials and optimized bi- or trimetallic nanostructures with desired dimensions and optimum amount of electrochemically active sites ensuring specificity and stability (Waheed et al., 2018).

5.3 Application in sustainable agriculture

Apart from the heavy metal detection and detection of soil and water contaminants, nano-biosensors can be used as a tool to precisely monitor and track the amount of water, fertilizers, etc. for smart agriculture. This can promote sustainable practices among farmers when it comes to optimum utilization of water and fertilizer use. Also, the application of smart nano-biosensors has played a significant role in detecting soil carbon content, phosphorus, nitrate, potassium, and urea residual content (Antonacci et al., 2018). The application of nano-based smart delivery systems can provide more efficient and targeted delivery of nutrients to the specific plant cells due to their physical properties such as size (Dar et al., 2020; Solanki et al., 2015).

A rapid and sensitive colorimetric method was developed by Mura et al. (2015) to identify the nitrate in water using cysteamine modified gold nanoparticles. The presence of nitrate in the water turns the colloidal suspension from red to gray. These results can be further verified by ultraviolet-visible measurement. The detection mechanism is based on the nanogold particles stabilized with citrate and modified with cysteamine. This allows the detection of nitrate in water on-site without neither any skilled personal nor any expensive equipment. Pan et al. (2016) has developed a stable solid-state electrochemical sensor for the determination of total nitrogen in soil samples. Deng et al. (2016) reported the biosensing capabilities of a colorimetric sensing system using gold nanoparticles horseradish peroxidase and tetramethylbenzidine for detection and quantification of urea and urease. The nano-sensors capable of detecting phosphate majorly in liquid samples such as electrochemical sensors (Cinti et al., 2016) could be also modified to use for detecting phosphate content of solid samples as well (Antonacci et al., 2018). Changes in the electrical resistance of graphene nanoparticles in different humidity environments were utilized to develop a graphene-based wearable nano-sensor that can monitor moisture evaporation from plant leaves (Oren et al., 2017).

Another revolutionary application of nano-biosensor in agriculture is in the field of early detection of various plant diseases such as viral and fungal infections and thereby manage and control plant health and yield (Li et al., 2020; Yusof and Isha, 2020). For example, Skottrup et al. (2008) had reported the potential application of antibody-based biosensors for plant pathogen detection such as tobacco mosaic virus and cowpea mosaic virus. Noble metal nanoparticles such as gold, silver, platinum, etc. as well as polymeric nanoparticles such as silicon, polyacetylene, etc. are being studied by researchers as a marker for the identification of plant diseases. The application of DNA-based nano-sensors utilizing various sensing mechanisms such as electrochemistry, SERS, colorimetry, optical sensing, etc. had been discussed by Li et al. (2020) and Yadav and Yadav (2018).

6 Conclusion

Biosensors have versatile applications in the field of toxicology makes them an easily adaptable tool for convenient and functional application in sustainable agriculture and environment systems. Advances in nano-biosensors have made tremendous improvements in the field of biosensors as ultralow size units exhibit novel properties and practical applications compared to their bulk counterparts. Consistent attempts to improve characteristics of nanomaterials and novel nanocomposites of these environmental biosensors with increased attention on the in situ and real-time monitoring of pollutants are still ongoing. An integrated approach with combined possibilities of nano-biosensors, robotics, and GPS systems can create smart and sustainable agricultural systems and better management of the environment.

References

Abdel-Karim, R., Reda, Y., Abdel-Fattah, A., 2020. Review—nanostructured materials-based nanosensors. J. Electrochem. Soc. 167 (3), 037554. https://doi.org/10.1149/1945-7111/ab67aa.

Ali, M.A., Jiang, H., Mahal, N.K., Weber, R.J., Kumar, R., Castellano, M.J., Dong, L., 2017. Microfluidic impedimetric sensor for soil nitrate detection using graphene oxide and conductive nanofibers enabled sensing interface. Sens. Actuators B 239, 1289–1299. https://doi.org/10.1016/j.snb.2016.09.101.

Antonacci, A., Arduini, F., Moscone, D., Palleschi, G., Scognamiglio, V., 2018. Nanostructured (bio)sensors for smart agriculture. Trends Anal. Chem. 98, 95–103. https://doi.org/10.1016/j.trac.2017.10.022.

Arlett, J.L., Myers, E.B., Roukes, M.L., 2011. Comparative advantages of mechanical biosensors. Nat. Nanotechnol. 6 (4), 203–215. https://doi.org/10.1038/nnano.2011.44.

Bakirhan, N.K., Topal, B.D., Ozcelikay, G., Karadurmus, L., Ozkan, S.A., 2020. Current advances in electrochemical biosensors and nanobiosensors. Crit. Rev. Anal. Chem., 1–16. https://doi.org/10.1080/10408347.2020.1809339.

Baron, R., Saffell, J., 2017. Amperometric gas sensors as a low-cost emerging technology platform for air quality monitoring applications: a review. ACS Sens. 2 (11), 1553–1566. https://doi.org/10.1021/acssensors.7b00620.

Chalklen, T., Jing, Q., Kar-Narayan, S., 2020. Biosensors based on mechanical and electrical detection techniques. Sensors 20 (19), 26–37. https://doi.org/10.3390/s20195605.

Chang, C.C., Chen, C.P., Wu, T.H., Yang, C.H., Lin, C.W., Chen, C.Y., 2019. Gold nanoparticle-based colorimetric strategies for chemical and biological sensing applications. Nanomaterials 9 (6), 861. https://doi.org/10.3390/nano9060861.

Chatterjee, B., Das, S.J., Anand, A., Sharma, T.K., 2020. Nanozymes and aptamer-based biosensing (Elsevier Enhanced Reader). Mater. Sci. Energy Technol. 3, 127–135. https://doi.org/10.1016/j.mset.2019.08.007.

Chen, X., Zhai, N., Snyder, J.H., Chen, Q., Liu, P., Jin, L., Zheng, Q., Lin, F., Zhou, H., 2015. Colorimetric detection of Hg^{2+} and Pb^{2+} based on peroxidase-like activity of graphene oxide–gold nanohybrids. Anal. Methods 7, 1951–1957. https://doi.org/10.1039/C4AY02801E.

Chiu, T.-C., Huang, C.-C., 2009. Aptamer-functionalized nano-biosensors. Sensors 9, 10356–10388. https://doi.org/10.3390/s91210356.

Cinti, S., Talarico, D., Palleschi, G., Moscone, D., Arduini, F., 2016. Novel reagentless paper-based screen-printed electrochemical sensor to detect phosphate. Anal. Chim. Acta 919, 78–84. https://doi.org/10.1016/j.aca.2016.03.011.

Damborský, P., Švitel, J., Katrlík, J., 2016. Optical biosensors. Essays Biochem. 60 (1), 91–100. https://doi.org/10.1042/EBC20150010.

Dar, F.A., Qazi, G., Pirzadah, T.B., 2020. Nano-biosensors: NextGen diagnostic tools in agriculture. In: Rehman, T.B.-H.K., Pirzadah (Eds.), Nanobiotechnology in Agriculture: An Approach Towards Sustainability. Springer International Publishing, pp. 129–144, https://doi.org/10.1007/978-3-030-39978-8_7.

Deng, H.-H., Hong, G.-L., Lin, F.-L., Liu, A.-L., Xia, X.-H., Chen, W., 2016. Colorimetric detection of urea, urease, and urease inhibitor based on the peroxidase-like activity of gold nanoparticles. Anal. Chim. Acta 915, 74–80. https://doi.org/10.1016/j.aca.2016.02.008.

Dhar, D., Roy, S., Nigam, V.K., 2019. Advances in protein/enzyme-based biosensors for the detection of pharmaceutical contaminants in the environment. In: Kaur Brar, S., Hegde, K., Pachapur, V.L. (Eds.), Tools, Techniques and Protocols for Monitoring Environmental Contaminants. Elsevier, pp. 207–229, https://doi.org/10.1016/B978-0-12-814679-8.00010-8 (Chapter 10).

di Tuoro, D., Portaccio, M., Lepore, M., Arduini, F., Moscone, D., Bencivenga, U., Mita, D.G., 2011. An acetylcholinesterase biosensor for determination of low concentrations of Paraoxon and Dichlorvos. New Biotechnol. 29 (1), 132–138. https://doi.org/10.1016/j.nbt.2011.04.011.

Dong, J., Fan, X., Qiao, F., Ai, S., Xin, H., 2013. A novel protocol for ultra-trace detection of pesticides: combined electrochemical reduction of Ellman's reagent with acetylcholinesterase inhibition. Anal. Chimica Acta 761, 78–83. https://doi.org/10.1016/j.aca.2012.11.042.

Dutta, K., Panda, S., 2019. Thermodynamic and charge transport studies for the detection of heavy metal ions in electrochemical sensors using a composite film of aniline, N-phenylglycine and graphene oxide. J. Electrochem. Soc. 166 (14), B1335–B1342.

Fang, Y., Bullock, H., Lee, S.A., Sekar, N., Eiteman, M.A., Whitman, W.B., Ramasamy, R.P., 2016. Detection of methyl salicylate using bi-enzyme electrochemical sensor consisting salicylate hydroxylase and tyrosinase. Biosens. Bioelectron. 85, 603–610. https://doi.org/10.1016/j.bios.2016.05.060.

Frost, M.S., Dempsey, M.J., Whitehead, D.E., 2015. Highly sensitive SERS detection of Pb2+ ions in aqueous media using citrate functionalised gold nanoparticles. Sensors Actuators B Chem. 221, 1003–1008. https://doi.org/10.1016/j.snb.2015.07.001.

Ghorbanpour, M., Bhargava, P., Varma, A., Choudhary, D.K., 2020. Biogenic Nano-Particles and Their Use in Agro-Ecosystems. Springer, Singapore, https://doi.org/10.1007/978-981-15-2985-6.

Hammond, J.L., Formisano, N., Estrela, P., Carrara, S., Tkac, J., 2016. Electrochemical biosensors and nanobiosensors. Essays Biochem. 60 (1), 69–80. https://doi.org/10.1042/EBC20150008.

Han, K., Liang, Z., Zhou, N., 2010. Design strategies for aptamer-based biosensors—sensors. Sensors 10, 4541–4557. https://doi.org/10.3390/s100504541.

Huang, X., Zhu, Y., Kianfar, E., 2021. Nano biosensors: properties, applications and electrochemical techniques. J. Mater. Res. Technol. 12, 1649–1672. https://doi.org/10.1016/j.jmrt.2021.03.048.

Kaushik, A., Solanki, P.R., Ansari, A.A., Sumana, G., Ahmad, S., Malhotra, B.D., 2009. Iron oxide-chitosan nanobiocomposite for urea sensor. Sensors Actuators B Chem. 138 (2), 572–580. https://doi.org/10.1016/j.snb.2009.02.005.

Lafleur, J.P., Senkbeil, S., Jensen, T.G., Kutter, J.P., 2012. Gold nanoparticle-based optical microfluidic sensors for analysis of environmental pollutants. Lab Chip 12 (22), 4651–4656. https://doi.org/10.1039/c2lc40543a.

Li, J., Zhu, J.-J., 2013. Quantum dots for fluorescent biosensing and bio-imaging applications. Analyst 138 (9), 2506–2515. https://doi.org/10.1039/C3AN36705C.

Li, M., Gou, H., Al-Ogaidi, I., Wu, N., 2013. Nanostructured sensors for detection of heavy metals: a review. ACS Sustain. Chem. Eng. 1, 713–723.

Li, Y., Ran, G., Lu, G., Ni, X., Liu, D., Sun, J., Xie, C., Yao, D., Bai, W., 2019. Highly sensitive label-free electrochemical aptasensor based on screen-printed electrode for detection of cadmium (II) ions. J. Electrochem. Soc. 166 (6), B449–B455. https://doi.org/10.1149/2.0991906jes.

Li, Z., Yu, T., Paul, R., Fan, J., Yang, Y., Wei, Q., 2020. Agricultural nanodiagnostics for plant diseases: recent advances and challenges. Nanoscale Adv. 2 (8), 3083–3094. https://doi.org/10.1039/c9na00724e.

Liu, Y., Deng, Y., Li, T., Chen, Z., Chen, H., Li, S., Liu, H., 2018. Aptamer-based electrochemical biosensor for mercury ions detection using AuNPs-modified glass carbon electrode. J. Biomed. Nanotechnol. 14 (12), 2156–2161. https://doi.org/10.1166/jbn.2018.2655.

Liu, X., Li, W.J., Li, L., Yang, Y., Mao, L.G., Peng, Z., 2014. A label-free electrochemical immunosensor based on gold nanoparticles for direct detection of atrazine. Sens. Actuators B Chem. 191, 408–414. https://doi.org/10.1016/j.snb.2013.10.033.

Liu, X., Liu, J., 2021. Biosensors and sensors for dopamine detection. View 2, 1–16. https://doi.org/10.1002/VIW.20200102.

Liu, J., Lv, G., Gu, W., Li, Z., Tang, A., Mei, L., 2017. A novel luminescence probe based on layered double hydroxides loaded with quantum dots for simultaneous detection of heavy metal ions in water. J. Mater. Chem. C 5 (20), 5024–5030. https://doi.org/10.1039/C7TC00935F.

Liu, B., Zhuang, J., Wei, G., 2020. Recent advances in the design of colorimetric sensors for environmental monitoring. Environ. Sci. Nano 7 (8), 2195–2213. https://doi.org/10.1039/d0en00449a.

Long, F., Zhu, A., Shi, H., 2013. Recent advances in optical biosensors for environmental monitoring and early warning. Sensors 13 (10), 13928–13948. https://doi.org/10.3390/s131013928.

Maghsoudi, A.S., Hassani, S., Mirnia, K., Abdollahi, M., 2021. Recent advances in nanotechnology-based biosensors development for detection of arsenic, lead, mercury, and cadmium. Int. J. Nanomedicine 16, 803–832. https://doi.org/10.2147/IJN.S294417.

Malekzad, H., Sahandi Zangabad, P., Mirshekari, H., Karimi, M., Hamblin, M.R., 2017. Noble metal nanoparticles in biosensors: recent studies and applications. Nanotechnol. Rev. 6 (3), 301–329. https://doi.org/10.1515/ntrev-2016-0014.

Malik, P., Katyal, V., Malik, V., Asatkar, A., Inwati, G., Mukherjee, T.K., 2013. Nanobiosensors: concepts and variations. ISRN Nanomater. 2013, 1–9. https://doi.org/10.1155/2013/327435.

Mandal, N., Adhikary, S., Rakshit, R., 2020. Nanobiosensors: recent developments in soil health assessment. In: Soil Analysis: Recent Trends and Applications. Springer, Singapore, https://doi.org/10.1007/978-981-15-2039-6_15.

Martinazzo, J., Marie De Cezaro, A., Nava, A., Rigo, A.A., Muenchen, D.K., Nava Brezolin, A., Manzoli, A., de Lima Leite, F., Steffens, C., Steffens, J., 2018. Pesticide detection in soil using biosensors and nanobiosensors PROGRAMA FUTURO CIENTISTA view project nanoneurobiophysics view project pesticide detection in soil using biosensors and nanobiosensors. Biointerface Res. Appl. Chem. https://doi.org/10.2016/Accepted.

McConnell, E.M., Nguyen, J., Li, Y., 2020. Aptamer-based biosensors for environmental monitoring. Front. Chem. 8, 434. https://doi.org/10.3389/fchem.2020.00434.

Min, J., Yea, C.H., El-Said, W.A., Choi, J.W., 2009. The application of cell based biosensor and biochip for environmental monitoring. In: Atmospheric and Biological Environmental Monitoring. Springer, Dordrecht, pp. 261–273, https://doi.org/10.1007/978-1-4020-9674-7_18.

Mondal, B., Ramlal, S., Lavu, P.S., Kingston, J., 2018. Highly sensitive colorimetric biosensor for staphylococcal enterotoxin B by a label-free aptamer and gold nanoparticles. Front. Microbiol. 9, 179. https://doi.org/10.3389/fmicb.2018.00179.

Mura, S., Greppi, G., Roggero, P.P., Musu, E., Pittalis, D., Carletti, A., Ghiglieri, G., Irudayaraj, J., 2015. Functionalized gold nanoparticles for the detection of nitrates in water. Int. J. Environ. Sci. Technol. 12 (3), 1021–1028. https://doi.org/10.1007/s13762-013-0494-7.

Naresh, V., Lee, N., 2021. A review on biosensors and recent development of nanostructured materials-enabled biosensors. Sensors 21 (4), 1109. https://doi.org/10.3390/s21041109.

Negi, N.P., Choephel, T., 2020. Biosensor: an approach towards a sustainable environment. In: Nanobiosensors for Agricultural, Medical and Environmental Applications. Springer, Singapore, pp. 42–62, https://doi.org/10.1007/978-981-15-8346-9.

Nikitin, M., Deych, K., Grevtseva, I., Girsova, N., Kuznetsova, M., Pridannikov, M., Dzhavakhiya, V., Statsyuk, N., Golikov, A., 2018. Preserved microarrays for simultaneous detection and identification of six fungal potato pathogens with the use of real-time PCR in matrix format. Biosensors (Basel) 8 (4), 129–147. https://doi.org/10.3390/bios8040129.

Oren, S., Ceylan, H., Schnable, P.S., Dong, L., 2017. Wearable electronics: high-resolution patterning and transferring of graphene-based nanomaterials onto tape toward roll-to-roll production of tape-based wearable sensors. Adv. Mater. Technol. 2 (12), 1770055. https://doi.org/10.1002/admt.201770055.

Ortiz-Tena, J.G., Ruhmann, B., Sieber, V., 2018. Colorimetric determination of sulfate via an enzyme cascade for high-throughput detection of sulfatase activity. Anal. Chem. 90, 2526–2533. https://doi.org/10.1021/acs.analchem.7b03719.

Pan, P., Miao, Z., Yanhua, L., Linan, Z., Haiyan, R., Pan, K., Linpei, P., 2016. Preparation and evaluation of a stable solid state ion selective electrode of polypyrrole/electrochemically reduced graphene/glassy carbon substrate for soil nitrate sensing. Int. J. Electrochem. Sci. 11 (6), 4779–4793. https://doi.org/10.20964/2016.06.7.

Piro, B., Shi, S., Reisberg, S., Noel, V., Anquetin, G., 2016. Comparison of electrochemical immunosensors and aptasensors for detection of small organic molecules in environment, food safety, clinical and public security—biosensors. Biosensors 6, 1–22. https://doi.org/10.3390/bios6010007.

Pohanka, M., 2018. Overview of piezoelectric biosensors, immunosensors and DNA sensors and their applications. Materials 11 (3), 448. https://doi.org/10.3390/ma11030448.

Prasad, S., 2014. Nanobiosensors: the future for diagnosis of disease? Nanobiosens. Dis. Diagn. 1. https://doi.org/10.2147/ndd.s39421.

Prasad, B.B., Jauhari, D., Tiwari, M.P., 2014. Doubly imprinted polymer nanofilm-modified electrochemical sensor for ultra-trace simultaneous analysis of glyphosate and glufosinate. Biosens. Bioelectron. 59, 81–88. https://doi.org/10.1016/j.bios.2014.03.019.

Pumera, M., Sánchez, S., Ichinose, I., Tang, J., 2007. Electrochemical nanobiosensors. Sensors Actuators B Chem. 123 (2), 1195–1205. https://doi.org/10.1016/j.snb.2006.11.016.

Purohit, B., Vernekar, P.R., Shetti, N.P., Chandra, P., 2020. Biosensor nanoengineering: design, operation, and implementation for biomolecular analysis. Sens. Int. 1, 100040. https://doi.org/10.1016/j.sintl.2020.100040.

Rocchitta, G., Spanu, A., Babudieri, S., Latte, G., Madeddu, G., Galleri, G., Nuvoli, S., Bagella, P., Demartis, M.I., Fiore, V., Manetti, R., Serra, P.A., 2016. Enzyme biosensors for biomedical applications: strategies for safeguarding analytical performances in biological fluids. Sensors 16 (6), 780. https://doi.org/10.3390/s16060780.

Saini, R.K., Bagri, L.P., Bajpai, A.K., 2017. Smart nanosensors for pesticide detection. In: New Pesticides and Soil Sensors. Academic Press, pp. 519–559, https://doi.org/10.1016/b978-0-12-804299-1.00015-1.

Salouti, M., Khadivi Derakhshan, F., 2020. Biosensors and nanobiosensors in environmental applications. In: Biogenic Nano-Particles and Their Use in Agro-Ecosystems. Springer, Singapore, https://doi.org/10.1007/978-981-15-2985-6_26.

Selvarajan, S., Alluri, N.R., Chandrasekhar, A., Kim, S.J., 2017. Unconventional active biosensor made of piezoelectric BaTiO3 nanoparticles for biomolecule detection. Sensors Actuators B Chem. 253, 1180–1187. https://doi.org/10.1016/j.snb.2017.07.159.

Seo, H.B., Gu, M.B., 2017. Aptamer-based sandwich-type biosensors. J. Biol. Eng. 11, 11–18. https://doi.org/10.1186/s13036-017-0054-7.

Shen, J., Zhou, T., Huang, R., 2019. Recent advances in electrochemiluminescence sensors for pathogenic bacteria detection. Micromachines 10 (8), 532. https://doi.org/10.3390/mi10080532.

Siddiquee, S., Rovina, K., Yusaf, N.A., Rodrigues, F.K., Suryani, S., 2014. Nanoparticle-enhanced electrochemical biosensor with DNA immobilization and hybridization of *Trichoderma harzianum* gene. Sens. Bio-Sens. Res. 2, 16–22.

Skottrup, P.D., Nicolaisen, M., Justesen, A.F., 2008. Towards on-site pathogen detection using antibody-based sensors. Biosens. Bioelectron. 24 (3), 339–348. https://doi.org/10.1016/j.bios.2008.06.045.

Solanki, P., Bhargava, A., Chhipa, H., Jain, N., Panwar, J., 2015. Nano-fertilizers and their smart delivery system. In: Nanotechnologies in Food and Agriculture. Springer, Cham, pp. 81–101, https://doi.org/10.1007/978-3-319-14024-7_4.

Syrek, K., Skolarczyk, M., Zych, M., Sołtys-Mróz, M., Sulka, G.D., 2019. A photoelectrochemical sensor based on anodic TiO2 for glucose determination. Sensors 19 (22), 4981. https://doi.org/10.3390/s19224981.

Tavakkoli Yaraki, M., Tan, Y.N., 2020. Recent advances in metallic nanobiosensors development: colorimetric, dynamic light scattering and fluorescence detection. Sens. Int. 1, 100049. https://doi.org/10.1016/j.sintl.2020.100049.

Ullah, N., Mansha, M., Khan, I., Qurashi, A., 2018. Nanomaterial-based optical chemical sensors for the detection of heavy metals in water: recent advances and challenges. TrAC Trends Anal. Chem. 100, 155–166. https://doi.org/10.1016/j.trac.2018.01.002.

Valdés-Ramírez, G., Cortina, M., Ramírez-Silva, M.T., Marty, J.L., 2008. Acetylcholinesterase-based biosensors for quantification of carbofuran, carbaryl, methylparaoxon, and dichlorvos in 5% acetonitrile. Anal. Bioanal. Chem. 392 (4), 699–707. https://doi.org/10.1007/s00216-008-2290-7.

Verma, N., Bhardwaj, A., 2015. Biosensor technology for pesticides—a review. Appl. Biochem. Biotechnol. 175 (6), 3093–3119. https://doi.org/10.1007/s12010-015-1489-2.

Waheed, A., Mansha, M., Ullah, N., 2018. Nanomaterials-based electrochemical detection of heavy metals in water: current status, challenges and future direction. TrAC Trends Anal. Chem. 105, 37–51. https://doi.org/10.1016/j.trac.2018.04.012.

Weerathunge, P., Ramanathan, R., Shukla, R., Sharma, T.K., Bansal, V., 2014. Aptamer-controlled reversible inhibition of gold nanozyme activity for pesticide sensing. Anal. Chem. 86, 11937–11941. https://doi.org/10.1021/ac5028726.

White, D.A., Buell, A.K., Dobson, C.M., Welland, M.E., Knowles, T.P., 2009. Biosensor-based label-free assays of amyloid growth. FEBS Lett. 583 (16), 2587–2592. https://doi.org/10.1016/j.febslet.2009.06.008.

Willner, M.R., Vikesland, P.J., 2018. Nanomaterial enabled sensors for environmental contaminants. J. Nanobiotechnol. 16 (1), 1–16. https://doi.org/10.1186/s12951-018-0419-1.

Wujcik, E.K., Wei, H., Zhang, X., Guo, J., Yan, X., Sutrave, N., Wei, S., Guo, Z., 2014. Antibody nanosensors: a detailed review. RSC Adv. 4 (82), 43725–43745. https://doi.org/10.1039/C4RA07119K.

Xia, N., Wang, Q., Liu, L., 2015. Nanomaterials-based optical techniques for the detection of acetylcholinesterase and pesticides. Sensors 15 (1), 499–514. https://doi.org/10.3390/s150100499.

Xu, L., Shoaie, N., Jahanpeyma, F., Zhao, J., Azimzadeh, M., Al-Jamal, K.T., 2020. Optical, electrochemical and electrical (nano)biosensors for detection of exosomes: a comprehensive overview. Biosens. Bioelectron. 161, 112222. https://doi.org/10.1016/j.bios.2020.112222.

Yadav, R., Kushwah, V., Gaur, M.S., Bhadauria, S., Berlina, A.N., Zherdev, A.V., Dzantiev, B.B., 2020. Electrochemical aptamer biosensor for As^{3+} based on apta deep trapped Ag-Au alloy nanoparticles-impregnated glassy carbon electrode. Int. J. Environ. Anal. Chem. 100 (6), 623–634. https://doi.org/10.1080/03067319.2019.1638371.

Yadav, A., Yadav, K., 2018. Nanoparticle-based plant disease management: tools for sustainable agriculture. In: Nanobiotechnology Applications in Plant Protection. Springer, Cham, pp. 29–61, https://doi.org/10.1007/978-3-319-91161-8_2.

Yang, Y., Asiri, A.M., Du, D., Lin, Y., 2014. Acetylcholinesterase biosensor based on a gold nanoparticle-polypyrrole-reduced graphene oxide nanocomposite modified electrode for the amperometric detection of organophosphorus pesticides. Analyst 139 (12), 3055–3060. https://doi.org/10.1039/c4an00068d.

Yoo, H., Jo, H., Soo-Oh, S., 2020. Detection and beyond: challenges and advances in aptamer-based biosensors. Mater. Adv. 1, 2663–2687. https://doi.org/10.1039/d0ma00639d.

Yusof, N.A., Isha, A., 2020. Nanosensors for early detection of plant diseases. In: Husen, A., Jawaid, M. (Eds.), Nanomaterials for Agriculture and Forestry Applications. Elsevier, pp. 407–419, https://doi.org/10.1016/B978-0-12-817852-2.00016-0 (Chapter 16).

Zamfir, L.G., Puiu, M., Bala, C., 2020. Advances in electrochemical impedance spectroscopy detection of endocrine disruptors. Sensors 20 (22), 6443. https://doi.org/10.3390/s20226443.

Zhang, R., Chen, W., 2014. Nitrogen-doped carbon quantum dots: facile synthesis and application as a "turn-off" fluorescent probe for detection of Hg2+ ions. Biosens. Bioelectron. 55, 83–90. https://doi.org/10.1016/j.bios.2013.11.074.

Zhang, C., Wang, L., Tu, Z., Sun, X., He, Q., Lei, Z., Xu, C., Liu, Y., Zhang, X., Yang, J., 2014. Organophosphorus pesticides detection using broad-specific single-stranded DNA based fluorescence polarization aptamer assay. Biosens. Bioelectron. 55, 216–219. https://doi.org/10.1016/j.bios.2013.12.020.

Zhang, Y.-L., Wang, L., Zhang, H.-C., Liu, Y., Wang, H.-Y., Kang, Z.-H., Lee, S.-T., 2013. Graphitic carbon quantum dots as a fluorescent sensing platform for highly efficient detection of Fe3+ ions. RSC Adv. 3 (11), 3733–3738. https://doi.org/10.1039/C3RA23410J.

Zhang, B., Wei, C., 2018. Highly sensitive and selective detection of Pb^{2+} using a turn-on fluorescent aptamer DNA silver nanoclusters sensor. Talanta 182, 125–130. https://doi.org/10.1016/j.talanta.2018.01.061.

Zhao, V.X.T., Wong, T.I., Zheng, X.T., Tan, Y.N., Zhou, X., 2020. Colorimetric biosensors for point-of-care virus detections. Mater. Sci. Energy Technol. 3, 237–249. https://doi.org/10.1016/j.mset.2019.10.002.

Zhu, C., Yang, G., Li, H., Du, D., Lin, Y., 2015. Electrochemical sensors and biosensors based on nanomaterials and nanostructures. Anal. Chem. 87 (1), 230–249. https://doi.org/10.1021/ac5039863.

Zhu, Y.C., Mei, L.P., Ruan, Y.F., Zhang, N., Zhao, W.W., Xu, J.J., Chen, H.Y., 2019. Enzyme-based biosensors and their applications. In: Advances in Enzyme Technology. Elsevier, pp. 201–223, https://doi.org/10.1016/B978-0-444-64114-4.00008-X.

Ziegler, J.M., Andoni, I., Choi, E.J., Fang, L., Flores-Zuleta, H., Humphrey, N.J., Kim, D.H., Shin, J., Youn, H., Penner, R.M., 2021. Sensors based upon nanowires, nanotubes, and nanoribbons: 2016-2020. Anal. Chem. 93 (1), 124–166. https://doi.org/10.1021/acs.analchem.0c04476.

Index

Note: Page numbers followed by *f* indicate figures and *t* indicate tables.